Reflector Antennas

OTHER IEEE PRESS BOOKS

Phase-Locked Loops & Their Application, *Edited by W. C. Lindsey and M. K. Simon*
Digital Signal Computers and Processors, *Edited by A. C. Salazar*
Systems Engineering: Methodology and Applications, *Edited by A. P. Sage*
Modern Crystal and Mechanical Filters, *Edited by D. F. Sheahan and R. A. Johnson*
Electrical Noise: Fundamentals and Sources, *Edited by M. S. Gupta*
Computer Methods in Image Analysis, *Edited by J. K. Aggarwal, R. O. Duda, and A. Rosenfeld*
Microprocessors: Fundamentals and Applications, *Edited by W. C. Lin*
Machine Recognition of Patterns, *Edited by A. K. Agrawala*
Turning Points in American Electrical History, *Edited by J. E. Brittain*
Charge-Coupled Devices: Technology and Applications, *Edited by R. Melen and D. Buss*
Spread Spectrum Techniques, *Edited by R. C. Dixon*
Electronic Switching: Central Office Systems of the World, *Edited by A. E. Joel, Jr.*
Electromagnetic Horn Antennas, *Edited by A. W. Love*
Waveform Quantization and Coding, *Edited by N. S. Jayant*
Communication Satellite Systems: An Overview of the Technology, *Edited by R. G. Gould and Y. F. Lum*
Literature Survey of Communication Satellite Systems and Technology, *Edited by J. H. W. Unger*
Solar Cells, *Edited by C. E. Backus*
Computer Networking, *Edited by R. P. Blanc and I. W. Cotton*
Communications Channels: Characterization and Behavior, *Edited by B. Goldberg*
Large-Scale Networks: Theory and Design, *Edited by F. T. Boesch*
Optical Fiber Technology, *Edited by D. Gloge*
Selected Papers in Digital Signal Processing, II, *Edited by the Digital Signal Processing Committee*
A Guide for Better Technical Presentations, *Edited by R. M. Woelfle*
Career Managment: A Guide to Combating Obsolescence, *Edited by H. G. Kaufman*
Energy and Man: Technical and Social Aspects of Energy, *Edited by M. G. Morgan*
Magnetic Bubble Technology: Integrated-Circuit Magnetics for Digital Storage and Processing, *Edited by H. Chang*
Frequency Synthesis: Techniques and Applications, *Edited by J. Gorski-Popiel*
Literature in Digital Processing: Author and Permuted Title Index (Revised and Expanded Edition), *Edited by H. D. Helms, J. F. Kaiser, and L. R. Rabiner*
Data Communications via Fading Channels, *Edited by K. Brayer*
Nonlinear Networks: Theory and Analysis, *Edited by A. N. Willson, Jr.*
Computer Communications, *Edited by P. E. Green, Jr. and R. W. Lucky*
Stability of Large Electric Power Systems, *Edited by R. T. Byerly and E. W. Kimbark*
Automatic Test Equipment: Hardware, Software, and Management, *Edited by F. Liguori*
Key Papers in the Development of Coding Theory, *Edited by E. R. Berlekamp*
Technology and Social Institutions, *Edited by K. Chen*
Key Papers in the Development of Information Theory, *Edited by D. Slepian*
Computer-Aided Filter Design, *Edited by G. Szentirmai*
Laser Devices and Applications, *Edited by I. P. Kaminow and A. E. Siegman*
Integrated Optics, *Edited by D. Marcuse*
Laser Theory, *Edited by F. S. Barnes*
Digital Signal Processing, *Edited by L. R. Rabiner and C. M. Rader*
Minicomputers: Hardware, Software, and Applications, *Edited by J. D. Schoeffler and R. H. Temple*
Semiconductor Memories, *Edited by D. A. Hodges*
Power Semiconductor Applications, Volume II: Equipment and Systems, *Edited by J. D. Harnden, Jr. and F. B. Golden*
Power Semiconductor Applications, Volume I: General Considerations, *Edited by J. D. Harnden, Jr, and F. B. Golden*
A Practical Guide to Minicomputer Applications, *Edited by F. F. Coury*
Active Inductorless Filters, *Edited by S. K. Mitra*
Clearing the Air: The Impact of the Clean Air Act on Technology, *Edited by J. C. Redmond, J. C. Cook, and A. A. J. Hoffman*

Reflector Antennas

Edited by

A. W. Love

Member of the Technical Staff
Space Division
Rockwell International

A volume in the IEEE PRESS Selected Reprint Series, prepared under the sponsorship of the IEEE Antennas and Propagation Society.

IEEE PRESS

The Institute of Electrical and Electronics Engineers, Inc. New York

PRINTED IN THE UNITED STATES OF AMERICA

IEEE International Standard Book Numbers: Clothbound: 0-87942-103-7
Paperbound: 0-87942-104-5

Library of Congress Catalog Card Number 77-94519

Sole Worldwide Distributor (Exclusive of the IEEE):

JOHN WILEY & SONS, INC.
605 Third Ave.
New York, NY 10016

Wiley Order Numbers: Clothbound: 0-471-04605-1
Paperbound: 0-471-04606-X

Contents

"One cannot escape the feeling that these mathematical formulae have an independent existence and an intelligence of their own, that they are wiser than we are, wiser even than their discoverers, that we get more out of them than was originally put into them."

—Heinrich Hertz
1857–1894

Part I
Introduction: Papers of General Interest

The real introduction to this volume of reprints dealing with reflector antennas is to be found in the first paper, one which grew out of an oral presentation that I was privileged to give at the 1975 URSI meeting in Boulder, Colorado. Later, in a repetition to members of the Los Angeles Chapter of the IEEE Antennas and Propagation Society, I commented on the historically ancient use of reflectors in optics. Thereupon, Professor N. G. Alexopoulos of UCLA pointed to the enduring legend that Archimedes set fire to the Roman invaders' fleet during the siege of Syracuse in 212 B.C. by using burning mirrors to focus the sun's rays on the wooden ships. Could it have been, perhaps, a touch of national pride that impelled the good professor to hail Archimedes as the father of the array antenna? Be that as it may, reflectors, as antennas, are only as old as the radio art itself. Nevertheless, there is a wealth of literature on the subject, a fact which made the process of selecting papers for this book a lengthy and difficult task.

To keep the book to the size deemed most effective by the IEEE Press Editorial Board (circa 400 pages) has resulted in the unfortunate exclusion of a number of excellent papers. One such, by Clarricoats and Poulton [1], appeared during preparation of this volume, but its sheer length, 35 pages, precluded its inclusion. A review paper, it treats in considerable detail that class of antennas comprised of circularly symmetric reflectors with axially symmetric feed systems. Admittedly, there are topics which have been omitted altogether. Cylindrical and conical reflectors, for example, have found only limited use and therefore have been left out. The subject of monopulse tracking antennas has likewise been omitted, not due to any limited usage, of course, but because much of the material dealing with this topic is in the form of reports and proprietary documents. To a great extent, this also seems to be true of the much newer field of satellite-borne antennas for the creation of specially contoured beams. These generally use an oversized reflector and a multiplicity of feeds designed to provide a beam intercept having a specified outline, say that of a country or a subcontinent as seen from geostationary orbit.

The book itself is divided into nine parts, the first being introductory in nature and including historically significant papers that are quite general in character. Parts II, III, and V develop the theory of radiation and pattern formation in symmetrical, front-fed reflectors. Parts IV, VI, and IX do likewise for Cassegrainian and offset (i.e., unsymmetrical) systems and for spherical mirrors. Aberrations due to feed displacement and to phase errors caused by reflector surface roughness form the subject matter for Parts VII and VIII with the former section branching, appropriately, into multiple beam formation by means of lateral feed displacement. For the reader who wishes to delve more deeply into any of these topics, bibliographies have been provided for each of the parts. They will be found at the end of each part.

In order to mitigate, in some degree, the loss of those papers that had to be excluded, I have tried to make the bibliographies comprehensive. This is particularly true for Part I. Because of the very broad and general nature of this part, it has seemed wise to divide its bibliography into seven different categories, none of which fits directly into the subject matter of the remaining eight parts.

As for Part I itself, the second paper is of considerable historical importance. In it, Cutler gives one of the earliest published analyses of reflector antenna requirements, including discussions of aperture efficiency, polarization, shadowing, and feed design. The paper by Jones examines aperture field requirements in detail and introduces the concept of a plane wave feed, consisting of crossed electric and magnetic dipoles, capable of producing a purely linearly polarized field in the aperture. In the last paper, Koffman again examines this kind of feed, termed a Huygens source, and shows that it permits the use of a reflector formed by parallel conducting slats in lieu of a solid surface. In the interesting paper by Carter, it is shown that the phase center of a symmetrical reflector antenna usually lies between the paraboloid vertex and the focus, but that its location is generally quite different in the E and H planes.

Some highlights in reflector antenna development[1]

A. W. Love

Space Division, D/193 SK86, Rockwell International, 12214 Lakewood Boulevard, Downey, California 90241

(Received February 25, 1976.)

Reflector antennas have been used since the radio pioneering era of Lodge, Hertz, and Marconi, but it took the exigent demands of radar in World War II to stimulate a real development in the reflector art. Subsequent interest in the science of radio astronomy and the inception of microwave ground communication links were responsible for a burgeoning growth in the field, so that in the 1940s and 1950s the design principles and requirements for prime focus fed systems were well established. Cassegrain, or secondary focus systems, and horn reflectors came into prominence in the early 1960s with the advent of satellite tracking and communication networks. The desire to maximize the gain, or the gain-temperature ratio, then led to development of sophisticated techniques for properly shaping the illumination over the reflector aperture in order to maximize efficiency and minimize spillover, among them being the shaping of the sub-reflector in Cassegrain systems and the use of multimode and hybrid mode feed horns. Not all reflector antennas utilize paraboloidal surfaces. Some recent developments in line source feeds make the spherical reflector attractive for scanning applications and the conical reflector for deployable, space-borne antennas. The large 1000-foot diameter reflector at Arecibo is a well known example of the former. Although some extremely large spaceborne reflector antennas have been proposed and studied, the largest now in use appears to be the unfurlable 30-foot reflector carried by ATS-6. Finally, some gain comparisons are given for a few of the (electrically) largest reflectors that have been built both for radio astronomy and for space communications. If some milestones in reflector development have been overlooked it is due to the limitations inherent in a review paper.

INTRODUCTION

The reflectors that are of most interest in the antenna field are all derived from the conic sections. The geometrical properties of these figures have been known for more than 2000 years and have been exploited in optics for centuries. The five basic conic sections shown in Figure 1 are all described by the same polar equation,

$$r/f = (1 + e)/(1 + e \cos \theta) \qquad (1)$$

where f is focal length and the pole is at one focus. They differ only in respect to the eccentricity, e, as noted in the figure. Reflecting surfaces are generated by translation of the curves, or by rotation around the focal axis to generate a figure of revolution. Such reflectors are truly wideband devices, capable in principle of operation from radio to optical frequencies.

[1] Based on an invited paper presented at the 1975 USNC/URSI Meeting, Boulder, Colorado.

EARLY HISTORY

There is little doubt that the reflector antenna was born in the year 1888 in the laboratory of Heinrich Hertz, who experimentally demonstrated the existence of the electromagnetic waves that had been predicted theoretically by James Clerk Maxwell some fifteen years earlier. In his experiments Hertz used a cylindrical parabolic mirror of zinc, like that in Figure 2, with a spark-gap excited dipole placed on the focal line. A similar dipole-mirror combination served as a receiver and Hertz was able to demonstrate the generation, propagation, reception, and detection of electromagnetic radiation at a wavelength of 66 cm. His work stimulated a host of scientists toward further investigation; among them we find familiar names, Lodge, Fleming, Marconi, and Michelson, as well as less familiar ones, Trouton, Righi, Bose, and Branly. A fascinating account of the clever researches of these Hertzians in the period up to 1900 has been given by *Ramsey* [1958], who estimated that Hertz's 1.2 by 2 m reflector generated a radiation pattern having half-power beamwidths of about 80° by 35°. Realiz-

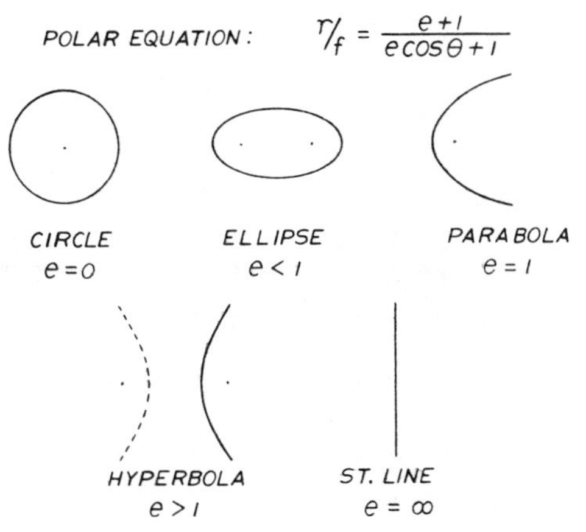

Fig. 1. The conic sections.

ing that almost one half of the dipole's radiated energy was uncollimated, we can infer that the gain could not have exceeded 8.5 db, corresponding to an aperture efficiency of 10%.

Microwave optics, a burgeoning field today, was born in the Hertzian era; indeed, the term "quasi-optical" was coined by Lodge to describe many of the experimental techniques in use. Those early experimenters, working with what we would regard as the crudest of equipment, displayed remarkable ingenuity in devising wire-grid polarizers, "cut-off" metallic gratings, artificial dielectrics and even dual-prism directional couplers. They investigated such phenomena as total internal reflection, double refraction and the Brewster angle at wavelengths of 10 and 3 cm. Ramsey gives an absorbing account

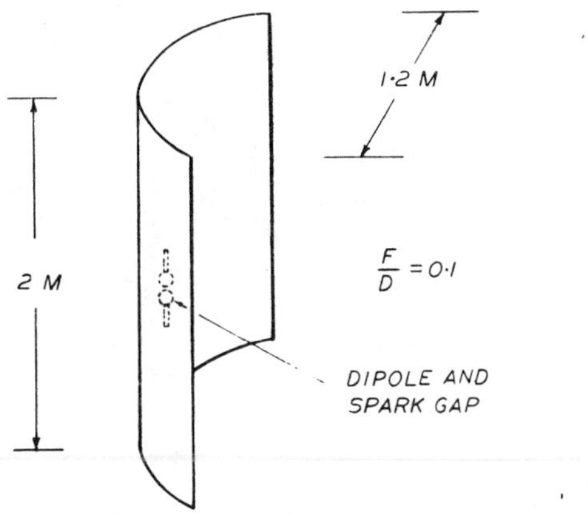

Fig. 2. The first reflector antenna: Hertz's cylindrical parabola, 1888.

of a lecture-demonstration by the Indian physicist J. Chunder Bose, given at the Royal Institution in London in 1897. He set up a microwave spectrometer that made use of the first horn antenna (he called it a collecting funnel) and that also incorporated plane and cylindrical mirrors, a dielectric prism, and, following Lodge's use of hollow pipes, a hollow waveguide radiator. All this at wavelengths down to 5 mm! With the apparatus he obtained the value 1.734 for the refractive index of sulphur, and he investigated artificial dielectrics, even going so far as to create macroscopic models of left-handed and right-handed sugar molecules by using twisted fibers of jute to produce rotation of the plane of polarization.

THE PERIOD FROM 1900 TO WORLD WAR II

The year 1900 marked the beginning of an eclipse in microwave optical researches. The waning interest in this field was replaced by a waxing interest in much lower frequencies, stimulated by Marconi's experiments which showed that long wavelengths could be used for long distance communication. Consequently, nothing of interest or importance in reflector development came about between 1900 and 1930. Indeed, as late as the 1940s, *Terman* [1943] wrote that "There is almost no information published in English on parabolic antennas." He did, however, cite a few references to German and French work of the 1930s. It was in that decade that two events of historical importance occurred. The first was the discovery, in 1931 by *Jansky* [1933], of extraterrestrial radio emission, marking the genesis of the science of radio astronomy. The second occurred in 1937 when the first large paraboloidal reflector was constructed and used as a radio telescope antenna by *Reber* [1944]. The reflector was 9.6 m in diameter and the wavelength 1.9 m. But apart from Reber's observations, carried on until 1944, radio astronomy remained a dormant science that was not revived until after the end of the war.

THE WORLD WAR II YEARS

The war, of course, was responsible for a vast and intensive development effort in all branches of electronics in general and in microwave physics in particular. The exigent demands of radar led to the perfection of the klystron oscillator and the invention of the high power magnetron and, of course, stimulated a concentrated effort to develop a coherent and unified theory of microwave antennas (as opposed to lower frequency, wire antennas) founded on a solid base of physical optics and electromagnetic theory. The results of that effort were eminently successful, we know, but the information that was generated, coming as it did from many different laboratories, both government and industrial in the USA and the Commonwealth nations, was necessarily scattered and of a clandestinely restricted nature. Fortunately we do not have to search the musty report files of these laboratories, for the important information was published in the postwar years in that marvelous distillation called the Radiation Laboratory Series, compiled by the Massachusetts Institute of Technology. The subject of microwave antennas was well covered in volume 12, the now classic text by *Silver* [1949], in which we find references to the contributions of L. J. Chu, E. U. Condon, C. C. Cutler, R. C. Spencer, and L. C. Van Atta, to name but a few of the American workers.

THE POSTWAR RISE OF RADIO ASTRONOMY

The immediate postwar period saw a rebirth of radio astronomy, for many former wartime radar scientists now had time to inquire into the intriguing discoveries of Jansky and Reber and to pursue some of the investigations that had been denied them during the war. A number of laboratories and radio telescopes were set up, notably at Jodrell Bank in England and in Sydney, Australia, where the author had the privilege of working under Dr. E. G. Bowen and the late Dr. J. L. Pawsey at the Radiophysics Laboratory of the Commonwealth Scientific and Industrial Research Organization. It was at Jodrell Bank that the first really large reflectors were built and put into use. The first, in 1947, was the transit telescope, a fixed, zenith pointing, paraboloid of diameter 218 feet and focal length 126 feet. The reflecting surface was a grid of unidirectional wires, 8 inches apart, supported by 24 radial steel cables that passed over 23.3-foot-high steel posts at the rim, and constrained to approximate a parabola by tie-downs at 27.3-foot intervals. The feed consisted of a pair of folded dipoles with reflectors, supported by an inclined mast. Limited beam steering could be achieved by pulling on the mast's guy wires so as to displace the feed from the focal point. The gain, as reported by *Brown and Lovell* [1957], was a little more than 31 db at 100 MHz; leakage through the wire reflector caused about 1 db of loss.

The success of this instrument demonstrated the great value of a large collecting aperture and, at the same time, pointed up the need for full steerability. Thus it became the forerunner of the first truly large, fully steerable reflector antenna, the 250-foot-diameter paraboloid at Jodrell Bank that was begun in 1952 and completed in 1957. It is a "deep dish" reflector, having a focal ratio of 0.25, that is provided with an altazimuth mount of the wheel and track variety. An excellent account of the radio astronomical considerations underlying its design is given in the book by *Brown and Lovell* [1957]. Not surprisingly, they failed to anticipate one of its major uses, that of tracking Russian launched satellites, a function which it performed nobly for a good many years.

THE TWENTY YEARS FROM 1945–1965

Clearly, the reflector art has not grown by quantum leaps, but rather in a methodical, progressive way and so it is that in the two decades following World War II we can do no more than point to a few highlights. Thus, for example, tolerance theory probably had its beginning in the work of *Spencer* [1949], who investigated the effects of aperture phase errors on the gain. A definitive treatment of reflector surface roughness effects on gain was first given by *Ruze* [1952] and Figure 3 displays one important result. It shows that for

$$G = \eta\left[\frac{\pi D}{\lambda}\right]^2 \cdot \exp\left[-\left(\frac{4\pi\sigma}{D}\right)^2 \cdot \left(\frac{D}{\lambda}\right)^2\right]$$

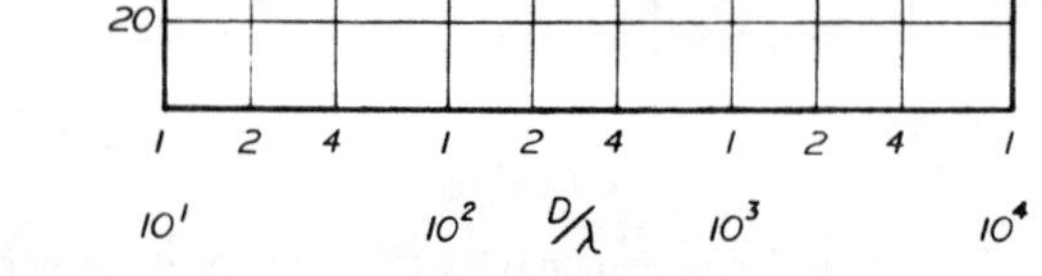

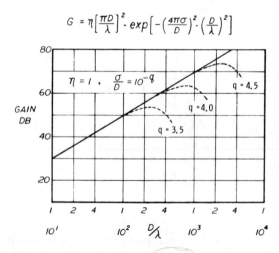

Fig. 3. The effect of reflector roughness on antenna gain.

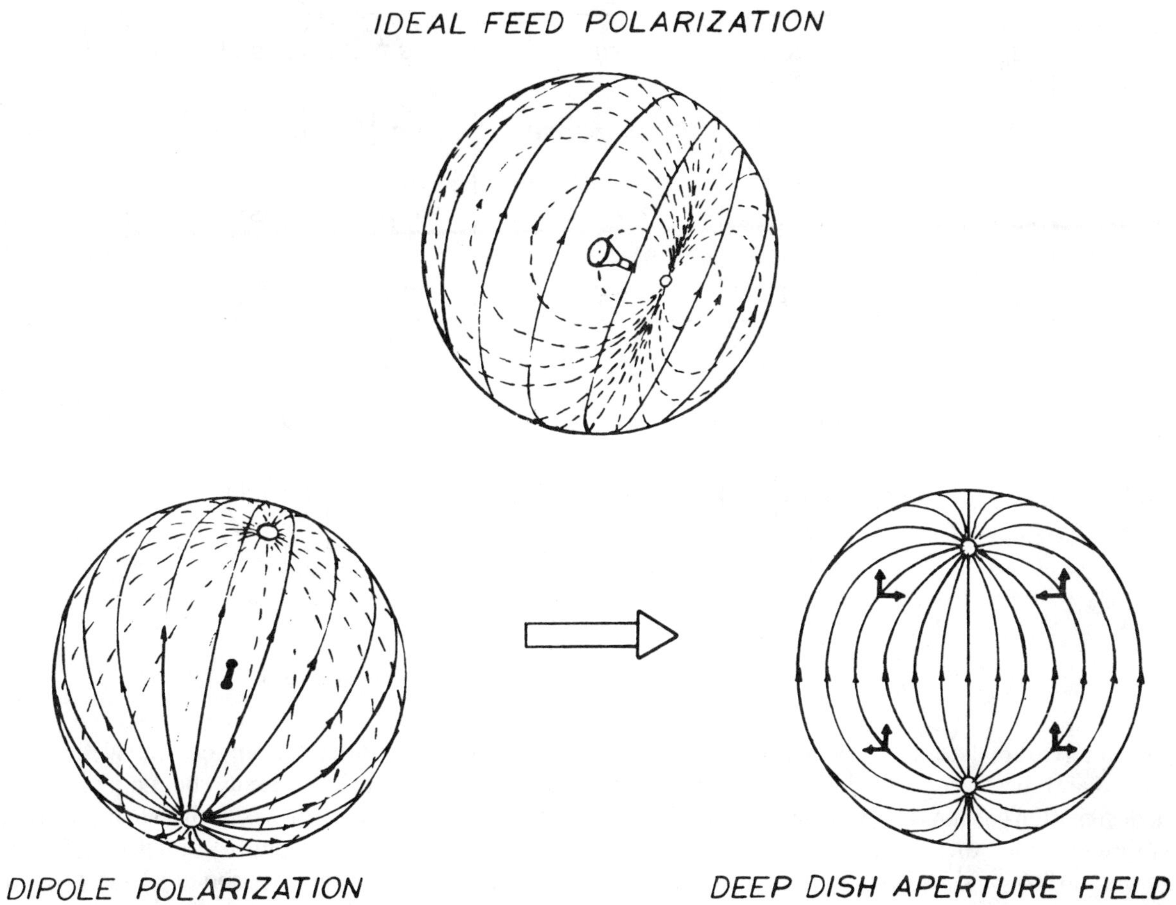

Fig. 4. Polarization characteristics of reflector antennas.

any reflector antenna there is a wavelength at which the gain reaches a maximum. This wavelength depends on σ, the rms deviation of the reflector surface from the ideal paraboloid,

$$\lambda_{max} = 4\pi\sigma \quad (2)$$

provided that the roughness is randomly distributed in a gaussian fashion, with correlation interval large compared to the wavelength. With aperture efficiency η the gain is

$$G = \eta\,(\pi D/\lambda)^2 \exp\left[-(4\pi\sigma/\lambda)^2\right] \quad (3)$$

When (2) is substituted into (3) the maximum gain is given by

$$G_{max} = 20q - 16.3 + 10\log_{10}\eta \quad \text{db} \quad (4)$$

where q is an index of smoothness, defined by

$$\sigma/D = 10^{-q} \quad (5)$$

D being the reflectors' aperture diameter. We shall have more to say about the index q a little later. Additional important contributions to tolerance theory were made by *Robieux* [1956], *Bracewell* [1961], and more recently by *Vu* [1969].

One of the first published papers dealing with the polarization characteristics of reflector antennas was that of *Cutler* [1947] who showed qualitatively that the ideal feed should radiate a spherical wave with the polarization characteristics shown in the upper part of Figure 4, if the reflector as a whole is to radiate with linear polarization in a fixed direction. He examined the polarization characteristics of the then-popular dipole feed (lower left of the figure) and showed that it would give rise to cross-polarized components in the reflected field, as sketched at the lower right of Figure 4. He further observed that the cross-polarized aperture field components would give rise to four minor lobes

E_θ	$A\cos\theta\cos\phi$	$B\cos\phi$	$\cos\phi[B + A\cos\theta]$
E_ϕ	$A\sin\phi$	$B\sin\phi\cos\theta$	$\sin\phi[A + B\cos\theta]$
Z_0H_θ	$A\sin\phi$	$B\sin\phi\cos\theta$	$\sin\phi[A + B\cos\theta]$
Z_0H_ϕ	$A\cos\theta\cos\phi$	$B\cos\phi$	$\cos\phi[B + A\cos\theta]$

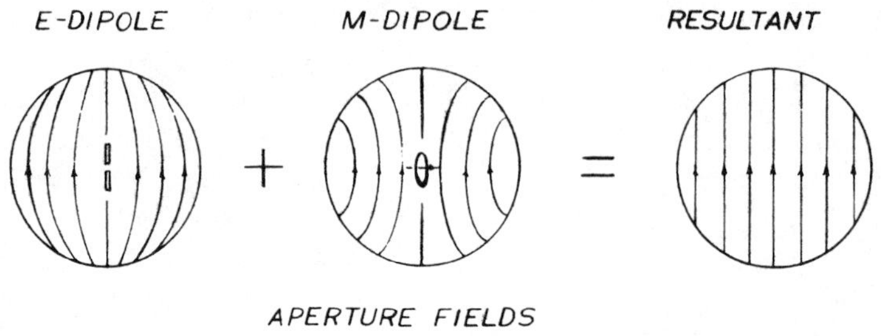

Fig. 5. The Huygen's source with ideal feed polarization characteristics.

in the intercardinal planes. For many years these cross-polar lobes were called Condon lobes, but that designation has now almost disappeared. Figure 5 is drawn from the work of *Jones* [1954] and *Koffman* [1966] and shows how an electric dipole and a magnetic dipole may be combined to produce a unidirectional Huygens source having ideal polarization characteristics for feeding a paraboloid. This occurs when the dipole strengths are such that $A = B$ (see Figure 5), in which case all field components vary in exactly the same way with the polar angle θ, namely, as $\cos^2(\theta/2)$. It is interesting, as Koffman noted, that a combination of crossed electric and magnetic dipoles can be adjusted to eliminate cross-polarized radiation in any reflector formed as a surface of revolution from a conic section. It is only necessary to impose the condition $A = eB$, where e is the eccentricity of the conic section and A and B are the electric and magnetic dipole strengths.

No discussion of reflector antennas would be complete without a reference to that family which is derived from the 17th century optical telescope devised by the Abbé Cassegrain, including variants due to Gregory and to Newton. Cassegrain antennas were being experimentally developed at least as early as the mid 1950s, and the whole family has been well described by *Hannan* [1961], who appears to have introduced the concept of equivalent parabolas. This concept can easily be understood by reference to Figure 6, in which the virtual and real foci are indicated by F_V and F_R, respectively. The two focal lengths of the hyperboloidal sub-reflector are f_1 and f_2, and their ratio determines the magnification, m, of the system. Hannan showed that the equivalent parabola has a focal length that is m times that of the primary reflector. Two-reflector antenna systems have been analyzed from the point

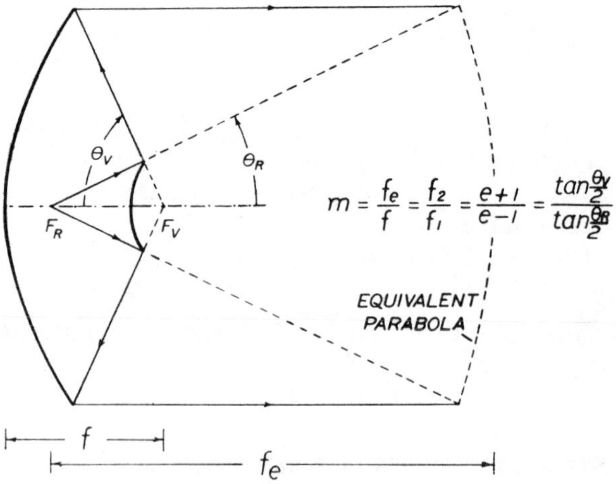

Fig. 6. The Cassegrain reflector antenna.

of view of geometrical optics by *Kinber* [1962], while *Morgan* [1964] has considered generalized Cassegrain systems. In a relatively recent paper [*Gniss and Ries*, 1970], a warning has been issued concerning the danger in using the equivalent-parabola concept when the feed is displaced, either axially or transversely, from the Cassegrain focus. Cassegrain systems suffer more from aperture blocking than do prime focus fed systems because the sub-reflector blocks the plane wave emanating from the main reflector while the feed horn blocks the spherical wave coming from the sub-reflector. For this reason Cassegrain antennas are usually considered only for those applications requiring a half-power beamwidth of less than about one degree.

THE PAST DECADE: MAXIMIZING THE GAIN

About 1960 the need arose to track space probes; along with the development of communications satellites this meant that large antennas were no longer the exclusive province of radio astronomers. 26 and 64 m diameter reflectors were built for the Goldstone deep-space communications facility, the latter being dedicated in 1966, while military and commercial satellite communications ground stations using 18 to 30 m diameter dishes became manifold. Such large antennas were, and are, costly. It was estimated [*Potter et al.*, 1966] that the cost of a single large reflector antenna was proportional to its diameter raised to the power 2.78. Thus it became imperative to increase the aperture efficiency and to reduce feed spillover so as to maximize the gain/temperature ratio (G/T) for these large, expensive structures. Effort in this direction began on several fronts early in the 1960s. *Galindo* [1963] showed that with a two-reflector system it is possible to achieve arbitrary phase and amplitude distributions over the main aperture. The uniform phase, constant amplitude case that leads to maximum aperture efficiency is, of course, included as a special case. *Williams* [1965] has described a modified Cassegrain system to which this approach was elegantly applied, leading to an improvement of about 25% in aperture efficiency. Its operation can be explained quite simply in terms of ray optics. Starting with a conventional Cassegrain system with prescribed feed pattern, the hyperboloidal sub-reflector's surface is deformed in such a way as to increase the ray density radially outward from the axis. When properly done a good approximation to a uniform amplitude distribution across the main aperture is realized, but only at the expense of a nonuniform phase distribution. However, the phase error can be corrected by a relatively minor change in the main reflector contour without significantly affecting the amplitude taper.

By 1960 the horn had, for the most part, replaced the dipole as a reflector feed. Nevertheless the dominant mode horn's characteristics are far from ideal for this purpose, chiefly because its principal E and H plane radiation patterns are quite different. *Potter* [1963] devised a clever solution to this problem in the case of the TE_{11} mode-excited conical horn. He noted that a step discontinuity in diameter near the throat of the horn would cause some of the dominant mode to be converted to the higher-order TM_{11} mode. He further showed that the correct combination of these two modes at the horn aperture would lead to a radiation pattern having almost identical patterns in the E and H planes, and that the normally high E plane sidelobes would be suppressed to a very low level. The lower part of Figure 7 indicates how the electric fields of the TE_{11} and TM_{11} modes combine in the aperture, while the upper portion shows the far field E plane radiation patterns of the individual modes and of the combined modes. *Jensen* [1963] applied the concept to a square pyramidal horn, in which the dominant mode is the TE_{10}, and showed that the correct higher mode in this case is a hybrid mixture of TE_{12} and TM_{12} modes. *Cohn* [1970] then described a method of generating and controlling these hybrid modes by changing the flare angle at appropriate points in the horn. *Potter and Ludwig* [1963] extended the concept to include additional higher modes (TE_{12}, TE_{13}, and TM_{12}) in the conical horn for purposes of beam shaping and showed how to obtain a feed pattern that more nearly approximates a uniform illumination over a reflector aperture. Multimode horns are not broadband devices, however, because the various modes propagate with different velocities. Hence a change in frequency disturbs the phase relationships between the modes at the aperture plane.

This difficulty is overcome in the corrugated horn which seems to have been conceived in the USA and in Australia almost at the same time, about 1964. In this horn, grooves are cut to a depth of between $\lambda/4$ and $\lambda/2$ in the horn walls and their effect is to cause the tangential magnetic field to

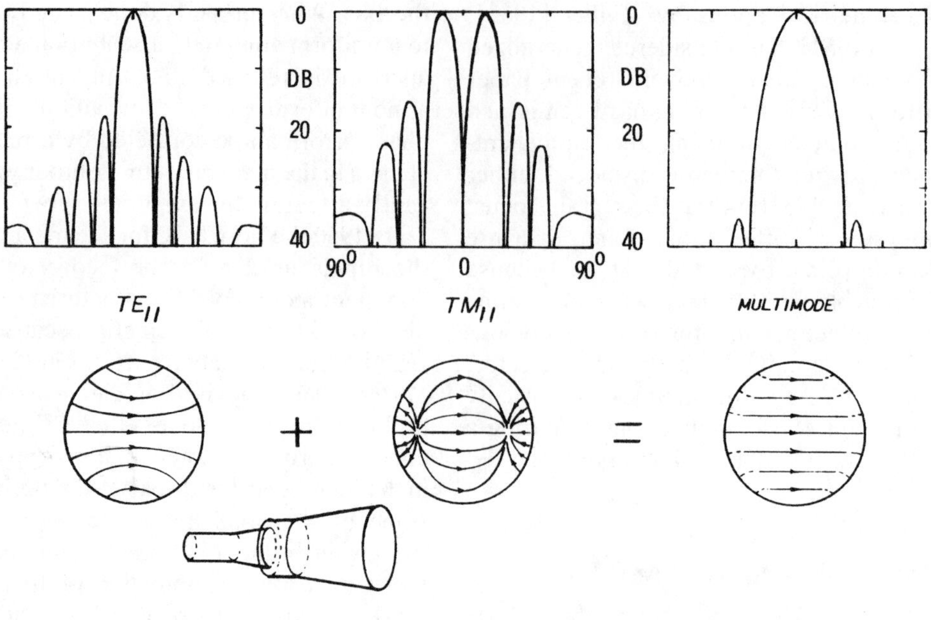

Fig. 7. The dual-mode conical horn.

vanish, or at least to assume a small value, on the walls. With essentially identical boundary conditions on both E and H, a hybrid mode called HE_{11} is created that consists of TE_{11} and TM_{11} components that travel with the same velocity. In the USA, *Kay* [1964] used grooved walls in a wide flare angle horn and called it the scalar feed because its properties were largely independent of polarization. A sketch of this feed horn is shown at the left of Figure 8. The scalar horn, due to the large flare angle, has a phase error in the aperture that exceeds 180°. For this reason it should be thought of as propagating a spherical wave which continues to expand after it leaves the aperture; its phase center is in the throat.

The events that led Australian workers to devise the corrugated feed horn have been described in a short note by *Minnett and Thomas* [1972]; an analytical description of the synthesis of the hybrid HE_{11} mode was given in an earlier paper by the same authors [*Minnett and Thomas*, 1966]. Beam-shaping, as with the multimode horn, can be achieved by adding higher-order hybrid modes (HE_{12}, etc.) but only at the expense of increased frequency sensitivity and loss in bandwidth. A corrugated waveguide feed is sketched at the right in Figure 8 in which the HE_{11} and higher-order hybrid modes generated at the step can be supported. Because it radiates a pattern with the polarization properties of a Huygens source it results in a very low level of cross-polar radiation in reflector antennas. The striking success of this horn as a feed has been an important factor leading to improved performance in reflector systems and has inspired a great deal of investigation in a number of countries. We would be remiss if we did not mention also the contributions of *Rumsey* [1966] and *Lawrie and Peters* [1966] in the USA, *Jeuken* [1969] in the Netherlands, *Narasimhan and Rao*, [1970] in India, and *Clarricoats and Saha* [1971] in the UK.

OTHER FORMS OF REFLECTOR ANTENNAS

To this point our discussions have been concerned entirely with paraboloidal and hyperboloidal reflec-

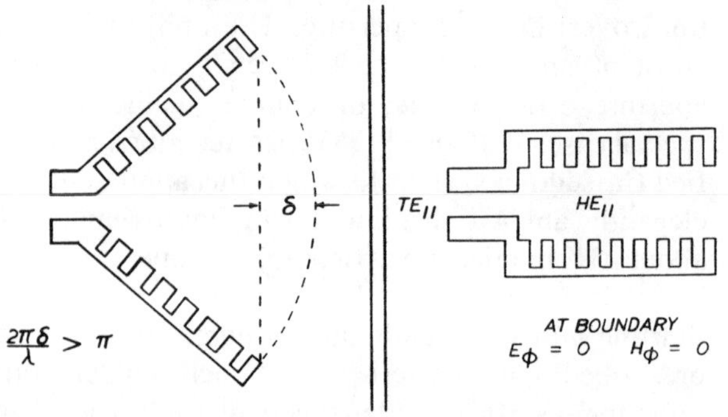

Fig. 8. The scalar horn and the corrugated feed.

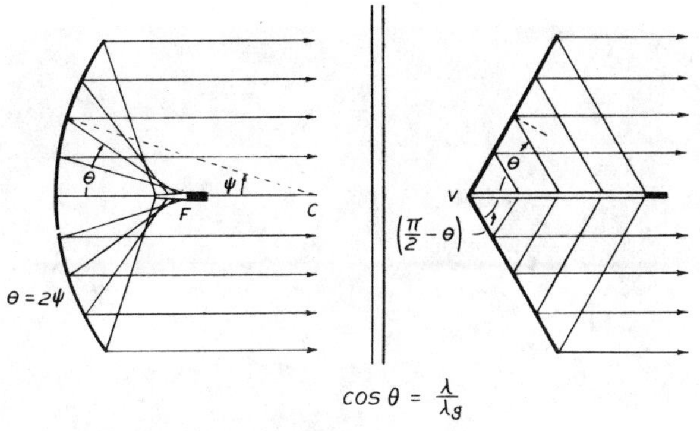

Fig. 9. Spherical and conical reflectors with line source feeds.

tors, representative of only two of the five conic sections. It is hardly necessary to remark that plane mirrors have found great utility in the antenna art, and that ellipsoidal mirrors are used whenever the need arises to focus the antenna's radiation pattern to a small region in the Fresnel zone, where the range is appreciably less than D^2/λ. Even this does not exhaust the useful possibilities of the conic sections, for reflector antennas can also be made that utilize the sphere, and even the cone itself, as is indicated in Figure 9. The conical reflector at the right is shown fed by a line source in which radiation is everywhere emitted at a constant angle θ to the axis. If the line source is a leaky waveguide then this condition will obtain when the propagation constant in the guide is a constant, given by

$$\beta = k \cos \theta \tag{6}$$

The geometry of the cone has obvious structural advantages in space applications when it is necessary to deploy a large antenna. A five-foot-diameter conical reflector with a waveguide line source feed has been successfully built and tested at 9 GHz (J. J. Gustincic, personal communication, 1969). The approach has two shortcomings: bandwidth is limited because the radiation angle θ is frequency dependent, and ohmic loss may be excessive due to the length of the waveguide.

The way in which a line source may be used to correct spherical aberration when used with a spherical reflector is illustrated at the left in Figure 9. Again, a leaky waveguide may be used, but in this case the radiation angle θ must vary along the guide as indicated in the figure, in accordance with the relation $\theta = 2\psi$ where ψ is measured at the center of the sphere, as shown. Equation (6) then shows that the propagation constant must vary in a prescribed manner with distance along the waveguide. This can be done, for example, by appropriate tapering of the cross-sectional dimensions of the guide. The advantage of the spherical mirror is that beam scanning may be performed with a stationary reflector by moving the feed, a virtue that first seems to have been pointed out by *Ashmead and Pippard* [1946]. They proposed to avoid the problem of spherical aberration by confining illumination to the paraxial zone of the sphere, that is, to a region over which the sphere does not differ from a paraboloid by more than one eighth of a wavelength. This practice, however, leads to a large value of F/D and to poor area utilization. That a line source could be used to correct the spherical aberration was first suggested by *Spencer et al.* [1949]. This approach has been used successfully in the NAIC spherical reflector at Arecibo, Puerto Rico, operated by Cornell University [*LaLonde and Harris*, 1970; *LaLonde and Love*, 1972].

SOME NOTABLE MODERN REFLECTOR ANTENNAS

The Arecibo antenna is notable because it is the world's largest reflector: its mirror is in the form of a 70° cap of a sphere whose radius is 870 feet, and the resulting aperture diameter is 1000 feet, or 305 meters. Figure 10 is a photograph taken after a program to upgrade the antenna was completed in 1974 and it clearly shows the new surface, comprised of 38,778 thin sheet aluminum panels that now replace the original open wire-mesh reflector. The deviation from the ideal sphere has been held to less than 6 mm rms (a q index of 4.7), no mean feat when it is realized that the total reflecting area of the mirror is 19.8 acres. A continuous improvement program is being undertaken, utilizing laser surveying, with the goal of reducing surface roughness to 3 mm rms [*LaLonde*, 1974]. Figure 11 is a photograph that shows the scanning platform, carriage houses, and an assortment of feeds illustrative of the great versatility of this instrument. The feed in the foreground is a leaky cylindrical waveguide 96.6 feet long whose scale model development has been described by *Love* [1973]. It illuminates the full aperture at 430 MHz with 70% efficiency in the upgraded reflector. Recently a similar, but shorter, feed has been put into service at 1415 MHz that illuminates 700 feet of the aperture (L. M. LaLonde, personal communication, 1975), while a similar one is being designed

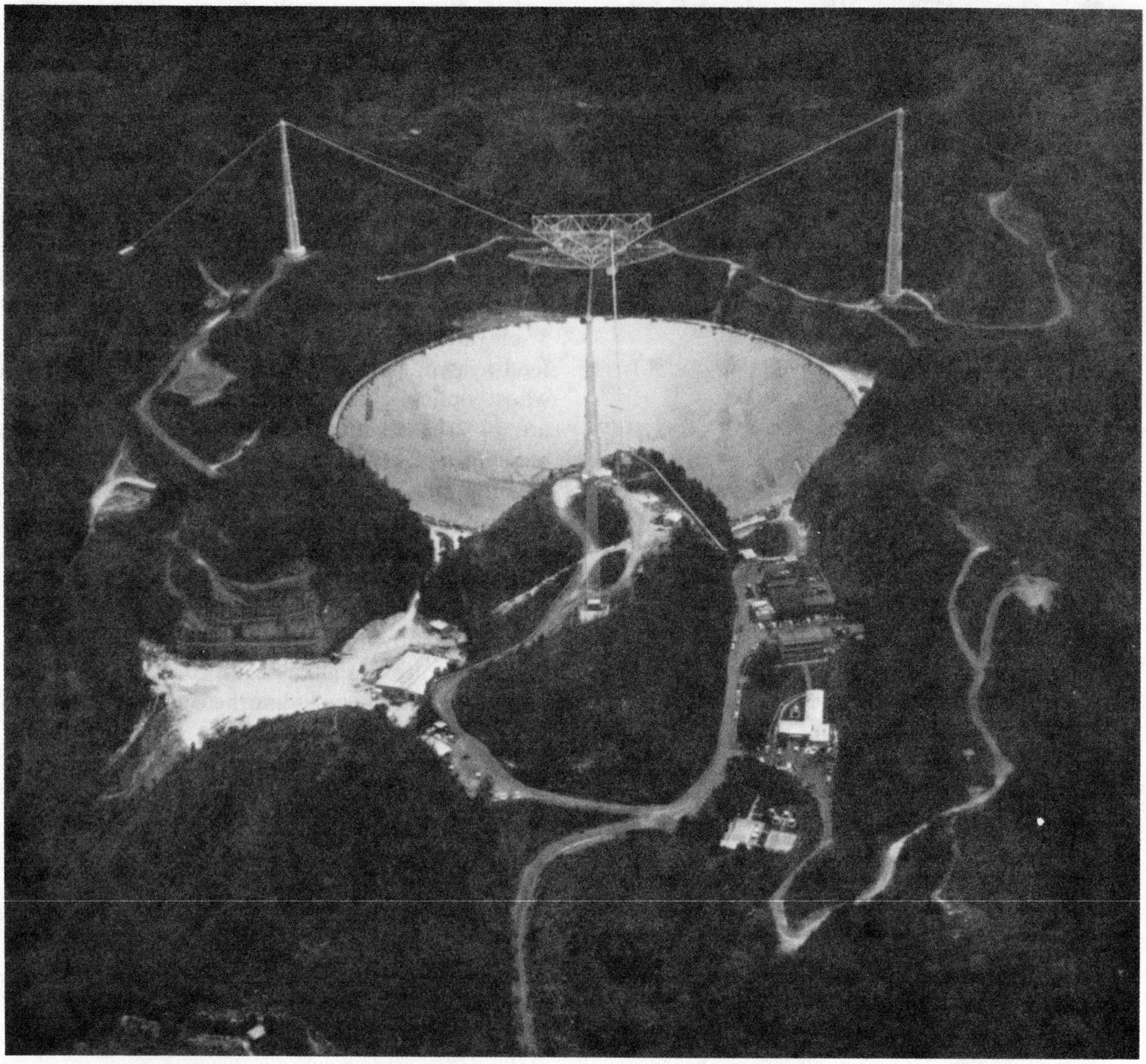

Fig. 10. The Arecibo 1000-foot spherical reflector.

to operate at 2380 MHz. The 1415 MHz feed realizes a gain of 67.8 db and can scan 11° away from the zenith in any direction without vignetting. It is interesting that the fixed reflector in this huge antenna is supported by cables and held to the proper contour by tensioned tie-downs in a manner not very different from the way in which the early transit telescope at Jodrell Bank was constructed.

Reflector antennas that would dwarf even the Arecibo instrument have been studied for use in the gravity-free environment of space. To date, however, the largest successful spaceborne reflector antenna is the Lockheed Missiles and Space Company's 30-foot-diameter "wrap-rib" reflector carried by the Applications Technology Satellite, ATS-6. This state-of-the-art unfurlable reflector is shown in its fully deployed state in Figure 12. It consists of three major sections: a hub, a set of 48 ribs, and a total of 48 reflective mesh panels. For launch purposes the ribs and mesh are wrapped around the hub, forming a very compact and tightly packed configuration. The 48 radial ribs are attached to the hub by means of hinges and are precision machined to conform to the desired paraboloid.

Fig. 11. Arecibo scan platform and feeds.

A copper-plated Dacron mesh is stretched between the ribs to create the reflective surface. In the stowed position the ribs are rotated on their hinge pins and are then tangentially wrapped around the hub, just as a flexible steel tape is stored on its spool; the mesh is carefully folded and packed between the ribs. A series of overlapping doors encloses the package and these are held in place by a steel cable. To unfurl the antenna a pyrotechnic device is used to cut the restraining steel cable. The spring-loaded doors then open and the stored energy in the wrapped ribs causes them to unwrap and to stretch the reflective mesh into position. This sequence takes only about two seconds in vacuum; the whole assembly weighs 60 kg. The mesh panels between the ribs depart somewhat from the desired paraboloidal surface, but the deviation is not more than about 0.8 mm. The measured gain of 55 db at 8.25 GHz indeed qualifies the ATS-6 reflector as an electrically large antenna.

Cost-effectiveness in nonmilitary satellite communications systems dictates the use of a relatively simple satellite antenna whose gain is limited by the need to provide a prescribed area coverage. High gain is therefore required of the antennas at the earth station terminals, and reflectors of up to 32 m in diameter are used. A good example of an antenna of this size is shown in Figure 13, through the courtesy of E-Systems, Inc., Dallas, Texas. It is part of the INTELSAT network, located in Portugal, and it operates in the 3.7 to 4.2 GHz downlink and 5.9 to 6.4 GHz uplink communications bands. When equipped with a cryogenically cooled, low-noise receiver this antenna achieves a gain/temperature ratio of 43 db at 4.0 GHz and 5° elevation angle. It employs a beam waveguide feed system [*Mizusawa and Kitsuregawa*, 1972], whose optical characteristics are illustrated in Figure 14. The advantage of this system is that the receiver and cryogenics can be housed in a fixed location beneath the reflector without incurring excessively high line losses. The altazimuth mount can cover 0° to 90° in elevation and ±170° in azimuth, with velocities of 0.002° to 0.3°/sec and with acceleration of 0.3°/sec^2. The tracking accuracy is 0.016° rms in 30 mph winds.

It would be unthinkable to conclude this brief discussion without including the world's largest fully

Fig. 12. The ATS-6 unfurlable 30-foot reflector.

steerable reflector antenna. This, of course, is the 100 m diameter radio telescope of the Max Planck Institute for Radioastronomy at Effelsberg, West Germany [*Hachenberg et al.*, 1973]. Photographs of this marvelous structure are shown in Figures 15 and 16. Its mechanical design incorporates a unique new principle, called homologous deformation, whereby deliberate, but controlled, elastic deflection under gravity is designed into the massive steelwork that supports the reflecting surface. With changing elevation angle, the deformations that take place are always such as to create a series of new paraboloidal surfaces having slightly displaced focal points. At a given elevation angle a computer commands a corresponding displacement of the feed so that defocussing does not occur. That the concept is successful is attested by the fact that the measured gain at 2.8 cm wavelength (which is nearly 78 db) does not vary by more than 1% as elevation angle changes from 30° to 90°. Both prime focus and Gregorian feed systems are used and some work has been done at $\lambda = 1.2$ cm, where half-power beamwidth is about 42 seconds of arc! Beyond 85 m in aperture diameter the reflecting surface is a wire mesh with 6 mm grid. Even under worst conditions the overall roughness does not exceed 0.65 mm rms. Therefore the q index (equation 5) is 5.1 for the innermost 85 m aperture. From (2) and (4) the resulting maximum realizable gain is 83 db at a wavelength of 8.2 mm, assuming a feed having 50% aperture efficiency. Such an estimate for the full 100 m aperture would be meaningless because of leakage loss through the outer mesh portion of the reflector.

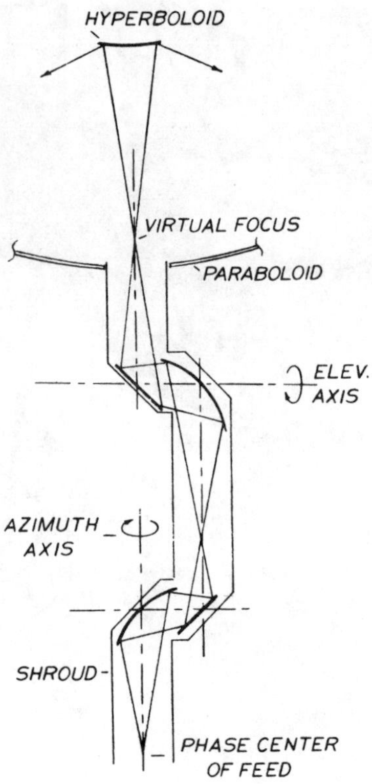

Fig. 14. Optics of beam waveguide feed system.

Fig. 13. 32-meter earth station antenna, Portugal.

CONCLUSION

This leads us to a concluding, brief comparison of the gains of a selected few of the largest extant reflector antennas. Figure 17 is a gain versus wavelength plot for five of these instruments that covers the frequency range from about 1 to 100 GHz. For the Effelsberg antenna the curve, at long wavelengths, corresponds to that for an aperture diameter of 100 m with a feed system yielding about 50% in aperture efficiency. At short wavelengths,

Fig. 15. The Effelsberg 100-meter radio telescope.

Fig. 16. The surface of the Effelsberg reflector.

however, the curve first begins to depart from linearity due to loss created by the outer mesh. Eventually the gain maximizes at about $\lambda = 1$ cm and then falls off in the manner appropriate to an aperture of 85 m with rms roughness of 0.65 mm. For the other cases shown in Figure 16 the dashed portions of the curves are extrapolations that have been carried beyond available experimental data by making use of (3). Thus, the Goldstone (California) reflector attains its maximum gain of about 78 db at around $\lambda = 2$ cm. The physically smaller reflectors at Krim (USSR) and at Kitt Peak (Arizona) reach their peak gains at much shorter wavelengths, approximately 3 and 2 mm, respectively.

Two curves are shown for the Arecibo spherical reflector. The one corresponding to the full 305 m (1000 foot) aperture is calculated for a feed efficiency of 70% and rms roughness of 6 mm. This curve is entirely dashed in Figure 17, indicating that it has been extrapolated with the aid of (3) beyond experimentally observed data at 430 and 611 MHz. The second curve is appropriate to the reduced aperture of 215 m (700 feet) for which the previously mentioned 1415 MHz line feed was designed. The measured gain of 68 db at this wavelength is indicated by the point enclosed within the square. If the current upgrading program is successful in reducing the rms roughness to 3 mm the effect will be to extend the linear portions of the two curves to peak gain levels that are 6 db

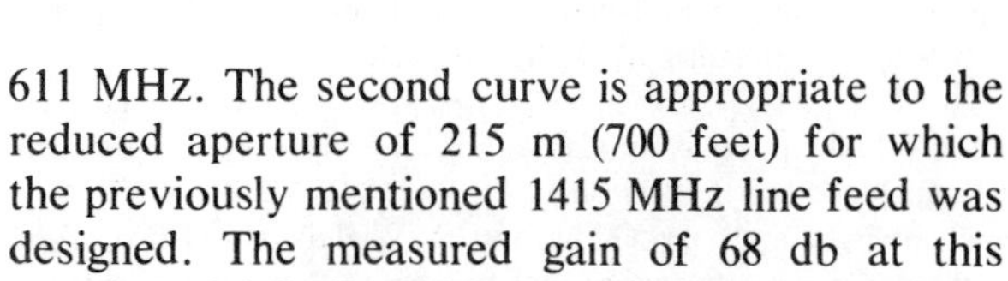

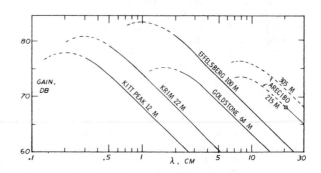

Fig. 17. Gains of some large reflector antennas.

higher than those shown in Figure 17. Whether any of this potential for increased gain can be realized is contingent upon the difficult task of creating efficient feed systems at higher frequencies.

Acknowledgments. The author is grateful to a number of persons for so promptly attending to his requests for photographs and information. His thanks go to L. M. LaLonde of Cornell University and the National Astronomy and Ionosphere Center; John B. Damonte, Lockheed Missiles and Space Company; George Moy, E-Systems, Inc., Garland Division; and finally R. Wielebinski, director of the Max Planck Institute for Radioastronomy, Bonn, FRG.

REFERENCES

Ashmead, J., and A. B. Pippard (1946), The use of spherical reflectors as microwave scanning aerials, *J. Inst. Elec. Eng., Part IIIA, 93*(4), 627.

Bracewell, R. N. (1961), Tolerance theory of large antennas, *IRE Trans. Antennas Propagat., AP-9*(1), 49–58.

Brown, R. H., and A. C. B. Lovell (1957), *The Exploration of Space by Radio,* 207 pp., John Wiley, New York.

Clarricoats, P. J. B., and P. K. Saha (1971), Propagation and radiation behaviour of corrugated feeds, *Proc. Inst. Elec. Eng., 118*(9), 1167–1186.

Cohn, S. B. (1970), Flare-angle changes in a horn as a means of pattern control, *Microwave J., 13*(10), 41–46.

Cutler, C. C. (1947), Parabolic antenna design for microwaves, *Proc. IRE, 35*(11), 1284–1294.

Galindo, V. (1963), Design of dual-reflector antennas with arbitrary phase and amplitude distributions, *IEEE Trans. Antennas Propagat., AP-12*(4), 403–408.

Gniss, H., and G. Ries (1970), Remarks on the concept of equivalent parabolas for Cassegrain antennas (in German with English abstract), *Electron. Lett., 6*(23), 737–739.

Hachenberg, O., B. H. Grahl, and R. Wielebinski (1973), The 100-meter radio telescope at Effelsberg, *Proc. IEEE, 61*(9), 1288–1295.

Hannan, P. W. (1961), Microwave antennas derived from the Cassegrain telescope, *IRE Trans. Antennas Propagat., AP-9*(2), 140–153.

Jansky, K. G. (1933), Electrical disturbances apparently of extraterrestrial origin, *Proc. IRE, 21*(10), 1387–1398.

Jensen, P. A. (1963), A low-noise multimode Cassegrain monopulse feed with polarization diversity, *NEREM Record 1963,* Vol. 5, pp. 94–95, Northeast Electronics Research and Engineering Meeting, Boston, Mass.

Jeuken, M. E. J. (1969), Experimental radiation pattern of the corrugated conical horn antenna with small flare angle, *Electron. Lett., 5*(20), 484–485.

Jones, E. M. T. (1954), Paraboloid reflector and hyperboloid lens antennas, *IRE Trans. Antennas Propagat., AP-2*(4), 119–127.

Kay, A. F. (1964), The scalar feed, *AFCRL Rep. No. 64-347 (AD No. 601609),* 42 pp., Air Force Cambridge Research Laboratories, Hanscom AFB, Mass. 01730.

Kinber, B. Ye. (1962), On two-reflector antennas, *Radio Eng. Electron. Phys.,* 7(6), 914–921.

Koffman, I. (1966), Feed polarization for parallel currents in reflectors generated by conic sections, *IEEE Trans. Antennas Propagat., AP-14*(1), 37–40.

LaLonde, L. M. (1974), The upgraded Arecibo Observatory, *Science, 186*(4160), 213–218.

LaLonde, L. M., and D. E. Harris (1970), A high-performance line source feed for the AIO spherical reflector, *IEEE Trans. Antennas Propagat., AP-18*(1), 41–48.

LaLonde, L. M., and A. W. Love (1972), A new high power dual polarized line feed for the Arecibo reflector, paper presented at the USNC/URSI Spring Meeting, Washington, DC.

Lawrie, R. E., and L. Peters, Jr. (1966) "Modifications of horn antennas for low sidelobe levels, *IEEE Trans. Antennas Propagat., AP-14*(5), 605–610.

Love, A. W. (1973), Scale model development of a high efficiency dual polarized line feed for the Arecibo spherical reflector, *IEEE Trans. Antennas Propagat., AP-21*(5), 628–639.

Minnett, H. C., and B. MacA. Thomas (1966), A method of synthesizing radiation patterns with axial symmetry, *IEEE Trans. Antennas Propagat., AP-14*(5), 654–656.

Minnett, H. C., and B. MacA. Thomas (1972), Propagation and radiation behaviour of corrugated feeds, *Proc. Inst. Elec. Eng., 119*(9), 1280.

Mizusawa, M., and T. Kitsuregawa (1972), A beam-waveguide feed having a symmetric beam for Cassegrain antennas, paper presented at the International IEEE G-AP Symposium, College of William and Mary, Williamsburg, Va.

Morgan, S. P. (1964), Some examples of generalized Cassegrainian and Gregorian antennas, *IEEE Trans. Antennas Propagat., AP-12*(6), 685–691.

Narasimhan, M. S., and B. V. Rao (1970), Hybrid modes in corrugated conical horns, *Electron. Lett., 6*(2), 32-34.

Potter, P. D. (1963), A new horn antenna with suppressed sidelobes and equal beamwidths, *Microwave J., 6*(6), 71–78.

Potter, P. D., and A. C. Ludwig (1963), Beamshaping by use of higher order modes in conical horns, *NEREM Record 1963,* Vol. 5, pp. 92–93, Northeast Electronics Research and Engineering Meeting, Boston, Mass.

Potter, P. D., W. D. Merrick, and A. C. Ludwig (1966), Big antenna systems for deep-space communications, *Astronautics and Aeronautics,* Oct., pp. 84–95.

Ramsey, J. F. (1958), Microwave antenna and waveguide techniques before 1900, *Proc. IRE, 46*(2), 405–415.

Reber, G. (1944), Cosmic static, *Astrophys. J., 100*(3), 279–287.

Robieux, J. (1956), Influence de la précision de fabrication d'une antenne sur ses performance, *Ann. Radio électricité, 11*(43), 29-56.

Rumsey, V. H. (1966), Horn antennas with uniform power patterns around their axes, *IEEE Trans. Antennas Propagat., AP-14*(5), 656–658.

Ruze, J. (1952), The effect of aperture errors on the antenna radiation pattern, *Nuovo Cimento Suppl., 9*(3), 364–380.

Silver, S. (1949), Microwave Antenna Theory and Design, 623 pp., McGraw-Hill, New York.

Spencer, R. C. (1949), A least square analysis of the effect of phase errors on antenna gain, AFCRC Rep. E5025, Air Force Cambridge Research Laboratories, Hanscom AFB, Mass. 01730.

Spencer, R. C., C. J. Sletten, and J. E. Walsh (1949), Correction of spherical aberration by a phased line source, in *Proceedings*

of the National Electronics Conference, Vol. 5, pp. 320–333.
Terman, J. E. (1943), *Radio Engineer's Handbook*, p. 837, McGraw Hill, New York.
Vu, T. B. (1969), Influence of correlation interval and illumination taper in antenna tolerance theory, *Proc. Inst. Elec. Eng.*, *116*(2), 195–202.
Williams, W. F. (1965), High efficiency antenna reflector, *Microwave J.*, *8*(7), 79–82.

Parabolic-Antenna Design for Microwaves*

C. C. CUTLER†, ASSOCIATE, I.R.E.

Summary—**This paper is intended to give fundamental relations and design criteria for parabolic radiators at microwave frequencies (i.e., wavelengths between 1 and 10 centimeters). The first part of the paper discusses the properties of the parabola which make it useful as a directional antenna, and the relation of phase polarization and amplitude of primary illumination to the over-all radiation characteristics. In the second part, the characteristics of practical feed systems for parabolic antennas are discussed.**

I. Introduction

The USE OF radio waves in the wavelength range between 1 and 10 centimeters has resulted in many innovations in directional-antenna design. In particular, it has become a common practice to focus microwave energy into a desired directional beam by the use of a metallic reflecting surface excited by radiation from a small, relatively nondirectional source. Where maximum directivity of the antenna is desired, the reflector shape is usually parabolic, with a primary source located at the focus and directed into the reflector area. The reflector may be a section of a surface formed by rotating a parabola about its axis (circular paraboloid), a parabolic cylinder, or a parabolic cylinder bounded by parallel conducting planes. Also, there is a choice of how much, and what part, of the parabolic curve is used for the reflector. It is the purpose of this paper to discuss the design of such antennas, particularly the paraboloid and associated wave-guide-feed radiators. In this discussion, where the subject matter applies to a general characteristic, the word parabola will refer to either a paraboloid, a parallel-plate parabola, or a parabolic cylinder. The word paraboloid will be used only where subject matter refers specifically to the circular paraboloid, which is the surface generated by rotating a parabolic curve about its axis.

II. Fundamental Relations

The following equations describe a parabolic curve and are useful in determining its properties. See Fig. 1 for the meaning of the symbols.

Cartesian co-ordinates:

$$y^2 = 4Fx. \tag{1}$$

Polar co-ordinates:

$$r = \frac{F}{\cos^2 \dfrac{\theta}{2}}. \tag{2}$$

Parametric equations:

$$y = 2F \tan \theta/2 \tag{3}$$

$$x = F \tan^2 \theta/2. \tag{4}$$

The properties of the parabola which make it particularly useful for focusing radiant energy into a directional beam are characterized by two ray considerations: First, any ray from the focus is reflected in a direction parallel to the axis of the parabola; and second, the distance traveled by any ray from the focus to the parabola and by reflection to a plane perpendicular to the parabola axis is independent of its path, and therefore such a plane represents a wave front of uniform phase.

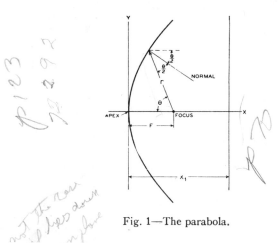

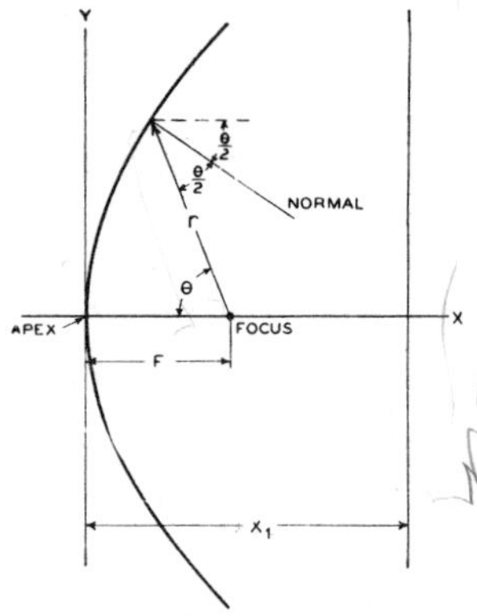

Fig. 1—The parabola.

The analysis of microwave antennas by consideration of geometrical rays may serve to give a crude picture, but generally it is necessary to use diffraction theory to obtain accurate results. In the discussion that follows both methods of attack are found useful.

III. Design Considerations

To obtain maximum efficiency from the paraboloid antenna requires a close control of amplitude, phase, and polarization of the field incident to the reflector. This puts rather strict requirements on the primary source of radiation, or the parabola "feed." In the first place, the feed must be small and of such configuration that it gives a spherical phase front; that is, from a distance it must appear as though the energy were radiated from a point. The amplitude of the radiation from the feed must be directed uniformly over a wide angle, to illuminate adequately the entire reflector area. Also, the field should be of such a nature that after reflection the waves will be properly polarized.

* Decimal classification: R326.8. Original manuscript received by the Institute, July 24, 1946; revised manuscript received, November 15, 1946.
† Bell Telephone Laboratories, Inc., New York, N. Y.

Reprinted from *Proc. IRE*, vol. 35, pp. 1284–1294, Nov. 1947.

A. Phase

The phase of the field radiated from an antenna depends on the electrical distance the wave has traveled to arrive at the point under consideration. This in itself is of no great significance, but if we measure the phase at all points in a field at a distance of several wavelengths from the source and connect points of equal phase we get a curve or surface representing the wave front from which we may draw certain conclusions. The direction of propagation of energy in the wave is perpendicular to the surfaces of constant phase. From one such surface we can project forward to find the destination of the wave, and we can project backward to locate the effective source and analyze its properties. On the basis of geometrical ray construction, we see that the deviation of such a surface from a sphere will cause a deviation of the wave front from a plane, after reflection from an ideal paraboloid. Similarly, we may find by projecting back that the apparent source is not a point, but is instead a line or some peculiar surface. Such an apparent source does not necessarily have a significant relation to the physical size and shape of the radiator, but it does give a basis for comparing various feeds, and often suggests methods of correction. If the phase front from a feed is not spherical, the phase in the aperture can be corrected by changing the shape of the reflector. For a small phase deviation Φ, the compensating correction to r of (2) is

$$\Delta r = \frac{\lambda}{2\pi} \frac{\Phi}{1 + \cos\theta} \qquad (5)$$

If the phase front is not spherical, or is not corrected for, the radiation pattern will be distorted and the gain reduced. The effect on the pattern depends upon a number of factors, so it is difficult to generalize. However, a widening of the main lobe at low levels, or a filling-in of the nulls between minor lobes, usually indicates deviations of phase.

It is a fallacy to attribute the limitation of directivity of a parabolic antenna to phase deviations because of the physical size of the feed. Both the half-wave doublet, with or without a reflector, and an open-ended wave guide give very good phase distributions in spite of their relatively large physical size. The ultimate limitation to the sharpness of the beam is the diffraction at the paraboloid aperture and is due to the limited size of the effective area in wavelengths.

B. Amplitude

To make effective use of the area of the paraboloidal reflector, the energy must be distributed over the surface with some degree of uniformity. However, it is important to avoid loss of energy by waves radiated from the feed which fail to strike the reflector. This energy is called "spill-over," and to obtain the optimum gain efficiency from a paraboloid it is necessary to design the combination of reflector and feed to compromise between the loss due to spill-over and the loss due to nonuniformity of illumination. There is a direct relationship between the directivity of the feed and the angle subtended at the focus by the paraboloid for optimum gain. Furthermore, if a circular section of a paraboloid is used, it is important that the feed should radiate energy with circular symmetry. With the assumption of circular symmetry of feed pattern and reflector, and ideal phase and polarization conditions, the relationship between the feed directivity and the subtended angle of the reflector may be determined as follows:

The amplitude of the field at a distant point on the paraboloid axis is the sum of the contributions from all elementary areas of a plane through the circular aperture of the reflector.

$$E_p \sim \int_0^{y_1} yV(y)dy \int_0^{2\pi} d\Psi = 2\pi \int_0^{y_1} yV(y)dy \qquad (6)$$

where

$V(y)$ = amplitude of the incident field at any point on the surface of the reflector

$= (1/F)\, U(\theta) \cos^2 \theta/2$ (from 2)

$U(\theta)$ = relative amplitude of field radiated from the feed where $U(\theta) = 1$ when $\theta = 0$

Ψ = the angle describing rotation about the axis

y_1 = radius of reflector.

Other symbols are indicated in Fig. 1. The power gain of the antenna is proportional to E_p^2.

$$C_a = \Omega \left[\int_0^{y_1} yU(\theta) \cos^2 \frac{\theta}{2}\, dy \right]^2 \qquad (7)$$

where the proportionality function Ω may be obtained by comparing this to the gain of a system consisting of a circular area illuminated by the same primary source at a great distance. For such a system the gain is

$$\lim_{y_1/F \to 0} G_a = G_h G_t \frac{\text{Area of circle of radius } y_1}{\text{Area of sphere of radius } F} \qquad (8)$$

where

G_h = gain of the primary source or feed illuminating the reflector

G_t = theoretical gain of uniformly excited circular area, which is[1]

$$G_t = \frac{4\pi A}{\lambda^2} = \left(\frac{2\pi y_1}{\lambda}\right)^2 \quad \text{if} \quad \frac{y_1}{\lambda} \gg 1. \qquad (9)$$

From (8),

$$\lim_{y_1/F \to 0} G_a = G_h \left(\frac{\pi y_1^2}{\lambda F}\right)^2. \qquad (10)$$

[1] J. C. Slater, "Microwave Transmission," McGraw-Hill Book Co., New York, N. Y., 1942; p. 260.

Also, from (7),

$$\lim_{y_1/F \to 0} G_a = \frac{1}{4} \Omega y_1^4. \tag{11}$$

Equating (10) and (11),

$$\Omega = 4G_h \left(\frac{\pi}{\lambda F} \right)^2 \tag{12}$$

and therefore, from (7),

$$G_a = 4G_h \left(\frac{\pi}{\lambda F} \right)^2 \left[\int_0^{y_1} y U(\theta) \cos^2 \frac{\theta}{2}\, dy \right]^2. \tag{13}$$

From (3),

$$y = 2F \tan \frac{\theta}{2}; \quad \text{and} \quad dy = \frac{F d\theta}{\cos^2 \frac{\theta}{2}} \tag{14}$$

and

$$G_a = 16G_h \left(\frac{\pi F}{\lambda} \right)^2 \left[\int_0^{\theta_1} U(\theta) \tan \frac{\theta}{2}\, d\theta \right]^2. \tag{15}$$

Now, G_h may be obtained from $U(\theta)$:

$$G_h = \frac{2}{\int_0^{\pi} [U(\theta)]^2 \sin \theta\, d\theta} \tag{16}$$

and

$$G_a = 32 \left(\frac{\pi F}{\lambda} \right)^2 \frac{\left[\int_0^{\theta_1} U(\theta) \tan \frac{\theta}{2}\, d\theta \right]^2}{\int_0^{\pi} [U(\theta)]^2 \sin \theta\, d\theta}. \tag{17}$$

The efficiency of a radiator is taken as the ratio of its gain to that of a uniformly illuminated aperture of the same area (see (9)).

$$\text{Efficiency} = 2 \cot^2 \frac{\theta_1}{2} \frac{\left[\int_0^{\theta_1} U(\theta) \tan \frac{\theta}{2}\, d\theta \right]^2}{\int_0^{\pi} [U(\theta)]^2 \sin \theta\, d\theta}. \tag{18}$$

With a given feed-radiation characteristic $U(\theta)$, the integrals can be evaluated by graphical methods, or by using a Fourier analysis, and the relationship of the subtended angle of the reflector to the efficiency can be obtained. Fig. 2 shows a typical plot of such a calculation for the radiation from a circular wave guide 0.84 wavelength in diameter whose radiation characteristic is shown in Fig. 3. The broad maximum of efficiency indicates that the subtended angle is not critical. It can be seen that the greatest efficiency is obtained with a reflector subtending an angle such that the radiation toward the edges is between 8 and 12 decibels below that at the center. In other words, the intensity of the energy radiated toward the edge of a parabolic reflector usually should be about one-tenth of the maximum intensity. It should be noted that the value "one-tenth" relates to the energy per unit solid angle in the primary

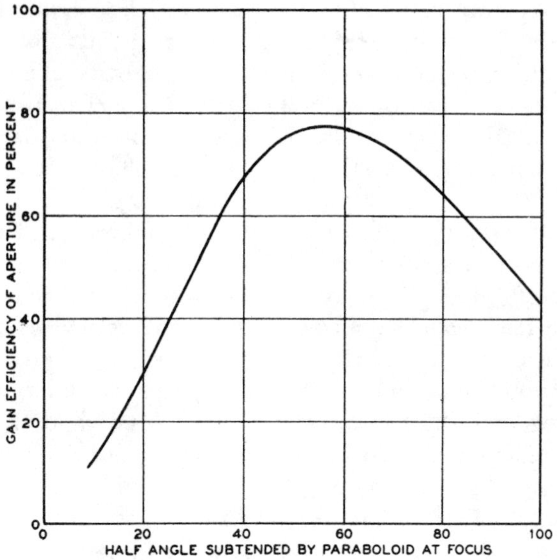

Fig. 2—Area efficiency of a paraboloid as a function of its proportions.

pattern, taken at a constant distance from the feed. The intensity at the edge of the reflector is further reduced because of the increased space attenuation in the longer path.

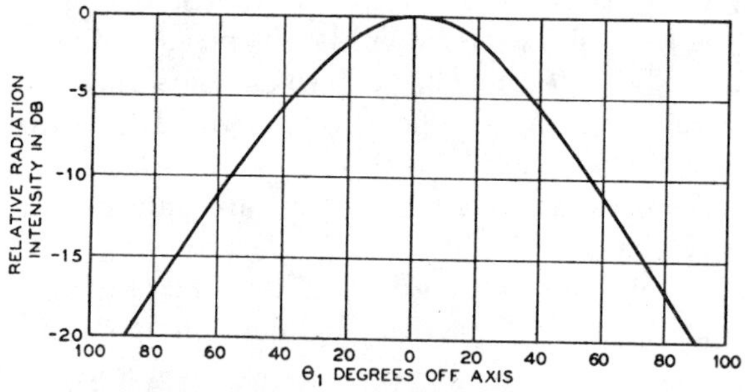

Fig. 3—Radiation pattern of 0.84-wavelength-diameter circular wave-guide aperture.

Calculations of the type given above indicate a theoretical gain efficiency of about 80 per cent for paraboloidal antennas, but because of defects in the phase and polarization characteristics and certain sources of interference to be discussed later, an efficiency much higher than 65 per cent is rarely obtained.

The effect of the amplitude distribution on the radiation pattern is very direct; but usually it is not an important factor in determining the desired illumination characteristic provided that sharp variations in intensity are avoided, and that the illumination is suitable from the point of view of gain. The diffraction pattern of a circular aperture, uniformly illuminated, has minor lobes 17 decibels below the maximum, and any tapering of illumination toward the edge of the aperture will reduce the lobes still further. A smooth reduction of intensity (of 10 decibels or more) towards the edge of the aperture results in a pattern with minor lobes 25 decibels or more below the major lobe. (This is usually less than minor lobes from other causes.) Where low side lobes are of paramount importance it may be desirable to use a deeper paraboloid, or, if beam sharpness is more important, a more shallow paraboloid than the gain criterion would indicate.

It should be noted here that the "spill-over" radiation mentioned above is the main source of wide-angle lobes from parabolic antennas, i.e., lobes at 90 degrees or further from the beam. The paraboloid is apt to be worse in this regard than other structures. This undesired radiation can be reduced (but only at the expense of gain or size) by using more directive feeds, or by extending the reflector sufficiently to intercept the energy.

C. Polarization

The radiation characteristic of the feed should be of such a nature that all the waves will be polarized in the same direction after reflection from the paraboloid surface. All field components which emerge from the aperture with polarization perpendicular to the average are wasted and contribute to minor-lobe radiation. This requirement will be satisfied if at any point the direction of polarization of the wave makes the same angle with a plane through the feed axis of symmetry and the point as does the average polarization of the feed. This may be seen by examination of such a wave after reflection, or, conversely, by imagining a polarized plane wave incident to the paraboloid, and analyzing the reflected wave over a sphere surrounding the focus. This required field is markedly different from that of a doublet, as may be seen by examining Figs. 4 and 5. The doublet has two poles (or points of indeterminate polarization and zero field strength) opposite each end of the antenna. The desired plane-polarized circularly symmetric feed has one pole directly behind the source. Lines indicating the desired polarization of the electric vector on a spherical surface around the feed, describe circles tangent to one another at a point on the surface of the sphere directly behind the feed. This specifies the polarization in all directions from the feed, but it is of significance only in the field radiated in the direction of the paraboloid surface.

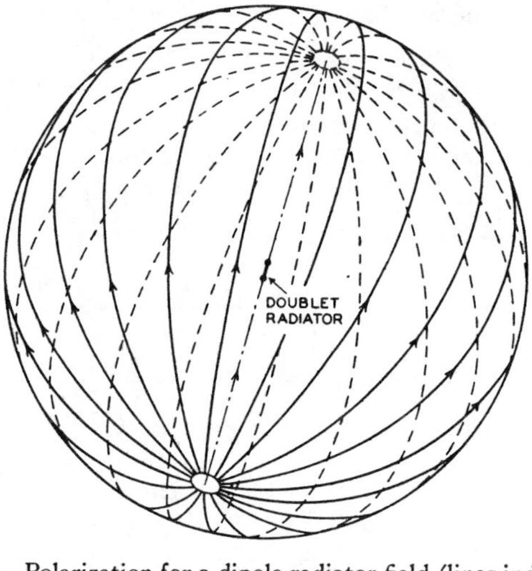

Fig. 4—Polarization for a dipole radiator field (lines indicate direction of E vector).

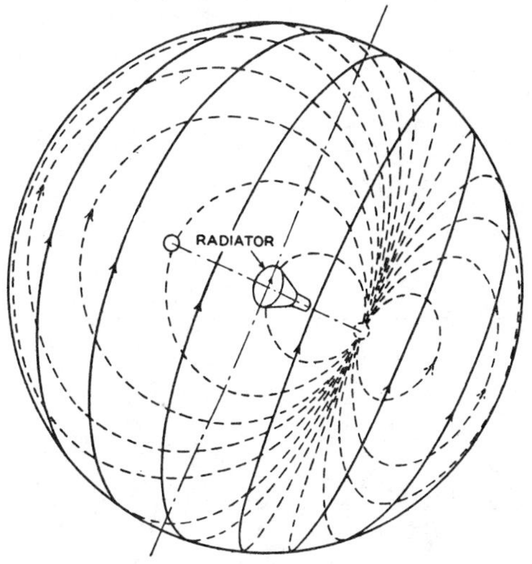

Fig. 5—Polarization of spherical wave front of field from an ideal paraboloidal feed (lines indicate direction of E vector).

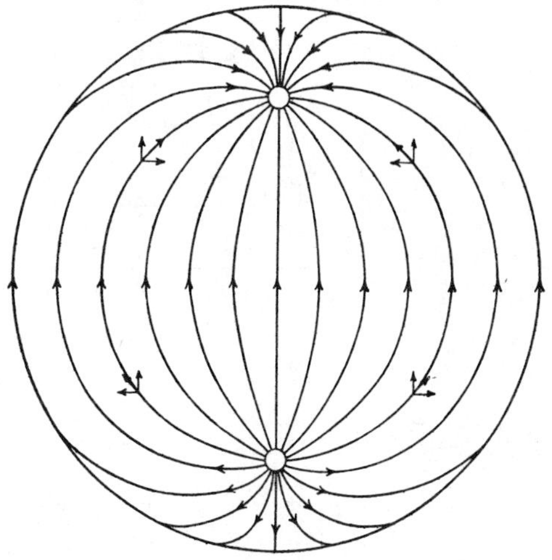

Fig. 6—Polarization of field in the aperture of a deep paraboloid fed from a dipole radiator.

If a feed having a poor polarization characteristic is used in a paraboloid, the resulting radiation pattern will contain regions where the polarization is perpendicular to that of the feed. Generally this energy is concentrated in four minor lobes, located in the quadrants between the plane of polarization and a perpendicular plane intersecting the axis of the paraboloid. For instance, consider a paraboloid excited by a feed having a polarization characteristic as shown in Fig. 4. After reflection from a deep paraboloid, the energy emerging through a plane across the aperture of the paraboloid will be polarized approximately as shown in Fig. 6. The component of field perpendicular to the feed polarization is called the "cross-polarized" field, and the resulting distant radiation, the "cross-polarized" radiation. The cross-polarized field for the case being considered has a maximum in each of the four quadrants of the reflector, as can be seen in Fig. 6. The resulting radiation pattern (Fig. 7) has cross-polarized lobes appearing in planes at 45 degrees to the axes of symmetry. Nearly all radiators have some cross-polarized radiation, but it is often undetected because of measurement techniques which discriminate against it.

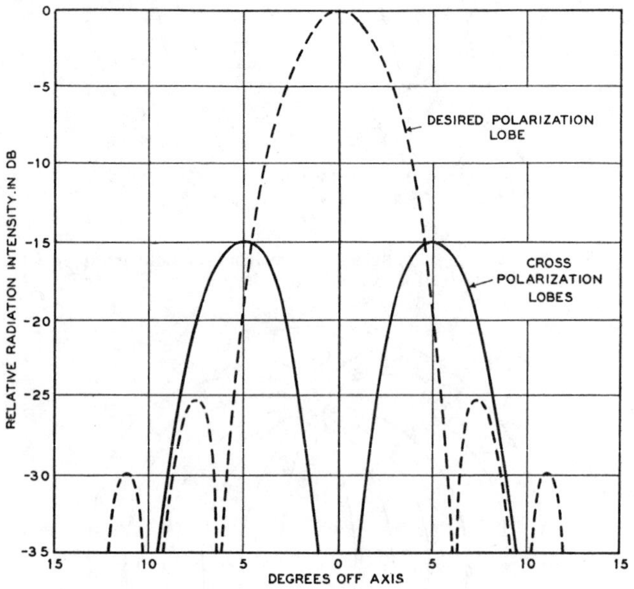

Fig. 7—Radiation pattern illustrating cross-polarized lobes.

D. Interference of Feed with Reflected Beam

The effect of the rearward radiation from a feed is indicated by a plot of gain against feed position. (See Fig. 8.) As the feed (wave guide in this case) is moved along the axis, the gain oscillates as the field radiated directly from the feed adds at various phases to that reflected from the paraboloid. To obtain optimum gain, the position of the feed should coincide with the focal point of the reflector. This requires that (for small antennas) the focal length shall be co-ordinated with the wavelength to assure proper phase of the rearward radiation from the feed.

The beam is deflected if the feed is moved laterally away from the focal point, and the defocusing loss does not take place as rapidly as it does for longitudinal motion. For the usual paraboloid proportions, by moving the feed the beam can be shifted about twice the half-power beam width with only ½ decibel loss in gain. The beam shift can be doubled if the feed is fixed and the reflector is tilted.

A defect in many parabolic radiator designs is the fact that the feed obstructs the path of the reflected field. This creates a region of low intensity (or shadow) at the

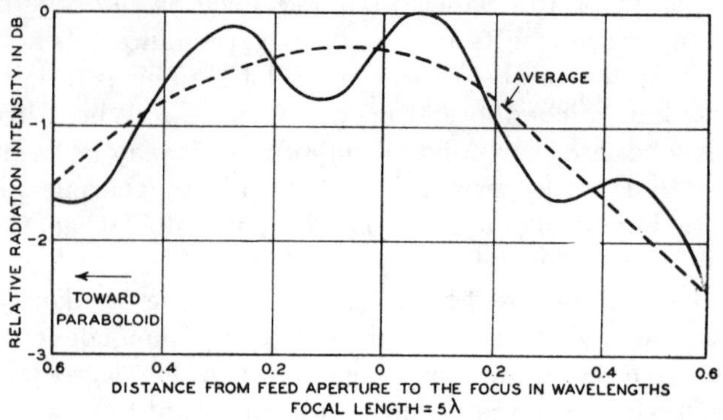

Fig. 8—Effect of rearward radiation on paraboloid gain as a function of axial feed position.

center of the aperture. The effect on the radiation pattern can be approximated by taking the difference of the radiation from the aperture and from the shadow area located at the feed position. This second hypothetical radiator is a small, relatively nondirectional source the same size and shape as the shadow, and acts on the resulting pattern to reduce the main lobe somewhat and raise the side lobes at least to the level of the radiation from the second source. This is illustrated in Fig. 9. The shadow effect may be reduced somewhat by "streamlining" the back of the feed by tapering it in the *E* plane.

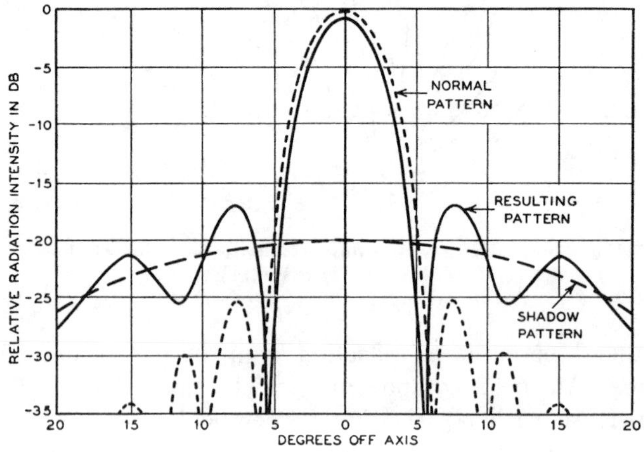

Fig. 9—Effects of a shadow on a paraboloid radiation pattern.

Another effect of the feed being in the path of the reflected wave is that some of the energy from the reflected wave returns to the feed system, producing an impedance mismatch. The absolute value of this impedance is fairly constant as a function of frequency or of feed position, but varies rapidly in phase because of the long round-trip path length of the reflected wave.

The mismatch may be corrected in the feed with an iris (or a stub line) over a narrow bandwidth, but inevitably this results in a more serious impedance mismatch at a frequency such that the focal length has changed by one-quarter wavelength. Moreover, if it is required that the feed be moved with respect to the reflector (in order to direct the beam) the impedance of the feed will change in phase and magnitude, making it impossible to match at a fixed point in the feed system. Thus, for conditions that require relative motion of the feed it may be desirable to match the feed to free space and tolerate the impedance change resulting from the reflector.

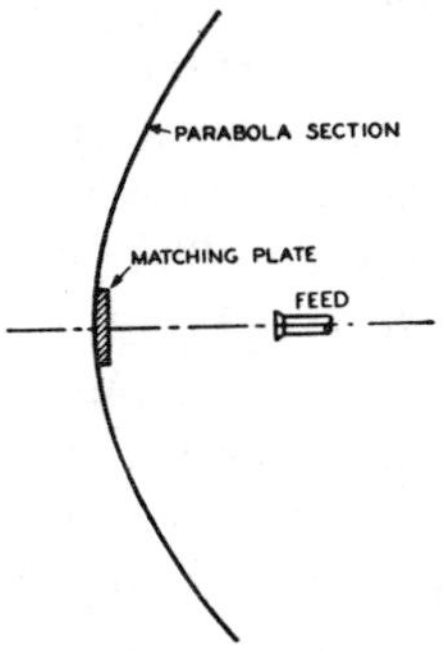

Fig. 10—Apex-matching plate for improving the impedance at the feed.

There are other ways of avoiding the effect of the reflector on the impedance of the feed. One method is to raise a portion of the reflecting surface to produce a reflected signal in the feed, equal and opposite to that received from the remainder of the reflector, thus canceling the reflected signal at the focus. This apex-matching plate is illustrated in Fig. 10. Since the two sources of reflection, namely, the undisturbed part of the reflector and the apex-matching plate, are at nearly the same distance from the focus, this impedance correction is effective over a very wide frequency band. Of course, the energy reflected from the raised surface is scattered widely and reduces the gain somewhat and increases the minor-lobe level of the radiation pattern.

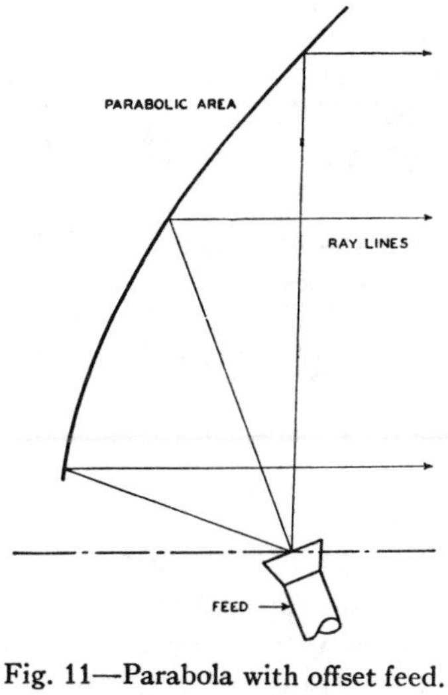

Fig. 11—Parabola with offset feed.

A method of avoiding the above-mentioned impedance problem, and also the shadow interference, is to use an off-set feed with a parabolic area (cylindrical or paraboloidal) located at one side of the apex, as shown in Fig. 11. In this type of antenna the feed is, for all practical purposes, clear of the reflected energy, and the bandwidth is limited only by the properties of the feed employed.

IV. Feed Systems

A. Half-Wave-Doublet Feeds

The earliest parabolic antennas evolved from attempts to increase the directivity of the half-wave doublet antenna by using sheet reflectors. However, as the art progressed and higher-gain antennas were required, it became apparent that the simple half-wave doublet was an ineffective source for exciting large parabolas.

The doublet antenna radiates uniformly in a plane perpendicular to its length. If a paraboloid is made to subtend a solid angle of 180 degrees at the focus, half of the energy will be radiated into space without striking the reflector. If this "lost energy" is properly phased with that of the reflected beam, it contributes to the gain of the antenna, as was discussed in connection with rearward radiation, and therefore the loss is not serious, provided that the aperture area is only a few square wavelengths. However, for large antennas most of the energy which does not strike the reflector is wasted. To reduce this loss and increase the radiating efficiency of the over-all system, it is necessary to direct most of the energy from the feed into the paraboloid.

The half-wave-doublet feed can be made more directive by using techniques familiar in wire antennas at lower frequencies, but the simple parasitically excited reflector appears to be the most practical. The reflector can be another doublet, a plane sheet, a half cylinder, or a hemisphere. The disk and the half cylinder appear to give best operation, but the other reflectors also have been used in specific applications.

The doublet antenna is at a disadvantage for feeding paraboloids in that the polarization characteristic is poor, as was discussed earlier. Beyond 90 degrees in the *E* plane, the polarization actually reverses, and extending the paraboloid beyond 90 degrees in this plane would result in a decrease in gain. To obtain minimum cross-polarization effects, and best distribution of illumination from a doublet radiator, a relatively shallow reflector should be used, as is indicated by the fact that the paraboloid which gives best efficiency with most of these feeds subtends only 140 degrees at the focus.

B. Wave-Guide Feeds

At centimeter wavelengths it is practical to feed the parabola with the radiation from an open-ended wave guide. The radiation characteristic of a wave-guide aperture is dependent upon the size and shape of the aperture and the mode or modes of propagation within the guide. Where a circular paraboloid is used, a circular $TE_{1,1}$ wave guide may be used for a feed, and indeed gives almost the ideal phase and polarization characteristics with suitable directivity. A fairly nondirective source for illuminating a deep paraboloid may be obtained by loading a small-diameter guide with a dielectric. A more directional source of illumination for a shallow paraboloid may be obtained by using a larger-diameter wave-guide aperture or by flaring the aperture into a small conical horn.

A rectangular $TE_{1,0}$ wave guide does not generally give a circularly symmetric radiation pattern, but it is suitable for feeding a paraboloidal section which is cut to subtend a wide angle in the E plane and a narrow angle in the H plane. A contour of uniform intensity in the pattern of such a feed is approximately elliptical, so the most efficient reflector area should be nearly elliptical, although sometimes it is mechanically more practical to use a rectangular shape. The directivity in the electric and magnetic planes can be controlled more or less independently by the corresponding aperture dimensions. The phase characteristic of a rectangular wave-guide aperture used as a radiator is very good, provided only the dominant mode is transmitted to the aperture. The measured polarization characteristic is usually deformed somewhat from the ideal, but it is still very good in the useful part of the amplitude pattern.

Where more directivity is required in the feed than can be obtained with a simple aperture, some form of wave-guide horn may be used. However, the phase characteristics of horn feeds should be examined carefully. In particular, the rectangular sectoral horn, which is often used for feeding elliptical paraboloid sections having large ratios of major-to-minor axes, has a poor phase characteristic. For instance, if the front of constant phase is measured for a sectoral horn of optimum flare,[2] it is found to be circular in the plane of flare, with the phase center near the apex of the angle of flare. In the other plane the phase front is also circular with the center at the horn aperture. Since these centers may be several wavelengths apart, the phase front is far from spherical, and may deviate a large fraction of a wavelength over the paraboloid surface. Acceptable operation is usually obtained for such a horn if the aperture is located at the focus, but better efficiency can be obtained by altering the shape of the reflector or by using a feed with more desirable phase characteristics.

There are many ways of obtaining a feed pattern for an elliptically cut paraboloid which will have more desirable phase characteristics than the sectoral horn with optimum flare. Some improvement may be had by using a flare angle somewhat smaller than that for optimum gain, but this requires a much longer horn for relatively small improvement. For reflectors with a major-to-minor-axis ratio of from 3 to 5, a "two-mode" or "box" type of horn has been found to give very good results.

The "two-mode" horn is simply a wide rectangular wave-guide aperture which is excited with both the $TE_{1,0}$ and the $TE_{3,0}$ modes of propagation. It can be shown that if both these modes are present in the aperture in the proper amplitude ratio and relative phase, the resulting aperture field approaches the desirable condition of uniform amplitude and phase. The two modes are set up in the guide by exciting it abruptly from a smaller guide which carries only the dominant ($TE_{1,0}$) mode. (See Fig. 12.) Since the large guide is wide enough to propagate the $TE_{1,0}$, $TE_{2,0}$, and the $TE_{3,0}$ modes, a discontinuity at the junction will tend to excite all three modes. However, if symmetry about a central plane is maintained, the $TE_{2,0}$ mode is not set up, and only the first and third modes are excited. At the junction, of course, many odd-order modes must be present in order to satisfy the boundary conditions presented by the walls of the guide, but if the larger guide is not wide enough to propagate modes higher than the third, they may for the sake of this discussion be ignored. At the junction the two modes add to approximately

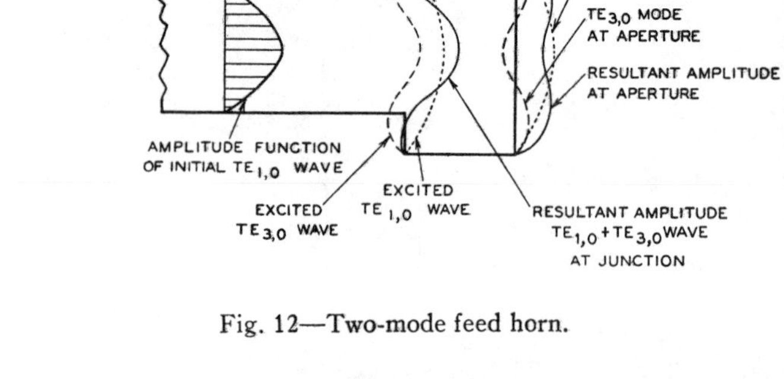

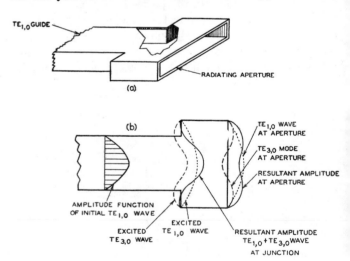

Fig. 12—Two-mode feed horn.

the field distribution of the original wave. However, since the two waves propagate with different velocities, their relative phases vary along the length of the guide, and at any other section the resultant amplitude will be different. When the length of the large guide is such that the relative phases of the two waves has changed by 180 degrees, they will add to give a uniphase field across the aperture with a nearly uniform amplitude. The superposition of these waves is shown in Fig. 12.

[2] S. A. Schelkunoff, "Electromagnetic Waves," D. Van Nostrand Co., Inc., New York, N.Y., 1943; pp. 363–365.

The relative amplitudes of the $TE_{1,0}$ and $TE_{3,0}$ waves can be obtained by making a Fourier analysis of the incident field at the junction, taking into consideration only the first two terms of the series. The field at the junction may be expressed as

$$f(x) = A_1 \cos x + A_3 \cos 3x + A_5 \cos 5x + \cdots + A_p \cos px \qquad (19)$$

where p is any odd integer. The constants A_p may be found from

$$A_p = \frac{4}{\pi}\int_0^{\pi/2} f'(x) \cos (px) dx \qquad (20)$$

where $f'(x)$ is the field incident to the junction, and is equal to $f(x)$. It may be taken as

$$f'(x) = \cos \frac{x}{b}\Bigg]_{x=0}^{x=(\pi/2)b}, \qquad (21)$$

and

$$f'(x) = 0\Bigg]_{x=(\pi/2)b}^{x=\pi/2}$$

where b is the ratio of the smaller to the larger guide widths. With values of 1 and 3 for p, the values of A_1 and A_3 may be obtained, and since higher-order waves are not propagated, the first two terms of the series (19) give the field propagated in the large guide.

In analyzing this field at any point it is necessary to know the relative amplitude of the two modes. This is given by

$$\frac{A_3}{A_1} = \frac{b^2 - 1}{9b^2 - 1}\,\frac{\cos \frac{3\pi}{2} b}{\cos \frac{\pi}{2} b} \qquad (22)$$

which is plotted as a function of b in Fig. 13.

To phase the two modes properly, the large guide must be of such a length as to cause 180 degrees relative phase shift between the two components. This requires a length equal to

$$\frac{L}{\lambda} = \frac{1}{2}\,\frac{\lambda_1\lambda_2}{\lambda_1 - \lambda_2} \qquad (23)$$

where λ_1 and λ_2 are the guide wavelengths for the two modes. Substituting for these, we get

$$\frac{L}{\lambda} = \frac{1}{\sqrt{4 - \left(\frac{3\lambda}{w}\right)^2} - \sqrt{4 - \left(\frac{\lambda}{w}\right)^2}} \qquad (24)$$

where λ is the air wavelength, and w the guide width.

The impedance presented to the guide by the two-mode feed horn has about the same magnitude as that of an open-ended wave guide, and can be easily matched over a fairly wide band of frequencies. For some applications it has been matched by adding a dielectric plate (of appropriate thickness) over the aperture, which also seals the guide from the weather. The impedance has

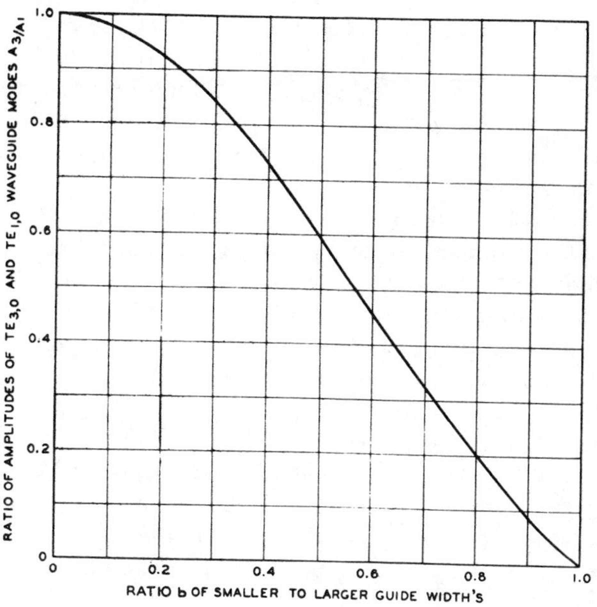

Fig. 13—Relative amplitudes of modes in a two-mode horn.

also been matched by exciting the horn directly from a wave-guide elbow, and matching both elbow and horn simultaneously at the corner. A horn of this type is show in the photograph, Fig. 14.

Fig. 14—Paraboloidal antenna with two-mode feed horn.

In spite of the fact that the wavelength appears critically in the dimensioning of this horn, it has been found to give acceptable operation over a bandwidth of at least 10 per cent.

Where still more directivity in the feed is desired, other types of antennas (such as lenses, and even small parabolas) may be used as paraboloid feeds.

C. Rear Wave-Guide Feeds

The wave-guide feeds discussed so far are all directive along the axis of the paraboloid and away from the feed mounting, and as a result they all require mechanical support and a source of power in front of the reflector. It is desirable, in most applications, to avoid the interference of the feeding guide and supporting structure with the reflected beam and, therefore, a number of rear feeds, or feeds supported and fed from the apex of the parabola, have been developed.

An early attack on this problem was to use a round wave-guide aperture, supported at the paraboloid apex and lying along the axis, opening toward the focus, with a metallic reflector to direct the energy back into the paraboloid. However, the simple geometric image picture of the field is not accurate, and we are not aware of any success in attempts to reshape the reflecting plate so as to produce a good over-all pattern. Finally, it was discovered that the phase front of this feed, instead of being spherical as desired, is toroidal with an apparent center in a ring lying between the circumference of the disk reflector and the wave guide, as shown in Fig. 15.

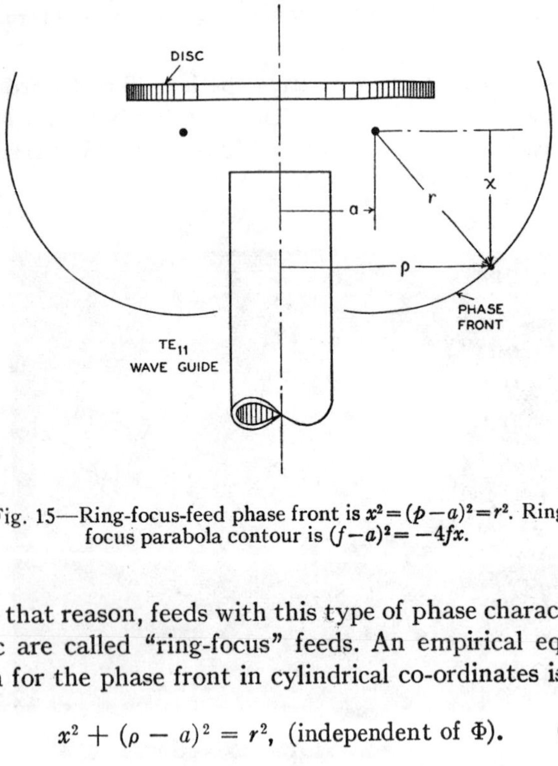

Fig. 15—Ring-focus-feed phase front is $x^2=(p-a)^2=r^2$. Ring-focus parabola contour is $(f-a)^2=-4fx$.

For that reason, feeds with this type of phase characteristic are called "ring-focus" feeds. An empirical equation for the phase front in cylindrical co-ordinates is:

$$x^2 + (\rho - a)^2 = r^2, \text{ (independent of } \Phi). \quad (25)$$

The reflecting surface for such a source should be

$$(\rho - a)^2 = - 4Fx, \text{ (independent of } \Phi). \quad (26)$$

which is the surface generated by a parabola rotated about a line parallel to its axis and displaced a distance a from the axis. Such a surface is called a "ring-focus" paraboloid.

The cirular symmetry of the amplitude characteristic of a ring-focus feed becomes somewhat better when a cup, instead of the disk, is used to direct the energy from the feed into the paraboloid, as shown in Fig. 16. The

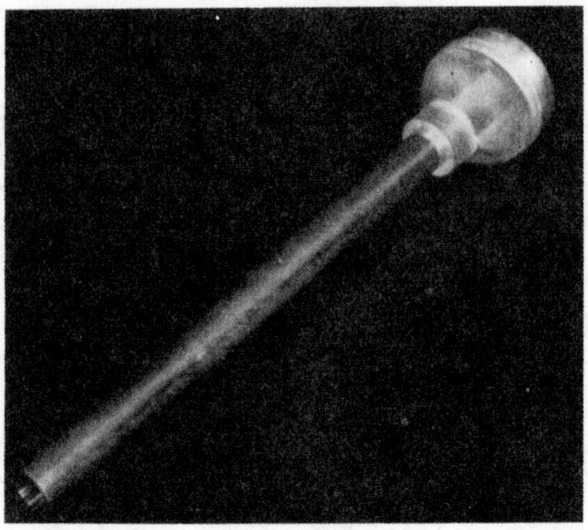

Fig. 16—Ring-focus feed.

amplitude characteristic of this feed is shown in Fig. 17. The distribution is more nearly uniform and covers a wider angle than that for other feeds. The optimum an-

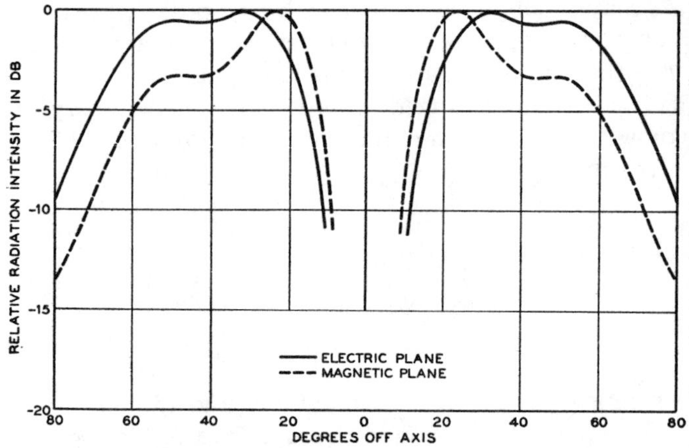

Fig. 17—Radiation pattern of ring-focus feed.

gle subtended by the paraboloid is about 160 degrees and consequently the reflector is much deeper than that required by most feeds.

A picture of the field in the aperture of the ring-focus feed may be obtained by sampling the energy with a small probe. If the rim of the cup is extended, the part of the structure outside of the feeding wave guide comprises a large-diameter coaxial line. Because of the size of the line a great many modes of propagation may be supported, but only two modes can be initiated because of symmetry in the system. They are analogous to the $TE_{1,1}$ and the $TE_{1,2}$ modes in a circular guide. These modes are sketched in Fig. 18. By adjusting the dimensions of a ring-focus feed, it has been possible to set up either of these modes, and combinations of the two, with various amplitudes and phases. Usually such combinations give very uneven amplitude and polarization distributions at the aperture, but when the dimensions are chosen to give the proper amplitudes and phases in the

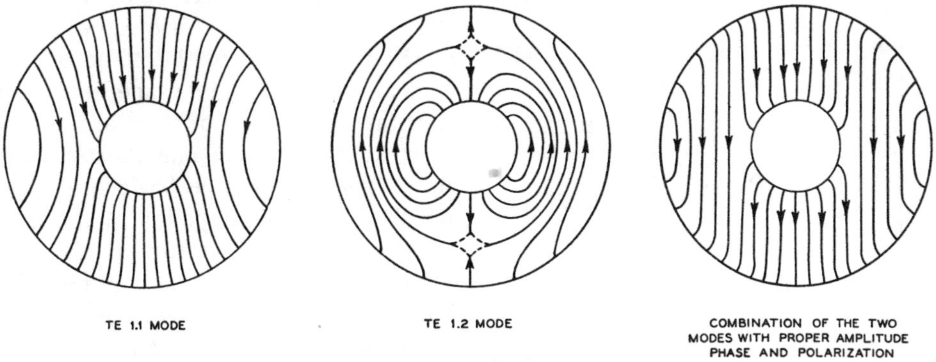

Fig. 18—$TE_{1,1}$ and $TE_{1,2}$ modes of propagation in coaxial, as combined in the ring-focus feed.

two waves, the amplitude at the aperture becomes very uniform and the lines of polarization are nearly parallel. It is this condition that accounts for the uniformity of the amplitude and phase characteristics of the ring-focus feed. Of course, there is not enough length of coaxial line to justify analyzing the field wholly in terms of these modes, but adding a length of line indicates that the fields present are closely related to these modes of propagation.

The impedance match of this feed is poor, and when matched by an iris in the wave guide the bandwidth is very narrow. Attempts to match impedance by changing the shape of the cup or disk have been unsuccessful because of the undesirable effects on the radiation pattern. Ring-focus antennas have given about 0.4 decibel more gain than the other paraboloid antennas described because of the closer approach to uniformity in the amplitude distribution of this type of feed.

Another method of feeding a paraboloid is by dividing the power in a rectangular guide, by providing two exit apertures symmetrically disposed with respect to the

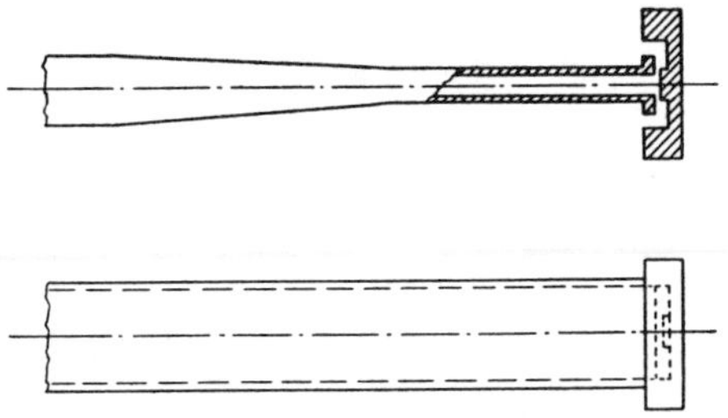

Fig. 19—Dual-aperture rear-feed horn.

median magnetic plane, as shown in Figs. 19 and 20. Here the radiation is from two apertures which, to give a smooth amplitude pattern, should be less than a half-wavelength apart. It is also important that the apertures should not be too close to the guide wall; therefore, the guide must not be wider than about one-quarter wavelength. The phase front on either side appears to be coming from the nearer aperture and its image reflected in the adjacent side of the wave guide, thus resulting in phases in the E and H planes at the edge of the reflector which differ by about 30 degrees. The polarization shows a tendency to depart from the ideal similarly to a "magnetic dipole" (a loop antenna or a narrow slot in a conducting plane), but this is not serious. This feed has several advantages over many other rear feeds in that the directivity is directly dependent on the dimensions of the aperture; the impedance may be

Fig. 20—Dual-aperture rear feeds.
(a), (b) Feed for circular paraboloids.
(c) Feed for elliptical paraboloids.
(d) Two-mode rear-feed horn.

matched by properly dimensioning the cavity; and the structure can be easily weatherproofed by incorporating windows across the apertures.

The directivity of the feed in the E plane can be controlled, within limits, by the separation of the slot from the wave-guide wall and somewhat by the width of the slot. The directivity in the H plane is roughly inversely proportional to the length of slot from a length of one-half to three half-wavelengths, as shown in Fig. 21. For greater directivity a slot longer than 3 half-wavelengths

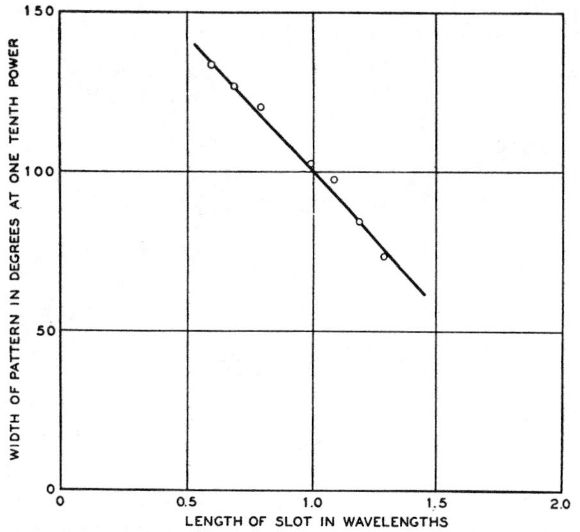

Fig. 21—H-plane directivity as a function of the length of the openings in the dual-aperture feed horn.

may be used applying the principles of the "two-mode" horn previously described and shown in Fig. 12. The rear-feed, dual-aperture, two-mode horn [Fig. 20(d)] has the junction between the wide and narrow guides at the cavity where the energy is divided between the two radiating apertures. Thus the transmission in the head of the horn consists of two modes ($TE_{1,0}$ and $TE_{3,0}$) and the length of the path necessary to properly phase the modes is built into the head.

If the wave guide is not changed as the slots are lengthened, it soon ceases to adequately shield the radiating openings from one another, with the result that the E-plane pattern becomes sharper. Where it is desired to have the E plane broad, the outside wall of the wave guide should be extended in width to at least equal the length of the slots, also shown in Fig. 20(d). The size of the flat surface in the plane of the apertures is not critical, but some advantage in uniformity of illumination (and reduction of rearward radiation) is obtained if it extends at least one-quarter wavelength from the slots. There are three basic dimensions of the cavity which affect the impedance, namely, the width and depth, and an indentation at the center of the cavity as shown in Fig. 19. The feed for a circular paraboloid having slots about 0.7-wavelength long and a circular cavity is shown in Figs. 20(a) and (b). In this case the diameter, the depth, and the indentation at the center of the cavity may be controlled to match almost any impedance condition. For horns with longer slots a rectangular cavity is usually used and the impedance is controlled by the depth and the indentation at the center of the cavity. The relation of the cavity dimensions to the impedance is not simple and usually requires an experimental determination for each new application.

Acknowledgment

The development of the ideas described in this paper cannot be attributed to any particular time and place, inasmuch as many of the most basic ones may be traced back over several decades. However, most of the recent development described resulted from war-project work during the last six years within the Bell Telephone Laboratories at Deal, N. J., and Holmdel, N. J.

Paraboloid Reflector and Hyperboloid Lens Antennas*

E. M. T. JONES†, ASSOCIATE, IRE

Summary—**A theoretical analysis of the radiating properties of the paraboloid reflector and the hyperboloid lens shows that low amplitude cross-polarized radiation and high gain factors can be obtained from a paraboloid reflector excited by a plane-wave source. Low amplitude, cross-polarized radiation can also be obtained from the hyperboloid lens with a plane-wave feed, but with a lower gain factor. It is found that the measured properties of the antennas agree reasonably well with the theoretical predictions. Also it is found experimentally that principal plane side lobes of the order of —40 db can be obtained with a short focal length hyperboloid lens.**

Introduction

THIS PAPER describes an investigation of the radiation properties of two perfectly focusing devices: the paraboloid reflector and the hyperboloid dielectric lens, excited at their foci by a short electric dipole, a short magnetic dipole, and a plane-wave source. The purpose of this investigation is to examine the radiation characteristics of these antennas, which, at least in the limit of infinitesimal wavelengths, have no phase aberrations in the aperture field. The analysis is divided into two parts: (1) the aperture fields of the two antennas are first computed, and, (2) the far-zone diffraction patterns are then determined. The power gain of the systems is obtained as a by-product of the diffraction pattern computation. Some experimental results are also discussed.

Paraboloid Reflector Aperture Distributions

The aperture distributions of the reflector are computed in two steps. First, the current induced on the surface of the reflector by radiation from the feed is determined. Next, the electric fields in the aperture, arising from these currents, are computed. The assumptions made in the derivation are:

a. The reflector is in the far-zone of the feed, so that only fields varying as the reciprocal of the distance from the feed to the reflector are significant.

b. The feed pattern is the same with the reflector in place as when it is absent.

c. Energy traveling in the region between the reflector aperture and the feed follows the straight line paths predicted by geometric optics, while the polarization of the aperture field is determined by the plane-wave boundary conditions at the reflector surface, namely, that the total tangential electric field in the incident and reflected waves must be zero.

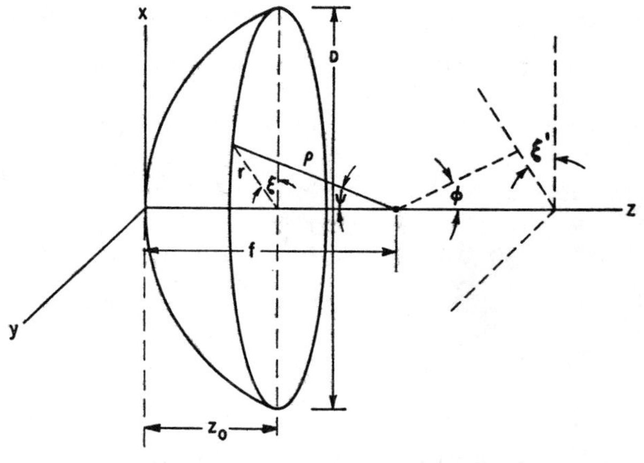

Fig. 1—Sketch of paraboloid reflector illustrating the notation used in the analysis.

A sketch of the parabola showing the notation appears in Fig. 1. The surface current $\overline{K}$ induced on the reflector by the feed at the focus is

$$\overline{K} = 2(\bar{n} \times \overline{H}) = \frac{2}{\eta}\bar{n} \times (\bar{\rho}_0 \times \overline{E})$$

$$= \frac{2}{\eta}\left[\bar{\rho}_0(\bar{n}\cdot\overline{E}) + \overline{E}\cos\frac{\psi}{2}\right], \tag{1}$$

where

$\overline{E}$ = far-zone electric field of the feed
$\overline{H}$ = far-zone magnetic field of the feed
η = 377 ohms

* Original manuscript received by the PGAP, November 11, 1953; revised manuscript received April 12, 1954.
† Stanford Research Institute, Stanford, Calif.

Reprinted from *IRE Trans. Antennas Propagat.*, vol. AP-2, pp. 119–127, July 1954.

and the outward normal $\bar{n}$ to the reflector surface is

$$\bar{n} = -\bar{x}\cos\xi\sin\frac{\psi}{2} - \bar{y}\sin\xi\sin\frac{\psi}{2} + \bar{z}\cos\frac{\psi}{2}, \qquad (2)$$

and the unit vector $\bar{\rho}_0$ in the ρ direction is

$$\bar{\rho}_0 = \bar{x}\cos\xi\sin\psi + \bar{y}\sin\xi\sin\psi - \bar{z}\cos\psi. \qquad (3)$$

In general, the surface current induced on the paraboloid will have components K_x, K_y, and K_z. For high-gain reflectors having apertures many wavelengths in diameter, it is only the K_x and K_y components of current that contribute to the principal part of the diffraction pattern centered about the z axis. Therefore, component K_z will be neglected in the rest of the discussion.

Short Electric-Dipole Feed

For a short electric-dipole feed lying along x axis, of height dx, and excited with a current I flowing in $-x$ direction, the far-zone components of electric field are

$$\overline{E} = \frac{j\eta I dx \epsilon^{-jk\rho}}{2\lambda\rho}\left[\bar{x}(\cos^2\psi\cos^2\xi + \sin^2\xi) - \bar{y}\frac{\sin 2\xi\sin^2\psi}{2} + \bar{z}\frac{\sin 2\psi\cos\xi}{2}\right]. \qquad (4)$$

The substitution of (2), (3), and (4) into (1) yields

$$\overline{K} = \frac{jIdx\epsilon^{-jk\rho}}{\lambda\rho}\cos\frac{\psi}{2}\left[\bar{x}(\cos\psi + \sin^2\xi(1-\cos\psi)) - \frac{\bar{y}}{2}\sin 2\xi(1-\cos\psi)\right]. \qquad (5)$$

The projected electric field in the aperture is given as

$$-\frac{\overline{K}\eta}{2\cos\frac{\psi}{2}}.$$ [1]

Making this substitution it is found that[2]

$$\overline{E}_a = -\frac{j\eta I dx\epsilon^{-jk(f+z_0)}}{4\lambda\rho}\{\bar{x}[(1+\cos\psi) - (1-\cos\psi)\cos 2\xi] - \bar{y}\sin 2\xi(1-\cos\psi)\}. \qquad (6)$$

It is seen that the aperture electric field has a unidirectional component along the x axis as well as a cross-polarized component parallel to the y axis. The cross-polarized component has the interesting symmetry property, in that it is oppositely directed in adjacent quadrants.

Fig. 2 shows the distribution of electric field in the aperture of a paraboloid reflector with an electric-dipole feed, together with characteristics of the patterns arising from such a distribution. It is seen that the principal polarization patterns measured in the E and H planes have the customary shape with their maxima on the polar axis. On the other hand, the cross-polarized lobes (which are often called Condon lobes) measured, in the planes at 45 degrees to the principal planes where the cross-polarization is a maximum, have minimum intensity on the polar axis and maximum intensity off axis. The position of maximum intensity corresponds approximately to the position of the first null in the principal polarization patterns. Because these cross-polarized lobes often have a magnitude considerably higher than the first side lobe of the principal polarization patterns, they can be troublesome in a radar application in which the target reflects energy polarized at right angles to that incident upon it.

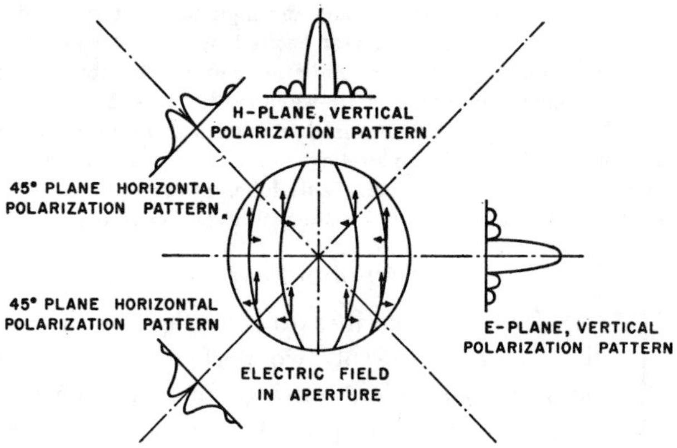

Fig. 2—Electric field in the paraboloid reflector aperture and resulting far-zone radiation patterns when the paraboloid is excited by a vertically oriented electric dipole.

Short Magnetic-Dipole Feed

For a short magnetic dipole lying along the y axis with a length dy and excited with a magnetic current M flowing in the $+y$ direction, the far-zone components of electric field are

$$\overline{E} = \frac{jMdy\epsilon^{-jk\rho}}{2\lambda\rho}[\bar{x}\cos\psi + \bar{z}\sin\psi\cos\xi]. \qquad (7)$$

Following the procedure outlined above, the projected field in the aperture is

$$\overline{E}_a = -\frac{jMdy\epsilon^{-jk(f+z_0)}}{4\lambda\rho}\{\bar{x}[(1+\cos\psi) + (1-\cos\psi)\cos 2\xi] + \bar{y}[\sin 2\xi(1-\cos\psi)]\}. \qquad (8)$$

Here again it is seen that the aperture electric field has a unidirectional component along the x axis but the cross-polarized component of electric field is oppositely directed. Fig. 3 shows a sketch of the aperture field.

Plane-Wave Feed

If the paraboloid could be excited by a combination of electric and magnetic dipoles, oriented at right angles to one another and having the proper values of electric and

[1] S. Silver, "Microwave Antenna Theory and Design," McGraw-Hill Book Co., New York, N. Y., p. 417; 1949.

[2] E. U. Condon, "Theory of Radiation from Paraboloid Reflectors," Westinghouse Research Report SR-105; September, 1941.

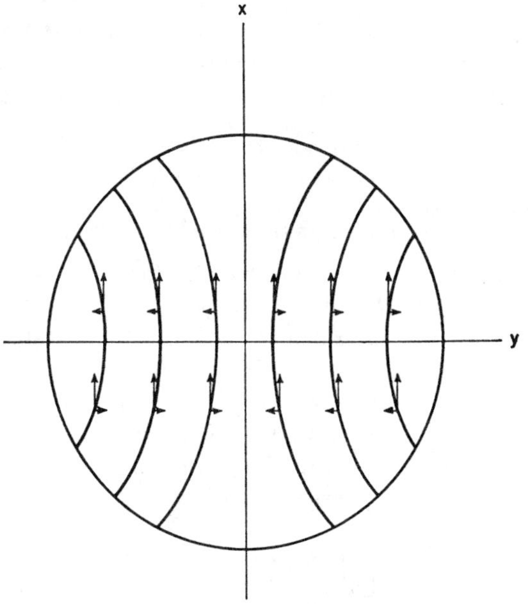

Fig. 3—Electric field in paraboloid reflector aperture when paraboloid is excited by a short magnetic dipole lying along y axis.

magnetic current, it is seen that the y directed component of the aperture field could be made to disappear. The proper value of the ratio of the magnetic to electric current is η which is the same as the ratio of the equivalent magnetic and electric current densities in a plane-wave source. The aperture field, $\overline{E}_a$, of a paraboloid excited by a plane wave-source with electric field intensity E_x and dimensions dx and dy small compared to a wavelength, is then seen to be from (6) and (8)

$$\overline{E}_a = -\bar{x}j\frac{E_x dx dy}{\lambda\rho}\frac{(1+\cos\psi)}{2}\epsilon^{-jk(f+z_0)}. \tag{9}$$

A sketch of this aperture field is shown in Fig. 4.

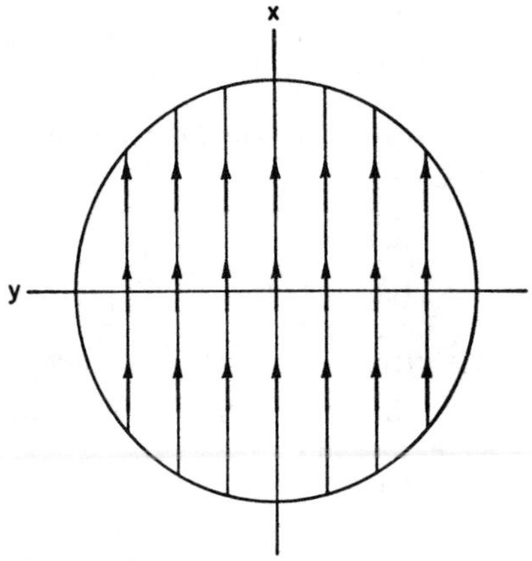

Fig. 4—Electric field in paraboloid reflector aperture when paraboloid is excited by a small plane-wave source polarized along x axis.

Hyperboloid Lens Aperture Distributions

Fig. 5 shows a sketch of the hyperboloid lens illustrating the notation used in the analysis. The aperture field of the hyperboloid lens of refractive index n is determined, subject to the conditions outlined previously, in a slightly different manner than the aperture field of the paraboloid reflector. First, the field incident on the lens from the feed is resolved into a component in the plane of incidence and one perpendicular to it. The amount of each of these components transmitted into the lens through surface 1 is then computed from the plane-wave boundary conditions at the dielectric-air interface, which are summarized in Fresnel's equations. Finally, the amount of energy transmitted through surface 2 is computed. The assumption is made, in deriving this aperture field, that the multiple transits of energy in the lens between the two surfaces are unimportant. These transits can be neglected because they give rise to small amplitude and phase variations arranged in concentric rings in the aperture, superimposed on the constant phase amplitude distribution produced by the single transit. In any lens of reasonable size they produce small amplitude diffraction patterns, split off at wide angles from the polar axis.[3,4]

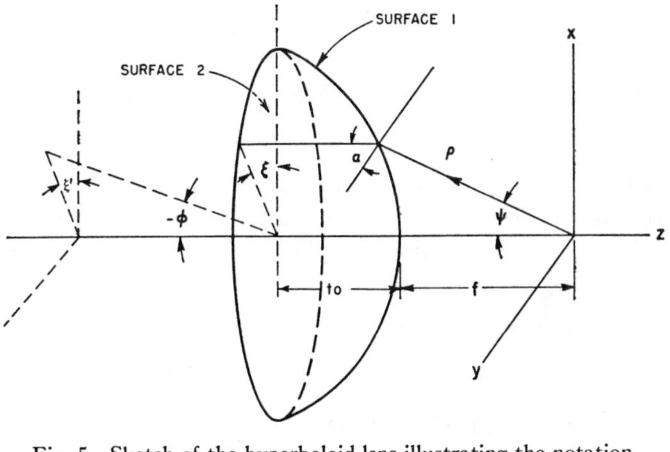

Fig. 5—Sketch of the hyperboloid lens illustrating the notation used in the analysis.

Electric-Dipole Feed

When the hyperboloid lens is excited by an electric dipole lying along the x axis, of length dx, excited by a current I following in the $-x$ direction, the electric field incident on the curved surface of the lens is

$$\overline{E} = j\frac{\eta I dx\epsilon^{-jk\rho}}{2\lambda\rho}[\bar{\psi}\cos\xi\cos\psi - \bar{\xi}\sin\xi] \tag{10}$$

where the $\bar{\psi}$ component lies in the plane of incidence, and the $\bar{\xi}$ component lies perpendicular to the plane of incidence. The electric field transmission coefficient, $T_{\parallel}$, for the field polarized in the plane of incidence is

$$T_{\parallel} = \frac{2\cos(\psi+\alpha)\sin\alpha}{\sin(\psi+2\alpha)\cos\psi}, \tag{11}$$

[3] J. Brown, "Effect of amplitude variations in aperture fields on side lobes," *Jour. Inst. Elect. Eng.*, vol. 97, part III, pp. 419–424; November, 1950.

[4] N. I. Korman, E. B. Herman, and I. R. Ford, "Analysis of microwave antenna side-lobes," *RCA Review*, vol. 13, pp. 323–334; September, 1952.

while the electric field transmission coefficient for the $T_\perp$ component of field, polarized perpendicular to the plane of incidence is

$$T_\perp = \frac{2\cos(\psi+\alpha)\sin\alpha}{\sin(\psi+2\alpha)}. \tag{12}$$

The transmitted electric field lying in the plane of incidence becomes

$$E_\parallel = j\frac{\eta I dx}{\lambda\rho}\epsilon^{-jk\rho}\left[\frac{\cos(\psi+\alpha)\sin\alpha\cos\xi}{\sin(\psi+\alpha)}\right], \tag{13}$$

while the transmitted field lying perpendicular to the plane of incidence becomes

$$E_\perp = -j\frac{\eta I dx\epsilon^{-jk\rho}}{\lambda\rho}\left[\frac{\cos(\psi+\alpha)\sin\alpha\sin\xi}{\sin(\psi+\alpha)}\right]. \tag{14}$$

After refraction at the curved surface of the lens, there will be no ϵ_z field. The ϵ_x- and ϵ_y-components of the field will be

$$\epsilon_x = E_\parallel\cos\xi - E_\perp\sin\xi, \tag{15}$$

and

$$\epsilon_y = E_\parallel\sin\xi + E_\perp\cos\xi.$$

These transverse components of field will all arrive in phase at the aperture after traveling their respective distances through the lens. The value of electric field just outside the dielectric will be

$$E_a = \bar{x}\frac{2j\eta I dx\epsilon^{-jk(f+nt_0)}}{\lambda\rho}\left[\frac{n}{(n+1)}\,\frac{\cos(\psi+\alpha)\sin\alpha}{\sin(\psi+2\alpha)}\right]. \tag{16}$$

Thus the hyperboloid lens excited by an electric dipole has no cross-polarized components of aperture field.

Magnetic-Dipole Feed

When the hyperboloid lens is excited by a magnetic dipole lying along the y axis with a length dy and excited by a magnetic current M flowing in the $+y$ direction, the field incident on the curved surface of lens becomes

$$\overline{E} = \frac{jMdy\epsilon^{-jk\rho}}{2\lambda\rho}[\bar{\psi}\cos\xi - \bar{\xi}\sin\xi\cos\psi]. \tag{17}$$

Tracing the field through the lens as before, now the aperture field has both a principal and transverse polarization component. The aperture field is

$$\overline{E}_a = \frac{jMdy\epsilon^{-jk(f+nt_0)}}{\lambda\rho}\,\frac{n}{n+1}\,\frac{\cos(\psi+\alpha)\sin\alpha}{\sin(\psi+2\alpha)\cos\psi}$$
$$\cdot\left[\bar{x}(1+\cos^2\psi+\sin^2\psi\cos 2\xi)+\bar{y}(\sin 2\xi\sin^2\psi)\right]. \tag{18}$$

Plane-Wave Feed

The aperture field of the lens, when it is excited by a plane-wave source with electric field intensity E_x, can be obtained in the manner outlined previously. The aperture field is

$$\overline{E}_a = j\frac{E_x dxdy}{\lambda\rho}\cdot\frac{n}{(n+1)}\cdot\frac{\cos(\psi+\alpha)\sin\alpha}{\sin(\psi+2\alpha)\cos\psi}\epsilon^{-jk(f+nt_0)}$$
$$\cdot\left\{\bar{x}\left[(1+\cos\psi)^2+\sin^2\psi\cos 2\xi\right]+\bar{y}\sin^2\psi\sin 2\xi\right\}, \tag{19}$$

and it is seen that here too, there is both a principal and cross-polarized component of aperture field.

Far-Zone Diffraction Patterns

The integral expressions for the far-zone field components, E_ϕ and $\overline{E}_{\xi'}$ of the foregoing aperture distributions, that have the ratio of tangential E to tangential H equal to the impedance of free space, can be set up in the standard fashion. For narrow radiated beams it is permissible to approximate the spherical co-ordinate far-zone fields by rectangular components E_x and E_y. The expressions for these components are

$$\overline{E} = j\frac{\epsilon^{-jkR}}{\lambda R}\left[\int_A \overline{E}_a\epsilon^{-jkr\sin\phi\cos(\xi-\xi')}rdrd\xi\right]. \tag{20}$$

Paraboloid Reflector with Electric-Dipole Feed

If the following substitutions are made in (6)

$$u = \frac{r}{2f}, \qquad \gamma = \frac{D}{4f},$$

$$F = \frac{j\eta I dx\epsilon^{-jk(f+z_0)}}{2\lambda},$$

$$\rho = \frac{2f}{1+\cos\psi}, \qquad \beta = 2kf\sin\phi,$$

the far-zone fields $\overline{E}$ of the paraboloid excited by an electric dipole become

$$\overline{E} = j\frac{\epsilon^{-jkR}}{\lambda R}4fF\int_0^{2\pi}\int_0^{\gamma}\left\{\bar{x}\left(-\frac{1}{(1+u^2)^2}+\frac{u^2\cos 2\xi}{(1+u^2)^2}\right)\right.$$
$$\left.+\bar{y}\left(\frac{u^2\sin 2\xi}{(1+u^2)^2}\right)\right\}\epsilon^{-j\beta u\cos(\xi-\xi')}udud\xi. \tag{21}$$

Integration with respect to the angular variable ξ can be carried out since

$$\int_0^{2\pi}\epsilon^{j\beta u\cos(\xi-\xi')}\begin{Bmatrix}\cos n\xi\\ \sin n\xi\end{Bmatrix}d\xi = 2\pi(j)^n\begin{Bmatrix}\cos n\xi'\\ \sin n\xi'\end{Bmatrix}J_n(\beta u).$$

Performing this integration (21) becomes

$$\overline{E} = -j\frac{\epsilon^{-jkR}}{\lambda R}8\pi fF\int_0^{\gamma}\left\{\bar{x}\left(\frac{uJ_0(\beta u)}{(1+u^2)^2}+\frac{u^3J_2(\beta u)\cos 2\xi'}{(1+u^2)^2}\right)\right.$$
$$\left.+\bar{y}\frac{u^3J_2(\beta u)\sin 2\xi'}{(1+u^2)^2}\right\}du. \tag{22}$$

This integral cannot be evaluated exactly because of the presence of the term $1/(1+u^2)^2$. However, if this term is approximated by a polynomial as

$$\frac{1}{(1+u^2)^2} = a_0 + a_2u^2 + \cdots + a_{2n}u^{2n}, \tag{23}$$

these integrals can be evaluated in terms of tabulated

TABLE I

COMPUTED PATTERN CHARACTERISTICS AND GAIN FACTOR OF PARABOLOIDS EXCITED BY A SHORT ELECTRIC DIPOLE

D-Wave Lengths	f/D	H-Plane Half-Power Beam-Width (degrees)	E-Plane Half-Power Beam-Width (degrees)	Position of Cross-Polarization Maximum	H-Plane First Side-Lobe Level (db)	E-Plane First Side-Lobe Level (db)	Cross-Polarized Lobe Max. Level (db)	Gain Factor
37.2	0.25	1.6	2.2	1.8	−16.5	−36.5	−15.8	0.41
37.2	0.30	1.6	2.0	1.8	−16.5	−32	−18.1	0.37
37.2	0.40	1.6	1.75	1.8	−17.2	−24.7	−22.2	0.32
37.2	0.46	1.6	1.75	1.8	−17.2	−22.9	−24.3	0.28
37.2	0.60	1.6	1.75	1.8	−17.4	−20	−28	0.19

functions, since repeated integration by parts puts each term in the form

$$\int x^{p+1}J_p(x)dx = x^{p+1}J_{p+1}(x). \tag{23a}$$

The coefficients a_{2n} of (23) are chosen here so that the polynomial is the least square approximation to the desired function.[5] It has been found that retaining only the first three coefficients that have values of 0.9823, −1.468, and 0.7445 respectively, the difference between the desired and approximating function is less than .02.

Performing the above manipulations, it is found that the far-zone field is

$$\begin{aligned}\overline{E} = -j\frac{\epsilon^{-ikR}}{\lambda R}8\pi fF\Bigg\{\bar{x}\Bigg[&(a_0\gamma^2 + a_2\gamma^4 + a_4\gamma^6)\frac{J_1(\beta\gamma)}{\beta\gamma}\\ &- 2(a_2\gamma^4 + 2a_4\gamma^6)\frac{J_2(\beta\gamma)}{(\beta\gamma)^2} + 8a_4\gamma^6\frac{J_3(\beta\gamma)}{(\beta\gamma)^3}\Bigg]\\ &+ \bar{x}\cos 2\xi'\Bigg[(a_0\beta^2\gamma^6 + a_2\beta^2\gamma^8 + a_4\beta^2\gamma^{10})\frac{J_3(\beta\gamma)}{(\beta\gamma)^3}\\ &- 2\beta^2(a_2\gamma^8 + 2a_4\gamma^{10})\frac{J_4(\beta\gamma)}{(\beta\gamma)^4} + 8a_4\beta^2\gamma^{10}\frac{J_5(\beta\gamma)}{(\beta\gamma)^5}\Bigg]\\ &+ \bar{y}\beta^2\sin 2\xi'\Bigg[(a_0\gamma^6 + a_2\gamma^8 + a_4\gamma^{10})\frac{J_3(\beta\gamma)}{(\beta\gamma)^3}\\ &- 2(a_2\gamma^8 + 2a_4\gamma^{10})\frac{J_4(\beta\gamma)}{(\beta\gamma)^4} + 8a_4\gamma^{10}\frac{J_5(\beta\gamma)}{(\beta\gamma)^5}\Bigg]\Bigg\}\end{aligned} \tag{24}$$

Neglecting direct radiation from the feed, the maximum power gain of the paraboloid excited by an electric dipole can be computed as follows from the definition of power gain.

$$G = 4\pi\frac{\text{Power radiated per unit solid angle by the aperture}}{\text{Total power radiated by the feed}}$$

$$= \frac{4\pi}{\lambda^2}\left(\frac{\pi D^2}{4}\right)\left\{6\left[\frac{a_0\gamma}{2} + \frac{a_2\gamma^3}{4} + \frac{5}{12}a_4\gamma^5\right]^2\right\}. \tag{25}$$

The term in braces, { }, in (25) is often called the gain factor. It has a maximum value of unity for a uniformly illuminated aperture which has constant phase, and lesser values for all other amplitude distributions which have constant phase.

Diffraction patterns have been computed from (24) for a series of paraboloids with a 37.2 wavelength aperture diameter and various focal lengths. The gain factor for these reflectors has been computed from (25). A typical pattern is shown in Fig. 6 for one of these paraboloids which has an f/D ratio of 0.25. Pertinent data about this pattern and the other computed patterns are summarized in Table I above.

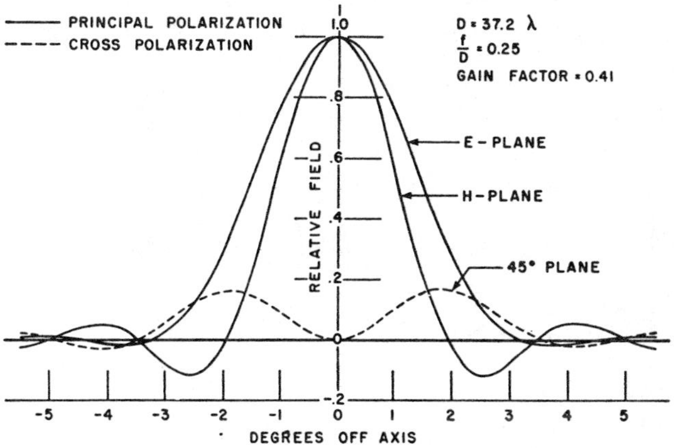

Fig. 6—Paraboloid diffraction patterns with electric dipole feed.

It is seen from Fig. 6, that when the feed lies in the aperture plane, the principal polarization pattern is wider in the *E*-plane than the *H*-plane and the side-lobe level is higher in the *H*-plane than the *E*-plane. This behavior is consistent with the fact that the aperture distribution is much more strongly tapered in the *E*-plane than the *H*-plane. The cross-polarized lobes measured in planes at 45 degrees to the principal planes, where they attain their maximum value, lie quite close to the polar axis and are higher than the first side lobes in either of the principal planes. The gain factor for this antenna, however, is only 0.41. Table I shows that paraboloids of the same aperture diameter but with longer focal lengths have similar *H*-plane patterns, but have *E*-plane patterns with narrower beam-widths and higher side-lobe levels. The cross-polarized lobes and gain factor, however, both decrease as the focal length increases.

[5] W. E. Milne, "Numerical Calculus," Princeton University Press, Princeton, N. J., chap. IX, 1949.

Paraboloid Reflector with Magnetic-Dipole Feed

The results for the paraboloid excited by a short electric dipole oriented along the x axis apply, with minor modifications, to a paraboloid excited by a short magnetic dipole lying along the y axis. Therefore, a detailed discussion will not be given here. However, the principal characteristics of a paraboloid excited by a magnetic dipole lying along the y axis are summarized below.

(1) The E-plane diffraction patterns of the paraboloid with magnetic-dipole feed correspond to the H-plane diffraction patterns of the paraboloid with electric-dipole feed.
(2) The H-plane diffraction patterns of the paraboloid with the magnetic-dipole feed correspond to the E-plane diffraction patterns of the paraboloid with electric-dipole feed.
(3) The cross-polarized patterns in both cases are the negative of one another.
(4) The gain of the two systems is the same.

Paraboloid Reflector with Plane-Wave Feed

When a paraboloid reflector is fed by a plane-wave feed with no variation of field intensity in the E-plane dimension B, and a half-sinusoidal variation in field intensity across the H-plane dimension A, the aperture field of the paraboloid becomes

$$\overline{E}_a = -\bar{x}j\,\frac{E_x AB(1+\cos\psi)^2}{2\pi\lambda f}\,\epsilon^{-jk(f+\rho_0)} \cdot \left[\frac{\sin\left(\frac{\pi B}{\lambda}\sin\psi\cos\xi\right)}{\frac{\pi B}{\lambda}\sin\psi\cos\xi}\cdot\frac{\cos\left(\frac{\pi A}{\lambda}\sin\psi\sin\xi\right)}{1-\left(\frac{2A}{\lambda}\sin\psi\sin\xi\right)^2}\right]. \quad (26)$$

If the dimension A, of the plane-wave feed is chosen to be about 1.42 B, the principal lobe of the E- and H-plane patterns of the feed are almost identical and the patterns in other planes are very similar. Using this approximation then, the paraboloid aperture field becomes

$$\overline{E}_a = -\bar{x}j\,\frac{E_x 1.42B^2(1+\cos\psi)^2}{2\pi\lambda f}\,\epsilon^{-jk(f+\rho_0)} \cdot \left[\frac{\sin\left(\frac{\pi B}{\lambda}\sin\psi\right)}{\frac{\pi B}{\lambda}\sin\psi}\right]. \quad (27)$$

Proceeding as in the case of the electric-dipole feed, it is easy to show that the far-zone field of the reflector with a plane-wave feed is

$$\overline{E} = \bar{x}\,\frac{E_x 5.68B^2\epsilon^{-jk(f+\rho_0+R)}}{\lambda^2 fR}\left[(b_0d^2+b_2d^4+b_4d^6)\frac{J_1(\beta d)}{\beta d} - 2(b_2d^4+2b_4d^6)\frac{J_2(\beta d)}{(\beta d)^2} + 8b_4d^6\frac{J_3(\beta d)}{(\beta d)^3}\right], \quad (28)$$

where $d=D/2$ and the coefficients b_0, b_2, and b_4 are the coefficients of the polynomial which approximates the aperture distribution normalized to a maximum value of unity.

It is noticed that the far-zone field has no cross-polarized radiation fields.

Similarly, the power gain of the reflector excited by the plane-wave source becomes

$$G = \left(\frac{4\pi}{\lambda^2}\right)\left(\frac{\pi D^2}{4}\right)\left\{\frac{45.4B^2}{\pi\lambda^2 f^2}\left[\frac{a_0 d}{2}+\frac{a_2 d^3}{4}+\frac{5}{12}a_4d^5\right]^2\right\} \quad (29)$$

where, as before, the factor in the braces, $\{\ \}$, is the gain factor.

The diffraction pattern of a paraboloid with a 37.2 wavelength diameter aperture, and f/D ratio of 0.46, and an edge illumination of -12 db is shown in Fig. 7. The shape of this principal polarization pattern is rotationally symmetric about the polar axis because of the rotationally symmetric properties assumed for the feed pattern.

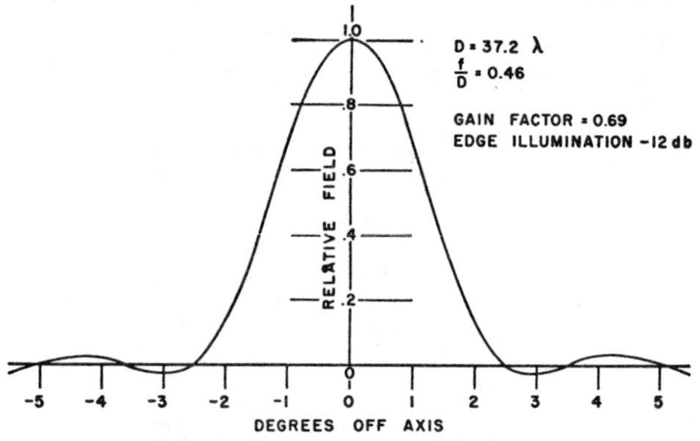

Fig. 7—Paraboloid diffraction pattern with plane-wave feed.

Diffraction patterns and gain factor of this paraboloid for various edge illuminations have been computed and are tabulated in Table II below.

TABLE II
PATTERN CHARACTERISTICS AND GAIN FACTOR OF A PARABOLOID EXCITED BY A PLANE-WAVE SOURCE
Aperture Diameter = 37.2 Wavelengths, $f/D = 0.46$

Edge Illumination (db)	Half-Power Beam-Width (degrees)	First Side-Lobe Level (db)	Gain Factor
− 8	1.74	−23.7	0.53
−10	1.80	−26	0.66
−12	1.82	−29.1	0.69
−16	1.96	−37.1	0.67
−20	2.06	−46	0.61
−30	2.36	−36.5	0.51

As the edge illumination decreases from -8 db, the width of the main beam increases as is expected. On the other hand, the first side-lobe level initially decreases, reaching a minimum for an edge illumination of about -20 db, and then increases. The gain factor is low for high edge illumination, where the aperture efficiency is high, but the spill-over loss is large. A maximum of gain is reached at an edge illumination of about -12 db. For still lower edge illuminations, the gain decreases because the aperture efficiency decreases.

Hyperboloid Lens with Plane-Wave Feed

The aperture field of a hyperboloid lens just outside the dielectric, fed by a plane-wave feed that has no variation of field intensity in the E-plane dimension B, and a half sinusoidal variation in field intensity across the H-plane dimension A, subject to the previous approximations, is

$$\overline{E}_a = j \frac{E_x 2.84 B^2}{\pi \lambda f} \epsilon^{-jk(f+nt_0)} \cdot \left\{ \frac{n \sin\left(\frac{\pi B}{\lambda}\right) \sin\psi \tan\psi \cos^2(\psi+\alpha)}{(n^2-1)\frac{\pi B}{\lambda} \sin\psi \sin(\psi+2\alpha)} \cdot \bar{x}\left[(1+\cos\psi)^2 + \sin^2\psi \cos 2\xi\right] + \bar{y} \sin^2\psi \sin 2\xi \right\}. \quad (30)$$

Eq. (29) can be written in terms of the approximating polynomials in r, the aperture radius, as

$$\overline{E}_a = j \frac{E_x 2.84 B^2 \epsilon^{-jk(f+nt_0)}}{\pi \lambda f} \{\bar{x}[f(r) + g(r) \cos 2\xi] + \bar{y} g(r) \sin 2\xi\}. \quad (31)$$

Proceeding as before, the far-zone field becomes

$$\overline{E} = -\frac{E_x 5.68 B^2 \epsilon^{-jk(f+nt_0+R)}}{\lambda^2 f R} \int_0^d \{\bar{x}[f(r) J_0(\beta r) - g(r) J_2(\beta r) \cos 2\xi'] - \bar{y} g(r) J_2(\beta r) \sin 2\xi'\} r dr. \quad (32)$$

It is found that in order to have the polynomials $f(r)$ and $g(r)$ approximate this aperture distribution closely, they must contain terms in r as

$$c_0 + c_1 r + c_2 r^2 + c_3 r^3.$$

Eq. (32) can be evaluated in terms of tabulated functions, since integration by parts puts each term either in the form of (23a) or[6]

$$\int_0^z (ax)^n J_n(ax) dx = 2^{n-1}\sqrt{\pi}\, \Gamma(n + \tfrac{1}{2}) z \cdot [J_{n+1}(az) S_n(az) - S_{n+1}(az) J_n(az)] + \frac{(az)^{n+1} J_n(az)}{(2n+1)a},$$

where $S_n(ax)$ is a Struve function.

[6] N. W. McLachlan and A. L. Meyers, "Integrals involving Bessel and Struve functions," *Phil. Mag.*, vol. 21, p. 437; February, 1936.

The power gain of the hyperboloid lens is computed in similar fashion to the computation of power gain of the paraboloid with plane-wave feed, in terms of the coefficients of the polynomial $f(r)$. The expressions for the diffraction field and power gain will not be written down since they are rather lengthy. However, some data obtained from these expressions will be presented. Fig. 8

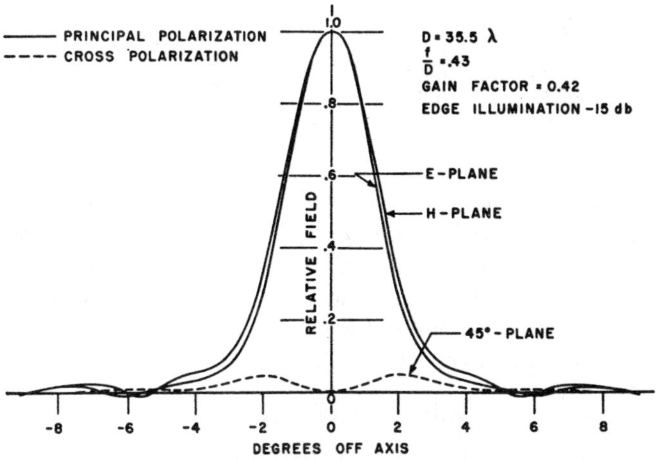

Fig. 8—Hyperboloid lens diffraction patterns with plane-wave feed.

shows the diffraction pattern of a hyperboloid lens, with a 35.5 wavelength diameter aperture, an f/D ratio of 0.41, and a refractive index of 1.57 excited by a plane-wave source to give a -15 db edge illumination at the curved surface of the lens. It is noticed that, although the illumination of the curved surface of the lens is symmetric about the polar axis, the E-plane pattern here is narrower than the H-plane pattern. The reason for this is because illumination in this plane is less tapered, since the transmission coefficient for the lens is greater in the E plane. The computed gain factor of the lens is a maximum for this edge illumination, at a value of 0.42, which is almost 2.2 db less than the 0.69 gain factor obtainable with the paraboloid of about the same f/D ratio. This decreased gain is due to the approximately 0.21 db reflection loss at each lens surface, coupled with the poor aperture efficiency and high spill-over loss of the lens. Table III, on the following page, summarizes the important pattern characteristics of the above lens and another lens, of refractive index 1.57 for various edge illuminations. This table shows that, for each of the lenses, the beam-widths in the principal planes are narrow for high-edge illumination, and wide for low-edge illumination. The gain factor, on the other hand, reaches a maximum for some intermediate illumination which represents the best compromise between high aperture efficiency and low spill-over loss. For each lens, the maximum level of the cross-polarized lobes decreases as the edge illumination is decreased, since the average amplitude of the cross-polarized aperture fields is less for low-edge illumination. The longer focal-length lens has lower cross-polarized lobes for a given edge illumination since it subtends a smaller angle at the feed point. The angular

TABLE III

COMPUTED PATTERN CHARACTERISTICS AND GAIN FACTOR OF TWO HYPERBOLOID LENSES EXCITED BY PLANE-WAVE SOURCES

D-Wave Lengths	f/D	Edge Illumination (db)	E-Plane Half-Power Beam Width (degrees)	H-Plane Half-Power Beam Width (degrees)	Max. Level of Cross-Polarization Lobe (db)	E-Plane First Side-Lobe Level (db)	H-Plane First Side-Lobe Level (db)	Gain Factor
35.5	0.43	−10	2.0	2.1	−31.4	−30.5	−28.4	0.35
35.5	0.43	−15	2.2	2.3	−32.0	−40	−46	0.42
35.5	0.43	−20	2.4	2.5	−32.8	−38	−34	0.38
35.5	0.43	−24	2.6	2.7	−34	−40	−30.5	0.31
35.5	0.77	−10	1.9	2.0	−38.7	−32.8	−40	0.53
35.5	0.77	−15	2.0	2.2	−39.2	−40	−45	0.53
35.5	0.77	−20	2.2	2.4	−40	−31.4	−40	0.47

position of the maximum of all the cross-polarized lobes for these lenses is about 2 degrees from the polar axis. There seems to be no systematic variation in the amplitude of the first side lobes in the principal plane patterns as edge illumination is varied, as there is in the case of the paraboloid reflector.

EXPERIMENTAL RESULTS

Experimental radiation patterns and power gain measurements have been made at an operating frequency of 35,000 mc for a paraboloid reflector fed with a small rectangular horn, and for two hyperboloid dielectric lenses fed by small rectangular horns. The dimensions of the reflector and lenses used in these experiments correspond to those used in the above theoretical computations. The paraboloid reflector used is a searchlight reflector with a contour accuracy of better than a hundredth of a wavelength. The rectangular feed horn is connected to the signal source behind the reflector by means of a waveguide bent in the H plane through a total angle of 180 degrees in two 90 degree steps. Thus, there is a small amount of aperture blocking caused by this waveguide, which is not taken into account in the theory. The hyperboloid lenses are machined from polystyrene; their contour accuracy also being on the order of a hundredth of a wavelength.

An experimental radiation pattern of the paraboloid reflector excited by a pyramidal-horn feed, which gives an edge illumination of −20 db in the E plane and −21.5 db in the H plane, is shown in Fig. 9. Here it is noticed the cross-polarized lobes are −25 db. It has been found that the level of these lobes decreases monotonically as the H-plane dimension of the horn is increased. Because the wave impedance of the mode in the horn approaches 377 ohms (the impedance of free space) as this dimension is increased, this behavior tends to confirm the theory that a paraboloid excited by a plane-wave source with a wave impedance of 377 ohms would have no cross-polarized lobes at all. The level of the first side lobe in the E plane is −28 db, while that in the H plane is −33 db. The H-plane pattern has a half-power beam-width of 2.1 degrees; while the E-plane pattern has a half-power beam-width of 2.0 degrees because the aperture illumination is more strongly tapered in this plane. Reference to Table II shows that the experimental half-power beam-widths agree closely with the theoretical values, while the side lobe levels in the two cases are quite different.

Experimental radiation patterns of a hyperboloid lens, with an f/D ratio of 0.42, excited by a horn feed that gives an edge illumination of −24 db in the E and H planes, are shown in Fig. 10. It is noticed that the main beam is relatively quite broad, but the side lobes are very low, the highest one being −41 db. The cross-polarized lobes are −29 db, and lie close to the polar axis (compared to the width of the main beam). The

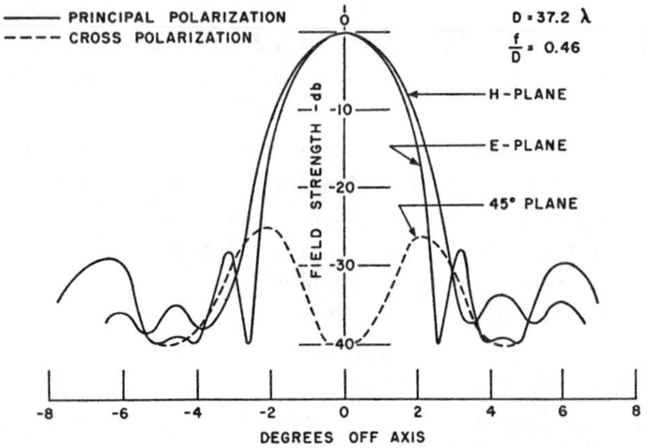

Fig. 9—Measured radiation patterns of a paraboloid reflector with a horn feed.

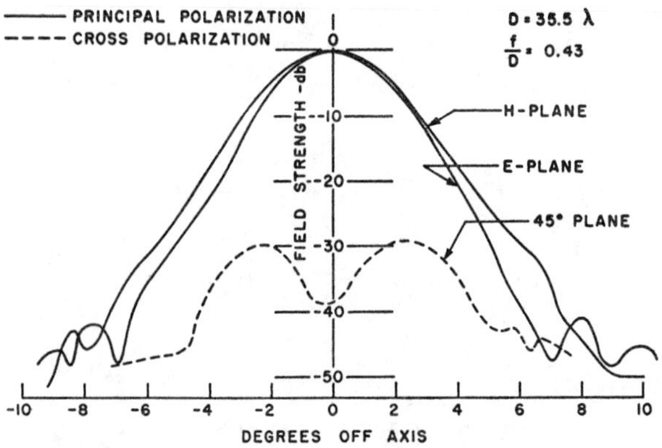

Fig. 10—Measured radiation patterns of hyperboloid lens with a horn feed.

TABLE IV

MEASURED PATTERN CHARACTERISTICS OF TWO HYPERBOLOID LENSES EXCITED BY PYRAMIDAL HORN FEEDS

D-Wave Lengths	f/D	Principal Plane Av. Illumination (db)	*E*-Plane Half-Power Beam-Width (degrees)	*H*-Plane Half-Power Beam-Width (degrees)	Max. Level of Cross-Polarization (db)	*E*-Plane First Side-Lobe Level (db)	*H*-Plane First Side-Lobe Level (db)	Gain Factor
35.5	0.42	−24	2.4	2.5	−29	−41	−42	0.26 ±0.013
35.5	0.77	−10	1.90	1.95	−28.5	−24.5	−27	0.42 ±0.021
35.5	0.77	−15	2.0	2.05	−31	−30	−30	0.41 ±0.021
35.5	0.77	−20	2.05	2.1	−35	−31.5	−31	0.390±0.020

measured gain factor of this lens is 0.26 which is low but consistent with the broad main beam. It is expected that the measured gain of the lenses should be about 8 per cent lower than the theoretical because the measured aperture efficiency of the experimental feed horns are only 75 per cent while the aperture efficiency of the plane-wave feeds used in the theoretical computation is 81 per cent. Allowing for this difference though, the experimental gain is still about 10 per cent less than theoretical. The additional discrepancy is probably accountable to absorption loss in the lens, and experimental error.

In Table IV, above, are summarized the results of measurements on the two hyperbolic lenses for various edge illuminations.

Comparison of the experimental measurements shown in Table IV with the theoretical results of Table III, for the lens with $f/D=0.77$, shows that the theory predicts the correct trend of the experimental results. The beamwidth of the experimental patterns is always narrower in the E plane than the H plane, and the principal plane beamwidths increase with decreasing edge illumination. However, the experimental beamwidths seem always to be narrower than theoretical. The measured gain factor is consistently about 20 per cent less than theoretical. The variation in amplitude of the measured cross-polarization lobes varies in the predicted fashion with edge illumination; however, the lobe values are always higher than theoretical. The measured side-lobe level in the principal planes, however, always decreases with decreasing-edge illumination.

CONCLUSION

It has been shown that low cross-polarized radiation lobes can be obtained from paraboloid reflectors when they are excited by small horns which approximate plane-wave sources. Low principal-polarization side lobes and low cross-polarized lobes can be obtained with a hyperboloid lens excited by a small horn, but the main beam is wide and the aperture efficiency is less than for paraboloids.

ACKNOWLEDGMENT

The work reported in this paper was sponsored by the Signal Corps Engineering Laboratory, Fort Monmouth, New Jersey.

Phase Centers of Microwave Antennas*

DAVID CARTER†

Summary—**This paper is concerned with the location of the phase centers of microwave antennas. The inadequacy of conventional aperture theory for the accurate description of phase centers is discussed. Formulas are developed and, for numerical indications, calculations are made for paraboloidal reflectors of different f/D ratios and a class of primary patterns which provide an approximate representation of a great many common feeds. The results are presented in graphical form to provide useful design information and show the dependence of principal E- and H-plane phase center location on feed and dish parameters. Contrary to the prediction of aperture theory, it is shown that the phase centers of axially symmetric antennas are not in the aperture plane, but they are dispersed about it.**

THERE IS a good deal of ambiguity on the location of the phase centers of microwave antennas. Different answers are obtained depending on the approach used. For some purposes, where antennas are rotated or used in interferometric systems, knowledge of phase center location is desirable and often necessary. For small antennas this information is rather easily measured in the laboratory. However, the large ranges required for the measurement of large aperture antennas at small wavelengths make such measurements exceedingly difficult. Reflections in the range, mechanical inaccuracies, etc., may well make accurate measurements impossible.

Theoretical calculations have been made to obtain phase distribution over the main lobe.[1] These calculations are based on the aperture distribution method of calculating far-field patterns[2–4] and show the phase and amplitude distributions in the far field for different assumed aperture-field distributions. While this theory satisfactorily predicts the far-field amplitude distributions for small angles off-axis, it cannot accurately locate the phase centers of real antennas.

The major difficulty is the fact that the total aperture-field distribution is generally not known and not used in the formulations referred to above. To obtain these formulations from information that might usually be known in microwave antennas, such as feed patterns (amplitude, phase and polarization) and reflector geometry, approximations are made which seriously affect the location of the phase center. In the first place the location of the "aperture" plane is somewhat arbitrary.[5] If one employs the usual methods of geometrical optics, for example, to obtain the aperture-phase and amplitude distributions from a point source feed pattern and a paraboloid reflector,[6] one obtains the same distributions on any two parallel planes which are perpendicular to the axis of the reflector. Calculations of the far-field patterns from these aperture distributions by means of the Fraunhofer field approximation,[7]

$$U_p \simeq \frac{je^{-jkR}}{\lambda R} \int_A F(\xi, \eta) e^{j(k_x \xi + k_y \eta)} d\xi d\eta,$$

then would be identical and one would obtain phase centers differing in location by the axial separation of these two arbitrarily chosen planes. The notation used here is the same as that of S. Silver.[4] (See Fig. 1.)

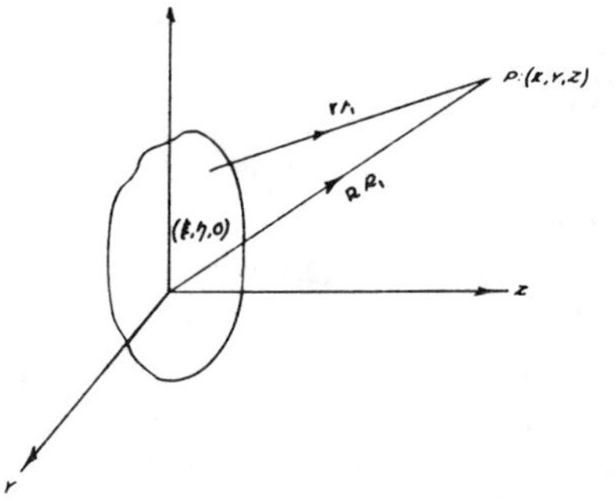

Fig. 1—Coordinate system. The origin is at the focal point of the reflector.

A second significant discrepancy occurs in the fact that there is no counterpart of the longitudinal component of reflector current in the aperture distribution.[8] This longitudinal component of current makes no significant contribution to the amplitude patterns at angles near the axis, and it has generally been neglected. However, it does contribute to the curvature of the far-field phase pattern and is, therefore, significant in the calculation of phase centers. For this reason, phase centers derived from the far-field patterns of representative aperture distributions,[9] will not coincide with the actual phase centers of the antennas represented by these aperture distributions. In fact it will be shown that paraboloids illuminated by point source feeds with axial symmetric patterns have principal plane phase centers which are not coincident and whose location depends on the feed and dish parameters. This is in contrast to the results of aperture theory which would pre-

* Manuscript received by the PGAP, November 10, 1955; revised manuscript received, July 30, 1956.

† San Jose State College, San Jose, Calif.

[1] C. C. Allen, "Radiation patterns for aperture antennas with nonlinear phase distributions," 1953 IRE CONVENTION RECORD, Part 2, pp. 9–12.

[2] R. C. Spencer, "Fourier Integral Methods of Pattern Analysis," R. L. Rep. No. 762-1; January 21, 1956.

[3] H. T. Friis and W. D. Lewis, "Radar antennas," *Bell Sys. Tech. J.*, vol. 26, pp. 232–246; April, 1947.

[4] Silver, S., "Microwave Antenna Theory and Design," McGraw-Hill Book Co., Inc., New York, N. Y., 1949.

[5] *Ibid.*, p. 158.

[6] *Ibid.*, p. 419.

[7] *Ibid.*, p. 173.

[8] *Ibid.*, p. 420.

[9] Allen, *op. cit.*, p. 9.

Reprinted from *IRE Trans. Antennas Propagat.*, vol. AP-4, pp. 597–600, Oct. 1956.

dict coincident phase centers located in the aperture plane in such cases.

This latter statement follows directly from the fact that the Fourier transform of an even function is real. Thus the Fraunhofer field of an aperture distribution, $F(\xi, \eta)$, is given by[10]

$$U_p = \frac{j}{\lambda R} e^{-jkR} \int_{-\infty}^{\infty} \int_{-\infty}^{\infty} u(\xi, \eta) e^{j(k_x \xi + k_y \eta)} d\xi d\eta$$

where $u(\xi, \eta)$ vanishes everywhere except inside the aperture area, wherein it coincides with the aperture distribution. Since the integral is real if $u(\xi, \eta)$ is a real (except for an arbitrary complex constant) function even with respect to both of its arguments, the phase of U_p will be independent of direction, (θ, ϕ), except for discrete points when U_p passes through zero making the phase change by π radians. In the case of the paraboloid illuminated by a point source feed with axial symmetric patterns, having negligible cross-polarization component of reflected field, the aperture distribution of the principal polarization component is axially symmetric and phase constant.[11] Therefore the corresponding aperture distribution representation, $u(\xi, \eta)$ possesses more than enough symmetry to make its Fourier transform real. Hence the equiphase surface in the region of the main lobe in the Fraunhofer field is given by kR = constant, which is a sphere with center at the origin of coordinates. Thus, for this example, the aperture distribution approximation predicts that there will be a single point phase center located at the origin which is taken at the center of the aperture plane.

To obtain a more accurate description of the phase center of such antennas, the phase distribution in the far field will be calculated from the current distribution over the reflector. Inclusion of the longitudinal component of reflector current will be seen to separate the principal E- and H-plane centers of phase. In addition this approach will locate these phase centers with respect to the vertex of the paraboloid. This will be done in terms of a primary feed gain parameter, and the paraboloids angular aperture to show the variation of these phase center locations with feed and dish parameters.

Choosing a class of primary patterns which provide an approximate representation of a great many common feeds, the axially symmetric point source primary feed gain function will be taken as (see Fig. 2)

$$G_p(\psi) = G(0) \cos^n \psi = G(0) \left(\frac{1 - x^2}{1 + x^2}\right)^n .$$

Then assuming as before that for most feeds of interest the cross polarization component of reflected field may be neglected in calculating the principal polarization diffraction patterns, it has been shown that

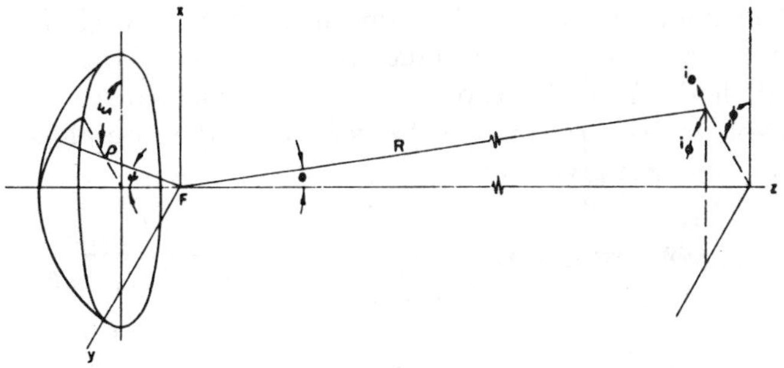

Fig. 2—Coordinate system. The aperture is in the $x=y$ plane.

principal E- and H-plane patterns of the radiation field of such a paraboloid antenna is given by[12]

$$\vec{E}(\theta) = C \frac{e^{-j(kR + 2kf \cos^2(\theta/2))}}{R} \cdot \begin{cases} \vec{i}_\theta [I_{z2} \sin\theta - I_{x1} \cos\theta + j(I_{z1} \sin\theta + I_{x2} \cos\theta)] & \text{in the } E\text{-plane} \\ \pm \vec{i}_\phi (I_{x1} - jI_{x2}) & \text{in the } H\text{-plane} \end{cases}$$

where

$$I_{x1} = \int_0^X \left(\frac{1 - x^2}{1 + x^2}\right)^{n/2} \frac{x}{1 + x^2} \cdot \cos\left(2kfx^2 \sin^2 \frac{\theta}{2}\right) J_0(2kfx \sin\theta) dx,$$

$$I_{x2} = \int_0^X \left(\frac{1 - x^2}{1 + x^2}\right)^{n/2} \frac{x}{1 + x^2} \cdot \sin\left(2kfx^2 \sin^2 \frac{\theta}{2}\right) J_0(2kfx \sin\theta) dx,$$

$$I_{z1} = \int_0^X \left(\frac{1 - x^2}{1 + x^2}\right)^{n/2} \frac{x^2}{1 + x^2} \cdot \cos\left(2kfx^2 \sin^2 \frac{\theta}{2}\right) J_1(2kfx \sin\theta) dx,$$

$$I_{z2} = \int_0^X \left(\frac{1 - x^2}{1 + x^2}\right)^{n/2} \frac{x^2}{1 + x^2} \cdot \sin\left(2kfx^2 \sin^2 \frac{\theta}{2}\right) J_1(2kfx \sin\theta) dx,$$

$$x = \tan \frac{\psi}{2}$$

and C is a constant equal to

$$-4j\omega\mu f \left[\frac{n+1}{\pi} \left(\frac{\epsilon}{\mu}\right)^{1/2} P_T\right]^{1/2} .$$

[10] Silver, *op. cit.*, p. 174.
[11] *Ibid.*, p. 419.
[12] D. Carter, "Wide angle radiation in pencil beam antennas," *J. Appl. Phys.*, vol. 26, pp. 645–652; June, 1955.

The notation here is the same as Carter's,[12] and the origin of coordinates is taken at the focus of the paraboloid. From this expression for the principal E- and H-plane patterns, it can be seen that the equiphase contours are given by

$$R = \begin{cases} \text{const.} - 2f\cos^2\dfrac{\theta}{2} + \dfrac{1}{k}\tan^{-1}\dfrac{I_{z1}\sin\theta + I_{x2}\cos\theta}{I_{z2}\sin\theta - I_{x1}\cos\theta} & \text{in the } E\text{-plane} \\ \text{const.} - 2f\cos^2\dfrac{\theta}{2} - \dfrac{1}{k}\tan^{-1}\dfrac{I_{x2}}{I_{x1}} & \text{in the } H\text{-plane} \end{cases}$$

Here one can see the effect of the longitudinal component of reflector current in separating the principal E- and H-plane phase centers. Thus if I_{z1} and I_{z2} were negligible the equiphase contour in the E plane would reduce to the same expression as obtained for the H plane. The fact that the longitudinal component of reflector current contributes to the principal E-plane equiphase contour and does not contribute in the H plane is not surprising. Consideration of the distribution of this component on the paraboloid as represented in Fig. 3 shows that it has odd symmetry about the H plane and the contributions from both sides of this plane cancel everywhere in the plane. This does not happen in the principal E plane because the longitudinal component of reflector current has even symmetry about that plane.

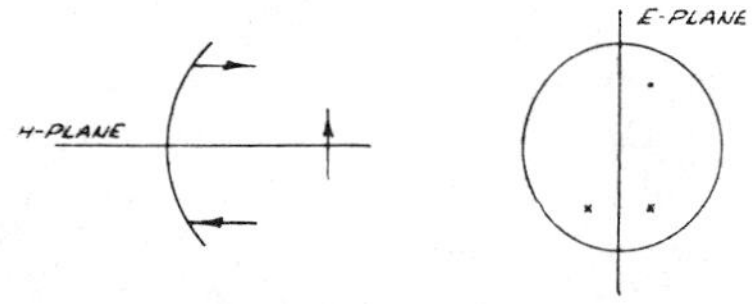

Fig. 3—Longitudinal component of reflector current. The distribution of this current component is represented by the horizontal arrows and the dots and crosses to illustrate the symmetry about the principal E and H planes. The vertical arrow at the focus indicates the polarization of the feed.

To obtain the phase centers, the radius of curvature of the equiphase contours will be evaluated on the axis of the main lobe. If $R(\theta)$ is the equiphase contour, the radius of curvature, ρ, is given by

$$\rho = R\,\frac{\left[1 + \dfrac{1}{R^2}\left(\dfrac{dR}{d\theta}\right)^2\right]^{3/2}}{1 + \dfrac{2}{R^2}\dfrac{dR}{d\theta} - \dfrac{1}{R}\dfrac{d^2R}{d\theta^2}}.$$

The phase patterns of axially symmetric antennas are axially symmetric and have continuous first derivatives. Therefore, on the axis, the first derivative of R with respect to θ vanishes and

$$\rho(0) = \frac{R(0)}{1 - \left(\dfrac{1}{R}\dfrac{d^2R}{d\theta^2}\right)_{\theta=0}} \simeq R(0) + \left[\frac{d^2R}{d\theta^2}\right]_{\theta=0}.$$

in the far zone. Hence the phase center lies behind the origin a distance equal to

$$\rho(0) - R(0) = \left[\frac{d^2R}{d\theta^2}\right]_{\theta=0}.$$

It will be convenient to express the phase center locations as fractions of the focal length in front of the vertex of the paraboloid. Calling this normalized distance ζ, it can be seen that

$$\zeta = \frac{f - [\rho(0) - R(0)]}{f} = 1 - \frac{1}{f}\left[\frac{d^2R}{d\theta^2}\right]_{\theta=0}.$$

Performing the somewhat laborious operations indicated it can be shown that for $\theta = 0$,

$$\frac{\partial I_{x2}}{\partial\theta} = I_{x2} = I_{z1} = I_{z2} = 0$$

$$\frac{\partial^2 I_{x2}}{\partial\theta^2} = \frac{\partial I_{z1}}{\partial\theta} = kf\int_0^X \left(\frac{1-x^2}{1+x^2}\right)^{n/2}\frac{x^3}{1+x^2}\,dx$$

and

$$\zeta = \frac{\displaystyle\int_0^X \left(\frac{1-x^2}{1+x^2}\right)^{n/2}\frac{x^3}{1+x^2}\,dx}{\displaystyle\int_0^X \left(\frac{1-x^2}{1+x^2}\right)^{n/2}\frac{x}{1+x^2}\,dx}\;X \quad \begin{matrix} 1 & \text{in the } H\text{-plane} \\ \text{or} & \\ 3 & \text{in the } E\text{-plane.} \end{matrix}$$

This last formula locates the phase centers as a function of the primary feed gain parameter, n, and the angular aperture of the paraboloid, $\Psi = 2\tan^{-1}x$. Choosing a range of the parameters n and Ψ to obtain representations of the most common configurations, the phase center locations were evaluated and plotted in Figs. 4 and 5. For purposes of comparison a plot of aperture plane location vs angular aperture was added to Fig. 5.

Contrary to the prediction of aperture theory, Figs. 4 and 5 indicate that the phase centers are not in the aperture plane, but they are dispersed about it. For the most part, they lie between the aperture plane and the vertex which is the region to the left of the aperture plane locus in Fig. 5. The H-plane phase centers always lie behind the aperture plane, near the vertex. This is also true for the E-plane phase centers in most cases. However, the axial component of reflector current, which does contribute in the E plane, pushes the phase centers further ahead of the vertex than their corresponding location in the principal H plane. The reason for this is that the reflector's curvature forces the axial component of current distribution to lie ahead of the vertex, producing a maximum of this component distribution near the aperture plane. In some cases of low primary feed gain and small angular aperture, Fig. 5 indicates that the principal E-plane phase centers are located in front of the aperture plane.

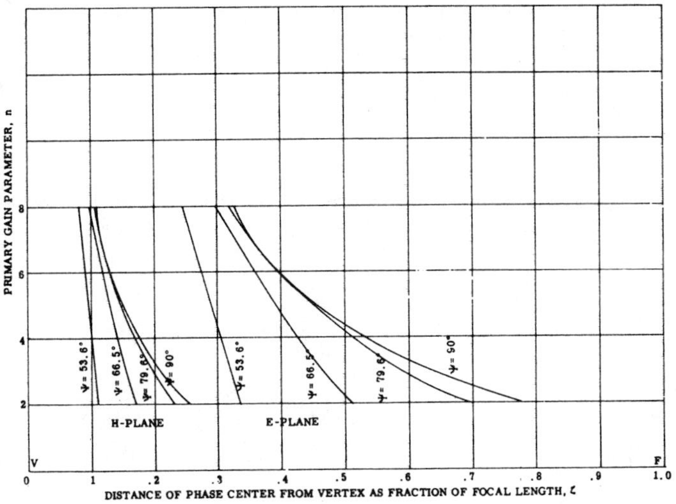

Fig. 4—Principal plane phase centers. These curves show the variation of the phase center locations with primary feed gain in the principal E and H planes of a paraboloid reflector fed by a primary having a gain function.

$$G_p = \begin{cases} 2(n+1)\cos^n \psi & \text{for } 0 \leqq \psi \leqq \dfrac{\pi}{2} \\ 0 & \text{for } \dfrac{\pi}{2} \leqq \psi \leqq \pi. \end{cases}$$

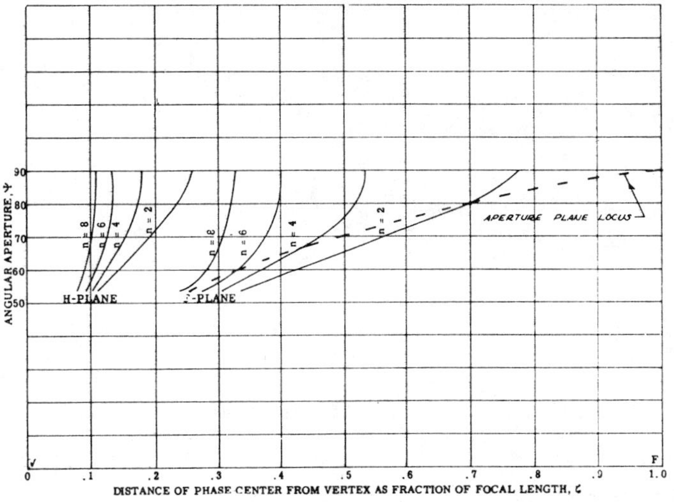

Fig. 5—Principal plane phase centers. These curves show the variation of the phase center locations with angular aperture in the principal E and H planes of a paraboloid reflector fed by a primary having a gain function,

$$G_p = \begin{cases} 2(n+1)\cos^n \psi & \text{for } 0 \leqq \psi \leqq \dfrac{\pi}{2} \\ 0 & \text{for } \dfrac{\pi}{2} \leqq \psi \leqq \pi. \end{cases}$$

Finally it should be noted from Figs. 4 and 5 that, in general, decreasing the paraboloid angular aperture and increasing primary feed gain moves both principal E- and H-plane phase centers towards the vertex of the paraboloid. This was to be expected since either of these two variations tends toward the limiting case of reflection by a plane mirror of a point source, a case in which image theory tells us that the phase center is located in the reflecting plane. Thus decreasing the angular aperture means either decreasing the paraboloid diameter for a fixed focal length or increasing the focal length for a fixed diameter. In the former case the aperture plane is moved back toward the vertex and in the latter case the vertex is moved forward toward the aperture plane, both variations effectively producing a flatter reflector. Similarly increasing primary feed gain increases the current in the zero curvature vertex region at the expense of the currents near the aperture plane, effectively producing a flatter current distribution. Since either of these variations produces a flatter current distribution located near the vertex, the phase centers should move together toward the vertex.

In conclusion, it should be stressed that these results apply strictly only for feeds which are axially symmetric point sources located at the focus of the paraboloid. This provides an approximation for a great many common feeds. However, the phase centers of the reflected radiation field can be varied and even pushed behind the dish by defocusing and separating the primary feed phase centers.

Feed Polarization for Parallel Currents in Reflectors Generated by Conic Sections

IRWIN KOFFMAN, MEMBER, IEEE

Abstract—The family of surfaces of revolution obtained from conic sections are the sphere, ellipsoid, paraboloid, hyperboloid, and plane. These "conic" surfaces find various uses as reflectors in antenna systems because of their focusing or imaging properties. When any of the "conic" reflectors are used, it is of interest to specify an "ideal" polarization of the incident field such that the currents induced in the reflector flow in parallel paths. This specification will permit the reflector to be formed by parallel conducting wires or slats.

This paper presents a method for determining the "ideal" feed polarization that should be incident upon a "conic" reflector. It is specified in terms of the polarization characteristics of elemental electric and magnetic dipoles. It will be shown that for any "conic" surface, the "ideal" polarization can always be specified in terms of crossed electric and magnetic dipoles whose relative intensities are related to the eccentricity of the "conic" reflector.

Introduction

THE POLARIZATION characteristics of a feed illuminating a reflector affects the direction of currents induced in the reflector. The direction of such currents determines the far-field polarization of the antenna and is of particular interest when the reflecting surface is polarization-sensitive. This paper derives the condition which guarantees that the induced currents flow in parallel paths in any of the surfaces of revolution formed by conic sections. These surfaces are the sphere, ellipsoid, paraboloid, hyperboloid, and plane. This is of practical interest when an efficient "conic" reflector is to be formed by parallel conducting wires or slats. It will be shown that for any "conic" surface, illuminated by a feed at the reflector focus, the "ideal" feed polarization for parallel current flow can always be simply specified in terms of crossed electric and magnetic dipoles in the proper ratio. The generalization requires that the relative intensity of the crossed dipole pair be related to the eccentricity (ϵ) of the "conic" surface [1].

It is well known [2], [3] that when the reflector is a paraboloid, a feed having the polarization characteristics of a pair of crossed electric and magnetic dipoles of equal intensity, and located at the focus, will induce currents in the reflector that are everywhere parallel. This is illustrated in Fig. 1, where the electric and magnetic dipole pair are represented by a short current element and a small loop, respectively. Such a feed, commonly called a Huygens source, is "ideal" in the sense that the reflector need not be solid but may be formed by closely-spaced parallel conductors; in addition, the far-field radiation from the reflector is free of cross polarization. The parallel conductor type of construction can always be used to form an efficient reflector if the feed causes the induced currents in an otherwise solid reflector to flow in parallel paths.

When the reflector is not a paraboloid, but is any member of the family of "conic" surfaces, the "ideal" polarization of the source may be expressed as a combination of electric and magnetic dipoles in the proper intensity ratio. These reflectors, shown in cross section in Fig. 2, all exhibit focusing or imaging properties and find use in various reflector systems.

The most common antenna reflector is, of course, the paraboloid; however, other types find various uses. For example, hyperboloid reflectors (either solid or formed by parallel conducting elements) are used in Cassegrain reflecting systems [4]. In this case, a feed located between hyperboloid and paraboloid reflectors is imaged at the focus of the paraboloid. Other reflector systems may use elliptical reflectors to refocus the feed radiation to a point. An example is the Gregorian system which

Manuscript received June 7, 1965; revised August 23, 1965.
The author is with Wheeler Laboratories, Inc., Smithtown, N. Y.

Reprinted from *IEEE Trans. Antennas Propagat.*, vol. AP-14, pp. 37–40, Jan. 1966.

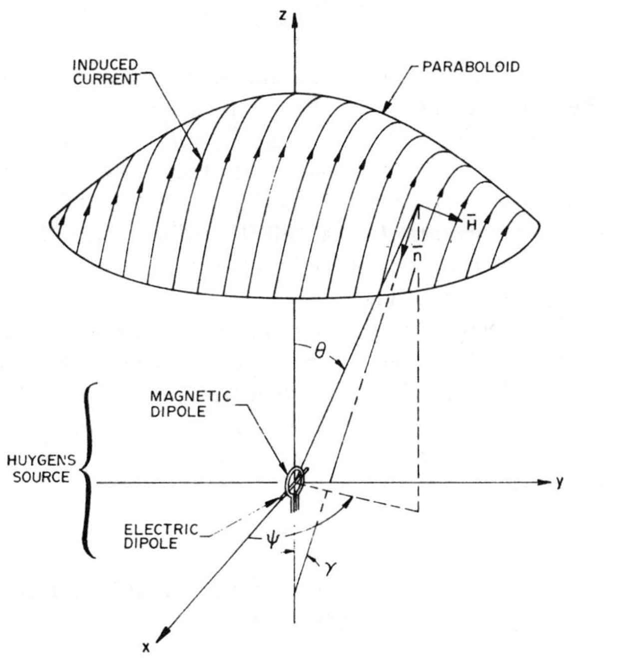

Fig. 1. Induced currents in a paraboloid excited by a Hugyens source.

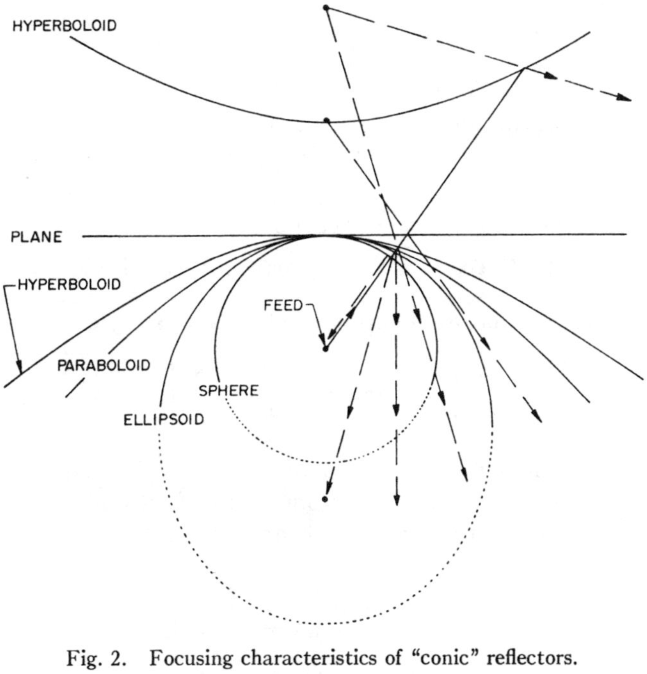

Fig. 2. Focusing characteristics of "conic" reflectors.

uses ellipsoid and paraboloid reflectors. Plane reflectors are commonly used with paraboloids in microwave radio relay systems.

For any "conic" reflector, it is of interest to determine the "ideal" feed which induces parallel currents so that it can be compared against an actual feed. In the case of a reflector formed by parallel wires or slats, the departure from the "ideal" determines the percentage of feed radiation that passes through the reflector.

Electric and Magnetic Dipoles

Electric and magnetic dipoles are elemental point source radiators that are convenient models which, in practice, cannot be perfectly realized. Their value as a concept is useful in the present discussion since they can provide a means for comparing the polarization characteristics of an actual feed against the ideal.

The electric dipole may be regarded as a short current element carrying a time-varying current. When oriented so that its dipole moment is in the $-x$ direction (Fig. 1), the E and H fields have components in the $\bar{\theta}$ and $\bar{\phi}$ directions as follows:

Electric Dipole:

$$E_\theta = A \cos\theta \cos\phi \tag{1}$$

$$E_\phi = A \sin\phi \tag{2}$$

$$H_\theta = \frac{A}{\eta} \sin\phi \tag{3}$$

$$H_\phi = \frac{A}{\eta} \cos\theta \cos\phi \tag{4}$$

where A is constant at a fixed radius from the element, and η is the characteristic impedance of free space.

Analogous to the electric dipole is the magnetic dipole. It may be regarded as a small current-carrying loop. When the magnetic dipole is oriented so that its dipole moment is in the $-y$ direction (Fig. 1), the E and H fields have components in the $\bar{\theta}$ and $\bar{\phi}$ direction as follows:

Magnetic Dipole:

$$E_\theta = B \cos\phi \tag{5}$$

$$E_\phi = B \sin\phi \cos\theta \tag{6}$$

$$H_\theta = \frac{B}{\eta} \sin\phi \cos\theta \tag{7}$$

$$H_\phi = \frac{B}{\eta} \cos\phi \tag{8}$$

where B is a constant.

A combination of crossed magnetic and electric dipoles gives rise to a field that is a linear combination of the electric and magnetic dipole fields. A special case is when each dipole field is of equal intensity. Such a radiator is commonly referred to as a Huygens source. In general, though, the E and H fields of the combination are given as follows:

$$E_\theta = B \cos\phi(1 + X \cos\theta) \tag{9}$$

$$E_\phi = B \sin\phi(X + \cos\theta) \tag{10}$$

$$H_\theta = \frac{B}{\eta} \sin\phi(X + \cos\theta) \tag{11}$$

$$H_\phi = \frac{B}{\eta} \cos\phi(1 + X \cos\theta) \tag{12}$$

where X is the relative strength of electric to magnetic dipole.

Figure 3 illustrates the polarization characteristics of the crossed electric and magnetic dipole pair for various

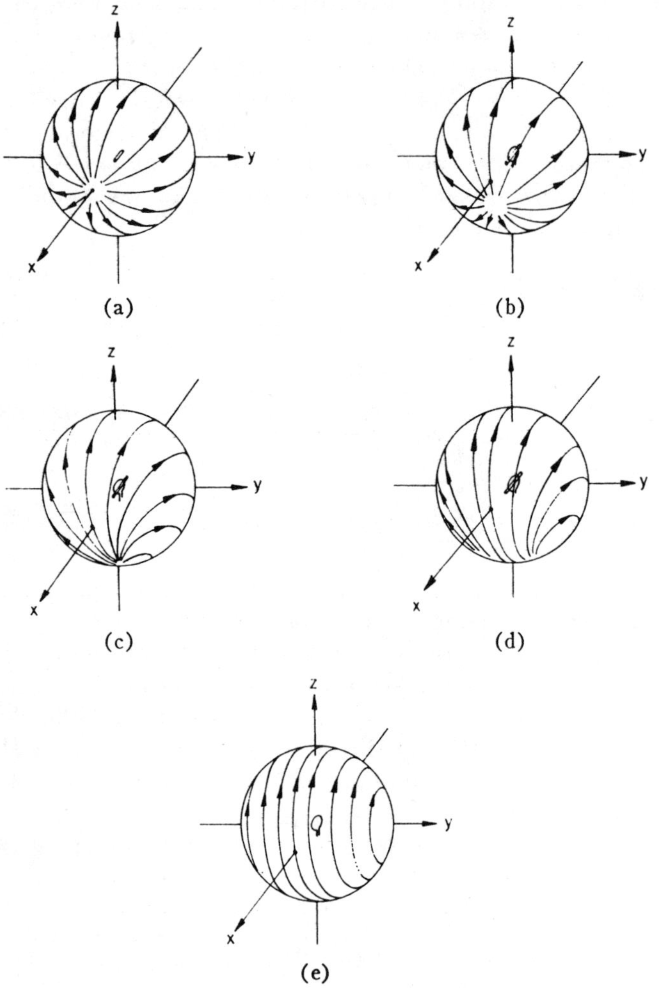

Fig. 3. Polarization of elemental radiators. (a) Electric dipole. (b) Electric dipole > magnetic dipole. (c) Electric dipole = magnetic dipole (Hugyens source). (d) Electric dipole < magnetic dipole. (e) Magnetic dipole.

intensity ratios of electric to magnetic dipole. These sketches show the electric field lines plotted on a sphere about the elemental radiators. Note that when the ratio is infinite, $X=\infty$, the radiator consists of only the electric dipole, Fig. 3(a). When $X=3$, the radiator is a Huygens source Fig. 3(c). When $X=0$, Fig. 3(e), the radiator consists of only the magnetic dipole. Of interest is the case, when $X<1$ (electric dipole intensity less than magnetic dipole), since this approximates the polarization characteristics of a horn-type radiator (excited by a TE mode) whose aperture dimensions are in the order of a wavelength. In the case when the horn aperture is very large, the polarization of the horn is similar to the Huygens source, Fig. 3(c). For very small apertures the polarization resembles the magnetic dipole case, Fig. 3(e).

"Ideal" Polarization

The "ideal" polarization characteristics of a radiator illuminating a reflector is deduced from the boundary conditions that must be satisfied at the surface of the reflector. These conditions relate the incident magnetic field to the induced currents as follows:

$$\bar{J} = 2(\bar{n} \times \bar{H}). \tag{13}$$

Note that $\bar{n}$ represents the unit normal at any point on the reflector and, therefore, defines the surface contour or type of conic reflector, while $\bar{H}$ represents the incident magnetic field and, therefore, relates to the polarization of the feed. By properly specifying the direction of the induced currents $\bar{J}$, the "ideal" polarization of a feed located at the focus of conic reflector can be determined by application of (13).

The specification which requires that $\bar{J}$ flow in parallel current paths everywhere in the reflector (Fig. 1) is deemed "ideal" since this condition permits the use of a reflector formed by simple parallel elements. For the geometry shown, the condition for parallel current flow requires that

$$J_y = 0. \tag{14}$$

Expressing the unit normal $\bar{n}$ and the incident $\bar{H}$ field in rectangular coordinates as follows:

$$\bar{n} = (-\sin\gamma)(\cos\phi)\bar{x} - (\sin\gamma)(\sin\phi)\bar{y} - (\cos\gamma)\bar{z} \tag{15}$$

where

$$\gamma = \arc\tan\left(\frac{\sin\theta}{\epsilon + \cos\theta}\right),$$

$$\epsilon = \text{eccentricity of "conic" surface} \tag{16}$$

and

$$\bar{H} = (H_\theta \cos\theta\cos\phi - H_\phi \sin\phi)\bar{x} + (H_\theta\cos\theta\sin\phi + H_\phi\cos\phi)\bar{y} - (H_\theta\sin\theta)\bar{z}, \tag{17}$$

it is found that application of (13) and (14) yields,

$$\frac{H_\theta}{H_\phi} = \tan\phi\left[\frac{\epsilon + \cos\theta}{1 + \epsilon\cos\theta}\right]. \tag{18}$$

Recognizing that (18) is the same as the following equation for the radiation from a pair of crossed electric and magnetic dipoles [see (11) and (12)],

$$\frac{H_\theta}{H_\phi} = \tan\phi\left[\frac{X + \cos\theta}{1 + X\cos\theta}\right], \tag{19}$$

it can be concluded that such a feed will induce currents in the desired direction. The only specification regards the ratio of electric to magnetic dipole strength. This means that the "ideal" polarization of the feed illuminating a "conic" reflector should be the same as the polarization characteristics of a pair of crossed electric and magnetic dipoles whose relative strength X is equal to the eccentricity ϵ of the reflector, i.e.,

$$\epsilon = X. \tag{20}$$

For example, if the reflector is a paraboloid, $\epsilon = 1 = X$, the "ideal" polarization as previously noted is that of a Huygens source.

Table I lists the "ideal" polarization characteristics of the feed required for each of the conic surfaces. These

TABLE I
"IDEAL" FEED FOR EACH TYPE OF "CONIC" REFLECTOR

Conic Surface	Eccentricity	Feed
Sphere	$\epsilon=0$	Magnetic Dipole (Electric Dipole=0)
Ellipsoid	$0<\epsilon<1$	Electric+Magnetic Dipole ($ED=\epsilon\ MD$)
Paraboloid	$\epsilon=1$	Electric+Magnetic Dipole ($ED=MD$)
Hyperboloid	$1<\epsilon<\infty$	Electric+Magnetic Dipole ($ED=\epsilon\ MD$)
Plane	$\epsilon=\infty$	Electric Dipole (Magnetic Dipole=0)

polarization characteristics are expressed in terms of the polarization characteristics of elemental electric and magnetic dipole radiators in the proper intensity ratio.

ACKNOWLEDGMENT

The author would like to thank P. W. Hannan of Wheeler Laboratories, Inc., for helpful discussions on this subject.

REFERENCES

[1] I. Koffman, "Reflectiving surfaces formed by wire grids," MEE Report, Polytechnic Institute of Brooklyn, Brooklyn, N. Y., June 1961.
[2] C. C. Cutler, "Parabolic-antenna design for microwaves," *Proc. IRE*, vol. 35, pp. 1284–1294, November 1947.
[3] E. M. T. Jones, "Paraboloid reflector and hyperboloid lens antennas," *IRE Trans. on Antennas and Propagation*, vol. AP-2, pp. 119–127, July 1954.
[4] P. W. Hannan, "Microwave antennas derived from the Cassegrain telescope," *IRE Trans. on Antennas and Propagation*, vol. AP-9, pp. 140–153, March 1961.

Bibliography for Part I

A. General Properties of Reflectors

[1] P. J. B. Clarricoats and G. T. Poulton, "High-efficiency microwave reflector antennas—A review," *Proc. IEEE*, vol. 65, pp. 1470–1504, Oct. 1977.
[2] K. S. Kelleher, "Relations concerning wave fronts and reflectors," *J. Appl. Phys.*, vol. 21, pp. 573–576, June 1950.
[3] C. J. Sletten, R. B. Mack, W. G. Mavroides, and H. M. Johnson, "Corrective line sources for paraboloids," *IRE Trans. Antennas Propagat.*, vol. AP-6, pp. 239–251, July 1958.
[4] M. K. Hu, "Fresnel region field distributions of circular apertures," *IRE Trans. Antennas Propagat.*, vol. AP-8, pp. 344–346, May 1960.
[5] R. T. Nash, "Beam efficiency limitations of large antennas," *IEEE Trans. Antennas Propagat.*, vol. AP-12, pp. 918–923, Dec. 1964.
[6] H. C. Ko, "Radio-telescope parameters," *IEEE Trans. Antennas Propagat.*, vol. AP-12, pp. 891–898, Dec. 1964.
[7] C. J. Sletten and P. Blacksmith, "The paraboloid mirror," *Appl. Opt.*, vol. 4, pp. 1239–1251, Oct. 1965.
[8] L. Peters and T. E. Kilcoyne, "Radiating mechanisms in a reflector antenna system," *IEEE Trans. Electromagn. Compat.*, vol. EMC-7, pp. 368–374, Dec. 1965.
[9] G. F. Koch, "Parabolantennen mit geringer Rauschtemperatur," *NTZ*, vol. 18, pp. 324–330, 1965.

B. Beam-Shaping Reflectors

[10] A. S. Dunbar, "Calculation of doubly curved reflectors for shaped beams," *Proc. IRE*, vol. 36, pp. 1289–1296, Oct. 1948.
[11] T. F. Carberry, "Analysis theory for the shaped-beam doubly curved reflector antenna," *IEEE Trans. Antennas Propagat.*, vol. AP-17, pp. 131–138, Mar. 1969.
[12] A. Brunner, "Possibilities of dimensioning doubly curved reflectors for azimuth-search radar antennas," *IEEE Trans. Antennas Propagat.*, vol. AP-19, pp. 52–57, Jan. 1971.
[13] C. F. Winter, "Dual vertical beam properties of doubly curved reflectors," *IEEE Trans. Antennas Propagat.*, vol. AP-19, pp. 174–180, Mar. 1971.
[14] A. P. Norris and B. S. Westcott, "Realisation of generalized far fields by reflector synthesis," *Electron. Lett.*, vol. 10, pp. 322–324, July 25, 1974.
[15] F. Brickell and B. S. Westcott, "Reflector design as an initial-value problem," *IEEE Trans. Antennas Propagat.*, vol. AP-24, pp. 531–533, July 1976.

C. Side and Back Radiation in Reflector Antennas

[16] L. B. Tartakovskii, "Side radiation from ideal paraboloid with circular aperture," *Radio Eng. Electron. Phys.*, vol. 4, pp. 14–28, June 1959.
[17] B. Ye Kinber, "Lateral radiation of parabolic antennas," *Radio Eng. Electron. Phys.*, vol. 6, pp. 481–492, Apr. 1961.
[18] H. N. Kritikos, "The extended aperture method for the determination of the shadow region radiation of parabolic reflectors," *IEEE Trans. Antennas Propagat.*, vol. AP-11, pp. 400–404, July 1963.
[19] L. Lewin, "Main reflector rim diffraction in back direction," *Proc. Inst. Elec. Eng.*, vol. 119, pp. 1100–1102, Aug. 1972.
[20] G. L. James and V. Kerdemelidis, "Selective reduction in back radiation from paraboloidal reflector antennas," *IEEE Trans. Antennas Propagat.*, vol. AP-21, pp. 886–887, Nov. 1973.
[21] C. M. Knop, "On the front to back ratio of a parabolic

dish antenna," *IEEE Trans. Antennas Propagat.*, vol. AP-24, pp. 109–111, Jan. 1976.

D. Miscellaneous Forms of Reflectors

[22] K. S. Kelleher, "A new microwave reflector," in *IRE Nat. Conv. Rec.*, vol. 1, part 2, 1953, pp. 56-57.

[23] R. C. Spencer, F. S. Holt, H. M. Johanson, and J. Sampson, "Double parabolic cylinder pencil-beam antenna," *IRE Trans. Antennas Propagat.*, vol. AP-3, pp. 4–8, Jan. 1955.

[24] G. D. Peeler and D. H. Archer, "A toroidal microwave reflector," in *IRE Nat. Conv. Rec.*, vol. 1, part 1, 1956, pp. 242–247.

[25] S. R. Jones and K. S. Kelleher, "A new low noise, high gain antenna," in *IEEE Int. Conv. Rec.*, vol. II, part 1, 1963, pp. 11–17.

[26] D. G. Berry, R. G. Malech, and W. A. Kennedy, "The reflectarray antenna," *IEEE Trans. Antennas Propagat.*, vol. AP-11, pp. 645–651, Nov. 1963.

[27] A. C. Schell, "The multiplate antenna," *IEEE Trans. Antennas Propagat.*, vol. AP-14, pp. 550–560, Sept. 1966.

[28] A. B. Crawford and R. H. Turrin, "A packaged antenna for short-hop microwave radio systems," *Bell Syst. Tech. J.*, vol. 48, pp. 1605–1622, July–Aug. 1969.

[29] L. G. Josefsson, "A broad-band twist reflector," *IEEE Trans. Antennas Propagat.*, vol. AP-19, pp. 552–554, July 1971.

[30] H. W. Ehrenspeck, "A new class of medium size high efficiency reflector antennas," *IEEE Trans. Antennas Propagat.*, vol. AP-22, pp. 329–332, Mar. 1974.

[31] C. Dragone, "An improved antenna for microwave radio systems consisting of two cylindrical reflectors and a corrugated horn," *Bell Syst. Tech. J.*, vol. 53, pp. 1351–1377, Sept. 1974.

[32] G. Hyde, R. W. Kreutel and L. V. Smith, "The unattended earth terminal multiple-beam torus antenna," *COMSAT Tech. Rev.*, vol. 4, Fall 1974.

E. Deployable and Erectable Reflectors

[33] J. A. Fager and R. Garriott, "Large-aperture expandable truss microwave antenna," *IEEE Trans. Antennas Propagat.*, vol. AP-17, pp. 452–458, July 1969.

[34] A. C. Ludwig, "A new geometry for unfurlable antennas," *Microwaves*, vol. 9, pp. 41–42, Nov. 1970.

[35] A. C. Ludwig, "Conical-reflector antennas," *IEEE Trans. Antennas Propagat.*, vol. AP-20, pp. 146–152, Mar. 1972.

[36] P. G. Ingerson and W. C. Wong, "The analysis of deployable umbrella parabolic reflectors," *IEEE Trans. Antennas Propagat.*, vol. AP-20, pp. 409–414, July 1972.

[37] W. A. Imbriale, P. G. Ingerson, and W. C. Wong, "Experimental verification of the analysis of umbrella parabolic reflectors," *IEEE Trans. Antennas Propagat.*, vol. AP-21, pp. 705–708, Sept. 1973.

[38] W. A. Imbriale and W. V. T. Rusch, "Scalar analysis of nonsymmetrically distorted umbrella reflector," *IEEE Trans. Antennas Propagat.*, vol. AP-22, pp. 112–114, Jan. 1974.

F. Reflector Fabrication

[39] J. W. Dawson, "28-ft liquid-spun radio reflector for millimeter wavelengths," *Proc. IRE*, vol. 50, p. 1541, June 1962.

[40] R. A. Semplak and R. H. Turrin, "Pressure-formed parabolic reflectors for millimeter wavelengths," *IEEE Trans. Antennas Propagat.*, vol. AP-16, pp. 762–764, Nov. 1968.

G. Radio Telescopes and Large Tracking Antennas

[41] J. W. Findlay, "Radio telescopes," *IEEE Trans. Antennas Propagat.*, vol. AP-12, pp. 853–864, Dec. 1964.

[42] C. W. Tolbert, A. W. Straiton, and L. C. Krause, "A 16-foot diameter millimeter wavelength antenna system, its characteristics and its applications," *IEEE Trans. Antennas Propagat.*, vol. AP-13, pp. 225–229, Mar. 1965.

[43] H. E. King, E. Jacobs, and J. M. Stacey," A 2.8 arc-min beamwidth antenna: Lunar eclipse observations at 3.2 mm," *IEEE Trans. Antennas Propagat.*, vol. AP-14, pp. 82–91, Jan. 1966.

[44] P. D. Potter, W. D. Merrick, and A. C. Ludwig, "Big antenna systems for deep-space communications," *Astronautics and Aeronautics*, vol. 4, pp. 84–95, Oct. 1966.

[45] G. S. Levy, D. A. Bathker, A. C. Ludwig, D. E. Neff, and B. L. Seidel, "Lunar range radiation patterns of a 210-foot antenna at *S*-band," *IEEE Trans. Antennas Propagat.*, vol. AP-15, pp. 311–313, Mar. 1967.

[46] J. R. Cogdell *et al.*, "High resolution millimeter reflector antennas," *IEEE Trans. Antennas Propagat.*, vol. AP-18, pp. 515–529, July 1970.

[47] M. L. Meeks and J. Ruze, "Evaluation of the Haystack antenna and radome," *IEEE Trans. Antennas Propagat.*, vol. AP-19, pp. 723–728, Nov. 1971.

[48] O. Hachenberg, B. H. Grahl, and R. Wielebinski, "The 100-meter radio telescope at Effelsberg," *Proc. IEEE*, vol. 61, pp. 1288–1295, Sept. 1973.

Part II
Focal Region Fields: Prime Focus Feed Requirements

That the descriptive phrase "Airy rings," used to describe the nature of Fraunhofer diffraction phenomena in the focal plane of a circular aperture, is familiar to microwave and antenna engineers is testimony to the fact that the general features of the focal plane pattern have been well known for nearly a century and a half. This description, stated mathematically as a $[2J_1(u)/u]^2$ distribution in intensity, is an incomplete and unsatisfactory approximation when applied to the aperture and focal region of a reflector formed from a paraboloid of revolution. In the first place, it says nothing about polarization effects because it is a scalar solution to a vector problem; in the second place, it is significantly in error when applied to reflectors having focal ratios in the range 0.25 to 0.5 such as are commonly used in microwave antennas.

The first paper in this part is an attempt by Watson at a rigorous analysis, based on Maxwell's equations of the electromagnetic field, to determine the field distribution in the focal plane of a paraboloidal reflector. Not surprisingly, the analysis is heavily mathematical and difficult to follow. It does, however, contribute significantly to an understanding of the problem for it shows that the fields near the focus may be calculated quite adequately from the radiation of the induced primary surface currents alone. The term "primary" simply means the usual $2(\bar{n} \times \bar{H})$ current at a point on the reflector surface. Much farther away from the focus it is necessary to consider the secondary current distribution and to include the effect of the reflector rim.

The lack of quantitative data in the Watson paper is remedied in the second one, an important contribution by Minnett and Thomas toward an understanding of the nature of the fields near the axis of a focusing reflector. Using only the induced primary surface currents due to an incident plane wave field, they show, in a physically satisfying way, that these fields can be represented by a spectrum of hybrid waves that are simply linear combinations of the familiar TE_{1n} and TM_{1n} mode fields appropriate to perfectly conducting circular pipes. One noteworthy fact stands out, however, and that is that it is not possible for a single pipe to satisfy, simultaneously, the boundary conditions for both the TE and the TM mode sets because of their different radial periodicities. Nevertheless, the boundary conditions can be satisfied in a single hollow tube if it has an anisotropic internal surface reactance. One way to achieve this condition is to cut annular slots in the inner surface of a round metal waveguide. When the slots are $\lambda/4$ in depth, the boundary conditions on E and H are forced to be the same and the required anisotropic surface reactance is obtained. This constitutes the genesis of corrugated waveguide feeds and forms the subject for the third paper in this part, by Thomas.

In the fourth paper, Vu considers the problem of obtaining the optimum combination of amplitude and phase of the hybrid modes in a round guide that will match, at the guide aperture, the hybrid mode field near the axis of a symmetrical reflector. A simple method is given whereby the approximate mode composition may be derived from the solution of a small set of linear, simultaneous equations.

Thomas returns in the fifth paper to turn the problem around. Thus, he first derives an expression for the pattern of a hybrid mode field radiating from the circular aperture of a corrugated pipe and he then determines the conditions that are necessary for axial symmetry and for pure linear polarization. With this waveguide radiator placed at the focus of a paraboloidal mirror, he next derives the aperture distribution for the reflector. Finally, he determines the gain and aperture efficiency in the secondary pattern as a function of the number of hybrid modes supported by the feed. As is customary, these hybrid waves are termed HE_{1n} or EH_{1n} modes according as the mode content factor γ is equal to either +1 or -1. When $\gamma = 1$ the TE and TM components that comprise the hybrid mode are in exactly the right phase and amplitude to create maximum field on the axis of the waveguide radiator, along with an axisymmetric pattern that is purely linearly polarized. The fact that $\gamma = -1$ for the EH_{1n} hybrid indicates that the TE and TM components are 180° out of phase, resulting in a null on the axis of the feed's radiation pattern. Thus, only the HE_{1n} hybrid mode feed may be used to obtain maximum efficiency and purely linear polarization in a paraboloidal reflector.

There is a considerable body of literature dealing with hybrid mode waveguide radiators and corrugated horns. The latter are often called scalar horns when the flare angle is relatively large. A great deal of this literature is concerned with the use of such horns as feeds in reflector antennas, and much of it could appropriately be presented in this volume. It has not been included, however, for two reasons. The first is that the book would have to be unduly lengthened. The second is that the most important and representative papers have already been collected in Part VI of the reprint volume *Electromagnetic Horn Antennas* (A. W. Love, Ed., IEEE Press, 1976). One that did not appear in that book has been included as the sixth paper here. It is a very terse presentation by Narasimhan and Malla in which approximate methods are used to determine the half-flare angle α_0 of a scalar feed horn which will yield a specified sidelobe level in reflectors of various f/D ratios. It is unfortunate that the authors failed to specify the aperture size or the slant length of the horn. The term "scalar horn" implies not only the use of corrugated walls, but also that the aperture size and flare angle are such as to assure at least 180° of phase error in its aperture. It may therefore be inferred that there is a lower limit to the aperture diameter given by $D/\lambda > \cot \alpha_0/2$. The use of a larger horn would be permissible and presumably would be beneficial in reducing spillover beyond the reflector rim.

The seventh paper, by Ghobrial, analyzes the copolar and crosspolar diffraction images in the focal plane of paraboloidal reflectors for both linearly and circularly polarized incident fields. He shows that the crosspolar field vanishes at the focus and along the principal axes, being concentrated in the quadrants at $\pm 45^\circ$ to these directions. The $\pm 45^\circ$ axes are directions of stationary polarization in the sense that the slope of the crosspolar pattern is zero at the focus.

Again it appears convenient to divide the bibliography into categories having somewhat different subject matter. For the benefit of the reader who does not have ready access to the above-mentioned reprint volume on horns, the bibliography includes all the relevant papers from that book.

The Field Distribution in the Focal Plane of a Paraboloidal Reflector

W. H. WATSON

Summary—The electromagnetic field in the focal plane of a finite axially-symmetrical paraboloidal reflector illuminated by a plane wave of arbitrary polarization incident nearly normally at the vertex, has been investigated. This paper describes the investigation, explains the method, and summarizes some of its results. The reflector is assumed to be of a focal length which is large compared with the wavelength of the incident wave, and the plane of the aperture is assumed to be about halfway between the focus and the vertex. The purpose is to view the diffraction pattern in the focal plane by treating the currents induced on the concave surface of the reflector so that the electromagnetic effect due to them on one part of the surface is taken into account at another by means of waves of current and charge on the paraboloid. The wave equation is derived by approximation from Maxwell's equations; this is feasible because the radii of curvature are large compared with the wavelength. The calculation proceeds by finding approximate solutions of the wave equation by the WKB method, and by noting the "turning point" of the waves at the same distance from the vertex as that of the geodesic corresponding to each pair of characteristic modes of the same harmonic order.

The modes are determined and normalized, and thereon the coupling of the incident field is computed. From this representation of the current and charge distribution the electromagnetic field in the focal plane is derived. The integrations required are made by asymptotic approximation using the method of stationary phase.

It is shown that in the vicinity of the focus the contribution from these secondary currents is very small compared with the field due to the primary currents $2(\vec{n}\times H^{\mathrm{inc}})$. The primary field is represented by Fourier-Bessel series the number of terms in which is substantially $[m_1 \sin \alpha+1/2]$ where $m_1=(4\pi/\lambda)\times$focal length, and $\alpha=$angle of incidence at the vertex. The representation that approximates to geometrical optics when m_1 is large enough does not emerge until the parameter $m_1^{1/2} \sin \alpha$ is greater than unity and involves $m_1^{1/2}$ times as many terms in the series. This does not occur until α is many times the angular half beamwidth of the main lobe of the reflector.

Manuscript received October 8, 1963; revised March 27, 1964. This investigation was supported entirely by the Independent Research Fund of the Lockheed Missiles and Space Company.

The author is with the Lockheed Missiles and Space Company, Sunnyvale, Calif.

Reprinted from *IEEE Trans. Antennas Propagat.*, vol. AP-12, pp. 561-569, Sept. 1964.

Introduction

THE AIM OF THIS investigation is to form a practical idea of the diffraction pattern produced in the vicinity of the focal plane of a finite axially-symmetric paraboloidal reflector when a monochromatic plane electromagnetic wave is incident at a small angle to the axis. The focal length (a) and the diameter (2R) of the reflector are assumed very large compared with the wavelength (λ). Accordingly, we shall be concerned with the form of the electromagnetic field distribution in amplitude, phase, and polarization that tends, in the limit of infinitely small wavelength, to the coma pattern derived from geometrical optics.

Since in actual practice the precise form of reflector may deviate from the ideal geometrical form, and also, since one can improvise special arrangments to weaken the effect of the fringe field [1] due to diffraction at the edge, the important point to be kept in mind is that the main effects to be considered arise from the distribution of currents and varying charges on the finite concave surface of the reflector.

There are two main ways in which one can approach determining the current distribution. One is to set out from the pair of integral equations satisfied by the components of the current density [2]. The main source of difficulty in this is to find a satisfactory way of representing explicitly the connection between currents at different places on the paraboloid. The alternative is to use the appropriate differential equation to represent electrical oscillations on the concave surface of the reflector, and to determine from the boundary condition at the surface the coupling of the incident wave to the various modes of oscillation. This latter method is used in the present work. The differential equation represents the propagation of waves on the surface of the paraboloid as the means by which currents at one place induce currents at another. The characteristic forms of the solution of the differential equation treated asymptotically in terms of the large ratio of the focal length to the wavelength become the basis of representing the incident field at the surface. Accordingly, we regard the current system set up on the paraboloid as the sum of the distribution of primary currents corresponding to the physical optics approximation and of the distribution of secondary currents due to the propagation over the surface of characteristic waves excited by the incident radiation.

Once the current and charge distributions have been effectively represented, the electromagnetic field near the focal plane due to them may be computed. It will be noted that the possibility of attempting to solve the electromagnetic diffraction problem for a finite reflector as a boundary value problem in three dimensions has been tacitly dismissed. The reason for this, quite apart from the mathematical difficulties, is that from the physical point of view it poses the problem in too sharp a form. In the boundary value problem, the condition at the rim of the reflector must enter intrinsically, whereas the aim here is to leave the effect indefinite and depend on the fact that oscillations are set up on the face of the finite reflector. The primary effect of altering in a minor way the geometry or physical structure of the reflector is to alter the current distribution on its surface; the field at a distance is secondary. This point of view is characteristic of the physical treatment of scattering; the important physical process occurs in the vicinity of the object scattering the waves.

In the calculations[1] the mathematical approximation has been exploited wherever possible on the ground that the purpose of the calulations is to improve physical understanding of the current distributions on the reflector and the fields due to them. The physical model could also be improved to take into account the differing effects on the various characteristic modes due to radiation by them, but this would involve calculations of a much larger order so it has not been attempted.

The Characteristic Modes

We shall use "parabolic" coordinates (ξ, η, ϕ) for treating the electromagnetic field in the neighborhood of the reflector surface using the convention of Magnus and Oberhettinger [3]. This convention differs from that adopted in mathematical works on the application of the confluent hypergeometric function to discussing electromagnetic waves in parabolic coordinates [4].

For the Cartesian coordinates we have

$$x = \xi\eta \cos \phi \qquad y = \xi\eta \sin \phi \qquad z = \tfrac{1}{2}(\xi^2 - \eta^2), \tag{1}$$

and we consider the electromagnetic field in the neighborhood of the paraboloid

$$\eta^2 = \eta_1{}^2 \equiv 2a \tag{2}$$

where a is the focal length of the reflector. In what follows we shall use $2a$ as the unit of length.

Since B_η, E_ξ, and E_ϕ vanish on the paraboloid, and since the normal derivatives of E_ξ and E_ϕ are small compared with gradients of E_η, B_ξ, and B_ϕ along the surface which has radii of curvature very large compared with the wavelength, we may approximate from Maxwell's equations in paraboloidal coordinates [3] as follows:

$$H_\xi = \frac{1}{h_3}\frac{\partial\psi}{\partial\phi}, \quad H_\phi = -\frac{1}{h_1{}^2}\frac{\partial}{\partial\xi}(h_1\psi), \quad E_\eta = j\omega\mu_0\psi \tag{3a}$$

where

$$h_1 = \sqrt{\xi^2 + \eta^2} \qquad h_3 = \xi\eta. \tag{3b}$$

The components of surface current and charge density are given by

$$\sigma = -j\omega\mu_0\kappa_0\psi, \qquad j_\xi = H_\phi, \qquad j_\phi = -H_\xi. \tag{3c}$$

[1] Presented in the original manuscript of this paper.

Conservation of electric charge (see Appendix I) on the surface yields

$$\frac{\partial\sigma}{\partial t} - \frac{1}{h_1h_3}\left[\frac{\partial}{\partial\phi}(h_1j_\phi) + \frac{\partial}{\partial\xi}(h_3j_\xi)\right] = -2j\omega\kappa_0E_\eta^{\text{inc}} \quad (4)$$

where E_η^{inc} is the normal component of the incident field.

The characteristic waves are solutions of the homogeneous equation

$$\frac{1}{h_1h_3}\left(\frac{\partial}{\partial\xi}\left(\frac{h_3}{h_1}\frac{\partial\psi}{\partial\xi}\right) + \frac{\partial}{\partial\phi}\left(\frac{h_1}{h_3}\frac{\partial\psi}{\partial\phi}\right)\right) + k^2\psi = 0, \quad (5)$$

the differential expression being the Beltrami parameter $\Delta_2\psi$ with respect to the paraboloid.

Let $\xi=\eta_1 t$ so $t=\tan\theta/2$, where θ is the angle between the axis and the radius vector from the focus to (ξ, ϕ). Noting that $k\eta_1^2=4\pi a/\lambda=m_1$ is a large number, we derive for the characteristic mode $\psi_m=t^{-1/2}e^{-im\phi}u_m(t)$ satisfying (5), the following equation governing $u_m(t)$:

$$\frac{d^2u_m}{dt^2} + \frac{m_1^2(t^2-\tau^2)(1+t^2)}{t}u_m = 0$$

where

$$m = m_1\tau. \quad (6)$$

Here we depend on the large size of m_1 to obtain asymptotic expressions for the u_m which are properly expressed in terms of the confluent hypergeometric functions. The WKB approximation to $u_m(t)$ is

$$\frac{t^{1/2}e^{i\Phi_m}}{m_1^{1/2}(t^2-\tau^2)^{1/4}(1+t^2)^{1/4}}$$

where

$$\Phi_m = \int \frac{m_1}{t}(t^2-\tau^2)^{1/2}(1+t^2)^{1/2}dt. \quad (7)$$

It is evident that Φ_m becomes imaginary when $t<\tau$. A simple calculation shows that the transition from real to imaginary values of Φ_m is accomplished by representing the wave functions as satisfying the equation for the Airy functions.

It can be seen that in the vicinity of $t=0$, the vertex of the paraboloid, $e^{i\Phi_m}\sim 0(e^{-m/2})$. It will be noted also that the Bessel function approximation to $u_m(t)$ for t small effectively ignores the curvature of the paraboloid.

Thus the characteristic mode of order m has the remarkable property of becoming evanescent when $t<\tau$. This is quite in accord with the geometrical interpretation of the WKB solution, for it can be proved that the turning point of the geodesic paths of the rays corresponding to the phase $\Phi_m \pm m\phi$ is $t=\tau$. The paraboloid thus acts as a filter which suppresses the high-order harmonics far from the vertex and progressively admits the lower-order harmonics closer and closer to the vertex as the order m is reduced. The standing waves of order m, *viz*, $t^{-1/2}\,u_m(t)\sin m\phi$ or $t^{-1/2}\,u_m(t)\cos m\phi$, are reflected in the neighborhood of $t=\tau$ and of $t=t_1$, the rim of the reflector.

For physical reasons it is clear that representation of the wave function based on (7) is inadequate to deal with the radiation by the currents on the paraboloidal surface. When $t\gg\tau$ the waves travel along the surface, and it is only as the wavelength on the surface starts to grow because $t\to\tau$ that there is effective radiation off the surface. Of course, there is a corresponding radiation associated with reflection at the rim, which leads to the diffraction dependent on the size of the reflector. The reflection of the characteristic waves in the vicinity of $t=\tau$ is associated with the diffraction pattern, in the limit of vanishing free-space wavelength, corresponding to the geometrical coma of the paraboloid reflector. Accordingly, we must look to the Airy equation as the basis of approximate representation of ψ_m.

In place of the independent variable t, we introduce s given by

$$s = \left(\frac{t^2-\tau^2}{1+\tau^2}\right)^{1/2} \quad \text{so} \quad t^2=\tau^2+(\tau^2+1)s^2=(\tau^2+1)(c^2+s^2)$$

where

$$c^2 = \frac{\tau^2}{\tau^2+1}. \quad (8)$$

The wave function $\psi_m(s, \phi)$ is now given by

$$\frac{(c^2+s^2)^{1/2}}{m_1^{1/2}(1+\tau^2)(1+s^2)^{1/4}}e^{\pm i\Phi_m \pm im\phi} \quad (9)$$

where, except for $\tau\to 0$, Φ_m is adequately represented by

$$\Phi_m = \tfrac{1}{3}m_1\beta s^3$$

where

$$\beta = \frac{2(\tau^2+1)^2}{\tau^2} \quad (10a)$$

and

$$\Phi_0 = m_1(s+\tfrac{1}{6}s^3). \quad (10b)$$

We now normalize the wave functions for the standing waves over the finite paraboloid and obtain the form

$$\psi_m^{sc} = G_m(s)\cos\left(\Phi_m - \frac{\pi}{4}\right)\begin{matrix}\sin\\ \cos\end{matrix}\, m\phi, \qquad s^2 \geq 0 \quad (11)$$

where

$$G_m = \frac{0.463m^{1/3}(c^2+s^2)^{1/2}}{m_1(1+\tau^2)(1+s^2)^{1/4}s^{3/2}} \qquad m \geq 1 \quad (12a)$$

$$G_0 = \frac{0.564}{t_1(1+s^2)^{1/4}s^{1/2}} \quad (12b)$$

$$\int_0^{t_1}\int_0^{2\pi} |\psi_m|^2(1+t^2)^{1/2}t\,dt\,d\phi = 1. \qquad (13)$$

The phase chosen for ψ_m in (11) is justified by treating the "turning point" in applying the WKB method to allow only the decaying exponential for s imaginary [5], [6].

The Excitation of Currents on the Reflector by a Plane Wave

The wave systems treated in the foregoing section are, of course, relevant to the estimation of the secondary effects that correspond to the induction of currents and charge at one place on the reflector by the field due to those at another place. The primary current system is equivalent to the tangential components of the magnetic field in the incident plane wave as follows:

$$J_t = 2H_\phi^{\text{inc}} \qquad J_\phi = -2H_t^{\text{inc}}. \qquad (14)$$

We shall imagine the plane electromagnetic wave to have propagation vector

$$\boldsymbol{k} = -k(\sin\alpha, 0, \cos\alpha)$$

so the direction of incidence is in the XZ plane, making the angle α with the negative direction of the Z axis. (See Fig 1.) The unit normal to the paraboloid at (t, ϕ) is

$$\left(\frac{-t\cos\phi}{\sqrt{1+t^2}}, \frac{-t\sin\phi}{\sqrt{1+t^2}}, \frac{1}{\sqrt{1+t^2}}\right).$$

If the electric force is the vector

$$\boldsymbol{E}_{\text{inc}} = [\lambda(0,1,0) + \mu(\cos\alpha, 0, -\sin\alpha)]e^{i\boldsymbol{k}\cdot\boldsymbol{r}} \qquad (15)$$

the normal component of E_{inc} at (t, ϕ) is

$$E_\eta = -\frac{(\lambda\sin\phi + \mu\cos\phi\cos\alpha)t + \mu\sin\alpha}{(1+t^2)^{1/2}}e^{i\boldsymbol{k}\cdot\boldsymbol{r}}. \qquad (16)$$

The complex numbers λ and μ are associated respectively with the electric force components $\perp$ and $\|$ to the plane of incidence at the vertex, and permit the representation of arbitrarily polarized incident radiation. We note that

$$\boldsymbol{k}\cdot\boldsymbol{r} = -m_1(t\sin\alpha\cos\phi + \tfrac{1}{2}t^2\cos\alpha) \qquad (17)$$

and expand $\exp(-im_1t\sin\alpha\cos\phi)$ in the series

$$\sum_{m=0}^{\infty}\frac{2e^{-im\pi/2}}{1+\delta_{0m}}J_m(q)\cos m\phi \qquad (18)$$

where

$$q = m_1t\sin\alpha, \qquad \delta_{0m}=0\ (m\neq 0), \qquad \delta_{00}=1.$$

Since

$$\boldsymbol{H}_{\text{inc}} = \frac{1}{\mu_0\omega}(\boldsymbol{k}\times\boldsymbol{E}_{\text{inc}})$$

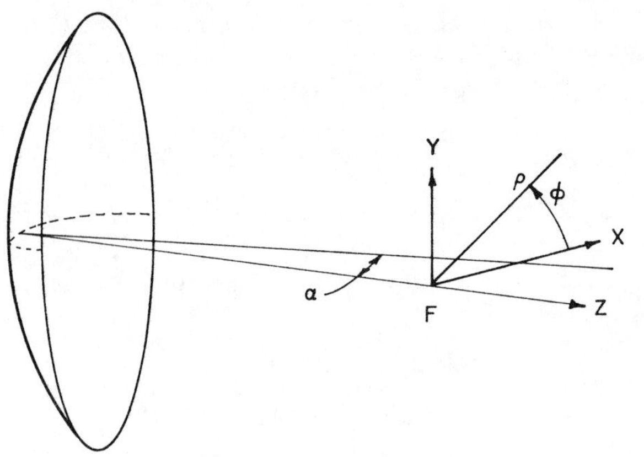

Fig. 1—Coordinate system.

we readily find H_t, H_ϕ, and by using (18) express the primary currents J_ϕ, J_t in series form

$$J_\phi = \frac{-e^{-im_1\cos\alpha t^2/2}}{\mu_0c(1+t^2)^{1/2}}\sum_{m=1}^{\infty}\{\lambda\cos\alpha\cos m\phi[e^{-i(m-1)\pi/2}J_{m-1}(q) + e^{-i(m+1)\pi/2}J_{m+1}(q)] - \mu\sin m\phi[e^{-i(m-1)\pi/2}J_{m-1}(q) - e^{-i(m+1)\pi/2}J_{m+1}(q)] - \lambda t\sin\alpha e^{-im\pi/2}J_m(q)\} \qquad (19a)$$

$$J_t = \frac{-e^{-im_1\cos\alpha t^2/2}}{\mu_0c(1+t^2)^{1/2}}\sum_{m=1}^{\infty}\{\lambda\cos\alpha\sin m\phi[e^{-i(m-1)\pi/2}J_{m-1}(q) - e^{-i(m+1)\pi/2}J_{m+1}(q)] + \mu\cos m\phi[e^{-i(m-1)\pi/2}J_{m-1}(q) + e^{-i(m+1)\pi/2}J_{m+1}(q)\}. \qquad (19b)$$

The secondary currents j_t and j_ϕ determined by (3) are derived from the solution of the inhomogeneous equation, (4). In Appendix II it is shown that it suffices to analyze $(1+t^2)E_\eta^{\text{inc}}$ in terms of the characteristic modes, *i.e.*,

$$(1+t^2)E_\eta^{\text{inc}} = \sum_{m=0}(\lambda A_m\psi_m^s + \mu B_m\psi_m^c) \qquad (20)$$

so

$$A_m = \iint E_\eta^{\text{inc}}\psi_m^s t(1+t^2)^{3/2}dt\,d\phi$$

$$B_m = \iint E_\eta^{\text{inc}}\psi_m^c t(1+t^2)^{3/2}dt\,d\phi. \qquad (21)$$

The integrals (21) are evaluated by finding asymptotic approximations to them. The procedure adopted relies on the saddle-point method in integration on the complex s plane.

Radiation by the Primary Current Distribution

The primary currents are expressed in Fourier series (19a, b) which we shall write

$$\mu_0cJ_\phi = \sum_{m=1}^{\infty}[\lambda\cos\alpha\cos m\phi P_m(t) - \mu\sin m\phi Q_m(t) - \lambda t\sin\alpha\sin m\phi R_m(t)]$$

$$\mu_0 c J_t = \sum_{m=1}^{\infty} [\lambda \cos\alpha \sin m\phi Q_m(t) + \mu \cos m\phi P_m(t)]$$

$$\mu_0 c J_0 = -\frac{1}{c}\sum_{m=1}^{\infty} [\lambda t \sin m\phi Q_m(t) + \mu t \cos\alpha \cos m\phi P_m(t)$$

$$+ 2\mu \sin\alpha \cos m\phi R_m(t)] - \frac{1}{c}\mu \sin\alpha R_0(t) \quad (22)$$

where

$$P_m(t) = [e^{-i(m-1)\pi/2} J_{m-1}(q) + e^{-i(m+1)\pi/2} J_{m+1}(q)] \cdot \frac{e^{-im_1 \cos\alpha t^2/2}}{(1+t^2)^{1/2}}$$

$$Q_m(t) = [e^{-i(m-1)\pi/2} J_{m-1}(q) - e^{-i(m+1)\pi/2} J_{m+1}(q)] \cdot \frac{e^{-im_1 \cos\alpha t^2/2}}{(1+t^2)^{1/2}}$$

$$R_m(t) = e^{-im\pi/2} J_m(q) \frac{e^{-im_1 \cos\alpha t^2/2}}{(1+t^2)^{1/2}} \cdot$$

At the point (ρ, ϕ, z) distant R from the point (ρ', ϕ', z') [or $(t, \phi', \frac{1}{2}t^2)$] on the paraboloidal surface, the components of the electromagnetic potential $(A_\rho, A_\phi, A_z, A_0)$ are given by the usual formula, for example,

$$A_\phi = \frac{\mu_0}{4\pi}\frac{m_1^3}{k^3}\iint J_\phi \frac{e^{ikR}}{R} t(t+t^2)^{1/2} dt d\phi, \quad (23)$$

in terms of

$$J_\rho = J_t(1+t^2)^{-1/2}, \quad J_\phi, \quad J_z = J_t t(1+t^2)^{-1/2}, \quad J_0.$$

In (23) we have taken into account the effect of using 2a as the unit of length in treating the geometry of the paraboloid.

Now we may expand e^{ikR}/R in cylindrical waves as follows, since we are interested in the field distribution in the focal plane not too far from the focus,

$$\frac{e^{ikR}}{R} = \sum_{n=0}^{\infty} \frac{2\cos n(\phi-\phi')}{1+\delta_{0n}} F_n(z-z', \rho, \rho') \quad (24)$$

where

$$R^2 = (z-z')^2 + \rho^2 + \rho'^2 - 2\rho\rho' \cos(\phi - \phi')$$

and

$$F_n(z-z', \rho, \rho')$$

$$= k\int_0^\infty e^{-|z-z'|k\sqrt{\sigma^2-1}} J_n(k\rho\sigma) J_n(k\rho'\sigma) \frac{\sigma d\sigma}{(\sigma^2-1)^{1/2}} \quad (25)$$

with appropriate conditions on $\arg(\sigma-1)$ and $\arg(\sigma+1)$ at $\sigma=0$ and as $\sigma\to\infty$ [10].

Substituting (25) in (23) and integrating with respect to ϕ' we have

$$A_\rho = -\frac{\mu_0 m_1^3}{2k^3}\sum_n \int_0^{t_1} F_n\{Q_n(t) \sin n\phi \,\lambda \cos\alpha$$

$$+ P_n(t) \cos n\phi \,\mu\} t dt \quad (26)$$

and corresponding expressions for A_ϕ, A_z, A_0.

An asymptotic representation of F_n may be derived by considering the appropriate contour integral of which (25) forms a part, there being a cut in the complex plane from $\sigma=-1$ to $\sigma=+1$. Provided that $k|z-z'|$ is large, as it is when the maximum value of t^2 is $\frac{1}{2}$, and that $z\sim a$, the quantity F_n is given by

$$F_n = \frac{kJ_n(k\rho)J_n(k\rho')}{|z-z'|} \cdot$$

In particular, for $z=a$ (focal plane),

$$F_n = \frac{2k}{m_1}\frac{J_n(k\rho)J_n(k\rho')}{1-t^2} \cdot \quad (27)$$

Thus, we write

$$A_\rho = -\frac{\mu_0 m_1^2}{k^2}\sum_{n=1}^{\infty} J_n(k\rho)\{\mu L_n \cos n\phi + \lambda \cos\alpha M_n \sin n\phi\}$$

$$A_\phi = -\frac{\mu_0 m_1^2}{k^2}\sum_{n=1}^{\infty} J_n(k\rho)\{\lambda \cos\alpha\, L_n \cos n\phi - \mu M_n \sin n\phi$$

$$- \lambda \sin\alpha T_n \sin n\phi\}$$

$$A_z = -\frac{\mu_0 m_1^2}{k^2}\sum_{n=1}^{\infty} J_n(k\rho)\{\mu N_n \cos n\phi$$

$$+ \lambda \cos\alpha S_n \sin n\phi\} \quad (28)$$

where

$$L_n = \int_0^{t_1} P_n(t) J_n(m_1 t) \frac{t dt}{1-t^2}$$

$$N_n = \int_0^{t_1} P_n(t) J_n(m_1 t) \frac{t^2 dt}{1-t^2}$$

$$M_n = \int_0^{t_1} Q_n(t) J_n(m_1 t) \frac{t dt}{1-t^2}$$

$$S_n = \int_0^{t_1} Q_n(t) J_n(m_1 t) \frac{t^2 dt}{1-t^2}$$

$$T_n = \int_0^{t_1} R_n(t) J_n(m_1 t) \frac{t^2 dt}{1-t^2} \cdot \quad (29)$$

These integrals may be approximated by the stationary phase method replacing the Bessel functions by their asymptotic expressions for large argument. This procedure will fail, however, unless the argument exceeds the order of the Bessel function. Without entering the elaborate details of calculation it is possible to state in principle what is involved, by introducing the critical parameter $m_1^{1/2} \sin\alpha = (4\pi a/\lambda)^{1/2} \sin\alpha$. When this quantity is much greater than 1,

$$L_n \simeq \frac{e^{im_1/2 \sec\alpha + i(n+1/4)\pi}}{m_1 \sin\alpha} H_n^{(1)}(m_1 \tan\alpha) \qquad M_n = 0$$

$$N_n \simeq \frac{-e^{im_1/2 \sec\alpha + i(n+1/4)\pi}}{m_1 \sin\alpha} H_n^{(2)}(m_1 \tan\alpha) \qquad S_n = 0$$

$$T_n \simeq \frac{e^{im_1/2 \sec\alpha + i(n+1/4)\pi}}{2m_1 \sin\alpha} H_n^{(1)}(m_1 \tan\alpha). \quad (30)$$

These approximations are relevant to the asymptotic approach to geometrical optics. Even for a large microwave paraboloidal reflector, where $m_1 \sim 4000$, the condition $m_1^{1/2} \sin \alpha \gg 1$ is not met until α is several degrees and the source of radiation lies well outside the main lobe of the reflector.

On the other hand, when $m_1^{1/2} \sin \alpha$ is not large enough to justify the foregoing approximation we find the saddle-point *cum* pole contributions to L_n, etc., depend on

$$C\left(\sqrt{\frac{m_1}{2}} \sin \alpha\right) \quad \text{and} \quad S\left(\sqrt{\frac{m_1}{2}} \sin \alpha\right)$$

(see Appendix III) where

$$C(x) = \int_0^x \cos t^2 dt \quad \text{and} \quad S(x) = \int_0^x \sin t^2 dt \tag{31}$$

are the Fresnel integrals. Further, the functions $P_n(t)$, $Q_n(t)$, $R_n(t)$ which appear in the integrands of (29) for L_n, etc., are dependent on Bessel functions of $m_1 t \sin \alpha$. Since $m_1 \sin \alpha$ is the turning point of the Bessel differential equation of order ν when $\nu = m_1 \sin \alpha$, it is evident, and is supported in detail by the formulas of Debye and Nicholson, that $J_\nu(m_1 \sin \alpha)$ decreases very rapidly with increasing ν as it passes this critical value [6]. Accordingly, we may obtain an approximate representation by cutting off all integrals for which $\nu > [m_1 \sin \alpha + \frac{1}{2}]$ or 2, whichever is the greater. *The field will then be represented by a finite series instead of an infinite series.*

It will be noticed that in the approximation (30), since $m_1^{1/2} \sin \alpha \gg 1$, the number of terms in the series representation is much greater than $m_1^{1/2}$. Let us consider some numerical values. Suppose $m_1 = 3820$, α = beam halfwidth, say 0.1°; then $[m_1 \sin \alpha + \frac{1}{2}] + 7$. This is the number of terms to be considered when the radiation is incident at an angle equal to one half of the angular diameter of the main beam of this reflector. At r beam halfwidths off axis, the number of terms entering is $O(7r)$. On the other hand, when the conditions for applying (30) are met, say $m_1^{1/2} \sin \alpha = 3$, $\alpha = 3°$, $n = 185$.

The Field Distribution in the Focal Plane

We consider first the field components given by (28) due to the primary current distribution. To compute these we must sum the series

$$1)\ \sum_{n=1}^{\infty} J_n(k\rho) \cos n\phi L_n \qquad 2)\ \sum_{n=1}^{\infty} J_n(k\rho) \cos n\phi N_n$$

$$3)\ \sum_{n=1}^{\infty} J_n(k\rho) \sin n\phi S_n \qquad 4)\ \sum_{n=1}^{\infty} J_n(k\rho) \sin n\phi T_n \tag{32}$$

in which ρ is the distance from the focus to the point at which the series represents the appropriate field component. It is apparent that the number of terms that should be considered in these series is $N = [m_1 \sin \alpha + \frac{1}{2}]$ or 2, whichever is the greater, but that it is $k\rho$ that effectively determines the number of terms to be summed in the series when $k\rho < N$. Thus when the field is to be computed for points not far from the focus, only the first few terms of the series are significant. Further, the analysis shows that the relative extent to which the finite diameter affects the form of the diffraction pattern in the vicinity of the focus is of the order m_1^{-1}, and is, therefore, negligible for a large reflector. On the other hand, the effect of increasing the small angle of incidence, α, is to increase the number of terms required in the series to represent the field. When α becomes sufficiently great to justify the use of the approximate forms (30), we can see the image of geometrical optics emerge, for, by the addition theorem for cylindrical functions, the series (32a) tends to $H_0^{(1)}(k\rho')$ where $\rho'^2 = \rho^2 + 4a^2 \tan^2\alpha + 4\rho a \tan \alpha \cos \phi$ [11].

The computation of the radiation by the secondary currents is tedious and will not be given here in any detail. Nevertheless, it is worth noting that the asymptotic approximations show that the main sources of radiation due to currents associated with ψ_n are at the turning point $t = \tau = n/m_1$ and at the rim $t = t_1$. We present the orders of magnitude of the dominant terms. For the current harmonic derived from ψ_n^s, the order of the Fourier-Bessel harmonic $J_n(k\rho) \cos n\phi$ of the field in the focal plane is $n^{31/12}/m_1^3 \sin \alpha$; whereas, for the current harmonic derived from ψ_n^c, the order is $n^{43/12}/m_1^3$ compared with the corresponding coefficient in the series representation of the field due to the primary currents. If these contributions to the focal field are to count, n must be a substantial fraction of m_1, but then the value of $J_n(k\rho)$ would be quite negligible for such high-order harmonics, as we have already noted. It is therefore evident that the conditions under which the effect of the finite size of the reflector and the effects of secondary currents on the reflector can be made evident in the diffraction pattern are that the angle of incidence of the incident radiation at the vertex be not small and that the field should be examined at a substantial distance from the focus.

Accordingly, we reach the following conclusions. For distances up to a few tens of wavelengths from the focus of a large reflector, the field may be calculated very well from the radiation due to the primary current system alone. The field components may be represented by Fourier-Bessel series; the number of terms in which increases as the incident radiation is directed at an increasing angle to the axis of the paraboloid. The number of terms is substantially $[m_1 \sin \alpha + \frac{1}{2}]$.

As α is still further increased and we proceed farther from the focus in representing the effective field, not only is the number of harmonics increased, but the field contributions due to the secondary currents and those arising from the finite size of the reflector must come into account. Eventually the diffraction equivalent of the coma pattern given by geometrical optics is developed. Whereas in the latter each point of the focal plane at the coma corresponds to one point of the reflector surface for given incident radiation and this involves an infinite series of harmonics, in the diffraction phenomenon the number of harmonics does not exceed exceed m_1.

As to the polarization of the field in the focal plane not too far from the focus, (28), (32) and the results of Appendix III may be used to show that circularly polarized incident radiation produces circularly polarized field in the focal plane.

Finally, it is obvious that even with the largest microwave reflectors, the number of harmonics required to represent the field in the focal plane due to distant sources lying within the main lobe of the reflector is small enough for practical numerical analysis. Provided that physical means for measuring and analyzing the field distribution can be achieved, it should be possible from this analysis to infer some of the structure of an extended source of radiation.

Appendix I

The expression which appears as the first term in (4) is the net inflow of surface current per unit area. Eq. (4) can be derived from the integral equations [12].

$$j(P) = \frac{1}{2\pi}\iint \vec{n}_P \times (\nabla_P \times \vec{J}\,(Q)e^{ikR}/R)\, dS_Q \tag{33}$$

where R is the distance between P and Q, and

$$\sigma(P) = -\frac{1}{1\pi}\iint (\vec{n}_P \cdot \nabla_P J_0\,(Q)e^{ikR}/R)\, dS_Q - \frac{j\omega\mu_0}{2\pi}\iint (\vec{n}_P \cdot \vec{J}\,(Q)e^{ikR}/R)\, dS_Q, \tag{34}$$

expressing $\vec{J}(Q)$ and $J_0(Q)$ in terms of the incident field. The calculation is simplified by noting that it is the singularity in the integrand that leads to the right-hand side of the inhomogeneous differential equation. The last term of (34) does not contribute to this.

Appendix II

Let us reduce the partial differential equation (4) to an ordinary one in ψ_m:

$$\frac{d^2\psi_m}{dt^2} + \frac{1}{t}\frac{d\psi_m}{dt} + \frac{(m_1^2t^2 - m^2)(1+t^2)}{t^2}\psi_m = -2j\omega k_0[(1+t^2)E_\eta^{\text{inc}}]_m = F_m(t). \tag{35}$$

$[(1+t^2)E_\eta^{\text{inc}}]_m$ is the coefficient of $e^{im\phi}$ in the Fourier expansion of $E_\eta^{\text{inc}}(1+t^2)$. Using the confluent hypergeometric function to construct the Green's function for the solution of (35) under the conditions ψ_m finite at $t=0$, $d\psi_m/dt=0$ when $t=t_1$ we obtain

$$g_m(t,u) = m_{K,m/2}(-im_1t^2)m_{K,m/2}(-im_1u^2) \quad 0 \le t \le u$$

$$g_m(t,u) = m_{K,m/2}(-im_1t^2)m_{K,m/2}(-im_1u^2) + f_m(t,u) \quad u \le t \le t_1 \tag{36}$$

where

$$f_m(t,u) = \frac{m_{K,m/2}(-im_1u^2)w_{K,m/2}(-im_1t^2) - m_{K,m/2}(-im_1t^2)w_{K,m/2}(-im_1u^2)}{m_{K,m/2}(-im_1u^2)w'_{K,m/2}(-im_1u^2) - m'_{K,m/2}(-im_1u^2)w_{K,m/2}(-im_1u^2)} \tag{37}$$

and for ψ_m,

$$\psi_m = \int_0^{t_1} F_m(u)m_{K,m/2}(-im_1u^2)m_{K,m/2}(-im_1t^2)u\,du + \int_0^t F_m(u)f_m(t,u)du. \tag{38}$$

In (38) we note that the second term on the right will contribute significantly to the value of $\psi_m(t)$ only if $t > m/m_1$, and, since the Wronski determinant is proportional to u^{-2} and the w functions for small argument depend on u^{1-m}, we see that this contribution to the value of $\psi_m(t)$ may be neglected for low-order modes. For higher-order modes, the phases of $F_m(u)$ and of the m and w functions vary so rapidly that the contribution of this second term is again negligible on the basis of stationary phase considerations.

Thus, to find the secondary currents we analyze the expression (16) using (18) for the normal component of the incident electric force in terms of the characteristic modes, *i.e.*,

$$(1+t^2)E_\eta = \sum_{m=0} (\lambda A_m\psi_m{}^s + \mu B_m\psi_m{}^c). \tag{39}$$

Appendix III

The integrals (29) are linear combinations of integrals of the form

$$I_{\nu r} = \int_0^{t_1} J_\nu(m_1t\sin\alpha)J_n(m_1t)\frac{t^r e^{-im_1\cos t\alpha^2/2 - i\nu\pi/2}}{(1-t^2)(1+t^2)^{1/2}} \tag{40}$$

where ν takes the values n, $n \pm 1$.

We write the four terms of the integrand in (40) in the form

$$\sum_{a_1,a_2} \frac{\text{i}f_\nu\,(a_1,a_2)}{2\pi m_1(\sin\alpha)^{1/2}(1-t^2)(1+t^2)^{1/2}} \tag{41}$$

where

$$f_\nu(a_1,a_2) = m_1t(a_1 + a_2\sin\alpha) - a_1(2n+1)\frac{\pi}{4} - a_2(2\nu+1)\frac{\pi}{4} - \frac{\nu\pi}{2} - \frac{1}{2}m_1t^2\cos\alpha.$$

The symbols a_1, a_2 take the possible values ± 1. The point of stationary phase is

$$t_s(a_1, a_2) = a_1 \sec \alpha + a_2 \tan \alpha \tag{42}$$

which is near one of the poles of the integrand, *viz.*, $t = a_1$. For $a_1 = +1$, $t_s(1, -1)$, and $t_s(1, +1)$ lie respectively to the left and right of the pole. Thus, to compute the contribution to the integral from $a_1 = +1$, we require three paths of integration: C^- from $t=0$ through $t_s(1, -1)$ to $\infty e^{-i\pi/4}$, C^+ from $t=0$ through $t_s(1, +1)$ to $\infty e^{-i\pi/4}$ and C_1 from t_1 to $\infty e^{+i\pi/4}$. This contribution is

$$\int_{C+} \frac{e^{if_\nu(+1,+1)} t^{r-1} dt}{4\pi m_1 (\sin\alpha)^{1/2}(1-t)(1+t^2)^{1/2}} + \int_{C-} \frac{e^{if_\nu(+1,-1)} t^{r-1} dt}{4\pi m_1 (\sin\alpha)^{1/2}(1-t)(1+t^2)^{1/2}} - \int_{C_1} \frac{(e^{if_\nu(+1,+1)} + e^{if_\nu(+1,-1)}) t^{r-1} dt}{2\pi m_1 (\sin\alpha)^{1/2}(1-t^2)(1+t^2)^{1/2}} + 2\pi i R_+ \tag{43}$$

where R_+ is the residue at the pole $t = +1$ of the first of the integrands.

The integrals corresponding to (43) for $a_1 = -1$ have also to be evaluated. In the resulting sum, to compute (41) the residue contributions from the two poles cancel since the contours encircle them in opposite senses, and the residues there are equal.

If $m_1^{1/2} \sin \alpha \gg 1$, we obtain good approximations to the sum of the integrals starting from $t=0$ by considering the leading term of the asymptotic expression given by the stationary phase method, leading to (30).

When $m_1^{1/2} \sin \alpha$ is not large enough to justify the foregoing approximation, we treat the integrals such as a first of (43) in the following way [8]. Let

$$s = (t-1)\sqrt{\tfrac{1}{2} m_1 \cos \alpha}$$

$$\begin{aligned} f_\nu(+1, +1) &= i m_1 (1 + \sin\alpha - \tfrac{1}{2}\cos\alpha) \\ &\quad + \frac{i m_1 (1 + \sin\alpha - \cos\alpha) s}{\sqrt{\tfrac{1}{2} m_1 \cos\alpha}} - is^2 \\ &\simeq i m_1 (\tfrac{1}{2} + \sin\alpha) + i\sqrt{2 m_1} \sin \alpha s \\ &\quad - is^2 + O(\alpha^2). \end{aligned} \tag{44}$$

We therefore consider the integral

$$I = \int_\gamma e^{as - is^2} \frac{ds}{s} \quad \text{where} \quad a = i\sqrt{2m_1} \sin \alpha. \tag{45}$$

The path γ makes $-\pi/4$ with the real axis of s so that $e^{-is^2} \to e^{-r^2}$. Since

$$\int_0^a e^{xs} ds = \frac{1}{s} e^{as} - \frac{1}{s},$$

$$I = \int_\gamma \int_0^a e^{xs - is^2} dx ds - \int_\gamma \frac{e^{-is^2}}{s} ds.$$

The second of these integrals is equal to $-i\pi$. We reverse the order of integration in the first and obtain

$$\begin{aligned} I &= i\pi + \int_0^a dx \int_\gamma e^{xs - is^2} ds \\ &= i\pi + \sqrt{\pi} \int_0^a e^{-(i/2)x^2} dx \\ &= i\pi + 2\sqrt{\pi} \int_0^{a/2} e^{-iy^2} dy \\ &= i\pi + \pi e^{i\pi/4} \operatorname{erf}(e^{i\pi/4}\sqrt{\tfrac{1}{2} m_1} \sin\alpha) \\ &= i\pi + \sqrt{2\pi} i \left\{ C\left(\sqrt{\frac{m_1}{2}} \sin\alpha\right) \right. \\ &\quad \left. - iS\left(\sqrt{\frac{m_1}{2}} \sin\alpha\right) \right\} \end{aligned} \tag{46}$$

where

$$C(x) = \int_0^x \cos t^2 dt \qquad S(x) = \int_0^x \sin t^2 dt.$$

To obtain the results for the integrals in which the exponent of the exponential factor of the integrand is $f(+1, -1)$ or $f(-1, +1)$ we change the sign of the argument of C and S and note that

$$C(-x) = -C(x) \qquad S(-x) = -S(x).$$

Thus we deduce the results corresponding to (30):

$$\begin{aligned} I_{n+1,1} = I_{n-1,1} &= \frac{e^{im_1(1/2 + \sin\alpha)}}{2\pi m_1 (2\sin\alpha)^{1/2}} \\ &\quad \cdot [\pi i + \sqrt{2\pi}\, i(C - iS)] e^{i(n+1)\pi/2} \\ I_{n,1} &= \frac{e^{im_1(1/2 - \sin\alpha)}}{2\pi m_1 (2\sin\alpha)^{1/2}} \\ &\quad \cdot [\pi i - \sqrt{2\pi}\, i(C - iS)] e^{-in\pi/2} \\ I_{n-1,2} = I_{n+1,2} &= \frac{e^{im_1(1/2 - \sin\alpha)}}{2\pi m_1 (2\sin\alpha)^{1/2}} \\ &\quad \cdot [\pi i - \sqrt{2\pi}\, i(C - iS)] e^{-in\pi/2} \\ I_{n,2} &= \frac{e^{im_1(1/2 + \sin\alpha)}}{2\pi m_1 (2\sin\alpha)^{1/2}} \\ &\quad \cdot [\pi i + \sqrt{2\pi}\, i(C - iS)] e^{i(n-1)\pi/2}. \end{aligned} \tag{47}$$

We now consider the integrals along the contour C_1. For I_ν, the contribution is

$$\frac{-e^{-im_1 \cos\alpha t^2/2}\left\{ e^{im_1 t_1}(e^{i(\nu - n + 1/2)\pi} - e^{-in\pi/2}) + e^{-im_1 t_1}(e^{i(\nu + n/2)\pi} + e^{i(n+1)\pi/2}) \right\}}{4\pi m_1^2 (\sin\alpha)^{1/2}(1 - t_1^2)(1 + t_1^2)^{1/2}(1 - t_1 \cos\alpha)} \tag{48}$$

in which we have neglected $\sin \alpha$ compared with $(1 - t_1 \cos \alpha)$. For $I_{\nu,2}$ we multiply the first term in $\{\ \}$ of (48) by t_1 and the second term by $-t_1$.

The integrals (29) are obtained from the combinations

$$L_n = I_{n-1,1} + I_{n+1,1} \qquad N_n = I_{n-1,2} + I_{n+1,2}$$
$$M_n = I_{n-1,1} - I_{n+1,1} \qquad S_n = I_{n-1,2} - I_{n+1,2}$$
$$T_n = I_{n,2}. \tag{49}$$

The "end-point" contributions derived from (48) and combined in accordance with (49) yield, on neglecting $\sin \alpha / 1 - t_1 \cos \alpha$,

$$L_n: \quad \frac{(e^{i\pi/2} + e^{i(n+1)\pi})e^{-i\pi/4}J_n(m_1t_1)e^{-im_1 \cos \alpha t_1^2/2}}{(2\pi \sin \alpha)^{1/2}m_1^{3/2}(1 - t_1^2)(1 + t_1^2)^{1/2}(1 - t_1 \cos \alpha)}$$

$$M_n: \quad 0$$

$$N_n: \quad \frac{t_1\{(e^{-i\pi/2} + e^{in\pi})e^{i\pi/4}J_n'(m_1t_1) + (e^{i\pi/2} + e^{in\pi})e^{-i\pi/4}J_n(m_1t_1)\}e^{-im_1 \cos \alpha t_1/2}}{(2\pi \sin \alpha)^{1/2}m_1^{3/2}(1 - t_1^2)(1 + t_1^2)^{1/2}(1 - t_1 \cos \alpha)}$$

$$S_n: \quad \frac{t_1\{(e^{-i\pi/2} - e^{in\pi})e^{i\pi/4}J_n'(m_1t_1) + (e^{in\pi} - e^{i\pi/2})e^{-i\pi/4}J_n(m_1t_1)\}e^{-im_1 \cos \alpha t_1^2/2}}{(2\pi \sin \alpha)^{1/2}m_1^{3/2}(1 - t_1^2)(1 + t_1^2)^{1/2}(1 - t_1 \cos \alpha)}$$

$$T_n: \quad \frac{t_1(e^{-i\pi/2} + e^{i(n-1)\pi})e^{i\pi/4}J_n'(m_1t_1)e^{-im_1 \cos \alpha t_1^2/2}}{(2\pi \sin \alpha)^{1/2}m_1^{3/2}(1 - t_1^2)(1 + t_1^2)^{1/2}(1 - t_1 \cos \alpha)} \cdot \tag{50}$$

The pole *cum* saddle-point contributions derived from (47) are

$$L_n: \quad \frac{e^{im_1(1/2 + \sin \alpha)}}{\pi m_1(\sin \alpha)^{1/2}}\{\pi + \sqrt{2\pi}(C - iS)\}e^{i\pi + in\pi/2}$$

$$M_n: \quad 0$$

$$N_n: \quad \frac{e^{im_1(1/2 - \sin \alpha)}}{\pi m_1(\sin \alpha)^{1/2}}\{\pi - \sqrt{2\pi}(C - iS)\}e^{i\pi/2 - in\pi/2}$$

$$S_n: \quad 0$$

$$T_n: \quad \frac{e^{im_1(1/2 + \sin \alpha)}}{\pi m_1(\sin \alpha)^{1/2}}\{\pi + \sqrt{2\pi}(C - iS)\}e^{in\pi/2} \tag{51}$$

Let us now return to consider the effect of taking into account the failure of the assumed asymptotic representation of the Bessel functions of large argument and order. We note that in (40) it is $J_\nu(m_1t \sin \alpha)$ that should be considered for ν approaching $m_1 \sin \alpha$, since the value of the function enters to determine the asymptotic value of the integral when t is in the neighborhood of $t = 1$. When ν exceeds $m_1t \sin \alpha$, this Bessel function is a slowly varying function of t and hence should no longer enter to determine the saddle point which will

therefore coincide to within $O(\alpha^2)$ with the pole of the integrand of (40) at $t = +1$ and $t = -1$. In (46) we put $\alpha = 0$ to compute the integral

$$\int_{c_+} \exp\left(ia_1\left(m_1t - \frac{2n+1}{4}\pi\right) - \frac{i\nu\pi}{2} - \frac{im_1 t^2}{2}\cos \alpha\right)$$
$$\cdot \frac{J_\nu(m_1 \sin \alpha)t^{\nu-1/2}dt}{2\pi m_1^{1/2}(1 - t)(1 + t^2)^{1/2}},$$

for example. [Cf. (C.4).]

Acknowledgment

The author acknowledges gratefully the kind interest of E. A. Blasi, and helpful discussions with N. A. Logan, A. S. Dunbar, and Professor V. H. Rumsey.

References

[1] B. Ye. Kinber, *Radioteckhnika i. Elektronika*, vol. 7, pp. 90–98; January, 1962.
[2] V. A. Fock, J. of Phys. U.S.S.R., vol. 10, pp. 130–136; 1946.
[3] W. Magnus and F. Oberhettinger, "Formulas and Theorems for the Functions of Mathematical Physics," Chelsea Publishing Co., New York, N. Y.; 1949.
[4] H. Buchholz, *Z. angew Math. Mech.*, vol. 23, pp. 47–58, 101–118; 1943.
[5] L. I. Schiff, "Quantum Mechanics," McGraw-Hill Book Co., Inc., 2nd ed., New York, N. Y., pp. 187–190; 1955.
[6] R. E. Langer, *Amer. Math. Soc. Bull.*, ser. 2, vol. 40, p. 545; 1934.
[7] M. Born and E. Wolf, "Principles of Optics," Pergamon Press, London, England; 1959.
[8] M. V. Cerrillo, "On the Evaluation of Integrals of the Type $f(\tau_1, \tau_2, \cdots, \tau_n)1/2\pi i\int F(s) \exp W(s, \tau_1, \tau_2, \cdots, \tau_n)ds$," Research Lab. of Electronics, M.I.T., Cambridge, Tech. Rept. No. 55: 2a; 1950.
[9] H. Jeffreys and B. S. Jeffreys, "Methods of Mathematical Physics," Third Ed., Cambridge Press, London, England; 1956.
[10] H. Buchholz, "Die Konfluente Hypergeometrische Funktion," Springer-Verlag, Berlin, Germany, p. 171; 1953.
[11] G. N. Watson, "Theory of Bessel Functions," Cambridge, p. 231; 1952.
[12] A. W. Maue, *Z. f. Phys.*, vol. 126, pp. 602–608; 1949.

Fields in the image space of symmetrical focusing reflectors

H. C. Minnett, B.Sc., B.E., Mem.I.E.E.E., and B. MacA. Thomas, B.E., M.Eng.Sc., Ph.D., C.Eng., M.I.E.E.

Synopsis

The fields scattered by circular symmetric reflector illuminated by a linearly polarised wave incident normally on the aperture are calculated from the induced surface currents. It is shown that the fields in the axial region can be represented by a spectrum of near-spherical hybrid waves propagating along the axis. For large microwave focusing reflectors, the wavefronts are effectively plane in the significant part of the image space. The axial wave fields are linear combinations of the TE_{1n} and TM_{1n} fields appropriate for circular metal pipes, but can be bounded only by anisotropic-reactance surfaces. Axial-wave theory is used to investigate the characteristics of the fields in the focal region of a paraboloidal reflector, when the incident wave is uniform and plane. For radiotelescope focal ratios, the image structure differs significantly from the classical Airy pattern, deduced by scalar analysis, of optical focusing systems. Energy vortexes circulating about the dark rings influence the efficiency obtainable from aperture-type feeds in the focal plane. Application of axial-wave analysis to spherical reflectors, and the synthesis of high-efficiency low-noise feeds, using hybrid-waves in corrugated guides, are described briefly.

List of principal symbols

$A(U)$, $B(U)$, $C(U)$ = parameters of total field at Q
a = radius of aperture in focal plane
c = $1/\sqrt{(\mu_0\epsilon_0)}$ = free-space velocity
D = diameter of paraboloid aperture
dS = element of reflector area at M
$\boldsymbol{E}$, $\boldsymbol{H}$ = field incident on reflector
E_σ, E_ϕ = components of incident field $\boldsymbol{E}$
$\boldsymbol{E}_s$, $\boldsymbol{H}_s$ = field scattered by reflector
E_β, E_δ = spherical components of scattered field at Q
E_ρ, E_ξ, E_z = cylindrical components of scattered field at Q
f = focal length of paraboloid
$g(s)$ = aperture illumination function of paraboloid
h = distance from origin O to reflector vertex V
$\boldsymbol{J}$ = surface current density
$J_n(u)$ = Bessel function of first kind, order n
j = $\sqrt{-1}$
k = $2\pi/\lambda$
k_0 = $\frac{k}{2}(1 - \cos\theta_0)$
$\bar{k}$ = $\frac{k}{2}(1 + \cos\theta_0)$
$\boldsymbol{n}$ = unit normal to reflector surface
P = power absorbed by aperture of radius a in focal plane
P_0 = power incident on paraboloid aperture
p = integer
q = $pa/\lambda f$
R, θ, ϕ = spherical co-ordinates of dS from the origin O
σ, ϕ, z = cylindrical co-ordinates of S from the origin O
r, β, δ = spherical co-ordinates of Q from the origin at M (Fig. 2)
r_0, β_0, 0 = spherical co-ordinates of axial point P from origin at M (Fig. 2)
$\boldsymbol{r}_1$ = unit vector from dS to point Q near axis
s = normalised radius of paraboloid aperture
$\boldsymbol{S}$ = Poynting vector
u = $k\rho\sin\theta$
U = $k\rho\sin\theta_0$
U_a = $ka\sin\theta_0$
Z_0 = $\sqrt{(\mu_0/\epsilon_0)}$ = impedance of free space
α = angle of incidence to $\boldsymbol{n}$
γ = dimensionless parameter specifying the ratio of the scattered field components at P
Δ = phase anomaly
ϵ_0, μ_0 = electric and magnetic constants of free space
ζ = r_0/R
η = aperture efficiency of paraboloid and feed
θ_0 = value of θ at reflector rim
κ = dimensionless parameter specifying the amplitude of the scattered field components at P
λ = free-space wavelength
ρ, ξ, z = cylindrical co-ordinates of Q from origin O
Φ = phase of fields at P
Ψ = $\{\exp(-jkr)\}/r$
ψ = angle between MP and z axis
ω = angular frequency of wave

Paper 5642 E, first received 1st November 1967 and in revised form 18th June 1968

Mr. Minnett and Dr. Thomas are with the Division of Radio Physics, Commonwealth Scientific & Industrial Research Organisation, Sydney, Australia

1 Introduction

The widespread use in recent years of paraboloidal reflector aerials has stimulated considerable interest in the development of improved feed systems. In particular, the high unit-area cost of very large paraboloidal reflectors has emphasised the need for feeds giving higher aperture efficiency and greater discrimination against noise radiation from the ground. Even modest improvements in these two factors can be equalled only by a very costly increase in aperture area.[1]

In radiotelescopes used for astronomical research, prime-focus feed systems are simple and flexible. The aperture efficiency, however, is usually below 65%, and ground radiation response can significantly degrade the performance of low-temperature receivers, as well as introducing spurious polarisation effects. Although the tapered response of such feeds does reduce the secondary sidelobe level, there are many applications in which this is not of primary concern, and their use then represents a substantial waste of expensive aperture area. Basically, the need is for a design technique permitting the synthesis of any desired response over the paraboloid aperture with rapid cutoff to very low levels beyond the rim. This would allow the designer to devise the optimum solution for a given application.

Investigation of this problem in connection with the Parkes 210ft radio telescope in 1963 suggested a theoretical solution, which has since been studied in detail. The initial approach was based on the concept of an *ideal* feed at the focus of an error-free paraboloid without aperture obstructions. With uniform, linearly polarised plane waves incident normally on

Reprinted with permission from *Proc. IEE*, vol. 115, pp. 1419–1430, Oct. 1968.

the paraboloid aperture, an ideal feed and matched load would absorb all the energy intercepted by the aperture.* The aperture efficiency would therefore be 100%. In the transmitting case, the ideal feed would, by reciprocity, radiate only in directions within the solid angle of the aperture, over which an identical outward-going wave would be established. Since the directional pattern of the feed must be the same in each case, the feed, when receiving, would not respond to noise generated by terrestrial sources outside this solid angle.

In this ideal system, the focal-region fields during reception are identical with those when transmitting with the direction of propagation reversed. Achieving a performance approaching the ideal depends on synthetising these fields, but the usual optical analysis of the focal region is inadequate for investigating this possibility. Thus Airy's well known equation[2] $2J_1(U)/U$, for the amplitude distribution in the focal plane of a circular lens, is valid only for the large focal ratios (f/D) common in optical systems. Although both amplitude and phase near the focus of a lens have been studied extensively at both optical[3] and microwave[4,5,6,7] frequencies, the scalar-wave theory used gives no information on the detailed configuration of the electromagnetic fields in the image. It is natural, however, for a radio engineer contemplating the dark rings of the Airy pattern to speculate on the arrangement of the fields there, and the possibility of inserting a cylindrical boundary to 'trap' the enclosed energy.

A vector solution was therefore sought for the fields in the focal region of a circular paraboloid.† It was discovered that the fields could be expressed as an infinite set of hybrid waves, each the sum of TE and TM component waves propagating along the reflector axis. This representation provides a very useful physical picture of the phenomena at the focus, and suggests applications beyond the scope of the original analysis. A study of the boundary conditions for the hybrid waves showed that they are the natural modes of certain corrugated waveguide structures, thus providing, in principle, a method of synthetising the fields required for an ideal feed.[11] As expected, the radiation fields of these modes have inherent symmetry about the axis, a property which may be exploited in radioastronomical polarisation measurements.[12,13]

In this paper, the original analytical technique has been generalised, to deal with the field structure in the axial region of any circularly symmetric focusing reflector. The theory is then used to study the fields in the focal region of a paraboloid. Finally, the application of the theory to other focusing systems is briefly discussed. The propagation of hybrid axial waves in corrugated waveguides, their radiation characteristics, and application to the synthesis of improved feed systems, will be considered in a separate paper.

2 Fields near the axis of a symmetrical reflector

2.1 General theory

Consider a linearly polarised plane wave incident normally on the circular aperture of an axially symmetric reflector (Fig. 1) having a shape $R = F(\theta)$, which tends to focus the scattered fields in the region of the axis. We require to calculate the field vector at a point $Q(\rho, \xi, z)$ in the image space.

The incident fields E, H induce current and charge distributions on the reflector surface, which will be assumed to be perfectly conducting. A line distribution of charge around the rim of the reflector, where the current density J is discontinuous, is also necessary to satisfy the equation of continuity. The fields radiated by these sources may be written in terms of J only:[14]

$$E_s = \frac{j}{4\pi\omega\epsilon_0}\int_S \{k^2 J\Psi + J\,.\,\nabla(\nabla\Psi)\}dS \quad . \quad . \quad (1)$$

$$H_s = \frac{1}{4\pi}\int_S (J \times \nabla\Psi)dS \quad . \quad . \quad . \quad . \quad . \quad . \quad . \quad (2)$$

where $\Psi = e^{-jkr}/r$ and the time-dependent factor has been taken as $\exp(j\omega t)$. The surface-charge field is exactly annulled

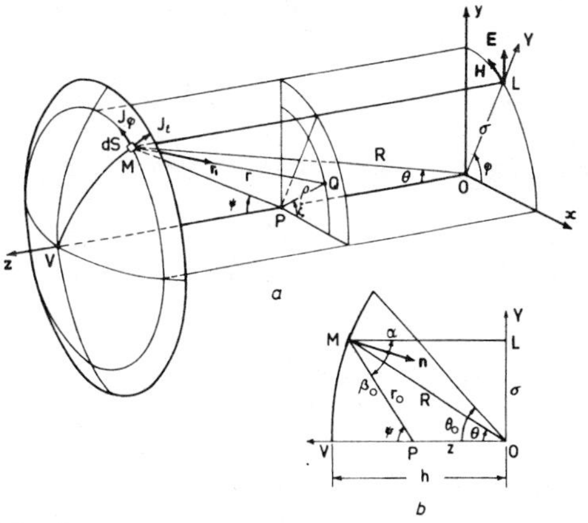

Fig. 1
Geometry of axially symmetric focusing reflector
a Co-ordinate system
b Plane of incidence

at all points by a component of the line-charge field, leaving a net contribution from the charges given by the second term of eqn. 1. The remaining field terms represent radiation by the current, which is related to the incident field by

$$J = 2(n \times H) \quad . \quad . \quad . \quad . \quad . \quad . \quad . \quad . \quad . \quad . \quad (3)$$

Eqn. 3 involves the usual assumption that the currents induced on dS are those that would be excited if dS were part of an infinite plane tangent at that point. This is justified because of the very large number of wavelengths in the radii of curvature of the reflectors to be considered.*

For distances r exceeding a few wavelengths, we can write

$$\nabla\Psi = -jk\Psi r_1$$

and $\quad J\,.\,\nabla(\nabla\Psi) = -k^2\Psi(J\,.\,r_1)r_1$

where r_1 is a unit vector directed from dS to the field point. Putting $\mu_0\omega = kZ_0$, the scattered fields become

$$\left.\begin{aligned} E_s &= -\frac{jkZ_0}{4\pi}\int_S \{J - (J\,.\,r_1)r_1\}\frac{e^{-jkr}}{r}dS \\ H_s &= \frac{jk}{4\pi}\int_S \{J \times r_1\}\frac{e^{-jkr}}{r}dS \end{aligned}\right\} \quad . \quad . \quad (4)$$

Thus, in the far zone, the longitudinal field due to J is cancelled by the field $(J\,.\,r_1)r_1$ due to the charges. Only components of J normal to r_1 contribute to the scattered fields which are therefore entirely transverse.

2.2 Components of the scattered field

The field components scattered by a surface element dS will now be determined in more specific form. The H field incident on dS at R, θ, ϕ resolves into cylindrical components

$$H_\sigma = -H\cos\phi,\ H_\phi = H\sin\phi$$

parallel and perpendicular respectively to the plane of incidence. If α is the angle of incidence to the normal n (Figs. 1 and 2), the components of surface current density, by substitution in eqn. 3, are

$$J_\phi = 2H\cos\alpha\cos\phi,\ J_t = -2H\sin\phi$$

where J_t is the radial component flowing towards the rim. The reflector current is polarised parallel to the yz plane, and its phase, which lags progressively from rim to vertex, is given by $e^{-jkR\cos\theta}$ with the plane xOy as phase reference.

* In this definition, supergain possibilities are excluded from consideration.
† Subsequently, it was found that an investigation of this problem had been undertaken previously by Robieux and Tocquec.[8] Their field expressions, however, are in error, because of inconsistency in the use of left- and right-handed co-ordinate systems. More recently, Kennaugh and Ott[9] have made an analysis based on the superposition of vector spherical waves, but have published only a brief communication on the field components along the principal axes. Watson[10] has discussed the more general case, when the incident wave arrives at a small angle to the reflector axis, and has particularly studied the exact distribution of the induced reflector currents. These generalisations introduced considerable analytical complications, and no specific results were presented.

* Watson[10] has shown that 'secondary' currents introduced by a more exact analysis have negligible influence

To compute the field scattered in any direction r_1, we take a set of polar co-ordinates r, β, δ, aligned as in Fig. 2. By resolving the current components normal to r_1, the fields of the spherical wave scattered by dS are found from eqn. 4 to be:

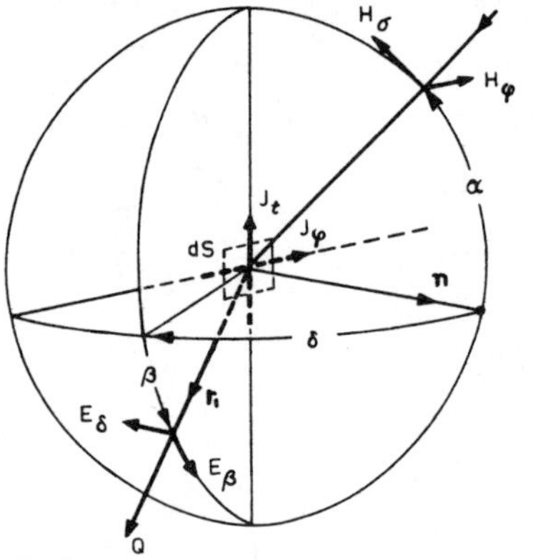

Fig. 2
Spherical co-ordinate system on surface element dS

$$\left.\begin{aligned} E_\beta &= \frac{jkE}{2\pi r}(\cos\beta\sin\phi + \cos\alpha\sin\delta\cos\phi)e^{-jkr}dS \\ E_\delta &= \frac{jkE}{2\pi r}(\cos\alpha\cos\delta\cos\phi)e^{-jkr}dS \\ H_\beta &= E_\delta/Z_0,\ H_\delta = -E_\beta/Z_0 \end{aligned}\right\} \quad (5)$$

2.3 Approximations near the axis

The fields of greatest interest in a focusing system are those near the axis, and this allows the equations to be simplified. In Fig. 1, P is a point (0, 0, z) where the perpendicular plane containing Q(ρ, ξ, z) cuts the axis, and MP $= r_0$. It is shown in Appendix 10.1 that the phase at Q relative to the reference plane xOy is

$$\Phi_Q = \Phi - k\{\rho\sin\psi\cos(\phi-\xi) - \frac{\rho^2}{2r_0}(1-\tfrac{1}{2}\sin^2\psi) + \frac{\rho^2}{4r_0}\sin^2\psi\cos 2(\phi-\xi)\} \quad (6)$$

where $\Phi = k(z + r_0\cos\psi + r_0)$

is the phase at P and ψ is the inclination of r_0 to the axis.
If ρ is restricted so that

$$\rho/\lambda \leqslant 0{\cdot}35\sqrt{(r_0/\lambda)} \quad . \quad . \quad . \quad . \quad (7)$$

the second-order terms in eqn. 6 may be neglected with an error of less than $\pi/8$ radians for any value of $(\phi-\xi)$ and ψ. Then

$$\Phi_Q = \Phi - k\rho\sin\psi\cos(\phi-\xi) \quad . \quad . \quad . \quad . \quad (8)$$

This is equivalent to assuming that the wavefront near P is a plane normal to the ray MP (Fig. 3). It is easily verified from this Figure that the phase difference between Q and P is $k(\mathrm{NP}) = k\rho\sin\psi\cos(\phi-\xi)$.

The boundary of the region of validity of the plane-wave approximation defined by eqn. 7 is shown in Fig. 4 by the heavy line. The size of this region in wavelengths is considerable for large microwave reflectors ($D/\lambda \gg 1$). For example, its diameter is about 11λ near the focus of the Parkes 210ft paraboloid (at $\lambda = 10$cm) and about 9λ near the paraxial focus of the Arecibo 1000ft spherical reflector (at $\lambda = 75$cm), as indicated in Fig. 4. These dimensions are ample for the study of image formation and feed development.

Since the radiation pattern of dS is very broad, we may also assume that the field amplitudes are constant over the section of the wavefront of interest. Hence the fields at Q (direction β, δ) are equal to those at N (direction β_0, 0), and, since NP $\ll r_0$, they are negligibly different in amplitude from those at P. This assumption is discussed in Appendix 10.2, where it is shown that, for $r_0/\lambda > 50$, the error is less than 5% in the region where the plane-wave approximation is applicable (Fig. 4).

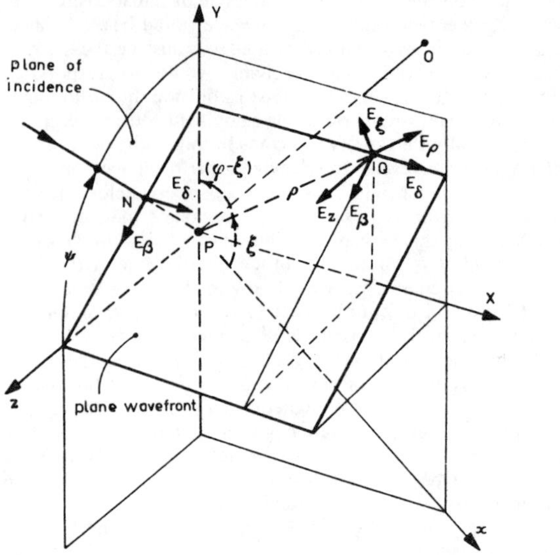

Fig. 3
Resolution of the scattered field components at a point Q near the reflector axis

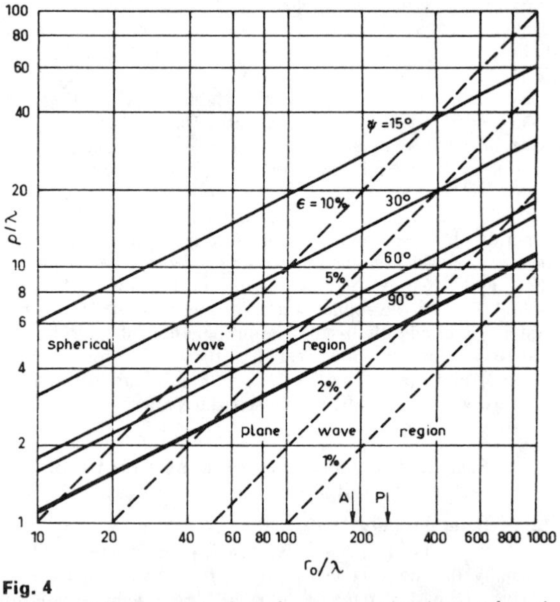

Fig. 4
Regions of the plane and spherical wave approximations, and maximum amplitude error ϵ at Q for a wave scattered from dS
Values of ρ/λ for Arecibo spherical reflector at $\lambda = 75$ cm (A), Parkes paraboloid at $\lambda = 10$ cm (P)

Accordingly, in eqn. 5, we can put $\delta = 0$, $\beta = \beta_0$ and $r = r_0$ in the amplitude term. In addition, since

$$dS = \sigma r_0 \sec(\psi - \alpha)d\psi d\phi \text{ and } \beta_0 = \psi - \alpha$$

eqn. 5 becomes

$$\left.\begin{aligned} E_\beta &= \frac{jkhE}{2\pi}\kappa\sin\phi e^{-j\{\Phi - k\rho\sin\psi\cos(\phi-\xi)\}}d\psi d\phi \\ E_\delta &= \frac{jkhE}{2\pi}\gamma\kappa\cos\phi e^{-j\{\Phi - k\rho\sin\psi\cos(\phi-\xi)\}}d\psi d\phi \end{aligned}\right\} \quad (9)$$

where κ and γ are the dimensionless parameters:

$$\kappa = \frac{\sigma}{h} = \frac{r_0\sin\psi}{h},\ \gamma = \frac{\cos\alpha}{\cos(\psi-\alpha)} \quad . \quad . \quad . \quad (10)$$

The fields at Q have now been expressed in terms of those at the axial point P, where the amplitude, polarisation and phase are determined by κ, γ and Φ, respectively. For a given position of P, these parameters are functions only of ψ, since both α and r_0 depend on ψ when the reflector shape has been specified.

2.4 Total scattered field

The total field at Q is the sum of the fields due to all the waves scattered from the reflector surface. For this calculation, the fields are best expressed in cylindrical components, which can be found readily from Fig. 3.

Thus, resolving the incident component E_β, which is polarised in the plane of incidence, gives

$$\left.\begin{aligned} E_\sigma &= -E_\beta \cos\psi \cos(\phi - \xi) \\ E_\xi &= -E_\beta \cos\psi \sin(\phi - \xi) \\ E_z &= +E_\beta \sin\psi \end{aligned}\right\} \quad (11)$$

Resolving E_δ, polarised normal to the plane of incidence, produces

$$\left.\begin{aligned} E_\rho &= +E_\delta \sin(\phi - \xi) \\ E_\xi &= -E_\delta \cos(\phi - \xi) \\ E_z &= 0 \end{aligned}\right\} \quad (12)$$

Substituting in eqn. 11 for E_β from eqn. 9, and integrating over the reflector surface, gives

$$\left.\begin{aligned} E_\rho &= \frac{-jkhE}{2\pi}\int_0^{\psi_0}\Big\{\kappa \cos\psi e^{-j\Phi} \int_0^{2\pi} \sin\phi \cos(\phi - \xi) e^{jk\rho \sin\psi \cos(\phi-\xi)} d\phi\Big\} d\psi \\ E_\xi &= \frac{-jkhE}{2\pi}\int_0^{\psi_0}\Big\{\kappa \cos\psi e^{-j\Phi} \int_0^{2\pi} \sin\phi \sin(\phi - \xi) e^{jk\rho \sin\psi \cos(\phi-\xi)} d\phi\Big\} d\psi \\ E_z &= \frac{jkhE}{2\pi}\int_0^{\psi_0}\Big\{\kappa \sin\psi e^{-j\Phi} \int_0^{2\pi} \sin\phi e^{jk\rho \sin\psi \cos(\phi-\xi)} d\phi\Big\} d\psi \end{aligned}\right\} \quad (13)$$

where ψ_0 is the semiangle subtended at P by the reflector aperture. A similar procedure shows that the magnetic-field components are

$$H_\rho = E_\xi / Z_0 \cos\psi, \; H_\xi = -E_\rho / Z_0 \cos\psi$$
$$H_z = 0$$

Likewise, the field produced by the normal component E_δ is found, by substitution from eqn. 9 in eqn. 12, to be:

$$\left.\begin{aligned} E_\rho &= \frac{jkhE}{2\pi}\int_0^{\psi_0}\Big\{\gamma\kappa e^{-j\Phi} \int_0^{2\pi} \cos\phi \sin(\phi - \xi) e^{jk\rho \sin\psi \cos(\phi-\xi)} d\phi\Big\} d\psi \\ E_\xi &= \frac{-jkhE}{2\pi}\int_0^{\psi_0}\Big\{\gamma\kappa e^{-j\Phi} \int_0^{2\pi} \cos\phi \cos(\phi - \xi) e^{jk\rho \sin\psi \cos(\phi-\xi)} d\phi\Big\} d\psi \\ E_z &= 0 \\ H_\rho &= E_\xi \cos\psi / Z_0, \; H_\xi = -E_\rho \cos\psi / Z_0 \\ H_z &= \frac{jkhE}{2\pi Z_0}\int_0^{\psi_0}\Big\{\gamma\kappa \sin\psi e^{-j\Phi} \int_0^{2\pi} \cos\phi e^{jk\rho \sin\psi \cos(\phi-\xi)} d\phi\Big\} d\psi \end{aligned}\right\} \quad (14)$$

The total scattered field at Q is obtained by adding corresponding components of eqns. 13 and 14.

3 Axial-wave representation

3.1 TE and TM component waves

A useful and physically informative description of these fields can be developed by considering eqns. 13 and 14 separately, and integrating initially only around an annular ring of the reflector.

Integration of eqn. 13 from $\phi = 0$ to $\phi = 2\pi$ represents the summation of all the scattered plane waves polarised in their plane of incidence. This integration gives a field at Q which is transverse-magnetic to the reflector axis:

$$\left.\begin{aligned} E_\rho &= -jkhE\kappa \cos\psi \sin\xi J_1'(u) e^{-j\Phi} d\psi \\ E_\xi &= -jkhE\kappa \cos\psi \cos\xi \frac{J_1(u)}{u} e^{-j\Phi} d\psi \\ E_z &= -khE\kappa \sin\psi \sin\xi J_1(u) e^{-j\Phi} d\psi \\ H_\rho &= E_\xi / Z_0 \cos\psi, \; H_\xi = -E_\rho / Z_0 \cos\psi \\ H_z &= 0 \end{aligned}\right\} \quad (15)$$

where $u = k\rho \sin\psi$ is the generalised radial distance. Since the phase Φ is independent of ρ and ξ, it is the same everywhere in the transverse plane containing Q and P. The field distribution in this plane is shown in Fig. 5*a*, together with the values of u, for which $J_1(u) = 0$.

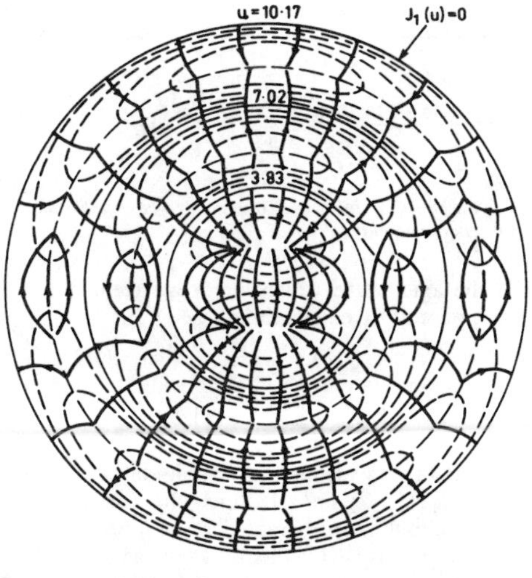

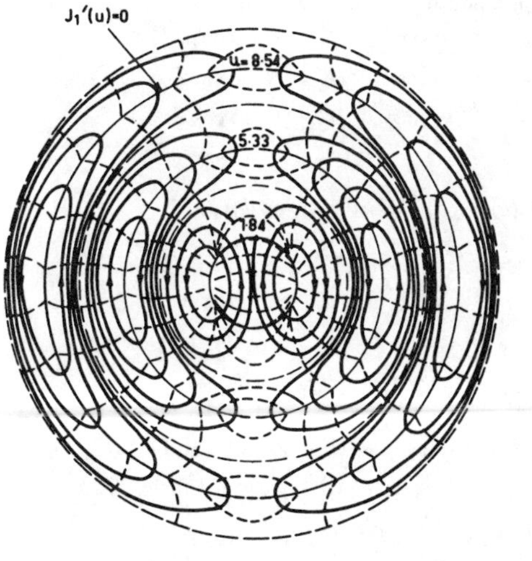

Fig. 5
Transverse field patterns for axial waves
a TM_{1n}
b TE_{1b}
—— E lines
---- H lines

Similarly, the ϕ integration of eqn. 14 sums all the scattered plane waves which are polarised normal to their plane of incidence. The result is a field transverse-electric to the reflector axis with components

$$\left.\begin{aligned} E_\rho &= -jkhE\gamma\kappa \sin\xi \frac{J_1(u)}{u} e^{-j\Phi} d\psi \\ E_\xi &= -jkhE\gamma\kappa \cos\xi J_1'(u) e^{-j\Phi} d\psi \\ E_z &= 0 \\ H_\rho &= E_\xi \cos\psi/Z_0,\ H_\xi = -E_\rho \cos\psi/Z_0 \\ H_z &= \frac{-khE}{Z_0}\gamma\kappa \sin\psi \cos\xi J_1(u) e^{-j\Phi} d\psi \end{aligned}\right\} \quad . \quad (16)$$

Fig. 5*b* shows this distribution, together with the values of u, for which $J_1'(u) = 0$.

These fields consist of component waves incident on P with the free-space velocity c from a cone of directions of semiangle ψ. In a time Δt, each wave component travels Δr_0 and P advances $\Delta z = -\Delta r_0 \sec\psi$ along the axis. Thus the patterns of Fig. 5 propagate along the axis away from the reflector with velocity $c \sec\psi$.

Eqns. 15 and 16 are TM_{1n} and TE_{1n} plane-wave solutions of Maxwell's equations in cylindrical co-ordinates with one circumferential periodicity, as a result of the linearly polarised excitation. There are n radial periodicities, when n is to be determined when the boundaries are specified. These solutions are appropriate for hollow cylindrical waveguides, and, if the walls have infinite electrical conductivity, the boundaries may be located on any of the circles shown in Fig. 5 by solid lines. On these circles, the boundary conditions are satisfied since $\boldsymbol{E}$ is everywhere normal and $\boldsymbol{H}$ everywhere tangential. The analysis in this Section is, in fact, the reverse of the decomposition of cylindrical waveguide modes into an infinite set of plane waves propagating at an angle ψ to the axis.[15]

The TE and TM axial wave fields, however, have different radial periodicities, and the normal boundary conditions given above cannot be satisfied at the same radius in each case. It is impossible, therefore, to bound the total field by a tube of infinitely conducting material, i.e. by an 'electric' wall; similarly, it cannot be bounded by a 'magnetic' wall requiring $\boldsymbol{H}$ to be everywhere normal.

3.2 Hybrid axial waves

The total field generated in a transverse plane by the annular ring $d\psi$ is the sum of the TM_{1n} and TE_{1n} axial waves. Since each of these waves travels at the same velocity along the axis, the combined axial wave will propagate as an entity at this velocity. Adding eqns. 15 and 16, the field components of the combined wave can be written

$$\left.\begin{aligned} E_\rho &= E_\rho(u)\sin\xi \quad & H_\rho &= H_\rho(u)\cos\xi \\ E_\xi &= E_\xi(u)\cos\xi \quad & H_\xi &= H_\xi(u)\sin\xi \\ E_z &= E_z(u)\sin\xi \quad & H_z &= H_z(u)\cos\xi \end{aligned}\right\} \quad . \quad . \quad (17a)$$

where

$$\left.\begin{aligned} E_\rho(u) &= -j\frac{khE}{2}\kappa\{(\gamma + \cos\psi)J_0(u) \\ &\qquad + (\gamma - \cos\psi)J_2(u)\} e^{-j\Phi} d\psi \\ E_\xi(u) &= -j\frac{khE}{2}\kappa\{(\gamma + \cos\psi)J_0(u) \\ &\qquad - (\gamma - \cos\psi)J_2(u)\} e^{-j\Phi} d\psi \\ E_z(u) &= -khE\kappa \sin\psi J_1(u) e^{-j\Phi} d\psi \\ H_\rho(u) &= -j\frac{khE}{2Z_0}\kappa\{(1 + \gamma\cos\psi)J_0(u) \\ &\qquad + (1 - \gamma\cos\psi)J_2(u)\} e^{-j\Phi} d\psi \\ H_\xi(u) &= j\frac{khE}{2Z_0}\kappa\{(1 + \gamma\cos\psi)J_0(u) \\ &\qquad - (1 - \gamma\cos\psi)J_2(u)\} e^{-j\Phi} d\psi \\ H_z(u) &= -\frac{khE}{Z_0}\kappa\gamma \sin\psi J_1(u) e^{-j\Phi} d\psi \end{aligned}\right\} \quad (17b)$$

This is a plane wave with both E_z and H_z components, and will therefore be called the HE_{1n} hybrid axial wave. We note from eqns. 15 and 16 that γ determines the ratio of the longitudinal fields of the TE and TM axial waves respectively:

$$\gamma = Z_0 H_z(u)/E_z(u)$$

The axial velocity $c \sec\psi$ and wavelength $\lambda \sec\psi$ are determined at a particular axial point by the semiangle ψ subtended by the generating zone. The field distribution over the wavefront depends only on ψ, and the details of the pattern, but not the basic form, change slowly as the wave propagates. However, over substantial sections of the axis, the pattern can be considered fixed. This will be illustrated in Section 4, when dealing with the focal region of a paraboloid.

3.3 Curvature of the wavefronts

It is of interest to consider the axial-wave phase fronts at distances from the axis exceeding the plane-wave approximation. A reference to eqn. 6 shows that a second-order approximation for the phase at Q is

$$\Phi_Q = \Phi - k\left\{\rho \sin\psi \cos(\phi - \xi) - \frac{\rho^2}{2r_0}(1 - \tfrac{1}{2}\sin^2\psi)\right\} \quad (18)$$

provided that

$$\rho/\lambda \leqslant 0{\cdot}5 \operatorname{cosec}\psi\sqrt{(r_0/\lambda)} \quad . \quad . \quad . \quad . \quad . \quad . \quad . \quad (19)$$

This restriction enables the last term in eqn. 6 to be neglected with an error not exceeding $\pi/8$, and is less severe than the plane-wave approximation, especially for low ψ values, as shown in Fig. 4.

The second-order phase term in eqn. 18 is independent of ϕ, so that, when the ϕ integrations of Section 3.1 are performed, the axial-wave phase term is modified from $\exp(-j\Phi)$ to $\exp[-j\{\Phi + k\rho^2(1 - \frac{1}{2}\sin^2\psi)/2r_0\}$. Thus, the phase at Q in the transverse plane lags behind that at P by $k\rho^2(1 - \frac{1}{2}\sin^2\psi)/2r_0$.

Let $Q'(\rho, \xi, z + \Delta z)$ be a point on the wavefront (Fig. 6)

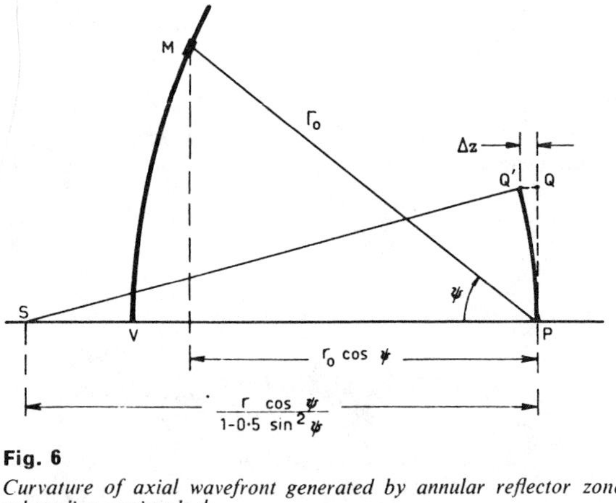

Fig. 6
Curvature of axial wavefront generated by annular reflector zone subtending semiangle ψ

of the axial wave which has a wavelength of $\lambda \sec\psi$. Then the phase at Q' equals the phase at P:

$$k\cos\psi(QQ') = k\rho^2(1 - \tfrac{1}{2}\sin^2\psi)/2r_0$$

Thus $$\rho^2 = 2r_0\left(\frac{\cos\psi}{1 - \frac{1}{2}\sin^2\psi}\right)\Delta z \quad . \quad . \quad . \quad . \quad . \quad . \quad . \quad (20)$$

so that the wavefront is paraboloidal with a focal length of $r_0\cos\psi/2(1 - \frac{1}{2}\sin^2\psi)$. Since the curvature is very small, this is closely approximated by a sphere with centre S at a distance $r_0\cos\psi/(1 - \frac{1}{2}\sin^2\psi)$ to the left of P. This is not a fixed phase centre. For $\psi = \pi/2$, P and S coincide in the plane of the generating annulus, but, as the axial wave recedes, S moves behind the vertex V. As $\psi \to 0$, S approaches a limiting position on V.

The first-order phase approximation obviously specifies the region surrounding the axes within which the curved axial wavefronts can be considered plane (to within $\lambda/16$). Beyond this boundary, the field patterns will still be essentially those already discussed, but adjusted to conform to the slightly curved wavefront.

3.4 The axial-wave spectrum

Each annular zone of the reflector generates an axial wave characterised by a particular value of ψ. The uniform plane wave incident on the aperture is therefore transformed by reflection into a spectrum of axial waves extending from $\psi = 0$ to $\psi = \psi_0$. In eqn. 17, κ specifies the amplitude spectrum of the waves, Φ the phase spectrum, and γ the spectrum of field patterns. The form of the spectrum in each case is determined by the reflector shape. The ψ integration in the total-field calculation (eqns. 13 and 14) is therefore equivalent to summing the field components of all the axial waves in the spectrum.

The above concepts are clearly not limited by the initial assumption of uniform plane-wave illumination. Substituting any linearly polarised, circularly symmetric distribution simply modifies the spectrum of axial waves produced. Thus, if the resulting aperture amplitude varies radially, the aperture spectrum will be weighted by a function of ψ. Similarly, the phase spectrum will be modified if there is a radial variation of phase produced, for example, by an illuminating source at a finite distance. If the source is close to the reflector,* curvature of the incident spherical wavefront changes α, and therefore the spectrum of γ.

The reflector shape is specified by the relation $R = F(\theta)$, and ψ can also be expressed as a function of θ. It is therefore convenient in computations to eliminate ψ and write the axial-wave fields (eqn. 17) in terms of the independent variable θ. The spectrum of waves then extends from $\theta = 0$, corresponding to the reflector vertex, to $\theta = \theta_0$, corresponding to the rim. Expressions for the quantities appearing in eqn. 17 (κ, γ, Φ, u and $\cos\psi$) are given in Appendix 10.3 as functions of θ. When substituting for $d\psi$ in the equation, it is convenient to redefine the amplitude spectrum as $\kappa(\theta)$, so that $\kappa(\theta)d\theta = kd\psi$ as shown in the Appendix.

4 Fields in the focal region of a paraboloid

The axial-wave theory developed above will be used in this Section to study the fields in the focal region of a paraboloid illuminated by a uniform plane wave.

4.1 Limits of the region

For a paraboloid with the origin of co-ordinates at the focus, $h = f$, $R = f\sec^2\theta/2$, $\alpha = \theta/2$, so that eqns. 53, 54 and 55 in Appendix 10.3 reduce to

$$k(\theta) = \frac{2\tan\theta/2}{\zeta^2}(1 - z/R) \quad . \quad . \quad . \quad . \quad . \quad . \quad (21)$$

$$\gamma = \frac{\zeta}{1 - z/R} \quad . \quad . \quad . \quad . \quad . \quad . \quad . \quad . \quad (22)$$

$$\Phi = kR(\cos\theta + \zeta) \quad . \quad . \quad . \quad . \quad . \quad . \quad (23)$$

where $$\zeta = \frac{r_0}{R} = \sqrt{\left\{1 - \frac{2z}{R}\cos\theta + \left(\frac{z}{R}\right)^2\right\}}$$

By confining attention to small distances z from the focus, ζ may be expanded binomially, to obtain

$$\zeta = 1 - \frac{z}{R}\cos\theta - \tfrac{1}{2}\left(\frac{z}{R}\right)^2(1 + \cos^2\theta) + \ldots \quad (24)$$

Neglecting terms of higher order than the first, the phase of an axial wave near the focus becomes

$$\Phi = k(2f - z\cos\theta) \quad . \quad . \quad . \quad . \quad . \quad . \quad . \quad . \quad (25)$$

showing that all axial waves in the spectrum become cophased at the focal plane $z = 0$. This is characteristic of any system with exact focusing.

At a distance z, the phase error due to neglect of the second-order term in eqn. 24 will not exceed $\pi/8$, provided

$$|z|/\lambda \leqslant \frac{0\cdot35\sec^2\theta/2}{\sqrt{(1 + \cos^2\theta)}}\sqrt{(f/\lambda)} \quad . \quad . \quad . \quad . \quad . \quad (26)$$

* For a distance of five aperture diameters, γ is increased by about 10% above the value for a plane incident wave

For the Parkes paraboloid, $f/D = 0\cdot41$ and θ ranges from 0 to 63°. When operating at $\lambda = 10$cm, $f/\lambda = 262$, and $|z|/\lambda$ varies from 4 to 7 over the spectrum of axial waves. This is quite adequate for present purposes, and eqns. 7 and 26 will be used to define the limits of the focal region considered. The size of this region in wavelengths increases with decreasing λ, and it is interesting to note that, for a mirror with $f = 10$cm at an optical λ of 5×10^{-5}cm, the region is 150λ in diameter and 300λ long.

Within the focal region, γ and $\kappa(\theta)$ will be assumed to have constant values equal to those in the focal plane:

$$\gamma = 1, \quad \kappa(\theta) = 2\tan\frac{\theta}{2}$$

It is easily verified from eqns. 22 and 21 that the errors due to this assumption are less than 1% and 2%, respectively. It may also be assumed, with an error not exceeding 2%, that $\psi = \theta$ throughout the focal region. Axial-wave velocity and wavelength are then constant and equal to $c\sec\theta$ and $\lambda\sec\theta$, respectively.

4.2 Properties of axial waves near focus

Substituting the expressions given above for $\kappa(\theta)$, γ and Φ into eqn. 17*b*, the axial-wave fields in the focal region are given by eqn. 17*a* with

$$\left.\begin{aligned}
E_\rho(u) &= -jkfE\sin\theta\left\{J_0(u) + \tan^2\frac{\theta}{2}J_2(u)\right\}e^{jkz\cos\theta}d\theta \\
E_\xi(u) &= -jkfE\sin\theta\left\{J_0(u) - \tan^2\frac{\theta}{2}J_2(u)\right\}e^{jkz\cos\theta}d\theta \\
E_z(u) &= -4kfE\sin^2\frac{\theta}{2}J_1(u)e^{jkz\cos\theta}d\theta
\end{aligned}\right\} \quad (27)$$

$$H_\rho(u) = \frac{+E_\rho(u)}{Z_0}, \quad H_\xi(u) = \frac{-E_\xi(u)}{Z_0}, \quad H_z(u) = \frac{+E_z(u)}{Z_0}$$

where $u = k\rho\sin\theta$, and the common phase factor e^{-2jkf} has been omitted.

The field distribution over the wavefront is therefore constant inside the region considered. It will also be noticed that the E and H fields are *identical* in form except for a rotation of 90° in ξ and, at corresponding points (ρ, ξ and ρ, $\xi + \pi/2$), the ratio of the E to the H component is equal to Z_0. It can be shown that this unique characteristic of hybrid-wave fields having $\gamma = 1$ ensures that their radiation patterns are axially symmetrical.

For axial waves at the lower end of the spectrum ($\theta < 0\cdot2$rad), the total transverse field is:

$$E_y = -jkfE\theta J_0(k\rho\theta)e^{jkz}d\theta \quad . \quad . \quad . \quad . \quad . \quad . \quad (28)$$

polarised parallel to the wave incident on the paraboloid. The magnetic field is orthogonal and given by $H_x = E_y/Z_0$. Thus the TM and TE components of Fig. 5 combine into a linearly polarised hybrid wave (Fig. 7*a*).

Where θ is not small, however, the transverse field distribution is rather more complicated, as in Fig. 7*b*, which has been computed for $\theta = 45°$. The crosspolarised field is zero only along the axes $\xi = 0$, $\pi/2$, and on the circles where $J_2(u) = 0$. The crosspolarisation and curvature of the field lines are particularly marked in the region where $J_0(u) = 0$. Also, the E and H plane fields are orthogonal only where the crosspolarisation is zero, and on the circle where $J_0(u) = 0$. For large θ, the field amplitude is no longer symmetrically distributed about the axis. This is shown by the amplitude contours for the transverse electric field $E(u)$ in Fig. 7*c*; the bright and dark rings are not quite circular, and the amplitude varies around the ring.

The distribution of power over the wavefront has a number of interesting features. The time-average power density in the axial direction is

$$\bar{S}_z = \tfrac{1}{2}\,\mathrm{Re}\{E_\rho H_\xi^* - E_\xi H_\rho^*\} \quad . \quad . \quad . \quad . \quad . \quad . \quad (29)$$

$$= -\frac{1}{2Z_0}\,\mathrm{Re}\{E_\rho(u)E_\xi^*(u)\}$$

using the hybrid-wave symmetry relations. Thus the power flow is symmetrical about the axis, although the field amplitudes are not. This is true for any θ and is a result of the equality of the TE and TM components when $\gamma = 1$.

Substitution from eqn. 27 gives

$$\bar{S}_z = -\frac{(kEf\sin\theta)^2}{2Z_0}\left\{J_0^2(u) - \tan^4\frac{\theta}{2}J_2^2(u)\right\} \quad . \quad . \quad (30)$$

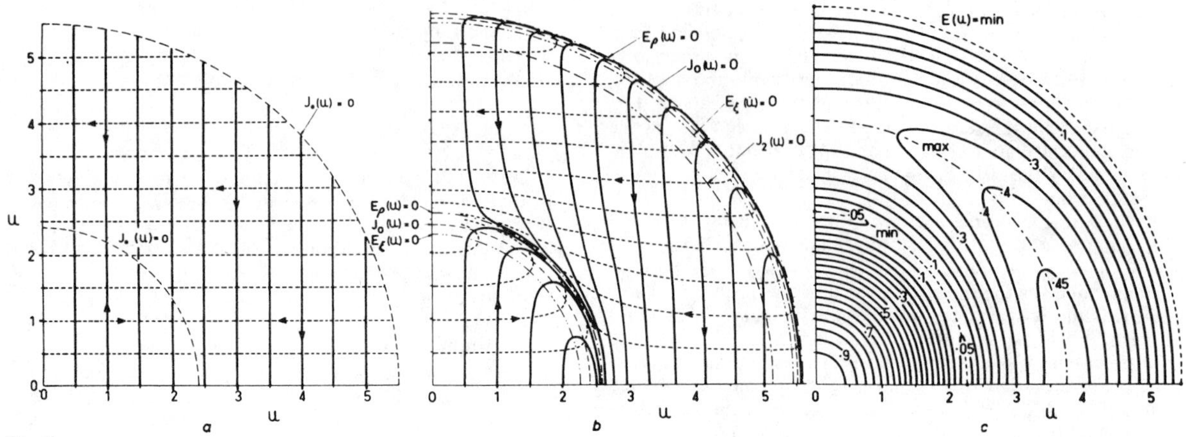

Fig. 7

Transverse field distribution of HE_{1n} hybrid axial wave in focal region

a θ small
b $\theta = 45°$
——— E lines
- - - - H lines
c Contours of amplitude $E(u)$, $\theta = 45°$

Power flow is therefore concentrated about the axis in concentric zones which, for small θ, are separated by 'dark' rings $J_0(u) = 0$ on which $\bar{S}_z = 0$. In the general case, however, the dark rings split into pairs defined by $J_0(u) = \pm\tan^2\frac{\theta}{2}J_2(u)$, on which $E_\xi(u) = 0$ and $E_\rho(u) = 0$ are respectively zero (Fig. 7*b*). Between each pair of dark rings, $\bar{S}_z$ reverses sign, so that power flows opposite to the direction of wave propagation. It is easily verified that the transverse components of the Poynting vector $\boldsymbol{S}$ averaged over a cycle are zero. The reverse power must therefore be supplied and absorbed by the fields at the ends of the focal region where different conditions apply (see Section 4.5).

The 3-dimensional structure of hybrid-wave fields exhibits some distinctive features. Consideration of Fig. 7*b* shows that the cylindrical surfaces, on which $E_\rho(u) = H_\rho(u) = 0$, divide the fields into annular cells, within which the interlinked closed loops of the $\boldsymbol{E}$ and $\boldsymbol{H}$ lines are entirely enclosed. This differs from the structure of TM and TE fields (Fig. 5), where the successive $\boldsymbol{E}$ cells are intermingled with the $\boldsymbol{H}$ cells.

4.3 Field variation along the axis

The variation of the field along the axis in the focal region is given by integrating eqn. 27 with $u = 0$. The electric field is parallel to the incident field and is

$$E_y(z) = -2k_0Ef\frac{\sin k_0z}{k_0z}e^{j(kz+\Delta)} \quad . \quad . \quad . \quad . \quad . \quad (31)$$

where $\quad k_0 = k\sin^2(\theta_0/2)$ and $\Delta = \frac{\pi}{2} - k_0z$

In this equation, Δ, which is usually termed the *phase anomaly* in optics,[3,5] measures the deviation of the phase of the resultant wavefront from that of a pure spherical wave converging on the focus. As shown in Fig. 8, discontinuities of π radians occur in Δ when $\sin k_0z/k_0z$ changes sign.

These results can be interpreted physically as interference patterns produced by the spectrum of axial waves travelling along the axis. Remote from the focal plane, the waves interfere destructively, and the resultant amplitude is small. In the focal region, the waves approach phase coherence and become precisely cophased in crossing the focal plane, after which the process reverses.

The $|E_y(z)|$ pattern is symmetrical about the focus because the axial-wave amplitudes remain nearly constant in the focal region. However, over a longer segment of the axis the interference pattern will be asymmetrical because of progressive changes in the waves as they propagate.

Eqn. 31 can be rewritten in the form:

$$E_y(z) = -2k_0Ef\frac{\sin k_0z}{k_0z}e^{j\left(\bar{k}z + \frac{\pi}{2}\right)} \quad . \quad . \quad . \quad . \quad (32)$$

where $\quad \bar{k} = k\cos^2(\theta_0/2) = \frac{k}{2}(1+\cos\theta_0)$

Since the phase constant of an individual axial wave is $k\cos\theta$, $\bar{k}$ is the average phase constant of the waves generated

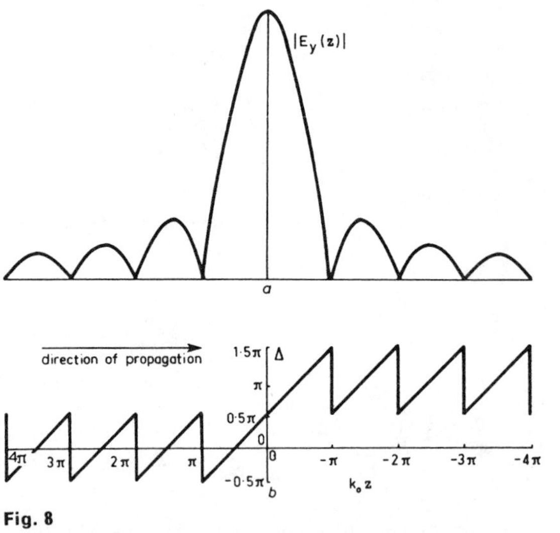

Fig. 8

Fields on the axis in focal region of paraboloid

a Amplitude $E_y(z)$
b Phase anomaly Δ

by the rim and vertex of the paraboloid. The resultant wavefronts have an axial wavelength of $\lambda\sec^2(\theta_0/2)$, and their velocity is $c\sec^2(\theta_0/2)$; they therefore exceed the free-space values.

Also

$$k_0 = \frac{k}{2}(1 - \cos\theta_0)$$

is the difference of the phase constants at the extremes of the spectrum, and this determines the axial period of the interference pattern. From eqn. 32, the zeros occur when

$$z/\lambda = \pm\frac{p}{2}\operatorname{cosec}^2\frac{\theta_0}{2}$$

where p is an integer. In radiotelescopes, they are much more closely spaced than in the usual optical image. Thus, for the Parkes paraboloid ($\theta_0 = 1 \cdot 1$ radians), $\frac{z}{\lambda} = \pm 1 \cdot 83p$ whereas for $f/D = 2 \cdot 5(\theta_0 = 0 \cdot 2\,\text{rad})$, $\frac{z}{\lambda} = \pm 50p$

4.4 Focal-plane distribution

To obtain the fields at a point off the axis, the spectrum of axial waves must be integrated, using the complete eqns. 27. The resulting fields may be written in the form

$$\left.\begin{array}{ll} E_\rho = E_\rho(U)\sin\xi & H_\rho = H_\rho(U)\cos\xi \\ E_\xi = E_\xi(U)\cos\xi & H_\xi = H_\xi(U)\sin\xi \\ E_z = E_z(U)\sin\xi & H_z = H_z(U)\cos\xi \end{array}\right\} \quad . \quad (33a)$$

$$\left.\begin{array}{l} \text{where } E_\rho(U) = -2jkfE\sin^2\frac{\theta_0}{2}\{A(U)+B(U)\} \\ E_\xi(U) = -2jkfE\sin^2\frac{\theta_0}{2}\{A(U)-B(U)\} \\ E_z(U) = -8kfE\sin^2\frac{\theta_0}{2}\{\dot{C}(U)\} \end{array}\right\} \quad . \quad (33b)$$

$$H_\rho(U) = E_\rho(U)/Z_0, \quad H_\xi(U) = -E_\xi(U)/Z_0, \quad H_z(U) = E_z(U)/Z_0 \quad . \quad . \quad (33c)$$

and $U = k\rho\sin\theta_0$. In these equations, the functions

$$\left.\begin{array}{l} A(U) = \frac{1}{2}\operatorname{cosec}^2\frac{\theta_0}{2}\int_0^{\theta_0}\sin\theta J_0(k\rho\sin\theta)e^{jkz\cos\theta}d\theta \\ B(U) = \frac{1}{2}\operatorname{cosec}^2\frac{\theta_0}{2}\int_0^{\theta_0}\sin\theta\tan^2\frac{\theta}{2}J_2(k\rho\sin\theta)e^{jkz\cos\theta}d\theta \\ C(U) = \frac{1}{2}\operatorname{cosec}^2\frac{\theta_0}{2}\int_0^{\theta_0}\sin^2\frac{\theta}{2}J_1(k\rho\sin\theta)e^{jkz\cos\theta}d\theta \end{array}\right\} \quad (34)$$

have been normalised to the values at the focus ($\rho = 0, z = 0$).

In general, $A(U)$, $B(U)$ and $C(U)$ are complex. In the focal plane, however, $z = 0$ and the functions are real. If, in addition, θ_0 is small ($\leqslant 0 \cdot 2$ rad),

$$A(U) = 2\frac{J_1(U)}{U}, \quad B(U) \approx 0, \quad C(U) = \frac{\theta_0}{2}\frac{J_2(U)}{U} \quad (35)$$

Substituting eqn. 35 into eqn. 33, and putting $\theta_0 = D/2f$, the amplitude of the focal plane field is

$$|E_y| = 2E\left(\frac{\pi D^2}{4f\lambda}\right)\frac{J_1(k\rho\theta_0)}{k\rho\theta_0} \quad . \quad . \quad . \quad . \quad . \quad . \quad (36)$$

with $|H_x| = |E_y|/Z_0$. Eqn. 36 corresponds to the scalar solution obtained by Airy for the distribution of light in the focal plane of a lens. The dark rings occur on the circles where $J_1(k\rho\theta_0) = 0$ and the polarisation of the transverse field is parallel to the incident field.

If the incident aperture field is nonuniform but varies radially as $g(s)$, where s is the normalised aperture radius, the amplitude spectrum of the axial waves must be weighted by $g(s)$. The focal-plane distribution then becomes

$$|E_y(q)| = E\left(\frac{D^2}{4f\lambda}\right)2\pi\int_0^\infty g(s)J_0(2\pi qs)s\,ds \quad . \quad . \quad . \quad (37)$$

where $q = \rho D/2\lambda f$ and it is understood that $g(s) = 0$ for $s > 1$. This equation expresses the well known fact that the aperture and focal-plane distributions are related by the Hankel transform pair:[16]

$$\left.\begin{array}{l} F(q) = 2\pi\int_0^\infty g(s)J_0(2\pi qs)s\,ds \\ g(s) = 2\pi\int_0^\infty F(q)J_0(2\pi qs)q\,dq \end{array}\right\} \quad . \quad . \quad . \quad . \quad . \quad (38)$$

Eqns. 36 and 37, however, are not valid for radiotelescopes, which usually have f/D values in the range $0 \cdot 25$–1. In this case, the complete eqns. 33 with $z = 0$ must be used. The functions $A(U)$ and $B(U)$ have been computed, and are shown in Fig. 9 for values of θ_0 equal to 5°, 30°, 63° and 90°. The $\theta_0 = 5°$ curves correspond to eqn. 35 and Airy's solution. As θ_0 increases, $A(U)$ departs from this value and $B(U)$ becomes appreciable.

Since eqns. 33, representing the sum of axial waves, are

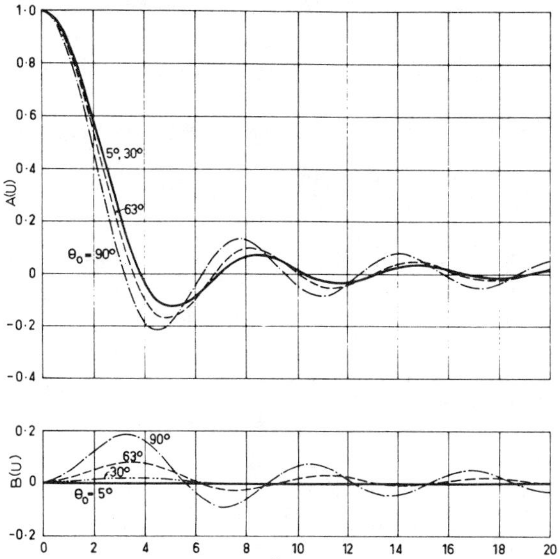

Fig. 9
Functions $A(U)$ and $B(U)$ for paraboloid subtending semiangle θ_0 at focus

similar in form to eqns. 27, it is to be expected that the total focal-plane fields will have many of the characteristics of the individual axial-wave components. This can be illustrated by the field distribution in the focal plane of the Parkes paraboloid ($\theta_0 = 63°$) computed from the curves of Fig. 9. Fig. 10a shows the $\boldsymbol{E}$ fields; the $\boldsymbol{H}$ fields, in accordance with the symmetry relations (eqn. 33c), are identical, but rotated 90° in ξ. For the same reason, the radiation pattern of the distribution is perfectly symmetrical about the axis. Since $E_\rho(U) \neq E_\xi(U)$ in general (see eqn. 33b), the $\boldsymbol{E}$ polarisation is not parallel to the incident field except along the axes $\xi = 0, \pi/2$, and on the circles for which $B(U) = 0$. Furthermore, it is orthogonal to the $\boldsymbol{H}$ lines only where the cross-polarisation is zero, and on the circle where $A(U) = 0$. As shown by the amplitude contours, the amplitude is not symmetrical for the transverse electric field $E(U)$, in Fig. 10b, about the axis. The bright and dark Airy rings are not circular, and display 2- and 4-cycle amplitude variations in ξ, respectively.

The phase centres of the spectrum of axial wavefronts arriving at the focus of the Parkes paraboloid all lie within $0.1f$ of the vertex, as is shown by putting $\psi = \theta$ and $r_0 = R = f\sec^2(\theta/2)$ in eqn. 20. The curvature of the image pattern of Fig. 10 is therefore very small, and, at $\lambda = 10$cm, the deviation from the focal plane does not amount to $\lambda/16$ until $\rho/\lambda = 5 \cdot 7$ (Fig. 4), so that nine Airy rings are enclosed. Image curvature becomes apparent, however, with long-focal-length apertures having moderate D/λ values.* At the other extreme, the image of optical focusing systems is physically so small that wavefront curvature is quite insignificant.

From the axial-wave viewpoint, the finite size of the central spot of a diffraction image is a consequence of fundamental laws limiting the size of the basic cells of the electromagnetic field structure. The image is the sum of axial waves, and cannot be smaller than the central cell of the wave at the high end of the spectrum. This wave is generated by the annular zone at the reflector rim, and has a diameter determined by $J_0(k\rho\sin\theta_0) = 0$, or $2\rho = 0 \cdot 76\lambda\operatorname{cosec}\theta_0$. As θ_0 approaches $\frac{\pi}{2}$ ($f/D = 0 \cdot 25$), the diameter decreases to a limit

* This is illustrated by the work of Farnell on 50cm-diameter lenses operating at $\lambda = 3 \cdot 22$cm.[7] If eqn. 22 is applied to the case of the lens having $f/D = 1 \cdot 26$ ($\theta_0 = 22 \cdot 6°$), the calculated deviation from the focal plane is $0 \cdot 4\lambda$ at $\rho/\lambda = 4 \cdot 1$, enclosing two Airy rings. This is in very good agreement with the measurements reported

of 0·76λ, and the length $\lambda \sec \theta_0$ becomes infinite. This is analogous to the cutoff diameter of the fundamental mode in a waveguide.

vortexes, the deduced energy flow bypasses the zero of the Airy pattern in a manner consistent with the Poynting-vector calculations of Fig. 12.

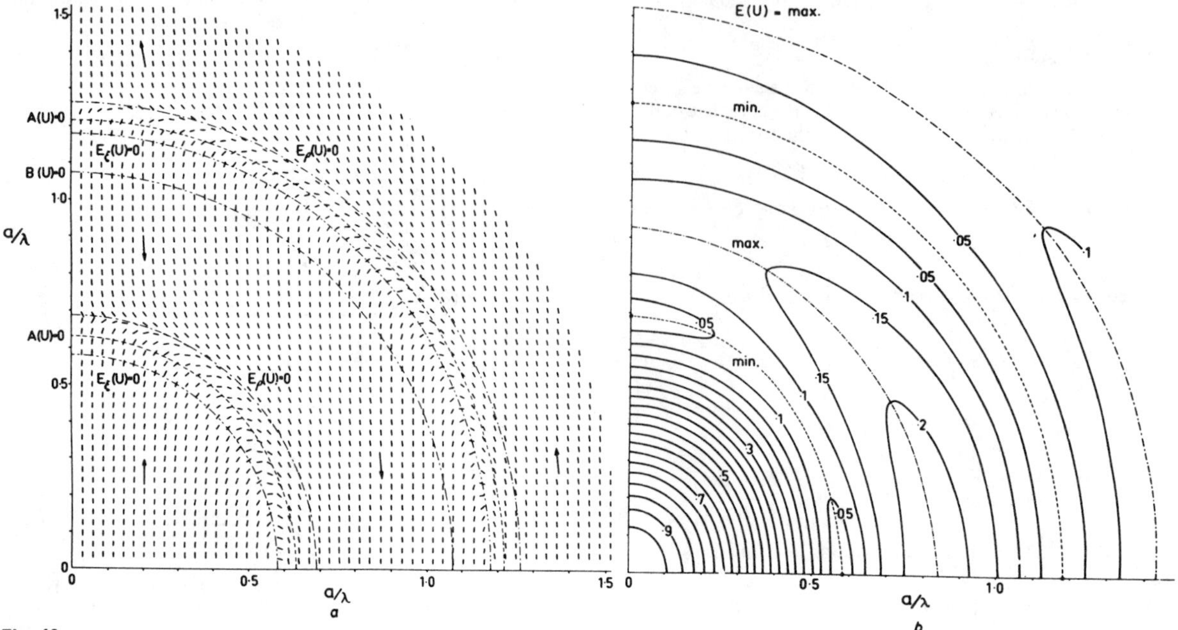

Fig. 10
Field distribution in focal plane of paraboloid with $\theta_0 = 63°$
a Polarisation of E field
b Contours of amplitude E(U)

4.5 Energy flow in the focal region

The average rate of energy flow in a given direction is determined by the real part of the component of the complex Poynting vector in that direction. Thus

$$\left.\begin{aligned} \bar{S}_z &= \frac{-2(kfE)^2}{Z_0}\sin^4\frac{\theta_0}{2}\{(A_r^2 + A_i^2) - (B_r^2 + B_i^2)\} \\ \bar{S}_\rho &= \frac{8(kfE)^2}{Z_0}\sin^4\frac{\theta_0}{2}\{(A_r - B_r)C_i - (A_i - B_i)C_r\} \\ \bar{S}_\xi &= 0 \end{aligned}\right\} \quad (39)$$

where the subscripts r and i denote real and imaginary parts, respectively, of the functions $A(U)$, $B(U)$ and $C(U)$. As is to be expected, the flow of energy is axially symmetrical.

Since the functions are real at the focal plane, the power flow is normal to the plane, and is equal to:

$$\bar{S}_z = \frac{-2(kfE)^2}{Z_0}\sin^4\frac{\theta_0}{2}\{A^2(U) - B^2(U)\} \quad . \quad . \quad (40)$$

The radial function $\{A^2(U) - B^2(U)\}$ is plotted in Fig. 11 and shows the progressive change in the Airy intensity pattern, which occurs when θ_0 increases through the values 5°, 30°, 63°, 90°. The central disc decreases slightly in diameter, and the surrounding rings become brighter. The dark rings split into pairs, given by $A(U) = \pm B(U)$, corresponding to $E_\xi(U) = 0$ and $E_\rho(U) = 0$, respectively, between which the energy flow across the focal plane is reversed. To investigate this reversal, the Poynting-vector components on each side of the focal plane have been computed from eqn. 39. Fig. 12 shows the lines of energy flow in a longitudinal plane near the first dark ring of the Parkes paraboloid ($U = 3 \cdot 25$). This indicates that vortexes of energy circulate about the dark ring where $E_\xi(U) = 0$. At the second dark ring, where $E_\rho(U) = 0$, the energy flow is entirely radial on each side of the focal plane, and the vortex loop just fails to close.

These results may be compared with calculations made by Farnell[6] of the image of a microwave lens. Because scalar diffraction theory was used, the direction of energy flow had to be obtained by assuming that it was normal to the phase front at each point. Although the f/D ratio of 2·2 is too long, and the scale of the diagram too small, to show energy

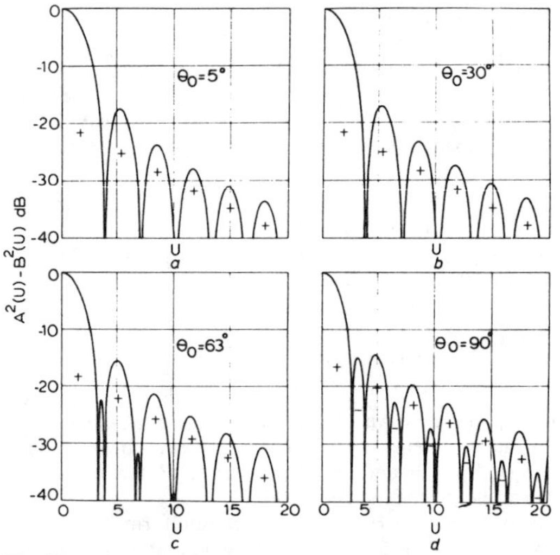

Fig. 11
Intensity functions in focal plane of paraboloid
a $\theta_0 = 5°$ *b* $\theta_0 = 30°$ *c* $\theta_0 = 63°$ *d* $\theta_0 = 90°$

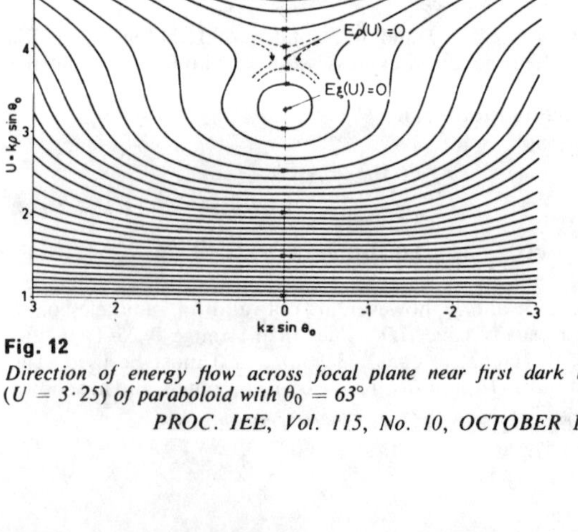

Fig. 12
Direction of energy flow across focal plane near first dark ring ($U = 3 \cdot 25$) of paraboloid with $\theta_0 = 63°$

5 Efficiency of aperture-type feeds

Consider a feed with a circular aperture centred on the focus, and suppose the field-distribution set up over the aperture, when transmitting, is the conjugate of the fields during reception, and either E or H is reversed. Such a feed will absorb all the energy of the received fields incident on its aperture and, if we exclude the possibility of other supergain distributions, will be optimum for the given aperture diameter.

We may now compute the extent to which an aperture feed, defined in this way, approaches the ideal discussed in Section 1 as the radius a is increased. The power P absorbed by the aperture is $\int_0^a \bar{S}_z 2\pi\rho d\rho$, and, since the power incident on the aperture of the paraboloid is $P_0 = 2\pi\left(f \tan\frac{\theta_0}{2}\right)^2 E^2/Z_0$, the aperture efficiency of the paraboloid and feed becomes

$$\eta = \frac{P}{P_0} = \tfrac{1}{2}\int_0^{U_a} \{A^2(U) - B^2(U)\}\, U dU \quad . \quad . \quad . \quad (41)$$

where $U_a = ka \sin\theta_0$.

For small θ_0, $U \approx ka\theta_0$ and, from eqn. 35, $A(U) = 2J_1(U)/U$, $B(U) = 0$. Eqn. 41 then becomes

$$\eta = 1 - J_0^2(U_a) - J_1^2(U_a) \quad . \quad . \quad . \quad . \quad . \quad . \quad . \quad (42)$$

which is Rayleigh's formula for the fraction of the light falling within a circle of radius a at the focus of a lens.[3]

Eqn. 42 is plotted in Fig. 13 for $\theta_0 = 5°$. Efficiency curves have also been numerically computed from the more general expression (eqn. 41) for θ_0 equal to 30°, 63°, 75° and 90°, and are given in the same Figure. For the higher values of θ_0, the efficiency increases more slowly with $ka \sin\theta_0$ than would be expected from the optical theory. This is because some energy from the central disc of the focal-plane pattern is redistributed into the surrounding rings. Moreover, for values of U_a corresponding to the zones of reversed power flow, the slope of the efficiency curve becomes negative.

Because of the redistribution of energy, the size a/λ of feed aperture required to achieve a given efficiency tends to increase with θ_0. The image size, however, decreases with θ_0, so that the two factors are opposed. As a result, the feed-aperture size required is a minimum in the middle range of θ_0 values (Fig. 14).

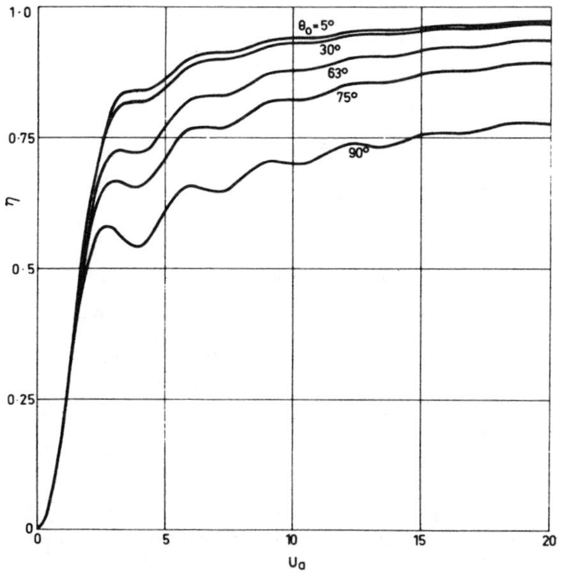

Fig. 13
Aperture efficiency η of paraboloid with semiangle θ_0 and optimum aperture-type feed of radius a

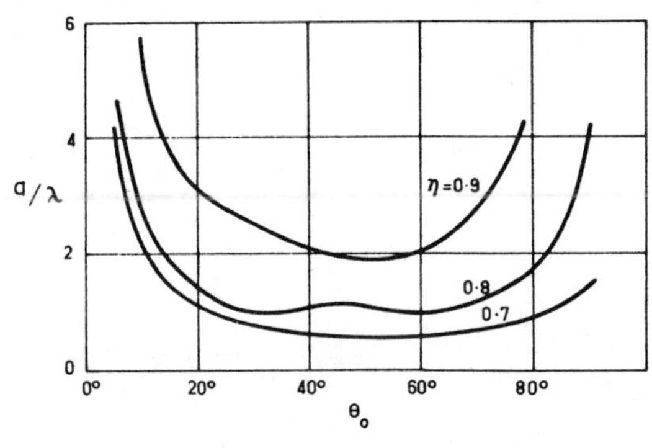

Fig. 14
Radius a of optimum aperture-type feed giving efficiency η with paraboloid having semiangle θ_0

6 Other applications of axial-wave theory

The theory developed in Sections 2 and 3 shows that the image fields of any symmetrical and symmetrically illuminated focusing reflector can be represented by a spectrum of axial waves. Consequently, any desired symmetrical field distribution over the reflector aperture with sharp cutoff at the rim can be synthetised if the spectrum of waves produced at an axial point can be generated over an aperture normal to the axis at that point.

The high-efficiency focal-plane feed for paraboloids considered in the preceding Section is one example. Other spectra could be selected to produce a compromise between efficiency and sidelobes, or to focus the paraboloid on a distant axial point instead of infinity. In practice, the desired synthesis will be approximated very closely if the feed aperture encloses most of the energy in the focal plane image.

It can be shown that the hybrid axial waves specified by eqn. 17 can propagate as discrete modes in a cylindrical waveguide having walls with anisotropic surface reactances. The latter determine γ and ψ, and can be produced physically with corrugated metal walls. In principle, therefore, the synthesis problem can be solved by generating the required spectrum of modes in a corrugated waveguide with an open end in the focal plane.

These concepts can be extended to other axially symmetric focusing systems. Because of their freedom from coma, spherical reflectors are attractive, but spherical aberration makes efficient feeding difficult. In terms of axial-wave theory, the phase spectrum of the waves generated is wide, and phase coincidence does not occur at any point of the axis. However, an axial-wave feed designed for the phase spectrum at a particular point on the axis will correct the aberration by phasing within the feed. High aperture efficiency can also be achieved by matching to the amplitude spectrum of the waves as in the case of the paraboloid. Owing to the lower concentration of energy near the axis, however, the feed aperture would need to be larger. Studies of the flow of energy in the region between the paraxial and marginal foci are therefore being made, using axial-wave theory to determine the optimum location for such a feed and its potential performance.

Other examples of symmetrical focusing systems, to which the axial-wave theory developed in this paper can be applied, are ellipsoidal reflectors and reflecting zone plates. It is also to be expected that many of the conclusions regarding the structure of image fields, which can be derived from the analysis of focusing reflectors, will be valid for circular lens systems.

7 Conclusions

The fields scattered by an axially symmetric reflector, when symmetrically illuminated by a linearly polarised wave incident normally on the aperture, have been shown to be equivalent near the axis to a spectrum of hybrid axial waves. The phase fronts of these waves are effectively plane within the image region of focusing reflectors having high values of D/λ.

This formulation of the fields provides a clear physical picture of the mechanism of image formation. By combining the appropriate spectrum of axial waves, the field components and energy flow in the image region may be calculated for a variety of symmetrical reflector shapes and aperture distributions. Since axial hybrid waves are the natural modes of propagation inside certain corrugated waveguide structures, the analysis suggests a method of synthetising any linearly polarised, symmetrical field distribution over the reflector aperture.

By applying these concepts to paraboloidal reflectors with f/D ratios commonly used in radiotelescopes, it has been shown that the image structure differs significantly from the classical Airy pattern, deduced by scalar analysis, of optical focusing systems. The distribution of the energy flowing across the focal plane has been investigated in order to evaluate the efficiency obtainable from aperture-type feeds of finite diameter. The results indicate that the diameter of the hybrid-wave feed required for a specified efficiency is a minimum over a range of f/D values centred on about $0{\cdot}6(\theta_0 = 50°)$.

8 Acknowledgments

This work was carried out as part of a study of the performance of the Parkes 210ft radio telescope, under NASA research grant NsG 240–62.

The authors are indebted to Dr. T. B. Vu for carrying out the computations of field structure and aperture efficiency.

9 References

1 RUZE, J.: 'Antenna cost, efficiency and system noise', *IEEE Trans.*, 1966, **AP–14**, p. 249
2 AIRY, G. B.: 'On the diffraction of an object glass with a circular aperture', *Trans. Camb. Phil. Soc.*, 1834, **5**, p. 283
3 BORN, M., and WOLF, E.: 'Principles of optics' (Pergamon, 1959), pp. 434–40
4 MATTHEWS, P. A., and CULLEN, A. L.: 'A study of the field distribution at an axial focus of a square microwave lens', *Proc. IEE*, 1956, **103** C, pp. 449–55
5 BACHYNSKI, M. P., and BEKEFI, G.: 'Study of optical diffraction images at microwave frequencies', *J. Opt. Soc. Amer.*, 1957, **47**, pp. 428–38
6 FARNELL, G. W.: 'Calculated intensity and phase distribution in the image space of a microwave lens', *Canad. J. Phys.*, 1957, **35**, pp. 777–83
7 FARNELL, G. W.: 'Measured phase distribution in the image space of a microwave lens', *ibid.*, 1958, **36**, pp. 935–43
8 ROBIEUX, J., and TOCQUEC, Y.: 'Etude théoretique de l'influence de la structure de la source premaire et des tolérances de fabrication sur les propriétés des antennes froides à grand gain', Space Radio Communication Symposium (Elsevier, 1962), pp. 319–62
9 KENNOUGH, E. M., and OTT, R. H.: 'Fields in the focal region of a parabolic receiving antenna', *IEEE Trans.*, 1964, **AP–12**, pp. 376–77
10 WATSON, W. H.: 'The field distribution in the focal plane of a paraboloidal reflector', *ibid.*, **AP–12**, 1964, pp. 561–69
11 MINNETT, H., and THOMAS, B.: 'Synthesis of improved feeds for large circular paraboloids', IEEE Conf. Publ. 21, 1966, pp. 262–66
12 MINNETT, H., and THOMAS, B. MACA.: 'A method of synthetising radiation patterns with axial symmetry', *IEEE Trans.*, 1966, **AP–14**, pp. 654–56
13 RUMSEY, V. H.: 'Horn antennas with uniform power patterns around their axis', *ibid.*, 1966, **AP–14**, pp. 656–58
14 SILVER, S.: 'Microwave antenna theory and design' (McGraw-Hill, 1949), pp. 146–49
15 PAGE, L., and ADAMS, N. I., JUN.: 'Electromagnetic waves in conducting tubes', *Phys. Rev.*, **52**, 1937, pp. 647–51
16 BRACEWELL, R.: 'The Fourier transform and its applications' (McGraw-Hill, 1965), p. 244

10 Appendixes

10.1 Phase near an axial point

In Figs. 1 and 3, QN is normal to MP, and $(\mathrm{QN})^2 = \rho^2 \sin^2(\phi - \xi) + \rho^2 \cos^2\psi \cos^2(\phi - \xi)$. The path length MQ $= r$ from dS to Q is thus:

$$r^2 = \{r_0 - \rho \sin\psi \cos(\phi - \xi)\}^2 + \mathrm{QN}^2$$

$$= r_0^2 - 2\rho r_0 \sin\psi \cos(\phi - \xi) + \rho^2 \quad . \quad . \quad . \quad (43)$$

Relative to the reference plane xOy, the phase of the field at Q is:

$$\Phi_Q = k(z + r_0 \cos\psi + r) \quad . \quad . \quad . \quad . \quad . \quad . \quad (44)$$

If eqn. 43 is expanded binomially, and terms of order $(\rho/r_0)^3$ and higher are ignored, eqn. 44 becomes

$$\Phi_Q = \Phi - k\left[\rho \sin\psi \cos(\phi - \xi) - \frac{\rho^2}{2r_0}\{1 - \sin^2\psi \cos^2(\phi - \xi)\}\right]$$

$$= \Phi - k\left\{\rho \sin\psi \cos(\phi - \xi) - \frac{\rho^2}{2r_0}(1 - \tfrac{1}{2}\sin^2\psi) + \frac{\rho^2}{4r_0}\sin^2\psi \cos 2(\phi - \xi)\right\} \quad . \quad . \quad . \quad (45)$$

where $\Phi = k(z + r_0 \cos\psi + r_0)$ is the phase at P.

10.2 Uniform amplitude approximation

It is assumed that, for points near the axis where $\rho \ll r_0$, the fields are constant over a section of wavefront from dS centred on the ray directed towards P. In this direction, $\delta = 0$ $\beta = \beta_0$. However, as Q departs from P, $|E_\delta|$ decreases as $\cos\delta$ (eqn. 5), and the vector assumes an inclination of δ to the wavefront. Resolved parallel to the wavefront, the true value is $\cos^2\delta$, where $\delta = \left(\frac{\rho}{r_0}\right)\sin(\phi - \xi)$. The maximum value of error, which occurs when $(\phi - \xi) = 90°$, is $\left(\frac{\rho}{r_0}\right)^2$; for $\left(\frac{\rho}{r_0}\right) < 0{\cdot}05$, the error is less than $\frac{1}{4}\%$.

The component $|E_\beta|$ varies with both β and δ, as shown by eqn. 5. For the general point Q where $\beta = \beta_0 + \Delta\beta$, and $\delta \neq 0$, the change in $|E_\beta|$ with respect to the value at P is:

$$|\Delta E_\beta| = -\Delta\beta \sin\beta_0 \sin\phi + \delta \cos\alpha \cos\phi$$

Now, $\Delta\beta \approx \left(\frac{\rho}{r_0}\right)\cos\psi \cos(\phi - \xi)$

$$\delta \approx \left(\frac{\rho}{r_0}\right)\sin(\phi - \xi)$$

The error, expressed as a fraction of the maximum value of $|E_\beta|$ with respect to the variation in ϕ, is:

$$\frac{|\Delta E_\beta|}{|E_\beta|} = -\frac{\rho}{r_0}\cos\psi \tan\beta_0 \sin\phi \cos(\phi - \xi) + \frac{\rho}{r_0}\frac{\cos\alpha}{\cos\beta_0}\cos\phi \sin(\phi - \xi)$$

$$= -\left(\frac{\rho}{r_0}\right)\{\cos\psi \tan\beta_0 \sin\phi \cos(\phi - \xi) - \gamma \cos\phi \sin(\phi - \xi)\} \quad . \quad . \quad . \quad . \quad (46)$$

Consideration of the range of values assumed by $\tan\beta_0 \cos\psi$ and γ in typical cases suggests that a value of unity is representative for both these factors. Eqn. 46 then becomes

$$\frac{|\Delta E_\beta|}{|E_\beta|} \approx -\left(\frac{\rho}{r_0}\right)\sin\xi \quad . \quad . \quad . \quad . \quad . \quad . \quad . \quad (47)$$

so that the maximum error $\epsilon = \frac{\rho}{r_0}$ occurs at $\xi = 90°$.

Boundaries beyond which ϵ exceeds 1, 2, 5 and 10% are shown by the broken lines in Fig. 4. These are intended only as a guide, and a separate estimate of errors should be made for each problem.

10.3 Axial-wave parameters

From Fig. 1*b*, $r_0^2 = R^2 + z^2 - 2zR\cos\theta$, so that

$$\zeta = \frac{r_0}{R} = \sqrt{\left\{1 - \frac{2z}{R}\cos\theta + \left(\frac{z}{R}\right)^2\right\}} \quad . \quad . \quad . \quad (48)$$

where $R = F(\theta)$ specifies the reflector shape.

If an element of surface at M subtends angles of $d\psi$ and $d\theta$ at P and O, respectively:

$$d\psi = \frac{1}{\zeta^2}\frac{\sin\theta \sec(\theta - \alpha)}{\sin\psi \sec(\psi - \alpha)}d\theta \quad . \quad . \quad . \quad . \quad . \quad (49)$$

Also $\cos(\psi - \theta) = (R - z\cos\theta)/r_0$

and $\sin(\psi - \theta) = z\sin\theta/r_0$

so that

$$\cos(\psi - \alpha) = \cos\{(\psi - \theta) + (\theta - \alpha)\}$$

$$= \frac{1}{\zeta}\left\{\cos(\theta - \alpha) - \frac{z}{R}\cos\alpha\right\} \quad . \quad . \quad (50)$$

Since

$$r_0 \sin \psi = R \sin \theta \quad . \quad . \quad . \quad . \quad . \quad . \quad . \quad . \quad . \quad . \quad (51)$$

$$\kappa = r_0 \sin \psi / h \text{ (from eqn. 10)}$$

$$= R \sin \theta / h \quad . \quad . \quad . \quad . \quad . \quad . \quad . \quad . \quad . \quad . \quad (52)$$

and $\kappa(\theta) = \kappa \dfrac{d\psi}{d\theta}$

$$= \frac{R \sin \theta}{h\zeta^2}\left\{1 - \frac{z}{R}\frac{\cos \alpha}{\cos(\theta - \alpha)}\right\} \quad . \quad . \quad . \quad (53)$$

using eqns. 49 and 50. Also

$$\gamma = \frac{\cos \alpha}{\cos(\psi - \alpha)} \text{ (from eqn. 10)}$$

$$= \frac{\zeta}{\dfrac{\cos(\theta - \alpha)}{\cos \alpha} - \dfrac{z}{R}} \quad . \quad . \quad . \quad . \quad . \quad . \quad . \quad . \quad . \quad . \quad (54)$$

by substituting from eqn. 50.

From eqn. 6:

$$\Phi = k\{z + r_0(1 + \cos \psi)\}$$

$$= k(R \cos \theta + r_0) \quad \text{(from Fig. 1}b\text{)}$$

$$= kR(\cos \theta + \zeta) \quad . \quad . \quad . \quad . \quad . \quad . \quad . \quad . \quad . \quad (55)$$

Using eqn. 51, the parameter $u = k\rho \sin \psi$ may be written

$$u = k\rho \sin \theta / \zeta \quad . \quad . \quad . \quad . \quad . \quad . \quad . \quad . \quad . \quad . \quad . \quad (56)$$

and sec ψ is given by

$$\sec \psi = \frac{\zeta}{(\cos \theta - z/R)} \quad . \quad . \quad . \quad . \quad . \quad . \quad . \quad . \quad . \quad (57)$$

When the reflector shape has been specified, the angle of incidence α is determined, and the above equations give $\kappa(\theta)$, γ, Φ, u and sec ψ as functions of θ and z.

Matching Focal-Region Fields with Hybrid Modes

Abstract—**A previous analysis of circularly symmetric reflectors has shown that the focal-region fields consist of an infinite spectrum of hybrid waves. This communication shows that these fields in the case of paraboloidal and spherical reflectors can be closely matched over the open end of a corrugated waveguide propagating a small number of hybrid modes.**

Introduction

The fields and energy flow in the focal regions of paraboloidal and spherical reflectors illuminated by a linearly polarized plane wave have been considered in [1] and [2], respectively. It was shown that the contributions of all the annular zones $d\theta$ of the reflector give rise to an infinite spectrum of hybrid waves propagating at different velocities along the axis in the focal region. For the paraboloid the waves arrive in phase across the focal plane. It has been shown [3] that these waves can propagate as discrete modes in a circumferentially corrugated waveguide where both the circumferential $\boldsymbol{E}$ and $\boldsymbol{H}$ fields at the waveguide boundary are zero. With the use of a transmission theorem due to Robieux [4] an expression is derived which enables the degree of matching of the focal region fields by a number of hybrid modes to be evaluated. The results obtained when applied to a paraboloid and a spherical reflector are then briefly discussed.

Transmission Coefficient

The fields to be matched are the focal region fields $\boldsymbol{E}_1,\boldsymbol{H}_1$ received on the waveguide aperture and the hybrid-mode fields $\boldsymbol{E}_2,\boldsymbol{H}_2$ existing there when the waveguide is transmitting. The voltage transmission coefficient as defined by Robieux is

$$T = \frac{1}{4}\int_s (\boldsymbol{E}_1 \times \boldsymbol{H}_2 + \boldsymbol{H}_1 \times \boldsymbol{E}_2)\, ds \tag{1}$$

where the power in the plane wave incident on the reflector and the total power of the hybrid modes in the waveguide are both unity. $\boldsymbol{E}_2,\boldsymbol{H}_2$ are assumed to be zero elsewhere on the closed surface s so that the integration extends only over the aperture. It is also assumed that $\boldsymbol{E}_1,\boldsymbol{H}_1$ are those existing in the absence of the waveguide and $\boldsymbol{E}_2,\boldsymbol{H}_2$ are the undisturbed waveguide fields. These approximations should become more valid as the aperture size is increased.

Where the fields have not been normalized it is convenient to write the power transmission coefficient (or efficiency) as in [5]:

$$\eta = \frac{|\int_s (\boldsymbol{E}_1 \times \boldsymbol{H}_2 + \boldsymbol{H}_1 \times \boldsymbol{E}_2)\, ds|^2}{16P_1P_2} \tag{2}$$

where P_1 and P_2 are the total powers contained in the two fields. In the ideal case, the incident fields $\boldsymbol{E}_1,\boldsymbol{H}_1$ would be matched across the aperture by their conjugates such that $\boldsymbol{E}_2 = \boldsymbol{E}_1^*$ and $\boldsymbol{H}_2 = \boldsymbol{H}_1^*$. Equation (2) then becomes

$$\eta_0 = \frac{\{\int_s \mathrm{Re}\, (\boldsymbol{E}_1 \times \boldsymbol{H}_1^*)\, ds\}^2}{4P_1P_2} \tag{3}$$

where

$$P_2 = \frac{1}{2}\int_s \mathrm{Re}\, (\boldsymbol{E}_2 \times \boldsymbol{H}_2^*)\, ds$$

$$= \frac{1}{2}\int_s \mathrm{Re}\, (\boldsymbol{E}_1 \times \boldsymbol{H}_1^*)\, ds.$$

Manuscript received July 24, 1969.

Substituting for P_2 in (3) yields

$$\eta_0 = \frac{1}{2P_1}\int_s \mathrm{Re}\, (\boldsymbol{E}_1 \times \boldsymbol{H}_1^*)\, ds \tag{4}$$

which is the ratio of the power incident on and absorbed by the aperture to the power incident on the reflector.

The fields $\boldsymbol{E}_1,\boldsymbol{H}_1$ are the sum of an infinite number of hybrid waves, whereas $\boldsymbol{E}_2,\boldsymbol{H}_2$, in practice, are the sum of the discrete set of hybrid modes able to propagate in the waveguide. The degree of matching which can be achieved will be considered for paraboloidal and spherical reflectors.

Fields in the Focal Region

The transverse fields in the focal plane of a paraboloid of half-angle θ_0, or in the focal region of a spherical reflector where θ_0 (the half-angle subtended by the rim of the dish at the center) is less than 30°, may be written in the form

$$E_\rho = E_\rho(U)\sin\xi, \quad H_\rho = H_\rho(U)\cos\xi$$
$$E_\xi = E_\xi(U)\cos\xi, \quad H_\xi = H_\xi(U)\sin\xi \tag{5a}$$

where

$$E_\rho(U) = \alpha k\{A(U) + B(U)\}$$
$$E_\xi(U) = \alpha k\{A(U) - B(U)\}$$
$$H_\rho(U) = E_\rho(U)/Z_0$$
$$H_\xi(U) = -E_\xi(U)/Z_0$$
$$U = k\rho\sin\theta_0$$
$$k = 2\pi/\lambda. \tag{5b}$$

$A(U)$ and $B(U)$ are integrals giving the summation of the hybrid-wave fields radiating from all the annular zones $d\theta$ of the reflector. [$A(U)$ and $B(U)$ are given in full in (34) of [1] for the paraboloidal reflector, and (4c) of [2] for the spherical reflector.] In addition, α' is equal to $2(fE)\sin^2(\theta_0/2)$ for the paraboloid (indicated by a single prime), and α'' is equal to $(RE)\sin^2\theta_0$ for the spherical reflector (indicated by a double prime). f and R are the focal length and radius of the paraboloidal and spherical reflectors, respectively. The power incident on the paraboloidal and spherical reflectors is given by

$$P_1' = (2\pi/Z_0)(fE)^2\tan^2(\theta_0/2)$$
$$P_1'' = (\pi/2Z_0)(RE)^2\sin^2\theta_0. \tag{6}$$

Hybrid-Mode Fields

It is convenient to write the HE_{1n} transverse field components in the following form [1]:

$$E_\rho = E_\rho(u_n)\sin\xi, \quad H_\rho = H_\rho(u_n)\cos\xi$$
$$E_\xi = E_\xi(u_n)\cos\xi, \quad H_\xi = H_\xi(u_n)\sin\xi \tag{7a}$$

where

$$E_\rho(u_n) = \beta k\{Q_n(C_n + D_n)\exp(j\Phi_n)\}$$
$$E_\xi(u_n) = \beta k\{Q_n(C_n - D_n)\exp(j\Phi_n)\}$$
$$H_\rho(u_n) = -E_\rho(u_n)/Z_0$$
$$H_\xi(u_n) = E_\xi(u_n)/Z_0 \tag{7b}$$
$$C_n = \sin\psi_n J_0(u_n)$$
$$D_n = \sin\psi_n\tan^2(\psi_n/2)J_2(u_n)$$
$$u_n = k\rho\sin\psi_n$$
$$\beta' = fE, \quad \beta'' = \tfrac{1}{2}RE. \tag{7c}$$

Reprinted from *IEEE Trans. Antennas Propagat.*, vol. AP-18, pp. 404-405, May 1970.

Q_n, Φ_n are the amplitude and phase of the nth mode, respectively. In the focal plane of the paraboloid the modes will be cophased ($\Phi_n = 0$), and for the spherical reflector the sign of the phase is opposite to that of the focal region fields to give a conjugate match. $\sin \psi_n = \lambda/\lambda_c$ (where λ_c is the cutoff wavelength) is determined from the boundary condition $E_\xi(\bar{u}_n) = 0$, $(\bar{u}_n = ka \sin \psi_n)$. In addition, since $H_\xi(u_n)$ is equal to zero, the longitudinal current is zero, ensuring minimum disturbance of the incident focal-region fields by the presence of the waveguide aperture.

The total aperture fields may be written

$$\sum^n E_\rho(u_n) = \beta k\{F(U) + G(U)\}$$

$$\sum^n E_\xi(u_n) = \beta k\{F(U) - G(U)\}$$

$$F(U) = \sum^n Q_n C_n \exp(j\Phi_n)$$

$$G(U) = \sum^n Q_n D_n \exp(j\Phi_n). \qquad (8)$$

The power flowing through the aperture is

$$P_2 = \frac{\pi}{Z_0}\int_0^a \mathrm{Re}\,\{\sum^n E_\rho(u_n) \sum E_\xi^*(u_n)\}\rho\, d\rho$$

$$= \frac{\pi\beta^2}{Z_0 \sin^2\theta_0}\int_0^{U_a} \{|F(U)|^2 - |G(U)|^2\}U\, dU$$

where $U_a = ka \sin \theta_0$, and (2) becomes

$$\eta = \frac{(\pi/Z_0)^2 \,|\int_0^a \{E_\rho(U) \sum^n E_\xi(u_n) + E_\xi(U) \sum^n E_\rho(u_n)\}\rho\, d\rho\,|^2}{4P_1P_2}$$

Substituting for P_1, P_2 yields

$$\eta = \delta \frac{|\int_0^{U_a} \{A(U)F(U) - B(U)G(U)\}U\, dU\,|^2}{\int_0^{U_a} \{|F(U)|^2 - |G(U)|^2\}U\, dU} \qquad (9)$$

where $\delta' = \frac{1}{2}$, $\delta'' = 2$. Equation (9) may be expressed in terms of real and imaginary components

$$\eta = \delta \frac{\begin{array}{l}|\int_0^{U_a} \{A_rF_r - B_rG_r + A_iF_i - B_iG_i \\ \qquad + j(A_rF_i - F_rA_i - B_rG_i + G_rB_i)\}U\, dU\,|^2\end{array}}{\int_0^{U_a} \{(F_r^2 + F_i^2) - (G_r^2 + G_i^2)\}U\, dU} \qquad (10)$$

where

$$A(U) = A_r - jA_i, \quad B(U) = B_r - jB_i$$

$$F(U) = F_r + jF_i, \quad G(U) = G_r + jG_i.$$

This result may be checked by noting that an exact conjugate match requires

$$F_r = A_r/\kappa, \quad F_i = A_i/\kappa, \quad G_r = B_r/\kappa, \quad G_i = B_i/\kappa \qquad (11)$$

where κ is a constant. It is easily verified that the imaginary term then vanishes from (10), which reduces to

$$\eta_0 = \delta \int_0^{U_a} \{(A_r^2 + A_i^2) - (B_r^2 + B_i^2)\}U\, dU. \qquad (12)$$

Equation (12) for paraboloidal and spherical reflectors is plotted in [1] and [2], respectively, where the results are derived by calculating the fraction of the received power P_1 incident on an aperture in the focal plane. This confirms that (10), for an exact conjugate match, reduces to the result predicted by (4).

Method of Matching

For matching in the focal plane of a paraboloid it appears desirable to arrange for the guide radius to correspond to a zero in the focal plane field $E_\xi(U)$, since the circumferential field at the edge of the corrugated guide, $E_\xi(\bar{u}_n)$ is equal to zero. Consequently, for a single-mode guide the radius is then automatically given by the position of the first zero in the Airy pattern, so that for n modes it is determined by the position of the nth zero. The relative amplitudes Q_n of the n modes must then be adjusted to maximize η using the closeness of match of the ρ and ξ components of the two aperture fields as a guide.

A rapid method of achieving this using the "vista" graphic display of a CDC 3600 computer was developed by Bao [6]. For the Parkes paraboloid ($\theta_0 = 63°$) he was able to achieve an almost perfect match resulting in efficiencies within 0.1 percent of η_0. From (12), or Fig. 13 of [1], this corresponds to efficiencies of 0.724, 0.828, 0.875, 0.901, and 0.919 for feed diameters capable of supporting one to five hybrid modes, respectively.

For the spherical reflector both the amplitudes Q_n and phases Φ_n of the modes must be varied to obtain an optimum match. As in the paraboloid case, the edge of the waveguide is placed at a null in the focal-region fields. The positions of these nulls for a spherical reflector with $\theta_0 = 20°$, $R/\lambda = 400$ have already been determined [2], and their normalized radius ka, and axial position z/R, are given in Table I. The optimization of the modes in this case was carried out using a computer program written by Warne, and the efficiencies obtained by using up to three hybrid modes show that η_0 calculated from (12) (see also [2]) can, in theory, be closely approached.

TABLE I

Number of Modes	z/R	ka	a/λ	η_0	η
1	0.51286	6.15	0.98	0.538	0.536
2	0.51725	11.90	1.90	0.796	0.782
3	0.52080	17.20	2.74	0.896	0.868

Two-mode hybrid feeds are in use at 6- and 13-cm wavelengths on the Parkes 210-foot radio telescope with a 10-percent increase in efficiency over that of a single hybrid-mode feed. In a spherical reflector the predicted improvement with such feeds would be 46 percent, and this has stimulated further work on the practical problem of exciting and controlling additional modes.

Acknowledgment

The author wishes to acknowledge valuable discussions with H. C. Minnett, who suggested the use of hybrid-mode feeds to correct the aberration of spherical reflectors, and the assistance of W. Warne in writing the computer program used to determine the matching efficiency for the spherical reflector.

B. MacA. Thomas
Radiophysics Lab.
CSIRO
Sydney, Australia

References

[1] H. C. Minnett and B. M. Thomas, "Fields in the image space of symmetrical focusing reflectors," *Proc. IEE* (London), vol. 115, pp. 1419–1430, October 1968.
[2] B. M. Thomas, H. C. Minnett, and V. T. Bao, "Fields in the focal region of a spherical reflector," *IEEE Trans. Antennas and Propagation* (Communications), vol. AP-17, pp. 229–231, March 1969.
[3] H. C. Minnett and B. M. Thomas, "A method of synthesizing radiation patterns with axial symmetry," *IEEE Trans. Antennas and Propagation* (Communications), vol. AP-14, pp. 654–656, September 1966.
[4] J. Robieux, "Lois générales de la liaison entre radiateurs d'ondes. Application aux ondes de surface et à la propagation," *Ann. Radioélec.*, vol. 14, pp. 187–229, July 1959.
[5] M. K. Hi, "Near-zone power transmission formulas," *IRE Natl. Conv. Rec.*, vol. 6, pt. 8, pp. 128–135, 1958.
[6] V. T. Bao, "Optimization of efficiency of deep reflectors," *IEEE Trans. Antennas and Propagation* (Communications), vol. AP-17, pp. 811–813, November 1969.

Optimisation of efficiency of reflector antennas: approximate method

Vu The Bao, B.E., Ph.D.

Abstract

A brief discussion of the theoretical approach to the problem of obtaining the maximum efficiency, for both paraboloidal- and spherical-reflector antennas, is presented. In both types of antenna, the problem reduces to obtaining the optimum combination of amplitude and phase of the hybrid modes propagating in the guide, so that, at the feed aperture, the resultant hybrid field closely matches that field which would be produced by a plane wave incident on the antenna. With shallow reflectors, a purely analytical method can be used, whereas graphical computer-aided optimisation techniques may be used to obtain the solution when the reflector is deep. The paper, however, presents a simple method of obtaining the approximate mode composition, which often proves to be satisfactory. As an illustration, results obtained by the approximate method are also compared with published results.

List of principal symbols

a = waveguide radius
c_m = relative amplitude of E_{1m}
E_T, E_{1m} = incident field and mth hybrid mode from guide (shallow reflector)
E_ρ, E_ρ^n = radial components of incident field and of nth hybrid mode from guide (deep reflector)
E_ξ, E_ξ^n = circumferential components of incident field and nth hybrid mode from guide (deep reflector)
$\mathrm{Im}(\rho), \mathrm{Re}(\rho)$ = imaginary and real parts of E_ξ in $\xi = 0$ plane
J_n = Bessel function of first kind, of order n
J_0' = first derivative of J_0
$j = \sqrt{(-1)}$
λ = wavelength
$k = \dfrac{2\pi}{\lambda}$ = wavenumber
m_i, γ_i = relative amplitude and phase of ith hybrid mode from guide (spherical)
R = radius of curvature of spherical reflector
$r_i = m_i \cos \gamma_i$
$s_i = m_i \sin \gamma_i$
T_i = relative amplitude of ith hybrid mode from guide (deep paraboloid)
$U = k\rho \sin \theta_0$
$U_l = k\rho_l \sin \theta_0$
$V = k\rho \sin \psi_0$
$V_l = k\rho_l \sin \psi_0$
$v = (k\rho \sin \psi)/\zeta$
z = distance from feed point to centre of the sphere
α_m = mth root of J_1
β_M = Mth root of J_0
ψ = angular position of point on sphere
$2\psi_0$ = angle subtended by spherical reflector at centre of sphere
Φ = phase error caused by spherical aberration
ρ, ξ = polar co-ordinates of point on feed aperture
ρ_l = point at which fields are exactly matched
ζ = distance from feed point to point on spherical reflector
θ = angular position of point on paraboloid
$2\theta_0$ = angle subtended by paraboloidal reflector at focus

1 Introduction

The widespread use of large-reflector antennas, capable of high-power transmission and low-noise reception in satellite communication and radioastronomy, has greatly stimulated research in this particular field. The problems can be grouped into two main categories: those related to the reflector, and those connected with the feed design. Although, in the past, much attention has been focused on the reflector design, the high cost of the reflector has slowly shifted the emphasis onto the design of an optimum feed, so that the antenna system can be efficiently exploited. To date, much effort has been spent on the search for a highly efficient feed, but the problem is far from being solved. This is particularly true of spherical reflectors, where attempts to overcome the effect of spherical aberration have not been very successful.

One of the most important theoretical treatments of the subject was described recently by Potter[1], who computed the maximum efficiency for a given feed size. However, it can be shown[2] that his results are too optimistic, and a more realistic way of predicting the maximum efficiency was given by Robieux,[3] or Minnett *et al.*[4] Basically, the efficiency is computed by assuming that the feed can absorb all the energy incident normally on its aperture. One can, without loss of generality, consider the receiving case where a plane wave, propagating in the direction of the axis of the antenna, is incident on the reflector. The reflected field has been calculated by Minnett *et al.*, and, in the focal region, the incident field has been shown to consist of a spectrum of axial hybrid waves. The maximum theoretical efficiency is then equal to the ratio of the energy incident on the feed aperture to the total energy received by the reflector; i.e. it is assumed that the incident field in the plane of the feed aperture is exactly matched, in the conjugate sense, by the resultant hybrid field from the feed, so that the incident energy is completely absorbed. Although the physical size of the feed rules out the possibility of achieving a complete absorption of the incident energy, an efficiency close to the maximum value can be obtained if the resultant hybrid field closely matches the incident field over the feed aperture. The problem therefore reduces to finding the required amplitude and phase combination to obtain the optimal match. The next step is to find a practical method of generating and controlling the hybrid modes so that the desired amplitude and phase relationship can be obtained. The latter problem, however, will be treated elsewhere. This paper only discusses a method of finding the appropriate mode compositions for both paraboloidal and spherical reflectors.

2 Historical background

If the reflector is shallow, the mathematical expressions of the fields are quite simple, and a purely analytical method can be used to obtain the optimum mode composition.[2] In fact, one can express the total incident field in the focal region as

$$E_T = A \frac{\{J_1(k\rho\theta_0)\}}{k\theta_0\rho} \quad \dots \dots \dots \dots \quad (1)$$

The co-ordinate system used in the derivation of this equation is shown in Fig. 1.

Paper 6028 E, first received 23rd July and in revised form 2nd October 1969
Dr. Vu is with the School of Electrical Engineering, University of New South Wales, PO Box 1, Kensington, NSW 2033, Australia

Reprinted with permission from *Proc. IEE*, vol. 117, pp. 30-34, Jan. 1970.

In addition, the hybrid waves from the guide, which are required to illuminate the reflector, are given by

$$E_{1m} = B_m J_0\left(\alpha_m \frac{\rho}{a}\right) \quad . \quad . \quad . \quad . \quad . \quad . \quad . \quad . \quad (2)$$

Thus eqns. 1 and 2 show that the fields are symmetrical about the axis of the antenna, and the problem is essentially

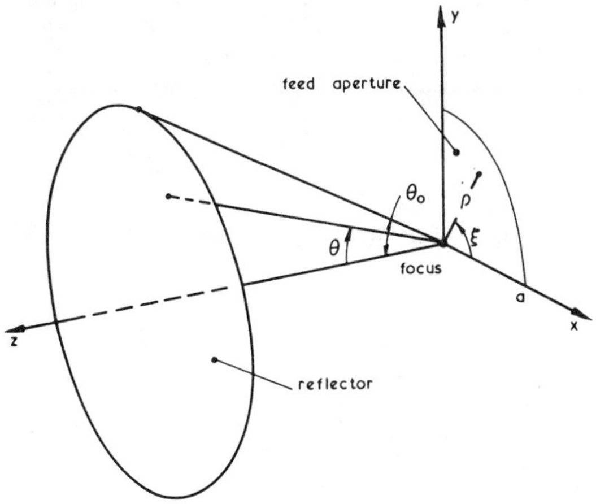

Fig. 1
Co-ordinate system for paraboloidal reflector

2-dimensional. As each E_{1m} vanishes at the wall of the guide, the first condition for a good match is that E_T also vanishes at $\rho = a$; i.e. the radius of the guide, designed to support a total number of M modes, is given by

$$a = \frac{\beta_M}{k\theta_0} \quad . \quad . \quad . \quad . \quad . \quad . \quad . \quad . \quad . \quad . \quad . \quad . \quad (3)$$

and the total field becomes

$$E_T = A\frac{\left\{J_1\left(\beta_M \frac{\rho}{a}\right)\right\}}{\beta_M \frac{\rho}{a}} \quad . \quad . \quad . \quad . \quad . \quad . \quad . \quad . \quad (4)$$

Therefore the amplitudes of the M hybrid modes, required to match the incident field, can be easily obtained by computing the M coefficients of the following series:[2]

$$\frac{J_1\left(\beta_M \frac{\rho}{a}\right)}{\beta_M \frac{\rho}{a}} = \sum_{m=1}^{M}\left\{c_m J_0\left(\alpha_m \frac{\rho}{a}\right)\right\}$$

$$\text{where } c_m = \frac{2}{\{\beta_M J_0'(\alpha_m)\}^2} \quad . \quad . \quad . \quad . \quad . \quad . \quad . \quad . \quad . \quad (5)$$

But $J_0'(z) = -J_1(z)$, and therefore c_m can be rewritten as

$$c_m = \frac{2}{\{\beta_M J_1(\alpha_m)\}^2} \quad . \quad . \quad . \quad . \quad . \quad . \quad . \quad . \quad . \quad (6)$$

With deep reflectors, however, the mathematical expressions of the fields become vastly more complicated so that other methods, such as graphic computer optimisation techniques,[5] may be more attractive. Unlike the fields in shallow reflectors, which retain their cylindrical symmetry about the antenna axis, the fields in deep reflectors are different in different planes passing through the axis. Thus one is faced with a 3-dimensional problem, since the combined hybrid field must match the incident field in all planes. As the solution to the problem involves a long process of trial and error, it is therefore clear that the restricted medium of numerical digits is not satisfactory for displaying the single-step results. Alternatively, active computer-graphics techniques enable real-time alteration of mode composition, and other parameters, to be effected by the designer. This facilitates the planning of the main program and results in efficient use of the computer memory. In addition, real-time interaction between the designer and the computer significantly reduces the total time involved in the matching process.

However, unless there is an approximate solution available, the search for the true optimum mode composition can be quite time-consuming. For this reason, it is desirable to find a simple method of obtaining an approximate solution which may be used as the starting point in the computer optimisation program. The following Section presents a simple graphical solution. It is significant that, if one is required to determine the optimum mode composition at one single frequency, this approximate method gives surprisingly good results. However, to obtain an insight into the bandwidth characteristics of the antenna, computer-aided optimisation techniques would still be required.

3 Approximate method for deep reflectors

It has been explained previously that the combined hybrid field must match the incident field in all planes passing through the antenna axis. Fortunately, experience with computer optimisation techniques[5] has shown that, if a good match can be achieved in a particular plane, the matching in all other planes will also be fairly satisfactory. Thus the matching problem can be solved in two steps. The total incident field, in the plane of the feed aperture, is first approximately matched in one plane by solving a set of N linear equations of N independent variables. The optimum match is then obtained with the aid of a high-speed digital computer. This Section is mainly concerned with the approximate solution.

3.1 Paraboloidal reflectors

To obtain the maximum efficiency for a given feed size, the feed must be positioned at the point where the energy density is highest. For a paraboloidal reflector, the feed aperture must therefore lie in the focal plane. The components of the incident field in this plane have been previously obtained by Minett *et al.*[4] In matching the fields, one is mainly concerned with the transverse components, which are given by

$$E_\rho = E(U)\sin\xi$$
$$E_\xi = E_\xi(U)\cos\xi \quad . \quad . \quad . \quad . \quad . \quad . \quad . \quad . \quad (7a)$$

where $E_\rho(U) = C(F_1 + F_2)$

$$E_\xi(U) = C(F_1 - F_2)$$

$$F_1 = \int_0^{\theta_0} \sin(\theta) J_0(k\rho \sin\theta)d\theta$$

$$F_2 = \int_0^{\theta_0} \sin(\theta)\tan^2\left(\frac{\theta}{2}\right) J_2(k\rho\sin\theta)d\theta \quad . \quad . \quad . \quad (7b)$$

C = a constant factor

Ideally, the corresponding components of the resultant hybrid wave, from the guide, must exactly match the incident field in every ξ plane at the aperture of the feed. In practice, however, the total number of hybrid modes which can propagate in the guide is determined by the guide's size. For a waveguide of radius a, the transverse components of the nth hybrid mode from the waveguide are

$$E_\rho^n = C(G_1 + G_2)$$
$$E_\xi^n = C(G_1 - G_2) \quad . \quad . \quad . \quad . \quad . \quad . \quad . \quad . \quad (8a)$$

where $G_1 = \sin(\theta_n) J_0(k\rho\sin\theta_n)$

$$G_2 = \sin(\theta_n)\tan^2\left(\frac{\theta_n}{2}\right) J_2(k\rho\sin\theta_n) \quad . \quad . \quad . \quad (8b)$$

θ_n is a function of ka and n, and for a circumferentially slotted waveguide it satisfies the following condition:

$$1 - \frac{u_n J_0(u_n)}{J_1(u_n)} = \cos\theta_n \quad . \quad . \quad . \quad . \quad . \quad . \quad . \quad . \quad (9)$$

As an illustration of the matching procedure, five hybrid modes from the waveguide are used to match the incident field. Fig. 2*a* shows the circumferential components E_ξ^n of the five hybrid modes. It is seen that all E_ξ^n vanish at the wall of the guide, as expected. Conversely, eqn. 1 shows that, in the $\xi = 0$ plane,

$$E_\rho = 0$$
$$E_\xi = E_\xi(U) \quad . \quad . \quad . \quad . \quad . \quad . \quad . \quad . \quad . \quad . \quad (10)$$

Thus if $E_\xi(U)$ is matched by correctly combining all E_ξ^n, the incident field will be matched in this plane. As in shallow reflectors, one should also choose the guide radius so that

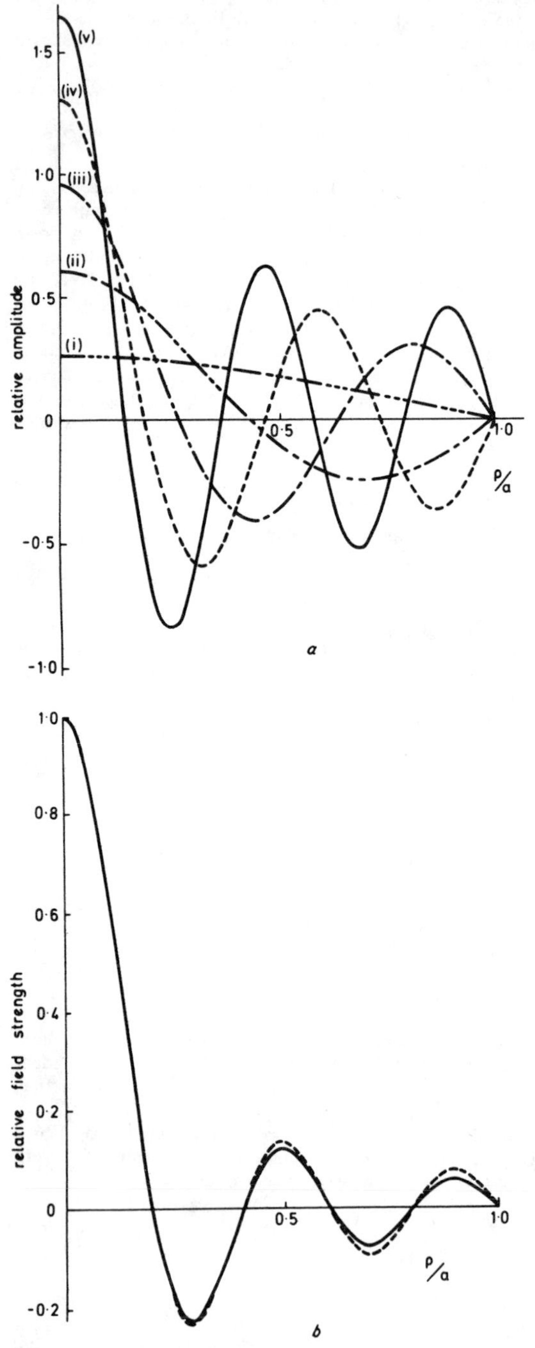

Fig. 2
Relative amplitudes of circumferential electric-field components for paraboloidal-reflector antenna when $a = 2{\cdot}9\lambda$
a Hybrid fields from the waveguide
(i) E_ξ^1 (ii) E_ξ^2
(iii) E_ξ^3 (iv) E_ξ^4
(v) E_ξ^5
b Matching of the incident field
———— incident field
- - - - - - approximate solution

$E_\xi(U)$ vanishes at $\rho = a$. In this example, the condition requires that $a = 2{\cdot}9\lambda$. A curve of $E_\xi(U)/\frac{\rho}{a}$, which satisfies this condition, is shown in Fig. 2*b*.

The problem therefore reduces to correctly combining the curves shown in Fig. 2*a*, so that the resultant curve closely fits the solid curve in Fig. 2*b*. Experience with computer-aided optimisation[5] indicates that a good fit may be obtained if the combined field *exactly* matches $E_\xi(U)$ at all the null points, as well as at the first maximum at $\rho = 0$; i.e. one has to determine the coefficient T_i so that the condition

$$E_\xi(U) = \sum_{i=1}^{5} T_i E_\xi^i \quad . \quad . \quad . \quad . \quad . \quad . \quad . \quad . \quad (11)$$

is satisfied at $\rho = 0$ and $\rho = \rho_l$. Eqn. 11 can be expressed in the matrix form as follows:

$$\begin{bmatrix} E(0) \\ E(U_1) \\ \cdot \\ \cdot \\ \cdot \\ E(U_4) \end{bmatrix} = \begin{bmatrix} x_1(0)\ x_2(0) \ldots & x_5(0) \\ x_1(\rho_1) & \cdot \\ \cdot & \cdot \\ \cdot & \cdot \\ \cdot & \cdot \\ x_1(\rho_4) \ldots & x_5(\rho_4) \end{bmatrix} \begin{bmatrix} T_1 \\ T_2 \\ \cdot \\ \cdot \\ \cdot \\ T_5 \end{bmatrix} \quad (12)$$

where, for simplicity, $x_i(\rho)$ denotes E_ξ^i. Thus the solution is obtained by solving a set of five linear equations of five independent variables. This is a standard problem and can be solved with ease. The calculated relative amplitudes T_i of the five modes are shown below.

HE_{11}	HE_{12}	HE_{13}	HE_{14}	HE_{15}
1·00	0·99	1·05	1·21	1·5

The combined field is also plotted against $E_\xi(U)$ in Fig. 2*b*. The approximate solution gives a reasonable match in the $\xi = 0$ plane.

3.2 Spherical reflectors

The problem of matching the transverse components of the electric field is more complex in spherical reflectors. The elementary modes of the infinite hybrid spectrum of the incident wave are not in phase with one another in the focusing region. As a result, both the radial and circumferential components of the incident field are complex quantities, and the problem of matching involves both amplitude and phase considerations; i.e. the hybrid modes from the guide must have correct amplitude and phase relative to one another if their resultant field is to match the incident field. In addition, a study of the energy flow[6] shows that the maximum efficiency obtainable, with a given feed size, is also dependent on the position of the feed relative to the paraxial focus. The transverse components of the incident field, at a given point near the paraxial focus, have the following forms:[6]

$$E_\rho = E_\rho(V) \sin \xi$$
$$E_\xi = E_\xi(V) \cos \xi \quad . \quad . \quad . \quad . \quad . \quad . \quad . \quad . \quad (13a)$$

where $E_\rho(V) = B(H_1 + H_2)$

$$E_\xi(V) = B(H_1 - H_2)$$

$$H_1 = \int_0^{\psi_0} K(\psi)(1 + \Gamma) J_0(v) e^{-j\Phi} d\psi$$

$$H_2 = \int_0^{\psi_0} K(\psi)(1 - \Gamma) J_2(v) e^{-j\Phi} d\psi$$

$$v = \frac{k\rho \sin \psi}{\zeta}$$

$$K(\psi) = \frac{\sin \psi \{1 - (z/R) \cos \psi\}}{\zeta}$$

$$\Gamma = \frac{\cos \psi - z/R}{\zeta}$$

$$\zeta = \sqrt{\left\{1 - \frac{2z\cos\psi}{R} + \left(\frac{z}{R}\right)^2\right\}}$$

$$\Phi = kR\left\{\cos(\psi) + \zeta + \frac{z}{R} - 2\right\} \quad . \quad . \quad . \quad (13b)$$

The system of reference axes is shown in Fig. 3. Although, at first glance, the problem seems to be extremely complicated,

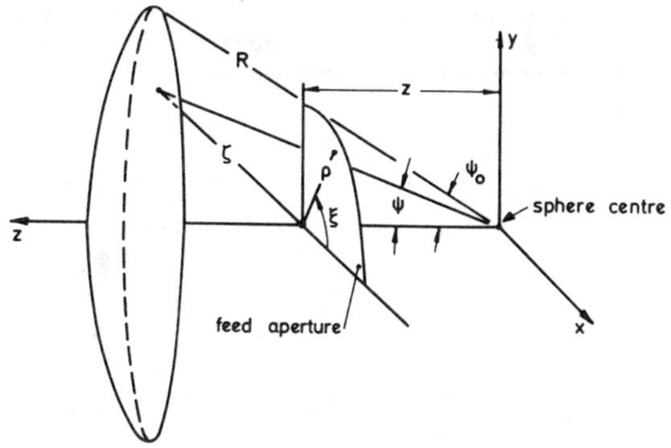

Fig. 3
Co-ordinate system for spherical reflector

a careful study of the field expressions shows that the nature of the problem is essentially the same as that of a paraboloid. Thus a similar approach to that described in Section 3.1 can be used. Eqn. 13*a* shows that, in the $\xi = 0$ plane,

$$E_\rho = 0$$

$$E_\xi = E_\xi(V) \quad . \quad . \quad . \quad . \quad . \quad . \quad . \quad . \quad . \quad . \quad . \quad (14)$$

so an approximate solution can be obtained if $E_\xi(V)$ is matched by correctly combining the N circumferential components E_ξ^n of the modes from the guide. The problem is essentially that of fitting two curves in a complex space. As in the Section 3.1, the first condition is that the radius of the guide should be chosen so that $E_\xi(V)$ vanishes at $\rho = a$. Once the guide radius is determined, the maximum number of propagating modes can be found. The next step is to determine the required amplitudes and phases of E_ξ^n so that $E_\xi(V)$ is exactly matched at some discrete points, the total number of which is equal to the total number of hybrid modes from the guide. Thus one has the following equation:

$$E_\xi(V_l) = \sum_{i=1}^{M} m_i x_i(\rho_l) \exp\{j^{(\gamma_i)}\} \quad . \quad . \quad . \quad . \quad . \quad (15)$$

where m_i and γ_i are the required amplitude and phase of the ith hybrid mode from the guide, and $x_i(\rho_l)$ denotes E_ξ^i at the discrete points $\rho = \rho_l$ where $E_\xi(V)$ is exactly matched.

However, a more convenient way of solving the problem is to match the real and imaginary parts separately. $E_\xi(V)$ can be written as

$$E_\xi(V) = R(\rho) + jI(\rho) \quad . \quad . \quad . \quad . \quad . \quad . \quad . \quad . \quad (16)$$

and eqn. 15 can be split into two sets of linear equations as follows:

$$\begin{bmatrix} R(0) \\ R(\rho_1) \\ \cdot \\ \cdot \\ \cdot \\ R(\rho_M) \end{bmatrix} = \begin{bmatrix} x_1(0) & x_2(0) \ldots & x_M(0) \\ x_1(\rho_1) & & \cdot \\ \cdot & & \cdot \\ \cdot & & \cdot \\ \cdot & & \cdot \\ x_1(\rho_M) \ldots & & x_M(\rho_M) \end{bmatrix} \begin{bmatrix} r_1 \\ r_2 \\ \cdot \\ \cdot \\ \cdot \\ r_M \end{bmatrix} \quad . \quad (17a)$$

and

$$\begin{bmatrix} I(0) \\ I(\rho_1) \\ \cdot \\ \cdot \\ \cdot \\ I(\rho_M) \end{bmatrix} = \begin{bmatrix} x_1(0) & x_2(0) \ldots & x_M(0) \\ \cdot & & \cdot \\ \cdot & & \cdot \\ \cdot & & \cdot \\ \cdot & & \cdot \\ x_1(\rho_M) \ldots & & x_M(\rho_M) \end{bmatrix} \begin{bmatrix} s_i \\ s_2 \\ \cdot \\ \cdot \\ \cdot \\ s_M \end{bmatrix} \quad . \quad (17b)$$

s_i and r_i are related to m_i and γ_i by the following relationships:

$$m_i^2 = r_i^2 + s_i^2$$

$$\gamma_i = \tan^{-1}\left(\frac{s_i}{r_i}\right) \quad . \quad . \quad . \quad . \quad . \quad . \quad . \quad . \quad . \quad . \quad (18)$$

Thus eqns. 17 are essentially the same as eqn. 12, and the two vectors s and r can be obtained without difficulty.

As an illustration, the incident field will be matched by using three hybrid modes from the guide. For comparison purposes, the same data as that used in published work[5] will be chosen; i.e. the feed is placed at a distance equal to $0 \cdot 5173R$ from the centre of the sphere of radius $R = 400\lambda$, and the angle subtended by the reflector *at the centre* of the sphere is 40°. To satisfy the condition

$$E_\xi|_{\rho=a} = 0 \quad . \quad . \quad . \quad . \quad . \quad . \quad . \quad . \quad . \quad . \quad . \quad (19)$$

the guide radius can be chosen to be

$$a = 1 \cdot 89\lambda \quad . \quad . \quad . \quad . \quad . \quad . \quad . \quad . \quad . \quad . \quad . \quad (20)$$

The real and imaginary parts of $E_\xi(V)$ are shown in Fig. 4. Although, with $a = 1 \cdot 89\lambda$, four hybrid modes can propagate

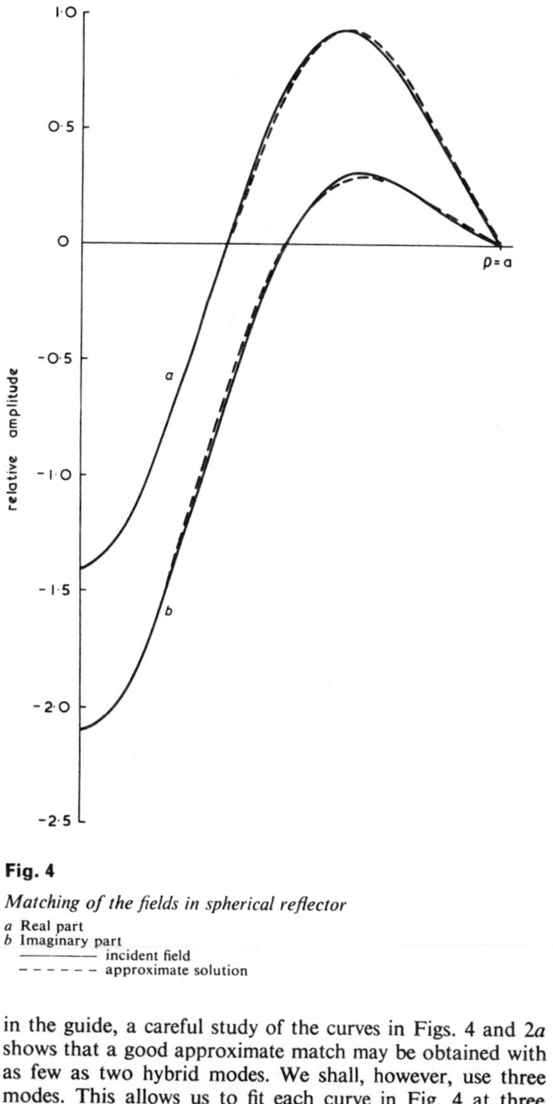

Fig. 4
Matching of the fields in spherical reflector
a Real part
b Imaginary part
——— incident field
– – – – approximate solution

in the guide, a careful study of the curves in Figs. 4 and 2*a* shows that a good approximate match may be obtained with as few as two hybrid modes. We shall, however, use three modes. This allows us to fit each curve in Fig. 4 at three points, apart from that point given by eqn. 19. It is important to note that, to obtain the maximum efficiency, a conjugate match is required. Thus the sign of the imaginary parts obtained for an absolute match must be reversed. The final results are shown in Table 1 and are virtually the same as previous results.

Table 1

COMPARISON OF RESULTS

Mode	HE_{11}	HE_{12}	HE_{13}
Relative amplitude . .	2·1	3·0	0·3
Relative phase, deg . .	0	−94	−63

The real and imaginary parts of the resultant hybrid field for the absolute match are shown in Fig. 4.

4 Conclusions

An approximate method of matching the fields in the focusing region of both spherical and paraboloidal reflector antennas has been presented. Despite its simplicity, this method gives results comparable to those obtained by more sophisticated methods which use computer-aided optimisation techniques. For paraboloidal reflectors, the problem reduces to solving a system of N linear equation of N variables. Similarly, for spherical reflectors, one can obtain the relative amplitudes and phases of the hybrid modes from the guide by solving two systems of linear equations. In the examples illustrated in Sections 3.1 and 3.2, it has been assumed that, in the $\xi = 0$ plane, E_ξ actually vanishes at $\rho = a$. However, to obtain an insight into the bandwidth characteristics of the antenna, it may be desirable to study the example where E_ξ is not equal to zero at $\rho = a$. The degree of *inherent* mismatch will depend on the amplitude of E_ξ at $\rho = a$, and there are no simple criteria for choosing the discrete points ρ_l at which E_ξ is to be exactly matched. Nevertheless, the simple method presented in this paper can be used to obtain an approximate solution which can then be used as the starting point in the computer-aided optimisation program.

5 References

1 POTTER, P. D.: 'Application of spherical wave theory to Cassegrainian-fed paraboloids', *IEEE Trans.*, 1967, **AP-15**, pp. 727–736

2 VU THE BAO: 'Optimisation of efficiency of shallow reflectors', *Electronics Lett.*, 1969, **5**, pp. 6–7

3 BROWN, G. M.: 'Space radio communication' (Elsevier, 1962), pp. 319–365

4 MINNETT, H. C., and THOMAS, B. M.: 'Fields in the image space of symmetrical focusing reflectors', *Proc. IEE*, 1968, **115**, (10), pp. 1419–1430

5 VU THE BAO: 'Optimisation of efficiency of deep reflectors', *IEEE Trans.*, 1969, **AP-17** (to be published)

6 THOMAS, B. M., MINNETT, H. C., and VU THE BAO: 'Fields in the focal region of a spherical reflector', *ibid.*, 1969, **AP-17**, pp. 229–232

Theoretical performance of prime-focus paraboloids using cylindrical hybrid-mode feeds

B. MacA. Thomas, B.E., M.Eng.Sc., Ph.D., Mem.I.E.E.E., C.Eng., M.I.E.E.

Indexing terms: *Antenna radiation patterns, Waveguide antennas*

Abstract

An expression is derived for the radiation pattern of a hybrid-mode field distributed over a circular aperture. The hybrid mode consists of a combination of TM_m and TE_m components (where $m \geqslant 1$) and is the type of wave which propagates in a cylindrical waveguide with an internal lossless anisotropic surface. The requirements for pattern symmetry and linear polarisation are then derived, and the form of the radiation pattern when the waveguide boundary consists of either circumferential or longitudinal slots is discussed. The advantages of using hybrid modes, instead of TM and TE modes, in a smooth pipe for optimising the performance of a paraboloid are considered. The principle of synthetising a desired paraboloid-aperture distribution by the superposition of the beams radiating from a hybrid-mode feed at the prime focus is discussed. The parameters of the hybrid modes in the waveguide, and its diameter, are optimised to obtain maximum aperture efficiency for a given number of modes by a field-matching technique at the feed aperture. Efficiencies for feeds supporting up to five hybrid modes are calculated for a paraboloid semiangle of 63°. These values are compared with those derived by field matching over the paraboloid aperture. The procedure for optimising the figure of merit of 1- and 2-hybrid-mode feeds is then considered, together with the effect of changing the mode-power ratio in the case of the 2-mode feed. Finally, the effect of frequency on the aperture efficiency and figure of merit of the 1- and 2-mode feeds is discussed.

List of principal symbols

a = radius of cylindrical waveguide
E_ρ, E_ξ, E_z = cylindrical components of waveguide fields
$E_\rho(u), E_\xi(u), E_z(u)$ = radial field variation of E_ρ, E_ξ, E_z
E_ϕ, E_ψ = components of radiated field at P
$E_\phi(v), E_\psi(v)$ = field variation of E_ϕ, E_ψ with respect to ψ only
$J_m(u)$ = Bessel function of first kind, order m
$k = 2\pi/\lambda$
$k_\rho = k \sin \psi_0$ = radial separation constant
$k_z = k \cos \psi_0$ = longitudinal propagation constant
ka = normalised feed radius at any frequency
$k_0 a$ = normalised feed radius at the 'centre' design frequency
L = length of corrugated waveguide
m = number of circumferential field variations
n = number of radial field variations
R, ψ, ϕ = spherical co-ordinates of a distant point P
T_R = receiver noise temperature
$u = k_\rho \rho = k\rho \sin \psi_0$
$u_a = k_\rho a = ka \sin \psi_0$
$v = ka \sin \psi$ = generalised radiation angle
X_ξ = normalised circumferential surface reactance
X_z = normalised longitudinal surface reactance
Z_0 = impedance of free space
γ = mode-content factor
δ = corrugated waveguide slotwidth/pitch ratio
η_0 = maximum efficiency
$\eta_a = \eta_s \eta_i$ = aperture efficiency
η_i = illumination efficiency
η_s = spillover efficiency
λ = free-space wavelength
$v = k\rho \sin \psi$
ρ, ξ, z = cylindrical co-ordinates of a point in the waveguide
Ψ = value of ψ at the paraboloid rim, with respect to the focus
ψ_0 = semiangle of cone of plane waves in cylindrical waveguide

1 Introduction

Analysis of the focal-region fields of an axially symmetric reflector illuminated by a linearly polarised plane wave has shown that they are equivalent to a continuous spectrum of plane hybrid waves propagating along the axis with different velocities.[1] Each hybrid wave is a linear combination of TE_{1n} and TM_{1n} fields. In the case of a paraboloidal reflector, all the hybrid waves arrive at the focal plane in phase.

The resultant field structure in the focal regions of both the paraboloidal[1] and spherical[2] reflectors has been considered in detail. With this information, the efficiencies have been calculated of 'ideal' feeds of radius a which absorb all the energy from the spectrum of hybrid waves over the aperture a.

It has also been shown that the hybrid waves are the natural modes of propagation in a circular waveguide when the walls have the values of anisotropic-surface reactances necessary to match the fields at the boundary.[3] For a given anisotropic surface, only a finite number of discrete hybrid modes can propagate in the waveguide. An expression has been derived which enables the calculation of the degree of matching between the focal-region fields of a paraboloidal or spherical reflector and the set of hybrid modes at the open end of a waveguide.[4] This approach to the optimisation of efficiency has been applied to particular cases for both the paraboloidal and spherical reflectors.[4,5]

In this paper, the optimisation of the aperture efficiency and the figure of merit of a paraboloid are considered mainly from the point of view of synthetising a desired reflector-aperture distribution by the superposition of the beams radiating from the hybrid-mode waveguide feed situated at the prime focus. An expression, given in Reference 3, for the theoretical radiation pattern of hybrid modes is derived, and radiation properties of the dominant and higher-order modes are discussed in detail. The effect of the values of the anisotropic-surface reactances on the shape of the patterns is briefly discussed.

Some experimental radiation patterns of 1- and 2-hybrid-mode feeds have already been published.[3,6,7] A detailed comparison of experimental results with the theory presented here will be considered in a later paper.

Paper 6524 E, first received 4th January and in revised form 12th July 1971
Dr. Thomas is with the Division of Radiophysics, Commonwealth Scientific & Industrial Research Organisation, Sydney, Australia

2 Hybrid-wave fields

2.1 Field equations

The most general expessions for the fields in a cylindrical waveguide with unspecified anisotropic-surface reactance require both TM_m and TE_m components to be propagating at the same velocity when $m \geqslant 1$. The resultant

Reprinted with permission from *Proc. IEE*, vol. 118, pp. 1539–1549, Nov. 1971.

cylindrical-field components (see Fig. 1 for the co-ordinate system) of the hybrid wave are:

$$\left.\begin{array}{ll} E_\rho = E_\rho(u)\sin m\xi & H_\rho = H_\rho(u)\cos m\xi \\ E_\xi = E_\xi(u)\cos m\xi & H_\xi = H_\xi(u)\sin m\xi \\ E_z = \dot{E}_z(u)\sin m\xi & E_z = H_z(u)\cos m\xi \end{array}\right\} \quad . \quad (1a)$$

where

$$E_\rho(u) = -jk\left\{\frac{k_z}{k}J_{m-1}(u) + \left(\gamma - \frac{k_z}{k}\right)\frac{mJ_m(u)}{u}\right\}$$

$$E_\xi(u) = -jk\left\{\gamma J_{m-1}(u) - \left(\gamma - \frac{k_z}{k}\right)\frac{mJ_m(u)}{u}\right\}$$

$$E_z(u) = k_\rho J_m(u) = \frac{u_a}{a}J_m(u)$$

$$H_\rho(u) = j\frac{k}{Z_0}\left\{\gamma\frac{k_z}{k}J_{m-1}(u) + \left(1 - \gamma\frac{k_z}{k}\right)\frac{mJ_m(u)}{u}\right\}$$

$$. \quad . \quad . \quad . \quad . \quad (1b)$$

$$H_\xi(u) = -j\frac{k}{Z_0}\left\{J_{m-1}(u) - \left(1 - \gamma\frac{k_z}{k}\right)\frac{mJ_m(u)}{u}\right\}$$

$$H_z(u) = -\frac{\gamma}{Z_0}k_\rho J_m(u) = -\frac{\gamma}{Z_0}\frac{u_a}{a}J_m(u)$$

and

$$u = k_\rho\rho = k\rho\left\{1 - \left(\frac{k_z}{k}\right)^2\right\}^{1/2}$$

$$u_a = k_\rho a$$

The mode-content factor γ is defined as the ratio of the longitudinal fields of the TE_m and TM_m components, respectively; i.e.

$$\gamma = -Z_0 H_z(u)/E_z(u)$$

When the TM_m and TE_m components are in phase, so that γ is positive, the hybrid mode is designated HE_m, and when the two components are out of phase, so that γ is negative, the

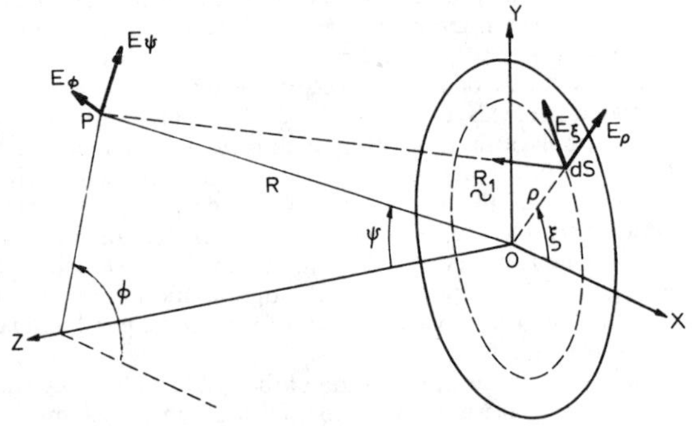

Fig. 1
Co-ordinate system for radiation pattern analysis

designation used is EH_m. For the particular case where $E_\xi(u_a)$ is zero at the boundary, the hybrid modes are designated HE_{mn} and EH_{mn} respectively, where n is the mode order and represents the number of zeros of $E_\xi(u)$ along a radius of the guide, excluding any on the axis.

When $\gamma = 0$, eqns. 1*b* reduce to those of the TM_m mode, and when $\gamma = \infty$, the TE_m mode field components are obtained by neglecting the terms not including γ. The modes of particular interest are the balanced $HE_1(\gamma = 1)$ modes, which have a predominantly linearly polarised field structure in the waveguide, and the $EH_1(\gamma = -1)$ modes, which have strong crosspolarised fields. It can readily be shown that both these modes can propagate in the same waveguide when the normalised circumferential and longitudinal surface reactances* satisfy the relation[3]

$$X_\xi X_z = -1 \quad . \quad . \quad . \quad . \quad . \quad . \quad . \quad . \quad . \quad . \quad (2)$$

* These reactances are defined, respectively, as

$$X_\xi = \frac{-jE_\xi(u_a)}{H_z(u_a)Z_0} \qquad X_z = \frac{jE_z(u_a)}{H_\xi(u_a)Z_0}$$

The longitudinal propagation constant of a waveguide mode is sometimes expressed in terms of the direction of the plane waves which constitute the mode.[8] For the circular waveguide, the Poynting vectors of the plane waves are aligned along a cone of semiangle ψ_0 with respect to the axis of the guide. The angle ψ_0 is given by $k_z = k\cos\psi_0$, and since $k^2 = k_z^2 + k_\rho^2$, $u_a = k_\rho a = ka\sin\psi_0$. Only when $\gamma = \pm 1$ are all the Poynting vectors of equal magnitude. In addition, the electric field of the plane waves making up the balanced hybrid mode then has no crosspolarised component.

2.2 Balanced hybrid modes in corrugated waveguide

Since the type of boundary used determines the field distribution across the aperture, and hence the shape of the radiation pattern, it is convenient to restrict the discussion of pattern synthesis to those waveguides having anisotropic surfaces which are readily realised in practice.

The anisotropic waveguide surface necessary to support a balanced hybrid wave is defined by eqn. 2; it can be approximated with metal corrugations, provided that a sufficient number per wavelength are used.[3] The edges of the flanges separating the slots force the electric field parallel to the slots to be zero at the boundary. The magnitude of the corresponding magnetic-field component, which is determined by the reactance of the slot, must also be zero for the balanced hybrid mode. This requires the slot to be at quarter-wave resonance.

The structure using circumferential slots when $E_\xi(u_a) = 0$ is the most convenient. For this case, the characteristic equation, determined from eqn. 1*b*, is

$$X_z = \frac{u_a J_m(u_a) J_m'(u_a)/ka}{\left\{m\dfrac{k_z}{k}\dfrac{J_m(u_a)}{u_a}\right\}^2 - \{J_m'(u_a)\}^2} \quad . \quad . \quad . \quad (3a)$$

Also, it can be shown that

$$\gamma = -\frac{m\dfrac{k_z}{k}J_m(u_a)}{u_a J_m'(u_a)} \quad . \quad . \quad . \quad . \quad . \quad . \quad . \quad . \quad (3b)$$

When a number of balanced HE_{1n} modes exist in a waveguide, it can readily be seen from eqn. 3*a* that u_a will be determined by the relation $J_0(u_a) \simeq 0$ when $k_z/k \to 1$, corresponding to the lowest-order mode. The value of u_a for the highest-order mode near cutoff ($k_z/k \to 0$) is determined by the relationship $J_1'(u_a) = 0$.†

Fast-wave solutions to the characteristic equations when $X_\xi = 0$ are shown in Fig. 2. Here the normalised longitudinal propagation constant for $X_z = \infty$ is plotted for the first

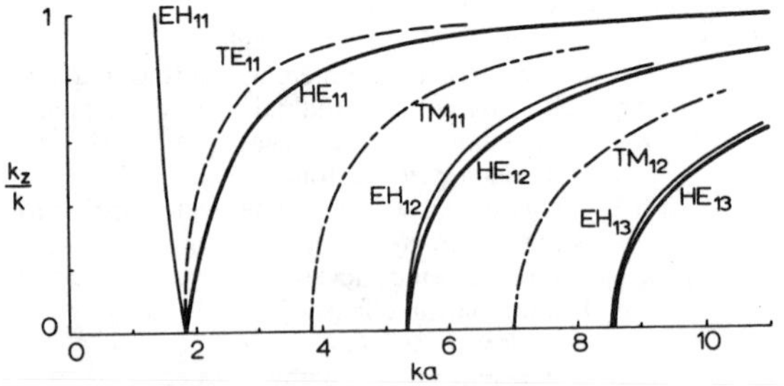

Fig. 2
Propagation characteristics for $X_\xi = 0$
For the HE_{1n} and EH_{1n} modes, $X_z = \infty$, and for the TM_{1n} and TE_{11} modes $X_z = 0$
(The TE_{12} and TE_{13} modes, not shown, fall between the corresponding HE_{1n} and EH_{1n} mode curves)

three balanced HE_{1n} modes together with the $EH_{1n}(\gamma = -1)$ modes. Superimposed are the solutions from a smooth metal pipe ($X_z = 0$). These are the TM_{1n} and TE_{11} modes, for which $J_1(u_a)$ and $J_1'(u_a)$, respectively, are zero. The curves for the higher-order TE_{1n} ($n \geqslant 2$) modes (not shown) are situated between the HE_{1n} and EH_{1n} curves.

As an alternative to circumferential corrugations, longitudinal slots one-quarter wavelength deep, where $X_z = 0$,

† The first roots of the expression $J_0(u_a) = 0$ are 2·405, 5·520 and 8·654, and for $J_1'(u_a) = 0$ they are 1·841, 5·331 and 8·536. These values apply to the HE_{11}, HE_{12} and HE_{13} modes, respectively. The values of the roots given by the two equations above converge as the order of the mode, n, increases

$X_\xi = \infty$ and therefore $E_z(u_a) = H_z(u_a) = 0$, could be used. The characteristic equation of the balanced mode is identical to that of the TM_{mn} mode, i.e. $J_m(u_a) = 0$. However, in the longitudinally corrugated waveguide, the TM_{mn} and TE_m components can propagate independently at the same velocity. This creates the practical problem of ensuring that both modes are excited with the same phase, and that the powers in each of the modes in the guide remain in the correct ratio in order to satisfy the balanced condition necessary for pattern symmetry. Even if the balanced condition could be maintained, it will be shown below that the radiation pattern of the mode has a null along the axis which may be undesirable for feeding a reflector. For these reasons, only the radiation from circumferentially corrugated waveguides will be considered in detail in the following Sections.

3 Radiation properties of cylindrical hybrid modes

3.1 Radiation from an annular ring

Before evaluating the expressions for the radiation pattern of a circular aperture containing hybrid-mode fields, it is instructive to consider the radiation from an annular aperture of normalised radius u, where the transverse-field components have circumferential variations given by eqn. 1a.

It is also convenient to write the far-field components E_ψ and E_ϕ at a distant point $P(R, \psi, \phi)$ (see Fig. 1) in the following form:

$$\left.\begin{aligned} E_\psi &= \frac{j^{m+1}}{2R} e^{-jkR} E_\psi(v) \sin m\phi \\ E_\phi &= \frac{j^{m+1}}{2R} e^{-jkR} E_\phi(v) \cos m\phi \end{aligned}\right\} \quad \dots \quad (4)$$

where $E_\psi(v)$ and $E_\phi(v)$ define the variation in the far field with respect to angle ψ only.

It is shown in Appendix 9.1 that the far-field components of the annular aperture of width $d\rho$ and radius ρ are as follows:

$$\left.\begin{aligned} E_\psi(v) &= jk\left[\{E_\xi(u) - Z_0 H_\rho(u)\cos\psi\}\frac{mJ_m(v)}{v} \right.\\ &\qquad \left. + \{E_\rho(u) + Z_0 H_\xi(u)\cos\psi\}J_m'(v)\right]\rho d\rho \\ E_\phi(v) &= jk\left[\{E_\rho(u)\cos\psi + Z_0 H_\xi(u)\}\frac{mJ_m(v)}{v} \right.\\ &\qquad \left. + \{E_\xi(u)\cos\psi - Z_0 H_\rho(u)\}J_m'(v)\right]\rho d\rho \end{aligned}\right\} \quad (5)$$

where $v = k\rho \sin\psi$.

When the fields at a radius ρ in the aperture are related by

$$\left.\begin{aligned} E_\rho(u) &= -Z_0 H_\rho(u) \\ E_\xi(u) &= Z_0 H_\xi(u) \end{aligned}\right\} \quad \dots \quad (6a)$$

the far-field components $E_\psi(v)$ and $E_\phi(v)$ at an angle ψ are equal. The magnitude of the resultant field vector, $|E| = (E_\psi^2 + E_\phi^2)^{1/2}$, is independent of the azimuthal angle ϕ, although the polarisation angle of this vector, $90° - (m-1)\phi$, remains constant only when $m = 1$. The radiation pattern is then linearly polarised and circularly symmetric. However, if the aperture fields are related by

$$\left.\begin{aligned} E_\rho(u) &= Z_0 H_\rho(u) \\ E_\xi(u) &= -Z_0 H_\xi(u) \end{aligned}\right\} \quad \dots \quad (6b)$$

then $E_\psi(v) = -E_\phi(v)$. Again, the magnitude of E is independent of ϕ, but the polarisation angle becomes $90° + (m+1)\phi$. The type of aperture field specified by eqn. 6b is therefore undesirable in a linearly polarised system.

In general, for $m = 1$, the magnitude of the linearly polarised component of the far field is given by

$$E_\psi(v)\sin^2\phi + E_\phi(v)\cos^2\phi$$

and the magnitude of the crosspolarised field is

$$\{E_\psi(v) - E_\phi(v)\}\sin\phi\cos\phi$$

Some of the basic radiation-pattern properties of the hybrid mode can be deduced directly from those of the annular ring, since the radiation pattern of the entire aperture is the superposition of the patterns of each of the elemental annuli comprising the complete aperture. Reference to eqns. 1b shows that for any value of u, the HE_m modes satisfy eqn. 6a when $\gamma = 1$, and that the EH_m modes satisfy eqn. 6b when $\gamma = -1$. Consequently, only the balanced HE_1 modes produce a linearly polarised, circularly symmetrical radiation pattern. If the EH_1 modes are present, the rotating polarisation angle $(90° - 2\phi)$ introduces asymmetry into the total radiation pattern. It is interesting to note that in the $\phi = \pm 45°$ planes, the radiated fields of the two types of mode are orthogonally crosspolarised. This property enables the radiation pattern of the $HE_1(\gamma = 1)$ modes to be measured separately from that of the $EH_1(\gamma = -1)$ modes. Another property of the $EH_1(\gamma = -1)$ mode is that the patterns of each of the annular rings, and hence the complete aperture, have a null in the axial direction.

3.2 Radiation of hybrid modes in a circular aperture

The assumptions in calculating the far-field radiation pattern of the hybrid-mode fields are similar to those used by Silver[9] for the TE_{mn} and TM_{mn} modes, except that the effect of any reflection at the aperture is neglected. Consequently, the fields at the aperture are assumed to be those of the undisturbed waveguide fields.

The far-field components $E_\psi(v)$ and $E_\phi(v)$ are evaluated by substituting eqn. 1b into eqn. 5, and summing the contributions from all the annular rings of width $d\rho$ across the circular aperture of radius a. Upon integration, it can be shown that these components take the following form:

$$E_\psi(v) = ka\left\{\gamma\left(1 + \frac{k_z}{k}\cos\psi\right)K_1 + \left(\frac{k_z}{k} + \cos\psi\right)K_2\right\}$$

$$E_\phi(v) = ka\left\{\left(1 + \frac{k_z}{k}\cos\psi\right)K_1 + \gamma\left(\frac{k_z}{k} + \cos\psi\right)K_2\right\} \quad \dots \quad (7)$$

where $K_1 = mJ_m(u_a)\dfrac{J_m(v)}{v}$

$$K_2 = \frac{J_m(u_a)J_m'(v) - \dfrac{v}{u_a}J_m(v)J_m'(u_a)}{1 - \left(\dfrac{v}{u_a}\right)^2}$$

and the generalised radiation angle $v = ka\sin\psi$. Eqn. 7 shows that $E_\psi(v) = E_\phi(v)$ when $\gamma = 1$. This was also deduced from the properties of an annular ring in Section 3.1. In particular, when $m = 1$, the radiation pattern is axially symmetrical and linearly polarised.

In the axial direction, where $v = 0$,

$$K_1 = K_2 = mJ_m(u_a)\left.\frac{J_m(v)}{v}\right|_{v\to 0}$$

The field strength in this direction is then dependent on $J_m(u_a)$, which is determined by the boundary conditions and mode order n.

The function K_1, when $m = 1$, represents a radiation pattern which has a main beam in the axial direction and sidelobes which rapidly decay in amplitude. On the other hand, the function K_2 mainly defines the shape of the main beam, and the sidelobe structure away from the axial direction. The direction ψ_0, given by $v = u_a$, corresponds to the maximum in the function K_2, which occurs when both numerator and denominator are zero. However, if the boundary conditions at the circumference of the waveguide force a field distribution across the aperture such that $J_1(u_a)$ is not zero, the function K_1 does not vanish and may change the position of the main beam from the value $\psi = \psi_0$ given by K_2. Consequently, the effect of K_1 on the resultant shape and position of the main beam will be greater for those beams closest to the axial direction.

Inspection of eqn. 7 also shows that the zeros in the radiation patterns of a series of modes never exactly coincide, except for the longitudinally slotted waveguide where

$J_m(u_a) = 0$. However, for balanced HE_{1n} modes in circumferentially corrugated and relatively large-diameter waveguides, the zeros almost coincide for the modes away from cutoff where the condition $J_0(u_a) \simeq 0$ holds (see Section 2.1).

Before discussing the radiation patterns of the HE_{1n} modes, it is instructive to consider briefly the principle of pattern synthesis as an extension of the concept of the decomposition of waveguide fields into a cone of plane waves of semiangle ψ_0 (see Section 2.1). On passing through the aperture, the cone of plane waves of the balanced hybrid mode gives rise to a symmetrical linearly polarised beam with a direction $\psi \simeq \psi_0$. In the limit as $a \to \infty$, the beam remains a cone of waves emerging from the waveguide at an angle ψ_0. However, as the radius of the guide is decreased, diffraction causes the width of the main lobe to increase. In addition, for a given waveguide size, k_z/k decreases (i.e. ψ_0 increases) with the order of the mode. Consequently, the main beam of the lowest-order mode, for which $k_z/k \simeq 1$ and $\psi_0 \to 0$, illuminates the central portion of the reflector. Each successive annular ring of the reflector is illuminated by each of the higher-order modes in turn.

To illustrate the above, Figs. 3*a* and *b* show the illumination of two consecutive elementary annular rings of a parabolic reflector by the HE_{17} and HE_{18} modes radiating from a waveguide with $ka = 50$ ($a \simeq 8\lambda$) and having circumferential slots ($X_\xi = 0$, $X_z = \infty$). The direction of the main beam occurs at an angle ψ_0. The high amplitude of the sidelobe in the axial direction ($\psi = 0$) is due to the K_1 term in eqn. 7 and accounts for half the field strength in this direction. Since the modes in Figs. 3*a* and *b* are not close to cutoff, the value of u_a is determined by the relation $J_0(u_a) \simeq 0$. It can be

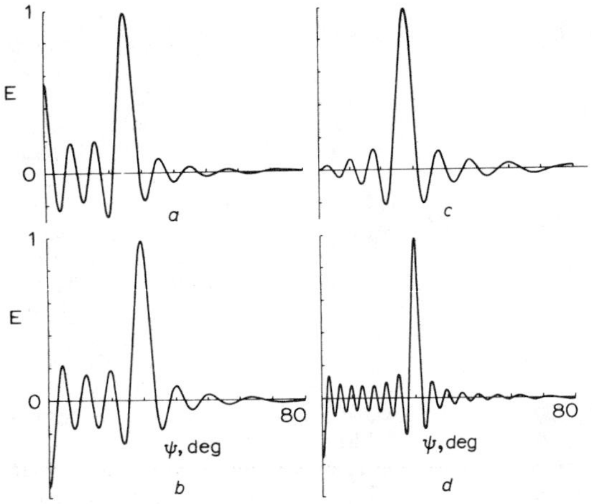

Fig. 3

Radiation patterns of the balanced HE_{1n} modes

(*a*) $n = 7$	$X_\xi = 0$	$X_z = \infty$	$ka = 50$	$\psi_0 = 25{\cdot}1°$
(*b*) $n = 8$	$X_\xi = 0$	$X_z = \infty$	$ka = 50$	$\psi_0 = 29{\cdot}1°$
(*c*) $n = 7\frac{1}{2}$	$X_\xi = \infty$	$X_z = 0$	$ka = 50$	$\psi_0 = 27{\cdot}1°$
(*d*) $n = 16$	$X_\xi = 0$	$X_z = \infty$	$ka = 100$	$\psi_0 = 29{\cdot}6°$

shown that the resulting value of u_a makes $J_1(u_a)$ a maximum. Hence the field in the axial direction, which is proportional to $J_1(u_a)$, is also a maximum for a given mode order. The number of lobes between the axial direction and the main beam, inclusive, is equal to the order of the mode. Consequently, for odd-order modes, the lobe in the axial direction has the same phase as the main beam, whereas for the even-order modes they are out of phase.

A radiation pattern with the main beam pointing midway between those produced by the HE_{17} and HE_{18} modes in a circumferentially corrugated waveguide is shown in Fig. 3*c*. In this case, the waveguide has longitudinal corrugations ($X_\xi = \infty$, $X_z = 0$) which require $H_z(u_a)$ to be zero at the boundary surface. $E_\xi(u_a)$ is a maximum at the surface corresponding to an extra half cycle of the field. It is useful to be able to label specifically such modes by generalising the usual notation, given in Section 2.1 for waveguides requiring $E_\xi(u_a)$ to be zero at the surface. This can be done if n represents half the number of zeros and maxima of $E_\xi(u)$ along a radius of the guide excluding any zero on the axis. Thus the mode under consideration is specified by $HE_{17\frac{1}{2}}$. Since $J_1(u_a) = 0$, the term K_1 in eqn. 7, and hence the radiated field in the axial direction, is zero. This case represents the other extreme to the patterns shown in Figs. 3*a* and *b*. To achieve a finite but nonmaximum axial field, both reactances of the waveguide wall must be nonzero.

For comparison with Fig. 3*b*, the effect of doubling the waveguide diameter, and also the mode order to keep ψ_0 fixed, is shown in Fig. 3*d*. Besides having twice the number of sidelobes, the beamwidth has been halved and the amplitude of the sidelobes slightly reduced.

The effect of reducing the diameter of the waveguide is shown in Fig. 4. This Figure shows the three hybrid modes

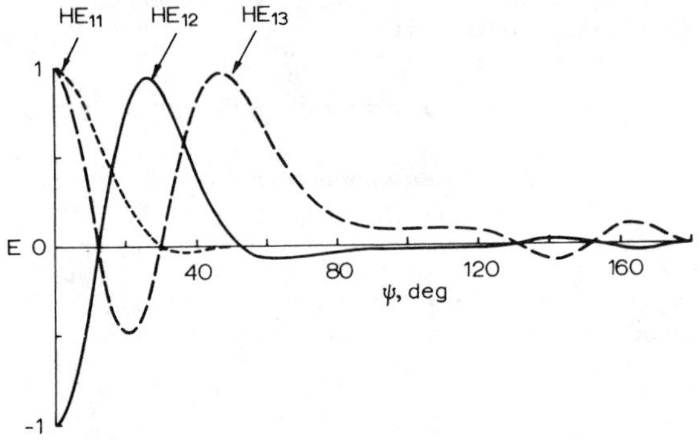

Fig. 4

Radiation patterns of the balanced HE_{11}, HE_{12}, and HE_{13} modes when $k_0a = 11{\cdot}0$

which can propagate in a waveguide with $X_\xi = 0$, $X_z = \infty$ and $ka = 11{\cdot}0$. Comparison with Figs. 3*a*, *b* and *d* shows that the field strength in the axial direction for the small-diameter waveguide has increased considerably, and is greater than that of the main beam.

4 Radiation patterns of TM and TE modes

Before considering the optimisation of the paraboloid performance using hybrid modes, the radiation properties of TM and TE modes in circular pipes, and their application to pattern synthesis, will be briefly summarised.

The boundary conditions in a metal pipe require that the tangential electric field be zero, i.e. $J_m(u_a)$ and $J'_m(u_a)$ are zero for the TM and TE modes, respectively. Eqn. 7 is then reduced to the following equations:

(*a*) TM_{mn} *mode*:

$$E_\psi(v) = ka\left(\frac{k_z}{k} + \cos\psi\right) K'_2$$

$$E_\phi(v) = 0 \qquad (8)$$

$$\text{where } K'_2 = \frac{v}{u_a}\,\frac{J'_m(u_a)J_m(v)}{1 - \left(\dfrac{u_a}{v}\right)^2}$$

(*b*) TE_{mn} *mode*:

$$E_\psi(v) = ka\left(1 + \frac{k_z}{k}\cos\psi\right) K_1$$

$$E_\phi(v) = ka\left(\frac{k_z}{k} + \cos\psi\right) K''_2 \qquad (9)$$

$$\text{where } K''_2 = J_m(u_a)\,\frac{J'_m(v)}{1 - \left(\dfrac{v}{u_a}\right)^2}$$

Eqns. 8 and 9 are in agreement with those derived by Silver.[9]

The TE_{11} mode is often used for illuminating a paraboloidal reflector. However, the beamwidth is greater in the *H*plane (E_ϕ) than in the *E*plane (E_ψ), and the crosspolarised energy is a maximum in the $\phi = \pm 45°$ planes. The radiation pattern of the TM_{11} mode is unsuitable by itself, since there is no energy in the *H*plane, and in the *E*plane the main lobe points off axis at an angle given by $\psi \simeq \psi_0$ (i.e. $v = u_a$) and there is a null at $\psi \simeq 0°$. However, the pattern symmetry and crosspolarisation characteristics of the TE_{11} mode can be improved by combining the TM_{11} and TE_{11} modes with the correct amplitude and phase in a single pipe.[10, 11]

Pattern synthesis using a number of TM_{1n} and TE_{1n} modes is relatively straightforward in principle, because, in the Eplane, both types of mode have nulls at angles defined by $J_1(v) = 0$, $(v \neq 0)$, except at the position of the $(n + 1)$th null of this function, where the main beam of the TM_{1n} mode is situated (see eqn. 8). Thus, at these angles, the amplitudes of each of the TM_{1n} modes can be adjusted independently so as to approximate a desired aperture distribution at these positions.[12, 13] In the Hplane, the field contributed by the TM_{1n} modes is zero. Also, in this plane, the main beam of each of the TE_{1n} modes occurs at the nth null of the function $J_1'(v) = 0$, which corresponds to a null in the patterns of the other TE_{1n} modes (see eqn. 9). Thus the amplitudes of each of the TE_{1n} modes may be adjusted independently to approximate a desired Hplane aperture distribution. Although the paraboloid-aperture distribution may be approximated closely at these discrete points, it does not necessarily follow that the resulting distribution will give maximum efficiency.

Exact pattern symmetry and linear polarisation can only be achieved with balanced HE_1 modes. In this case, only half the number of modes is necessary for pattern synthesis, compared with the approach described above using both TE_{1n} and TM_{1n} modes in a smooth metal pipe.

5 Pattern synthesis using hybrid modes

From the discussion in Section 3.2, it is clear that a given paraboloid-aperture distribution can be approximated by including all the modes in the corrugated waveguide for which $\psi_0 < \Psi$, the semiangle of the reflector. As the feed aperture is increased, the number of modes must also increase to ensure illumination of the reflector out to the rim, where $\psi = \Psi$. With a larger number of modes, there is the possibility of approximating very closely to any desired aperture distribution. In practice, however, the waveguide size and number of modes have had to be restricted up to the present because of the problem of mode control.

The method of pattern synthesis described in Section 4 for TE_{1n} and TM_{1n} modes in a metal pipe is more difficult for HE_{1n} modes radiating from circumferentially corrugated waveguides, because the nulls in the radiation patterns of the individual modes do not exactly coincide. In addition, the shape of the radiation pattern gives very little guidance as to the changes in the parameters necessary to improve the performance.

It is therefore desirable to devise a better technique to determine the waveguide size and relative mode powers for optimum paraboloid performance. The methods used to obtain either optimum aperture efficiency (η_a) or optimum figure of merit (f.m.) are discussed in the following Sections.

5.1 Optimum aperture efficiency

The aperture efficiency η_a of a reflector, with a given field distribution over the aperture, is equal to the ratio of gain of the reflector relative to the maximum value obtained when the distribution is uniform in amplitude, phase and polarisation. If this uniformity is not achieved, the total vector field produced at a distant, on-axis point by all the aperture elements is reduced, and $\eta_a < 1$. The aperture efficiency η_a may be expressed as the product $\eta_i \eta_p \eta_x \eta_s$ of component efficiencies. The first three of these take account of any departure from uniformity of the amplitude, phase and polarisation distributions, respectively, and η_s is the fraction of the radiated energy intercepted by the reflector.

When the feed consists of an aperture which radiates cophased hybrid waves and is located in the focal plane of the paraboloid, there are no phase or crosspolarisation losses ($\eta_p = \eta_x = 1$). Maximum aperture efficiency is then achieved when $\eta_i \eta_s$ is a maximum. The radiated field at the reflecting surface (and hence across the paraboloid aperture) is dependent on the f/D ratio of the reflector because of space attenuation, i.e. the field strength is proportional to $\cos^2 \psi/2$. Consequently, if radiation patterns at a fixed distance R are being considered, the ideal pattern for 100% aperture efficiency would have an amplitude distribution proportional to $\sec^2 \psi/2$ extending to $\pm\Psi$, and zero elsewhere. However, for a finite feed aperture, the approximation to this ideal distribution will have ripples across the top of the pattern and will taper more or less steeply towards the edge of the reflector, so that some energy spills over into the region $\psi > \Psi$. In general, if η_i is increased by improving the uniformity of the distribution within the range $\pm\Psi$, η_s will decrease. Conversely, less spillover requires the radiation pattern to taper more towards the rim so that η_i is reduced. Consequently, a compromise must be made between the two efficiencies η_i and η_s in order to achieve maximum η_a.

Expressions for the various components of aperture efficiency are given in Appendix 9.2 as functions of the feed radiation patterns. The aperture efficiency has been derived by comparing the match between the transmitting and receiving fields in the paraboloid aperture. The former are those produced by the feed when radiating, and the latter is the uniform, linearly polarised distribution due to a distant source. The expression for η_a agrees with that calculated from the secondary radiation pattern.[14] To optimise the aperture efficiency, η_a can be calculated from the feed radiation patterns for a range of feed diameters and mode powers. However, a more direct procedure results from the technique of focal-plane matching.[4]

For 100% aperture efficiency (or maximum possible gain), the aperture of the feed in the focal plane when radiating would have to contain the conjugate of the received focal-plane fields. For a close match between the continuous spectrum of hybrid waves arriving at the focal plane[1] and a finite set of modes at the feed aperture, it is necessary to choose the correct anisotropic-waveguide diameter and to optimise the relative powers of the waveguide modes. When the waveguide surface consists of circumferential corrugations, so that the field $E_\xi(u_a)$ at the edge of the guide is zero, the guide radius is chosen to correspond to a null in the incident focal-plane field.[4] For a single-hybrid-mode guide, the radius is given by the first zero, while for N modes it is determined by the position of the Nth zero in the focal-plane distribution.

In practice, the relative amplitudes of the modes necessary to approach the desired match for waveguide diameters chosen in the above manner may be determined by a trial-and-error computation, or more quickly by the use of a graphic display on a computer. The latter technique, in which the E_ρ and E_ξ components of both sets of fields are displayed and compared, has been investigated by Vu.[5] For the Parkes paraboloid, it was possible to converge rapidly to an almost perfect match to the incident fields.

Table 1 shows the aperture and spillover efficiencies computed for the Parkes paraboloid ($\Psi = 63°$) using prime-focus corrugated feeds carrying up to five hybrid modes. Note that these efficiencies are a measure of the feed performance and not overall aperture efficiency, since aperture blockage and surface errors have been neglected for the present purposes.

The values of η_a and η_s given in columns 7 and 8, respectively, have been calculated from the efficiency equations given in Appendix 9.2 by using the optimum waveguide radii (column 3), and mode powers (column 5), determined by the focal-plane matching technique. The relative powers of the modes were then varied in each case, and the values giving maximum η_a were found to agree with those obtained by focal-plane matching. The calculated aperture efficiencies also agreed with those computed from the focal-plane match (see column 6). The agreement is exact for the larger waveguide sizes, which is consistent with the fact that aperture-radiation theory improves in accuracy for larger values of ka. The values of η_a approach closely the maximum values η_0 in column 4. This shows that the focal-plane field distribution can be closely matched by a finite number of hybrid modes across the waveguide aperture.

Fig. 5a shows the theoretical radiation pattern of the 2-mode feed adjusted for optimum η_a, and also when the mode-power ratio is reduced from the optimum of 3·1 (5 dB) to 2·5 dB. Fig. 6 shows that the power ratio can vary over a wide range without significantly affecting η_a. This fact is useful in the practical design of 2-mode feeds, where a power

Table 1

PARABOLOID EFFICIENCIES, $\Psi = 63°$

Number of modes	ka	a/λ	η_0	Relative mode powers					Aperture efficiency η_a		Spillover efficiency η_s
				P_1	P_2	P_3	P_4	P_5	Focal-plane matching	Paraboloid aperture-plane matching	
1	3·65	0·58	0·725*	1·0					0·725	0·720	0·896
2	7·4	1·18	0·828	1·0	3·1				0·827	0·819	0·927
3	11·0	1·75	0·875	1·0	3·1	6·4			0·874	0·870	0·946
4	14·6	2·32	0·901	1·0	3·1	6·3	10·9		0·900	0·900	0·960
5	18·2	2·90	0·919	1·0	3·1	6·3	10·9	16·4	0·918	0·918	0·971

* For comparison, η_0 for a TE_{11} mode is 0·66

ratio as high as 5dB cannot be obtained from a 2-hybrid-mode exciter consisting of a step discontinuity between the small-diameter TE_{11}-mode input waveguide and the larger diameter corrugated waveguide.[15]

The radiation pattern of a 3-mode feed adjusted for optimum η_a is given in Fig. 5*b*. The individual mode patterns for this feed, given in Fig. 4, show that the field in the axial direction of the HE_{12} mode pattern is out of phase with that of the HE_{11} and HE_{13} modes, but that the main beams of the three modes are all in phase.

The radiation pattern of a 5-mode feed is given in Fig. 5*c*. A comparison of the radiation patterns of the 2-, 3-, and 5-mode feeds with that of the 1-hybrid-mode feed discussed in Section 5.2 (see Fig. 8*b*) shows the progressive improvement as the number of modes is increased. The patterns successively approximate more closely to the ideal $\sec^2 \psi/2$ distribution within the angle subtended by the reflector ($\Psi = \pm 63°$), and have a lower percentage spillover energy beyond the rim.

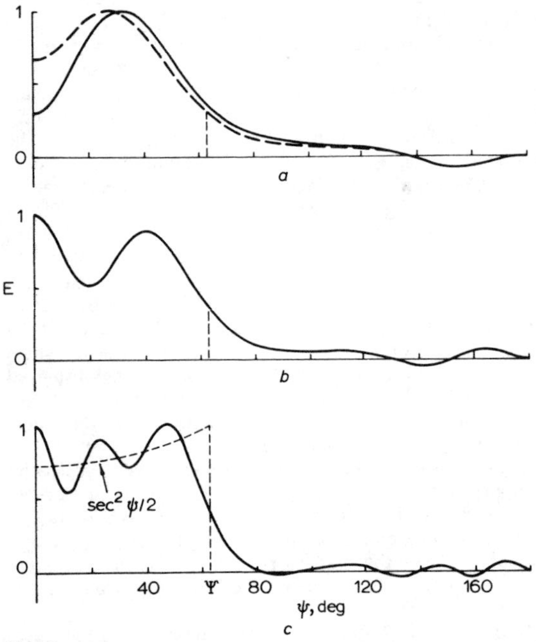

Fig. 5

Radiation patterns of hybrid-mode feeds designed for optimum η_a when $\Psi = 63°$

a Two modes, $k_0a = 7·4$
—— $P_{12}/P_{11} = 5$dB
- - - - $P_{12}/P_{11} = 2·5$dB
b Three modes, $k_0a = 11·0$
c Five modes, $k_0a = 18·2$

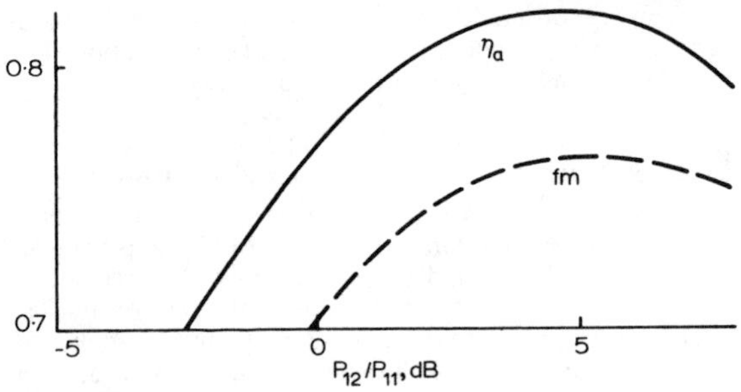

Fig. 6

Effect of P_{12}/P_{11} for a 2-mode feed

—— effect on η_a, with $k_0a = 7·4$ for maximum η_a
- - - - effect on f.m. ($T_R = 100$K), with $k_0a = 8·0$ for maximum f.m.

5.2 Optimum figure of merit

When low-noise receivers are employed, the parameter to be optimised is the figure of merit, which takes into account spillover as well as aperture efficiency. The figure of merit can be defined[16] as

$$\text{f.m.} = \frac{T_R \eta_a}{T}$$

where

T_R = receiver noise temperature, including components produced by losses in the transmission line connecting the feed to the receiver and the sky background noise, as seen in the main beam

T = total system noise temperature
$= T_R + (1 - \eta_s)T_S$

T_S = effective temperature of background as seen by the feed in the spillover region, taken as 150K (this figure assumes that half the spillover energy sees the ground at approximately 300K, and the other half sees the sky at 0K)[16]

Table 2

OPTIMUM FIGURE OF MERIT, $\Psi = 63°$

Number of modes	T_R	ka	Relative mode powers		f.m.	η_a	η_s
			P_1	P_2			
1	100	4·2	1·0		0·64	0·69	0·95
	50	4·5	1·0		0·60	0·66	0·97
2	100	8·0	1·0	3·1	0·765	0·80	0·97
	50	8·2	1·0	3·1	0·74	0·78	0·98

Once the parameters for optimum η_a have been established as above, a trial-and-error method can then be used to determine the optimum f.m. This is done by increasing the waveguide radius to reduce spillover and then computing the resultant radiation patterns for a range of mode powers. This is repeated for a range of waveguide diameters until the best result is obtained.

Table 2 gives the parameters and performances of 1- and 2-hybrid-mode feeds adjusted for maximum figure of merit, with $T_R = 100$K and 50K. The optimum ratio of 3·1 (5dB) of HE_{12} to HE_{11} mode powers for the 2-mode feed is the same for both receiver temperatures, and reference to Table 1 shows that the same power ratio is obtained when η_a is optimised. As with η_a, the power ratio can vary over a wide range without significantly affecting the f.m. (see Fig. 6). The effect of waveguide diameter on the f.m. is also not critical. For example, when $T_R = 100$K, the f.m. remains within 1% of the maximum value at $k_0a = 8{\cdot}0$ for ka between 7·7 and 8·2, and within 2% for ka between 7·5 and 8·4.

Fig. 7 shows the theoretical radiation pattern of the 2-mode feed adjusted for optimum f.m. ($T_R = 100$K), and when the

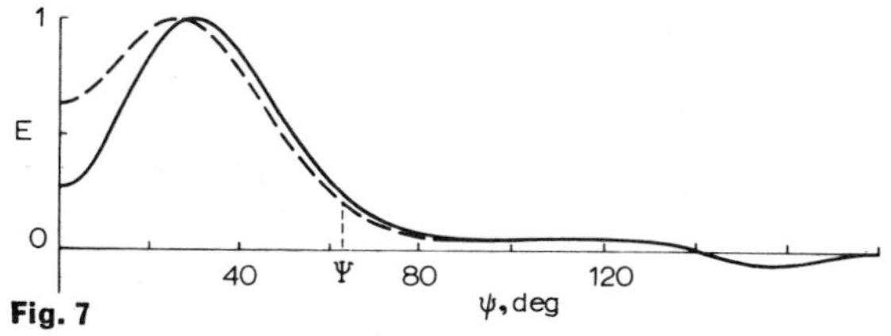

Fig. 7

Radiation pattern of a 2-mode feed designed for optimum f.m. ($T_R = 100K$) with $k_0a = 8{\cdot}0$

—— $P_{12}/P_{11} = 5$dB
---- $P_{12}/P_{11} = 2{\cdot}5$dB

mode-power ratio is reduced from 5 to 2·5dB. The Figure also shows that the edge illumination at $\psi = \Psi$, and hence the spillover, is less than that for the pattern giving optimum η_a in Fig. 5*a*.

5.3 Bandwidth properties

The effect of frequency on the performance of 1- and 2-hybrid-mode feeds intended for the illumination of a paraboloid with $\Psi = 63°$ will now be considered.

As the frequency of a circumferentially corrugated waveguide is reduced below slot resonance, where $X_z = \infty$, X_z becomes inductive and the mode-content factor γ less than 1. Above slot resonance, X_z is capacitive, and γ is greater than 1. The change in γ with frequency is least for a slotwidth/pitch ratio, δ, close to unity, and the change for a given fractional bandwidth increases as the waveguide diameter approaches the cutoff value. This is illustrated in Table 3, which shows how γ varies with frequency in 1- and 2-mode feeds with δ equal to 0·75 and 0·5. The relative half bandwidth on either side of the 'centre' frequency, f_0, is given by $\Delta f/f_0$, and at the band edges the performance as specified by either f.m. or η_a is identical.

The radiation patterns of the HE_{11}-mode feed designed for an optimum figure of merit, with $T_R = 100$K so that $k_0a = 4{\cdot}2$ at the slot-resonant or 'centre' frequency, are shown in Fig. 8 for $ka = 3{\cdot}4, 4{\cdot}2$ and $4{\cdot}8$, and $\delta = 0{\cdot}75$. At $ka = 3{\cdot}4$ and $4{\cdot}8$, the figure of merit has decreased by approximately 8% of the maximum value at $k_0a = 4{\cdot}2$ (see Fig. 9). The crosspolarisation efficiencies at $ka = 3{\cdot}4$ and $4{\cdot}8$ are 0·995 and 0·998, respectively. The edges of this band are indicated by dots in Fig. 9, the bandwidth being 34%.

Pattern asymmetry of a 2-hybrid-mode feed at a frequency away from slot resonance is mainly introduced by the HE_{12} mode, which is closest to cutoff, whereas the departure of γ from unity, and hence the pattern asymmetry of the HE_{11} mode, are small over the bandwidths obtainable (see Table 3).

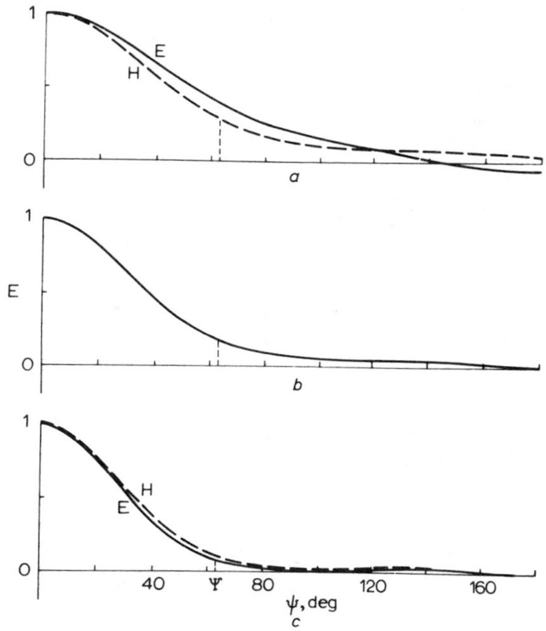

Fig. 8

Effect of $k(2\pi/\lambda)$ on the HE_{11}-mode feed patterns with $k_0a = 4{\cdot}2$ and $\delta = 0{\cdot}75$

(*a*) $ka = 3{\cdot}4$
(*b*) $k_0a = 4{\cdot}2$
(*c*) $ka = 4{\cdot}8$

However, the variations in the efficiency and figure of merit as a function of frequency depend not only on the extent to which ka deviates from the optimum value and γ departs from unity across the band, but also on the variation in the phase difference with frequency between the HE_{11} and HE_{12} modes at the feed aperture. In practice, the two modes are excited in phase at a step discontinuity, and it is necessary to provide a length of corrugated waveguide to bring the two modes back into phase at the aperture. The length of guide, L, required can be calculated from

$$\Delta\Phi = kL\left(\frac{k_z^{11}}{k} - \frac{k_z^{12}}{k}\right)$$

where k_z^{11}/k and k_z^{12}/k are the normalised axial propagation constants of the HE_{11} and HE_{12} modes, respectively, and $\Delta\Phi$ is the required phase difference between the two modes to be produced by L. Fig. 2 shows that, as the waveguide

Table 3

EFFECT OF FREQUENCY ON HYBRID-MODE PARAMETERS

Number of modes	k_0a	L_0/λ_0	δ	Lower band edge					Upper band edge				
				ka	$\Delta f/f_0$	γ_{11}	γ_{12}	$\Delta\Phi$	ka	$\Delta f/f_0$	$1/\gamma_{12}$	$1/\gamma_{11}$	$\Delta\Phi$
1	4·2		0·75 0·5	3·4	19%	0·54 0·36			4·8	15%	0·80 0·72		
2	8·0	4·45	0·75 0·5	7·6	5·0%	0·96 0·94	0·74 0·63	36° 41°	8·5	6·2%	0·95 0·92	0·72 0·61	−33° −35°
3	7·4	3·7	0·75 0·5	7·05	4·7%	0·95 0·93	0·70 0·59	37° 44°	7·88	6·5%	0·94 0·92	0·68 0·57	−38° −39°

radius a is reduced, the difference between the two propagation constants at slot resonance increases, so that the length L_0 necessary to bring the two modes into phase at the aperture decreases.

For a given waveguide radius and length, a change in frequency will cause $(k_z^{11}/k) - (k_z^{12}/k)$, and hence $\Delta\Phi$, to change. The change in the propagation constants is greater than that given by the difference between the curves for the HE_{11} and HE_{12} modes in Fig. 2, since γ departs from unity on both sides of slot resonance. To minimise the change, δ should be close to unity. Since $(k_z^{11}/k) - k_z^{12}/k)$ changes very rapidly near cutoff, the phase variation with frequency is greatest in this region, despite the shorter length of guide necessary to bring the two modes into phase at the centre frequency. For centre frequencies well above cutoff, however, the phase change over a given percentage bandwidth decreases only slowly as k_0a increases.

These properties are illustrated in Table 3, which shows that the 2-mode feed with k_0a equal to 8·0 has a length of $4{\cdot}45\lambda_0$ if the modes are excited in phase at the step, and that the phase shifts between the two modes at the aperture are approximately $\pm 34°$ at $ka = 7{\cdot}6$ and 8·5 and when $\delta = 0{\cdot}75$. When k_0a equals 7·4, the waveguide length is reduced to $3{\cdot}7\lambda_0$ and the phase shift at the band edges for the same percentage bandwidth increases by approximatey 3°. Note that the band edges are not situated symmetrically about the centre frequency, because the difference between the propagation constants, and hence the phase, changes more rapidly at the lower frequencies.

Fig. 10 shows the magnitude and phase of the radiation patterns of the feed with $k_0a = 8{\cdot}0$, at frequencies corresponding to $ka = 7{\cdot}6$ and 8·5. If the relative power of the HE_{12} mode is reduced from 5 to 2·5dB, the relative phase at $\psi = 0°$ decreases from 45° to 32°, and the magnitude of the dip in the radiation pattern decreases.

Fig. 11 shows how the frequency affects both η_a and f.m. when the power in the HE_{12} mode relative to the HE_{11} mode has been optimised (5 dB) to produce the maximum figure of merit for $T_R = 100$ K at $k_0a = 8{\cdot}0$ (curve fm_1). The change in the figure of merit produced by reducing the HE_{12} mode power by 2·5 dB is shown by curve fm_2. Fig. 11 also shows the effect of frequency for the waveguide size giving maximum

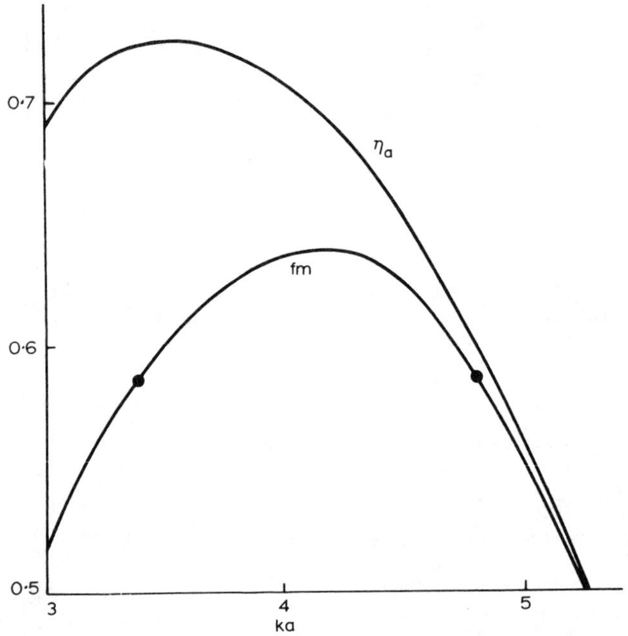

Fig. 9
Effect of k on η_a and f.m. for the HE_{11}-mode feed with $k_0a = 4{\cdot}2$ and $\delta = 0{\cdot}75$

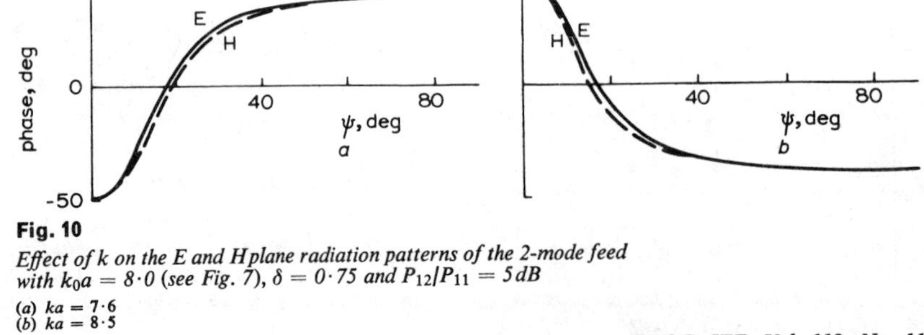

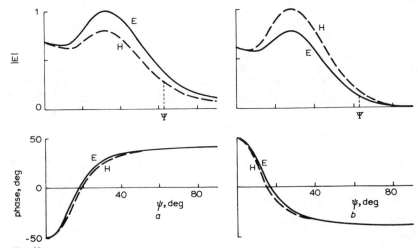

Fig. 10
Effect of k on the E and H plane radiation patterns of the 2-mode feed with $k_0a = 8{\cdot}0$ (see Fig. 7), $\delta = 0{\cdot}75$ and $P_{12}/P_{11} = 5$ dB

(a) $ka = 7{\cdot}6$
(b) $ka = 8{\cdot}5$

η_a ($k_0a = 7\cdot4$), where the relative HE_{12}mode power is optimum at 5 dB (curve η_{a1}), and also when this power ratio is reduced by 2·5 dB (curve η_{a2}). At the band edges, indicated by dots, the reduction in η_a for the feed $k_0a = 7\cdot4$, and in f.m. when $k_0a = 8\cdot0$, is 11%. The bandwidth is 11%. Fig. 11 also shows that the value of η_{a1} at $k_0a = 8\cdot0$ has only decreased by 2½% below the optimum value at $k_0a = 7\cdot4$.

mode powers and efficiencies obtained by the method of radiation-pattern synthesis agree closely with those obtained by matching the waveguide fields to the focal-plane fields; in addition, the values of the efficiencies approach the maximum values computed from the energy flow through an aperture of radius *a* centred on the focus. The figure of merit for the 1- and 2-hybrid-mode feeds can be optimised by increasing

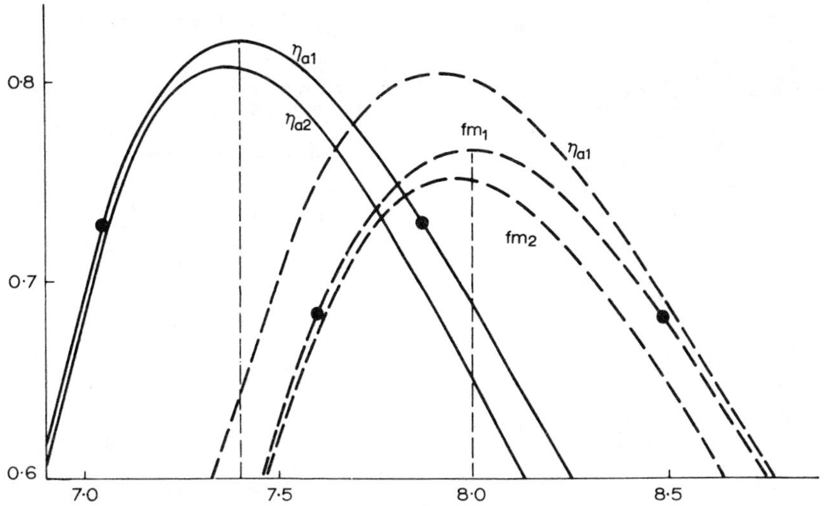

Fig. 11
Effect of k on the performance of the 2-mode feeds having $\delta = 0\cdot75$
—— designed for optimum $\eta_a(k_0a = 7\cdot4)$
- - - - designed for optimum f.m. ($k_0a = 8\cdot0$)
Subscript 1 signifies $P_{12}/P_{11} = 5$ dB; subscript 2 signifies $P_{12}/P_{11} = 2\cdot5$ dB

6 Conclusions

The theoretical expression for the radiation pattern of the undisturbed fields arriving at the open end of a circular waveguide indicates that a linearly polarised, circularly symmetric pattern can be achieved by using balanced HE_1 modes for which the mode-content factor γ is equal to 1. Although approximate pattern symmetry and linear polarisation can be obtained using TE_m and TM_m modes in a smooth metal pipe, exact symmetry is only achievable with balanced hybrid-mode fields.

The anisotropic reactive surface necessary to support balanced hybrid modes will also suport EH_1 modes with $\gamma = -1$, which can introduce both pattern asymmetry and crosspolarisation. Consequently, when linear polarisation is desired, the method used to excite the desired HE_1 modes must discriminate against the excitation of the unwanted EH_1 modes.

A corrugated waveguide consisting of circumferential slots appears to be the most practical type of structure for supporting balanced hybrid modes. Although the highest sidelobe in this case occurs in the axial direction, the sidelobes of two consecutive modes tend to cancel when they are in phase at the feed aperture. The radiation pattern of the balanced hybrid mode in a waveguide consisting of longitudinal slots has a null in the axial direction, and, in addition, the TE and TM components of the hybrid mode can propagate independently. Consequently this type of structure does not appear to be suitable for use as a feed.

The radiation pattern of a balanced hybrid mode has a main beam which radiates at an angle given by $\cos\psi_0 \simeq k_z/k$, where ψ_0 increases with the mode order. Consequently, each mode illuminates an annular ring of the paraboloidal reflector, so that any desired aperture distribution may be synthetised by suitably combining modes. Only half the number of distinct modes are required compared with the method of generating separate TM_{1n} and TE_{1n}modes in a smooth pipe.

The aperture efficiency of a paraboloid with a semiangle Ψ of 63° has been optimised using up to five balanced hybrid modes in a circumferentially corrugated waveguide. The the waveguide diameter, and for the 2-mode feed it is not necessary to change the mode-power ratio to achieve this (see Table 2). In addition, both the optimum values of the aperture efficiency and the figure of merit for a given receiver temperature are relatively insensitive to the mode-power ratio within fairly wide limits.

The maximum bandwidth is achieved by making the ratio of slotwidth to pitch of the circumferentially corrugated waveguide close to unity. Not only does this ensure the least change in γ with frequency, but it also minimises the phase shift with frequency between the two hybrid modes in a 2-mode feed. A single-hybrid-mode feed designed for a paraboloid with $\Psi = 63°$ and a receiver temperature $T_R = 100$ K, has a bandwidth of 34% for a reduction of 8% in the figure of merit at the band edges. The pattern asymmetry at these frequencies introduces negligible crosspolarisation. The bandwidth of the 2-mode feed is limited by the phase difference which exists between the two modes at the aperture of the feed. For feeds designed for maximum efficiency and also maximum figure of merit with $T_R = 100$ K, the bandwidth for a 11% reduction in the maximum values of these factors at the band edges is 11% in both cases.

Cylindrical hybrid-mode corrugated waveguides not only have application as prime-focus feeds for paraboloids, but also as feeds for spherical reflectors. Phillips and Clarricoats[17] have considered theoretically a spherical reflector with a Gregorian subreflector fed by an HE_{11}mode radiating from a cylindrical waveguide. Another approach to the efficient illumination of spherical reflectors is to use a multimode feed in the focal region where the amplitudes and phases of the hybrid modes are adjusted to obtain the best field match.[4]

7 Acknowledgments

The author desires to acknowledge many valuable discussions with H. C. Minnett, and also the assistance of D. N. Cooper in making available the computer programs for calculating the theoretical radiation patterns.

8 References

1 MINNETT, H. C., and THOMAS, B. MACA.: 'Fields in the image space of symmetrical focusing reflectors', *Proc. IEE*, 1968, **115**, (10), pp. 1419–1430
2 THOMAS, B. MACA., MINNETT, H. C., and VU, T. B.: 'Fields in the focal region of a spherical reflector', *IEEE Trans.*, 1969, **AP–17**, pp. 229–232
3 MINNETT, H. C., and THOMAS, B. MACA.: 'A method of synthesizing radiation patterns with axial symmetry', *ibid.*, 1966, **AP–14**, pp. 654–656
4 THOMAS, B. MACA.: 'Matching focal-region fields with hybrid modes', *ibid.*, 1970, **AP–18**, pp. 404–405
5 VU, T. B.: 'Optimisation of efficiency of deep reflectors', *ibid.*, 1969, **AP–17**, pp. 811–813
6 THOMAS, B. MACA.: 'Prime-focus one- and two-hybrid-mode feeds', *Electron. Lett.*, 1970, **6**, pp. 460–461
7 VU, T. B., and VU, Q. H.: 'Optimum feed for large radio telescopes: experimental results', *ibid.*, 1970, **6**, pp. 159–160
8 LAMONT, H. R. L.: 'Waveguides' (Methuen, 1953), chap. 4
9 SILVER, S.: 'Microwave antenna theory and design' (McGraw–Hill, 1949), chap. 10
10 POTTER, P. D.: 'A new horn with suppressed sidelobes and equal beamwidths', *Microwave J.*, 1963, **6**, (6), pp. 71–78
11 TURRIN, R. H.: 'Dual mode small-aperture antennas', *IEEE Trans.* 1967, **AP–15**, pp. 307–308
12 RUZE, J.: 'Circular aperture synthesis', *ibid.*, 1964, **AP–12**, pp. 691–694
13 LUDWIG, A. C.: 'Radiation pattern synthesis for circular-aperture horn antennas', *ibid.*, 1966, **AP–14**, pp. 434–440
14 LUDWIG, A. C.: 'Antenna feed efficiency' *in* JPL space programs summary IV, 1963, pp. 200–203
15 COOPER, D. N.: 'Complex propagation constants and the step discontinuity in corrugated cylindrical waveguide', *Electron. Lett.*, 1971, **7**, pp. 135–136
16 HERBISON-EVANS, D.: 'Optimum paraboloid aerial and feed design', *Proc. IEE*, 1968, **115**, (1), pp. 87–90
17 PHILLIPS, C. J. E., and CLARRICOATS, P. J. B.: 'Optimum design of a Gregorian-corrected spherical-reflector antenna', *ibid.*, 1970, **117**, (4), pp. 718–734
18 HARRINGTON, R. F.: 'Time-harmonic electromagnetic fields' (McGraw–Hill, 1961), chap. 3

9 Appendix

9.1 Radiation from a circular aperture

The far-field radiation components are determined from the magnetic and electric vector potentials of the equivalent electric and magnetic currents, respectively, in the aperture of the circular waveguide.

The aperture H and E fields are first expressed in terms of these equivalent electric and magnetic currents

$$J = n \times H$$

$$M = -n \times E$$

where the E and H fields are convenently written in the form shown in eqn. 1*a*. The cylindrical components of the equivalent aperture currents are then given by

$$J_\rho = -H_\xi(u) \sin m\xi \qquad M_\rho = E_\xi(u) \cos m\xi$$

$$J_\xi = H_\rho(u) \cos m\xi \qquad M_\xi = -E_\rho(u) \sin m\xi$$

To determine the magnetic and electric vector potentials, it is necessary to resolve these currents into an orthogonal set of Cartesian components, as follows:

$$J_x = J_\rho \cos \xi - J_\xi \sin \xi$$

$$J_y = J_\rho \sin \xi + J_\xi \cos \xi$$

A similar set of equations holds for M_x and M_y.

The field at a distant point P is determined from the magnetic and electric vector potentials A and F, respectively. For the magnetic vector potentials[18]

$$A_x = \frac{e^{-jkR}}{4\pi R} \int_S J_x e^{jk(\rho . R_1)} dS$$

$$A_y = \frac{e^{-jkR}}{4\pi R} \int J_y e^{jk(\rho . R_1)} dS$$

A similar set of equations exist for F_x and F_y in terms of M_x and M_y. The unit vector R_1 is directed from the element of area dS in the aperture to the distant point P. The phase of the field at the element with respect to the centre of the aperture is

$$k(\rho . R_1) = k\rho \sin \psi \cos (\phi - \xi)$$

It is usual to express the far field in spherical co-ordinates,[18] as

$$E_\psi = -jk(Z_0 A_\psi + F_\phi) \qquad H_\phi = \frac{E_\psi}{Z_0}$$

$$E_\phi = -jk(Z_0 A_\phi - F_\psi) \qquad H_\psi = -\frac{E_\phi}{Z_0}$$

where $A_\psi = (A_x \cos \phi + A_y \sin \phi) \cos \psi$

$$A_\phi = -A_x \sin \phi + A_y \cos \phi$$

and similarly for F_ψ and F_ϕ.

By making the appropriate substitutions, it can readily be shown that the fields at a distant point in terms of the fields existing in an element of aperture are

$$E_\psi = \frac{jk}{4\pi R} e^{-jkR}[\{E_\xi(u) - Z_0 H_\rho(u) \cos \psi\} \cos m\xi \sin(\phi - \xi) + \{E_\rho(u) + Z_0 H_\xi(u) \cos \psi\} \sin m\xi \cos (\phi - \xi)] e^{jk\rho \sin \psi \cos(\phi - \xi)} \rho \, d\rho \, d\xi$$

$$E_\phi = \frac{jk}{4\pi R} e^{-jkR}[- \{E_\rho(u) \cos \psi + Z_0 H_\xi(u)\} \sin m\xi \sin (\phi - \xi) + \{E_\xi(u) \cos \psi - Z_0 H_\rho(u)\} \cos m\xi \cos (\phi - \xi)\} e^{jk\rho \sin \psi \cos (\phi - \xi)} \rho \, d\rho \, d\xi$$

Integration with respect to ξ gives the distant field of an aperture consisting of an annular ring of width $d\rho$ and radius ρ. The result, which is given in eqns. 4 and 5, shows that the variations in field strength with respect to ϕ and ψ can be written as a product.

9.2 Paraboloid feed efficiency

The aperture efficiency of a paraboloidal reflector with no blockage or surface errors and fed by a cylindrical waveguide having fields with one circumferential variation has been considered by Ludwig.[14] The aperture efficiency η_a is dependent on the phase and amplitude distributions across the reflector aperture, as well as on the energy lost by the feed in the spillover region and in the crosspolarised direction. These factors are taken account of by the following efficiencies: η_p (phase loss), η_i (illumination), η_s (spillover) and η_x (crosspolarisation), respectively, and

$$\eta_a = \eta_s \eta_x \eta_p \eta_i$$

These factors are defined as follows:

$$\eta_s = \frac{\int_0^\Psi \{|A(\psi)|^2 + |B(\psi)|^2\} \sin \psi d\psi}{\int_0^\pi \{|A(\psi)|^2 + |B(\psi)|^2\} \sin \psi d\psi}$$

$$\eta_x = \frac{\int_0^\Psi \{|A(\psi)| + |B(\psi)\}^2 \sin \psi \, d\psi}{2\int_0^\Psi \{|A(\psi)|^2 + |B(\psi)|^2\} \sin \psi \, d\psi}$$

$$\eta_p = \frac{\left| \int_0^\Psi \{A(\psi) + B(\psi)\} \tan \frac{\psi}{2} d\psi \right|^2}{\left[\int_0^\Psi \{|A(\psi)| + |B(\psi)|\} \tan \frac{\psi}{2} d\psi \right]^2}$$

$$\eta_i = 2 \cot^2 \frac{\Psi}{2} \frac{\left[\int_0^\Psi \{|A(\psi) 1 + |B(\psi)|\} \tan \frac{\psi}{2} d\psi \right]^2}{\int_0^\Psi \{|A(\psi)| + |B(\psi)|\}^2 \sin \psi \, d\psi}$$

$$\eta_a = \cot^2 \frac{\Psi}{2} \frac{\left| \int_0^{\Psi} \{A(\psi) + B(\psi)\} \tan \frac{\psi}{2} d\psi \right|^2}{\int_0^{\pi} \{|A(\psi)|^2 + |B(\psi)|^2\} \sin \psi d\psi}$$

where $A(\psi) = |A(\psi)| e^{j\Phi_A(\psi)}$

$B(\psi) = |B(\psi)| e^{j\Phi_B(\psi)}$

are the E and H plane radiation patterns.

If all modes in the cylindrical waveguide are in phase so that $\Phi_A(\psi) = \Phi_B(\psi) = 0$, then $\eta_p = 1 \cdot 0$. Also, for the balanced hybrid modes where $A(\psi) = B(\psi)$, $\eta_x = 1 \cdot 0$. The efficiency equations are then reduced to

$$\eta_s = \frac{\int_0^{\Psi} A^2(\psi) \sin \psi d\psi}{\int_0^{\pi} A^2(\psi) \sin \psi d\psi}$$

$$\eta_i = 2 \cot^2 \frac{\Psi}{2} \frac{\left\{ \int_0^{\Psi} A(\psi) \tan \frac{\psi}{2} d\psi \right\}^2}{\int_0^{\Psi} A^2(\psi) \sin \psi d\psi}$$

$$\eta_a = \eta_i \eta_s$$

PARABOLOIDAL-REFLECTOR ILLUMINATION WITH CONICAL SCALAR HORNS

Indexing terms: Antenna radiation patterns, Reflector antennas, Antenna feeders

This letter describes a systematic procedure for calculating the half flare angle of a conical scalar horn, illuminating a paraboloidal reflector of known f/D ratio, with a view to realising a prescribed sidelobe level for the secondary radiation.

Recently, the conical scalar horn (otherwise called the corrugated conical horn) has been the subject of study for many workers because of its attractive features, such as a low-noise feed. This letter described a systematic procedure for calculating the half flare angle α_0 of the scalar feed, to illuminate a paraboloidal reflector of known f/D ratio (Fig. 1A) so as to realise a prescribed sidelobe level for the secondary radiation.

Computation of the primary-feed requirements for illuminating a paraboloidal dish with a given f/D ratio to obtain a desired sidelobe level* is well known.[1] The formulas

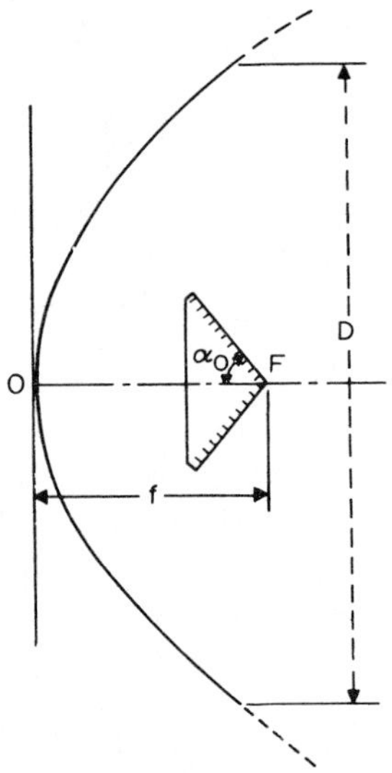

Fig. 1A *Geometry of paraboloid*

and other data necessary for the above computation are based on experimental investigations and a few on analytical results supported by measured data,[2,3] which implies that the design objective can be realised in practice to a good degree of accuracy.

The halfpower beamwidth (b.w.) of the primary feed for a prescribed f/D ratio and sidelobe radiation level of the paraboloid can be easily computed, following the procedure outlined by Blake.[1] When the scalar feed is employed as the primary feed, it is necessary to derive a formula relating the b.w. to α_0, so that the sidelobe-radiation level of the reflector can be controlled by varying α_0. To do this, an analytically simple asymptotic solution for the $\bar{E}$ and $\bar{H}$ fields over the aperture scalar horn of long axial length, described by one of the authors in previous letters,[4,5] is employed. Since the above-mentioned asymptotic solution forms the basis for the design calculations to be described, it was felt necessary to compare the magnitudes of the tangential electric fields corresponding to the exact[6] and simpler solutions over the aperture for the HE_{11} mode when α_0 is large ($0 \leqslant \alpha_0 \leqslant 90°$). Calculated values of $|\bar{E}_t(\theta)|/|\bar{E}_t(0)|$ corresponding to the exact[7] and asymptotic solutions were compared for a horn with $\alpha_0 = 90°$. The comparison revealed that the simpler solution for $\bar{E}_t$ is accurate enough for all practical purposes.† Therefore, when the feed supports the HE_{11} mode, the aperture field is given by

$$\bar{E}_t = -(a_{11}/r)\, Z_0\, B_n(kr)\, J_0\left(\frac{2\cdot405}{\alpha_0}\theta\right) e^{j\phi}(\bar{a}_\theta + j\bar{a}_\phi) \quad . \quad . \quad (1)$$

where the notation used by one of the authors in a previous letter[5] has been used. From eqn. 1, one obtains the following expression for the -3 dB level of $|\bar{E}_t|$ over the aperture:

$$\theta_1 = 0\cdot4685\alpha_0 \quad . \quad . \quad . \quad . \quad . \quad . \quad . \quad . \quad . \quad . \quad (2)$$

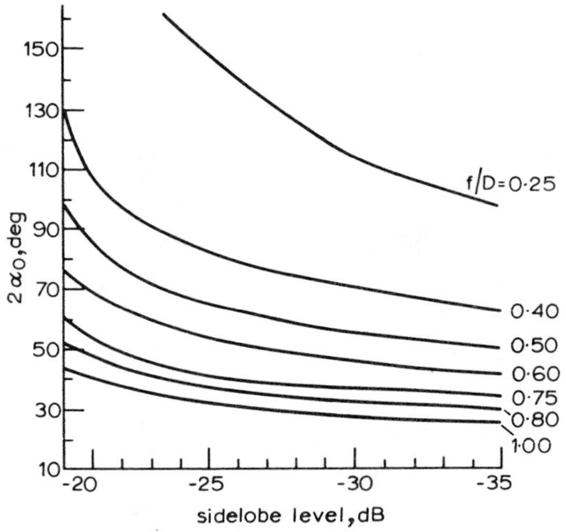

Fig. 2 *Influence of α_0 and f/D ratio on sidelobe level*

In eqn. 2, θ_1 is the polar angle (measured from the axis of symmetry FO in Fig. 1A) at which $|\bar{E}_t(\theta_1)| = (1/\sqrt{2})|\bar{E}_t(0)|$.

It has been pointed out by Clarricoats and Saha[8] that the aperture field of a scalar horn (for the HE_{11} mode) for small values of θ resembles the Gaussian function. Hence the diffracted far field should also resemble the aperture field $\bar{E}_t(\theta)$ (at least up to the -3 dB level), because of the well known transform properties of the Gaussian function. (It is also possible to explain this behaviour physically from a ray-optical point of view.) Hence the halfpower beamwidth of a scalar feed supporting the HE_{11} mode is approximately given by

$$\text{b.w.} = 2\theta_1 = 0\cdot9370\alpha_0 \quad . \quad . \quad . \quad . \quad . \quad . \quad . \quad . \quad (3)$$

Computed values of b.w., using eqn. 3, have been compared with the measured values [9,10] in Fig. 1B for several values of α_0. The comparison justifies the validity of eqn. 3.

Following the analytical procedure outlined by Blake,[1] and using the available data relating to the taper and sidelobes for circular apertures,[2,3] the value of α_0 necessary for achieving a prescribed sidelobe radiation for the paraboloid is calculated with the aid of eqn. 3, using the f/D ratio as a parameter. The calculated results are presented in a graphical form in Fig. 2, which may be used as a guide in the design of scalar feeds for paraboloidal-reflector illumination.

* The sidelobe level in question corresponds to the first lobe of the secondary radiation

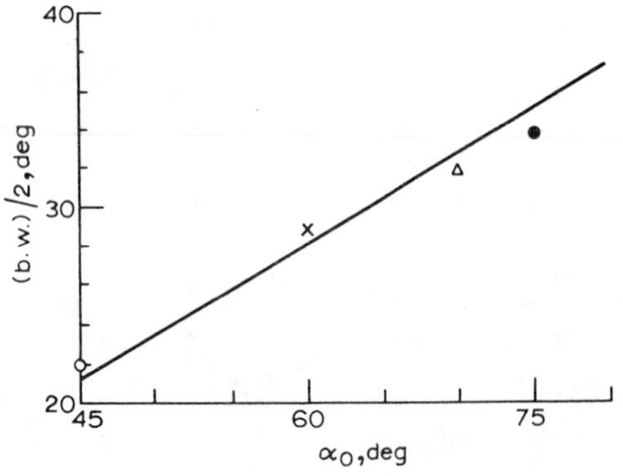

Fig. 1B *Half-power beamwidths of scalar horns*

——— calculated
○○○ $\alpha_0 = 45°$ $kr_0 = 39\cdot8$ (Jeuken)
××× $\alpha_0 = 60°$ $kr_0 = 17\cdot8$ (Booker)
△△△ $\alpha_0 = 70°$ $kr_0 = 17\cdot8$ (Booker)
●●● $\alpha_0 = 75°$ $kr_0 = 14\cdot9$ (Jeuken)

† A similar comparison was made for values of $\alpha_0 > 90°$ and a good correlation between the two solutions was noted

Reprinted with permission from *Electron. Lett.*, vol. 8, pp. 111-112, Mar. 9, 1972.

The authors are thankful to Mr. Govindarajan, of the Department of Electrical & Electronic Engineering, Birla Institute of Technology & Science, Pilani, for his help in carrying out the numerical computations involved in this work.

M. S. NARASIMHAN
Y. B. MALLA

Department of Electrical Engineering
Indian Institute of Technology
PO IIT, Madras-36, India

9th February 1972

References

1 BLAKE, L. V.: 'Antennas' (Wiley, 1966), chap. 6
2 ADAMS, R. J., and KELLEHER, K. S.: 'Pattern calculation for antennas of elliptical aperture', *Proc. Inst. Radio Eng.*, 1950, **38**, p. 1052
3 JASIK, H.: 'Antenna engineering handbook' (McGraw-Hill, 1961), chap. 12
4 NARASIMHAN, M. S., and RAO, B. V.: 'Hybrid modes in corrugated conical horns', *Electron. Lett.*, 1970, **6**, pp. 32–34
5 NARASIMHAN, M. S., and RAO, B. V.: 'Diffraction by wide-flare-angle corrugated conical horns', *ibid.*, 1970, **6**, pp. 469–471
6 CLARRICOATS, P. J. B.: 'Analysis of spherical hybrid modes in corrugated conical horn', *ibid.*, 1969, **5**, pp. 189–190
7 SAHA, P. K.: 'Tables of associated Legendre functions for scalar feed applications'. Queen Mary College, London, Department of Electrical & Electronic Engineering research report 01/1969
8 CLARRICOATS, P. J. B., and SAHA, P. K.: 'Propagation and radiation behaviour of corrugated feeds', *Proc. IEE*, 1971, **118**, (9), pp. 1177–1186
9 JEUKEN, M. E. J., and LAMBRECHTSE, C. W.: 'Small corrugated conical-horn antenna with wide flare angle', *Electron. Lett.*, 1969, **5**, pp. 489–490
10 BOOKER, D. D., and MCINNES, P. A.: 'Computer-predicted performance of corrugated conical feeds using experimental primary-radiation patterns', *ibid.*, 1970, **6**, pp. 18–20

Co-Polar and Cross-Polar Diffraction Images in the Focal Plane of Paraboloidal Reflectors: A Comparison Between Linear and Circular Polarization

S. I. GHOBRIAL, MEMBER, IEEE

Abstract—Equations for the co-polar and cross-polar fields in the focal plane of paraboloidal reflectors when excited by a plane wave are derived for linear and circular polarization. It is shown that for linear polarization the co-polar diffraction image consists of an elliptically shaped bright region followed by dark and bright zones. On the other hand, the cross-polar image consists essentially of four bright spots separated by a dark cross. These theoretical findings were confirmed by measurements. With circular polarization the co-polar diffraction image comprises a bright central circle followed by dark and bright rings, whereas the cross-polar image consists of a dark central circle followed by bright and dark rings. It is also shown that the field distribution in the image space is affected to a great extent by the angular semi-aperture ψ'. Graphs of the variation of the image characteristics with ψ' are also given.

I. Introduction

RECENT investigations in the study of reflector antennas utilizing two orthogonal polarizations revealed that high polarization discrimination can be achieved if the fields in the focal plane of the reflector when excited by a plane wave are orthogonal to that of the primary feed when operated in the transmit mode [1], [2]. A knowledge of the polarization nature of the field in the focal plane is therefore useful for the design of primary feeds with high cross-polar discrimination. The polarization characteristics of focal plane fields in the case of paraboloid excited by a linearly polarized wave was studied by a number of authors [3]–[5]. However, the case of circular polarization remains to be investigated. In this work equations for co-polar and cross-polar focal plane fields will be derived when a circularly polarized wave is exciting the reflector. It will be shown that while in the case of linear polarization the co-polar diffraction image consists of an almost elliptical bright region followed by dark and bright zones, with circular polarization the co-polar diffraction image consists of a bright circular region followed by dark and bright rings. This distribution is very similar to the well-known Airy distribution. On the other hand, the cross-polar diffraction image consists of a dark circle followed by bright and dark rings. This is very different from the distribution arising from linear polarization where the cross-polar image consists of four bright zones separated by a dark cross [6]. It is also shown that the deviation of the co-polar image from the Airy distribution decreases monotonically with ψ', the angular semi-aperture (in the case of linear polarization). As ψ' approaches zero, the co-polar field distribution approaches the Airy distribution where as the cross-polar image vanishes completely.

Manuscript received September 2, 1975; revised January 31, 1976.
S. I. Ghobrial is with the University of Khartoum, Khartoum, Sudan.

II. Analysis—Focal Plane Fields

A. Linear Polarization

Consider a plane wave linearly polarized along the x-axis incident on the reflector: $\boldsymbol{e}_i = \boldsymbol{i}|e| + \boldsymbol{j}0 + \boldsymbol{k}0$. On reflection the incident vector $\boldsymbol{e}_i$ changes direction and the components of the field in the focal plane can be obtained from the current distribution method as (Appendix I):

$$E_x(r,\zeta) = A + B \cos 2\zeta \tag{1}$$

$$E_y(r,\zeta) = B \sin 2\zeta \tag{2}$$

Reprinted from *IEEE Trans. Antennas Propagat.*, vol. AP-24, pp. 418–424, July 1976.

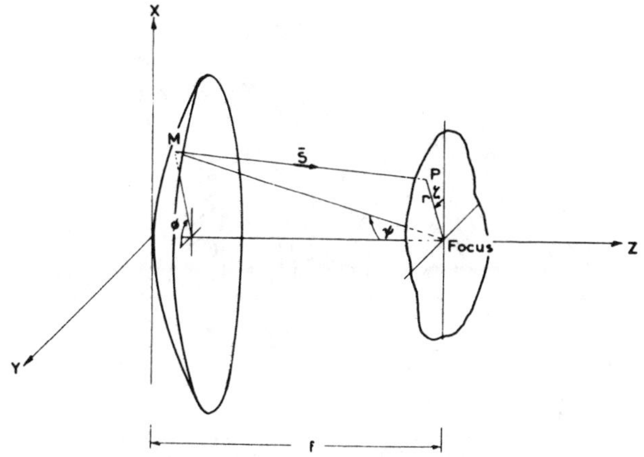

Fig. 1. Coordinate systems.

where

$$A = K \int_0^{\psi'} J_0(\beta \sin \psi) \sin \psi \, d\psi \quad (3)$$

$$B = K \int_0^{\psi'} \frac{1 - \cos \psi}{1 + \cos \psi} J_2(\beta \sin \psi) \sin \psi \, d\psi \quad (4)$$

and

β	$2\pi r/\lambda$;
$J_0(\beta \sin \psi)$ and $J_2(\beta \sin \psi)$	Bessel functions of the first kind;
λ	wavelength of the incident wave;
ψ, r, and ζ	as defined in Fig. 1.

On the other hand, if the incident wave is polarized along the y-axis, i.e., $\boldsymbol{e}_i = \boldsymbol{i}0 + \boldsymbol{j}|e| + \boldsymbol{k}0$, then the following expressions are obtained:

$$E_x(r,\zeta) = B \sin 2\zeta \quad (5)$$

$$E_y(r,\zeta) = A - B \cos 2\zeta. \quad (6)$$

B. Circular Polarization

If the polarization of the incident wave is circular, then we may express this as the sum of two linearly polarized waves in time quadrature. Thus the field incident on the reflector can be expressed symbolically as

$$\boldsymbol{e}_i = |e|(\boldsymbol{i} + \boldsymbol{j}e^{j\pi/2}). \quad (7)$$

The x and y field components in the focal plane can now be obtained by using (1), (2), (5), and (6) in conjunction with (7); thus the following results are obtained:

$$E_x = A + B \cos 2\zeta + jB \sin 2\zeta = |E_x| \exp (j\phi_x) \quad (8)$$

$$E_y = B \sin 2\zeta + j(A - B \cos 2\zeta) = |E_y| \exp (j\phi_y) \quad (9)$$

and the field in the focal plane can be written as

$$|E_x| \exp (j\phi_x)\boldsymbol{i} + |E_y| \exp (j\phi_y)\boldsymbol{j}.$$

From (8) and (9) it is seen that in the focal plane polarization is elliptical, indicating that cross-polarization has taken place as expected. The cross-polar component in this case consists of a circularly polarized wave but with its sense of rotation opposite to that of the incident wave. In order to determine the relative magnitude of the cross-polarized field we need to express the focal plane field as a sum of two circularly polarized waves: one right hand circular and the other left hand circular. Now, rewriting the expression for the field in the focal plane,

$$\boldsymbol{E} = |E_x| \exp (j\phi_x)\boldsymbol{i} + |E_y| \exp (j\phi_y)\boldsymbol{j}$$
$$= |E_x| \exp (j\phi_x)[\boldsymbol{i} + \boldsymbol{j}m \exp (j\phi')] \quad (10)$$

where

$$m = |E_y/E_x| \quad \text{and} \quad \phi' = \phi_y - \phi_x$$

expressing $\boldsymbol{E}$ as the sum of two circularly polarized waves, we have

$$\boldsymbol{E} = c[\boldsymbol{i} + \boldsymbol{j} \exp (j\pi/2)] + d[\boldsymbol{i} - \boldsymbol{j} \exp (j\pi/2)] \quad (11)$$

where c and d are complex quantities whose magnitudes are given by

$$|c| = |E_x/2|(1 + m^2 + 2m \sin \phi')^{1/2} \quad (12a)$$

$$|d| = |E_x/2|(1 + m^2 - 2m \sin \phi')^{1/2}. \quad (12b)$$

In (12) $|c|$ and $|d|$ represent the magnitudes of the co-polar and cross-polar fields, respectively.

From (8) and (9)

$$|E_x| = (A^2 + B^2 + 2AB \cos 2\zeta)^{1/2}$$

$$|E_y| = (A^2 + B^2 - 2AB \cos 2\zeta)^{1/2}.$$

Therefore,

$$m = [(A^2 + B^2 - 2AB \cos 2\zeta)/(A^2 + B^2 + 2AB \cos 2\zeta)]^{1/2}$$

and

$$\phi' = \tan^{-1} \frac{A - B \cos 2\zeta}{B \sin 2\zeta} - \tan^{-1} \frac{B \sin 2\zeta}{A + B \cos 2\zeta}$$

from which

$$\sin \phi' = \frac{A^2 - B^2}{[(A^2 + B^2 - 2AB \cos 2\zeta)(A^2 + B^2 + 2AB \cos 2\zeta)]^{1/2}}.$$

Substituting for $|E_x|$, m, and $\sin \phi'$ in (12) we get

$$|E_{\text{co-pol}}| = |c| = A = \kappa \int_0^{\psi'} J_0(\beta \sin \psi) \sin \psi \, d\psi \quad (13)$$

which quantity is independent of ζ. This clearly shows that the co-polar field distribution in the focal plane is a function of the distance between the point at which the field is to be calculated and the focus only. Therefore, the co-polar diffraction image in the focal plane is circularly symmetrical. Similarly, the cross-polar field distribution is obtained as

$$|E_{\text{cross-pol}}| = |d| = B$$
$$= \kappa \int_0^{\psi'} \frac{1 - \cos \psi}{1 + \cos \psi} J_2(\beta \sin \psi) \sin \psi \, d\psi. \quad (14)$$

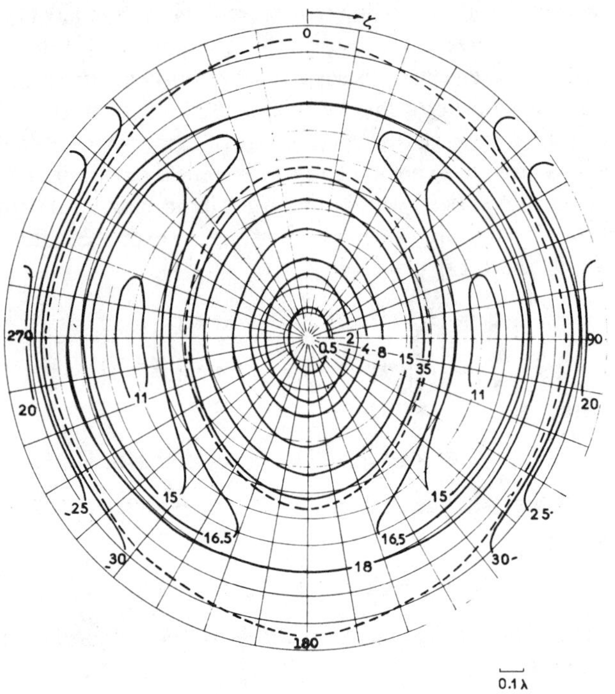

NUMBERS (IN dB) ARE: MATCHED POL. PEAK/MATCHED POL.

Fig. 2. Co-polar contours in the focal plane of an 80° reflector (linear polarization).

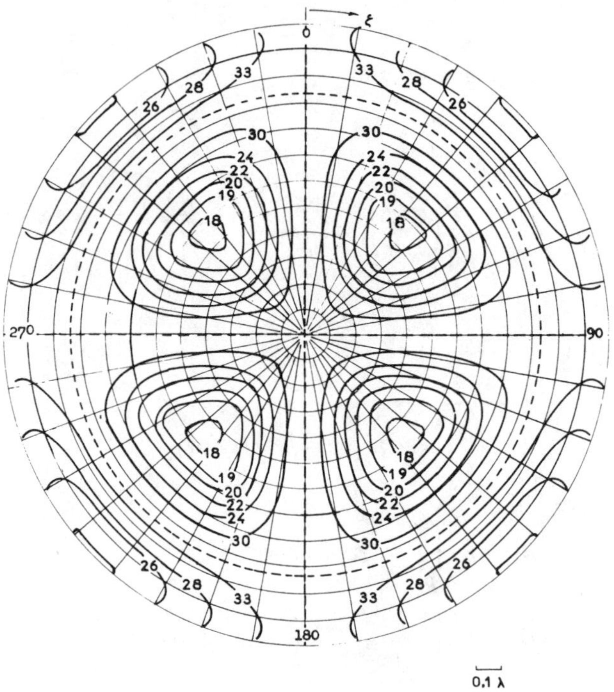

NUMBERS (IN dB) ARE: MATCHED POL. PEAK/CROSS POL.

Fig. 3. Cross-polar contours in the focal plane of an 80° reflector (linear polarization).

This is again independent of ζ and therefore the cross-polar diffraction image is also circularly symmetrical about the focus.

Now comparing (1) and (13), which give the co-polar field distribution for linear and circular polarizations, respectively, it is seen that (1) reduces to (13) if $\zeta = \pm 45°$, $\pm 135°$; i.e., in the focal plane the field distribution is the same for linear and circular polarizations along the radial lines defined by $\zeta = (2n + 1)45°$, $n = 0,1,\cdots$. Also the peak value of the co-polar field is the same for both modes of polarization being $K(1 - \cos \psi')$.

III. Computations

Using (1), (2), (13), and (14) contours of constant co-polar and cross-polar focal plane fields were plotted for both linear and circular polarizations. These are shown in Figs. 2–5. An angular semi-aperture of 80° was assumed in these calculations. The following features are to be noted.

1) The linear polarization co-polar diffraction image consists of an almost elliptically shaped region with its major axis along the direction of the co-polar field. For a constant value of r the variation of the co-polar field with ζ is harmonic superimposed on a constant as suggested by the expression $A + B \cos 2\zeta$.

2) The linear polarization cross-polar diffraction image consists of four bright zones separated by a dark cross. Also the field changes sense whenever the principal axes are crossed; i.e., cross-polar fields are in anti-phase in adjacent quadrants. Moreover, the variation of field intensity with ζ is harmonic and is completely determined by $\cos 2\zeta$ as seen from (2).

3) For circular polarization circular symmetry is maintained in both co-polar and cross-polar images. Also the variation of field intensity with r is identical to that of linear polarization along the radius vector defined by $\zeta = (2n + 1)45°$.

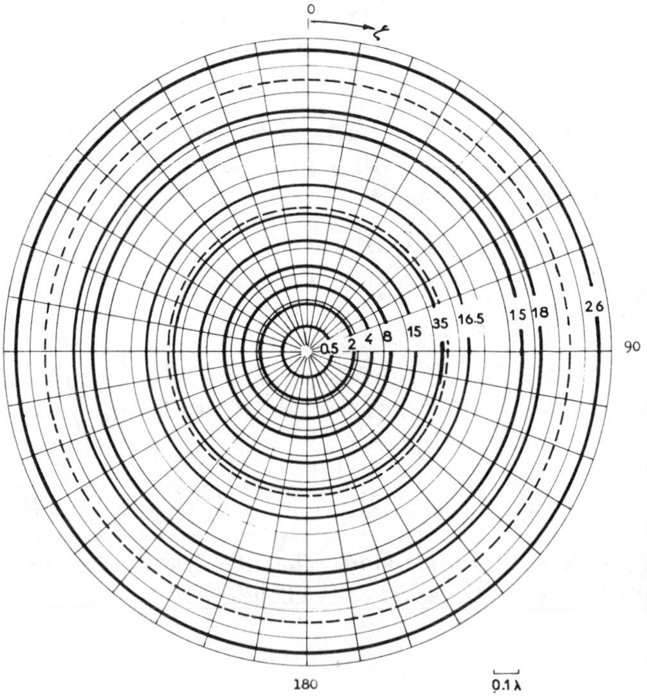

NUMBERS (IN dB) ARE: CO-POLAR PEAK/CO-POLAR.

Fig. 4. Co-polar contours in the focal plane of an 80° reflector (circular polarization).

Figs. 6 and 7 give the variation of co-polar and cross-polar field intensities with r along the lines $\zeta = 0°$, 45°, and 90°. The case of $\zeta = 45°$ is of special interest since it gives the variation of field with r for circular polarization

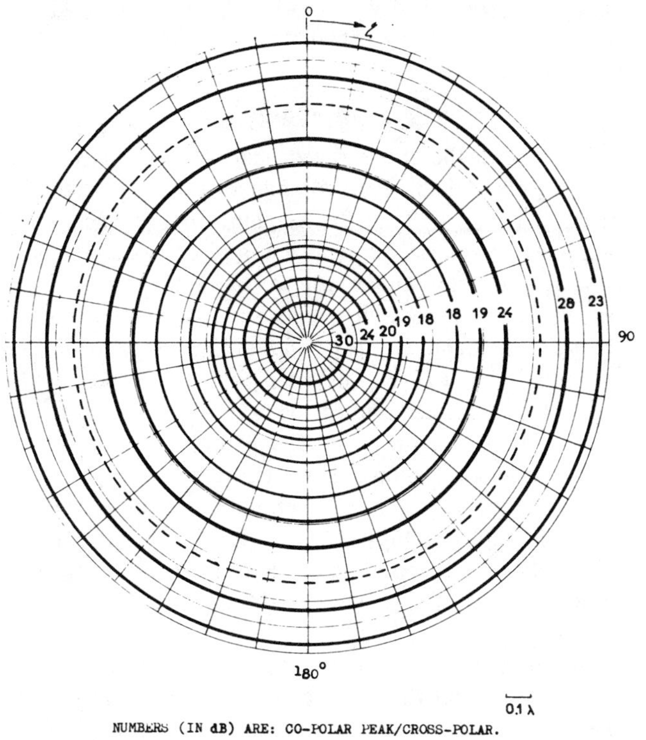

Fig. 5. Cross-polar contours in the focal plane of an 80° reflector (circular polarization).

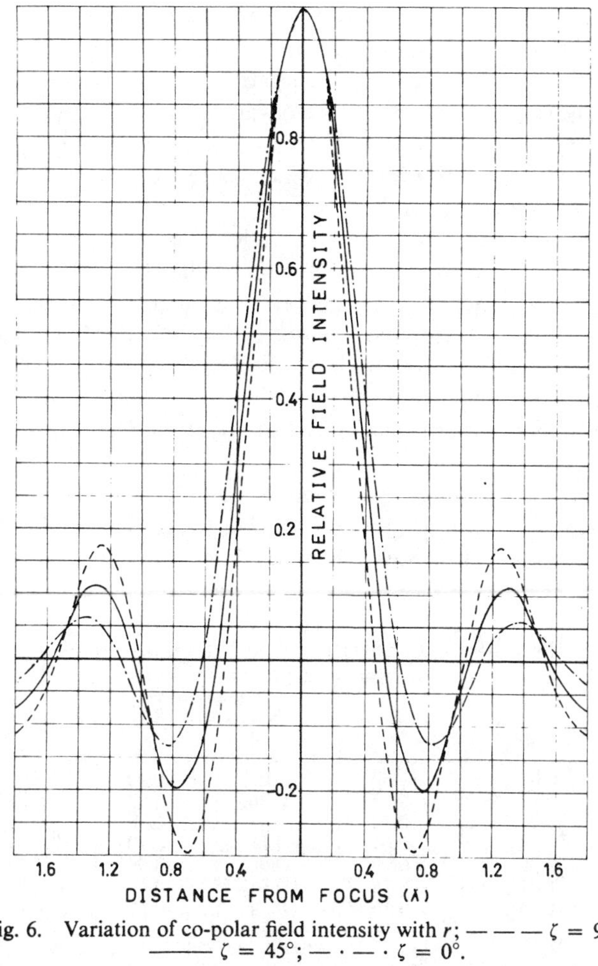

Fig. 6. Variation of co-polar field intensity with r; — — — $\zeta = 90°$; ——— $\zeta = 45°$; — · — · $\zeta = 0°$.

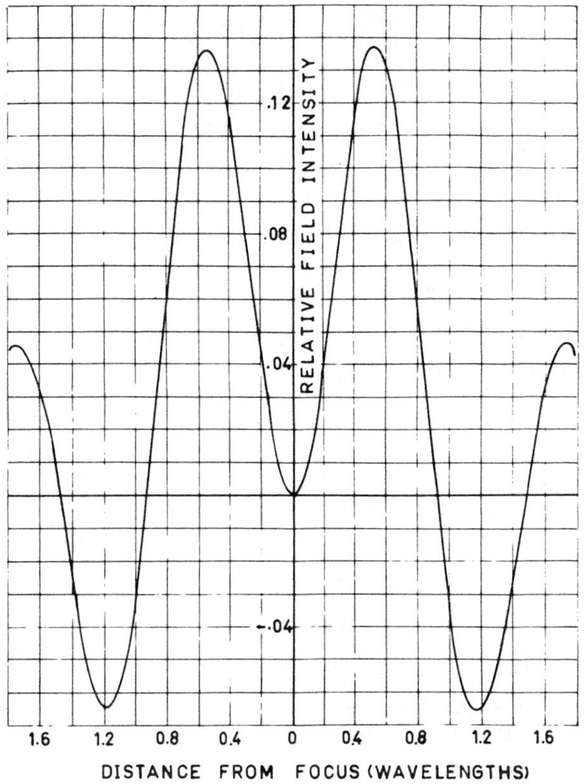

Fig. 7. Cross-polar field intensity variation with r along $\zeta = 45°$

for all ζ. From Fig. 6 it is seen that the three curves intersect at $r = 0.94\lambda$ and $r = 1.48\lambda$. These values of r are the zeros of the function B,

$$B = \kappa \int_0^{\psi'} \frac{1 - \cos \psi}{1 + \cos \psi} J_2(\beta \sin \psi) \sin \psi \, d\psi.$$

Thus for the above mentioned values for r the co-polar image has circular symmetry where as the cross-polar image has nulls.

The cross-polar field variation with r is shown in Fig. 7 for $\zeta = \pm(2n + 1)45°$, $n = 0,1$. Along the radius $\zeta = 0°$ and $\zeta = 90°$ the cross-polar field vanishes as seen from Fig. 3. It is worth noting that Figs. 6 and 7 are true for paraboloids of 80° angular semi-aperture and of any diameter. Changing the diameter affects the absolute value of the field intensity but has no effect on the structure of the diffraction image. However, changing the angular semi-aperture ψ' (this amounts to changing the f/D ratio) affects the diffraction image. This is seen to be true by considering the upper limit of integration of (3) and (4). Figs. (8) and (9) give the variation of some of the characteristics of the co-polar and cross-polar images with ψ'. In Fig. 8 the characteristics of the diffraction image for linear polarization are shown. The solid curves give these characteristics along $\zeta = 45°$. These also represent the characteristics of circular polarization images. It is seen that as ψ' approaches 0°, the curves for $\zeta = 0°$ and $\zeta = 90°$ come closer to those of $\zeta = 45°$. It is also worth noting that as ψ' becomes smaller the position of the first co-polar null moves away from the focus. In fact as ψ' approaches zero the first co-polar null moves indefinitely away from the focus. This is not un-

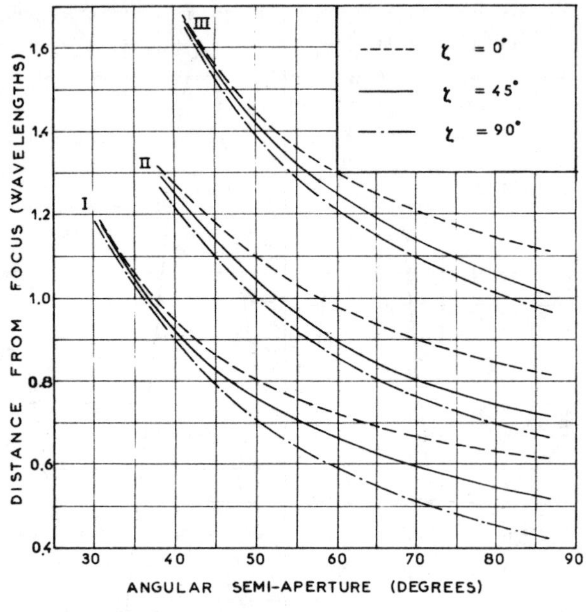

Fig. 8. Co-polar diffraction image characteristics as a function of the angular semi-aperture.

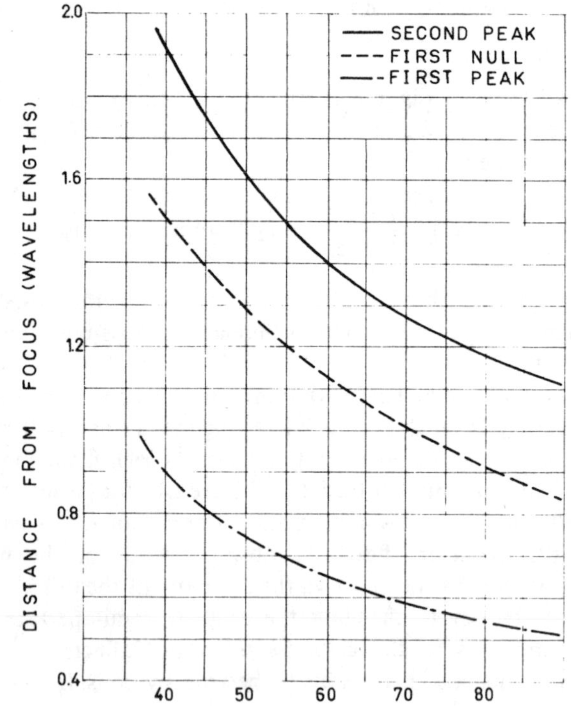

Fig. 9. Cross-polar diffraction image characteristics as a function of the angular semi-aperture.

expected since for $\psi' = 0$ the paraboloid acts no longer as a lens. Also from Fig. 9 it is seen that the position of the first and second cross-polar peaks move away from the focus as ψ' decreases. In the limiting case as ψ' tends to zero, the position of the peaks move indefinitely away from the focus. This is expected since when ψ' is zero the paraboloidal surface transforms to a plane reflector parallel to the incident wave. Such a surface introduces no depolarization.

IV. Measurements

To confirm the above theoretical findings measurements of focal plane fields were conducted using a paraboloid of 31 cm diameter (9.7λ at the frequency used). The transmitting linearly polarized antenna was placed at a far distance from the reflector under test so that a plane wave should be received. The receiving reflector was oriented such that the plane incident wave is normal to the axis of the reflector. A small dipole was used as the pickup probe. Across this a diode was connected and the demodulated voltage was then fed to a meter via thin wires. It was found that this arrangement introduces but little distortion in the focal plane fields. The outcome of these measurements are presented in the form of graphs in Figs. 10 and 11. Theoretical curves are also included for comparison. It is seen that the agreement between theory and measurements is good.

In Fig. 10(a) and (b), the variation of co-polar field with r, the distance from the focus for $\zeta = 0°$ and $\zeta = 90°$ is shown. It is clear that the agreement is good, in particular the positions of maxima and nulls are accurately predicted by theory. In Fig. 10(c) the variation of cross-polar field intensity as a function of the distance r is given for two values of ζ, namely 20° and 40°. Theoretical curves are also given. It is to be observed that the agreement is not as good as in the case of co-polar fields. This may be attributed to the inevitable misorientation of the pickup probe. Needless to mention that unlike co-polar measurements, small errors in misorientation lead to serious errors in measured cross-polar. The deviation between theory and measurements is also seen to increase with r. This is due to edge diffraction effects which were not taken care of in the present theory.

Table I gives relative amplitude of co-polar field as a function of r for $\zeta = \pm 40°$ and $\pm 20°$. From this table one may conclude that the field is symmetrical about the $\zeta = 0°$ axis. Measurements along $\zeta = \pi \pm 40°$ and $\zeta = \pi \pm 20°$ were found to yield similar results to those for $\zeta = \pm 40°$ and $\zeta = \pm 20°$, respectively. This demonstrates that the diffraction pattern is symmetrical about the focus as theory predicts.

Fig. 11 shows the variation of co-polar and cross-polar fields with ζ (and for $r = 0.4\lambda$). Measured results are represented by crosses. Theoretical graphs of the functions

TABLE I
Measured Co-Polar Amplitude Relative to that at the Focus

r (Wavelengths)	Measured Relative Amplitude			
	$\zeta = +40°$	$\zeta = -40°$	$\zeta = +20°$	$\zeta = -20°$
0.00	1.00	1.00	1.00	1.00
0.17	0.85	0.86	0.88	0.83
0.34	0.37	0.42	0.58	0.56
0.68	0.04	0.11	0.10	0.11
0.85	−0.06	−0.06	−0.04	−0.03
1.02	−0.15	−0.16	−0.05	−0.05
1.36	0.16	0.14	0.15	0.13

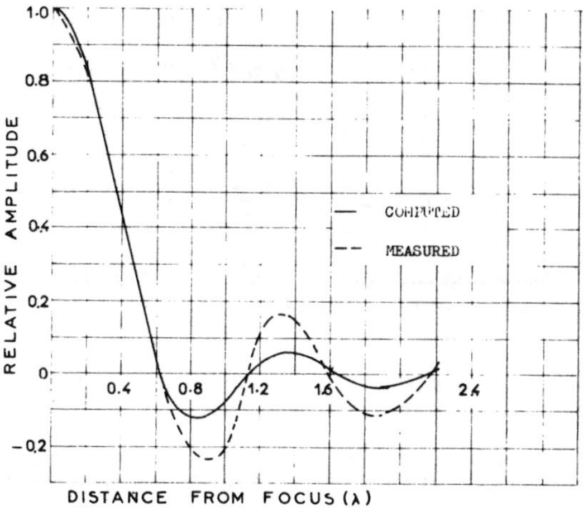

(a)

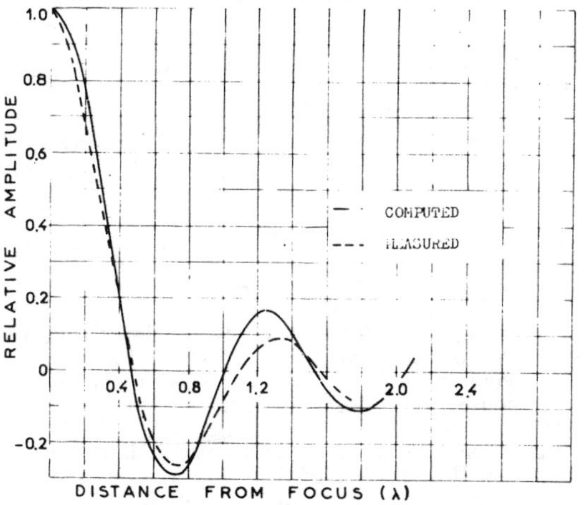

(b)

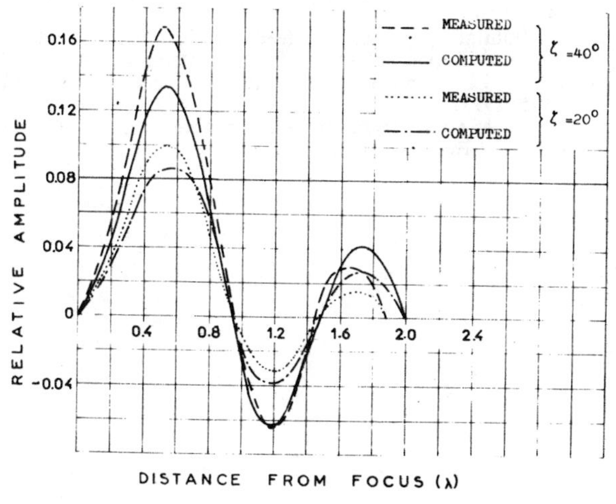

(c)

Fig. 10. Measured and computed diffraction patterns in the focal plane.

$A + B \cos 2\zeta$ and $B \sin 2\zeta$ are given as well. In plotting these theoretical curves the values of A and B were determined from measured results; i.e., A was obtained as the value of the co-polar at $\zeta = 45°$, B was then determined as the difference between the co-polar level at $\zeta = 0°$ and $\zeta = 45°$. This allows comparison of measured field dependence on ζ with theory while eliminating r dependent errors.

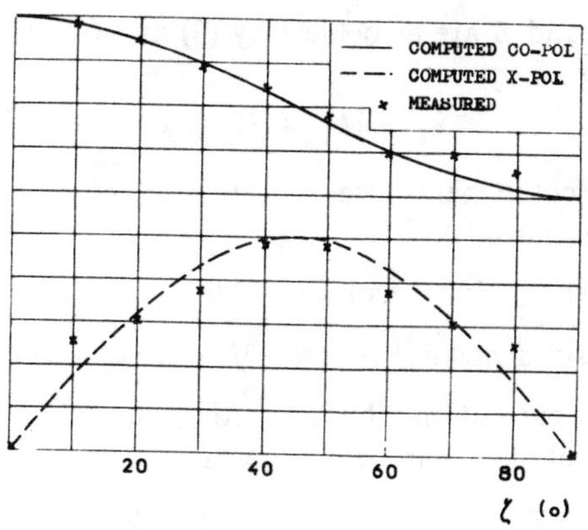

Fig. 11. Measured and computed variation of co-polar and cross-polar fields with ζ.

It is seen that agreement is excellent and one can conclude that measurements to confirm the theoretical findings presented in the previous sections.

To investigate the effect of the pickup probe back lobe on the above presented results measurements were conducted in the absence of the reflector. It was found out that the back lobe had a level of less than −27 dB relative to the co-polar at the focus, the back lobe being measured with the probe oriented to match the polarization of the incident plane wave. With the probe cross-polarized the back lobe had a level better than −50 dB. It was therefore concluded that the effect of back lobe on focal plane measurements was significant only when measuring field near the nulls of the co-polar diffraction image. In the case of cross-polar measurements the effect of the pickup probe back lobe was negligible. The error introduced by the above effect may, however, be isolated by interpolation.

V. Conclusions

The polarization nature of the field in the focal plane of paraboloidal reflectors was characterized. It was shown that for linear polarization the co-polar field distribution consists of almost elliptically shaped contours with maximum intensity at the focus. The cross-polar field is concentrated at the four quadrants formed by the principal axes. The cross-polar field vanishes at the focus and along the principal axes.

With circular polarization circular symmetry is maintained for co-polar and cross-polar fields. The field intensity variation with r being identical to that of the case of linear polarization along $\zeta = (2n + 1)45°$, $n = 0,1$. The effect of angular semi-aperture on focal plane fields was also studied and the results were given in the form of graphs. Practical measurements confirmed theoretical findings.

Appendix

To show that in the focal plane of a reflector excited by a plane wave linearly polarized along the x-axis, the x and y field components are given by

$$E_x(r,\zeta) = A + B \cos 2\zeta$$

$$E_y(r,\zeta) = B \sin 2\zeta$$

where A and B are as defined by (3) and (4). The incident field is given by

$$\boldsymbol{e}_i = |E|\boldsymbol{i} + \boldsymbol{j}0 + \boldsymbol{k}0. \tag{A1}$$

The field components after reflection are obtained from the reflecting matrix as

$$\boldsymbol{e}_r = |E|[[\sin^2\phi(1 - \cos\psi) - 1]\boldsymbol{i} + \sin\phi\cos\phi(1 - \cos\psi)\boldsymbol{j} - \boldsymbol{k}\sin\psi\sin\phi]. \tag{A2}$$

Using the current distribution method the field at any point in the focal plane can be written as

$$\boldsymbol{E} = C\iint_s [\boldsymbol{J} - (\boldsymbol{J}\cdot\boldsymbol{S}_r)\boldsymbol{S}_r]\exp\left(-j\frac{2\pi}{\lambda}(\rho + d)\right)\frac{1}{\rho}\,dS \tag{A3}$$

where

- C a constant;
- $\boldsymbol{J}$ current distribution density;
- $\boldsymbol{S}_r$ unit vector along MP (Fig. 1);
- d distance from the point M to a reference plane;
- ρ distance MP.

In terms of r, ζ, ϕ, and ψ the distance ρ is approximately[1] given by the relation

$$\rho \approx \frac{2f}{1 + \cos\psi} - \frac{r\sin\psi\sin(\phi + \zeta)}{1 + \cos\psi}. \tag{A4}$$

Substituting for $\boldsymbol{J} = \boldsymbol{n}\times(\boldsymbol{S}_r\times\boldsymbol{e}_r)$ where $\boldsymbol{n}$ is a unit vector normal to the reflecting surface, we get

$$\boldsymbol{E} = C\iint_s \boldsymbol{e}_r(\boldsymbol{n}\cdot\boldsymbol{S}_r)\exp\left(-j\frac{2\pi}{\lambda}(\rho + d)\right)\frac{1}{\rho}\,dS. \tag{A5}$$

[1] The validity of this approximation is subject to the constraint $r \ll 2f$.

But

$$dS = \rho^2 \sin\psi \sec(\psi/2)\,d\psi\,d\phi$$

$$\boldsymbol{n}\cdot\boldsymbol{S}_r = \cos(\psi/2)$$

$$d + \rho = 2f - \frac{r\sin\psi\sin(\phi + \zeta)}{1 + \cos\psi}.$$

On substituting the above results in (A5), the following expression is readily obtained,

$$\boldsymbol{E} = C\int_0^{2\pi}\int_0^{\psi'} \boldsymbol{e}_r \frac{\sin\psi}{1 + \cos\psi} \cdot \exp\left(j\frac{2\pi}{\lambda}r\sin\psi\sin(\phi + \zeta)\right)d\psi\,d\phi. \tag{A6}$$

Now, using (A2) and (A6) and using the identity

$$\int_0^{2\pi}\begin{Bmatrix}\cos n\phi\\ \sin n\phi\end{Bmatrix}\exp(j\beta\cos(\phi - \phi'))\,d\phi = 2\pi(j)^n\begin{Bmatrix}\cos n\phi'\\ \sin n\phi'\end{Bmatrix}J_n(\beta)$$

the required results are easily obtained.

References

[1] S. I. Ghobrial, "Some data for the design of low cross-polarization feeds," *Electron. Lett.*, vol. 9, pp. 465–466, 1973.

[2] G. T. Poulton and B. Claydon, "Determination of cross-polarization by use of focal-region fields," *Electron. Lett.*, vol. 9, pp. 568–569, 1973.

[3] B. Richard and E. Wolf, "Electromagnetic diffraction in optical systems. Pt. II—Structure of the image in an aplanatic system," *Proc. R. Soc. A.*, vol. 253, pp. 358–379, 1959.

[4] H. Minnet and B. Thomas, "Fields in the image space of symmetrical focusing reflectors," *Proc. Inst. Elec. Eng.*, vol. 115, pp. 1419–1430, 1968.

[5] S. I. Ghobrial, "A suggested feed for the reduction of cross-polarization in paraboloidal reflector antennas," Postgraduate School of Elec. and Electron. Eng., University of Bradford, Rep. 145, 1973.

[6] ——, "Cross-polarization in reflector antennas," Ph.D. thesis, University of Bradford, 1972.

Bibliography for Part II

A. Focal Region Fields

[1] E. M. Kennaugh and R. H. Ott, "Fields in the focal region of a parabolic receiving antenna," *IEEE Trans. Antennas Propagat.*, vol. AP-12, pp. 376–377, May 1964.

[2] Vu The Bao, "Optimisation of efficiency of shallow reflectors," *Electron. Lett.*, vol. 5, pp. 6–7, Jan. 9, 1969.

[3] D. J. Bem, "Electric field distribution in the focal region of an offset paraboloid," *Proc. Inst. Elec. Eng.*, vol. 116, pp. 679–684, May 1969.

[4] A. W. Rudge, "Focal plane field distribution of parabolic reflectors," *Electron. Lett.*, vol. 5, pp. 510–512, Oct. 16, 1969.

[5] v. H. Gniss and G. Ries, "Field pattern near the focus of paraboloid reflectors of low *f/D* ratio," *Arch. Elek. Übertragung.*, vol. 23, pp. 481–488, Oct. 1969.

[6] Vu The Bao, "Optimization of efficiency of deep reflectors," *IEEE Trans. Antennas Propagat.*, vol. AP-17, pp. 811–813, Nov. 1969.

[7] A. K. Sen, Y. Chassé, and M. Rouillard, "Measurement of the electromagnetic field intensity in the focal region of a wide-angle spherical reflector," *IEEE Trans. Antennas Propagat.*, vol. AP-19, pp. 426–430, May 1971.

[8] M. Landry and Y. Chassé, "Measurement of electromagnetic field intensity in focal region of wide-angle paraboloid reflector," *IEEE Trans. Antennas Propagat.*, vol. AP-19, pp. 539–543, July 1971.

[9] G. Poulton, "Image region fields for a stepped reflector," *Electron. Lett.*, vol 7, pp. 650–651, Oct. 21, 1971.

[10] J. Ruze, "Axial astigmatic fields in the focal region," *IEEE Trans. Antennas Propagat.*, vol. AP-23, pp. 734–735, Sept. 1975.

B. Simple Feed Systems

[11] B. Berkowitz, "Antennas fed by horns," *Proc. IRE*, vol. 41, pp. 1761–1765, Dec. 1953.

[12] J. W. Crompton, "On the optimum illumination taper for the objective of a microwave aerial," *Proc. Inst. Elec. Eng.*, vol. 101, part III, pp. 371–382, 1954.

[13] H. E. Green, "Paraboloidal reflector aerial with a helical feed," *Proc. IRE (Aust.)*, pp. 71–83, Feb. 1960.

[14] J. K. Shimizu, "Octave bandwidth feed horn for paraboloid," *IRE Trans. Antennas Propagat.*, vol. AP-9, pp. 223–224, Mar. 1961.

[15] J. Y. Wong, "A dual polarization feed horn for a parabolic reflector," *Microwave J.*, vol. 5, pp. 188–191, Sept. 1962.

[16] A. T. Moffet, "A novel duplex feed," *IEEE Trans. Antennas Propagat.*, vol. AP-12, p. 132, Jan. 1964.

[17] D. Herbison-Evans, "Optimum paraboloid aerial and feed design," *Proc. Inst. Elec. Eng.*, vol. 115, pp. 87–90, Jan. 1968.

[18] K. L. Walton and V. C. Sundberg, "Constant-beamwidth antenna development," *IEEE Trans. Antennas Propagat.*, vol. AP-16, pp. 510–513, Sept. 1968.

[19] A. W. Rudge and M. J. Withers, "Design of flared-horn primary feeds for parabolic reflector antennas," *Proc. Inst. Elec. Eng.*, vol. 117, pp. 1741–1749, Sept. 1970.

[20] G. W. Ewell, "Polarization-transforming antenna feed horns," *IEEE Trans. Antennas Propagat.*, vol. AP-19, pp. 681–682, Sept. 1971.

[21] R. Wohlleben, H. Mattes and O. Lochner, "Simple small primary feed for large opening angles and high aperture efficiency," *Electron. Lett.*, vol. 8, pp. 474–476, Sept. 21, 1972.

[22] R. W. Silberberg, "The paradisc antenna—A novel technique to improve the axial ratio of a circularly polarized high gain antenna system," *IEEE Trans. Antennas Propagat.*, vol. AP-21, pp. 108–110, Jan. 1973.

[23] J. H. Cowan, "Dual-band reflector feed element for frequency reuse applications," *Electron. Lett.*, vol. 9, pp. 596–597, Dec. 13, 1973.

[24] W. M. Truman and C. A. Balanis, "Optimum design of horn feeds for reflector antennas," *IEEE Trans. Antennas Propagat.*, vol. AP-22, pp. 585–586, July 1974.

[25] A. Kumar, "Experimental study of a dielectric rod enclosed by a waveguide for use as a feed," *Electron. Lett.*, vol. 12, pp. 666–668, Dec. 9, 1976.

[26] L. Shafai, "Broadening of primary-feed patterns by small E-plane slots", *Electron. Lett.*, vol. 13, pp. 102–103, Feb. 17, 1977.

C. Multimode Horn Feeds

[27] P. D. Potter, "A new horn antenna with suppressed sidelobes and equal beamwidths," *Microwave J.*, vol. 6, pp. 71–78, June 1963.

[28] R. H. Turrin, "Dual mode small aperture antennas," *IEEE Trans. Antennas Propagat.*, vol. AP-15, pp. 307–308, Mar. 1967.

[29] R. W. Gruner, "A 4- and 6-GHz, prime focus, CP feed with circular pattern symmetry," in *IEEE AP-S Symp. Dig.*, June 1974, pp. 72–74.

[30] G. F. Koch, "Coaxial feeds for high aperture efficiency and low spillover of paraboloidal reflector antennas," *IEEE Trans. Antennas Propagat.*, vol. AP-21, pp. 164–169, Mar. 1973.

D. Corrugated Horn Feeds

[31] A. F. Kay, "The scalar feed," AFCRL Rep. 64-347, AD601609, Mar. 1964.

[32] A. J. Simmons and A. F. Kay, "The scalar feed—A high performance feed for large paraboloid reflectors," in *Design and Construction of Large Steerable Aerials*, IEEE Conf. Publ. 21, pp. 213–217, 1966.

[33] C. M. Knop and H. J. Wiesenfarth, "On the radiation from an open-ended corrugated pipe carrying the HE_{11}

mode," *IEEE Trans. Antennas Propagat.*, vol. AP-20, pp. 644–648, Sept. 1972.
[34] H. C. Minnett and B. MacA. Thomas, "Propagation and radiation behaviour of corrugated feeds," *Proc. Inst. Elec. Eng.*, vol. 119, p. 1280, Sept. 1972.
[35] —, "A method of synthesizing radiation patterns with axial symmetry," *IEEE Trans. Antennas Propagat.*, vol. AP-14, pp. 654–656, Sept. 1966.
[36] T. B. Vu and Q. H. Vu, "Optimum feed for large radio-telescopes: Experimental results," *Electron. Lett.*, vol. 6, pp. 159–160, Mar. 19, 1970.
[37] B. MacA. Thomas, "Prime focus one- and two-hybrid mode feeds," *Electron. Lett.*, vol. 6, pp. 460–461, July 23, 1970.
[38] T. B. Vu and N. V. Hien, "A new type of high-performance monopulse feed," *IEEE Trans. Antennas Propagat.*, vol. AP-21, pp. 855–857, Nov. 1973.
[39] P. J. B. Clarricoats and P. K. Saha, "Propagation and radiation behaviour of corrugated feeds," *Proc. Inst. Elec. Eng.*, vol. 118, parts I and II, pp. 1167–1186, Sept. 1971.
[40] M. E. J. Jeuken, "Experimental radiation pattern of the corrugated conical horn with small flare angle," *Electron. Lett.*, vol. 5, pp. 484–486, Oct. 2, 1969.
[41] M. E. J. Jeuken and C. W. Lambrechtse, "Small corrugated conical horn antenna with wide flare angle," *Electron. Lett.*, vol. 5, pp. 489–490, Oct. 2, 1969.
[42] J. K. M. Jansen, M. E. J. Jeuken, and C. W. Lambrechtse, "The scalar feed," *Arch. Elek. Übertragung.*, vol. 26, pp. 22–30, Jan. 1972.
[43] M. S. Narasimhan and B. V. Rao, "Hybrid modes in corrugated conical horns," *Electron. Lett.*, vol. 6, pp. 32–34, Jan. 22, 1970.
[44] —, "Diffraction by wide-flare-angle corrugated conical horns," *Electron. Lett.*, vol. 6, pp. 469–471, July 23, 1970.
[45] M. S. Narasimhan, "Corrugated conical horn as a space feed for phased-array illumination," *IEEE Trans. Antennas Propagat.*, vol. AP-22, pp. 720–722, Sept. 1974.
[46] R. Price, "High performance corrugated feed horn for the unattended earth terminal," *COMSAT Tech. Rev.*, vol. 4, pp. 283–302, Fall 1974.

Part III
Radiation Pattern Analysis of Reflectors

The first paper, by Tsai *et al.*, is concerned with the radiation pattern analysis of two-dimensional reflector surfaces. Nevertheless, it is of importance to the antenna designer who must, in practice, deal with three-dimensional structures. The reason is that it compares two analytical techniques for pattern prediction, establishes their ranges of validity, and indicates the area in which they overlap. Thus, the practical designer is given guidelines to enable him to select the method best suited to his situation. The two techniques are the geometrical theory of diffraction (GTD) and the integral equation formulation (IEF) for the induced current.

In the paper by James and Kerdemelidis, edge diffraction methods using GTD are modified by the use of equivalent edge currents derived from cylindrical wave diffraction, and by the inclusion of a slope-wave correction term. When applied to a paraboloidal reflector, the results are comparable to those obtained from the physical optics approach in the forward direction, apart from the main beam region. In the shadow region, however, their method gives results that are in considerably better agreement with experiment than those of physical optics.

Another comparison between various analytical techniques is given in the brief third paper by Rusch, examples being given for both paraboloidal and hyperboloidal reflectors. This is followed by a GTD treatment of diffraction by axially symmetric reflectors, given by Rusch and Sorensen.

The final paper, by Kaufman, Croswell, and Jowers, is a practical demonstration of how numerical methods can be used for the rapid calculation of reflector antenna radiation patterns. The reflector is defined by the intersection of a feed illumination cone with any surface of revolution. Symmetry about the axis of revolution is not required, so that offset-fed geometries can be handled just as well as symmetrical cases.

A Comparison of Geometrical Theory of Diffraction and Integral Equation Formulation for Analysis of Reflector Antennas

LEONARD L. TSAI, MEMBER, IEEE, DONALD R. WILTON, MEMBER, IEEE, MICHAEL G. HARRISON, STUDENT MEMBER, IEEE, AND EDDY H. WRIGHT, STUDENT MEMBER, IEEE

Abstract—Four two-dimensional reflector antennas, the strip, the corner, the trough, and the inverted trough, are analyzed independently by both the geometrical theory of diffraction (GTD) and integral equation formulation (IEF). A comparison of the radiation patterns calculated by these two methods shows that for intermediate-sized structures (0.5–5λ), the agreement between the two methods is excellent. (Comparison with measured patterns is given where these are available in the literature.) Extensive calculations are also carried out to determine the ranges of validity of the methods, i.e., exactly how small a structure GTD can treat and how large a reflector IEF can adequately handle. A qualitative assessment of the relative merits of the two methods is also presented.

I. Introduction

IN RECENT YEARS much interest has been shown in the use of numerical methods applied to problems in electromagnetics. These methods typically involve an adaptation to numerical processing of standard classical approaches, e.g., integral equations [1], [2], variational and perturbational principles [3]–[5], geometrical diffraction analysis [6]–[8], and function-theoretic techniques [9], [10]. In problems where more than one method might be applicable, few user-oriented guidelines exist for deciding on an approach, although some comparisons of various methods and discussions of the relative merits of a particular method do appear in the literature [11]–[14]. It is the purpose of this paper to discuss and compare the well-known integral equation formulation (IEF) and the geometrical theory of diffraction (GTD) as applied to a particular class of reflector antennas.

In the IEF an integral equation is formulated which contains the induced current distribution on the antenna or scatterer as the unknown quantity. Once the current distribution is obtained, all the important characterizing properties (scattering or radiation patterns, scattering cross section, impedance, etc.) may be easily determined. The integral equation, which is usually an exact formulation, may then be reduced to an approximately equivalent matrix system by the so-called method of moments [1], and the resultant matrix inverted by computer to obtain the solution. The solution of the matrix system takes the form of a set of coefficients for an expansion of the current. It is generally recognized that the larger the configuration, the more degrees of freedom the current may assume and the larger the matrix that must be inverted. Therefore, an upper limit exists on the size of bodies that may be handled because of computer requirements.

The GTD is an approximate asymptotic method which treats diffraction as a localized phenomenon, and allows one to obtain the scattered fields directly from purely geometrical considerations. Using simple ray tracing, one includes contributions to the scattered field due to geometrical optics reflections, as well as diffracted fields which appear to emanate from edges and corners. In order to take into account the interactions of several edges and corners, higher order or multiple diffraction terms must also be included in some cases. Generally, all the calculations involve the evaluation of relatively simple functions, and, of course, no functional equations need to be solved. Since the method is an asymptotic method, however, it may be expected to give more accurate results as the frequency of excitation is increased. Therefore, in contrast to the IEF, the GTD method works best for larger bodies.

The following two-dimensional reflector antenna configurations are considered here: the strip reflector, the corner reflector, the inverted trough reflector, and the trough reflector. Investigated extensively are the patterns of these structures with line source excitation by both GTD and IEF methods. The currents available from IEF are also considered. Comparison with measured patterns is shown where these are available in the literature. The relative flexibility and difficulty in implementing the two methods as a function of the geometry is also discussed.

II. Integral Equation Formulation

The integral equation solved in each of the cases considered has the form [1]

$$\frac{-\omega\mu}{4}\int_{\text{reflector}} J_z(\bar{\rho}')H_0^{(2)}(k\,|\,\bar{\rho}-\bar{\rho}'\,|)\,dl' = -E_z^i(\bar{\rho}),\qquad \bar{\rho}\text{ on the reflector} \quad (1)$$

Manuscript received December 20, 1971; revised June 9, 1972. Portions of this paper were presented at the 1971 IEEE International Group on Antennas and Propagation Symposium, Los Angeles, Calif. This work was supported in part by the NSF under Grant GU/3833.

The authors are with the Department of Electrical Engineering, University of Mississippi, University, Miss. 38677.

Reprinted from *IEEE Trans. Antennas Propagat.*, vol. AP-20, pp. 705-712, Nov. 1972.

where k is the wavenumber of free space, J_z is the z component of electric surface current density, $H_0^{(2)}$ is the zeroth-order Hankel function of the second kind, and E_z^i is the z component of the incident electric field on the reflector surface. [An $\exp(j\omega t)$ time dependence is assumed and suppressed throughout for convenience.] In all cases considered here, $E_z^i(\bar{\rho}) = H_0^{(2)}(k\rho)$ which corresponds to the field of an electric line source located at the coordinate origin and carrying a current $I = -4/\omega\mu$. The term on the left-hand side of (1) represents the scattered fields due to the induced surface current J_z and (1) simply states that the scattered tangential electric field must equal the negative of the incident tangential electric field on the surface of the (perfectly conducting) reflector. To obtain the induced current in (1), the well-known method of moments is employed [1]. We first divide the reflector surface denoted by the arc length variable s, into $N - 1$ segments, denoting the end points of the segments by s_n, $n = 1,2,\cdots,N$. Then J_z is expanded in a set of triangular basis functions:

$$J_z = \sum_{n=1}^{N} \alpha_n T_n(s) \tag{2}$$

where

$$T_n(s) = \begin{cases} \dfrac{s - s_{n-1}}{s_n - s_{n-1}}, & s_{n-1} \le s \le s_n \\[2ex] \dfrac{s_{n+1} - s}{s_{n+1} - s_n}, & s_n \le s \le s_{n+1} \\[2ex] 0, & \text{elsewhere.} \end{cases}$$

Using this substitution and requiring (1) to hold only at the end points of the segments results in the linear system of equations,

$$\sum_{n=1}^{N} l_{mn}\alpha_n = g_m, \qquad m = 1,2,\cdots,N \tag{3}$$

where

$$l_{mn} = \frac{-\omega\mu}{4} \int_{s_{n-1}}^{s_n} T_n(s') H_0^{(2)}(k\,|\bar{\rho}_m - \bar{\rho}'|)\, ds' \tag{4}$$

$$g_m = -E_z^i(\bar{\rho}_m) \tag{5}$$

and $\bar{\rho}_m$ denotes the radius vector from the line source to the mth point on the reflector surface. Equation (3) constitutes a matrix equation for determining the coefficients α_n, which in turn give the surface current density according to (2). The far-field pattern of the reflector and line-source configuration is then given by

$$F(\theta) = 1 - \frac{\omega\mu}{4} \int_{\text{reflector}} J_z(\bar{\rho}') \exp[jk\rho' \cos(\phi' - \theta)]\, ds' \tag{6}$$

where $\bar{\rho}' = (\rho',\phi')$ is the polar coordinate representation of the vector from the illuminating line source to a point on the reflector and θ is the pattern observation angle. The pattern function $F(\theta)$ is normalized in (6) so as to give a unity pattern function for the line source when no reflector is present.

The actual evaluation of each element l_{mn} in (4) makes use of an approximate three-point rectangular quadrature whenever $m \neq n$. Whenever $m = n$, the Hankel function in (4) is singular and we use the small argument approximation for $H_0^{(2)}$ to perform the integration analytically [1].

At the edges and corners of the models considered, singularities in the surface current arise. If one desires to obtain an accurate representation for the currents near the singularities, the current must be sampled at smaller intervals. One scheme for doing this first recognizes that the order of the singularity at an edge or corner is known *a priori* [15]. With this approximation for J_z near the edges, assume that we make a change of variables $n = n(s')$ such that

$$|J_z|\, ds' = |J_z| \frac{ds'}{dn}\, dn \tag{7}$$

and require that $|J_z|\, ds'/dn =$ constant. The justification for this somewhat arbitrary requirement is that for equal increments in the new variable n, the contribution $|J_z|\, ds'$ to the area under the curve $|J_z(s')|$ will be constant. Thus the larger the value of current, the smaller we make the corresponding increment ds' so as to keep $|J_z|\, ds'$ constant. Let us apply this idea to the strip of length $2a$, located on the x axis, $-a \le x \le a$. The current is known to have a singularity which varies inversely as the square root of distance from the edges. We thus choose an approximate variation for $|J_z|$ according to

$$|J_z| \simeq \frac{1}{(a - x)^{1/2}} + \frac{1}{(a + x)^{1/2}}, \qquad -a \le x \le a. \tag{8}$$

This choice has the correct square root singularity near $x = \pm a$ and is relatively slowly varying in between as one would expect for the current near the center of the strip. Setting

$$|J_z| \frac{dx}{dn} = \text{constant} \tag{9}$$

and integrating the resulting differential equation, one obtains

$$x = \pm \left\{ a^2 - \left[\frac{2a - (K_1 n + K_2)^2}{2} \right]^2 \right\}^{1/2} \tag{10}$$

where K_1 and K_2 are arbitrary constants. It is now convenient to choose n to be the segment index (also denoted n) in (2) when n is an integer. Applying the conditions $x = -a$ when $n = 1$ and $x = +a$ when $n = N$, we may find the constants K_1 and K_2 resulting in the segmentation scheme,

$$x_n = \mp a\left\{1 - \left[1 - \left(\frac{2n - N - 1}{N - 1}\right)^2\right]^2\right\}^{1/2}$$

where the upper sign is used for $n = 1,2,\cdots,[N/2]$ and the lower sign for $n = [N/2] + 1,\cdots,N$. ($[N/2]$ is the largest integer not exceeding $N/2$.) In a similar manner, this scheme can also be extended to any corner or edge configuration. The segmentation scheme used for the corner reflector will be seen in Fig. 7, where the current values are given at the ends of the segments.

III. Geometrical Theory of Diffraction Analysis

The basic idea of the GTD method is that the line-source field illuminates, and in turn scatters from the antenna, causing single, and then multiple diffractions to emanate from the edges. Both the geometrical optics rays and diffracted rays may be shadowed by the antenna structure, and hence contribute to the total pattern only in their respective regions of applicability. The formulation in this paper follows standard GTD techniques [6]–[8], with cylindrical wave diffraction functions [8], [16] used to compute the diffracted waves. Because the primary purpose is a comparison between GTD and IEF, for brevity, only the necessary steps will be summarized (i.e., the equations to follow will merely indicate the various rays which contribute to the pattern and from whence they arise). However, for the four reflector geometries considered, each of which requires a separate formulation differing in detail and complexity, the following pattern contributions are all properly included: the direct geometrical optics ray, the various reflected geometrical optics rays, single and double diffraction rays from all exterior edges, and the pertinent reflected single and double diffraction contributions.

For the strip reflector, the total pattern $R_T(\theta)$, is computed by

$$R_T(\theta) = [R_1^{(1)} + R_1^{(2)}]e^{-jd_1} + [R_2^{(1)} + R_2^{(2)}]e^{-jd_2} + R_{\text{direct}} + [R_{\text{refl}}]e^{-jd_3} \quad (12)$$

with the various contributions as depicted in Fig. 1(a). The notation on diffracted rays are as exemplified by $R_1^{(2)}$ denoting the doubly diffracted ray (superscript), from edge 1 (subscript). (Explicit details for their computation are given in Appendix I). R_{direct} and R_{refl} are the geometrical optics rays, and the exponential factors arise from referring the phases to a common origin at the line source. Note that although not explicitly shown, all terms in (12) are functions of the antenna geometry and the angle θ.

For the 90° corner reflector, more terms may contribute, as depicted in Fig. 1(b). The pattern is given by

$$R_T(\theta) = [R_1^{(1)} + R_1^{(2)}]e^{-jd_1} + [R_2^{(1)} + R_2^{(2)}]e^{-jd_2} + R_3^{(2)}e^{-jd_3} + R_{\text{direct}} + R_{rg}^{(1)}e^{-jd_4} + R_{rg}^{(2)}e^{-jd_5} + R_{1rd}e^{-jd_6} + R_{2rd}e^{-jd_7}.$$

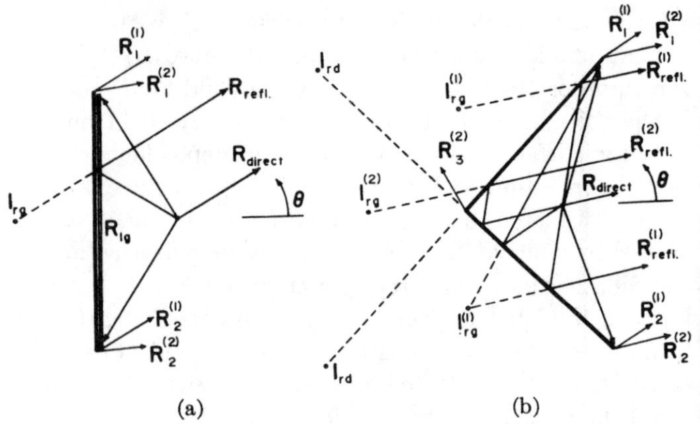

Fig. 1. (a) Ray diagram for strip reflector. (b) Ray diagram for 90° corner reflector.

The added terms $R_{rg}^{(1)}$ and $R_{rg}^{(2)}$ refer to the reflected geometrical optics rays which may undergo either one or two reflections before emerging from the corner reflector. For ease in computing the phase delay factors, as well as determining the ranges in θ where $R_{rg}^{(1)}$ and $R_{rg}^{(2)}$ exist, it is useful to view them as emanating from the images $I_{rg}^{(1)}$ and $I_{rg}^{(2)}$. In a similar manner, the reflected diffraction contributions, R_{1rd} from edge 1, and R_{2rd} from edge 2 (both singly and doubly diffracted rays), may be seen to radiate from the images I_{rd}. It should also be noted that the singly diffracted contributions, $R_1^{(1)}$ and $R_2^{(1)}$, arise from direct incidence from the line source, as well as from reflected geometrical optics incidence (from $I_{rg}^{(1)}$); and similarly, the singly diffracted terms will reflect from the corner, be reincident on the same edge, and give double diffraction (as, for example, $R_1^{(1)}$ from I_{rd} giving a contribution to $R_1^{(2)}$). This analysis is quite similar to that in [7] for a horn, and is an extension of that in [17] for a corner reflector where only single diffraction is used in the computation.

For the inverted trough reflector depicted in Fig. 2(a), the analysis is quite similar to that of the strip reflector, with the pattern given by

$$R_T(\theta) = [R_1^{(1)} + R_1^{(2)}]e^{-jd_1} + [R_2^{(1)} + R_2^{(2)}]e^{-jd_2} + R_{\text{direct}} + R_{\text{refl}}e^{-jd_r} + R_3^{(2)}e^{-jd_3} + R_4^{(2)}e^{-jd_4}. \quad (14)$$

The additional terms here are $R_3^{(2)}$ and $R_4^{(2)}$, the double diffraction from the rear edges. The multiple reflections these terms may subsequently undergo contribute only to the backlobes, and are assumed negligible. It may be noted that in the case of an electric line source, $R_3^{(2)}$ and $R_4^{(2)}$ are zero, hence (14) differs from (12) only in that a different wedge angle 90° is required to compute $R_1^{(1)}$, $R_1^{(2)}$, $R_2^{(1)}$, $R_2^{(2)}$. In fact, for the TM polarization, rays tangent to a conducting surface are always zero; for the TE polarization these terms do contribute and they are therefore retained for generality.

The analysis for the trough reflector illustrated in Fig. 2(b) becomes somewhat more complicated, primarily because the various multiple reflections must now be dealt

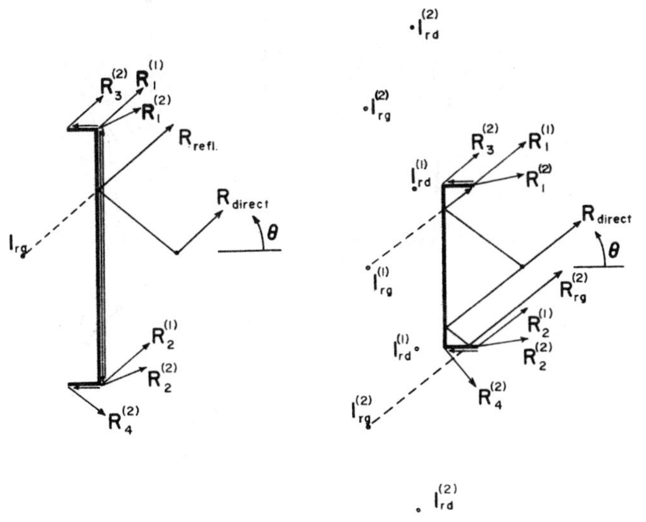

Fig. 2. (a) Ray diagram for inverted trough reflector. (b) Ray diagram for trough reflector.

with. For simplicity, the primary line source is assumed to lie outside the fenced edges. The pattern is given by

$$
\begin{aligned}
R_T(\theta) = {} & [R_1^{(1)} + R_1^{(2)}]e^{-jd_1} + [R_2^{(1)} + R_2^{(2)}]e^{-jd_2} \\
& + R_3^{(2)}e^{-jd_3} + R_4^{(2)}e^{-jd_4} + R_{\text{direct}} \\
& + R_{rg}^{(1)}e^{-jd_5} + R_{rg}^{(2)}e^{-jd_6} + R_{1rd}^{(1)}e^{-jd_7} \\
& + R_{2rd}^{(1)}e^{-jd_8} + R_{2rd}^{(2)}e^{-jd_9} + R_{1rd}^{(2)}e^{-jd_{10}}. \qquad (15)
\end{aligned}
$$

Both single and double diffractions from edges 1 and 2 are included. Double diffractions from edges 3 and 4 are also included. The analysis here is similar to that for the corner reflector and the arguments previously given for reflections and shadowing apply here. The reflected geometrical optics rays $R_{rg}^{(1)}$ and $R_{rg}^{(2)}$ are from $I_{rg}^{(1)}$ and $I_{rg}^{(2)}$. The various reflected diffracted rays $R_{1rd}^{(1)}$, $R_{1rd}^{(2)}$, $R_{2rd}^{(1)}$, $R_{2rd}^{(2)}$ (subscript denotes edge and superscript denotes the number of reflections) from edges 1 and 2 appear to emanate from $I_{rd}^{(1)}$ and $I_{rd}^{(2)}$ as shown. It should be noted that while only two reflections occur in the reflected geometrical optics ray, i.e., $I_{rg}^{(2)}$ is sufficient, an infinite number of reflections can take place for the reflected diffraction rays, i.e., $I_{rd}^{(n)}$ with $n \to \infty$. We include only up to $I_{rd}^{(2)}$ for convenience. (Inclusion of higher terms becomes tedious, especially if one notes how many different mechanisms there are, and only slight discontinuities are noticed in limited regions of the pattern as will be discussed.) It should be pointed out finally, that although (15) appears to merely contain more terms than (12), the individual terms are now more complicated. For example, $R_1^{(1)}$ in (15) alludes to single diffraction from the incident field of three sources, $I_{rg}^{(1)}$, $I_{rg}^{(2)}$, and the feeding line source, while the corresponding $R_1^{(1)}$ in (12) has only two incidences.

IV. Results and Discussions

Radiation patterns are computed for the four antenna structures using both IEF and GTD. Results are first given for the intermediate sizes to demonstrate that good

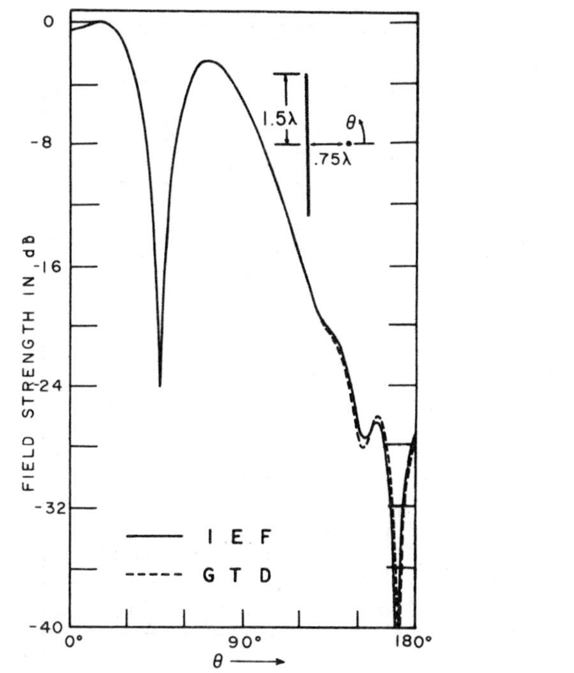

Fig. 3. Pattern comparison for strip reflector of length 1.5λ excited by electric line source at 0.75λ.

agreement can be obtained. The questions of convergence and range of applicability of each method will then be dealt with.

The radiation pattern for a 1.5λ strip reflector with electric line source excitation is given in Fig. 3. As can be seen, the agreement between GTD and IEF for this simple structure is excellent in the forward region, in fact the patterns overlap one another, and *slight* field strength deviations occur only in the deep backlobes, some 30 dB down from the main beam. Fig. 4 gives the calculated patterns compared to measured results [18] for a 90° corner reflector with length equal to 1λ. A modified reflector with possible applications for counterpoise antennas is the inverted trough reflector given in Fig. 5. Again, excellent agreement between theories (and measurement) are obtained in Figs. 4 and 5 for the electric line source case. It may be noted in passing, that the pattern in Fig. 5 deviated only slightly from that in Fig. 3, except in the backlobes, thus indicating the trailing edges have but little effect on antenna performance. The leading edges on the trough reflector given by Fig. 6, on the other hand, do influence the pattern because significant multiple reflections occur. The discontinuities in the GTD pattern near 90° are due to the presence of higher order reflected diffraction terms which were not included in the formulation. However, despite these anomalies, the overall agreement between GTD and IEF is still quite good for this structure. Thus through Figs. 3 to 6, it is shown that for intermediate size structures, close agreement in pattern can be obtained between the two methods.

Although only comparison of magnitudes has been given, both methods give phase information. An assessment of the agreement in phase is typified by consideration of the results for the strip reflector of Fig. 3. In the forward regions, $\theta < 135°$, less than two degrees differ-

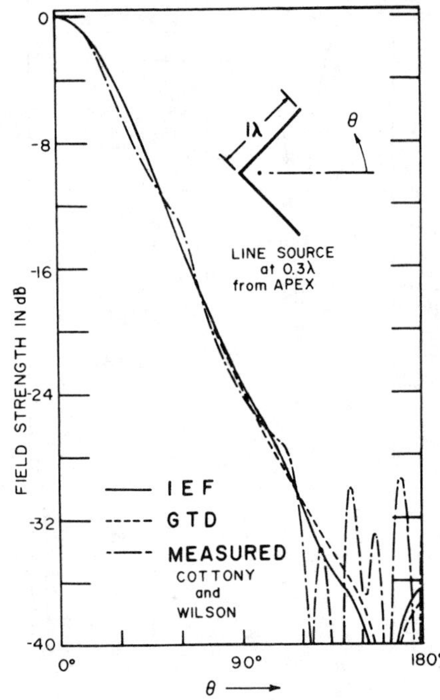

Fig. 4. Pattern comparison for 90° corner reflector of length 1λ with electric line source excitation.

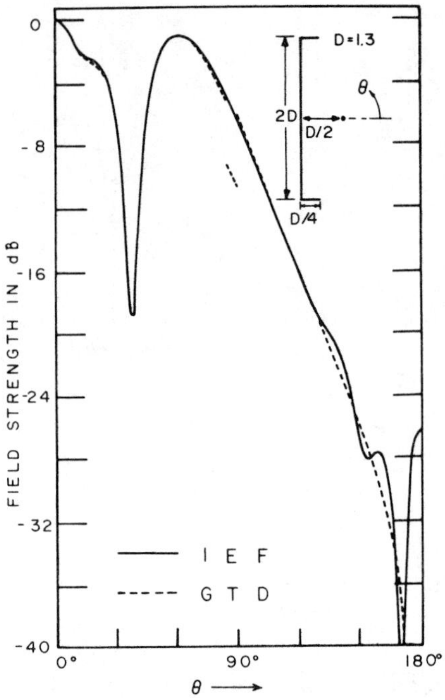

Fig. 6. Pattern comparison for trough reflector with electric line source excitation.

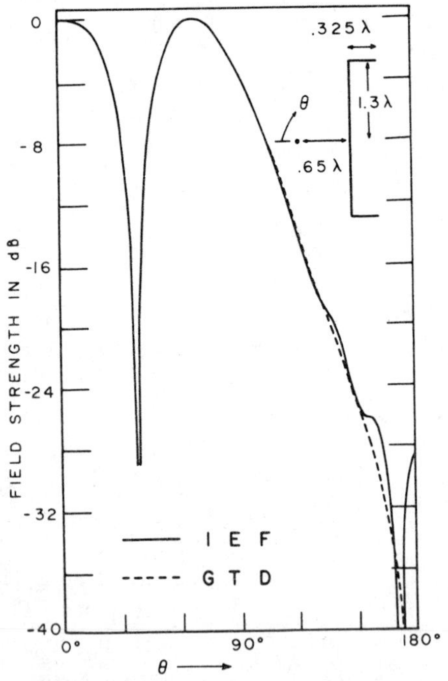

Fig. 5. Pattern comparison for inverted trough reflector with electric line-source excitation.

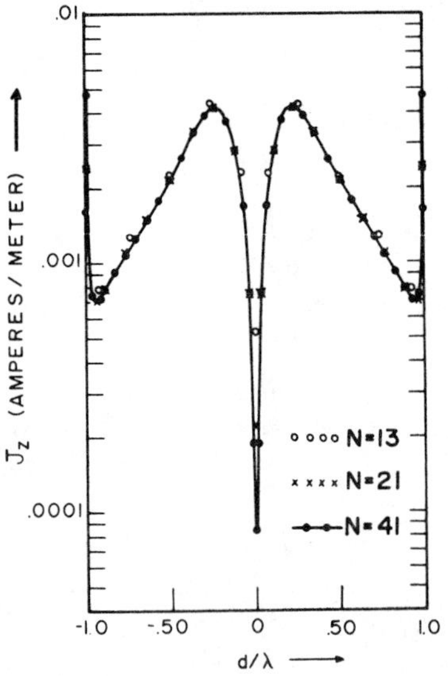

Fig. 7. Convergence of IEF current distribution for 90° corner reflector of length 1λ (electric line source).

ence is found between the phase patterns by GTD and IEF. In the deep backlobes, where larger differences may be expected, the agreement is still better than 10°.

One of the important questions in a moment method solution is its numerical convergence. Towards this end we examine the current distribution (Fig. 7) for the same 1λ corner reflector originally given in Fig. 4, as the number of sample points in the IEF solution is increased, i.e., $N = 13,21,41$. It can be readily seen that current converges, except near edge singularities. Hence, it may be concluded that 5 or 6 triangles per wavelength give good IEF results for the current distribution for this size corner reflector. As expected, patterns computed from these currents were found to be even more rapidly convergent and differed only in the pattern values below 30 dB. For larger structures, more sample points, of course, would be desirable, but due to the finite memory size of computers and cumulative computational errors, there necessarily exists some finite upper limit on how large a matrix one can practically invert. We, therefore, choose $N = 41$ as the standard sampling scheme and all other IEF results in this paper use this matrix size.

The pattern of a 57° corner reflector has also been calculated by IEF and similar close agreement with measurement [18] was found. This calculation was done to illustrate the versatility of IEF to changes in geometry of the structure. If a GTD solution for the 57° corner reflector was desired, then a much more general and difficult geometrical optics image scheme [17] would be necessary. On the other hand, the same formulation, and hence, one basic computer program with minor modifications for the various geometries, has been used by IEF for all the structures, while four different formulations were necessary for GTD for the four previous cases.

To give an assessment of the mutual regions of validity of the two methods, i.e., how small a structure will GTD accurately treat, and how large a structure can IEF practically handle, we show the panoramic series on the strip reflector and corner reflector in Figs. 8 and 9 as the structure size increases. In Fig. 8, the electric line source location is fixed at 0.75λ and the strip half length L is increased from 0.05λ to 0.25λ in the top three figures. For the small sizes here, of course, IEF gives very accurate results and these may be used as a standard. It is noted that GTD gives quite satisfactory patterns for L as small as 0.10λ, and even for $L = 0.05\lambda$, the mainlobe discrepancy is less than 0.2 dB. (The inaccuracies in GTD evidently arise from a breakdown of the localized field assumption and its inability to properly treat significant higher order diffraction contributions.) In the bottom three figures of Fig. 8, large strips with L ranging from 5.0λ to 10.0λ are given. Here, GTD results would be the standard as supported further by the close agreement between the 10.0λ and the infinite ground plane case. It is evident that IEF gives accurate patterns for $L = 5.0\lambda$, but backlobe degradation begins to occur for $L = 7.5\lambda$, with total pattern breakdown at $L = 10.0\lambda$. Since the matrix size is fixed (41×41 for $N = 41$) for IEF, the insight that may be gained from this comparison is not how large a structure one can treat, since a larger matrix can always be used, but how many sample points per wavelength will give accurate IEF patterns. Recalling also that a nonuniform sampling scheme is used, we conclude that an expansion using 3 triangles per wavelength where the current is not singular will give quite satisfactory patterns. (Near edge singularities more sample points will be required.)

Fig. 9 gives a similar comparison for the geometrically more complex corner reflector. It is seen that GTD patterns exhibit slight discontinuities for L small, with the result that good agreement is obtained for $L = 0.25\lambda$ (contrasted with $L = 0.10\lambda$ for the strip). Hence, as geometrical complexity increases, the lower size limit for accurate GTD patterns apparently also increases. For IEF patterns, the criterion of a minimum of 3 triangles per wavelength apparently still holds.

With these results, an overall assessment of the relative advantages and disadvantages of the two methods may be drawn. For IEF: 1) the method is versatile for changes in geometry, as evidenced by the fact that one formulation was used for all structures treated; 2) computation time is dependent on the size of the structure (proportional to N^3), hence the method is size limited; and 3) the method is accurate for small and intermediate sizes but becomes inaccurate for large structures. In general, an expansion using 3 triangles per wavelength gives accurate patterns, but for the intrinsically more difficult problem of current distribution, up to 5 triangles per wavelength is needed for convergence. For GTD: 1) the method is geometry dependent in that each structure requires a different treatment, and hence individual computer programs. In general, the computation time increases and the lower size limit also increases as the geometry becomes more complex; 2) for an individual geometric shape, however, the computation time is independent of size, hence the method is not size dependent; 3) no matrix inversion is required and the method provides more physical insight; 4) it is flexible for polarization changes (only a ± sign change in the diffraction function for the magnetic line source); and (5) the method is accurate for intermediate and large sizes, but not applicable for

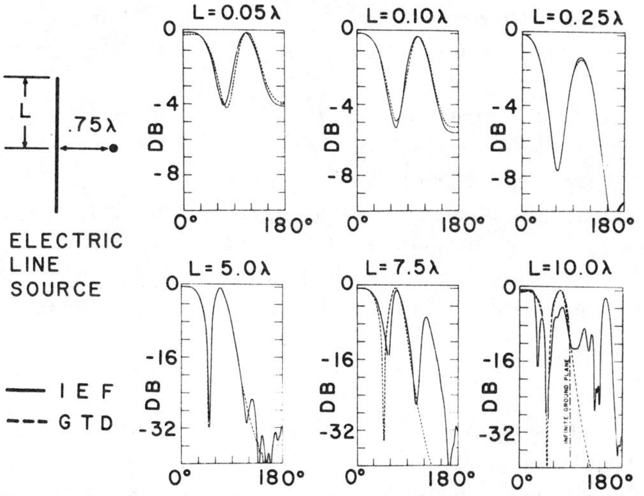

Fig. 8. Comparison of strip reflector patterns as function of reflector half-length L (electric line source 0.75λ in front of reflector).

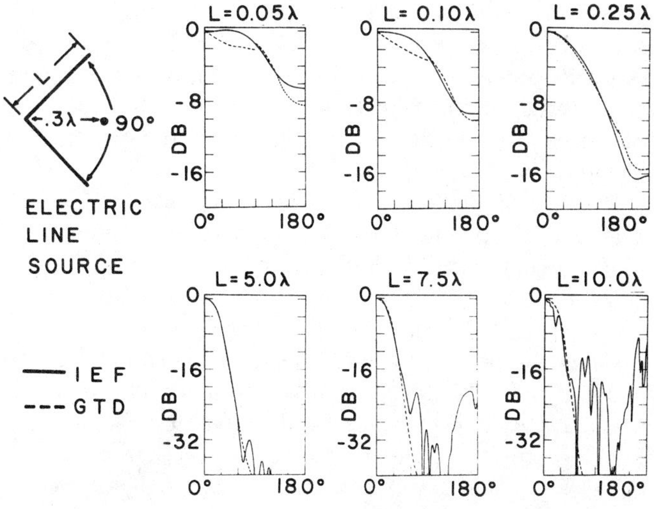

Fig. 9. Comparison of 90° corner reflector patterns as function of length of side L (electric line source centered at 0.30λ from the apex).

very small structures. Examples of lower size limits are 0.10λ for the strip reflector, and 0.25λ for the corner reflector.

A final comparison between the two methods is the computer time needed. In general, the maximum running times of the two methods were found to be of comparable magnitudes, though each one may require less for some cases, as for examples GTD for the strip, and IEF using smaller N for small sizes. This upper limit for the cases we considered was less than 4 minutes on the IBM 360/40.

V. Conclusion

In this paper IEF and GTD are demonstrated to be useful numerical methods for the analysis of various two-dimensional reflector antennas. The principle extension in this presentation has been an attempt to determine the limits of practical applicability of each method.

Results of this analysis indicate that there is considerable overlap in the ranges of structure sizes for which each method gives good results. For many bodies, this fact will provide a useful way of checking the accuracy of patterns obtained analytically. One will note that as the complexity of the structure increases, the overlapping region decreases.

For two of the structures where measured results are available, the two methods show excellent agreement with measurements except for the deep backlobes. The small variations are well within the range of deviations to be expected from an actual experimental model.

Using the results presented here, one should be able to make a judicious choice as to which of the two methods would be easier to apply and also give accurate results. For structures similar to the ones considered such a knowledge could result in a substantial savings of either programming or computer time.

In order to aid in determining the applicability of each method, the major advantages of each are summarized here:

Integral Equation Formulation (IEF):

1) Very accurate for small structures.
2) Applicable for a wide range of geometric configurations.
3) More information (current, impedance, etc.) readily available.

Geometrical Theory of Diffraction (GTD):

1) Accurate for large structures.
2) No large matrices to store or invert.
3) Easily adaptable to changes in polarization.

From the results and trends observed in this study, it is found that the two methods support as well as complement each other over a broad spectrum of reflector sizes. Furthermore, one may expect that the considerations seen for two-dimensional structures should also hold for the study of more practical three-dimensional reflectors.

Appendix I

For the computation of the diffracted rays from the strip reflector, the incident line source field is normalized hence the geometrical optics rays will have unity magnitude. The singly diffracted ray from edge 1 is given using standard GTD techniques [7], [8] as

$$R_1^{(1)}(\theta) = V_B(r_0,\psi - \psi_0) \pm V_B(r_0,\psi + \psi_0) \qquad (16)$$

where r_0 and ψ_0 are the source distance and incidence angle, and ψ is the observation angle dependent on θ. The $+$ sign applies for magnetic line sources (TE polarization) and the $-$ sign is for electric line sources (TM polarization). The diffraction function V_B is that applicable for line source incidence [16], given by

$$V_B(r,\phi) = I_{-\pi}(r,\phi) + I_{+\pi}(r,\phi)$$

with

$$I_{\pm\pi}(r,\phi) \approx \frac{\exp[-j(kr + \pi/4)]}{jn(2\pi)^{1/2}} a^{1/2} \cot\left(\frac{\pi \pm \phi}{2n}\right) \cdot \exp(jkra) \int_{(kra)^{1/2}}^{\infty} \exp(-j\tau^2)\, d\tau + \text{higher order terms} \qquad (17)$$

where $a = 1 + \cos(\phi - 2n\pi N)$, and N is a positive or negative integer or zero which most nearly satisfies

$$2n\pi N - \phi = \begin{cases} -\pi, & \text{for } I_{-\pi} \\ +\pi, & \text{for } I_{+\pi}. \end{cases}$$

The doubly diffracted ray from edge 2 is then found, making a localized uniform cylindrical wave assumption on the single diffraction wave, as

$$R_2^{(2)}(\theta) = R_{1G}[V_B(l,\psi - 0) \pm V_B(l,\psi + 0)] \qquad (18)$$

where R_{1G} corresponds to $R_1^{(1)}$ $(\theta = \pi/2)$. Higher order diffraction contributions may be computed similarly. Generally, the contribution of these terms is small and furthermore they are derived by assuming that the higher order diffracted fields are uniform cylindrical waves. This assumption may yield erroneous diffracted fields for the higher order diffraction terms. There are also several other ways of computing the multiple diffracted rays in GTD. The details for these methods are in [19], [20].

Acknowledgment

The authors are indebted to the reviewer for a comment which lead to the correction of a hitherto undetected programming error in the backlobes of the GTD patterns.

References

[1] R. F. Harrington, *Field Computation by Moment Methods.* New York: Macmillan, 1968, ch. 3.
[2] J. Van Bladel, *Electromagnetic Fields.* New York: McGraw-Hill, 1964.
[3] R. E. Collin, *Field Theory of Guided Waves.* New York: McGraw-Hill, 1960.
[4] N. Marcuvitz, *Waveguide Handbook* (M.I.T. Rad. Lab. Series). Boston, Mass.: Boston Tech., 1951.

[5] J. Schwinger and D. Saxon, *Discontinuities in Waveguides, Notes on Lectures by J. Schwinger.* New York: Gordon and Breach, 1958.
[6] J. B. Keller, "Geometrical theory of diffraction," *J. Opt. Soc. Amer.*, vol. 52, no. 2, pp. 116–130, Feb. 1962.
[7] J. S. Yu, R. C. Rudduck, and L. Peters, Jr., "Comprehensive analysis for *E*-plane of horn antennas by edge diffraction theory," *IEEE Trans. Antennas Propagat.*, vol. AP-14, pp. 138–149, Mar. 1966.
[8] R. C. Rudduck and L. L. Tsai, "Application of wedge diffraction and wave interaction methods to antenna theory," Short Course on Microwave Optics, Ohio State Univ., Columbus, 1969.
[9] L. A. Weinstein, *The Theory of Diffraction and the Factorization Method.* Boulder, Colo.: Golem Press, 1969.
[10] R. Mittra and S. W. Lee, *Analytical Techniques in the Theory of Guided Waves.* New York: Macmillan, 1971.
[11] J. Freeland and R. G. Kouyoumjian, "An evaluation of several approximate methods for analyzing the scattering from strips (plates)," in *1966 URSI–PGAP Fall Meeting Dig.*, p. 90, Dec. 1966.
[12] A. R. Neureuther *et al.*, "A comparison of numerical methods for thin wire antennas," in *1968 Fall URSI Meeting Dig.*
[13] E. K. Miller, G. J. Burke, and E. S. Selden, "Accuracy-modeling guidelines for integral-equation evaluation on thin-wire scattering structures," *IEEE Trans. Antennas Propagat.* (Commun.), vol. AP-19, pp. 534–536, July 1971.
[14] T. B. A. Senior and P. L. E. Uslenghi, "Comparison between Keller's and Ufimtsev's theories for the strip," *IEEE Trans. Antennas Propagat.* (Commun.), vol. AP-19, pp. 557–558, July 1971.
[15] D. S. Jones, *The Theory of Electromagnetism.* London, England: Pergamon, 1964, pp. 566–569.
[16] D. L. Hutchens and R. G. Kouyoumjian, "A new asymptotic solution to the diffraction by a wedge," *1967 URSI Spring Meeting Dig.*, pp. 154–155.
[17] Y. Obha, "On the radiation patterns of a corner reflector finite in width," *IEEE Trans. Antennas Propagat.*, vol. AP-11, pp. 127–132, Mar. 1963.
[18] A. C. Wilson and H. V. Cottony, "Radiation patterns of finite size corner-reflector antennas," *IRE Trans. Antennas Propagat.*, vol. AP-9, pp. 144–157, Mar. 1960.
[19] S. W. Lee, "A ray theory of diffraction by open-ended waveguides, I. Fields in waveguides," *J. Math. Phys.*, vol. 11, pp. 2830–2850, 1970.
[20] H. Y. Yee, L. B. Felsen, and J. B. Keller, "Ray theory of reflection from the open end of a waveguide," *SIAM J. Appl. Math*, vol. 16, pp. 268–300, Mar. 1968.

Reflector Antenna Radiation Pattern Analysis by Equivalent Edge Currents

GRAEME L. JAMES AND VASSILIOS KERDEMELIDIS

Abstract—**Equivalent edge currents, derived from the edge diffraction theory for a half-plane, are used to obtain the radiation patterns of a paraboloidal reflector antenna when illuminated by a source at the focus. Cylindrical wave diffraction coefficients are used. The method avoids infinities at caustics and shadow boundaries thus giving solutions which are finite everywhere. A slope-wave equivalent current correction term is applied when the illumination is tapered towards the edge of the reflector. Comparisons are given with the physical optics approach and experimental results.**

Introduction

METHODS of evaluating the field scattered from a reflector antenna fall into three categories: aperture field, induced current, and edge diffraction methods. Aperture field methods [1, pp. 158–162] give accurate solutions up to a few beamwidths from the forward axis. The induced current approach is tractable if an estimate of the current distribution on the reflector is made such as the physical optics approximation [1, pp. 144–149]. This provides radiation fields in close agreement with experimental results in the forward direction. The physical optics approximation is unreliable, however, in the shadow region because the currents on the back of the reflector are neglected. Edge diffraction methods using the geometrical theory of diffraction (GTD) [2] give valid results in the shadow region since the currents on either side of the edge are accounted for implicitly. Such methods are finding increasing applications to reflector antennas [3]–[7]. The two main limitations to the straightforward application of the GTD, are the infinities in the fields in certain directions (along shadow boundaries and at caustics of the diffracted rays), and the cases where the incident field tapers to zero at the reflector edge. If the source is at a known finite distance from the edge, then cylindrical wave diffraction coefficients [8] can be used to yield finite solutions in the shadow boundary regions.

In this paper equivalent edge currents derived from cylindrical wave diffraction are used to overcome the remaining limitations, i.e., caustics, and the cases where the incident field goes to zero at the reflector edge. In addition, a slope-wave correction term is added to account for the general case of non-uniform incidence at the edge, e.g., the case when the illumination is tapered.

Manuscript received April 24, 1972; revised September 5, 1972.
The authors are with the Department of Electrical Engineering, University of Canterbury, Christchurch 1, New Zealand.

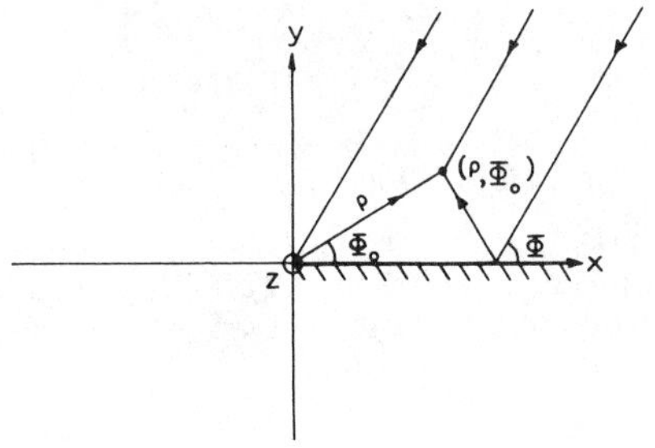

Fig. 1. Edge diffraction at half-plane.

Equivalent Current Method

The equivalent current concept was used with notable success by Millar [9]–[11] to solve for the axial and aperture fields in the case of an aperture in a plane screen. Recently, it has been used in other edge diffraction problems [12]. The approach here differs from [9]–[12] in that the equivalent currents, having been derived from cylindrical wave diffraction, are finite everywhere, and the radiation patterns of the currents are taken into account. (The previous applications were for fields in specific directions where the current elements were of fixed strength.)

To derive the equivalent edge currents for a half-plane, we begin with the diffracted field, in the (ρ,Φ,z) cylindrical polar coordinate system, due to a plane wave polarized in the z direction incident at an angle Φ upon a half-plane (Fig. 1), given from [8] as

$$u_{a \atop f} = V_D(\rho,\Phi_0 - \Phi) \mp V_D(\rho,\Phi_0 + \Phi) \tag{1}$$

where

$$V_D(\rho,\varphi) = \frac{-\exp(j\pi/4)}{\pi^{1/2}} \frac{|\cos(\varphi/2)|}{\cos(\varphi/2)} \cdot \exp(jk\rho\cos\varphi)F(s) \tag{2}$$

$$F(s) = \int_s^\infty \exp(-jt^2)\,dt \tag{3}$$

$$s = [k\rho(1+\cos\varphi)]^{1/2} \tag{4}$$

u_a (or u_f) is the scalar potential for electric (or magnetic) polarization in the z direction, k is the wavenumber, and the monochromatic time dependence $\exp(j\omega t)$ is

Reprinted from *IEEE Trans. Antennas Propagat.*, vol. AP-21, pp. 19–24, Jan. 1973.

suppressed. For s large,

$$V_D(\rho,\varphi) \sim \frac{-\exp(-j\pi/4)}{2(2\pi k)^{1/2}\cos(\varphi/2)} \frac{\exp(-jk\rho)}{\rho^{1/2}}. \qquad (5)$$

By reciprocity with the problem of plane wave diffraction in Fig. 1, the far field resulting from a z directed line source near the edge at (ρ_0,Φ_0), illustrated in Fig. 1 by reversing the direction of the arrows and replacing ρ with ρ_0, is given by [8]:

$$u_{\substack{a\\f}} = [V_D(\rho_0,\Phi - \Phi_0) \mp V_D(\rho_0,\Phi + \Phi_0)] \frac{\exp(-jk\rho)}{\rho^{1/2}} \qquad (6)$$

where the property $V_D(\rho,\varphi) = V_D(\rho,-\varphi)$ has been used.

Equation (6) has the form of a line source situated along the edge of the half-plane. Equating u_a (or u_f) in (6) with the far field from an infinitely long z directed filament of electric current I^e (or magnetic current I^m) along the z axis, given by [13], we get

$$I^{\substack{e\\m}} = \exp\left(\frac{j\pi}{4}\right) 2(2\pi k)^{1/2} \cdot [V_D(\rho_0,\Phi - \Phi_0) \mp V_D(\rho_0,\Phi + \Phi_0)]. \qquad (7)$$

For s large in (2), it follows that

$$I^{\substack{e\\m}} \sim -\left[\frac{1}{\cos\dfrac{\Phi - \Phi_0}{2}} \mp \frac{1}{\cos\dfrac{\Phi + \Phi_0}{2}}\right] \frac{\exp(-jk\rho_0)}{(\rho_0)^{1/2}}. \qquad (8)$$

Equation (7) gives the equivalent edge currents of the electric and magnetic diffraction far fields for the problem of a line source near the edge of a half-plane. The asymptotic value of (7), given by (8), is applicable everywhere, except in the region of the shadow boundaries. In this region it tends to infinity and (7) must be used.

Radiation Fields from Paraboloidal Reflectors Using Equivalent Edge Currents

The results of the previous section can be applied to radiation pattern analysis of reflector type antennas. It is assumed that in an element on the reflector edge, equivalent currents are given by (7) for a correspondingly oriented half-plane. The magnitudes of these currents are proportional to the incident electromagnetic field upon the element. Integration of the current elements around the reflector edge yields the diffracted fields. Addition of these fields to the reflected geometrical optics field and the direct radiation from the source, where it is not blocked by the reflector, gives the total radiation field. The application of this method to a circularly symmetrical paraboloidal reflector antenna will now be given.

Consider the paraboloidal reflector of Fig. 2 with the source at the focus. If the reflector is assumed to be in the far field of the source, then the electric field incident at the edge, $\boldsymbol{E}_e$, is given by

$$\boldsymbol{E}_e = [f_1(\xi,\psi_e)\hat{a}_\psi + f_2(\xi,\psi_e)\hat{a}_\xi] \frac{\exp(-jk\tau_e)}{\tau_e} \qquad (9)$$

with the symbols as defined in Fig. 2.

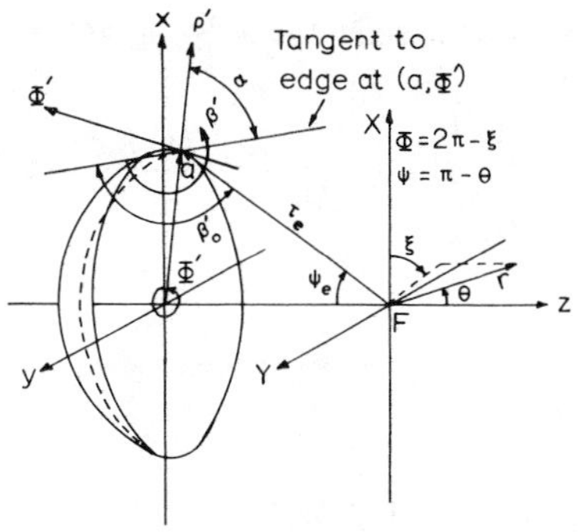

Fig. 2. Coordinates used for paraboloid.

The cross section of the reflector through (a,Φ') is shown in Fig. 2, where α is the angle between the tangent to the reflector edge at (a,Φ') and the ρ' axis, and β_0' is the angle of incidence of $\boldsymbol{E}_e$ to this tangent. For a symmetrical reflector, the quantities α, β_0', ψ_e, τ_e, are independent of Φ'. Initially, assume that the cross section at (a,Φ') represents a two-dimensional problem with a line source at the focus F. Then, from (9), the amplitude of the source in the direction of the edge is given by

$$\boldsymbol{E}_e = [f_1(2\pi - \Phi',\psi_e)\hat{a}_\psi + f_2(2\pi - \Phi',\psi_e)\hat{a}_\xi] \frac{1}{(\tau_e)^{1/2}}. \qquad (10)$$

With this amplitude term, the elemental equivalent electric (magnetic) source current $dJ'(d\boldsymbol{M}')$ at (a,Φ') for the diffracted field becomes

$$d\boldsymbol{J}' = \hat{a}_{\Phi'} \frac{\epsilon_2' f_2(2\pi - \Phi',\psi_e)}{j\omega\mu(\tau_e)^{1/2}} I^{e'} a\, d\Phi' \qquad (11)$$

$$d\boldsymbol{M}' = \hat{a}_{\Phi'} \frac{\epsilon_2' f_1(2\pi - \Phi',\psi_e)}{jk(\tau_e)^{1/2}} I^{m'} a\, d\Phi' \qquad (12)$$

where

$$\epsilon_2' = \begin{cases} 0, & \text{for } \beta' \le \alpha \\ 1, & \text{for } \beta' > \alpha \end{cases}$$

and β' is the angle measured from the tangent to the edge at (a,Φ') in the $\rho' - z$ plane. This term accounts for the blocking effect of the reflector on the far-field radiation from the equivalent currents.

Quantities $I^{e'}, I^{m'}$ are functions of the angle β'. Angles β' and θ are related by

$$\beta' = \begin{cases} \dfrac{\pi}{2} + \alpha + \theta, & \text{for } \dfrac{-\pi}{2} \le (\Phi - \Phi') < \dfrac{\pi}{2} \qquad (13) \\[2ex] \dfrac{\pi}{2} + \alpha - \theta, & \text{for } \dfrac{\pi}{2} \le (\Phi - \Phi') < \dfrac{3\pi}{2} \qquad (14) \end{cases}$$

where $0 \le \beta' \le 2\pi$, and θ is the angle measured from the z axis as shown in Fig. 2.

Writing $I^{\substack{e\\m}}(\pi/2 + \alpha + \theta)$ as $I^{\substack{e\\m}+}$, $I^{\substack{e\\m}}(\pi/2 + \alpha - \theta)$ as $I^{\substack{e\\m}-}$, and $(\Phi - \Phi')$ as φ, then integration of the current elements around the reflector edge gives the diffracted fields, $E_\theta{}^d, E_\Phi{}^d$, as

$$\left.\begin{matrix} E_\theta{}^d \\ \\ E_\Phi{}^d \end{matrix}\right\} = \frac{a}{(\tau_e)^{1/2}} \frac{\exp(-jkr)}{4\pi r} \left\{ \left[\epsilon_2{}^+ I^{e+} \int_{-\pi/2}^{\pi/2} f_2(2\pi + \varphi - \Phi, \psi_e) \begin{matrix} \cos\theta \sin\varphi \\ \\ \cos\varphi \end{matrix} \exp(j\gamma\cos\varphi)\, d\varphi \right.\right.$$

$$\left. + \epsilon_2{}^- I^{e-} \int_{\pi/2}^{3\pi/2} f_2(2\pi + \varphi - \Phi, \psi_e) \begin{matrix} \cos\theta \sin\varphi \\ \\ \cos\varphi \end{matrix} \exp(j\gamma\cos\varphi)\, d\varphi \right]$$

$$\pm \left[\epsilon_2{}^+ I^{m+} \int_{-\pi/2}^{\pi/2} f_1(2\pi + \varphi - \Phi, \psi_e) \begin{matrix} \cos\varphi \\ \\ \cos\theta \sin\varphi \end{matrix} \exp(j\gamma\cos\varphi)\, d\varphi \right.$$

$$\left.\left. + \epsilon_2{}^- I^{m-} \int_{\pi/2}^{3\pi/2} f_1(2\pi + \varphi - \Phi, \psi_e) \begin{matrix} \cos\varphi \\ \\ \cos\theta \sin\varphi \end{matrix} \exp(j\gamma\cos\varphi)\, d\varphi \right]\right\} \tag{15}$$

where $\gamma = ka \sin\theta$, and

$$\epsilon_2{}^\pm = \begin{cases} 0, & \text{for } \beta^\pm \leq \alpha, \quad \beta^\pm = \pi/2 + \alpha \pm \theta \\ 1, & \text{for } \beta^\pm > \alpha. \end{cases}$$

If the phase variations of f_1 and f_2 are small with respect to φ, then for γ large, direct application of the method of stationary phase [14] to (15) yields asymptotic expressions similar to those frequently used in the analysis of reflector antennas by edge diffraction theory, e.g., [5]. This method fails, however, in the axial regions of the reflector radiation pattern and in cases where the incident field tends towards zero at the stationary phase points on the reflector edge for the plane under consideration. For these cases it is necessary to evaluate the integrals in (15). The analytical solution is given in the Appendix.

In many practical applications the source is linearly polarized, so that (9) becomes

$$\boldsymbol{E}_e = [f_1(\psi_e) \sin\xi \hat{a}_\psi + f_2(\psi_e) \cos\xi \hat{a}_\xi] \frac{\exp(-jk\tau_e)}{\tau_e} \tag{16}$$

and the solution of (15) given in the Appendix by (35) becomes

$$\left.\begin{matrix} E_\theta{}^d \\ \\ E_\Phi{}^d \end{matrix}\right\} = \frac{a}{(\tau_e)^{1/2}} \frac{\exp(-jkr)}{8r} \begin{matrix} \sin\Phi \\ \\ \cos\Phi \end{matrix} \left[\left(I^{e+} f_2(\psi_e) \begin{matrix} G_A(\gamma)\cos\theta \\ \\ G_B(\gamma) \end{matrix} - I^{m+} f_1(\psi_e) \begin{matrix} G_B(\gamma) \\ \\ G_A(\gamma)\cos\theta \end{matrix} \right) \epsilon_2{}^+ \right.$$

$$\left. + \left(I^{e-} f_2(\psi_e) \begin{matrix} G_C(\gamma)\cos\theta \\ \\ G_D(\gamma) \end{matrix} - I^{m-} f_1(\psi_e) \begin{matrix} G_D(\gamma) \\ \\ G_C(\gamma)\cos\theta \end{matrix} \right) \epsilon_2{}^- \right] \tag{17}$$

where

$$G_{\substack{A\\C}}(\gamma) = J_0(\gamma) + J_2(\gamma) \pm \frac{j8}{\pi} \sum_{m=1,3,\cdots}^{\infty} \frac{m^2 - 2}{m(m^2 - 4)} J_m(\gamma)$$

$$G_{\substack{B\\D}}(\gamma) = J_0(\gamma) - J_2(\gamma) \mp \frac{j16}{\pi} \sum_{m=1,3,\cdots}^{\infty} \frac{1}{m(m^2 - 4)} J_m(\gamma) \tag{18}$$

and $J_m(\gamma)$ is the Bessel function of order m and argument γ. Equation (17) gives the diffracted fields for a linearly polarized source at the focus.

Slope-Wave Diffraction

In the formulation of the equivalent currents it was assumed that the half-plane was illuminated by a uniform plane wave. If, however, the incident field upon the reflector is tapered towards the edge, the previous theory does not predict the consequent modification to the diffracted field. Ahluwalia *et al.* [15] have considered the case of a nonuniform wave incident upon a screen. For a half-plane illuminated by a nonuniform plane wave (Fig. 1), their result for the field reduces to

$$u_{\substack{a\\f}} z(0) - \frac{1}{jk} \frac{\delta}{\delta\Phi_0} u_{\substack{a\\f}} \left.\frac{\delta z(0)}{\delta\Phi}\right|_{\Phi=\Phi_0} + [\text{higher order terms}] \tag{19}$$

where $z(0)$ is the value of the incident field at the edge, and u_a is the asymptotic form of (1) given by (5). The second term in (19) has the same form as the corresponding term in the slope-wave diffraction analysis of Rudduck and Wu [16] and agrees with the approach taken by Keller [2] for the special cases when $\Phi_0 = 0,\pi$. This correction term is proportional to the derivative of the incident field and is called the slope-wave diffraction term.

Applying the slope-wave diffraction to a nonuniform line source at a distance ρ_0 from the edge, as represented in Fig. 1 with the arrows reversed and ρ replaced with ρ_0, the second term in (19) becomes

$$-\frac{1}{jk\rho_0}\frac{\delta}{\delta\Phi_0}u_{a\,f}\frac{\delta h(\Phi)}{\delta\Phi}\bigg|_{\Phi=\Phi_0}$$

where $h(\Phi)$ is the radiation pattern of the line source. Defining the slope-wave equivalent currents, I^{se},I^{sm} as

$$I^{\substack{se\\sm}} = -\frac{1}{jk\rho_0}\frac{\delta}{\delta\Phi_0}I^{\substack{e\\m}} \tag{20}$$

then from (2) and (7)

$$I^{\substack{se\\sm}} = -4(2k)^{1/2}[V_{SD}(\rho_0,\Phi-\Phi_0) \pm V_{SD}(\rho_0,\Phi+\Phi_0)] \tag{21}$$

where

$$V_{SD}(\rho_0,\varphi) = \frac{|\cos(\varphi/2)|}{\operatorname{cosec}(\varphi/2)}\exp(jk\rho_0)\cdot\cos\varphi\left[\frac{\exp(-js^2)}{2s} - jF(s)\right]. \tag{22}$$

Quantities $F(s),s$ are defined in (3),(4), respectively. For s large, (21) becomes

$$I^{\substack{se\\sm}} \sim -\frac{1}{j2k\rho_0}\left[\frac{\sin\dfrac{\Phi-\Phi_0}{2}}{\cos^2\dfrac{\Phi-\Phi_0}{2}} \pm \frac{\sin\dfrac{\Phi+\Phi_0}{2}}{\cos^2\dfrac{\Phi+\Phi_0}{2}}\right]\frac{\exp(-jk\rho_0)}{(\rho_0)^{1/2}}. \tag{23}$$

Applying the slope-wave equivalent currents to the solution for the linearly polarized source, given by (17), yields

$$\left.\begin{matrix}E_\theta{}^d\\E_\Phi{}^d\end{matrix}\right\} = \frac{a}{(\tau_e)^{1/2}}\frac{\exp(-jkr)}{8r}\begin{matrix}\sin\Phi\\ \cos\Phi\end{matrix}\left\{\left[(I^{e+}f_2(\psi_e) + I^{se+}f_2'(\psi_e))\begin{matrix}G_A(\gamma)\cos\theta\\G_B(\gamma)\end{matrix} - (I^{m+}f_1(\psi_e) + I^{sm+}f_1'(\psi_e))\begin{matrix}G_B(\gamma)\\G_A(\gamma)\cos\theta\end{matrix}\right]\epsilon_2{}^+\right.$$
$$\left.+\left[(I^{e-}f_2(\psi_e) + I^{se-}f_2'(\psi_e))\begin{matrix}G_C(\gamma)\cos\theta\\G_D(\gamma)\end{matrix} - (I^{m-}f_1(\psi_e) + I^{sm-}f_1'(\psi_e))\begin{matrix}G_D(\gamma)\\G_C(\gamma)\cos\theta\end{matrix}\right]\epsilon_2{}^-\right\} \tag{24}$$

where

$$f_1'(\psi_e) = \frac{\delta}{\delta\psi}f_1(\psi_e)\bigg|_{\psi=\psi_0}; \qquad f_2'(\psi_e) = \frac{\delta}{\delta\psi}f_2(\psi_e)\bigg|_{\psi=\psi_0}.$$

For γ large, by the method of stationary phase, the asymptotic value of (24) is given by

$$E_\theta{}^d \sim -\frac{\exp(-jkr)}{4\pi r}\left[\frac{2\pi a^2}{\gamma\tau_e}\right]^{1/2}\sin\Phi[\epsilon_2{}^+(I^{m+}f_1(\psi_e) + I^{sm+}f_1'(\psi_e))\exp[j(\gamma-\pi/4)] + \epsilon_2{}^-(I^{m-}f_1(\psi_e) + I^{sm-}f_1'(\psi_e))\exp[-j(\gamma-\pi/4)]] \tag{25}$$

$$E_\Phi{}^d \sim \frac{\exp(-jkr)}{4\pi r}\left[\frac{2\pi a^2}{\gamma\tau_e}\right]^{1/2}\cos\Phi[\epsilon_2{}^+(I^{e+}f_2(\psi_e) + I^{se+}f_2'(\psi_e))\exp[j(\gamma-\pi/4)] + \epsilon_2{}^-(I^{e-}f_2(\psi_e) + I^{se-}f_2'(\psi_e))\exp[-j(\gamma-\pi/4)]]. \tag{26}$$

Equations (24)–(26) give the diffracted fields, including the slope-wave term, for a linearly polarized source at the focus.

Experimental and Theoretical Results

In the edge diffraction method, the field scattered from the reflector is given by the sum of the fields due to the equivalent currents and the geometrical optics term which, for a paraboloidal reflector, exists only along the forward axis. The scattered field for the H plane of a focal plane ($F/D = 0.25$) paraboloidal reflector illuminated by a short dipole at the focus, is given in Fig. 3, where in (16) we have $f_1(\psi) = \cos\psi$, $f_2(\psi) = 1$ for the normalized radiation pattern of the dipole, and $f_2'(\psi_e) = f_1(\psi_e) = 0$, $f_1'(\psi_e) = -1$ in (24). Fig. 3 shows that in the shadow region, considerable discrepancy exists between the equivalent current and physical optics methods. In forward directions the agreement is good except in the region of the axis. On the axis itself the edge diffraction field is given essentially by the predominant geometrical optics term and this has the same value as the physical optics field.

The H-plane pattern calculated from (24) could have been computed from the stationary phase formula of (26) with good accuracy except in the axial regions. For the E plane without the slope-wave term, stationary phase methods fail since the incident field from the source is zero at the stationary phase points on the reflector edge. With the slope-wave term included, (25) gives an accurate answer only in the lateral region of the radiation pattern. The E-plane patterns of Fig. 4, calculated from

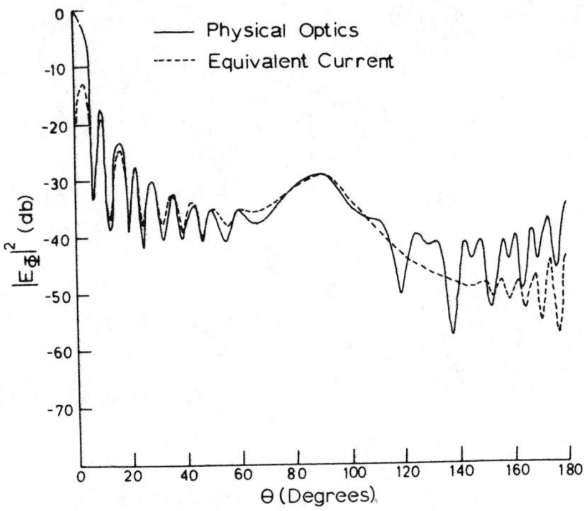

Fig. 3. H-plane field scattered from paraboloid illuminated by short dipole; $F/D = 0.25$, $D = 10\lambda$.

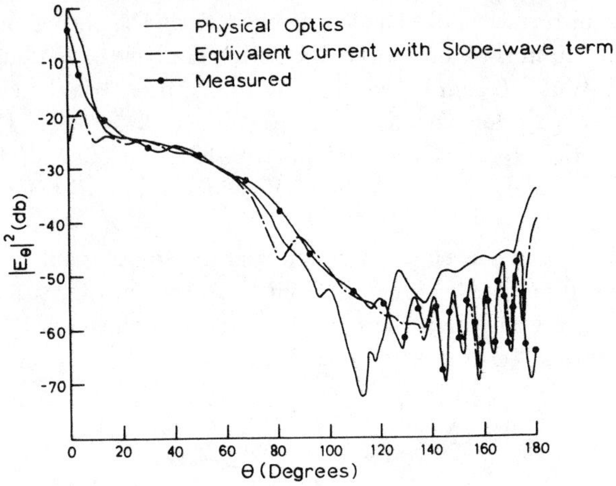

Fig. 6. Total E-plane field for paraboloid illuminated by short dipole; $F/D = 0.25$, $D = 10\lambda$.

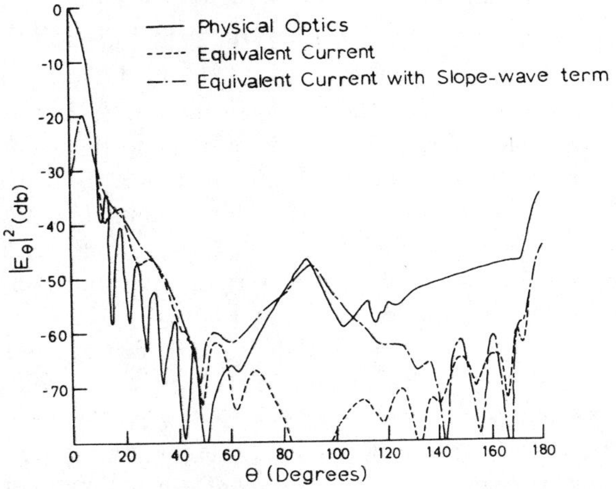

Fig. 4. E-plane field scattered from paraboloid illuminated by short dipole; $F/D = 0.25$, $D = 10\lambda$.

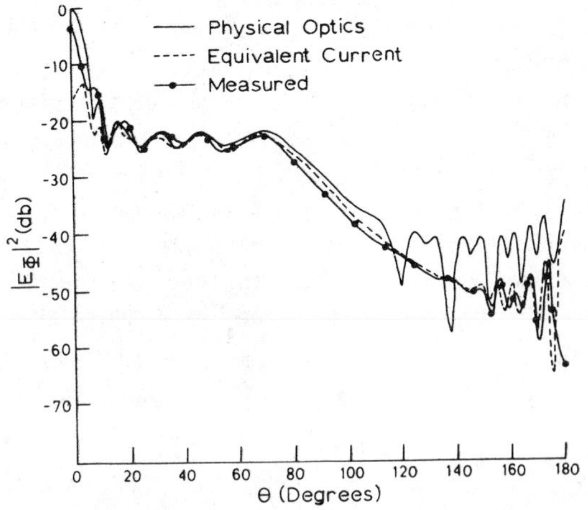

Fig. 5. Total H-plane field for paraboloid illuminated by short dipole; $F/D = 0.25$, $D = 10\lambda$.

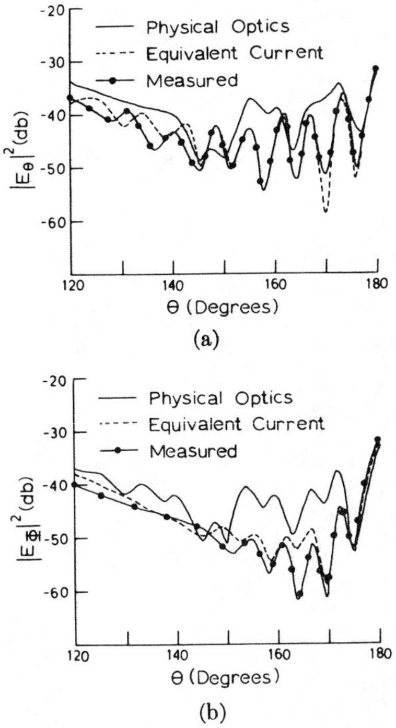

Fig. 7. Shadow region fields for paraboloid illuminated by small horn; $F/D = 0.43$, $D = 10\lambda$. (a) E plane. (b) H plane.

(24), indicate the importance of the slope-wave term in the lateral region.

Fig. 5 gives the measured H-plane field for a focal plane reflector illuminated by a half-wave dipole at its focus. The superiority of the equivalent current method in the shadow region is evident. The dip in the measured pattern on the back axis was caused by the feeding arrangement at the back of the reflector.

The field on the axis of a physical dipole is finite, consequently, $f_1(\psi_e)$ is not zero as assumed in Fig. 4. Although $f_1(\psi_e)$ is small, its contribution to the E-plane radiation is significant since it is the value of the incident field at the stationary phase points on the edge of the reflector for this plane. Fig. 6 shows that with $f_1(\psi_e)$ nonzero, the shadow region radiation in the E plane is significantly different from the ideal results cited in Fig. 4.

As another example, theory and experiment are compared for a front-horn-fed reflector, where the back of the reflector is clear of obstacles. The results are given in Fig. 7 for the shadow region of the principal planes.

Conclusion

The equivalent current method gives results comparable to the physical optics approach in the forward direction except in the region of the main beam. In the shadow region, the equivalent current method yields results in better agreement with the measured values than does the physical optics approach. It is in this region, where other methods fail, that the edge diffraction approach given in this paper is especially useful.

Appendix

Analytical Solution of General Equation for Diffraction Far Field

The analytical solution of (15) in the text is given here. In (15) the integrals are of the form

$$\Xi(\theta) = \int_{\theta}^{\pi+\theta} g(\varphi) \exp(j\gamma\cos\varphi)\, d\varphi. \tag{27}$$

Expressing $g(\varphi)$ in a Fourier series gives

$$g(\varphi) = \sum_{n=-\infty}^{\infty} G_n \exp(jn\varphi) \tag{28}$$

where

$$G_n = \frac{1}{2\pi}\int_0^{2\pi} g(\varphi)\exp(-jn\varphi)\, d\varphi \tag{29}$$

then

$$\Xi(\theta) = \sum_{n=-\infty}^{\infty} G_n K_n(\theta) \tag{30}$$

where

$$K_n(\theta) = \int_{\theta}^{\pi+\theta} \exp\{j[n\varphi + \gamma\cos\varphi]\}\, d\varphi. \tag{31}$$

Expressing the second exponential term of (31) in a Bessel function series gives

$$K_n(\theta) = \int_{\theta}^{\pi+\theta} \exp(jn\varphi) \cdot \left[J_0(\gamma) + 2\sum_{m=1}^{\infty} j^m J_m(\gamma)\cos m\varphi\right] d\varphi. \tag{32}$$

Evaluating the integral in (32) yields

$$K_n(\theta) = \pi\Big[J_0(\gamma)L(n,\theta) + \sum_{m=1}^{\infty} j^m J_m(\gamma) \cdot \big(L(n+m,\theta) + L(n-m,\theta)\big)\Big] \tag{33}$$

where

$$L(p,\theta) = \exp\left[jp\left(\theta + \frac{\pi}{2}\right)\right]\frac{\sin(p\pi/2)}{(p\pi/2)}. \tag{34}$$

Applying (30) to (15) in the text yields

$$\left.\begin{matrix} E_\theta^d \\ E_\Phi^d \end{matrix}\right\} = \frac{a}{(\tau_e)^{1/2}}\frac{\exp(-jkr)}{4\pi r} \cdot \sum_{n=-\infty}^{\infty}\left[K_n\left(-\frac{\pi}{2}\right)\left(I^{e+}\begin{matrix} a_n\cos\theta \\ c_n \end{matrix} \pm I^{m+}\begin{matrix} b_n \\ d_n\cos\theta \end{matrix}\right)\epsilon_2^+ + K_n\left(\frac{\pi}{2}\right)\left(I^{e-}\begin{matrix} a_n\cos\theta \\ c_n \end{matrix} \pm I^{m-}\begin{matrix} b_n \\ d_n\cos\theta \end{matrix}\right)\epsilon_2^-\right] \tag{35}$$

where

$$\left.\begin{matrix} a_n \\ c_n \end{matrix}\right\} = \frac{1}{2\pi}\int_0^{2\pi} f_2(2\pi+\varphi-\Phi,\psi_e)\begin{matrix}\sin\varphi \\ \cos\varphi\end{matrix}\exp(-jn\varphi)\, d\varphi$$

$$\left.\begin{matrix} b_n \\ d_n \end{matrix}\right\} = \frac{1}{2\pi}\int_0^{2\pi} f_1(2\pi+\varphi-\Phi,\psi_e)\begin{matrix}\cos\varphi \\ \sin\varphi\end{matrix}\exp(-jn\varphi)\, d\varphi. \tag{36}$$

References

[1] S. Silver, Ed., *Microwave Antenna Theory and Design.* New York: McGraw-Hill, 1949, pp. 144–149, 158–162.
[2] J. B. Keller, "Geometrical theory of diffraction," *J. Opt. Soc. Amer.*, vol. 52, pp. 116–132, Feb. 1962.
[3] W. V. T. Rusch and P. D. Potter, *Analysis of Reflector Antennas.* New York: Academic Press, 1970, pp. 51–57.
[4] B. E. Kinber, "The role of diffraction at the edges of a paraboloid in fringe radiation," *Radio Eng. Electron Phys.* (*USSR*), vol. 7, pp. 79–86, 1962.
[5] ——, "Lateral radiation of parabolic antennas," *Radio Eng. Electron Phys.* (*USSR*), vol. 6, pp. 481–492, 1961.
[6] S. Pogorzelski, "Diffraction by the edge of an antenna reflector," *Rozpr. Elecktrotech.*, vol. 13, pp. 451–475, 1967.
[7] V. P. Narbut and K. S. Khmel'nitskaya, "Polarization structure of radiation from axisymmetric reflector antennas," *Radio Eng. Electron. Phys.* (*USSR*), vol. 15, pp. 1786–1796, 1970.
[8] R. C. Rudduck, "Application of wedge diffraction to antenna theory," NASA Grant NSG-448, Dep. 1691-13, June 30, 1965.
[9] R. F. Millar, "An approximate theory of the diffraction of an electromagnetic wave by an aperture in a plane screen," *Proc. Inst. Elec. Eng.*, vol. 103C, pp. 177–185, 1955.
[10] ——, "The diffraction of an electromagnetic wave by a circular aperture," *Proc. Inst. Elec. Eng.*, vol. 104C, pp. 87–95, 1956.
[11] ——, "The diffraction of an electromagnetic wave by a large aperture," *Proc. Inst. Elec. Eng.*, vol. 104C, pp. 240–250, 1956.
[12] C. E. Ryan and L. Peters, "Evaluation of edge diffracted fields including equivalent currents for the caustic regions," *IEEE Trans. Antennas Propagat.*, vol. AP-17, pp. 292–299, May 1969.
[13] R. F. Harrington, *Time Harmonic Electromagnetic Fields.* New York: McGraw-Hill, 1961, p. 224.
[14] A. Erdelyi, *Asymptotic Expansions.* New York: Dover, 1956, p. 51.
[15] D. S. Ahluwalia, R. M. Lewis, and J. Boersma, "Uniform asymptotic theory of diffraction by a plane screen," *SIAM J. Appl. Math.*, vol. 16, pp. 783–807, 1968.
[16] R. C. Rudduck and D. C. F. Wu, "Slope diffraction analysis of T. E. M. parallel plate guide radiation patterns," *IEEE Trans. Antennas Propagat.*, vol. AP-17, pp. 797–799, Nov. 1969.

A Comparison of Geometrical and Integral Fields From High-Frequency Reflectors

W. V. T. RUSCH

Abstract—Various comparisons are made between geometrical optics, integrated physical optics, asymptotic physical optics, the geometrical theory of diffraction, and experimental data for fields from high-frequency paraboloidal and hyperboloidal reflectors.

Physical optics (PO), whereby the free-space dyadic Green's function is integrated over the geometrical-optics (GO) current distribution, is commonly used to analyze high-frequency reflectors, particularly, focusing reflectors [1]. Geometrical representations of the fields in terms of reflected or diffracted rays become possible when the field integrals possess stationary points which geometrically satisfy the generalized Fermat's principle. The most familiar representations of this type are GO [2] and the geometrical theory of diffraction (GTD) [3]. An additional stationary point exists when the observation point lies within the shadow "cast" by the reflector when it is illuminated by a source. In the limit $k \to \infty$ the contribution from this stationary point exactly cancels the source field in the shadow [4]. It is the purpose of this letter to present some comparisons between integral and geometrical fields for both convex and concave high-frequency reflectors.

The focused paraboloid with a point-source feed does not have a geometrical-optics stationary point. However, Kouyoumjian has shown that a combination of singly edge-diffracted rays for wide angles, equivalent ring currents for the rear axial caustic, and PO for the main-beam region will provide an accurate representation of the field which compares favorably with experimental data [5]. Fig. 1 is a typical segment of the calculated H-plane radiation pattern of a paraboloid for which $F/D = 0.4$, $D/\lambda = 25$, and the feed has a symmetric $\cos\theta$ pattern. The dashed curve is the PO result integrated over the illuminated front of the reflector. The solid curve represents the PO result evaluated asymptotically with saddle-point theory, producing equivalent PO geometrically edge-diffracted rays. The circles represent the Kouyoumjian GTD results. The two edge-diffracted results are virtually indistinguishable, and except for a shift of about 0.2° in the position of the maxima, the agreement with the integral PO result (which requires approximately 10^2 longer to evaluate each field point) is excellent.

GO, on the other hand, can be used to compute the radar cross section of a paraboloidal reflector. For an axially incident plane wave the result is $4\pi F^2$. For other angles of incidence the normalized cross section is plotted in Fig. 2 as a function of the angle of incidence. Plotted also are the integrated PO results for 10 and 31.6 wavelength reflectors. The integral result produces a nonzero cross section in the region beyond the reflection boundary.

The original Keller formulation of GTD [6] was applied to hyperboloids as early as 1967 [7]. The results compared favorably with

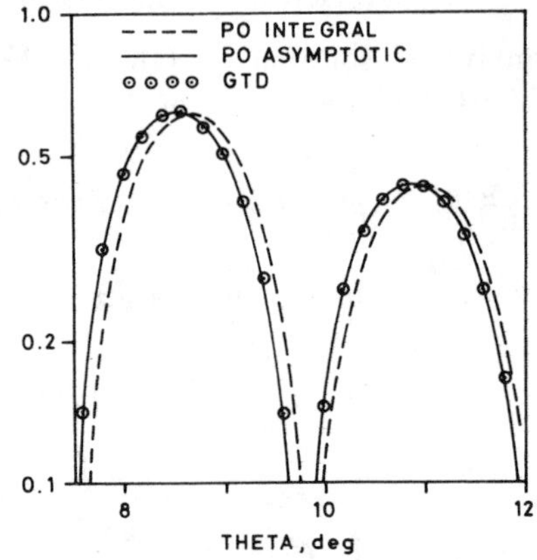

Fig. 1. Segment of paraboloid H-plane pattern for $\cos\theta$ feed, $F/D = 0.4$, $D/\lambda = 25.0$.

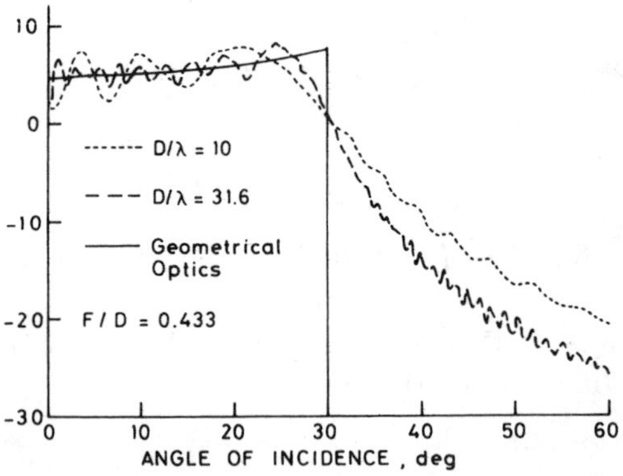

Fig. 2. Normalized radar cross section of a paraboloid.

the PO results, except for the singularities at shadow and reflection boundaries. The Kouyoumjian formulation with transition functions eliminated these singularities and yielded an accurate, rapidly computed radiation pattern. Typical amplitude and phase results [8] are plotted in Fig. 3 which compares GO, PO, GTD (Keller), and GTD (Kouyoumjian). The two GTD versions consists of the GO ray plus two singly edge-diffracted rays, with the Keller result exhibiting the singularity near the reflection boundary. In general, the PO and Kouyoumjian GTD curves agree closely in magnitude and phase, both in the illuminated and shadowed regions. The PO oscillations are somewhat larger than the GTD oscillations, indicative of the fact that two different edge current densities were assumed for the two results. In fact, in the H-plane the E vector is tangent to the edge at the point of edge-diffraction, and the actual current density becomes infinite at the edge for this polarization. Of particular significance is the agreement between the PO and Kouyoumjian GTD results in the vicinity of the reflection boundary. Similar results are found for the E-plane.

Manuscript received March 28, 1974; revised June 5, 1974.
The author is with the Electromagnetics Institute, Technical University of Denmark, DK-2800 Lyngby, Denmark, on leave from the Department of Electrical Engineering, University of Southern California, Los Angeles, Calif. 90007.

Reprinted from *Proc. IEEE*, vol. 62, pp. 1603–1604, Nov. 1974.

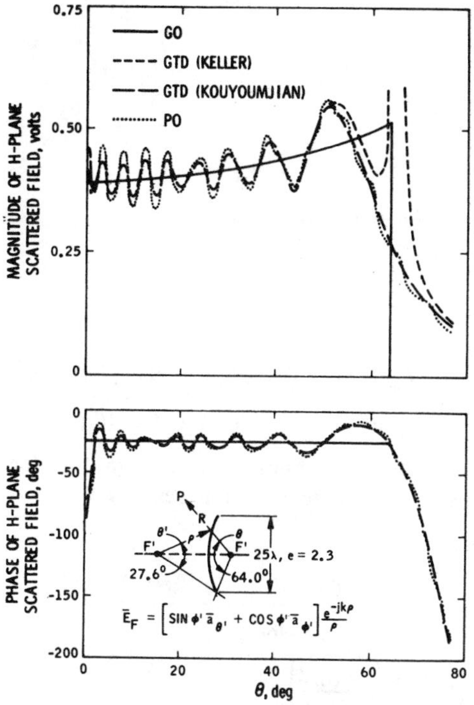

Fig. 3. Comparison of *H*-plane fields for hyperboloid with feed at external focus.

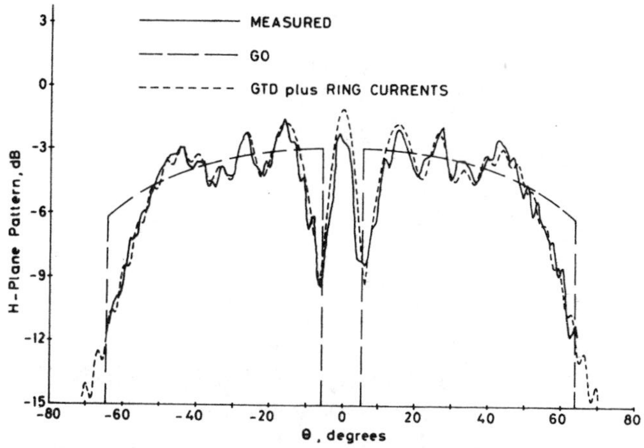

Fig. 4. Comparison of experimental and calculated GTD results for 11.7-wavelength hyperboloid with corrugated horn feed at external focus.

Fig. 4 presents a comparison of the calculated and measured scattered *H*-plane patterns from an 11.7-wavelength hyperboloid illuminated by a 10-GHz corrugated horn located at the hyperboloid's external focus [9]. The GO result includes the effects of sizable blocking by the horn. The GTD results include two edge rays from the reflector and one singly diffracted ray from the leading edge of the horn on the field-point side of the axis. For field points near the axis it is necessary to include contributions from equivalent ring currents around both the reflector and horn aperture rims. The GTD and experimental results agree within the experimental uncertainties involved, indicating the usefulness of the GTD procedures for such relatively complicated horn–reflector combinations.

References

[1] W. V. T. Rusch and P. D. Potter, *Analysis of Reflector Antennas.* New York: Academic Press, 1970.
[2] F. S. Holt, "Application of geometrical optics to the design and analysis of microwave antennas," A. F. Cambridge Res. Lab., Bedford, Mass., Rep. AFCRL-67-0501, Sept. 1967.
[3] R. G. Kouyoumjian, "The geometrical theory of application and applications," in *Numerical Techniques for Antennas and Electromagnetics* (short course notes), vol. VI. Univ. Southern California, June 25–29, 1973.
[4] W. V. T. Rusch, "Antenna notes," NB 84, vol. II, Electromagnetics Inst., Tech. Univ. Denmark, DK-2800 Lyngby, Denmark.
[5] P. A. J. Ratnasiri, R. G. Kouyoumjian, and P. H. Pathak, "The wide angle side-lobes of reflector antennas," Elec. Sci. Lab., The Ohio State Univ., Columbus, Ohio, Rep. 2183-1, 1970.
[6] J. B. Keller, "Diffraction by an Aperture," *J. Appl. Phys.*, vol. 28, pp. 426–444, 1957.
[7] W. V. T. Rusch, "Edge diffraction from truncated paraboloids and hyperboloids," Jet Propulsion Lab., Pasadena, Calif., JPL Tech. Rep. 32-1113, June 1, 1967.
[8] W. V. T. Rusch, Jet Propulsion Lab., Pasadena, Calif., Memo 3335-72-089, Sec. 333, August 21, 1972.
[9] O. Sørensen, "Analysis of defocused cassegrain antennas using geometrical theory of diffraction," Master's thesis, Lab. of Electromagnetic Theory, Tech. Univ. Denmark, Lyngby, Denmark, Fall 1973.

The Geometrical Theory of Diffraction for Axially Symmetric Reflectors

W. V. T. RUSCH, SENIOR MEMBER, IEEE, AND O. SØRENSEN

***Abstract*—The geometrical theory of diffraction (GTD) (cf. [1], for example) may be applied advantageously to many axially symmetric reflector antenna geometries. The material in this communication presents analytical, computational, and experimental results for commonly encountered reflector geometries, both to illustrate the general principles and to present a compact summary of generally applicable formulas.**

I. Axially Symmetric Reflector for Vector Spherical-Wave Point Source on Axis

If a point-source feed is located on the axis of an axially symmetric reflector, then only two singly edge-diffracted rays are possible for a distant field point $P(R,\theta,\phi)$, if $\theta \neq 0,\pi$ (Fig. 1). These rays are "diffracted" from Q_E^+, where the plane ϕ intersects the edge on the *same* side of the Z axis as the field point, and from Q_E^-, where the plane $\phi + \pi$ intersects the edge diametrically opposite Q_E^+ on the *opposite* side of the Z axis from the field point.

The geometry of the edge-diffracted ray from Q_E^- is shown in Fig. 2. (The surface is assumed to be convex.) The ray from the feed to Q_E^- defines θ'_{edge}, and the corresponding extreme geometrically reflected ray, when extended back to the Z axis, defines θ_{edge}. These two angles may then be used to define two intermediate angles used in the edge-diffraction notation of J. B. Keller [2]:

$$\delta_t = \frac{\theta_{\text{edge}} - \theta'_{\text{edge}}}{2} \tag{1a}$$

$$\alpha = -\frac{\theta_{\text{edge}} + \theta'_{\text{edge}}}{2}. \tag{1b}$$

Let the total singly edge-diffracted field be defined by

$$E_d = [E_{d\theta}(\theta,\phi)a_\theta + E_{d\phi}(\theta,\phi)a_\phi]\,\frac{e^{-jkR}}{R} \tag{2}$$

where the angles and unit vectors for both sets of feeds are defined in the sense indicated in Fig. 1. Then the components of E_d can be decomposed into contributions from both Q_E^+ and Q_E^-

$$E_{d\theta}(\theta,\phi) = E_{d\theta}^+(\theta,\phi) + E_{d\theta}^-(\theta,\phi) \tag{3a}$$

$$E_{d\phi}(\theta,\phi) = E_{d\phi}^+(\theta,\phi) + E_{d\phi}^-(\theta,\phi). \tag{3b}$$

Considering first the diffracted field from Q_E^- (because no transition functions are required for these rays)

$$E_{d\theta}^-(\theta,\phi) = \left\{-\frac{E_{f\theta}(\pi - \theta'_{\text{edge}}, \phi + \pi)e^{-jk\rho_{\text{edge}}}}{\rho_{\text{edge}}}\right\} \cdot \left\{e^{j(\pi/2)}\sqrt{\frac{D_{\text{refl}}/2}{\sin\theta}}\right\} \cdot \{\exp[jk[(2c + Z_{\text{edge}})\cos\theta - (D_{\text{refl}}/2)\sin\theta]]\}D_h^- \tag{4a}$$

$$E_{d\phi}^-(\theta,\phi) = \left\{-\frac{E_{f\phi}(\pi - \theta'_{\text{edge}}, \phi + \pi)e^{-jk\rho_{\text{edge}}}}{\rho_{\text{edge}}}\right\} \cdot \left\{e^{j(\pi/2)}\sqrt{\frac{D_{\text{refl}}/2}{\sin\theta}}\right\} \cdot \{\exp[jk[(2c + Z_{\text{edge}})\cos\theta - (D_{\text{refl}}/2)\sin\theta]]\}D_s^- \tag{4b}$$

where $E_{f\theta}$ and $E_{f\phi}$ are the θ and ϕ components of the incident field, ρ_{edge} is the distance from feed to Q_E^-, Z_{edge} is the Z coordinate of $Q_E^-(Z_{\text{edge}} < 0)$, D_{refl} is the reflector diameter, $Z = -2c$ is an arbitrary phase reference point on the negative Z axis, and the diffraction coefficients are given by

$$D_{\substack{s\\h}}^- = \begin{cases} \left\{-\dfrac{e^{-j(\pi/4)}}{2\sqrt{2\pi k}}\left[\dfrac{1}{\cos(\theta + \delta_t + \alpha)/2} \mp \dfrac{1}{\sin(\theta + \delta_t - \alpha)/2}\right]\right\}, & 0 < \theta \le \dfrac{\pi}{2} - \delta_t \\ 0, & \dfrac{\pi}{2} - \delta_t < \theta < \dfrac{\pi}{2} \\ \left\{-\dfrac{e^{-j(\pi/4)}}{2\sqrt{2\pi k}}\left[-\dfrac{1}{\cos(\theta + \delta_t + \alpha)/2} \pm \dfrac{1}{\sin(\theta + \delta_t - \alpha)/2}\right]\right\}, & \dfrac{\pi}{2} \le \theta < \pi \end{cases} \tag{5}$$

where the upper signs are taken for the soft (s) coefficient and the lower signs are taken for the hard (h) coefficient. Notice that

Manuscript received July 9, 1974; revised December 30, 1974.
W. V. T. Rusch is with the Electromagnetics Institute, Technical University of Denmark, Lyngby, Denmark, on leave from the Department of Electrical Engineering, University of Southern California, Los Angeles, Calif. 90007.
O. Sørensen is with the Electromagnetics Institute, Technical University of Denmark, Lyngby, Denmark.

Reprinted from *IEEE Trans. Antennas Propagat.*, vol. AP-23, pp. 414–419, May 1975.

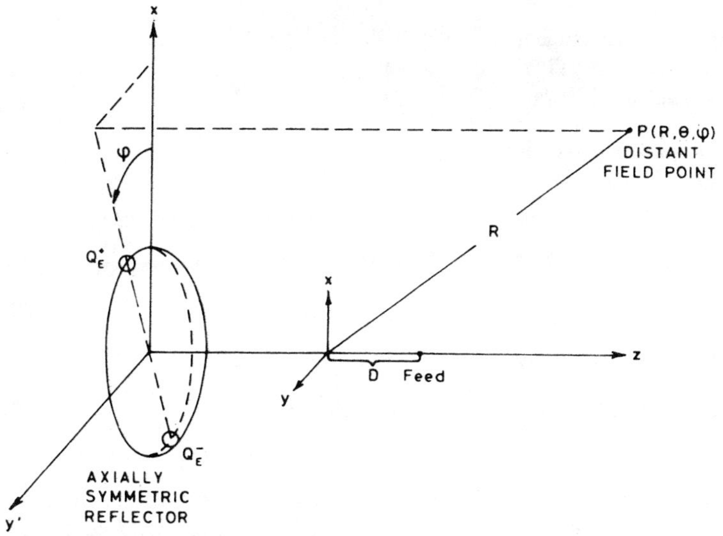

Fig. 1. Geometry of edge-diffraction points for axially symmetric reflector with feed on axis.

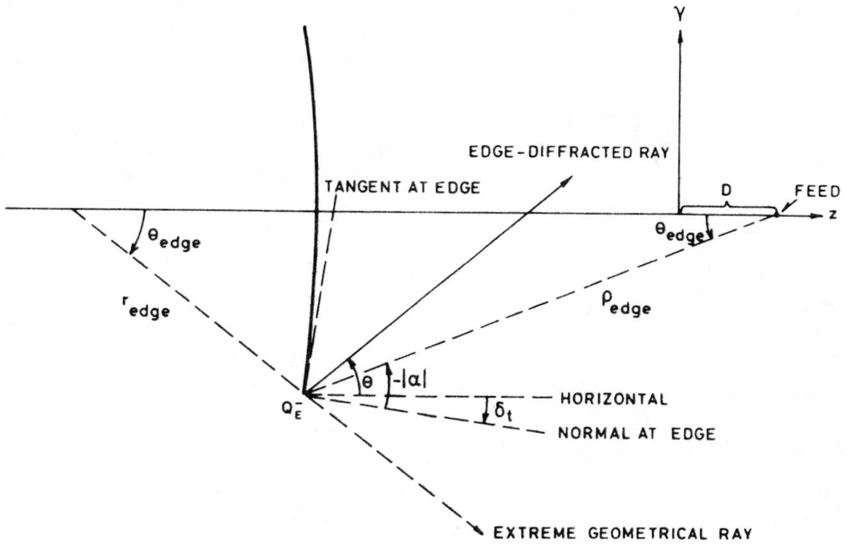

Fig. 2. Geometry of ray diffracted from Q_e^-.

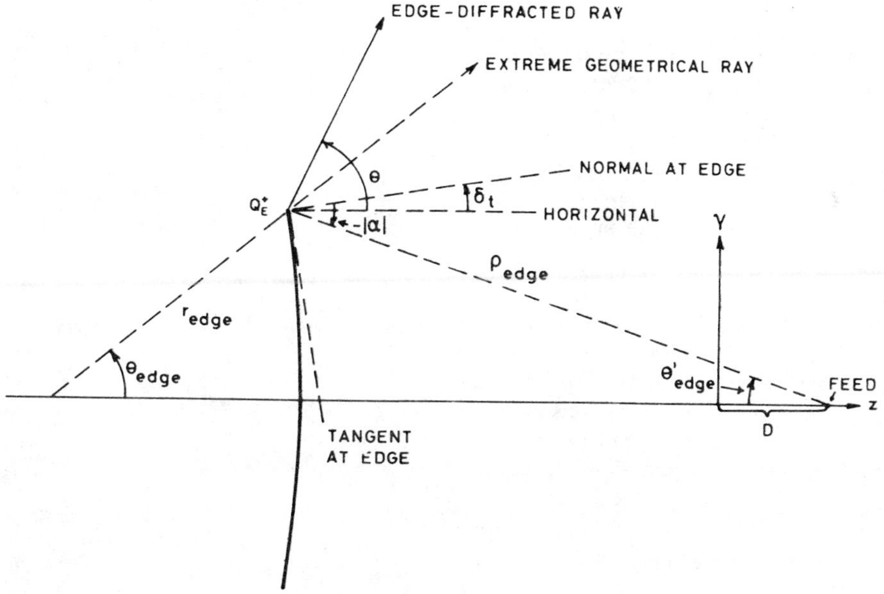

Fig. 3. Geometry of ray diffracted from Q_e^+.

D_s^- is zero in the direction of the surface tangent, $\theta = \pi/2 - \delta$, but D_h^- is not. Slightly modified but principally the same diffraction coefficients for singly edge-diffracted rays may be derived for concave surfaces.

The geometry of the ray diffracted from Q_E^+ is shown in Fig. 3. The fields of this ray are given by

$$E_{d\theta}^+(\theta,\phi) = \left\{\frac{E_{f\theta}(\pi - \theta'_{\text{edge}},\phi)e^{-jk\rho_{\text{edge}}}}{\rho_{\text{edge}}}\right\}\left\{\sqrt{\frac{D_{\text{refl}}/2}{\sin\theta}}\right\} \cdot \{\exp[jk[(2c + Z_{\text{edge}})\cos\theta + (D_{\text{refl}}/2)\sin\theta]]\}D_h^+ \tag{6a}$$

$$E_{d\phi}^+(\theta,\phi) = \left\{\frac{E_{f\phi}(\pi - \theta'_{\text{edge}},\phi)e^{-jk\rho_{\text{edge}}}}{\rho_{\text{edge}}}\right\}\left\{\sqrt{\frac{D_{\text{refl}}/2}{\sin\theta}}\right\} \cdot \{\exp[jk[(2c + Z_{\text{edge}})\cos\theta + (D_{\text{refl}}/2)\sin\theta]]\}D_s^+ \tag{6b}$$

where [1]

$$D_{\substack{s\\h}}^+ = \left\{-\frac{e^{-j(\pi/4)}}{2\sqrt{2\pi k}}\left[\frac{F[kL^i a(\theta - \delta_t - \alpha)]}{\cos(\theta - \delta_t - \alpha)/2} \pm \frac{F[kL^r a(\pi + \theta - \delta_t + \alpha)]}{\sin(\theta - \delta_t + \alpha)/2}\right]\right\} \tag{7}$$

$$a(x) \equiv 2\cos^2\left(\frac{x}{2}\right) \tag{8a}$$

$$F(kLa) = 2j\sqrt{kLa}\, e^{jkLa}\int_{\sqrt{kLa}}^{\infty} e^{-j\tau^2}\, d\tau. \tag{8b}$$

The transition functions $F(kLa)$ remove the singularities at the shadow boundary ($\theta = \pi - \theta'_{\text{edge}}$) and at the reflection boundary ($\theta = \theta_{\text{edge}}$). Furthermore,

$$L^i = \rho_{\text{edge}} \tag{9a}$$

$$L^r = \frac{\beta_1^r\beta_2^r\sin\theta_{\text{edge}}}{(D_{\text{refl}}/2)} \tag{9b}$$

where β_1^r and β_2^r are the principal radii of curvature of the reflected geometrical ray at Q_E^+ and $(D_{\text{refl}}/2)/\sin\theta_{\text{edge}}$ is the caustic distance evaluated in the direction of the reflection boundary.

For the point source on the Z axis illuminating the axially symmetric reflector, the resulting single edge-diffracted rays have a caustic at all points on the Z axis. Thus all ray path-lengths $\overline{FQ_EP}$ are the same if both F and P lie on the axis of symmetry. Ray optical solutions fail in the neighborhood of these caustics, as, for example, the $\sin\theta$ denominator in the square-root factor of (4) and (6) vanishes. Alternative solutions near caustics are canonical solutions [2], asymptotic solutions [3], [4], and integral solutions [1], [5]. The latter technique consists of replacing the edge-diffracted rays with contributions from equivalent electric and magnetic ring currents lying along the edge:

$$I_{e\phi} \cong -\frac{E_\phi D_s(\phi,\phi';\pi/2)}{\eta}\sqrt{\frac{8\pi}{k}}\, e^{-j\pi/4} \tag{10a}$$

$$I_{m\phi} = -H_\phi \eta D_h(\phi,\phi';\pi/2)\sqrt{\frac{8\pi}{k}}\, e^{-j\pi/4} \tag{10b}$$

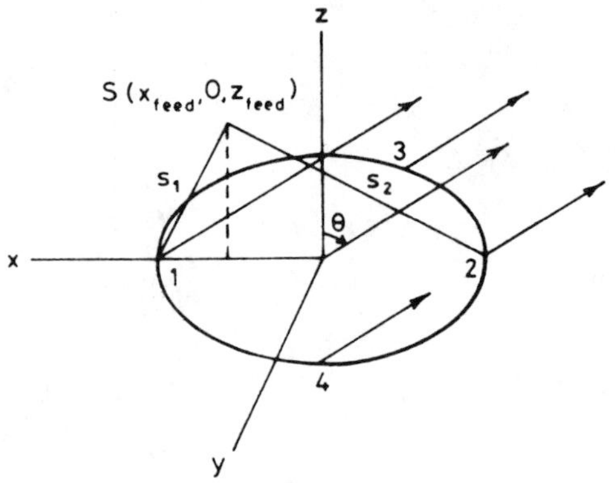

Fig. 4. Geometry of four possible edge rays.

where E_ϕ and H_ϕ are the edge-tangential incident electric and magnetic fields, $D_s(\phi,\phi';\pi/2)$ and $D_h(\phi,\phi';\pi/2)$ are the soft and hard edge-diffraction coefficients, and ϕ,ϕ' are the angles of diffraction and incidence in the "Ray-fixed" coordinate systems defined by Kouyoumjian [1]. For an arbitrary field point ϕ is different at various points on the rim. However, for a field point on axis ϕ is constant (at great distances on axis $\phi = \pi/2 - \delta_t$. Furthermore, for this symmetric case $\phi' = \pi/2 + \alpha$).

The most commonly encountered feed function is that of a far-field, $m = 1$, spherical-wave source for which the tangential fields incident on the edge are

$$E_\phi = d_1(\theta_s)\cos\phi_s\,\frac{e^{-jk\rho_{\text{edge}}}}{\rho_{\text{edge}}} \tag{11a}$$

$$H_\phi = \frac{1}{\eta}\, a_1(\theta_s)\sin\phi_s\,\frac{e^{-jk\rho_{\text{edge}}}}{\rho_{\text{edge}}}. \tag{11b}$$

The easily computed ring current fields then become as follows.

$\theta = 0$:

$$E_{\text{ring}} = a_y\frac{(D_{\text{refl}}/2)}{4}\exp[jk(2c + Z_{\text{edge}})]\frac{e^{-jkR}}{R} \cdot \left\{-\left[d_1(\pi - \theta'_{\text{edge}})\frac{e^{-jk\rho_{\text{edge}}}}{\rho_{\text{edge}}}\right] \cdot \left[\frac{1}{\cos(\theta'_{\text{edge}}/2)} - \frac{1}{\sin(\theta_{\text{edge}}/2)}\right] - \left[a_1(\pi - \theta'_{\text{edge}})\frac{e^{-jk\rho_{\text{edge}}}}{\rho_{\text{edge}}}\right] \cdot \left[\frac{1}{\cos(\theta'_{\text{edge}}/2)} + \frac{1}{\sin(\theta_{\text{edge}}/2)}\right]\right\}. \tag{12a}$$

$\theta = \pi$:

$$E_{\text{ring}} = a_y\frac{(D_{\text{refl}}/2)}{4}\exp[-jk(2c + Z_{\text{edge}})]\frac{e^{-jkR}}{R} \cdot \left\{-\left[d_1(\pi - \theta'_{\text{edge}})\frac{e^{-jk\rho_{\text{edge}}}}{\rho_{\text{edge}}}\right] \cdot \left[\frac{1}{\cos(\theta_{\text{edge}}/2)} - \frac{1}{\sin(\theta'_{\text{edge}}/2)}\right] - \left[a_1(\pi - \theta'_{\text{edge}})\frac{e^{-jk\rho_{\text{edge}}}}{\rho_{\text{edge}}}\right] \cdot \left[\frac{1}{\cos(\theta_{\text{edge}}/2)} + \frac{1}{\sin(\theta'_{\text{edge}}/2)}\right]\right\}. \tag{12b}$$

θ small but nonzero:

$$E_{\text{ring}} \cong \frac{(D_{\text{refl}}/2)}{4} \exp\,[jk(2c + Z_{\text{edge}}) \cos\theta]\, \frac{e^{-jk\rho_{\text{edge}}}}{\rho_{\text{edge}}} \frac{e^{-jkR}}{R}$$

$$\cdot \left\{ \boldsymbol{a}_\theta \left[-d_1(\pi - \theta'_{\text{edge}}) \left[\frac{1}{\cos(\theta'_{\text{edge}}/2)} - \frac{1}{\sin(\theta_{\text{edge}}/2)} \right] [J_0(\beta) + J_2(\beta)] \cos\theta \right.\right.$$

$$- a_1(\pi - \theta'_{\text{edge}}) \left[\frac{1}{\cos(\theta'_{\text{edge}}/2)} + \frac{1}{\sin(\theta_{\text{edge}}/2)} \right]$$

$$\left. \cdot\ [J_0(\beta) - J_2(\beta)] \right] \sin\phi$$

$$+ \boldsymbol{a}_\phi \left[-d_1(\pi - \theta'_{\text{edge}}) \left[\frac{1}{\cos(\theta'_{\text{edge}}/2)} - \frac{1}{\sin(\theta_{\text{edge}}/2)} \right] [J_0(\beta) - J_2(\beta)] \right.$$

$$- a_1(\pi - \theta'_{\text{edge}}) \left[\frac{1}{\cos(\theta'_{\text{edge}}/2)} + \frac{1}{\sin(\theta_{\text{edge}}/2)} \right]$$

$$\left.\left. \cdot\ [J_0(\beta) + J_2(\beta)] \cos\theta \right] \cos\phi \right\} \qquad (12c)$$

where $\beta = k(D_{\text{refl}}/2) \sin\theta$. Equation (12c) is an approximation in that the changes in D_s and D_h in equations (10a) and (10b) at various points on the rim have been assumed to be second order for relatively small values of θ.

II. Axially Symmetric Reflectors for Vector Spherical-Wave Point Source Off Axis

If a point-source feed is located off the axis of the axially symmetric reflector, Fermat's principle reveals that there are a maximum of four possible diffraction points around the rim of the reflector [6]. The situation becomes somewhat less complicated when the point-source feed, the reflector axis, and the observation point all lie in the same plane (taken, for example, to be the $X - Z$ plane in Fig. 4). Under these conditions, diffraction points 1 and 2 are possible for all possible positions of the distant field point. However, if the field point is on the opposite side of the axis from the source point, then two additional diffraction points 3 and 4 are also possible in the range of θ values, $\theta_1 \leq \theta \leq \theta_2$, [7] where

$$\theta_1 = \sin^{-1}\left[\frac{X_{\text{feed}}}{\sqrt{(D_{\text{refl}}/2 + X_{\text{feed}})^2 + (Z_{\text{edge}} - Z_{\text{feed}})^2}}\right] \qquad (13a)$$

$$\theta_2 = \sin^{-1}\left[\frac{X_{\text{feed}}}{\sqrt{(D_{\text{refl}}/2 + X_{\text{feed}})^2 + (Z_{\text{edge}} - Z_{\text{feed}})^2}}\right] \qquad (13b)$$

θ_1 corresponds to a caustic at infinity form point 2, while θ_2 corresponds to a caustic at infinity from point 1. Thus, as θ increases from θ_1 to θ_2, the additional roots 3 and 4 move from 2 to 1. In practice, for typical reflector geometries the range of values from θ_1 to θ_2 is sufficiently small so that the entire range is covered by radiation from the equivalent ring sources to avoid unrealistically large predicted fields. An example of a multicaustic region is for values of θ between θ_1 and θ_2 in the plane of scan as considered previously. Under these conditions it is necessary to account for the variations of E_ϕ, H_ϕ, D_s, and D_h by evaluating the radiation integrals numerically. Examples of typical results are shown in Fig. 5 for a laterally defocused

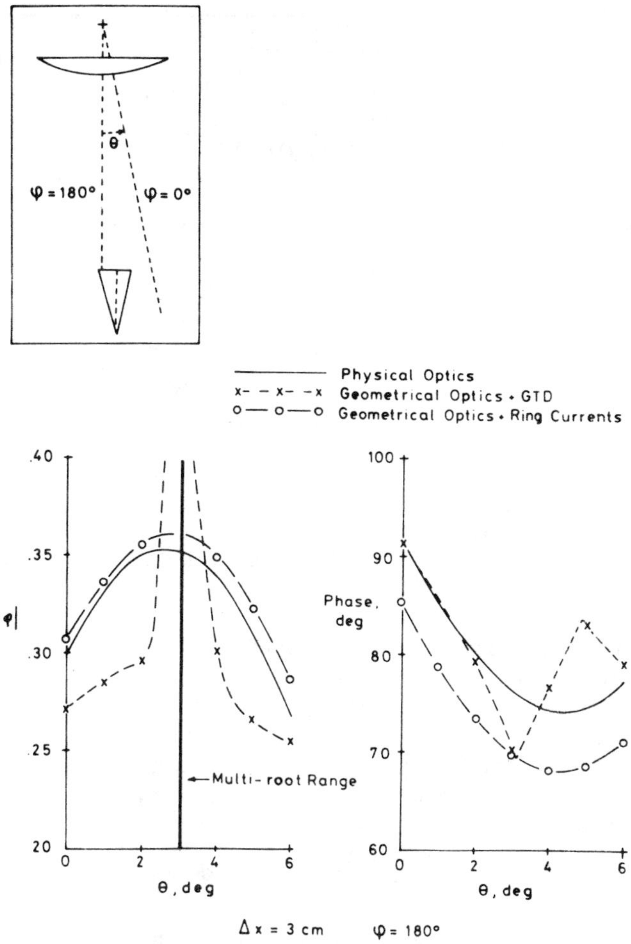

Fig. 5. Comparison of PO, GTD, and ring-current fields in multicaustic directions.

hyperboloid. On the opposite side of the axis in the plane of scan the multiroot range lies between 3.01 and 3.11 deg. However, the simple geometrical theory of diffraction (GTD) two-edge ray description of the field is seen to become unrealistically large outside the multiroot range. The ring-current contribution, on the other hand, is well behaved and very close to the physical optics results over a relatively wide angular range.

III. Surface Diffracted Rays

First-order surface rays will ordinarily not be directly excited by the feed in a reflecting antenna system. However, they may be excited by tangential edge-diffracted rays. An example may be seen in Fig. 2 where an edge-diffracted ray from Q_E^- in the direction $\theta = \pi/2 - \delta_t$ will tangentially graze the surface. Only the tangential H-field component will be nonzero in this direction, but the excited surface rays are necessary to provide a continuous field. In most instances, however, the edge-diffracted rays provide sufficiently accurate surface rays.

IV. Applications

A. Radiation from Prime-Focus Paraboloid

The Kouyoumjian group has used a combination of a single edge-diffracted GTD, equivalent ring-currents, and physical optics to compute the complete radiation pattern of a prime focus paraboloid [8], [9]. The calculated pattern was compared with the experimental results of Afifi, which were measured for a half-paraboloid on a groundplane with a monopole feed at the focus [10]. This experimental arrangement virtually eliminated aperture

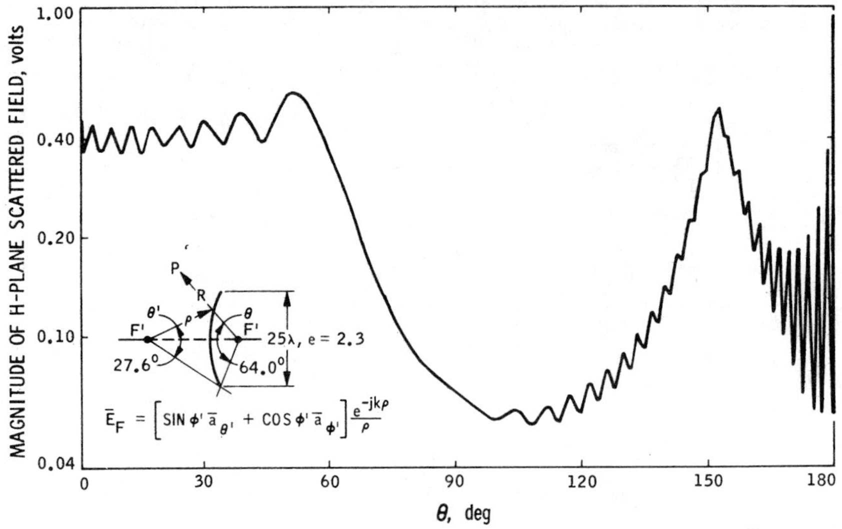

Fig. 6. Complete GTD *H*-plane pattern of hyperboloid with feed at external focus.

blocking. The vertical polarization eliminated the possibility of exciting surface-diffracted rays. The deep reflector ($F/D = 0.25$) eliminated rear spillover. The agreement between the calculated and measured patterns was very good between 0 deg and 130 deg. Beyond 130 deg the agreement was also quite good, considering that at such low signal levels part of the discrepancy was due to measurement error.

B. Radiation from Hyperboloid with Spherical-Wave Feed at Its External Focus

Results have been obtained for a "focused" hyperboloid using physical optics, geometrical optics, GTD (Keller), and GTD (Kouyoumjian) [11]. The geometrical-optics results are discontinuous at the reflection boundary. The physical optics and Kouyoumjian GTD results agree closely in magnitude and phase, both in the illuminated and shadowed regions as well as in the vicinity of the reflection boundary.

Fig. 6 is a complete *H*-plane pattern based on GTD. The fields at the reflection boundary at 64 deg and the shadow boundary at 152.4 deg are finite and continuous, and the two axial caustics are computed using the equivalent ring sources. This technique yields slightly different results at the rear axial caustic than the caustic correction factor of Keller [2], which did not yield *E*-plane–*H*-plane continuous fields on axis.

Results were also reported in [11] that compare the calculated and measured scattered *H*-plane patterns from a hyperboloid illuminated by a corrugated horn with its phase center at the hyperboloid's external focus [12]. The GTD results included two edge rays from the reflector and one from the leading edge of the horn on the field-point side of the axis. Equivalent ring currents around both the reflector and horn rims were used for field points near the axis. These GTD results agreed with the measured patterns within the experimental uncertainties.

V. Evaluation of Focused Equivalent Paraboloid Using GTD

The techniques of GTD provide a computational procedure to obtain RF performance data that hitherto was excessively costly in computer time. A prominent example of this is the dual-reflector antenna. Hitherto, it was only possible to analyze dual-reflector systems by integrating over both reflectors. The subreflector in a system of this type generally creates a shaped, rather than a focused, beam and consequently is readily amenable to GTD determination of the scattered field. This rapidly determined scattered field then provides the illumination function for a physical optics integration over the large focused primary mirror.

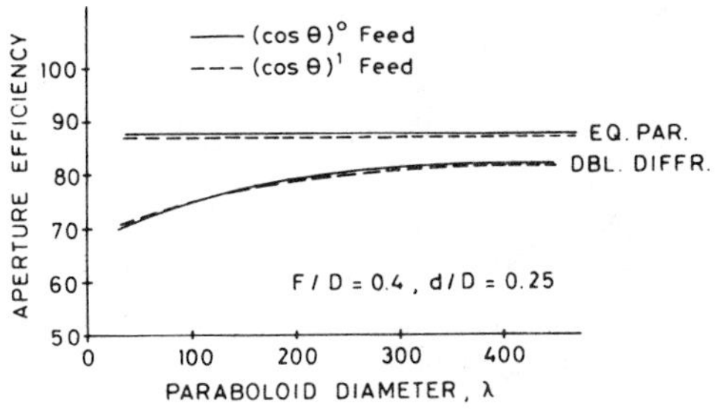

Fig. 7. Comparison of aperture efficiency values using GTD and equivalent paraboloid.

The equivalent paraboloid is said to provide an accurate technique for the calculation of performance characteristics for Cassegrain and Gregorian systems [13]. GTD provides an accurate and relatively economical technique to verify this assertation quantitatively. For example, in Fig. 7, the aperture efficiency (exclusive of spillover) is calculated for a Cassegrain system with a paraboloid $F/D = 0.4$ and subreflector diameter/main reflector diameter ratio $= 0.25$. Diffraction from the subreflector causes the illumination of the main reflector to be tapered. However, this taper is not included in the equivalent paraboloid calculation, and its predicted efficiency is significantly higher, even for subdish diameters as large as 100 wavelengths. Since the edge illumination of the equivalent paraboloid is higher than the actual edge illumination, it generally predicts higher wide sidelobes than actually exist [14].

References

[1] R. G. Kouyoumjian and P. H. Pathak, "A uniform geometrical theory of diffraction for an edge in a perfectly conducting surface," *Proc. IEEE*, vol. 62, pp. 1448–1461, Nov. 1974.

[2] J. B. Keller, "Diffraction by an aperture," *J. Appl. Phys.*, vol 28, pp. 426–444, 1957.

[3] I. Kay and J. B. Keller, "Asymptotic evaluation of the field at a caustic," *J. Appl. Phys.*, vol. 25, pp. 876–883, July 1954.

[4] D. Ludwig, "Uniform asymptotic expansions at a caustic," *Commun. Pure Appl. Math.*, vol. 19, pp. 215–250, May 1966.

[5] G. L. James and V. Kerdemelidis, "Reflector antenna radiation pattern analysis by equivalent edge currents," *IEEE Trans. Antennas Propagat.*, vol. AP-21, pp. 19–24, Jan. 1973.
[6] F. Molinet and L. Saltiel, "High frequency radiation pattern prediction for satellite antennas," Lab. Central de Télécommunications, ESTEC Contract 1820/72HP, July 1973.
[7] H. Bach, K. Pontoppidan and, L. Solymar, "High frequency radiation pattern prediction for satellite antennas," Lab. Electromagnetic Theory, Tech. Univ. Denmark, Lyngby, ESTEC Contract 1821/72HP, Dec. 1973.
[8] R. G. Kouyoumjian and P. A. J. Ratnasiri, "The calculation of the complete pattern of a reflector antenna," in *Proc. Int. Electronics Conf.*, 1969, pp. 152–153.
[9] P. A. J. Ratnasiri, R. G. Kouyoumjian, and P. H. Pathak, "The wide angle sidelobes of reflector antennas," Electro Sci. Lab. Ohio State Univ., Columbus, Rep. 2183-1, 1970.
[10] M. S. Afifi, "Radiation from the paraboloid of revolution," in *Electromagnetic Wave Theory*, part 2, J. Brown, Ed. New York: Pergamon, 1967, pp. 669–687.
[11] W. V. T. Rusch, "A comparison of geometrical and integral fields from high-frequency reflectors," *Proc. IEEE* (Lett.), vol. 62, pp. 1603–1604, Nov. 1974.
[12] O. Sørensen, "Analysis of defocused Cassegrain antennas using geometrical theory of diffraction," Master's thesis, Lab. Electromagnetic Theory, Tech. Univ. Denmark, Lyngby, Fall 1973.
[13] W. C. Wong, "On the equivalent parabola technique to predict the performance characteristics of a Cassegrainian system with an offset feed," *IEEE Trans. Antennas Propagat.*, vol. AP-21, pp. 335–339, May 1973.
[14] R. Booth, private communication.

Analysis of the Radiation Patterns of Reflector Antennas

J. F. KAUFFMAN, MEMBER, IEEE, WILLIAM F. CROSWELL, SENIOR MEMBER, IEEE, AND LEONARD J. JOWERS

Abstract—The development and application of a numerical technique for the rapid calculation of the far-field radiation patterns of a reflector antenna from either a measured or computed feed pattern are reported. The reflector is defined by the intersection of a cone with any surface of revolution or an offset sector of any surface of revolution. The feed is assumed to be linearly polarized and can have an arbitrary location. Both the copolarized and the cross polarized reflector radiation patterns are computed. Calculations using the technique compare closely with measured radiation patterns of a waveguide-fed offset parabolic reflector. The unique features of this technique are the freedom from restrictive feed assumptions and the numerical methods used in preparing the aperture plane electric field data for integration.

INTRODUCTION

THE MOTIVATION for this work was the development of a technique for the rapid computation of the radiation patterns of reflector antennas from the measured radiation patterns of multiple feeds. For communications satellite applications the feeds could be connected and phased to produce the required shaped beams [1]; in radiometric applications each feed, or set of feeds, might be operated at widely separated frequencies for the remote sensing of the physical properties of the earth's surface. Due to the blockage and scattering from such multiple feed systems for the conventional parabola, there is strong interest in the offset reflector geometry. The feeds may or may not be on focus depending upon the particular configuration. For the case of the feed on focus, Chu and Turrin [2] have published measurements of the beam displacement and cross polarization of the offset reflector along with calculations assuming a symmetrical feed pattern.

Briefly, the calculation reported here is formulated in the following manner. The equations of geometrical optics are used to calculate the reflected electric field using the radiation patterns of the feed and the parameters defining the reflector surface. Also obtained using geometrical optics are the direction of the reflected ray, and the point of intersection of the reflected ray with the aperture plane. These fields comprise the aperture distribution which is integrated over the aperture plane to yield the far-field radiation pattern.

The case of the on-axis-fed full paraboloid is the only one where the projected rays defined on ϕ = constant cuts of the feed radiation pattern, strike the aperture plane at points defining radial lines emanating from the origin. For other feed positions these data points will not lie on constant coordinate lines in any convenient coordinate system. Moreover, the point configuration changes each time the feed position and/or reflector geometry is changed. This is illustrated by comparing Fig. 5 later in the paper. In this paper an algorithm is presented for ordering these aperture data points in a rectangular coordinate system prior to the numerical integration. This coordinate system is used for all antenna configurations.

The most commonly used alternative formulation is the current distribution method [3], where the surface current J_s on the reflector is taken to be $2(\hat{n} \times H_i)$. H_i is the incident magnetic field calculated from the feed pattern using geometrical optics, and $\hat{n}$ is the unit normal to the reflector surface. This current is then integrated over the reflector

Manuscript received November 21, 1974; revised September 4, 1975.
J. F. Kauffman is with the Department of Electrical Engineering, North Carolina State University, Raleigh, NC 27607.
W. F. Croswell is with the NASA Langley Research Center, Hampton, VA 23665.
L. J. Jowers is with LTV Aerospace Corporation, Hampton, VA 23665.

Reprinted from *IEEE Trans. Antennas Propagat.*, vol. AP-24, pp. 53–65, Jan. 1976.

surface to yield the far-field radiation pattern. Common to both methods of analysis are the following. a) The surface current on the shadow side of the reflector is assumed to be zero. b) The discontinuity in the surface current at the edge of the reflector is neglected. c) Direct radiation from the feed and aperture blockage are not included. These approximations commonly restrict the accuracy of the calculations using either formulation to the main beam and the close in sidelobes [4].

The advantage of the method reported here is that the integration over the aperture plane can be performed with equal ease for any feed position and any feed pattern, whereas the integration over the reflector surface is time consuming, and becomes difficult when the feed is placed off-axis or when the feed radiation pattern has no symmetry which can be used to simplify the formulation.

FORMULATION

The equations here are developed for a paraboloid of revolution or any offset sector of a paraboloid of revolution. Three coordinate systems are needed: 1) a feed-centered system (x^0,y^0,z^0) as shown in Fig. 1 in which the radiation pattern of the feed has been measured, or in which the feed pattern is defined analytically; 2) a focus-centered system (x,y,z) also shown in Fig. 1 in which the equation defining the reflector is written. (The aperture plane is taken to be the $(x,y,0)$ plane: the electric field on this plane is calculated in this coordinate system); and 3) an (x_1,y_1,z_1) coordinate system shown in Fig. 2 which is defined by the transformation $x_1 = -z, y_1 = y, z_1 = x$. These coordinates are used to calculate the far-field pattern. The polar axes are the z axis and the z_1 axis in the (x,y,z) system and the (x_1,y_1,z_1) system, respectively. However, the latter system is oriented so the y_1z_1 plane is the aperture plane, and the x_1 axis is the boresight axis. In this orientation, a linear probe moved along great circle cuts will receive separable E_θ or E_ϕ components of the field, thereby simplifying the comparison of calculated and measured patterns, and allowing a clearer interpretation of polarized components of the field.

The parameters defining an offset sector are shown in Fig. 3. The other symbols shown on Fig. 1 are defined as follows.

1) $\hat{s}_i$ is a unit vector in the direction of an arbitrary ray incident on the reflector. Its direction is defined by the spherical coordinates (θ,ϕ) in the feed-centered system.

2) R is the distance from the phase center of the feed to the point at which the incident ray strikes the reflector.

3) $\hat{n}$ is the unit normal to the reflector surface.

4) $\hat{s}_r$ is a vector in the direction of the reflected ray.

5) d is the distance from the point of reflection (x_0,y_0,z_0) to the aperture plane.

The electric field incident on the reflector [5] is

$$\bar{E}_i = \frac{\left[\hat{s}_i \times \left(\begin{matrix}\hat{x}^0 \\ \hat{y}^0\end{matrix} \times \hat{s}_i\right)\right]}{\left|\hat{s}_i \times \left(\begin{matrix}\hat{x}^0 \\ \hat{y}^0\end{matrix} \times \hat{s}_i\right)\right|} \frac{\sqrt{P(\theta,\phi)}}{R} \tag{1}$$

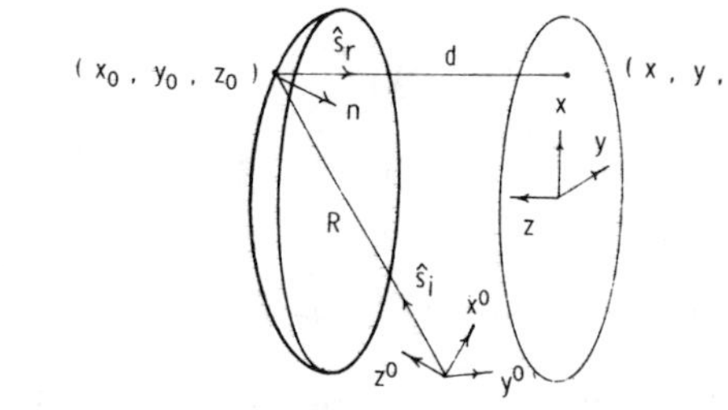

Fig. 1. Description of coordinate system.

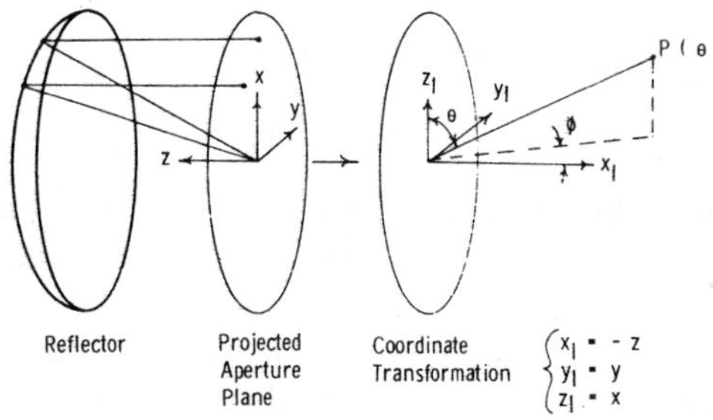

Fig. 2. Coordinate transformation of aperture fields.

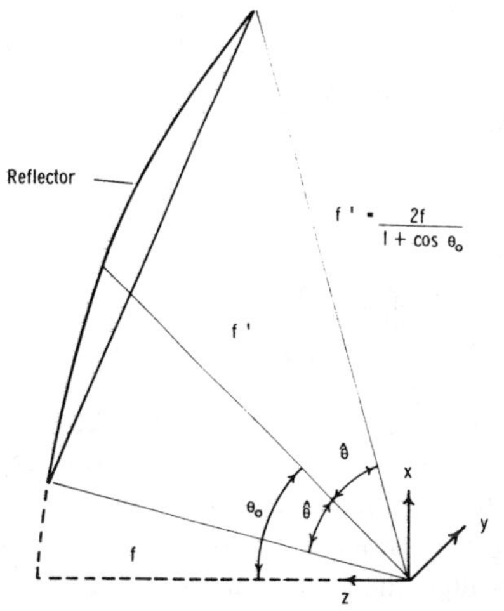

Fig. 3. Definition of parameters for offset reflector.

where $\hat{x}^0$ and $\hat{y}^0$ are unit vectors in feed-centered coordinates; the unit vector chosen depends on the polarization of the feed. The feed radiation pattern is $P(\theta,\phi)$ and $k = 2\pi/\lambda$. Equation (1) does not include the phase of $\bar{E}_i$ because the phase of the electric field over the aperture plane is calculated separately. $P(\theta,\phi)$ is the pattern measured by a linear probe aligned for maximum reception on boresight and kept in that orientation as the test antenna is rotated. Then, applying the boundary conditions at the reflector surface,

$$\bar{E}_r = 2(\hat{n} \cdot \bar{E}_i)\hat{n} - \bar{E}_i. \tag{2}$$

The distribution of electric field over the aperture plane can now be written as

$$E_{rx}e^{-j\psi} \text{ or } E_{ry}e^{-j\psi} \tag{3}$$

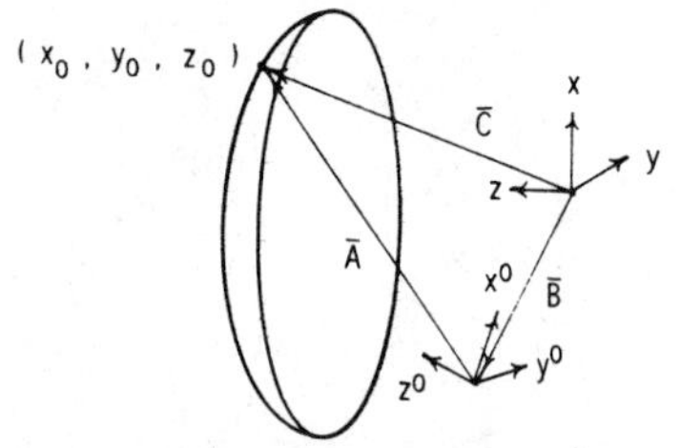

Fig. 4. Definition of distances, vectors, and transformations.

where

$$\psi = k(R + d) + \Phi(\theta,\phi) \quad (4)$$

$$d = \sqrt{(x - x_0)^2 + (y - y_0)^2 + z_0^2} \quad (5)$$

and $\Phi(\theta,\phi)$ is the phase pattern of the feed with respect to the origin of the feed coordinates. The x and y subscripts in (3) refer to the appropriate tangential component of the aperture field. The component which will be used in calculating the far-field pattern depends on the feed polarization and on whether the copolarized or the cross polarized secondary pattern is desired. (See (20) and (21).)

The distance R and the point (x_0,y_0,z_0) are needed in (4) and (5). The calculation of these quantities is described in terms of the vectors $\bar{A}$, $\bar{B}$, and $\bar{C}$ shown in Fig. 4. The components of $\bar{C}$ are the coordinates of the point of intersection in focus-centered coordinates.

$$\bar{C} = \bar{A} + \bar{B} = \hat{x}x_0 + \hat{y}y_0 + \hat{z}z_0 \quad (6)$$

where

$$\bar{A} = T\bar{A}^0 \quad (7)$$

$$\bar{B} = \hat{x}x' + \hat{y}y' + \hat{z}z' \quad (8)$$

and $\hat{x},\hat{y},\hat{z}$ are unit vectors in the focus-centered coordinates. The vectors $\bar{A}$ and $\bar{A}^0$ describe an arbitrary ray from the feed in the focus-centered and feed-centered coordinates, respectively. The transformation T rotates the feed coordinates through the Euler angles (δ,γ) to orient the feed coordinates parallel to the focal coordinates:

$$\bar{A}^0 = \hat{x}^0 R \sin\theta\cos\phi + \hat{y}^0 R \sin\theta\sin\phi + \hat{z}^0 R\cos\theta \quad (9)$$

$$T = \begin{bmatrix} \cos\delta & \sin\delta\sin\gamma & -\sin\delta\cos\gamma \\ 0 & \cos\gamma & \sin\gamma \\ \sin\delta & -\cos\delta\sin\gamma & \cos\delta\cos\gamma \end{bmatrix}. \quad (10)$$

The angles (δ,γ) are defined as positive when the rotation is counterclockwise as seen looking in the negative direction along the axis of rotation. The angle δ refers to rotation about the y axis, and the angle γ refers to rotation about the x axis. The components of the vector $\bar{B}$ define the origin of the feed coordinates in the focal-centered coordinate system.

The complete expressions for the components of $\bar{C}$ are

$$x_0 = R(\sin\theta\cos\phi\cos\delta + \sin\theta\sin\phi\sin\delta\sin\gamma - \cos\theta\sin\delta\cos\gamma) + x' \quad (11)$$

$$y_0 = R(\sin\theta\sin\phi\cos\gamma + \cos\theta\sin\gamma) + y' \quad (12)$$

$$z_0 = R(\sin\theta\cos\phi\sin\delta - \sin\theta\sin\phi\cos\delta\sin\gamma + \cos\theta\cos\delta\cos\gamma) + z'. \quad (13)$$

Equations (11), (12), and (13) together with the equation for the reflector

$$x_0^2 + y_0^2 = 4f(f - z_0) \quad (14)$$

can be solved simultaneously for x_0,y_0,z_0, and R.

The coordinates of a data point on the aperture plane are found from the intersection of the reflected ray with the $z = 0$ plane. The vector in the direction of the reflected ray is [5]

$$\bar{s}_r = \hat{s}_i - 2(\hat{n}\cdot\hat{s}_i)\hat{n}. \quad (15)$$

The parametric equations of a line along the reflected ray are

$$x = x_0 + h\cos\alpha_x$$

$$y = y_0 + h\cos\alpha_y$$

$$z = z_0 + h\cos\alpha_z \quad (16)$$

where $\cos\alpha_x$, etc., are the direction cosines of the reflected ray given by

$$\cos\alpha_x = \frac{\hat{x}\cdot\bar{s}_r}{|\bar{s}_r|}$$

$$\cos\alpha_y = \frac{\hat{y}\cdot\bar{s}_r}{|\bar{s}_r|}$$

$$\cos\alpha_z = \frac{\hat{z}\cdot\bar{s}_r}{|\bar{s}_r|}. \quad (17)$$

We want the intersection of this line with the $z = 0$ plane, thus, the parameter h must be

$$h = -z_0\frac{|\bar{s}_r|}{s_{rz}} \quad (18)$$

and the coordinates of the point, (x,y) are

$$x = x_0 - z_0\frac{s_{rx}}{s_{rz}}$$

$$y = y_0 - z_0\frac{s_{ry}}{s_{rz}} \quad (19)$$

where s_{rx}, s_{ry}, s_{rz} are the x, y, and z components of the vector $\bar{s}_r$.

In order to calculate the secondary radiation pattern, we transform the aperture data to the (x_1,y_1,z_1) coordinate system discussed in the preceding and integrate numerically over the aperture. Fig. 3 illustrates the transformation. The integrals to be evaluated are

$$E_\theta = \int_{S_1} E_{z1}\cos\phi e^{-j\psi}e^{jk[y_1\sin\theta\sin\phi + z_1\cos\theta]}\,dy_1\,dz_1 \quad (20)$$

and

$$E_\phi = \int_{S_1}[Ey_1\sin\theta + E_{z1}\cos\theta\sin\phi]e^{-j\psi} \cdot e^{jk[y_1\sin\theta\sin\phi + z_1\cos\theta]}\,dy_1\,dz_1 \quad (21)$$

where E_{y1} and E_{z1} are the tangential components of the aperture electric field, S_1 is the projected area of the aperture on the aperture plane, and (θ,ϕ) are the usual spherical coordinate angles defined with respect to the (x_1,y_1,z_1) coordinates. The term $\pm(j2ke^{-jkr_0})/(4\pi r_0)$, which is constant on the surface of a sphere of radius r_0, has been deleted from (20) and (21). Equations (20) and (21) are derived by defining $\bar{E}$ in terms of the electric vector potential $\bar{F}$, where $\bar{F}$ is the solution to the vector wave equation with magnetic current as the source term:

$$\bar{E} = -\frac{1}{\varepsilon}\nabla \times \bar{F}$$

$$\bar{F} = \varepsilon \int_{S_1} \frac{\bar{M}e^{-jkr}}{4\pi r}\, dS_1.$$

The magnetic currents are defined by the tangential electric field in the integration plane. The numerical techniques used to evaluate (20) and (21) are discussed in the following section.

Equations (20) and (21) give the θ and ϕ components of the electric field in the Fraunhofer region. The power received by a linear probe oriented along the $\hat{\theta}$ unit vector on the surface of a sphere of radius r_0 will be

$$P_\theta(\theta,\phi) = \frac{|E_\theta|^2}{2\eta}. \tag{22}$$

Similarly, if the linear probe is oriented along the ϕ unit vector, the received power will be

$$P_\phi(\theta,\phi) = \frac{|E_\phi|^2}{2\eta}. \tag{23}$$

P_θ is defined to be the copolarized radiation pattern, and P_ϕ is the cross polarized radiation pattern when the feed is linearly polarized along the x^0 axis of the feed-centered system. If the feed is linearly polarized along the y^0 axis, P_ϕ is the copolarized radiation pattern, and P_θ is the cross polarized radiation pattern. Other definitions of copolarization and cross polarization are possible, but these appear to be the most convenient for applications in which comparison of calculated and measured patterns is important [6].

Numerical Considerations

The numerical processing is modularized in five distinct functions. The functions, and the names of the program modules in which they are performed, are listed here:

1) input of program parameters (INPUT);
2) calculation of aperture plane data (ADIST);
3) sorting of aperture plane data to facilitate numerical integration (SORT);
4) calculation of edge data on the projected aperture (EDGES);
5) integration over the projected aperture to obtain radiation patterns (FARFLD).

A detailed description of the input parameters and of the programming aspects of the calculation is given in the companion Computer Program Descriptions section of this Transactions [7].

The quantization of the calculated aperture plane electric field is part of the second function listed above. This is done to align the calculated aperture field values so that the diffraction integrals, (20) and (21), can be performed in rectangular coordinates regardless of feed orientation. Comparison of Figs. 5(a) and 5(b) illustrates the position dependence of the calculated aperture field points upon the orientation of the feed; comparison of Figs. 5(b) and 5(d) demonstrates the effect of quantization. Numerically, quantization, (quantizing by grid lines), is accomplished by choosing y coordinate values for the $y =$ constant lines. Typically, these QUANTIZING GRID lines may be spaced a half-wavelength apart. The y coordinate of each calculated aperture field point is compared to the y values of the grid lines, and the field value is assigned to the closest $y =$ constant grid line.

The inputting of feed patterns which are either calculated or measured on $\phi =$ constant great circle cuts gives calculated fields at points on roughly radial lines in the aperture plane. Field points are thus dense near the center, and sparse near the edge of the projected aperture. (See Fig. 5(b).) A thinning procedure, based on the distance between points, is used to dispose of unneeded points in the dense area. (Compare Figs. 5(b) and 5(c).) The minimum desired distance between adjacent points is defined as the DENSITY GATE. This density gate is used as the criterion for eliminating redundant points. Thinning is carried out before quantization, and both functions are performed in the program module ADIST.

Field values are calculated by interpolation along the constant y-lines in places where points are sparse. The number of interpolations performed between existing points is called the INTERPOLATION NUMBER. Field values on the edge of the projected aperture on each constant y-line are needed for interpolation. These values are found by calculating the illumination for a reflector which is larger than actually desired, and using the points outside the desired projected aperture for interpolation to an edge point. The points outside the projected edge are then discarded. It is possible that this technique will also fail to produce an edge point. In this case, the edge point in the preceding constant y-line is assigned. The edge points are computed in the module named EDGES; the interpolations are performed in FARFLD.

The integrations performed to evaluate (20) and (21) utilize the generalized trapezoidal rule. The examples discussed in the following section were computed on a CDC 6600 machine. To get a rough idea of the computing time required, consider an antenna with the parameters shown below Table I. This aperture is approximately 28 λ in diameter. Peripheral processing time was 899 s, and computing time was 467 s for a total of 324 far-field points; this is approximately 1.4 s per point. For a 113 λ diameter reflector with a -10 dB amplitude taper and uniform phase the computing time was 278 s for 202 far-

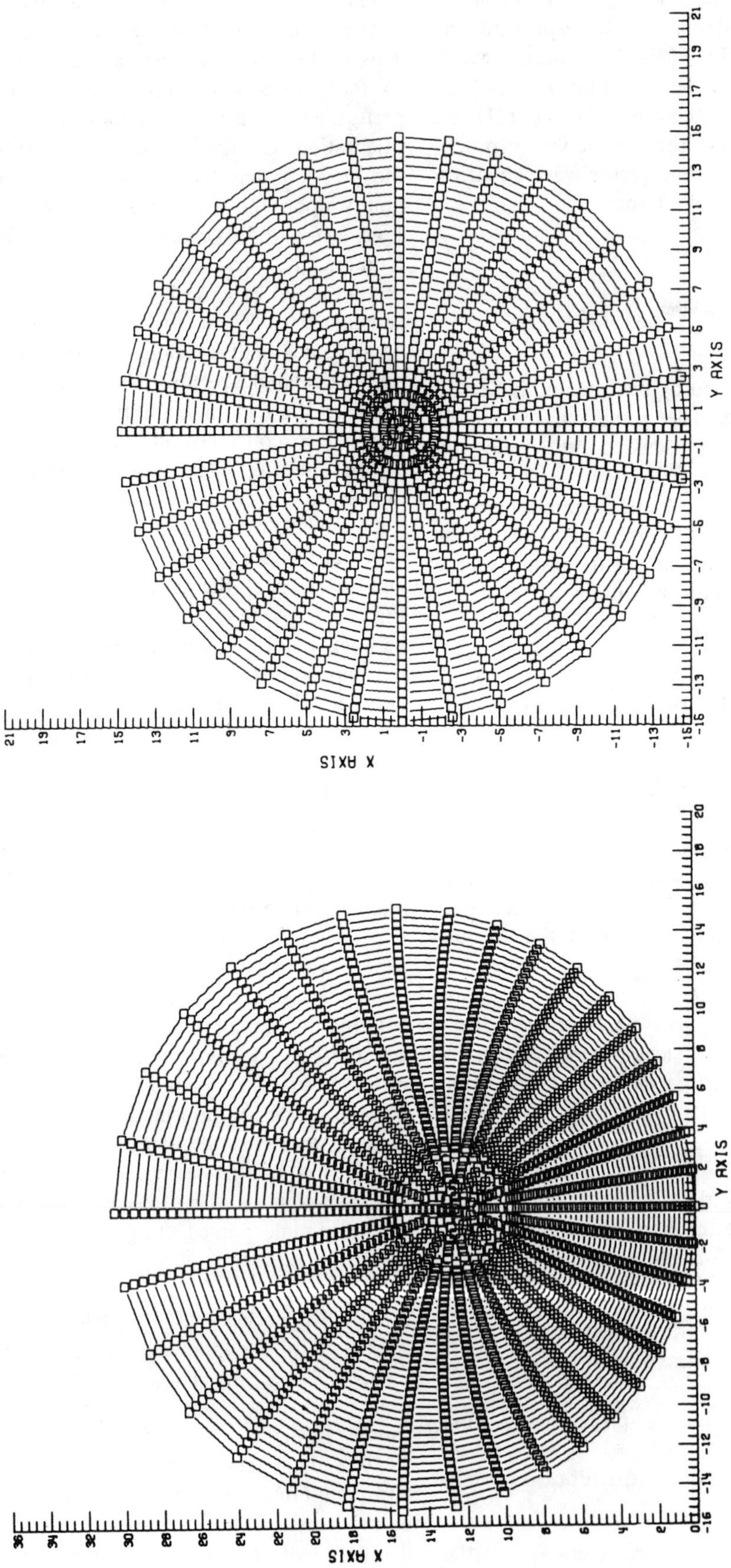

(a)

(b)

Fig. 5. Position dependence of calculated aperture field points upon location of feed. (a) Offset parabola. (b) Direct-fed parabola. (c) Direct-fed parabola with aperture data thinned. (d) Direct-fed parabola with aperture data thinned and quantized.

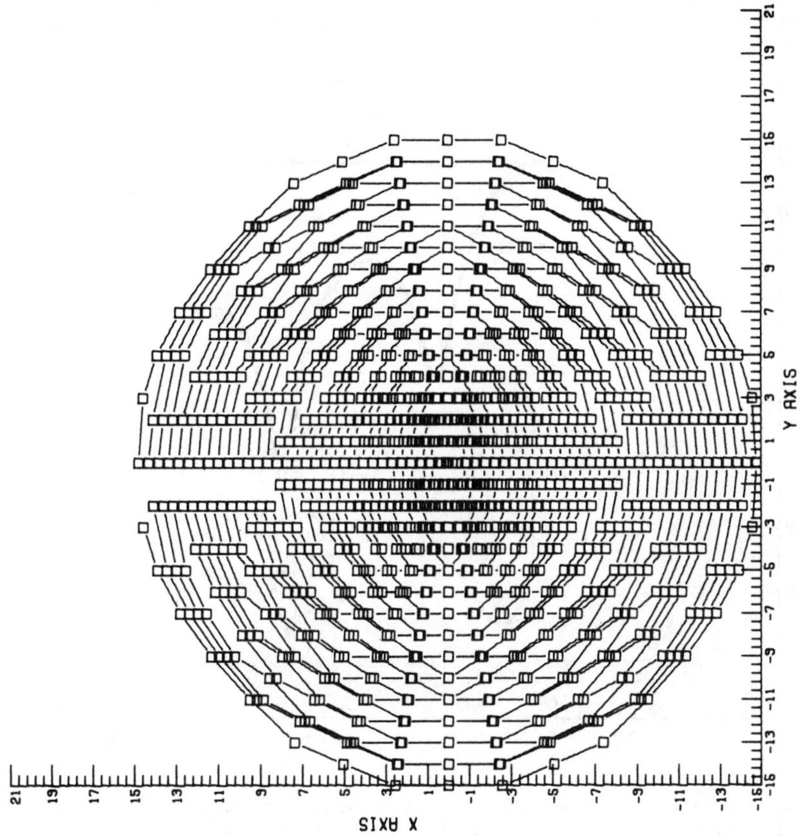

(d)

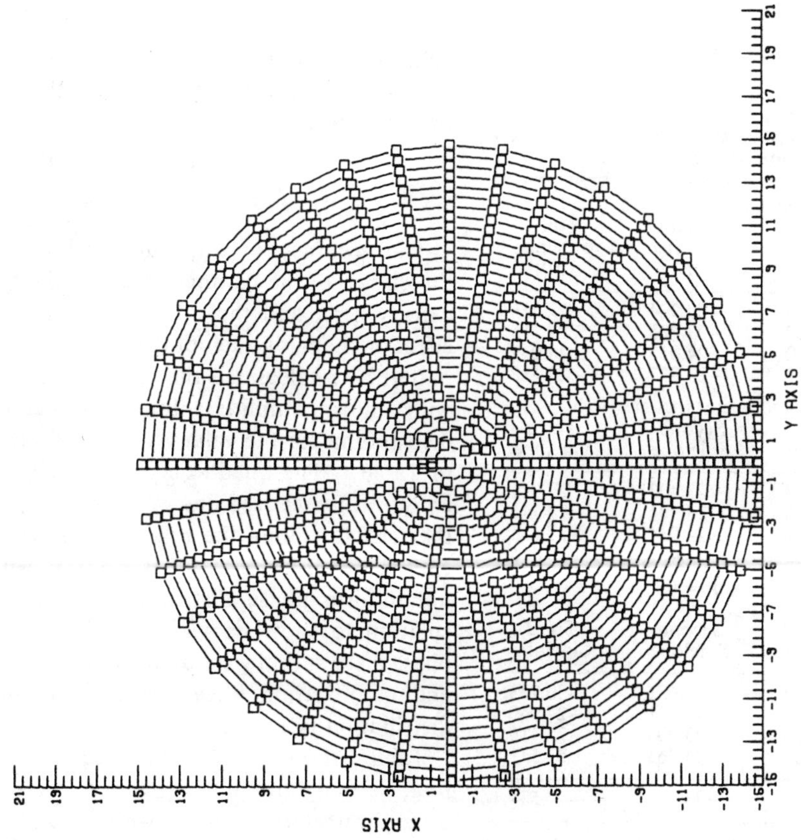

(c)

Fig. 5. Continued.

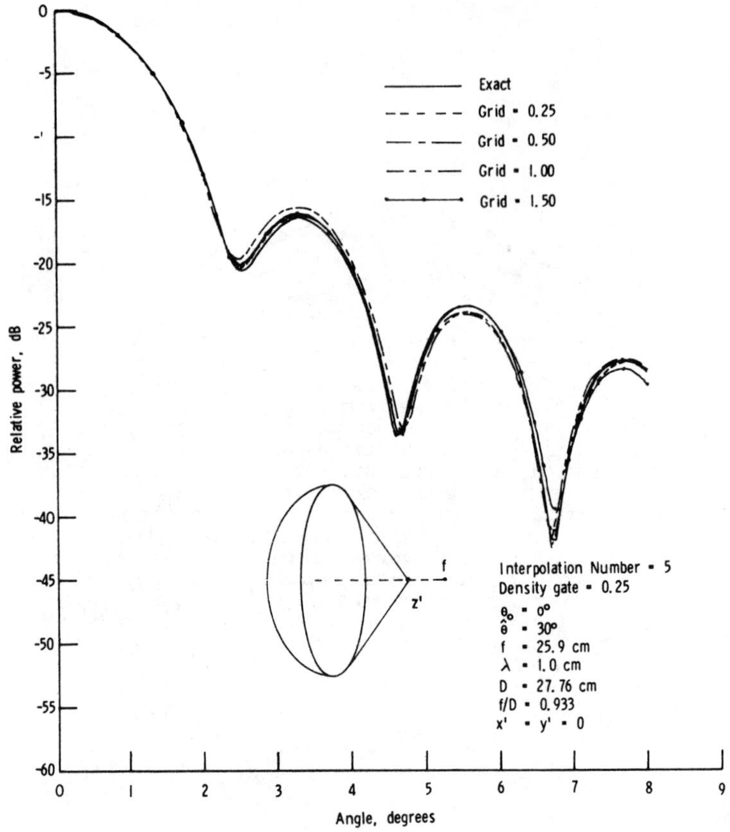

Fig. 6. Radiation patterns of reflector with uniform amplitude distribution and quadratic phase distribution computed as function of grid parameter. Quadratic phase distribution is obtained by moving feed along z' axis.

TABLE I
VARIATION OF BEAMWIDTH AND SIDELOBE WITH GRID, DENSITY GATE, AND INTERPOLATION NUMBER

GRID = 0.5, INTERPOLATION NUMBER = 5				
DENSITY GATE	½ 3 dB Beamwidth	First Sidelobe (dB)	Second Sidelobe (dB)	Third Sidelobe (dB)
0.25	1.06	−16.208	−23.904	−27.216
0.50	1.06	−16.209	−23.905	−27.715
1.00	1.06	−16.250	−24.013	−28.047
1.50	1.07	−16.263	−23.927	−28.418
DENSITY GATE = 0.25, INTERPOLATION NUMBER = 5				
GRID	½ 3 dB Beamwidth	First Sidelobe	Second Sidelobe	Third Sidelobe
0.25	1.06	−16.445	−23.403	−27.911
0.50	1.06	−16.208	−23.904	−27.716
1.00	1.06	−15.586	−24.034	−27.635
1.50	1.06	−16.033	−23.403	−28.415
DENSITY GATE = 0.25, GRID = 0.5				
INTERPOLATION NUMBER	½ 3 dB Beamwidth	First Sidelobe	Second Sidelobe	Third Sidelobe
2	1.06	−16.844	−25.683	−30.733
4	1.06	−16.272	−24.071	−27.940
5	1.06	−16.208	−23.904	−27.716
7	1.06	−16.153	−23.763	−27.531
Exact	1.06	−16.423	−23.272	−27.584

Note: $x' = y' = 0$, $z' = 1$ cm, $\theta_0 = 0°$, $\hat{\theta} = 30°$, $f = 25.9$ cm, $D = 27.76$ cm, $f/D = 0.933$, $\lambda = 1$ cm.

field points, or about 1.4 s per point. Note that the computing time required by this algorithm is quite insensitive to aperture size.

RESULTS

The accuracy of this numerical technique depends on the choice of the parameters called the quantizing grid, the density gate, and the interpolation number. To study the effect of changing each of these parameters on the radiation pattern, we calculated the radiation pattern of an aperture with uniform amplitude illumination and a quadratic phase distribution for the values of the various parameters shown in Table I. The quadratic phase distribution was obtained by offsetting the feed one wavelength from the focus toward the reflector along the reflector axis. This check case was chosen because the result is well known, and because the nonuniform phase distribution causes errors in the calculation to show up which might not be detected if the phase distribution were uniform. For example, interpolation and quantization have no effect on phase if the phase distribution is a constant.

Table I shows the variation in 3 dB half-beamwidth, and first, second, and third sidelobe levels when each of the three parameters are varied with the other two held constant. The exact values are given at the bottom of the table. By comparing with the exact result one can see that the values of DENSITY GATE = 0.25λ, QUANTIZING GRID = 0.23λ, and

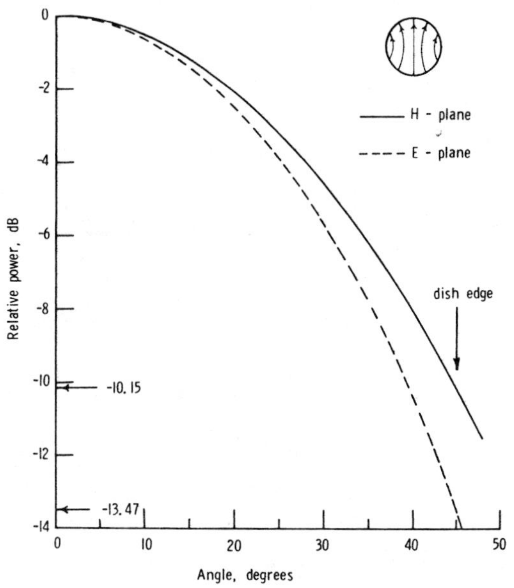

Fig. 7. Radiation patterns of open end circular waveguide, TE_{11} mode.

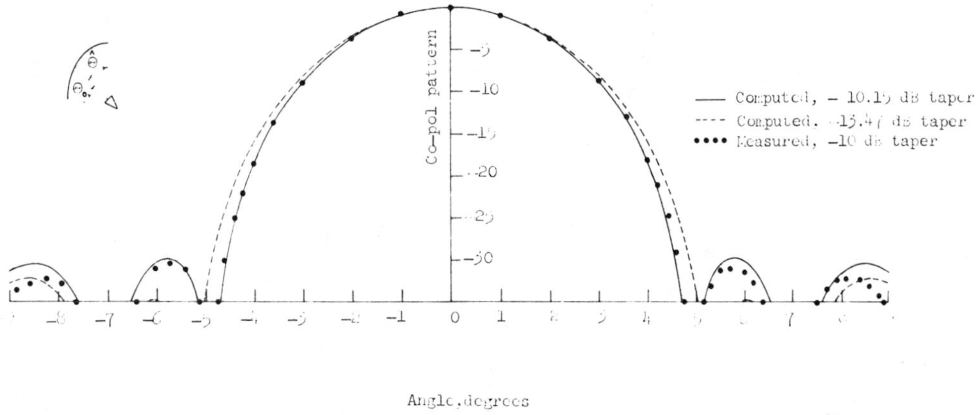

Fig. 8. Calculated and measured H-plane patterns of offset reflector, QUANTIZING GRID = 0.5, INTERPOLATION NUMBER = 5, DENSITY GATE = 0.25, $x' = y' = z' = 0$, $\theta = \theta_0 = 45°$, $f = 15.24$ cm, $D = 60.96$ cm, $d = 30.48$ cm, $\lambda = 1.62$ cm, $f/D = 0.25$.

INTERPOLATION NUMBER = 5 give the best results. All of the calculations discussed here were made using these values. To better show the variation, the radiation patterns are plotted in Fig. 6 using the data from the center section of Table I.

Another check case for which calculations were made is a 45° offset reflector. Measured and computed data for this case have been reported by Chu and Turrin [2]. They used a dual mode horn, which provided a -10 dB, circularly symmetric amplitude taper, as their feed. For our calculations, we used a circularly symmetric feed pattern with a nominal -10 dB taper generated from the H-plane pattern of an open ended circular waveguide. This feed pattern is shown in Fig. 7. The calculated patterns and the measured data of Chu and Turrin [2] are shown together in Figs. 8, 9, and 10. The complete H-plane ($\phi = 90°, 270°$) patterns are shown in Fig. 8. Both the copolarized and cross polarized patterns for $\phi = 90°$ are shown in Fig. 9. The calculated copolarized and cross polarized E-plane ($\phi = 0°$) patterns are shown in Fig. 10. The cross polarized field should be zero in the E-plane due to symmetry [2]. We calculated peak cross polarization levels of about -58 dB for the E-plane, which gives a good indication of the accuracy of the calculation.

Another check was provided by an experimental study of the radiation patterns of an offset parabolic reflector with the feed defocused along an axis near the offset focal axis [8]. The offset focal axis is defined by the angle θ_0 in Fig. 3. The parameters which define the antenna used in this study are

focal length	7.87 in
frequency	35 GHz
λ	0.337 in
aperture diameter	9.84 in $= 29\ \lambda$
feed polarization	$\hat{y}$
θ_0	37°
$\hat{\theta}$	31.15°
θ_f	40°.

θ_f defines the axis along which the feed was positioned. The feed was a dual mode horn [9] designed to provide a -15 dB amplitude taper at 34°. The feed was positioned along the

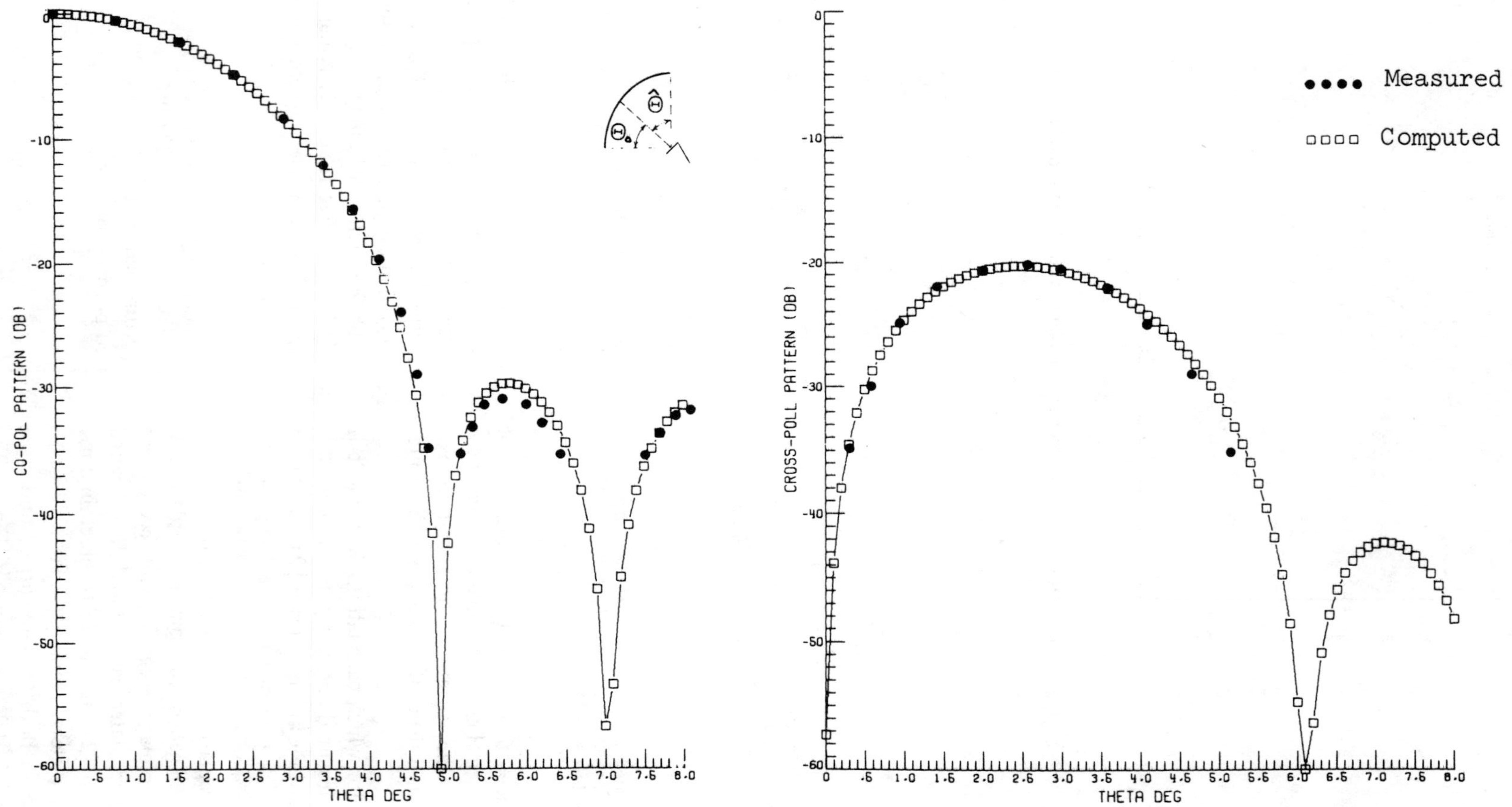

Fig. 9. Calculated and measured H-plane copolarized and cross polarized patterns for offset fed reflector, QUANTIZING GRID = 0.5, DENSITY GATE = 0.25, $x' = y' = z' = 0$, $\theta = \theta_0 = 45°$, $f = 15.24$ cm, $D = 60.96$ cm, $d = 30.48$ cm, $\lambda = 1.62$ cm, $f/D = 0.25$.

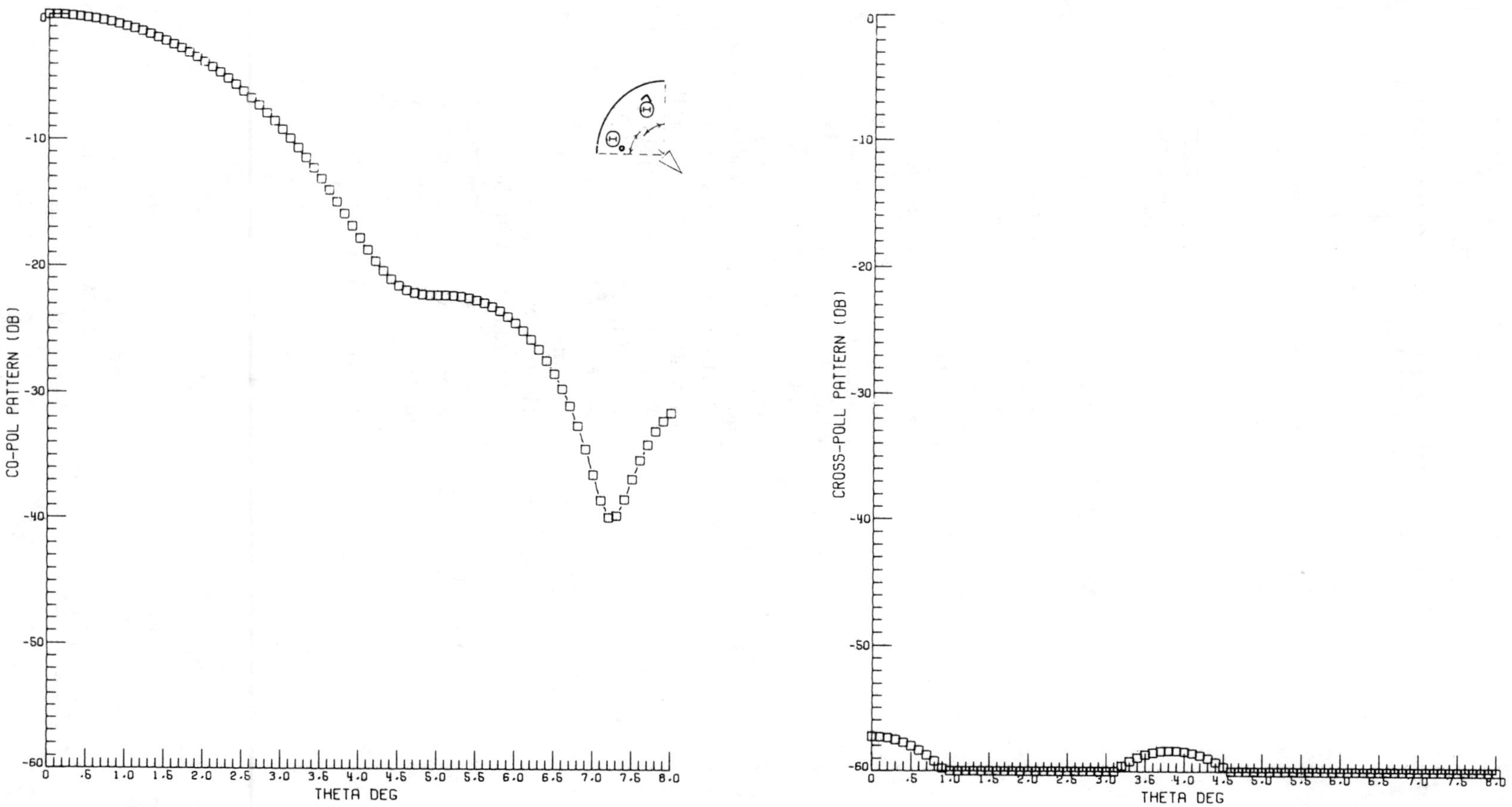

Fig. 10. Calculated E-plane copolarized and cross polarized patterns for offset fed reflector, QUANTIZING GRID = 0.5, DENSITY GATE = 0.25, $x' = y' = z' = 0$, $\theta = \theta_0 = 45°$, $f = 15.24$ cm, $D = 60.96$ cm, $d = 30.48$ cm, $\lambda = 1.62$ cm, $f/D = 0.25$.

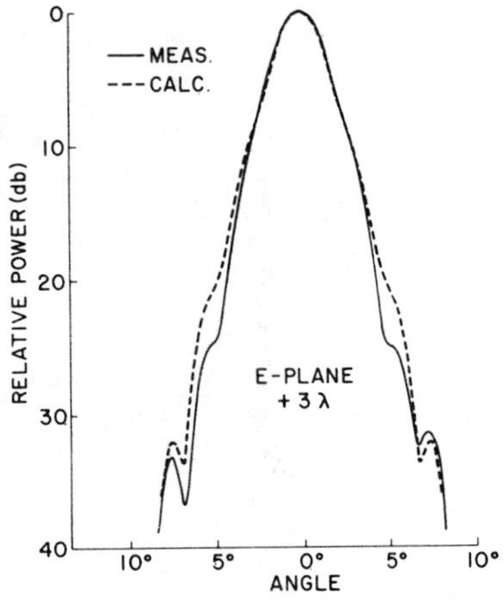

Fig. 11. *E*-plane radiation pattern. Feed offset 3λ toward reflector along $\theta_f = 40°$ axis.

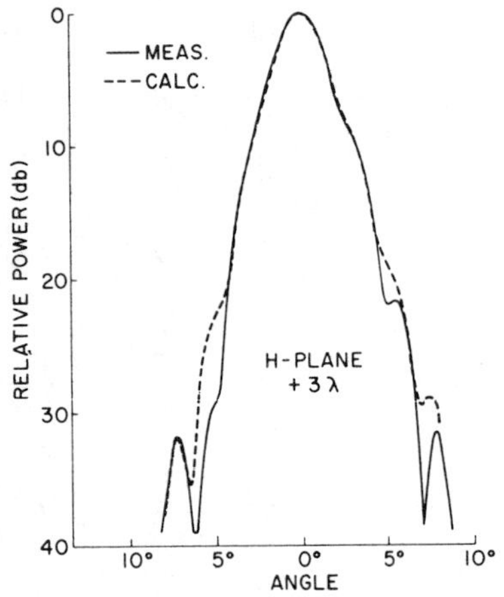

Fig. 12. *H*-plane radiation pattern. Feed offset 3λ toward reflector along $\theta_f = 40°$ axis.

TABLE II

MEASURED BEAMWIDTHS, SIDELOBE AND CROSS POLARIZATION LEVELS FOR FEED OFFSETS OF 0, 3, 5, 7λ TOWARD REFLECTOR ALONG $\theta_f = 40°$ AXIS

*	MEASURED					
	Beamwidths				**	***
	-3dB	-10dB	-20dB	-30dB		
+7λ						
E-PL	8.7	14.2	19.4	24.6	–	-23.8
H-PL	7.7	13.6	17.8	24.3	-27	-33
+5λ						
E-PL	5.5	9.0	14.2	17.2	–	-24
H-PL	5.9	9.9	13.1	16.7	-28	-33
+3λ						
E-PL	3.0	5.7	8.5	12.5	–	-24.4
H-PL	2.6	6.0	8.8	12	-29	-33
0						
E-PL	2.5	4.1	5.4	6.3	–	-22.8
H-PL	2.3	4.1	5.3	7	-22.8	-36

* Pattern cut and feed pos
** Highest sidelobe beyond -30dB BW
*** Max. cross-pol. level

TABLE III

CALCULATED BEAMWIDTHS, SIDELOBE AND CROSS POLARIZATION LEVELS FOR FEED OFFSETS OF 0, 3, 5, 7λ TOWARD REFLECTOR ALONG $\theta_f = 40°$ AXIS

*	CALCULATED					
	Beamwidths				**	***
	-3dB	-10dB	-20dB	-30dB		
+7λ						
E-PL	9.6	14.3	19.6	24.5	-29.7	-23
H-PL	9.1	14.5	20.1	26.3	-29.8	-73.8
+5λ						
E-PL	5.6	10.3	15.0	20.6	-30.6	-24
H-PL	6.0	10.3	15.0	18.7	-28.2	-75.3
+3λ						
E-PL	2.7	5.8	8.9	12.6	–	-24.8
H-PL	2.7	6.1	9.4	12.9	-32.3	-80
0						
E-PL	2.3	4.1	5.2	5.8	–	-24.6
H-PL	2.3	4.0	5.2	5.8	-26.2	-81.3

* Pattern cut and feed pos
** Highest sidelobe beyond -30dB BW
*** Max. cross-pol. level

θ_f axis by two orthogonal microscope drives. The measured and calculated *E*- and *H*-plane patterns for the feed 3 λ from the focus toward the reflector along the θ_f axis are shown in Figs. 11 and 12. The measured and calculated pattern characteristics for several other feed positions are tabulated in Tables II and III. Note the agreement between the measured and calculated results. Also, keep in mind that the feed causes more and more aperture blockage as it is moved toward the reflector along the θ_f axis—an effect which is not accounted for in the algorithm. There are several other observations which are worthy of mention.

1) The *E*- and *H*-plane beamwidths remain approximately equal as the feed is moved along the θ_f axis.

2) The sidelobe levels outside the ridged main beam maintain about the same level as the feed is defocused.

3) Maximum cross polarization diminishes as the feed is moved off focus.

4) The main beam is similar in shape to that of a corrugated horn with the same quadratic phase error.

These observations point to the use of an antenna of this type for radiometer applications.

A check against calculations made by another method was provided by work done at Aerojet Electrosystems Company [10]. The antenna for which the calculations were made is an offset parabolic reflector with a scalar horn feed. Its parameters are

feed polarization	$\hat{y}$
frequency	4.99 GHz
aperture diameter	39.37 in
focal length	31.5 in
θ_0	37°
$\hat{\theta}$	31°

feed phase center position $x' = 3.966$ in, $y' = 0$, $z' = 1.854$ in, feed pointing angle $\delta = -40°$ (40° above the horizontal).

The feed pattern is circularly symmetric with a −15 dB amplitude taper at 34°. The *H*-plane pattern is shown in Fig. 13, the *E*-plane pattern in Fig. 14, and the cross

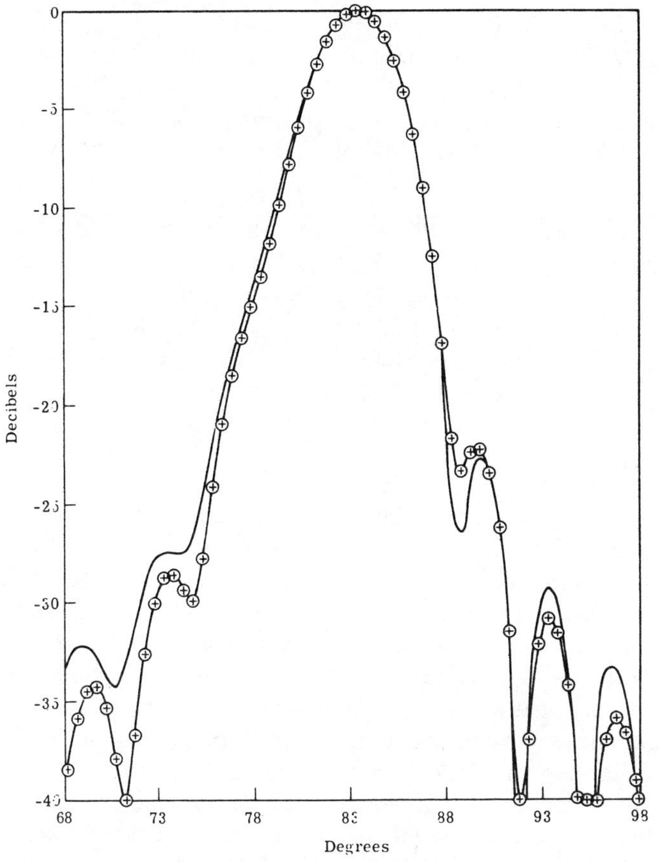

Fig. 13. Comparison of *H*-plane calculated patterns. ——— Rusch's [10] calculations. ⊕——⊕ Our calculations.

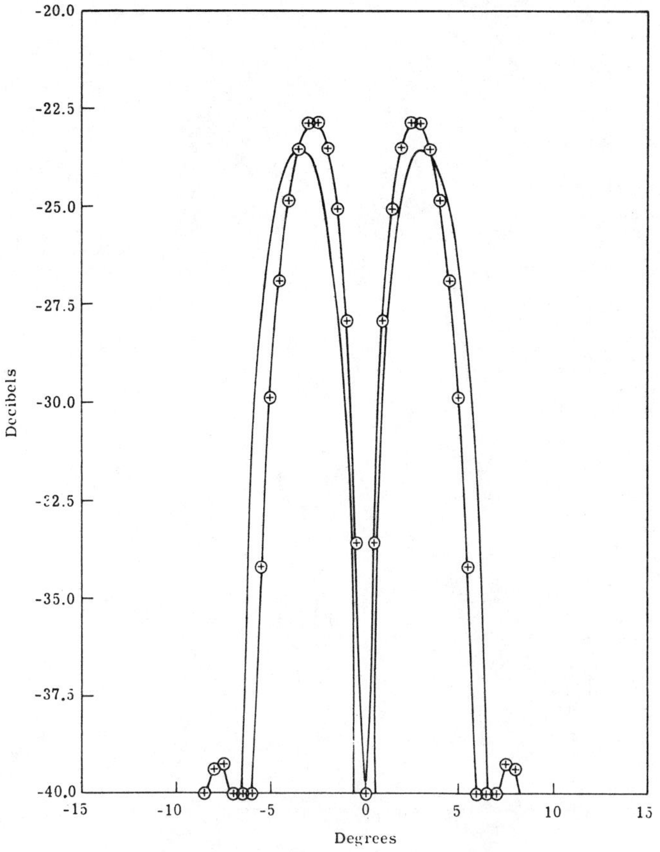

Fig. 15. Comparison of calculated cross polarized patterns. ——— Rusch's [10] calculations. ⊕——⊕ Our calculations.

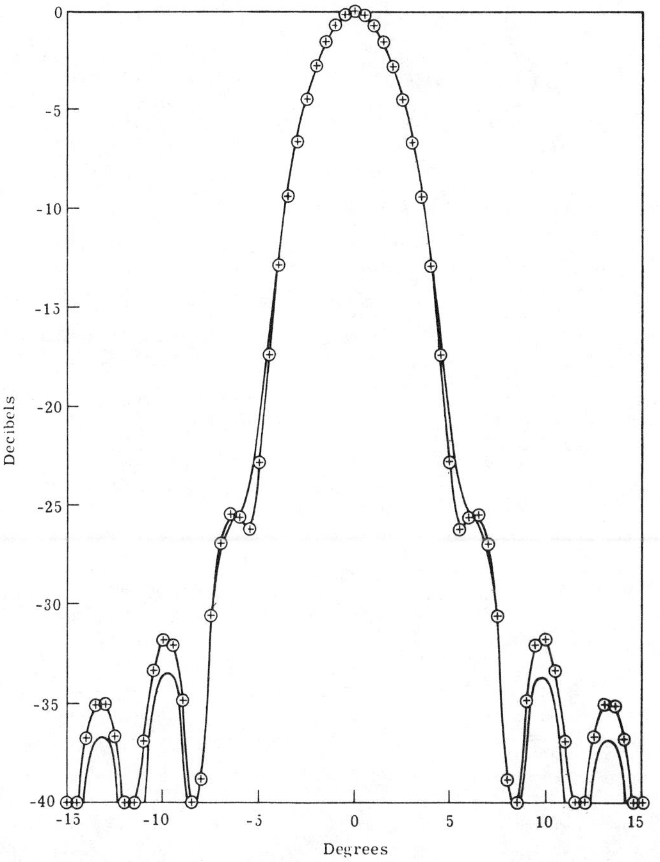

Fig. 14. Comparison of *E*-plane calculated patterns. ——— Rusch's [10] calculations. ⊕——⊕ Our calculations.

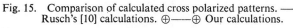

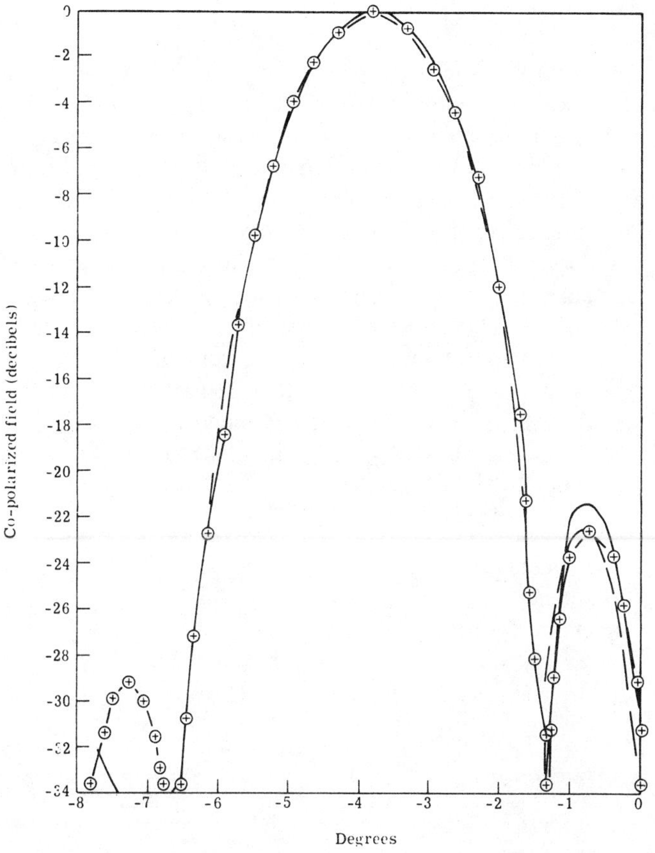

Fig. 16. Comparison of measured and calculated *E*-plane radiation patterns. ——— Rudge's [11] measurements. ⊕——⊕ Rudge's [11] calculations. — — — Our calculations.

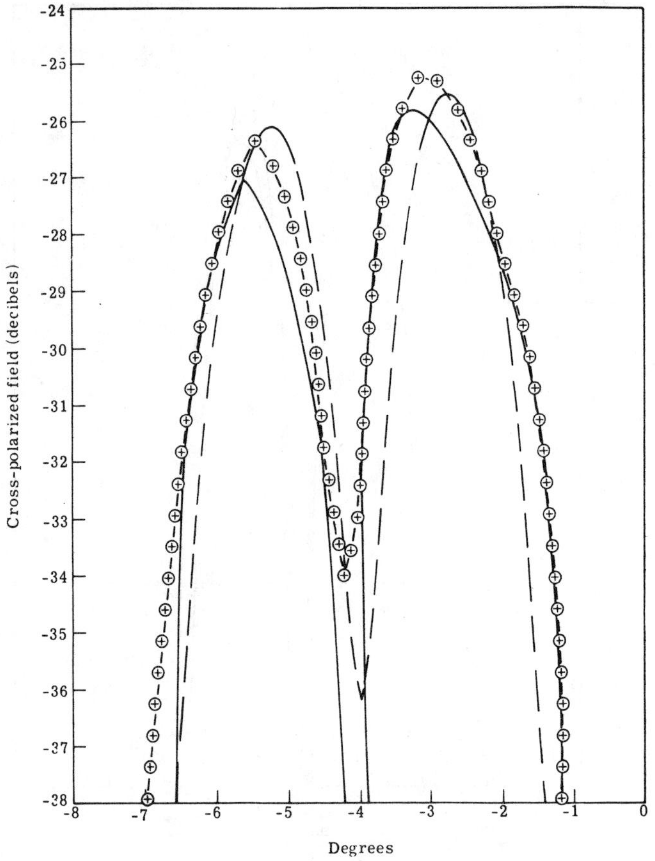

Fig. 17. Comparison of measured and calculated cross polarized patterns. ——— Rudge's [11] measurements. ⊕——⊕ Rudge's [11] calculations. — — — Our calculations.

polarization pattern (*E*-plane) in Fig. 15. (The cross polarization is zero in the *H*-plane.) The solid line represents the calculations made by Rusch [10]; the circled crosses represent the calculations made using the algorithm described in this paper. The computation performed in [10] uses geometrical optics to find the currents on the illuminated portion of the reflector and then integrates the far-field dyadic Green's function over the resulting current distribution. The agreement between the calculations made with these two different algorithms is quite good.

The final case which will be discussed provides a check against both the measured and computed results of Rudge [11]. The antenna is an offset parabolic reflector fed by a 1.57λ by 2.14λ rectangular horn. The horn provides a -10 dB amplitude taper at 30°. The other parameters are

feed polarization	$\hat{y}$
frequency	30 GHz
aperture diameter	36.2 cm
focal length	30.5 cm
θ_0	35°
$\hat{\theta}$	30°

feed phase center position $x' = 0$, $y' = -2.5$ cm, $z' = 0$ feed pointing angle $\delta = -35°$ (35° above the horizontal).

The copolarized patterns in the plane of the feed offset (*E*-plane) are shown in Fig. 16. The cross polarized patterns for the same plane are shown in Fig. 17. In both figures the solid line shows Rudge's measurements, the circled crosses show his calculation, and the dashed line shows the calculations made with the algorithm discussed here. The agreement among the three curves is good.

CONCLUSIONS

A numerical technique for computing the radiation patterns of reflector antennas with arbitrarily located feeds has been developed. An accuracy versus cost tradeoff can be made with a particular application in mind by adjusting the parameters of the algorithm, e.g., DENSITY GATE, QUANTIZING GRID, and INTERPOLATION NUMBER. It is recommended that for feed positions resulting in large phase and amplitude asymmetries in the aperture plane that the values of these parameters be investigated to insure accuracy. An advantage of the method is that it provides an easily visualized orderly sequential procedure for making pattern calculations. The calculations which have been made show good agreement with a variety of calculated and measured results.

REFERENCES

[1] K. G. Schroeder, "Pattern control for geostationary communication satellite antennas," Nat. Telecomm. Conf., Nov. 1973.
[2] T. S. Chu and R. H. Turrin, "Depolarization properties of offset reflector antennas," *IEEE Trans. Antennas Propagat.*, vol. AP-21, pp. 339–345, May 1973.
[3] S. Silver, *Microwave Antenna Theory and Design.* New York: McGraw-Hill, 1949.
[4] W. V. T. Rusch and P. D. Potter, *Analysis of Reflector Antennas.* New York: Academic, 1970, pp. 92–93.
[5] R. E. Collin and F. J. Zucker, *Antenna Theory Part II.* New York: McGraw-Hill, 1969.
[6] J. S. Hollis, T. S. Lyon, and L. Clayton, Jr., *Microwave Antenna Measurements.* Scientific-Atlanta, 1970, pp. 5–11, 5–12.
[7] L. J. Jowers, W. F. Croswell, and J. F. Kauffman, "Radiation patterns of paraboloid antennas," to be published in *IEEE Trans. Antennas Propagat.*
[8] J. F. Kauffman and W. F. Croswell, "Off focus characteristics of the offset fed parabola," AP-S Symp. Digest, June 2, 1975, pp. 358–361.
[9] M. C. Bailey, "The development of an *L*-band radiometer dual-mode horn," *IEEE Trans. Antennas Propagat.* (Commun.), vol. AP-23, May 1975, pp. 439–441,
[10] W. V. T. Rusch, "A study on a scanning multifrequency reflector antenna," Aerojet Electrosystems Co., Final IR&D Rep., AESC FY 1973, Rep. 8714-46-1, Part I.
[11] A. W. Rudge, "Multiple-beam antennas: offset reflectors with offset feeds," *IEEE Trans. Antennas Propagat.*, vol. AP-23, May 1975, pp. 317–322,

Bibliography for Part III

[1] J. B. Keller, "Geometrical theory of diffraction," *J. Opt. Soc. Amer.*, vol. 52, pp. 116–130, Feb. 1962.

[2] W. V. T. Rusch, "Physical-optics diffraction coefficients for a paraboloid," *Electron. Lett.*, vol. 10, pp. 358–360, Aug. 22, 1974.

[3] C. M. Knop, "An extension of Rusch's asymptotic physical optics diffraction theory of a paraboloid antenna," *IEEE Trans. Antennas Propagat.*, vol. AP-23, pp. 741-743, Sept. 1975.

[4] C. A. Mentzer and L. Peters, "A GTD analysis of the far-out sidelobes of Cassegrain antennas," *IEEE Trans. Antennas Propagat.*, vol. AP-23, pp. 702–709, Sept. 1975.

[5] M. Safak, "Calculation of radiation patterns of paraboloidal reflectors by high frequency asymptotic techniques," *Electron. Lett.*, vol. 12, pp. 229-231, Apr. 29, 1976.

Part IV
Cassegrain and Dual Reflector Systems

The two-reflector combination named after the inventor of the similarly configured astronomical telescope began to come into prominence when the need arose for large ground-based antennas for satellite tracking and communications. Because the hyperboloidal subreflector must, of necessity, be at least a few wavelengths in diameter, the loss in gain due to shadowing can be intolerably high unless the main reflector is many wavelengths in diameter. For this reason, the Cassegrain geometry is not normally attractive until the gain is required to exceed about 40 dB. The first comprehensive published treatment of this arrangement, that of Hannan, forms the first paper in this part. In it, he introduces the concepts of equivalent parabola and virtual focus, and he discusses means for minimizing, or even eliminating, subreflector blocking. Hannan's use of geometrical optics (GO) gives reasonably good results because Cassegrain antennas are almost always electrically large. GO, however, is inadequate for predicting fine detail and for determining spillover and noise temperature. In the second and third papers, analytical techniques are developed that overcome these deficiencies in the ray optics treatment. Rusch, for example, calculates the scattered field of the subreflector by integrating the induced current density over its front surface. He is then able to evaluate the spillover, i.e., that part of the field scattered by the subreflector that cannot be collimated by the main reflector. Potter, using the powerful technique of spherical wave expansion, is able not only to analyze spillover and determine how to reduce it, but also to generate more efficient aperture distributions and obtain improved subreflector impedance matching.

Wong, in the fourth paper, attempts to answer the question of the validity of the equivalent parabola concept in predicting the performance of a Cassegrain system in which the feed is laterally offset from the optical axis. For an example in which the main reflector is 253 wavelengths in diameter and the magnification factor is 6 ($f/D = 2$ for the equivalent parabola), he concludes that use of the concept produces negligible error for beam squints up to four beamwidths. There may be some conflict between this result and that obtained by Gniss and Ries [1]. By deriving series expansions for the aperture phase, they find "severe differences between the Cassegrain system and the prime focus-fed equivalent paraboloid for larger radial or axial defocusing." Their brief paper, in German, has not been included here.

A modified version of the Cassegrain antenna that is of some practical importance utilizes the dielguide feed [2], in which a cone of dielectric material extends from the feed horn to the subreflector and effectively supports the latter. A design procedure for determining the required subreflector profile is given in the fifth paper by Clarricoats, Salema, and Lim. The first two authors of that paper have published a detailed treatment of radiation by dielectric horns and their use as reflector feeds in a long, two-part paper [3] which may also be found in Part VII of the IEEE Press reprint book *Electromagnetic Horn Antennas*.

The final trio of papers in this part are concerned with Cassegrain, or two-reflector systems, in which the reflector surfaces are modified in order to achieve some specified aperture distribution. One of the first attempts to improve aperture efficiency by means of this technique was described in a short note by Green [4], who observed that the main aperture distribution could be made more nearly uniform by suitably shaping the subreflector. Since this, perforce, changes the phase distribution, he proposed a slight reshaping of the main reflector in order to correct the error. In the first paper of the trio, Galindo expands on this concept by creating a general synthesis technique, similar to that of Kinber [5], whereby an arbitrary phase and amplitude distribution may be generated in the aperture of the second reflector. Williams, in his paper, shows how the technique may be applied to obtain a near uniform aperture distribution, and therefore nearly maximum gain, as Green had done originally. In the last of the trio, Collins presents experimental results which show that shaping of the subreflector alone can yield overall aperture efficiencies ranging from 70 to 80 percent. The main reflector retains a paraboloidal shape, but its focal length is altered slightly in order to compensate for the phase change caused by the nonhyperboloidal subreflector.

Microwave Antennas Derived from the Cassegrain Telescope*

PETER W. HANNAN†, SENIOR MEMBER, IRE

Summary—**A microwave antenna can be designed in the form of two reflecting dishes and a feed, based on the principle of the Cassegrain optical telescope. There are a variety of shapes and sizes available, all described by the same set of equations. The essential performance of a Cassegrain double-reflector system may be easily analyzed by means of the equivalent-parabola single-reflector concept.**

Techniques are available for reducing the aperture blocking by the sub dish of the Cassegrain system: one method minimizes the blocking by optimizing the geometry of the feed and sub dish; other methods avoid the blocking by means of polarization-twisting schemes. The former method yields good performance in a simple Cassegrain antenna when the beamwidth is about 1° or less. The latter methods are available for any application not requiring polarization diversity, and an optimized set of polarization-operative surfaces has been developed for these twisting Cassegrain antennas.

Experimental results, presented for practical antennas of both types, illustrate the feasibility of these principles. A number of unusual benefits have been obtained in the various Cassegrain antenna designs, and additional interesting features remain to be exploited.

I. Introduction

FOR THE DESIGN of an optical telescope, the Cassegrain double-reflector system has often been utilized [1]–[4]. Compared with the single-reflector type, it achieves a high magnification with a short focal length, and allows a convenient rear location for the observer.

Recently, a number of microwave antennas have been developed which employ double-reflector systems similar to that of the Cassegrain telescope. Each of these antennas has achieved one or more particular benefits not obtainable with the ordinary single-reflector type. While the various designs may differ from each other to a considerable degree, there are certain basic features which are common to all.

It is the purpose of this paper to outline the design principles and essential properties of the Cassegrain antenna, and to discuss its advantages and limitations. Some of the techniques available for minimizing its limitations are described, and experimental results illustrating the practical nature of two particular designs are presented. Finally, a number of interesting applications for the Cassegrain antenna are mentioned.

II. Telescope vs Antenna

A Cassegrain telescope consists of two mirrors and an observing optical instrument, as indicated in Fig. 1. The primary mirror, which is a large concave mirror in the rear, collects the incoming light and reflects it toward the secondary mirror, which is a small convex mirror out in front. The secondary mirror then reflects the light back through a hole in the center of the primary mirror. When the incoming rays of light are parallel to the telescope axis, the final bundle of light rays is focused toward a point; at this location the observer places his eye or his camera.

The basic microwave antenna derived from the Cassegrain telescope is shown in Fig. 2. The microwave reflectors, which will be called the main dish and the sub dish, respectively, have surfaces similar in shape to those of the telescope. The microwave feed is a small antenna which, together with a transmitter or receiver, replaces the optical instrument of the telescope.

Analysis of the operation of a Cassegrain antenna system may be performed with the same semi-optical approximation commonly employed with an ordinary single-dish antenna. Usually the feed is sufficiently small so that the wave radiated by the feed can be described by the far-field pattern of the feed before reach-

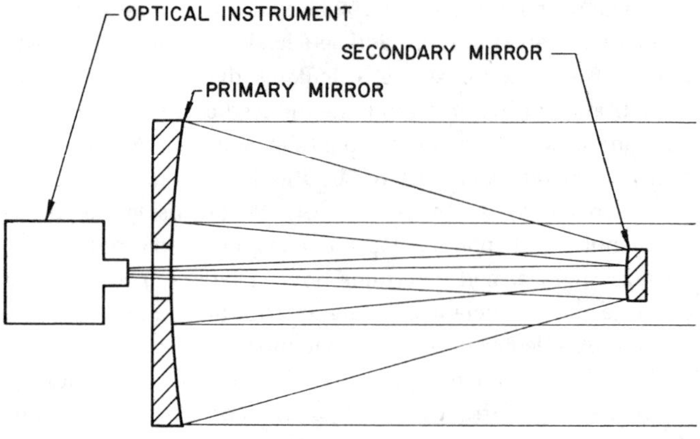

Fig. 1—Cassegrain telescope.

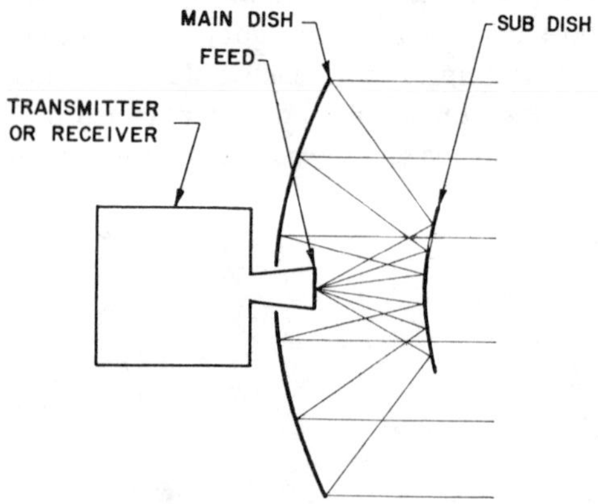

Fig. 2—Cassegrain antenna.

* Received by the PGAP, March 12, 1960. Revised manuscript received, July 5, 1960. Presented at the URSI-IRE Spring Meeting, Washington, D. C., May 2–5, 1960.

† Wheeler Labs., Smithtown, N. Y.

Reprinted from *IRE Trans. Antennas Propagat.*, vol. AP-9, pp. 140–153, Mar. 1961.

ing the sub dish, and the wave incident on the sub dish appears to travel along the rays originating from a point centered on the feed. The sub dish, which must be large enough to intercept the useful portion of the feed radiation, ordinarily reflects this wave essentially according to ray optics. On reaching the main dish, the wave is again reflected according to ray optics; and because of the geometry of the antenna elements, the rays emerge parallel and the wavefront has the flat shape which is usually desired. The amplitude of the emergent wave across the aperture has a taper which is determined by the radiation pattern of the feed, modified by the additional tapering effect of the antenna geometry. The far-field pattern of the antenna is, of course, a diffraction pattern whose characteristics depend on the amplitude taper of the emergent wave.

III. Geometry

The geometry of the Cassegrain system is simple and well-known, but it is helpful to have at hand those formulas describing the dish contours in terms of the significant antenna parameters. The classical Cassegrain geometry, shown in Fig. 3, employs a parabolic contour for the main dish and a hyperbolic contour for the sub dish. One of the two foci of the hyperbola is the real focal point of the system, and is located at the center of the feed; the other is a virtual focal point which is located at the focus of the parabola. As a result, all parts of a wave originating at the real focal point, and then reflected from both surfaces, travel equal distances to a plane in front of the antenna.

To completely describe a Cassegrain system, four fixed parameters are required, two for each dish. Since seven parameters are shown in Fig. 3, three are dependent on the other four, and three equations exist which describe this dependency. In the case of the main dish, the relationship is

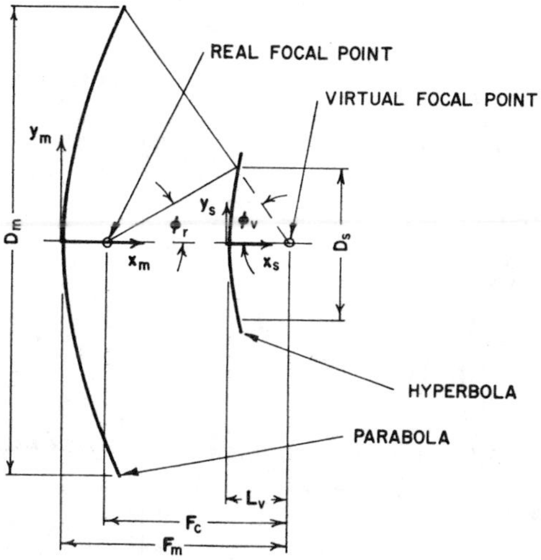

Fig. 3—Geometry of Cassegrain system.

$$\tan \tfrac{1}{2}\phi_v = \pm \frac{1}{4}\frac{D_m}{F_m}. \tag{1}$$

As will be discussed later, the positive sign in the above formula applies to the Cassegrain forms, and the negative sign to the Gregorian forms. In the case of the sub dish, the relationships are

$$\frac{1}{\tan \phi_v} + \frac{1}{\tan \phi_r} = 2\frac{F_c}{D_s}, \tag{2}$$

$$1 - \frac{\sin \frac{1}{2}(\phi_v - \phi_r)}{\sin \frac{1}{2}(\phi_v + \phi_r)} = 2\frac{L_v}{F_c}. \tag{3}$$

In a typical case, the parameters D_m, F_m, F_c, and ϕ_r might be determined by considerations of antenna performance and space limitations; ϕ_v, D_s, and L_v would then be calculated. It is interesting to note that a value for the parameter ϕ_r, which determines the beamwidth required of the feed radiation, may be specified independently of the ratio F_m/D_m, which determines the shape of the main dish.

The contour of the main dish is given by the equation

$$x_m = \frac{y_m^2}{4F_m}. \tag{4}$$

The contour of the sub dish is given by the equation

$$x_s = a\left[\sqrt{1 + \left(\frac{y_s}{b}\right)^2} - 1\right] \tag{5}$$

where

$$e = \frac{\sin \frac{1}{2}(\phi_v + \phi_r)}{\sin \frac{1}{2}(\phi_v - \phi_r)},$$

$$a = \frac{F_c}{2e} \qquad b = a\sqrt{e^2 - 1}.$$

The quantities e, a, and b are the parameters of the hyperbola: e is eccentricity, a is half the transverse axis, and b is half the conjugate axis.

So far, only the geometry of the classical Cassegrain system has been considered. However, the system may easily be extended to include a variety of forms, all obeying the basic formulas presented above. In Fig. 4, two series are shown in which the curvature of the sub dish is modified from the classical convex shape to a flat, and finally a concave shape. As this is done, the diameter of the sub dish increases. The first series shows the case in which the main dish is held invariant; this yields a progressive increase in the required feed beamwidth, and a progressive decrease in the axial dimension of the antenna. In the second series the feed beamwidth is held invariant; in this case, the main dish becomes progressively flatter, and the axial dimension of the antenna progressively increases.

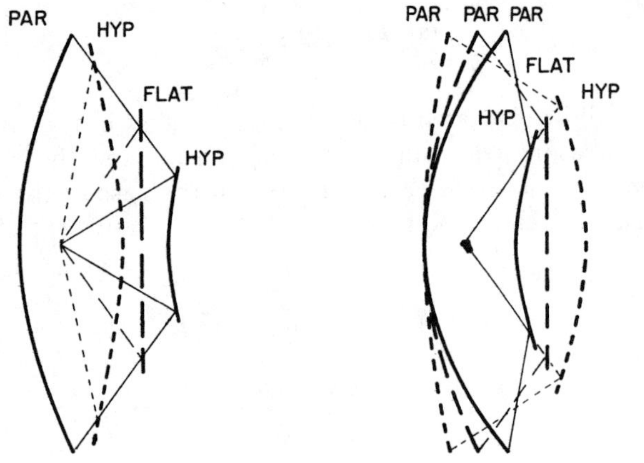

(a) Invariant main dish. (b) Invariant feed.

Fig. 4—Cassegrain modifications.

In Fig. 5, a series is presented in which the beamwidth of the feed is progressively increased while the overall dimensions of the antenna are held fixed. The range of values of some of the parameters previously mentioned are indicated alongside the sketches, together with the distinguishing characteristic of each case. (In each sketch an additional dish and some additional rays are shown in dashed lines, and one column has an additional parameter F_e/F_m; these will be discussed in the next section of this paper, and should be disregarded at this point.) The first three cases are similar to those shown in Fig. 4. In the fourth case, the main dish has degenerated to a flat contour and the sub dish has degenerated to a parabolic contour; here, the flat main dish may be placed at any distance from the sub dish, out to the region where the ray-optical approximation begins to fail. The final case carries the progression to the ridiculous extreme of a concave elliptical sub dish and a convex parabolic main dish, with the former being larger than the latter. It should be mentioned that in the two cases having one flat dish, the formulas presented before, while valid, are overly complicated and contain indeterminate factors; since the focusing is accomplished entirely by the curved parabolic surface, it is preferable to employ the simple formulas for an antenna having a single parabolic dish.

A further extension of the Cassegrain system is shown in Fig. 6. Here, the focal point of the main dish moves to a region between the two dishes, and the contour of the sub dish becomes concave elliptical. In the first of the two cases shown, the system is identical with that of the Gregorian telescope; however, both cases obey the formulas given previously for the Cassegrain system, if the proper values are employed. The ranges allowable for some of the parameters are indicated in the figure. In addition, the negative sign must be employed in (1) so as to maintain a positive F_m with the negative ϕ_v which occurs in the Gregorian forms. The first case, or classical Gregorian, is drawn so as to have the same over-all size and the same feed beamwidth as the classical

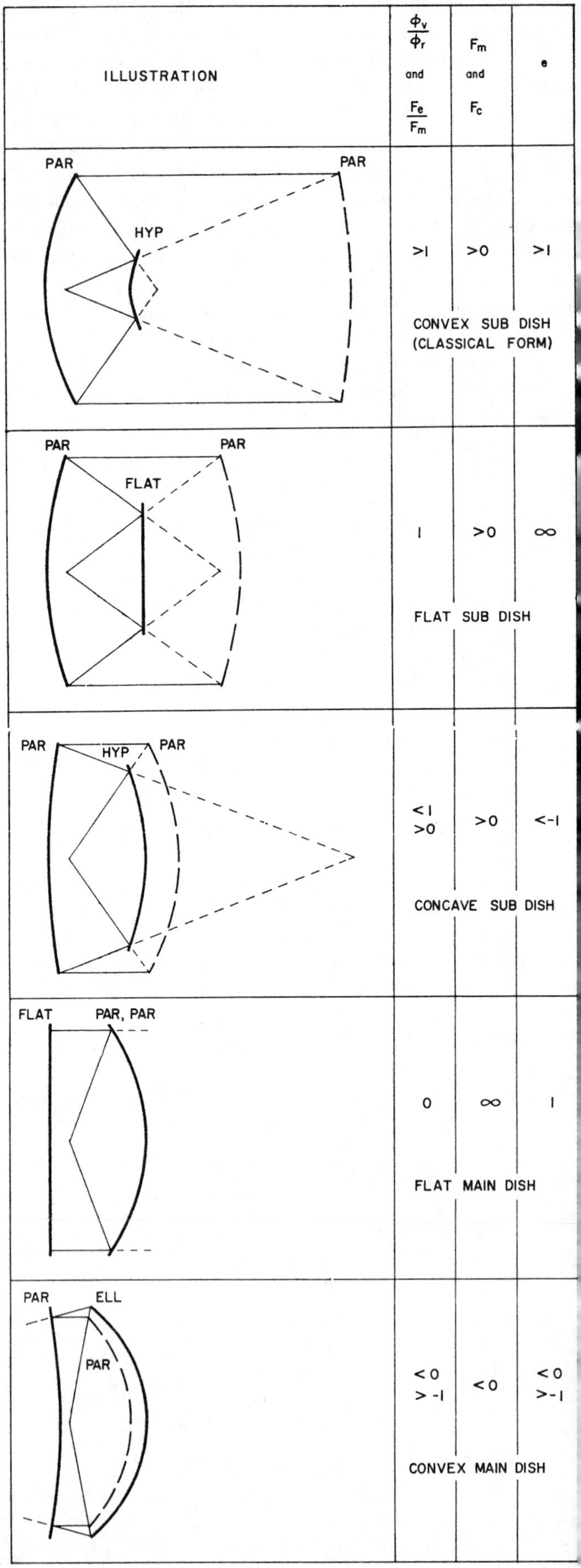

Fig. 5—Series of Cassegrain forms.

ILLUSTRATION	$-\frac{\phi_v}{\phi_r}$ and $\frac{F_e}{F_m}$	F_c	e
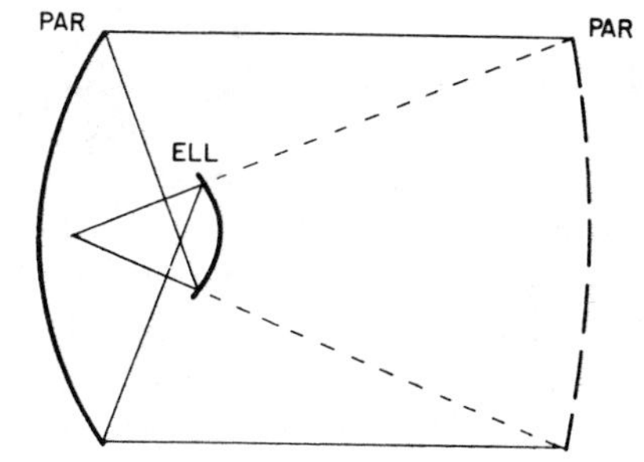	>1	>0	>0 <1
	FEED IN REAR (CLASSICAL FORM)		
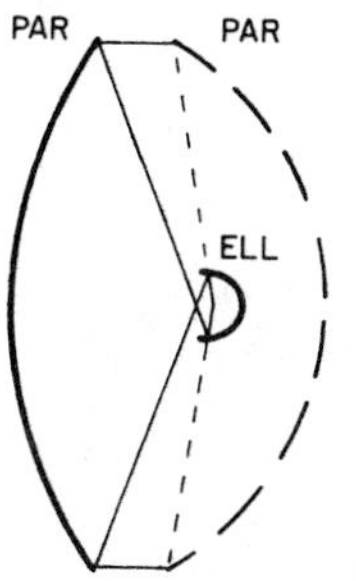	<1	<0	>0 <1
	FEED IN FRONT OF MAIN DISH FOCUS		

Fig. 6—Series of Gregorian forms.

Cassegrain in the first case of the previous figure. Under these conditions, the Gregorian form requires a shorter focal length for the main dish. In the second of the Gregorian forms shown, the feed has been moved to a location between the main dish focus and the sub dish, with the main dish kept the same as in the first case. This form would have several major disadvantages that would make it unattractive in most antenna applications.

All of the above-mentioned forms are members of the same family, which might be called the Cassegrain family. In every case, incoming rays collected by the main dish are focused toward a point. It should be mentioned that a further extension of the Cassegrain system can be made by modifying the contours of both dishes in such a way that incoming rays collected by the main dish are not focused exactly toward a point, while the final bundle of incoming rays, after reflection from the sub dish, remain focused toward a point. Although this may be a useful technique for achieving certain kinds of performance [5]–[7], it is beyond the scope of this paper.

IV. Equivalence Concepts

A. Virtual Feed

One concept which is helpful in understanding and predicting the essential performance of a Cassegrain antenna is that of a virtual feed. As shown in Fig. 7, the combination of real feed and sub dish is considered as

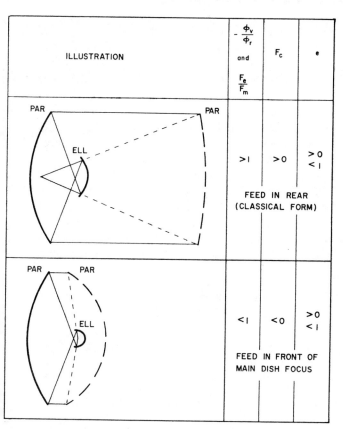

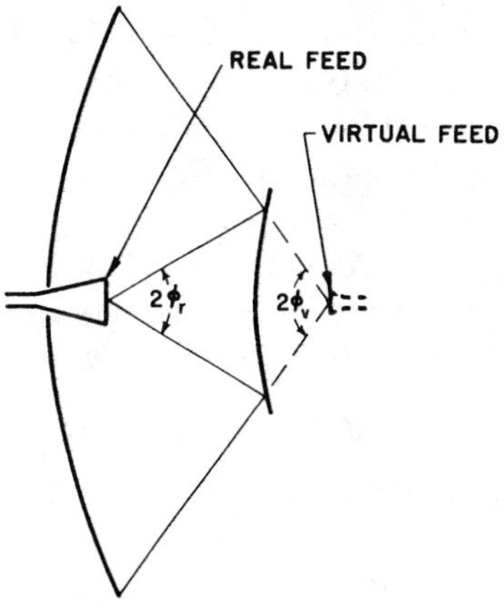

Fig. 7—Virtual-feed concept.

being replaced by a virtual feed at the focal point of the main dish. Thus the antenna becomes an ordinary single-dish design, having the same main dish but a different feed.

If both the real and virtual feeds had dimensions much larger than a wavelength, the configuration of the virtual feed could be determined by finding the optical image of the real feed in the sub dish. This condition seldom exists for a microwave antenna; however, if only the effective apertures of the feeds are considered, it is found that the imaging process yields approximately the correct results. For the classical Cassegrain configurtion shown in Fig. 7, the virtual feed has an effective aperture smaller than that of the real feed, and has a correspondingly broader beamwidth. The beamwidth increase is, of course, the result of the convex curvature of the sub dish; the ratio of virtual-feed to real-feed beamwidth is indicated by the quantity ϕ_v/ϕ_r.

For the various Cassegrain modifications, the range of values that the quantity ϕ_v/ϕ_r may achieve is given in the first column of Figs. 5 and 6, and from this the relative sizes of the effective apertures of the real and virtual feeds may be inferred. When the sub dish is flat, the virtual and real feeds are, of course, identical. For the Cassegrain system having a concave sub dish, the virtual feed has a beamwidth smaller than that of the real feed, and has a larger effective aperture. However in the classical Gregorian form, the concave sub dish results in an effective aperture of the virtual feed which is smaller than that of the real feed, just as in the classical Cassegrain system.

There are several situations in the design of microwave antennas in which the ability to obtain a different effective aperture of the virtual feed from that of the real feed is quite helpful. One such case occurs with a monopulse antenna, where it is difficult to reduce the overall size of the feed aperture to a wavelength or less, while maintaining efficient and wideband performance. On the other hand, a large feed aperture ordinarily requires a long focal length for effective utilization of the main aperture, thereby increasing the size of the antenna structure. This problem may be solved by means of the classical Cassegrain system of Fig. 7, which can incorporate a large feed while employing a short focal length for the main dish. Actually, the axial dimension of such an antenna is often less than the main focal length, because the virtual feed is beyond the sub dish. In addition, of course, there are no waveguide components required in this forward region.

B. Equivalent Parabola

The concept of a virtual feed furnishes a useful qualitative means for analyzing a Cassegrain antenna, but, in general, it is not convenient for an accurate quantitative analysis. In addition, the virtual feed assumes ridiculous proportions for certain of the Cassegrain configurations. A second concept, that of the equivalent parabola, overcomes these limitations.

As shown in Fig. 8, the combination of main dish and sub dish is considered as being replaced by an equivalent focusing surface, drawn with dashed lines in the figure, at a certain distance from the real focal point. The properties of this focusing element can be determined from a study of the "principal surface" of the Cassegrain system. This surface [8] is defined here as the locus of intersection of incoming rays parallel to the antenna axis with the extension of the corresponding rays converging toward the real focal point, as indicated in Fig. 8. It happens that for the Cassegrain system, the "principal surface" has a parabolic contour, and the focal length of this parabola exactly equals the distance from its vertex to the real focal point. As a result, this surface could be employed as a reflecting dish which would focus an incoming plane wave toward the real focal point in exactly the same manner as does the combination of main dish and sub dish. (Actually, the plane wave would have to be incident from the opposite direction; this is of no significance in the principles of this concept.) Thus the antenna again becomes an ordinary single-dish design, but this time having the same feed and a different main dish.

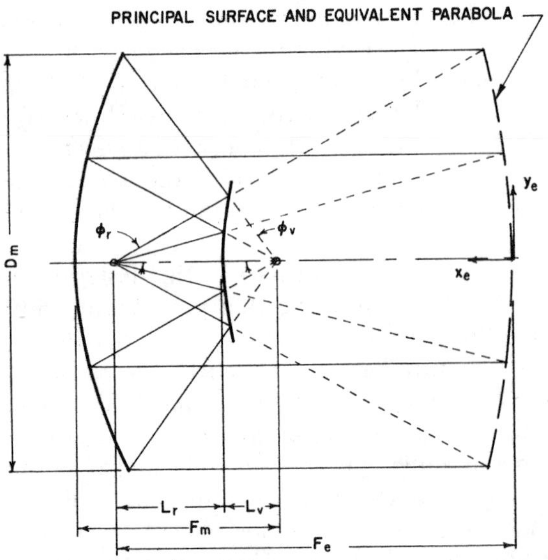

Fig. 8—Equivalent-parabola concept.

It should be mentioned that the equivalent parabola is based on simple ray analysis, rather than on an exact analysis of the wave action. This ray approximation is made throughout the paper, and is accurate enough for most purposes except when the sub dish is only a few wavelengths in diameter. When the wave analysis is necessary, consideration would have to be given to the Fresnel diffraction pattern formed at the main dish after reflection of the feed radiation by the sub dish.

The following equations provide the relationship between the equivalent parabola, the antenna parameters shown in Fig. 8, and some of the parameters previously mentioned:

$$\frac{1}{4}\frac{D_m}{F_e} = \tan \tfrac{1}{2}\phi_r , \tag{6}$$

$$x_e = \frac{y_e^2}{4F_e}, \tag{7}$$

$$\pm\frac{F_e}{F_m} = \frac{\tan \frac{1}{2}\phi_v}{\tan \frac{1}{2}\phi_r} = \frac{L_r}{L_v} = \frac{e+1}{e-1} . \tag{8}$$

In (8), the positive sign applies to the Cassegrain forms, and the negative sign to the Gregorian forms. Eqs. (6) and (7) describe the equivalent parabola itself, in terms of its *equivalent focal length*, F_e. Eq. (8) presents the various alternate expressions for the quantity F_e/F_m the ratio of equivalent focal length to focal length of the main dish. It is evident that with the classical Cassegrain system, the equivalent focal length is greater than the focal length of the main dish.

As might be expected, the equivalent-parabola concept also applies to the extended Cassegrain forms, and to the Gregorian forms, as well. The equivalent parabola for each of these cases is indicated by the dashed curves in Figs. 5 and 6, and the range of values for the quantity F_e/F_m is indicated in the first column of these figures. When the sub dish is flat, the equivalent focal length equals the focal length of the main dish. For the Cassegrain system having a concave sub dish, the equivalent focal length is shorter than that of the main dish. For the case of a flat main dish, the equivalent parabola is identical with the sub dish. In the classical Gregorian form, the equivalent focal length is greater than that of the main dish, as is also the case in the classical Cassegrain system.

In describing the magnifying properties of a Cassegrain optical telescope, it has become customary to employ the concept of the equivalent focal length [1], [2], [4]. It has also been recognized that the coma aberration of a Cassegrain telescope is the same as that of a telescope having a single parabolic mirror of focal length equal to the equivalent focal length of the Cassegrain [2]. These two aspects are readily explainable in terms of the equivalent parabola. It may be noted that since the Cassegrain optical telescope has an equivalent focal length greater than that of its large mirror, it has greater magnifying power and reduced coma compared with that obtained with only the single large mirror.

In the case of a microwave antenna, the equivalent-parabola concept yields properties similar to those mentioned above. The effective aperture of the feed should be such that the equivalent parabola is properly illuminated: when the equivalent focal length is greater than the focal length of the main dish, the optimum feed aperture is larger than that which would be optimum for a single-dish antenna having the same focal length as the main dish. This result is analogous to the magnifying properties [1]–[3] of the optical telescope; it also corresponds to the result obtained with the virtual-feed concept. Indeed, the ratio F_e/F_m is sometimes called the *magnification*. This is a valid approximation when applied to the relative sizes of the real and virtual feeds or images. However, it should not be confused with the magnification of an optical telescope containing an eyepiece, in which the term usually applies to the relative sizes of the image and the object.

As regards coma aberration, the equivalent-parabola concept yields the same results as in the optical case, when the off-axis beam angle is small. However, when this angle becomes appreciable, as may be necessary in a microwave antenna, the feed is sufficiently offset from the dish axes so that the principal surface is no longer closely approximated by the original equivalent parabola; therefore, the wide-angle coma may differ considerably from that calculated by the equivalent-parabola concept. Of course, other aberrations may become appreciable at the same time.

There are some significant uses of the equivalent-parabola concept in the microwave antenna which appear to have no application in the optical telescope. One such case involves the determination of amplitude taper across the main aperture of the antenna. For an ordinary single-dish antenna, the illumination is determined by the radiation pattern of the feed, modified by a "space-attenuation" characteristic which is a simple function of the F/D ratio [9]. For a Cassegrain antenna, exactly the same process is applicable, with the F/D ratio now being the ratio of equivalent focal length to main-dish diameter, F_e/D_m. In other words, the illumination is exactly the same as that which would exist across a single dish having the equivalent focal length and being illuminated with the same feed. When the equivalent focal length is greater than the diameter, the "space-attenuation" characteristic modifies the feed radiation only slightly; with a practical feed, such an antenna can have high efficiency even though it may have a physically short axial length.

V. Reduction of Aperture Blocking

The principal limitation on the application of the historical Cassegrain system to microwave antennas is the blocking of the main aperture by the sub dish [10], [11]. This problem has not been serious with optical telescopes because the requirements on characteristics of the diffraction pattern have not been severe, and because, for the relatively short wavelength of light, the size of the small reflector can be made very much less than that of the large reflector. With a microwave antenna, neither of these conditions ordinarily exists.

The presence of an opaque sub dish in the main aperture of the antenna creates a "hole" in the illumination which causes decreased gain and increased sidelobe levels. To analyze this effect, the resulting illumination may be resolved into two components [9], the original illumination plus a negative center, or "hole," as shown in Fig. 9(a). The resulting antenna pattern, shown in Fig. 9(b), can be determined by adding together the two pattern components, the original pattern plus a broad, low, negative pattern radiated by the "hole."

Although the above method facilitates an exact calculation of the shadowing effect for any case, it is instructive to apply the method to a particular simple case

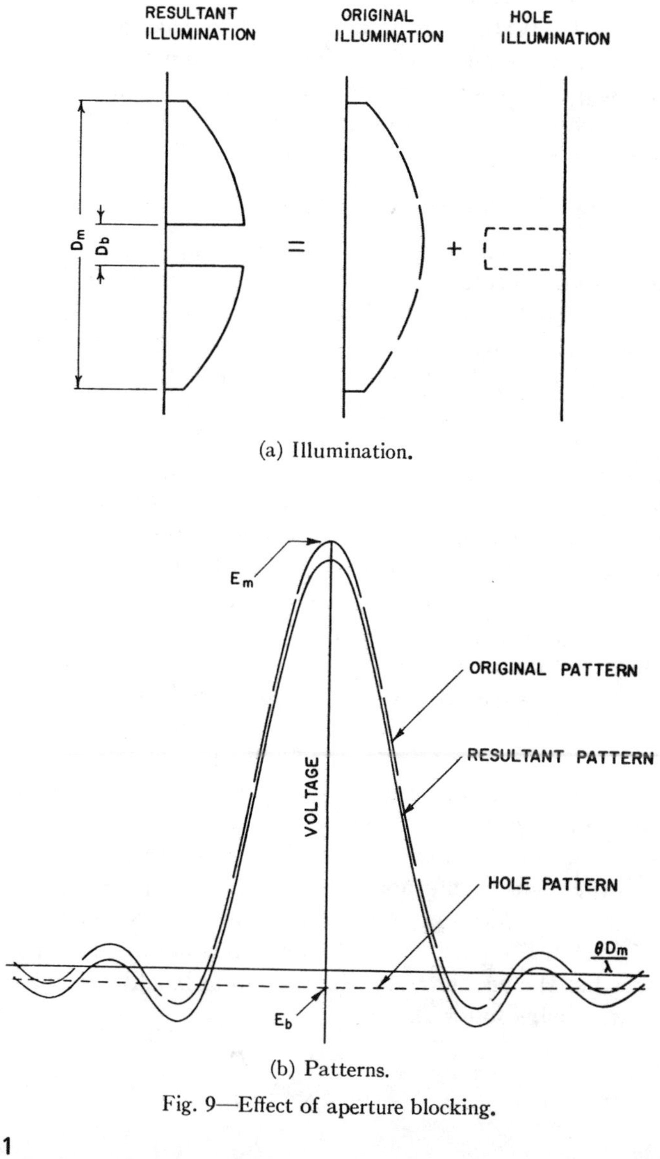

(a) Illumination.

(b) Patterns.

Fig. 9—Effect of aperture blocking.

which approximates many practical cases. If the main aperture is circular, and is assumed to have a completely tapered parabolic illumination, a small circular obstacle in the center of the aperture will create a "hole" pattern whose peak voltage relative to the peak voltage of the original pattern is

$$\frac{E_b}{E_m} = 2\left(\frac{D_b}{D_m}\right)^2. \tag{9}$$

where D_b is the diameter of the blocked portion of the aperture. This relative voltage is then subtracted from unity to yield the resultant relative peak voltage, and is added to the relative level of the first sidelobe to yield the resultant relative level.

The illumination hole is not the only effect created by the presence of an obstacle in the main aperture; the power which strikes the obstacle must also be accounted for. Usually this power reradiates and contributes an additional component to the sidelobes. For a particular sub dish and antenna configuration, it is often a straightforward process to estimate the amplitude pattern of this radiation. However the manner in which it combines with the original pattern is more complicated, and is likely to vary radically with a change of frequency. A further consideration of this effect is beyond the scope of this paper, and, even though it may sometimes be an important one, the effect will be neglected henceforth.

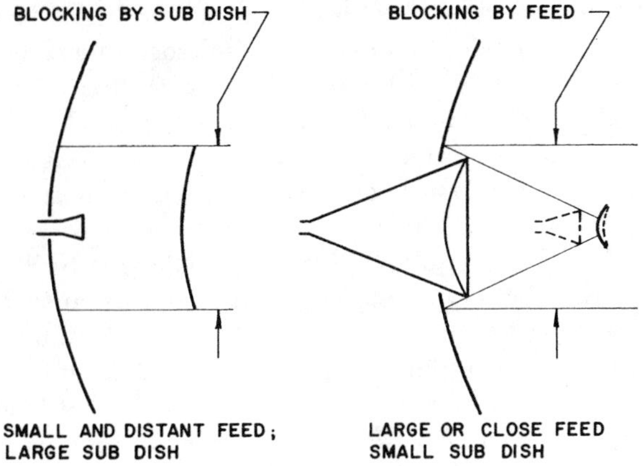

Fig. 10—Types of aperture blocking.

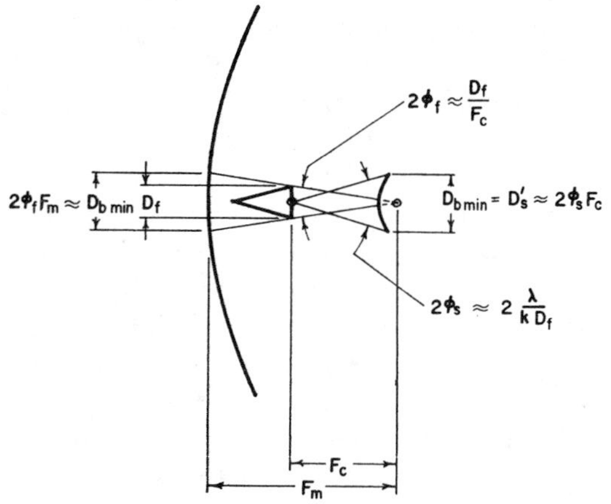

Fig. 11—Condition for minimum blocking by sub dish and feed.

A. Minimum Blocking with Simple Cassegrain

In order to determine the degree of aperture blocking to be expected in a Cassegrain antenna having an ordinary reflecting sub dish, it is necessary to consider those factors which influence the size of the sub dish. Essentially, the minimum size of the sub dish is determined by the directivity of the feed, and the distance between the feed and the sub dish. By making the feed more directive, or by decreasing its distance to the sub dish, the size of the sub dish may be reduced without incurring a loss caused by spillover of the feed radiation beyond the edge of the sub dish. However, as indicated in Fig. 10, a continuation of this process can eventually result in the feed itself creating a shadow in the main illumination which is greater than that created by the sub dish. It is evident that there is some intermediate condition in which neither the sub-dish nor the feed shadow predominates, and which would yield the least amount of aperture blocking; this may be termed the *minimum-blocking condition.*

In Fig. 11, the minimum blocking condition is shown, together with some approximate equations describing the basic relations between certain parameters. By combining these equations, a relationship is obtained which specifies the geometry for the minimum-blocking condition; it is as follows:

$$\frac{F_c}{F_m} \approx \frac{1}{2}\,\frac{kD_f^2}{F_c\lambda} \approx \frac{D_f}{D_s'} \tag{10}$$

where D_s' is the physical or blocking diameter of the sub dish, D_f is the physical or blocking diameter of the feed aperture, and k is the ratio of the effective feed-aperture diameter to its blocking diameter.[1] This approximate relationship assumes that the angles ϕ_s and ϕ_f, shown in Fig. 11, are small; and that the sub dish is much closer to the focus of the main dish than it is to the feed (F_e/F_m much larger than one). It also assumes that ray optics can describe the feed shadow; this is a good approximation when the feed is far from the sub dish. Within these limitations, minimum aperture blocking is obtained for a practical case in which there is essentially no spillover of the main lobe of the feed pattern past the edge of the sub dish. Although shown for the classical Cassegrain system, the above approximate analysis also applies for the classical Gregorian system [13].

It can be seen that the minimum-blocking condition

[1] Ordinarily k is slightly less than one; however, where a cluster of many feeds is employed to obtain a cluster of antenna beams, k can become quite small.

is not limited to a particular set of antenna dimensions, but includes a series ranging from the case of a feed located near the vertex of the main dish and having a diameter about equal to that of the sub dish, to the case of a feed located far in front of the main dish and having a diameter much smaller than the sub dish. In the former case, the feed should be focused approximately toward the focal point of the main dish in order that the illumination of the main aperture be characterized by a Fraunhofer diffraction pattern rather than a Fresnel pattern. In the latter case this is not necessary, but the feed must, of course, be excited by a length of transmission line and supported in its extended location; in an extreme form the latter case resembles a single-dish antenna with a splash-plate feed, although the principle of operation is quite different.

The diameter of the aperture blocking for the minimum-blocking condition is given by the following approximate equation:

$$D_{b\,\min} \approx \sqrt{\frac{2}{k} F_m \lambda} \tag{11}$$

where the limitations are the same as those mentioned previously. This equation also assumes that the total amount of aperture blocking is no greater than either of the two equal and coincident shadows; actually the blocking would be somewhat greater, particularly for the case of a small feed located close to the sub dish. It should also be mentioned that the approximation given in (11) implies that a classical Gregorian would have slightly less blocking than a classical Cassegrain of the same axial dimension, because of the shorter F_m of the former; however, a more exact formulation would show that just the reverse is true. The significant fact to note from (11) is that the minimum blocking diameter can be computed before determining the feed size and location, these latter dimensions finally being related by (10).

It is of interest to express (11) in some alternate approximate forms which more clearly illustrate the basic relationships;

$$\left(\frac{D_{b\min}}{D_m}\right)^2 \approx \frac{2}{k}\frac{\lambda}{D_m}\frac{F_m}{D_m} \approx \frac{\pi}{2k} 2\theta_{p/2}\frac{F_m}{D_m} \approx \frac{\pi}{2k}\frac{2\theta_{p/2}}{2\phi_v} \tag{12}$$

where $2\theta_{p/2}$ is the approximate half-power beamwidth of the antenna pattern in radians, and $2\theta_v$ is the approximate included angle formed by the main dish at the virtual feed in radians. (Actually the first and second forms of (12) are almost exactly equal when the main aperture is circular and has a completely tapered parabolic illumination, and the second and third forms are equal when $2\phi_v$ is small.) It is apparent from (12) that an antenna with a narrow beamwidth can have less relative aperture blocking than one with a wide beamwidth. This might be expected on the basis that the optical case, which has a very narrow beamwidth, has the capability for very small relative aperture blocking. Also apparent is the desirability of a small F/D ratio for the main dish, and an efficient feed aperture (k approaching one).

As an example, consider an antenna which is to have a pencil beam of one-degree half-power beamwidth, $F_m/D_m=0.3$, $k=0.7$, and which is to be optimized for the minimum-blocking condition. The second form of (12) yields a value of about .012 for $(D_{b\min}/D_m)^2$, which may then be applied in (9) to yield a value of about .024 for E_b/E_m. The aperture blocking in this antenna would therefore reduce the gain by about $\frac{1}{4}$ db and would increase a -23 db sidelobe[2] to about -20.5 db. This effect might be acceptable for some applications, but not for others. Thus a one-degree beamwidth might be considered as a rough boundary above which the simple Cassegrain design, even though optimized, would be unattractive.

B. Twisting Cassegrains for Least Blocking

The preceding discussion of a minimum-blocking design has assumed that, similar to an optical telescope, operation in all polarizations is required. However, many microwave antennas need operate in only one polarization, and in this event a considerable reduction of aperture blocking is possible. Fig. 12 presents one scheme for accomplishing this, by means of a polarization-twisting technique which avoids the sub-dish shadowing.

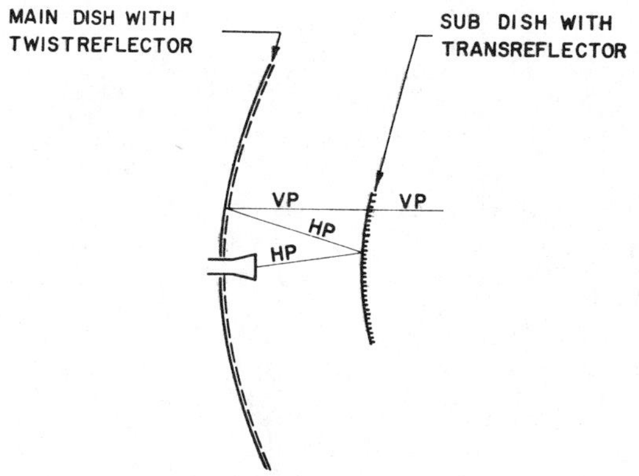

Fig. 12—Polarization twist for non-blocking sub dish.

In this scheme as shown, the sub dish comprises a horizontal grating, called a transreflector, which reflects a horizontally-polarized wave radiated by the feed. The main dish incorporates a surface design, called a twistreflector, which twists the horizontally-polarized wave to a vertically-polarized one as it reflects the wave back. The portion of this wave which is now incident on the

[2] The -23 db figure is a value which is typical for the first-sidelobe level when the illumination is tapered to about 11 db for maximum gain with a circular aperture.

sub dish is transmitted through unaffected, because the sub dish is transparent to a vertically-polarized wave. Thus there is no blocking by the sub dish at all. The feed does, of course, create aperture blocking; however, its size can be made quite small, and the blocking can be comparable with that of an ordinary single-dish design. In this scheme, therefore, it is advantageous to use a large sub dish with a small feed.

While the above process is theoretically perfect in the cardinal regions of the antenna aperture, the three-dimensional geometry of the system is such that there can be some loss into cross-polarized radiation toward the outer portion of the intercardinal regions. It is beyond the scope of this paper to consider this effect in detail, but some general comments can be made. One part of this effect occurs with the wave radiated from the sub dish to the main dish; for any but the most extreme Cassegrain forms, the loss here is usually so small as to be negligible. The other part occurs with the wave radiated from the feed to the sub dish, and the results are dependent on the polarization characteristics of the feed. There is often a moderate amount of loss here; however it is usually not greatly different from the loss in an ordinary single-dish antenna caused by the same effect.

Another scheme for reducing aperture blocking is indicated in Fig. 13; here, a polarization-twisting technique is employed to render the feed invisible. Two configurations are possible, both involving a sub dish which incorporates a twistreflector. In one case, a vertically-polarized feed is located behind the main dish, and the central portion of the main dish includes a transreflector having a horizontal grating. The feed radiates through the transreflector toward the sub dish, the sub dish reflects this wave and twists its polarization to horizontal, and the horizontally-polarized wave is then completely reflected by the main dish. In the other case, the feed is composed of thin horizontal elements and is located out in front of a simple main dish. When the feed radiates toward the sub dish, the vertically-polarized wave returned by the sub dish passes through the feed unaffected, is completely reflected by the main dish, and again passes through the feed. In both of these configurations, it is evident that the sub dish causes aperture blocking but the feed does not. Consequently, the feed may be greatly enlarged so that its increased directivity allows the sub dish to become quite small. As mentioned previously, when the feed becomes equal to or larger than the sub dish, the phase front across the feed aperture should be curved so as to focus the feed toward the vicinity of the main dish focus. It should also be mentioned that when this condition exists, the simple geometry of the Cassegrain system and the equivalence concepts no longer apply; however the basic operation of the antenna remains similar.

Of the two basic polarization-twisting schemes, the one having a twistreflecting main dish is of general applicability to many antenna developments [14]–[17],

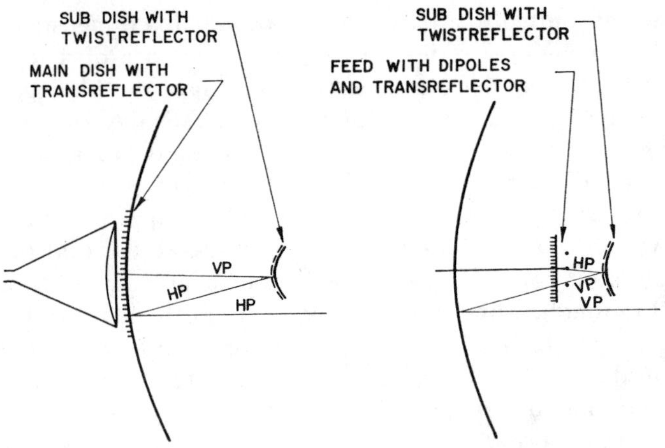

Fig. 13—Polarization twist for non-blocking feed.

while the one having a twistreflecting sub dish is useful in some special circumstances. For example, the former may be efficiently employed with any of the Cassegrain extensions shown in Fig. 5, and with the classical Gregorian form, as well. On the other hand, the latter should be limited to those forms in which the sub dish is small compared with the main dish. In either case it is essential, of course, to have suitable designs for the twistreflector and transreflector. One particular technique [16] involving thin metal wires embedded in fiberglass skins, has proven most satisfactory. While it is not the purpose of this paper to discuss these designs in detail, a brief description is in order as an indication of their practical nature.

For the transreflector design, a grating of thin wires, closely-spaced compared with a wavelength, has the property of being essentially a perfect reflector for parallel polarization, and being essentially invisible to perpendicular polarization. The cross-section of a practical structure which incorporates a quarter-wave sandwich support is indicated in Fig. 14(a). The wires may be placed all in one skin, or else they may be divided equally between the two skins as shown; either technique usually yields about the same result.

For the twistreflector design, a grating of metal wires oriented at 45° to the incident polarization may be placed in front of a reflecting surface. When the spacing between the grating and the reflecting surface is about three-eighths of a wavelength, and the grating is designed to allow about one-half of the parallel-polarized power to pass through, the twistreflector operates over a broad frequency band and over a wide range of incidence angles. The cross-section of a practical structure is shown in Fig. 14(b).

It is perhaps interesting to note in passing, that there are a number of uses for a twistreflector in addition to those already discussed. One such use occurs in an ordinary single-reflector antenna during transmission, when it is desired to prevent any of the wave reflected by the dish from getting back into the feed. This can be accomplished with a twistreflector on the dish [12].

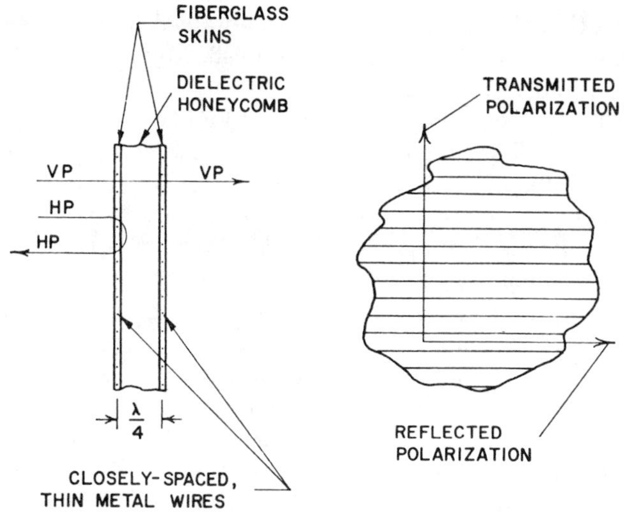

(a) Transreflector.

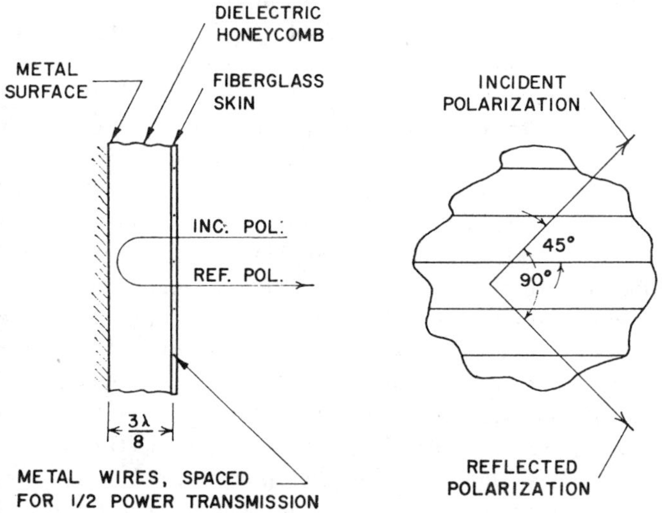

(b) Optimum twistreflector.

Fig. 14—Designs for polarization-operative surfaces.

Another use applies to a simple Cassegrain, in which it is desired to eliminate what may sometimes be an appreciable reflection by the sub dish back into the feed. Incorporation of a twistreflector on the sub dish achieves this result, and may create other advantages as well.

VI. Experimental Results

A number of Cassegrain antennas have been designed at Wheeler Laboratories, and their performance has been highly satisfactory. By way of illustration, the radiation patterns of two different antennas are presented; these two were designed for the Bell Telephone Laboratories on Army Ordnance projects.

One design is shown by the photograph in Fig. 15. It employs simple reflecting surfaces, and its geometry approaches that of the minimum-blocking configuration indicated in Fig. 11. Actually, as may be seen from the picture, the sub dish is appreciably larger than the feed shadow, and the blocking area is greater than the minimum possible by a factor of almost 3. The antenna is

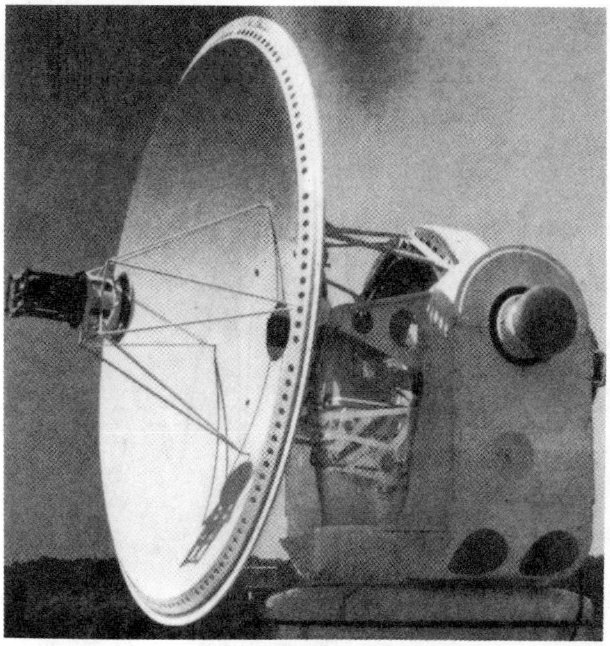

Fig. 15—Photograph of a simple Cassegrain.

shown in location on the roof of the antenna-development facility of Wheeler Laboratories at Smithtown, Long Island. Also visible in the picture are the precision mount for the antenna, and the versatile positioning devices for the feed and sub dish; these items were provided by the Bell Telephone Laboratories.

The radiation pattern and efficiency[3] of this Cassegrain antenna are shown by the solid curve in Fig. 16(a). The half-power beamwidth is 0.6 degrees; this is narrow enough so that the simple Cassegrain system without twisting is adequate for the intended application, even without complete optimization of the aperture blocking. As may be seen, the efficiency[3] of this Cassegrain is fairly high in spite of aperture blocking by both the sub dish and its rigid supporting system. This is probably a result of the efficient aperture utilization obtained with a long equivalent focal length. As expected, the near sidelobes are raised several db by the sub dish blocking; however this effect would also have occurred in the intended application had an ordinary single-reflector antenna been employed.

The dashed curve in Fig. 16(a) shows the radiation pattern which is obtained when the antenna is intentionally defocused by moving the sub dish a small distance toward the main dish. If desired, this technique might be used to provide a variable beamwidth. Alternatively, by moving the sub dish away from the main dish the antenna can be focused toward a point nearer than the far field; this would permit a greater concentration of power at such a point than could otherwise be obtained. Of course these focusing techniques are also

[3] The "efficiency" is used here as the ratio of measured gain to the gain which would be obtained if the same main aperture were uniformly illuminated, with no spillover or other losses.

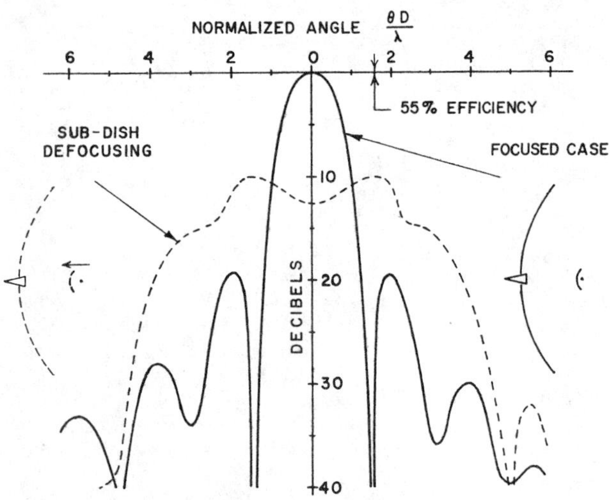

(a) Focused and defocused cases.

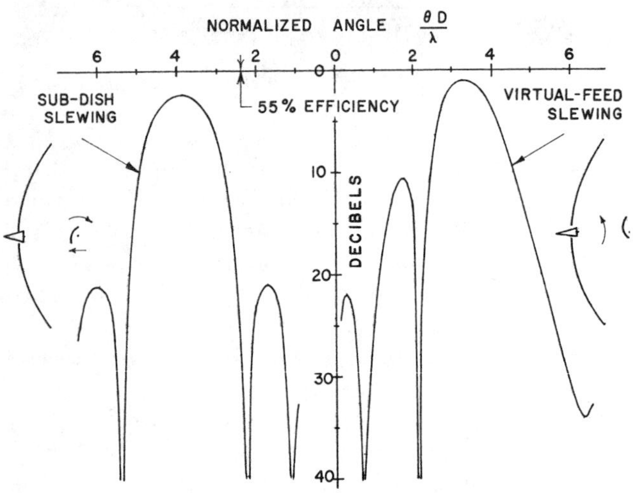

(b) Off-axis scanning methods.

Fig. 16—Radiation patterns of a simple Cassegrain.

available in the ordinary single-reflector antenna, as well as most other types. With this Cassegrain, however, it is possible to perform these operations by motion of a relatively small, passive device.

In Fig. 16(b), the patterns of this same antenna are shown for two cases in which the beam is scanned off axis by approximately three beamwidths. The pattern on the left is obtained by a movement of only the sub dish; the motion involves a substantial tilt about the main dish focus, plus a small axial motion to regain the focused condition in the plane of scan. It is evident that no appreciable coma is introduced by this process, since the pattern remains quite symmetrical. On the other hand, there is a considerable amount of astigmatism created; the pattern in the plane normal to the scan plane, not shown here, is quite broad. This defect is partly responsible for the decrease of gain which is apparent. The results presented above are typical of a Cassegrain system with a large F_e/F_m. It is perhaps instructive to mention that similar effects are obtained with this system when the scanning is accomplished by offsetting the feed and refocusing.

The pattern on the right side of Fig. 16(b) is obtained by tilting both the feed and the sub dish as a unit about the vertex of the main dish. This is equivalent to rotating the virtual feed about the same point. As can be expected, this results in a substantial degree of coma distortion, about the same as would be obtained by offsetting the feed in an ordinary single-reflector antenna of the same main focal length.

The other antenna design chosen as an example incorporates a twist-reflecting main dish and a transreflecting sub dish, such that the sub dish creates no aperture blocking, as illustrated in Fig. 12. The equivalent focal length of this antenna is designed to be just long enough so that a monopulse feed system can be employed in a size just large enough to utilize a simple cluster of four horns as the feed. The complicated monopulse plumbing is located in a convenient region behind the antenna.

The patterns of this antenna are shown in Fig. 17 for the sum and one difference channel. Also indicated are the computed points, determined from a knowledge of the feed pattern, and including a contribution from aperture blocking by the feed. The close correspondence between the two patterns is evident. Similar good agreement exists in the other properties of this antenna. The efficiency[3] of 54 per cent in the sum pattern is rather high for a monopulse system; this is a result of the inherent advantage of a Cassegrain system with a long equivalent focal length and very small aperture blocking. All of the above results confirm the nearly lossless behavior of the polarization-twisting technique and the surface designs of Fig. 14. While the twisting type of antenna requires additional effort during design and construction of the polarization-operative surfaces, it has proven practical to build in large quantities, and has yielded the expected good performance in the field.

VII. Benefits of Cassegrain Systems

In concluding the discussion of Cassegrain optics applied to microwave antennas, it is appropriate to outline some of the benefits obtainable. Perhaps most important is the ability to place the feed in a convenient position, while utilizing reflectors as the focusing elements. The rear location and forward direction for the feed are most desirable in various applications involving complicated feeds and associated plumbing.

One example of this advantage occurs in the case of an antenna intended for low-noise operation, as illustrated in Fig. 18. At present, a low-noise receiver is likely to be bulky and require a number of auxiliary connections as well as occasional adjustments; it is therefore inconvenient to mount it close to the feed out in front of a single-reflector antenna. Yet this is often done, because the attenuation in a waveguide from the

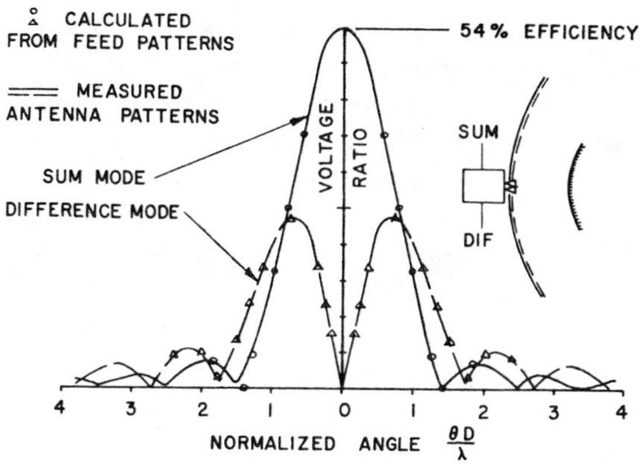

Fig. 17—Radiation patterns of a twisting Cassegrain.

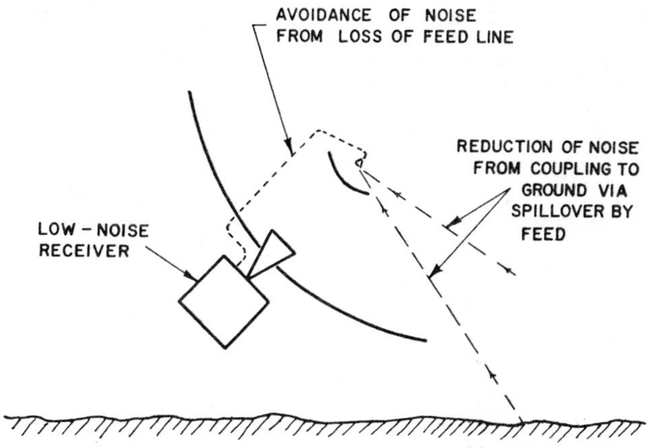

Fig. 18—Simple Cassegrain antenna for low-noise applications.

feed back to a receiver located behind the dish would introduce an excessive amount of noise power. The Cassegrain system furnishes the opportunity to avoid most of these difficulties.

There is another benefit obtainable with the Cassegrain system in a low-noise application. With the ordinary single-reflector antenna, there is usually a considerable amount of wide-angle sidelobe response caused by spillover radiation from the small feed out in front. This may introduce a very substantial amount of noise power into the antenna, by coupling to the radiation from the warm ground. In the case of a Cassegrain antenna, spillover radiation from the virtual feed can be very much less. This is because of the essentially ray-optic behavior of reflection from the sub dish, which results from its relatively large diameter in wavelengths. There remains to be considered, of course, spillover from the real feed past the edge of the sub dish. Although the total amount of this spillover power may be comparable with that in a single-reflector antenna, it is likely to be confined to direction relatively close to the antenna axis. There is also to be considered the sidelobe radiation created by the aperture blocking by the sub dish; here again, this is usually appreciable only in forward directions. As a result of these directional properties, the spillover and aperture blocking couple to the ground only when the antenna is pointed at a low elevation angle. In comparison, the ordinary single-reflector antenna is likely to have appreciable coupling to the ground even at high elevation angles.

It is possible in the case of a polarization-twisting Cassegrain system to reduce even this relatively narrow-angle sidelobe response. Fig. 19 illustrates this effect, for the scheme which involves a twistreflector at the main dish and a transreflector at the sub dish. Since the feed is horizontally polarized, it is essentially isolated from any vertically-polarized source, such as the normal ground reflection of the incoming wave, or one component of thermal ground radiation. If the transreflector is extended from the sub dish to the main dish in

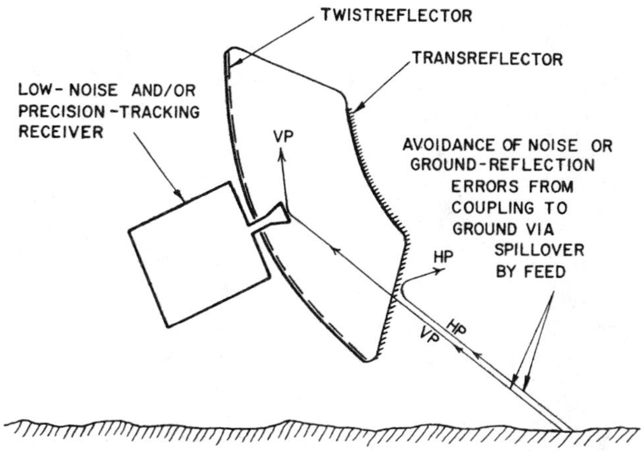

Fig. 19—Twisting Cassegrain antenna for further reduction of coupling to ground.

the lower portion of the antenna, isolation may also be achieved for the other polarization. Such an antenna, then, has effectively only those sidelobes which would be inherent in the illumination distribution of its main aperture. As a result, the antenna could provide accurate tracking of a target, as well as low-noise performance, down to elevation angles determined only by the decay rate of the inherent sidelobes.

In continuing the outline of benefits obtainable with a Cassegrain system, mention can be made of the ability to obtain an equivalent focal length much greater than the physical length; as discussed previously in this paper, various advantages may be obtained in this way. A third aspect is the capability for scanning or broadening the beam by moving one of the antenna surfaces. One case involving a small moving sub dish has been described here; there have also been designs utilizing a moving flat main dish for wide-angle scanning [15], [17].

The existence of two dishes and two focal points in the Cassegrain system gives rise to interesting methods for incorporating the separate functions of two antennas into one structure. On the left side of Fig. 20 a simple scheme is shown which provides a full-size plus a re-

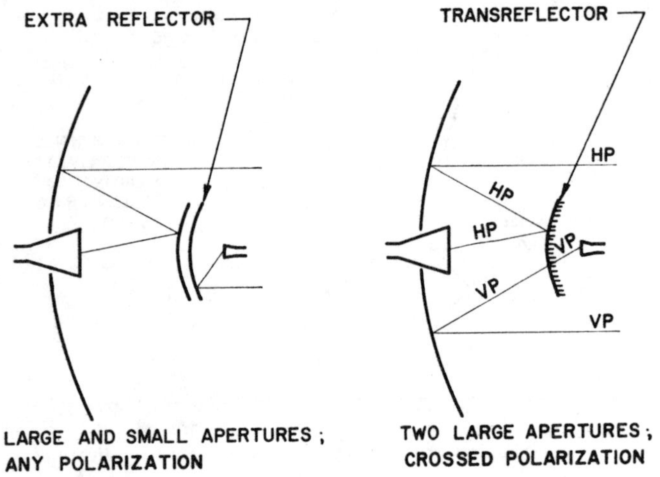

Fig. 20—Dual antennas, blocking sub dishes.

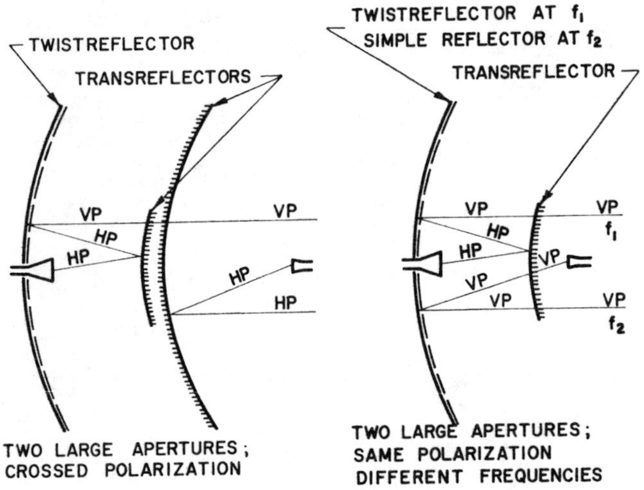

Fig. 21—Dual antennas, non-blocking sub dishes.

duced-size antenna combination, while on the right side there is illustrated a scheme for obtaining two full-size antennas having crossed polarizations. In both of these cases, the sub dish causes some blocking of one of the two apertures. On the left side of Fig. 21 an alternate scheme is shown for obtaining two full-size antennas having crossed polarizations; here, the polarization-twisting process eliminates any blocking by the sub dish. On the right side of Fig. 21, a scheme is indicated whereby two full-size antennas are obtained having the same polarization but operating at different frequencies, without any blocking by the sub dish. This case relies on a surface design for the main dish which is a twist-reflector at one frequency and an ordinary reflector at the other frequency.

VIII. Conclusion

To summarize the discussion of the principles and features of Cassegrain antennas, it has been shown that a simple set of formulas describe a number of forms, and that the essential performance can be calculated by means of simple equivalence concepts. The basic defect, aperture blocking by the sub dish, can be minimized or virtually eliminated by certain techniques. The Cassegrain system has proven both practical and advantageous in a number of operational antennas, and the tested performance has agreed closely with the computed predictions. A variety of benefits are obtainable with the Cassegrain system, and it provides a highly versatile form of microwave antenna capable of achieving good performance in a number of unusual applications. The second reflecting surface which is available in this system provides an extra degree of freedom to the antenna designer for application to his particular problem.

IX. Symbols

D_m = effective diameter of circular main dish (to edge rays).
D_s = effective diameter of circular sub dish (to edge rays).
D_s' = blocking diameter of sub dish.
D_f = diameter of feed.
D_b = diameter of aperture blocking.
$D_{b\,\min}$ = diameter of aperture blocking for minimum-blocking geometry.
F_m = focal length of main dish.
F_c = distance between foci of sub dish.
F_e = equivalent focal length of Cassegrain system.
L_v = distance from virtual focus (or main dish focus) to sub dish.
L_r = distance from real focus (or feed) to sub dish.
ϕ_v = angle between axis and edge ray, at virtual focus.
ϕ_r = angle between axis and edge ray, at real focus.
$2\phi_s$ = included angle between rays from real focus to physical edges of sub dish, in radians.
$2\phi_f$ = included angle between rays from virtual focus to edges of feed, in radians.
e = eccentricity of conic section.
a = transverse half-axis of conic section.
b = conjugate half-axis of conic section.
x_m, y_m = coordinates of main dish (axial, radial).
x_s, y_s = coordinates of sub dish (axial, radial).
x_e, y_e = coordinates of equivalent parabola (axial, radial).
λ = wavelength.
f = frequency.
k = ratio of effective diameter to blocking diameter of the feed.
θ = antenna pattern angle, in radians.
$2\theta_{p/2}$ = half-power beamwidth of antenna, in radians.
E = pattern voltage.
E_b = peak voltage of the supplemental negative pattern caused by aperture blocking.
E_m = peak voltage of the pattern of the main aperture without blocking.

X. Acknowledgment

The principles and geometry of the double-reflector systems described in this paper are based entirely on the principles developed for the optical telescope. This work originated in the 17th century, with particular forms being attributed to Gregory, Newton, and Cassegrain. It seems very probable that the simple concept of an equivalent parabola has long been known to those involved in the theory of optical reflecting telescopes; however the writer, having only a limited acquaintance with the optical literature, has not found any reference to this.

The existence of a minimum-blocking design was suggested by H. A. Wheeler, for the case of a feed located at the vertex of the main dish. This suggestion led to the general case of a feed located anywhere, and the basic relationship of minimum blocking as a function of beamwidth.

During the design of a radar antenna for the Bell Telephone Laboratories, the system involving a transreflecting sub dish and twistreflecting main dish was conceived [16] as the solution to the problem at hand. Later it was learned that the basic concept had already occurred to C. A. Cochrane [14] of Elliott Brothers, London. The scheme involving a twistreflecting sub dish and a transreflecting main dish or feed was devised in connection with the design of another antenna for the Bell Telephone Laboratories.

The use of thin wires in fiberglass for the polarization-operative surfaces was worked out in cooperation with K. B. Woodard of the Bell Telephone Laboratories. The particular set of dimensions which achieve wideband, wide-angle twistreflector performance was derived by H. Jasik, as a consultant to Wheeler Laboratories.

Of the four dual-antenna schemes mentioned in Section VII, it is of interest to note that the first has also been perceived and utilized by the Ryan Aeronautical Co., and the second by both Sperry Gyroscope Co., and Melpar, Inc.

The writer would like to express his appreciation for the support of Wheeler Laboratories in the preparation of this paper. He would also like to acknowledge the permission of the Bell Telephone Laboratories and the Army Ordnance Corps for the inclusion of experimental results obtained on certain antennas developed for them, and the helpful cooperation of R. L. Mattingly of the Bell Telephone Laboratories in this respect.

XI. Bibliography

[1] A. G. Ingalls, "Amateur Telescope Making," Scientific American, Inc., New York, N. Y., vol. 1, pp. 62–65, 215–218, 444–453; 1953.

[2] J. B. Sidgewick, "Amateur Astronomer's Handbook," Faber and Faber, Ltd., London, Eng., pp. 161–165; 1955.

[3] L. C. Martin, "Technical Optics," Sir Isaac Pitman and Sons, Ltd., London, Eng., vol. 2, pp. 75–78; 1954.

[4] D. O. Woodbury, "The Glass Giant of Palomar," Dodd Mead and Co., New York, N. Y.; 1957.

[5] W. M. Cady, M. B. Karelitz, and L. A. Turner, "Radar Scanners and Radomes," M.I.T. Rad. Lab. Ser., McGraw-Hill Book Co., Inc., New York, N. Y., vol. 26, pp. 55–61; 1948.

[6] W. Rotman, "A Study of Microwave Double-Layer Pillboxes, Part II—Multiple-Reflector Systems," AF Cambridge Res. Ctr., Bedford, Mass., Rept. No. TR-56-101; January, 1956.

[7] A. K. Head, "A new form for a giant radio telescope," *Nature*, vol. 179, pp. 692–693; April 6, 1957.

[8] F. A. Jenkins and H. E. White, "Fundamentals of Optics," McGraw-Hill Book Co., Inc., New York, N. Y., p. 156; 1957.

[9] C. C. Cutler, "Parabolic-antenna design for microwaves," Proc. IRE, vol. 35, pp. 1285–1286, 1288; November, 1947.

[10] K. S. Kelleher, "Microwave optics at naval research laboratory," *Proc. Symp. on Microwave Optics*, McGill University, Montreal, Canada, vol. 2, Paper No. 34; June, 1953.

[11] B. Woodward, "The Cassegrain Antenna," Advertisement by Airborne Instruments Lab., Proc. IRE, vol. 46, p. 2A; March, 1958.

[12] S. Silver, "Microwave Antenna Theory and Design," M.I.T. Rad. Lab. Series, McGraw-Hill Book. Co., Inc., New York, N. Y., vol. 12, pp. 190–192, 447–448; 1949.

[13] Advertisement by D. S. Kennedy and Co., *Aviation Week*, vol. 71, p. 4; July 27, 1959.

[14] C. A. Cochrane, "Improvements in or Relating to High Frequency Radio Aerials," British Patent No. 700,868, February, 1952–December, 1953; "High Frequency Radio Aerials," U. S. Patent No. 2,736,895, February, 1952–February, 1956.

[15] P. F. Mariner and C. A. Cochrane, "Improvements in or Relating to High Frequency Radio Aerials," British Patent No. 716,939; August, 1953–October, 1954.

[16] Wheeler Laboratories reports available through ASTIA: WL No. 658, ASTIA No. AD 116479, January, 1955; and WL No. 666, ASTIA No. AD 306000, April, 1955.

[17] R. W. Martin and L. Schwartzman, "A Rapid Wide Angle Scanning Antenna with Minimum Beam Distortion," *Proc. 1958 East Coast Conf. on Aeronautical and Navigational Electronics*, Baltimore, Md., pp. 47–51.

Scattering from a Hyperboloidal Reflector in a Cassegrainian Feed System*

W. V. T. RUSCH†, MEMBER, IEEE

Summary—**The scattered field from a hyperboloidal reflector is calculated by integrating the induced current density over the front of the hyperboloid. The resulting integral expressions for the fields possess a stationary term which, when evaluated, yields the geometrical ray-optics approximation to the scattering problem. The complete field, including diffraction effects, may be obtained by numerical evaluation of the integrals. The formulas are applied to a hyperboloid illuminated by an idealized, sharply cut off uniform feed pattern. Characteristic diffraction phenomena are reduced with increasing D/λ until the geometrical ray-optics result is obtained in the limit of vanishing wavelength. Theoretical field patterns are also obtained for a horn-fed hyperboloidal subreflector in a Cassegrainian feed system; they indicate that for moderately large hyperboloidal reflectors spillover may be reduced to an acceptable level, but there is a tendency toward increased forward spillover. The results of 9600-Mc model tests compare favorably with the theoretical patterns.**

* Received September 9, 1962; revised manuscript received February 13, 1963. This paper presents the results of one phase of research carried out at the Jet Propulsion Laboratory, California Institute of Technology, under Contract No. NAS 7-100, sponsored by the National Aeronautics and Space Administration.

† Electrical Engineering Department, University of Southern California, Los Angeles. Consultant to Jet Propulsion Laboratory, Pasadena, Calif.

I. Introduction

TWO-REFLECTOR Cassegrainian antenna systems have recently been developed to reduce spillover and increase aperture efficiency of paraboloid antennas [1]–[6]. These systems have numerous mechanical and electrical advantages over the conventional focal-point feed horn for large paraboloid antennas as used in ground stations for space communications. The Cassegrainian design was originally used in optical telescopes. At radio frequencies, however, a theoretical analysis of the scattering from the hyperboloidal subreflector may not be carried out using the ray-tracing techniques of geometrical optics, because spillover and back-lobe diffraction effects are unexplained by optical approximations. Inasmuch as these diffraction effects play an important role in the determination of equivalent antenna noise temperature, the techniques of vector diffraction theory must be employed in a useful analysis of the problem. In the limit of vanishing wavelength, the diffraction theory results

Reprinted from *IEEE Trans. Antennas Propagat.*, vol. AP-11, pp. 414–421, July 1963.

become identical with the geometrical ray-optics approximations.

The analysis that follows will consider the total radiation from a point feed illuminating a hyperboloid. The results are of sufficient generality to include arbitrary directivity and polarization of the feed. A complete analysis of the two-reflector problem (including the main paraboloidal reflector) can be made by using the field scattered from the hyperboloid as the illumination function in a similar analysis of the paraboloid problem [7].

II. Derivation of the Fields from the Current Distribution

The axially symmetric reflector (Fig. 1) is designed to convert a spherical wave emerging from the origin O into a spherical wave emerging from F. The methods of ray optics reveal that a hyperboloidal surface, with foci at O and F, is necessary to reflect a wave in the prescribed manner [1]. The polar equation for the hyperboloid is

$$\rho = \frac{-ep}{1 + e\cos\theta'} \qquad \theta_0 \leq \theta' \leq \pi \tag{1}$$

where

$$e \equiv \frac{c}{a} \qquad (e > 1) \tag{2a}$$

$$p \equiv c\left(1 - \frac{1}{e^2}\right). \tag{2b}$$

The outward surface normal from the front of the reflector is

$$\boldsymbol{n} = \frac{(1 + e\cos\theta')\boldsymbol{a}_\rho - e\sin\theta'\boldsymbol{a}_{\theta'}}{m(\theta')} \tag{3}$$

where

$$m(\theta') \equiv [(1 + e\cos\theta')^2 + (e\sin\theta')^2]^{1/2}. \tag{4}$$

The differential surface element on the front of the reflector is

$$dS = -\frac{\rho^2 \sin\theta' m(\theta') d\theta' d\phi'}{(1 + e\cos\theta')} \tag{5}$$

where ϕ' is the azimuthal coordinate (measured about the z axis) of a point on the reflector surface. The induced current distribution on the illuminated front of the reflector can be calculated by assuming that at every point the primary field is reflected as an infinite plane wave from an infinite plane tangent at the point of incidence. (The radius of curvature of the reflector must be much larger than the wavelength for this assumption to be valid.) The current in the "shadow" region on the back of the reflector is assumed to make negligible contribution to the field.

The electric field radiated from the primary feed at 0 may be described by

$$\boldsymbol{E}_{pf} = AP(\theta)\frac{e^{ikR}}{R}\mathbf{e}(\theta, \phi). \tag{6}$$

The unit vector $\mathbf{e}(\theta, \phi)$ describes the polarization of the primary field; $P(\theta)$ is the pattern factor of the primary field which is assumed to be axially symmetric. By integrating the induced surface current distribution over the reflector surface, it is possible to compute the field scattered from the reflector [8] as follows:

$$\boldsymbol{E}_s(\theta, \phi) = \left(\frac{iA}{\lambda}\right)\frac{e^{ikR}}{R} \cdot \int_S P(\theta')\frac{e^{ik\rho(1-\boldsymbol{a}_\rho\cdot\boldsymbol{a}R)}}{\rho}[\boldsymbol{n}\times\boldsymbol{h}(\theta', \phi')]_{\text{trans}} dS(\theta', \phi') \tag{7}$$

where

$$\boldsymbol{h}(\theta', \phi') = \boldsymbol{a}_\rho \times \mathbf{e}(\theta', \phi') \tag{8}$$

and only the transverse components of $\boldsymbol{n}\times\boldsymbol{h}(\theta', \phi')$ are involved in the integration.

If the primary feed is polarized in the x direction,

$$\begin{aligned}\boldsymbol{a}_\theta\cdot\boldsymbol{E}_s = {} & \left(\frac{iAep}{\lambda}\right)\left(\frac{e^{ikR}}{R}\right)\int_{\theta_0}^{\pi}\frac{P(\theta')\sin\theta' e^{i\psi_1}}{(1 + e\cos\theta')^2} \\ & \cdot\{-(1 + e\cos\theta')\cos\theta\cos\phi I_1(\phi) \\ & + \sin\theta\sin\theta' I_2(\phi) \\ & + \cos\theta(1 + e)(1 + \cos\theta')I_3(\phi)\}d\theta'\end{aligned} \tag{9a}$$

$$\begin{aligned}\boldsymbol{a}_\phi\cdot\boldsymbol{E}_s = {} & \left(\frac{iAep}{\lambda}\right)\left(\frac{e^{ikR}}{R}\right)\int_{\theta_0}^{\pi}\frac{P(\theta')\sin\theta' e^{i\psi_1}}{(1 + e\cos\theta')^2} \\ & \cdot\{\sin\phi(1 + e\cos\theta')I_1(\phi) \\ & + (1 + e)(1 + \cos\theta')I_4(\phi)\}d\theta'\end{aligned} \tag{9b}$$

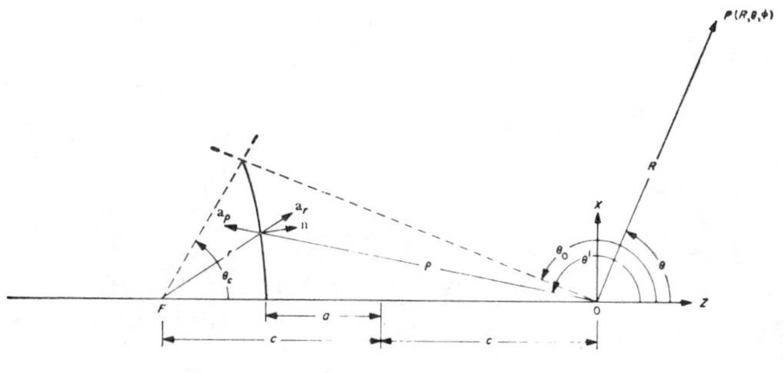

Fig. 1—Geometry of hyperboloidal reflector.

where

$$I_1(\phi) = \int_0^{2\pi} e^{i\psi_2 \cos(\phi'-\phi)} d\phi' \tag{10a}$$

$$I_2(\phi) = \int_0^{2\pi} \cos\phi' e^{i\psi_2 \cos(\phi'-\phi)} d\phi' \tag{10b}$$

$$I_3(\phi) = \int_0^{2\pi} \cos\phi' \cos(\phi'-\phi) e^{i\psi_2 \cos(\phi'-\phi)} d\phi' \tag{10c}$$

$$I_4(\phi) = \int_0^{2\pi} \cos\phi' \sin(\phi'-\phi) e^{i\psi_2 \cos(\phi'-\phi)} d\phi' \tag{10d}$$

and

$$\psi_1(\theta, \theta') = \frac{\gamma(\cos\theta' \cos\theta - 1)}{1 + e\cos\theta'} \geq 0 \tag{11a}$$

$$\psi_2(\theta, \theta') = \frac{\gamma \sin\theta \sin\theta'}{1 + e\cos\theta'} \leq 0 \tag{11b}$$

$$\gamma \equiv kep. \tag{11c}$$

Using the formula

$$e^{i\psi_2 \cos(\phi'-\phi)} = J_0(\psi_2) + 2\sum_{n=1}^{\infty} (i)^n J_n(\psi_2) \cos n(\phi'-\phi) \tag{12}$$

it is possible to reduce the integrals over ϕ' to simple combinations of trigonometric and Bessel functions, leaving the expressions for the scattered field as integrations over θ'.

The total radiation field is then obtained by a vectorial combination of (6) and (7). The E-plane field $E_\theta = (\boldsymbol{E}_{pf} + \boldsymbol{E}_s) \cdot \boldsymbol{a}_\theta$ is

$$E_\theta(\theta, \phi) = \left(iA\gamma \frac{e^{ikR}}{R} \cos\phi \right) I_\theta(\theta) \tag{13}$$

and the H-plane field $E_\phi = (\boldsymbol{E}_{pf} + \boldsymbol{E}_s) \cdot \boldsymbol{a}_\phi$ is

$$E_\phi(\theta, \phi) = \left(iA\gamma \frac{e^{ikR}}{R} \sin\phi \right) I_\phi(\theta) \tag{14}$$

where

$$I_\theta(\theta) = \frac{iP(\theta)}{\gamma} + \int_{\theta_0}^{\pi} \frac{\sin\theta' P(\theta') e^{i\psi_1}}{(1 + e\cos\theta')^2} \cdot \left[-\cos\theta \left\{ (1 - \cos\theta') \frac{1-e}{2} J_0(\psi_2) + (1 + \cos\theta') \frac{1+e}{2} J_2(\psi_2) \right\} + i \sin\theta \sin\theta' J_1(\psi_2) \right] d\theta' \tag{15}$$

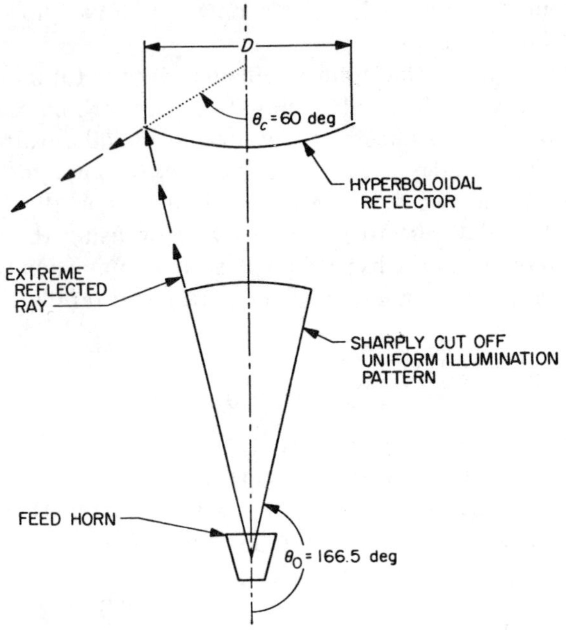

Fig. 2—Geometrical parameters and illumination pattern for fields plotted in Fig. 3.

and

$$I_\phi(\theta) = \frac{iP(\theta)}{\gamma} + \int_{\theta_0}^{\pi} \frac{\sin\theta' P(\theta') e^{i\psi_1}}{(1 + e\cos\theta')^2} \cdot \left[\frac{1-e}{2} (1 - \cos\theta') J_0(\psi_2) - \frac{1-e}{2} (1 + \cos\theta') J_2(\psi_2) \right] d\theta'. \tag{16}$$

III. Numerical Evaluation of the Fields

Evaluation of the theoretical E- and H-plane patterns, *i.e.*, $|I_\theta(\theta)|$ and $|I_\phi(\theta)|$, has been carried out on the IBM 7090 computer. For a primary field that is constant within the solid angle subtended by the hyperboloidal reflector and zero elsewhere (Fig. 2), the resulting H-plane patterns have been plotted in Figs. 3(a), 3(b), and 3(c). (Because of the axially symmetric nature of the excitation, the E-plane results are very similar and have not been included.) The reflector parameters for the three figures are: edge angle $\theta_0 = 166.5°$; extreme reflector geometrical ray $\theta_c = 60°$; eccentricity $e = 1.5$; and reflector diameters of 16.2λ, 32.4λ, 48.6λ. Superimposed on the calculated field patterns are the approximate ray-optics results (see Appendix II). In this limit of vanishing wavelength, the field is confined entirely within the region $0 \leq \theta \leq \theta_c$, and comparison with the integral results clearly indicates the contribution of diffractive effects.

Each of the three field patterns in Fig. 3 is characterized by the same six features: an approximately isotropic region (with minor oscillations) from 0° to about 50°; a region of approximately monotonic decrease

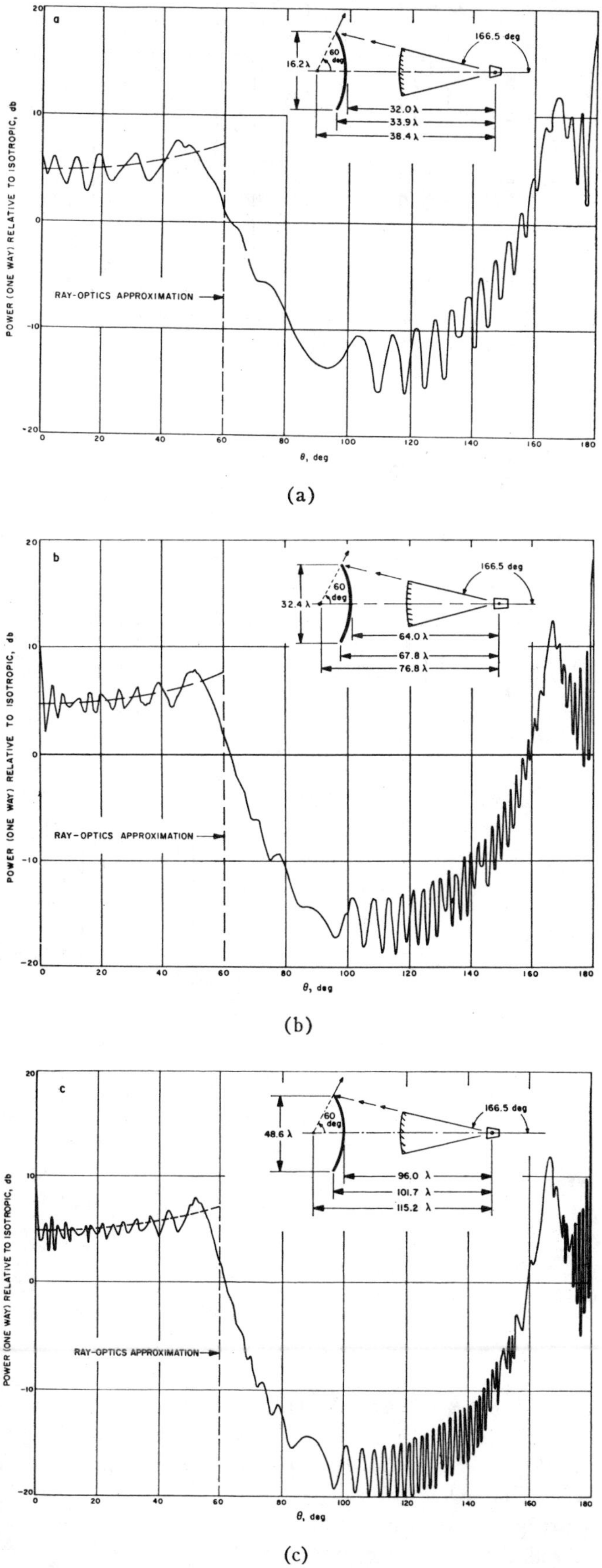

Fig. 3—*H*-plane radiation pattern from hyperboloidal reflector with uniform illumination.

from 50° to about 70°; a highly oscillatory region from 70° to θ_0; a local maximum at θ_0; an oscillatory region in the geometrical shadow from θ_0 to 180°; and a maximum at 180° nearly equal to the value of the unperturbed primary field that would exist in that direction if the reflector were not present. An increase of the diameter/wavelength ratio clearly has the effects of increasing the spatial frequency of oscillation in the oscillatory regions, decreasing the amplitude of the oscillations in the "isotropic" region, and decreasing the sidelobe level outside of the "isotropic" region. Consequently, the exact integral expressions more closely approach the geometrical optics approximation as the diameter/wavelength ratio increases.

Each of the six pertinent features indicated above is a well-known phenomenon from physical optics [9]–[15]. The field in the "isotropic" region consists almost entirely of the reflected ray expected from geometrical optics. The relatively minor oscillations superimposed on the geometrically reflected field are caused by interference among diffractive contributions which are not dominant in this region. From 50° to 70° (including the extreme reflected geometrical ray at 60°), a monotonic decay dominates the field, although very slight interference effects can be distinguished. The steepness of this decay determines the amount of rearward-directed spillover in a Cassegrainian antenna feed system. Similar monotonic decays are encountered in numerous diffraction problems [9], [10]. In the regions from 70° to θ_0 (where the direct and geometrically reflected fields are excluded) and from θ_0 to 180° (in the geometrical shadow), interference among diffractive contributions causes a highly oscillatory field pattern. The local maximum at θ_0, the boundary of the geometrical shadow, has also appeared in similar types of diffraction problems [11]. However, this local maximum apparently arises from a discontinuity of the illumination function or its derivatives in the penumbral region, because it is not present if a continuous illumination function is used (see Fig. 8). The peak at 180°, a well-known diffraction effect, was first predicted by Poisson [12] in the nineteenth century and subsequently verified by numerous experimental observations [13], [14]. This peak is the evolute of the rim of the hyperboloid as determined by the axially symmetric illumination [15].

If the combination of feed horn and hyperboloidal reflector of Fig. 2 illuminates a paraboloidal reflector with an angular semi-diameter of 60°, then the field from 0° to 60° illuminates the paraboloid; the field from 60° to 90° is directed beyond the edge of the paraboloid into the rear hemisphere (spillover); and the field from 90° to 180° is directed into the forward hemisphere (forward spillover). These three regions are indicated in Fig. 4. Estimates of the percentages of the total power directed into these three regions have been obtained by numerical integration and are plotted in Figs. 5(a), 5(b), and

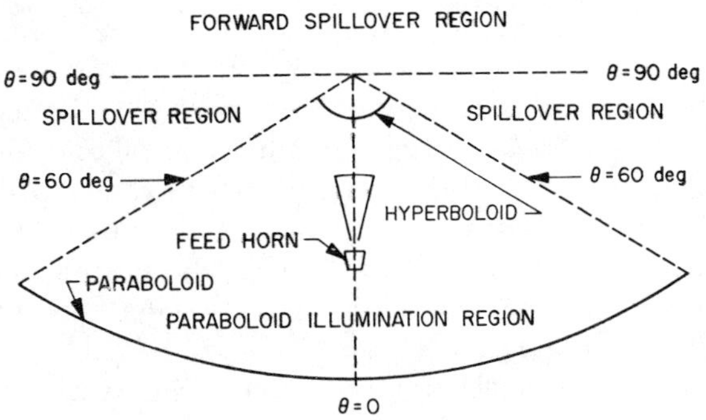

Fig. 4—Geometry of power distribution regions.

5(c) as functions of D/λ. (The scale of Fig. 5(b) is expanded.) Clearly, even for a hyperboloid diameter as large as 50 wavelengths, only 79 per cent of the total power is directed toward the paraboloid, with a correspondingly large spillover (5.2 per cent) and forward spillover (16 per cent). However, these values are not of immediate practical interest inasmuch as the idealized uniform, sharply cut off feed-horn pattern assumed for these calculations is not physically realizable.

IV. Experimental Results

The radiation pattern from a hyperboloidal reflector illuminated by a conventional feed horn has been measured at 9600 Mc. The feed horn, reflector, and mount are shown in Fig. 6. The reflector parameters for the experiment were: diameter 7.8λ; eccentricity $e=1.51$, edge angle $\theta_0=166.5°$; extreme reflected geometrical ray $\theta_c=60°$. The E-plane, H-plane, and both 45°-plane directivity patterns of the feed horn were measured and averaged (in magnitude) to yield a composite feed pattern (Fig. 7) that was numerically inserted into the calculations.

The field integrals [(15) and (16)] are based upon the assumption of axially symmetric excitation. The actual excitation was very nearly axially symmetric in magnitude, but it was found that the E- and H-plane phase centers of the feed horn did not coincide. This phase variation of the feed pattern was not included in the integrations. Consequently, a small degree of variation is expected between the experimental and the idealized theoretical results.

The measured H-plane pattern and the corresponding theoretical pattern are plotted in Fig. 8. Agreement is relatively good, except in the region from 0° to 20° and at wide angles from 80° to 145°. At small angles, however, the experimental pattern is distorted by feed-horn blockage, which was not included in the calculations. Since currents on the back of the reflector were not included in the integrations, close agreement is not expected for the back lobes. However, the extremely large theoretical back radiation is also exhibited by the measured patterns. Agreement is good in the region of greatest importance, from 20° to 80°.

A numerical integration of the theoretical pattern in Fig. 8 was carried out to estimate the power distribution

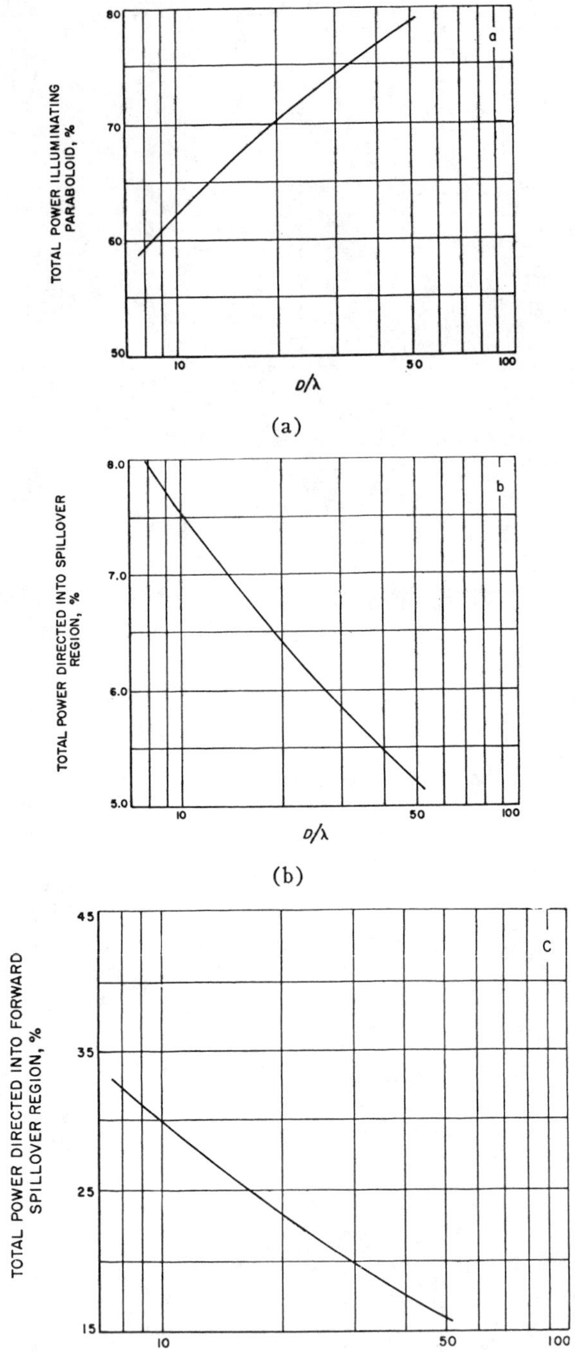

Fig. 5—(a) Percentage paraboloid illumination vs D/λ for uniform feed. (b) Percentage spillover vs D/λ for uniform feed. (c) Percentage forward spillover vs D/λ for uniform feed.

percentages (*cf.*, Section III). The results of this integration were: 73.8 per cent paraboloid illumination, 2.1 per cent spillover, and 24.1 per cent forward spillover. A well-designed low-noise feed horn may have a power distribution of approximately 98 per cent paraboloid illumination, 1–2 per cent spillover, and less than 1 per cent forward spillover for a paraboloidal semi-diameter of 60° ($f/D=0.42$) [16]. Consequently, if the hyperboloid for Fig. 8 were used as the subreflector of a Cassegrainian system, a somewhat increased antenna tem-

Fig. 6—Feed system for 9600-Mc hyperboloid radiation-pattern measurements.

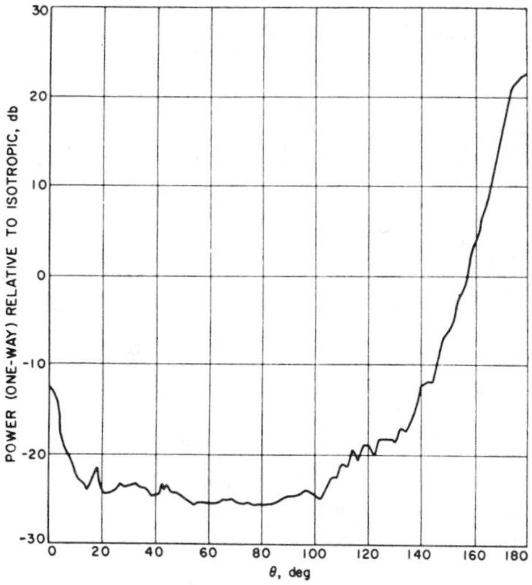

Fig. 7—Composite feed pattern.

perature due to spillover and increased near-sidelobe level would be expected.

The experiment was repeated with the same geometric parameters but with an increased reflector diameter of about 19.5λ. The measured H-plane pattern and corresponding theoretical pattern are plotted in Fig. 9. To avoid confusion of the two highly oscillatory patterns, only the envelope of the experimental pattern is plotted. Again, the feed-horn blocking is evident at small angles, and some variation is noticed in the wide-angle sidelobes and back lobes. Agreement is good, however, from about 10° to 135°. Power distribution integrations reveal 83.8 per cent paraboloid illumination, 1.4 per cent spillover, and 14.8 per cent forward spillover. The increased diameter has reduced spillover to an acceptable level, although the near-sidelobe level would still not compare favorably with a conventional low-noise, focal-point feed horn.

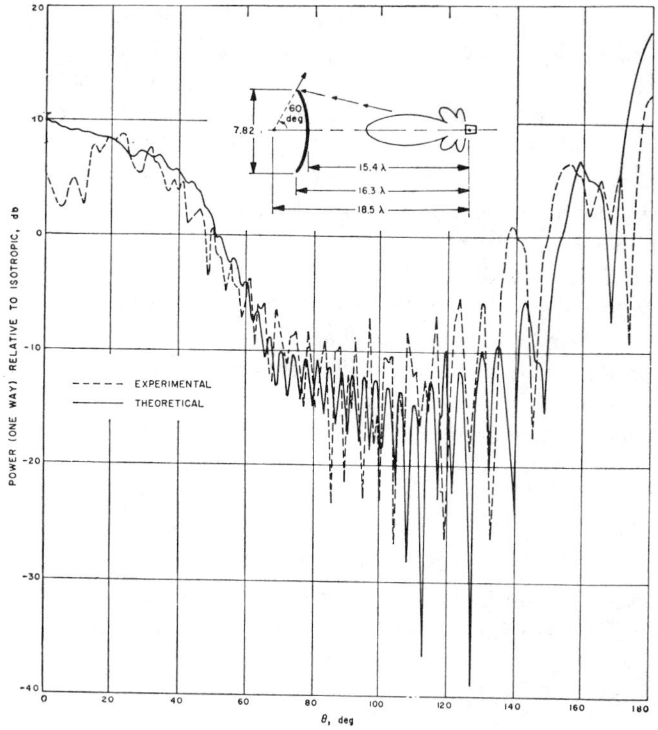

Fig. 8—Comparison of experimental and theoretical H-plane radiation pattern from 7.8 -wavelength hyperboloidal reflector with horn illumination.

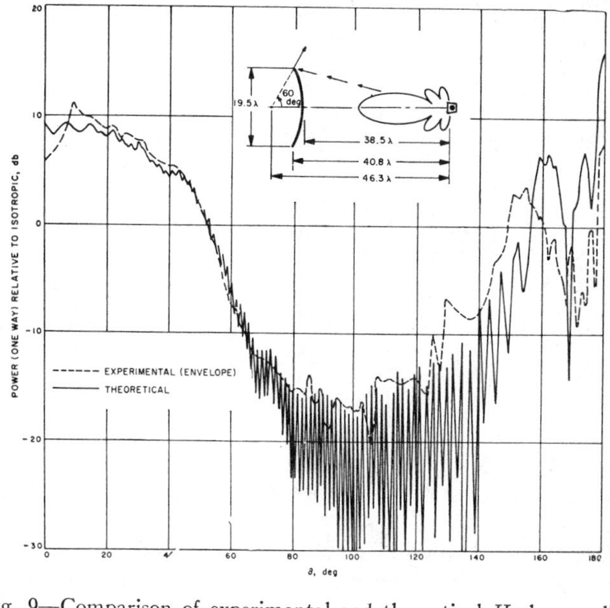

Fig. 9—Comparison of experimental and theoretical H-plane radiation pattern from 19.5-wavelength hyperboloidal reflector with horn illumination.

V. Discussion

The approximations of geometrical ray optics are inadequate to describe spillover and back radiation from a hyperboloidal reflector. However, a quantitative description of such phenomena is contained in the field integrals [(15) and (16)]. These equations may be used in the design of reflector feed systems employing hyperboloidal surfaces.

It is evident from the results of the previous sections that spillover from a Cassegrainian-fed paraboloid may be reduced to an acceptable level with moderately large hyperboloidal reflectors, but there is a tendency toward relatively high forward spillover. This situation cannot be resolved by indefinite increase of the hyperboloid diameter, since aperture blocking will eventually become prohibitive. For low-noise antenna systems it may be necessary to select a more suitable reflecting shape. It has already been discovered [4] that a conical flange extension to the hyperboloid will reduce spillover and produce a more uniform illumination pattern. Further reduction of forward spillover may be accomplished with a more suitable feed-horn pattern [17].

ACKNOWLEDGMENT

The authors wishes to indicate his gratitude for many valuable comments and suggestions by P. D. Potter on the experimental aspects of the problem. He would also like to thank W. Jungmeyer, who programmed the numerical evaluation of the field integrals.

APPENDIX I

SADDLE-POINT EVALUATION OF THE INTEGRALS

For large γ (*i.e.*, the linear dimensions of the surface $\gg\lambda$), the Bessel functions may be replaced by their asymptotic values. Then,

$$E_\phi(\theta, \phi) \simeq - AP(\theta)\frac{e^{ikR}}{R}\sin\phi + iA\gamma\frac{e^{ikR}}{R}\sin\phi \cdot \left\{ e^{-i(\pi/4)}\int_{\theta_0}^{\pi} F(\theta, \theta')e^{i(\psi_1+\psi_2)}d\theta' + e^{i(\pi/4)}\int_{\theta_0}^{\pi} F(\theta, \theta')e^{i(\psi_1-\psi_2)}d\theta' \right\} \tag{17}$$

where

$$F(\theta, \theta') = \frac{P(\theta')\sin\theta'}{(1 + e\cos\theta')(2\pi\psi_2)^{1/2}}. \tag{18}$$

Both integrals have a stationary point when $\theta=\theta'$. This stationary term exactly cancels the primary feed field in the shadow region behind the reflector. However, the integral with the exponential factor $e^{i(\psi_1+\psi_2)}$ has an additional stationary point, where

$$\frac{\sin\theta}{e + \cos\theta} = \frac{\sin\theta_s'}{e + \cos\theta_s'}. \tag{19}$$

It can be shown that $\theta' = \theta_s'$ corresponds to the ray-optics approximation for a ray reflected in the θ-direction (Fig. 10).

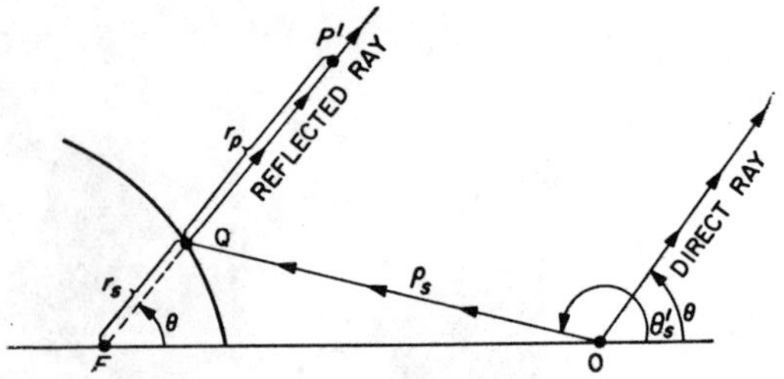

Fig. 10—Ray-optics geometry.

Using the method of stationary phase, the total field for a value of θ in the forward direction may be approximated by

$$E_\phi(\theta, \phi) = \underbrace{AP(\theta_s')\frac{\sin\theta_s'}{\sin\theta}\sin\phi\frac{e^{ikR}}{R}e^{i\gamma\psi(\theta,\theta_s')}}_{\text{stationary term}} \underbrace{- AP(\theta)\frac{e^{ikR}}{R}\sin\phi}_{\text{direct field}} \tag{20}$$

where

$$\psi(\theta, \theta_s') = \frac{\cos\theta\cos\theta_s' + \sin\theta\sin\theta_s' - 1}{(1 + e\cos\theta_s')}. \tag{21}$$

APPENDIX II

DERIVATION OF THE FIELDS USING THE APPROXIMATIONS OF RAY OPTICS

At P' (Fig. 10), the ϕ component of the field due to reflection from the hyperboloid is

$$[E_\phi(P')]_{\text{ref 1}} = \frac{E_0 e^{ikr_p}}{(r_s + r_p)}. \tag{22}$$

At Q, the point of reflection, the total field is zero. Hence

$$[E_\phi(Q)]_{\text{ref 1}} + [E_\phi(Q)]_{\text{incident}} = 0 \tag{23}$$

so that

$$\frac{E_0}{r_s} = A\sin\phi\frac{e^{ik\rho_s}}{\rho_s}P(\theta_s') \tag{24}$$

and

$$E_0 = A\sin\phi\frac{r_s}{\rho_s}e^{ik\rho_s}P(\theta_s') = A\sin\phi\frac{\sin\theta_s'}{\sin\theta}e^{ik\rho_s}P(\theta_s'). \tag{25}$$

As P' approaches P, the field point, $r_p \to R - \rho_s\cos(\theta_s' - \theta)$, and

$$[E_\phi(P)]_{\text{ref 1}} \simeq AP(\theta_s')\sin\phi\frac{\sin\theta_s'}{\sin\theta}\frac{e^{ikR}}{R}e^{ik\rho_s[1-\cos(\theta_s'-\theta)]}. \tag{26}$$

Using (1), (11), (21), and (26), it can easily be shown that

$$[E_\phi(P)]_{\text{refl}} = AP(\theta_s')\sin\phi \frac{\sin\theta_s'}{\sin\theta}\frac{e^{ikR}}{R}e^{i\gamma\psi(\theta,\theta_s')}. \quad (27)$$

Consequently, the stationary term in the field expression corresponds exactly to the high-frequency ray-optics approximation.

Nomenclature

- a length of semitransverse axis of hyperboloid
- A amplitude of primary field
- c semidistance between hyperboloid foci
- $dS(\theta', \phi')$ differential surface element on front of reflector
- e eccentricity of hyperboloid (c/a)
- $\mathbf{e}(\theta, \phi)$ unit vector describing polarization of primary electric field
- $\mathbf{E}$ total radiation field of system
- $\mathbf{E}_{pf}$ primary field incident on reflector
- $\mathbf{E}_s$ field scattered from reflector
- $\mathbf{h}(\theta, \phi)$ unit vector describing polarization of primary magnetic field
- k free-space propagation constant $(2\pi/\lambda)$
- $\mathbf{n}$ unit outward surface normal from front of reflector
- O, F foci of hyperboloid
- p $c[1-(1/e^2)]$
- $P(\theta)$ pattern factor of primary field
- r distance from secondary focus to point on reflector
- r_s distance from secondary focus to point of geometrical reflection
- r_p distance from point of geometrical reflection to field point
- R, θ, ϕ polar coordinates of field point P, with associated unit vectors $\mathbf{a}_R$, $\mathbf{a}_\theta$, $\mathbf{a}_\phi$
- γ $2\pi e(p/\lambda)$
- θ_c direction of extremum ray
- θ_0 polar angle of reflector edge
- θ_s' polar angle of point of geometrical reflection
- ρ, θ', ϕ' polar coordinates of point on reflector surface, with associated unit vectors $\mathbf{a}_\rho$, $\mathbf{a}_{\theta'}$, $\mathbf{a}_{\phi'}$.

References

[1] P. Foldes and S. G. Komlos, "Theoretical and experimental study of wide-band paraboloid antennas with central reflector feed," *RCA Rev.*, vol. XXI, pp. 94–116; March, 1960.

[2] P. W. Hannan, "Microwave antennas derived from the Cassegrain telescope," IRE Trans. on Antennas and Propagation, vol. AP-9, pp. 140–153; March, 1961.

[3] P. D. Potter, "Aperture efficiency of large paraboloidal antennas as a function of their feed system radiation characteristics," IEEE Trans on Antennas and Propagation, vol. AP-11; May 1963.

[4] ——, "Unique feed system improves space antennas," *Electronics*, vol. 35, pp. 36–40; June 22, 1962.

[5] ——, "The application of the Cassegrainian principle to ground antennas for space communications," IRE Trans. on Space Electronics and Telemetry, vol. SET-8, pp. 154–158; June, 1962.

[6] T. Sato and C. R. Stelzried, "An operational 960-Mc maser system for deep-space tracking missions," IRE Trans. on Space Electronics and Telemetry, vol. SET-8, pp. 164–170; June, 1962.

[7] W. V. T. Rusch, "Analytical Study of Wide-Angle and Back Lobe Paraboloidal Antennas," Jet Propulsion Lab., Pasadena, Calif., Research Summary 36-8, pp. 35–39; May 1, 1961.

[8] S. Silver, "Microwave Antenna Theory and Design," M.I.T. Rad. Lab. Ser., McGraw-Hill Book Co., Inc., New York, N. Y., vol. 12, p. 149; 1949.

[9] M. Born and E. Wolf, "Principles of Optics," Pergamon Press, London, England, chs. VIII and XI; 1959.

[10] R. W. P. King and T. T. Wu, "The Scattering and Diffraction of Waves," Harvard University Press, Cambridge, Mass., ch. 4; 1959.

[11] L. B. Tartakovskii, "Side radiation from ideal paraboloid with circular aperature," *Radiotekh. Elektron.*, vol. 4, pp. 920–929; month, 1959.

[12] J. Poisson, "Extrait d'une Lettre de M. Poisson à M. Fresnel," in "Annales de Chimie et de Physique," vol. 22, pp. 270–280; 1823.

[13] D. F. J. Arago, "Oeuvres Completes," Gide et J. Baudry, Paris, France, vol. VII, p. 1; 1858.

[14] J. Coulson and G. G. Becknell, "An extension of the principle of the diffraction evolute and some of its structural detail," *Phys. Rev.*, vol. 20, p. 607; 1922.

[15] J. B. Ketter, "A geometrical theory of diffraction," *Proc. Symp. Appl. Math.*, McGraw-Hill Book Co., Inc., New York, N. Y., vol. 8, pp. 27–52; 1958.

[16] D. Schuster, Jet Propulsion Lab., Pasadena, Calif., private communication; April 6, 1961.

[17] P. D. Potter, "A New Horn Antenna with Suppressed Side Lobes and Equal Beam Widths," Jet Propulsion Lab., Pasadena, Calif., Tech. Rept. No. 32-354; 1963.

Application of Spherical Wave Theory to Cassegrainian-Fed Paraboloids

PHILIP D. POTTER, MEMBER, IEEE

Abstract—**In this paper the spherical wave expansion technique is used to analyze the general properties of paraboloidal antenna feed systems. As a result, fundamental aperture efficiency and noise temperature limitations are established as quantitative functions of antenna wavelength size. A boundary-value solution is found for synthesis of ideal subreflector shapes in a Cassegrain-type feed system. The resulting surface is found to reduce the classical hyperboloid in the limit of zero wavelength. Applications of this synthesis technique to high performance feed systems and subreflector matching are discussed. Finally, an interesting quantitative cross-check between vector spherical wave and vector diffraction theories is obtained.**

I. Introduction

THE two-reflector antenna system invented for optical frequency use by Cassegrain has achieved widespread interest and usage in the microwave antenna field. Because Cassegrainian systems are typically used in electrically large antenna installations, geometric optics techniques have been generally used for design and analysis.[1]–[4] Some performance parameters are not quantitatively predictable from geometric optics, however, and as a result a number of diffraction analyses have been published.[5]–[8]

Although the optically-derived subreflector is the simplest and most practical for many applications, other surfaces may offer performance advantages in terms of aperture efficiency and spillover.[9]–[15] This paper considers a synthesis technique which proceeds along different lines from those previously presented. The feed system is defined such as to produce a spherical wavefront at infinity within a specified angular region. The amplitude radiation pattern is defined as a best-fit to a specified "ideal" pattern, within the constraints of finite feed system and main reflector sizes. A spherical wave expansion technique is then used to derive a subreflector contour and feed horn radiation pattern which yields the required feed system radiation patterns. The question of physically realizing the resulting required feed horn radiation patterns has been discussed in detail by Ludwig[16] and is not analyzed here.

Manuscript received December 30, 1966; revised May 31, 1967. This paper presents the results of one phase of research carried out at the Jet Propulsion Laboratory, California Institute of Technology, under Contract NAS 7-100, sponsored by NASA.

The author is with the Jet Propulsion Laboratory, Pasadena, Calif.

Reprinted from *IEEE Trans. Antennas Propagat.*, vol. AP-15, pp. 727–736, Nov. 1967.

II. Spherical Waves

As a useful idealization, the region between the main paraboloidal reflector and the feed system may be considered to be a source-free, homogeneous, isotropic medium with zero conductivity. Under these conditions, the electric and magnetic fields are completely described by the homogeneous vector wave equations

$$\nabla(\nabla \cdot \boldsymbol{E}) - \nabla \times (\nabla \times \boldsymbol{E}) + k^2\boldsymbol{E} = 0 \tag{1}$$

$$\nabla(\nabla \cdot \boldsymbol{H}) - \nabla \times (\nabla \times \boldsymbol{H}) + k^2\boldsymbol{H} = 0 \tag{2}$$

where

$\boldsymbol{E}$ = electric field
$\boldsymbol{H}$ = magnetic field
k = free space propagation constant.

The solutions to (1) and (2) in spherical coordinates are discussed in detail by Stratton.[17] Three families of independent vector solutions exist, designated $\boldsymbol{M}_{mn}$, $\boldsymbol{N}_{mn}$, and $\boldsymbol{L}_{mn}$. The last of these represents a family of plane waves and need not be considered further. The $\boldsymbol{M}_{mn}$ and $\boldsymbol{N}_{mn}$, however, describe the complex $\boldsymbol{E}$ and $\boldsymbol{H}$ fields of families of transverse electric (TE) and transverse magnetic (TM) waves. The field components are as follows:

TE_{mn} waves

$$\boldsymbol{E}_{mn} = jZ_0\boldsymbol{M}_{mn} \tag{3}$$

$$\boldsymbol{H}_{mn} = \boldsymbol{N}_{mn} \tag{4}$$

TM_{mn} waves

$$\boldsymbol{E}_{mn} = \boldsymbol{N}_{mn} \tag{5}$$

$$\boldsymbol{H}_{mn} = \frac{j}{Z_0}\boldsymbol{M}_{mn} \tag{6}$$

where

$$Z_0 = \text{free space impedance.}$$

It can be shown[18] that in a source-free region between two concentric spheres (one of these may have infinite radius), only TE and TM waves can exist, therefore (3) to (6) constitute a complete solution to possible fields in the region between a feed system and the main reflector. This set of waves is directly analogous to the set of modes which may exist in a waveguide; a complete knowledge of mode strengths at one point in space uniquely determines the fields (in amplitude, phase, and direction) elsewhere.

In describing the feed system fields, a right-handed polar coordinate system is used, as shown in Fig. 1. The vector functions $\boldsymbol{M}_{mn}$ and $\boldsymbol{N}_{mn}$ are given in terms of the polar angle ψ, the azimuthal angle ξ, the radius ρ, and the associated unit vectors. The assumed time dependence $e^{j\omega t}$ will be suppressed throughout this paper. Expressions for $\boldsymbol{M}_{mn}$ and $\boldsymbol{N}_{mn}$ are as follows:

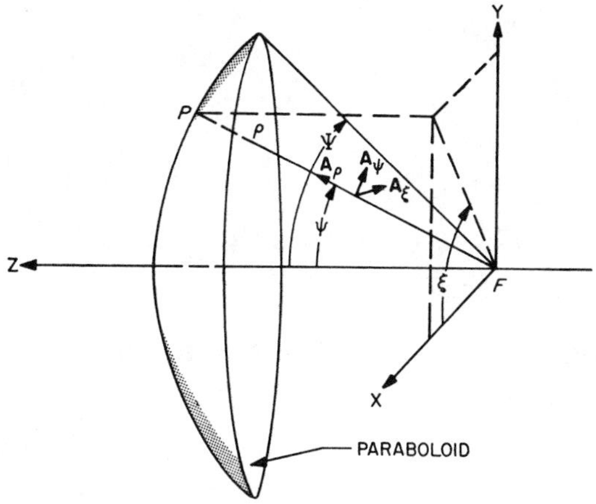

Fig. 1. Paraboloid geometry.

$$\boldsymbol{M}_{mn} = \mp \frac{m}{\sin\psi} Z_n(k\rho)P_n^m(\cos\psi)\begin{bmatrix}\sin\\ \cos\end{bmatrix} m\xi \boldsymbol{a}_\psi - Z_n(k\rho)\frac{d}{d\psi}[P_n^m(\cos\psi)]\begin{bmatrix}\cos\\ \sin\end{bmatrix} m\xi \boldsymbol{a}_\xi \tag{7}$$

$$N_{mn} = \frac{1}{k\rho}\frac{d}{d\rho}[\rho Z_n(k\rho)]\frac{d}{d\psi}[P_n^m(\cos\psi)]\begin{bmatrix}\cos\\ \sin\end{bmatrix} m\xi \boldsymbol{a}_\psi \mp \frac{m}{k\rho\sin\psi}\frac{d}{d\rho}[\rho Z_n(k\rho)]P_n^m(\cos\psi)\begin{bmatrix}\sin\\ \cos\end{bmatrix} m\xi \boldsymbol{a}_\xi + \frac{n(n+1)}{k\rho} Z_n(k\rho)P_n^m(\cos\psi)\begin{bmatrix}\cos\\ \sin\end{bmatrix} m\xi \boldsymbol{a}_\rho \tag{8}$$

where P_n^m $(\cos\psi)$ is the associated Legendre polynomial of the first kind[19] and $Z_n(k\rho)$ is the spherical Bessel function.[17] For the problem under consideration, the spherical waves must satisfy the radiation condition at infinity and hence the spherical Bessel functions are restricted to be spherical Hankel functions $h_n^{(2)}(k\rho)$. Tables of all of the involved functions are available.[20]–[22]

A. Restriction of Wave Order

The spherical waves given by (3) to (8) possess the orthogonality properties familiar in guided waves; a member of a set of these waves, excited by the feed system, maintains independence throughout free space with a constant total propagation power. Although each wave is a separable function of ψ, ξ, and ρ, this is not generally true of a sum of these waves; the sum may change character completely as a function of ρ. Of interest is the asymptotic behavior of the radial functions $h_n^{(2)}(k\rho)$:

Far-field region, $k\rho \gtrsim n^2$

$$h_n^{(2)}(k\rho) \approx (j)^{(n+1)}\frac{e^{-jk\rho}}{k\rho} \tag{9}$$

$$\frac{1}{k\rho}\frac{d}{d\rho}[\rho h_n^{(2)}(k\rho)] \approx (j)^n\frac{e^{-jk\rho}}{k\rho} \tag{10}$$

Fresnel region, $n \lesssim k\rho \gtrsim n^2$

$$h_n^{(2)}(k\rho) \approx (j)^{(n+1)} \frac{e^{-jkf_1(\rho,n)}}{k\rho} \tag{11}$$

$$\frac{1}{k\rho}\frac{d}{d\rho}\left[\rho h_n^{(2)}(k\rho)\right] \approx (j)^n \frac{e^{-jkf_2(\rho,n)}}{k\rho} \tag{12}$$

Near-field region, $0 \gtrsim k\rho \lesssim n$

$$h_n^{(2)}(k\rho) = f_3(\rho, n) \tag{13}$$

$$\frac{1}{k\rho}\frac{d}{d\rho}\left[\rho h_n^{(2)}(k\rho)\right] = f_4(\rho, n) \tag{14}$$

where $f_1(\rho, n)$, $f_2(\rho, n)$, $f_3(\rho, n)$, and $f_4(\rho, n)$ indicate complicated dependence upon ρ and n.

In the far-field region, all of the spherical waves suffer inverse-distance amplitude decay and all remain essentially in phase synchronism. In the Fresnel region, the waves have essentially this same amplitude dependence, but they do not maintain phase synchronism as a function of $k\rho$. In the near field, the waves behave nonsimply in both amplitude and phase.

Numerical investigation of the Hankel functions shows that the quantities $f_3(\rho, n)$ and $f_4(\rho, n)$ in (13) and (14) assume very large values for $k\rho \ll n$. This fact implies a possible supergain situation which has been investigated by several authors.[23]–[25] For large-wavelength two-reflector antenna systems, a different restriction upon N is more significant. This latter restriction, qualitatively indicated by (11) and (12), arises from a requirement that the paraboloidal main reflector be in the far-field region of the subreflector scattered fields to within some specified phase error tolerance. In Appendix I it is shown that the far-field restriction may be quantitatively expressed as follows:

$$N \leq B\sqrt{kf} = B\left[2\pi\left(\frac{f}{D}\right)\left(\frac{D}{\lambda}\right)\right]^{1/2} \tag{15}$$

where

B = a constant
f = paraboloid focal length
D = paraboloid diameter
λ = wavelength of operation.

Choice of the constant B will be a function of the wave amplitudes versus n and of the desired phase error at the paraboloid. Numerical investigation shows that the phase error effect is typically small for $B=1$ and severe for $B=2$.

The azimuthal index m is restricted to one since only these waves will radiate axially from the paraboloid.[8] Thus the general feed system radiation path may be expressed as a sum of TE_{1n} and TM_{1n} waves, where $1 \leq n \leq N$. Waves of the form TE_{10} and TM_{10} are not possible because of the properties of the associated Legendre polynomials which require that

$$n \geq m. \tag{16}$$

B. Feed System Expansion in Spherical Waves

The general expression for the feed system radiation pattern $\boldsymbol{E}_f(\psi, \xi, \rho)$ is given as follows:

$$\boldsymbol{E}_f(\psi, \xi, \rho) = \sum_{n=1}^{n=N} \left[A_{TE_{1n}}\boldsymbol{E}_{TE_{1n}} + A_{TM_{1n}}\boldsymbol{E}_{TM_{1n}}\right] \tag{17}$$

where

$A_{TE_{1n}}, A_{TM_{1n}}$ = complex wave coefficients.

$\boldsymbol{E}_{TE_{1n}}$ and $\boldsymbol{E}_{TM_{1n}}$ are given by (3) to (8) with obvious notation changes.

In the far field, the asymptotic expressions (9) and (10) may be used to simplify, yielding

$$\boldsymbol{E}_{TE_{1n}} = \frac{-(j)^n Z_0 e^{-jk\rho}}{k\rho}\left[\frac{P_n^1(\cos\psi)}{\sin\psi}\sin\xi\boldsymbol{a}_\psi + \frac{dP_n^1(\cos\psi)}{d\psi}\cos\xi\boldsymbol{a}_\xi\right] \tag{18}$$

$$\boldsymbol{E}_{TM_{1n}} = \frac{+(j)^n e^{-jk\rho}}{k\rho}\left[\frac{dP_n^1(\cos\psi)}{d\psi}\sin\xi\boldsymbol{a}_\psi + \frac{P_n^1(\cos\psi)}{\sin\psi}\cos\xi\boldsymbol{a}_\xi\right]. \tag{19}$$

A class of idealized, azimuthally-symmetric feed system patterns is chosen which eliminates aperture cross polarization[8] and also eliminates the feed system backlobe (see Appendix B).

$$\boldsymbol{E}_f(\psi, \xi, \rho) \stackrel{\Delta}{=} \frac{e^{-jk\rho}}{k\rho}(\sin\xi\boldsymbol{a}_\psi + \cos\xi\boldsymbol{a}_\xi)F(\psi) \tag{20}$$

where $F(\psi)$ is the desired polar pattern.

Standard techniques are applied to (17) to (20) (see Stratton[17]) to obtain the wave coefficients

$$A_{TE_{1n}} = \frac{-(-j)^n(2n+1)}{2Z_0 n^2(n+1)^2}\int_0^\pi F(\psi) \cdot\left[\frac{P_n^1(\cos\psi)}{\sin\psi} + \frac{dP_n^1(\cos\psi)}{d\psi}\right]\sin\psi d\psi \tag{21}$$

$$A_{TM_{1n}} = -Z_0 A_{TE_{1n}}. \tag{22}$$

For numerical use, it is convenient to define certain normalized functions as follows:

$$f_n(\psi) \stackrel{\Delta}{=} \frac{1}{n(n+1)}\left[\frac{P_n^1(\cos\psi)}{\sin\psi} + \frac{dP_n^1(\psi)}{d\psi}\right] \tag{23}$$

$$A_n \stackrel{\Delta}{=} \frac{Z_0 n(n+1)}{-(-j)^n} A_{TE_{1n}} = \frac{2n+1}{2}\int_0^\pi F(\psi)f_n(\psi)\sin\psi d\psi. \tag{24}$$

From (17) through (24), the feed system far-field radiation

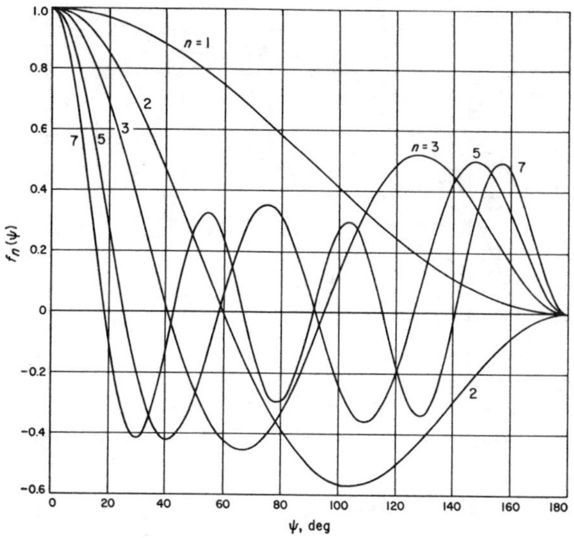

Fig. 2. Elemental wave functions $fn(\psi)$.

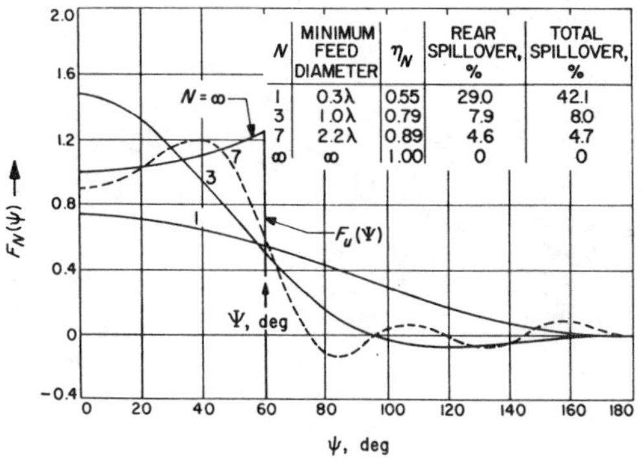

Fig. 3. Synthesis of ideal illumination pattern.

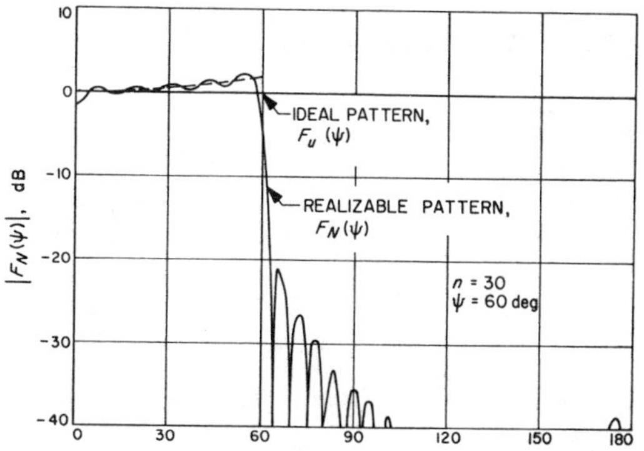

Fig. 4. Spherical wave fit to ideal illumination pattern.

patterns are given as

$$E_{fN}(\psi, \xi, \rho) = \frac{e^{-jk\rho}}{k\rho}(\sin \xi a_\psi + \cos \xi a_\xi)F_N(\psi) \tag{25}$$

$$F_N(\psi) = \sum_{n=1}^{n=N} A_n f_n(\psi) \tag{26}$$

where $F_N(\psi)$ is the spherical wave fit to the desired generalized polar pattern $F(\psi)$. A plot of the first few of the elemental wave functions, $f_n(\psi)$, is shown in Fig. 2.

C. Uniform Illumination Case

As a special case of considerable practical importance, $F(\psi)$ may be chosen such as to provide uniform aperture illumination over a specified angular region and no radiation elsewhere.

$$F_u(\psi) \triangleq \sec^2\left(\frac{\psi}{2}\right), \quad \psi_1 \le \psi \le \psi_0$$

$$F_u(\psi) \triangleq 0, \qquad 0 \le \psi < \psi_1 \text{ and } \psi_0 < \psi \le \pi. \tag{27}$$

For this case, the integration in (24) may be performed in closed form, resulting in a simple expression for A_n:

$$A_n(\psi_0, \psi_1) = \frac{2n+1}{n(n+1)}\left[\tan\left(\frac{\psi_0}{2}\right)P_n^1(\cos\psi_0) - \tan\left(\frac{\psi_1}{2}\right)P_n^1(\cos\psi_1)\right]. \tag{28}$$

If ψ_1 is selected to be zero, then the predicted aperture efficiency asymptotically approaches unity as N becomes large; similarly, the total spillover approaches zero as N becomes large.

Fig. 3 shows the spherical wave fits to (27) which are obtained for small N; Fig. 4 shows the excellent fit which is achieved for $N=30$. In the following section, the variation of performance with N and antenna size is quantitatively investigated.

III. Fundamental Performance Limitations

This section parallels in part independent work by personnel of the Hughes Aircraft Company;[27] the results are included here because of their relationship to the synthesis procedure described in the following section.

A. Aperture Efficiency

The system aperture efficiency η is given by[4]

$$\eta = \frac{2\cot^2\left(\frac{\psi}{2}\right)\left|\int_0^\psi F_N(\psi)\tan\left(\frac{\psi}{2}\right)d\psi\right|^2}{\int_0^\pi |F_N(\psi)|^2 \sin\psi d\psi} \tag{29}$$

where

ψ = edge angle of the paraboloid.

With use of (23) through (29), a general closed-form expression may be obtained for aperture efficiency

$$\eta = \frac{\left|\sum_{n=1}^{N} \frac{A_n}{n(n+1)} P_n^1(\cos\psi)\right|^2}{\sum_{n=1}^{N} \frac{A_n^2}{2n+1}}. \tag{30}$$

In general, numerical integration will be necessary to determine the wave coefficients A_n. For the important special case of a fit to uniform illumination (see Section II-C)

$$\eta = \frac{\left| \sum_{n=1}^{N} \frac{2n+1}{n^2(n+1)^2} P_n^1(\cos\psi) P_n^1(\cos\psi_0) \right|^2}{\sum_{n=1}^{N} \frac{2n+1}{n^2(n+1)^2} [P_n^1(\cos\psi_0)]^2} \tag{31}$$

where

ψ_0 = the pattern cutoff angle in the ideal pattern $F(\psi)$.

It is readily shown that a choice of $\psi = \psi_0$ will maximize η; this choice results in a compact form for (31)

$$\eta_{\max} = \sum_{n=1}^{N} \frac{2n+1}{n^2(n+1)^2} [P_n^1(\cos\psi)]^2$$

$$= 2 \sum_{n=1}^{N} \frac{[\overline{P}_n^1(\cos\psi)]^2}{n(n+1)} \tag{32}$$

where

$\overline{P}_n^1(\cos\psi)$ = *normalized* associated Legendre polynomial.[21]

In order to establish fundamental aperture efficiency limitations, it is necessary to relate N to antenna wavelength size by use of (15), which may be expressed

$$N = B\left[\frac{\pi}{2} \operatorname{ctn}\left(\frac{\psi}{2}\right)\left(\frac{D}{\lambda}\right)\right]^{1/2}. \tag{33}$$

Fig. 5 is a plot of (32) for selected values of aperture half angle ψ; smooth curves have been established between the integral values of n because these integral values do not appear to have special physical significance. As indicated by Fig. 2, a minimum n is required to efficiently synthesize a pattern for a given ψ. As shown in Fig. 5, the aperture efficiency rises rapidly for N values smaller than this critical value.

Fig. 6 is a plot of (33) for the case of $B=1$ and various aperture half angles. Combining these curves with those of Fig. 5, a plot of aperture efficiency versus aperture angular size may be constructed as shown in Fig. 7. The conflicting ψ dependences in the two previous figures largely compensate, with a result that fundamental aperture efficiency limitations are largely independent of f/D value. The degree to which these high performance levels are achievable in practice depends primarily on the state-of-the-art in feed horn design[16] and upon other practical considerations such as subreflector support design.[11] At present, overall efficiency is constrained by these considerations to about 70 percent. The subreflector blockage effect may be made negligibly small, since the spherical wave constraint is far-field distance rather than subreflector size (see Section II-A).

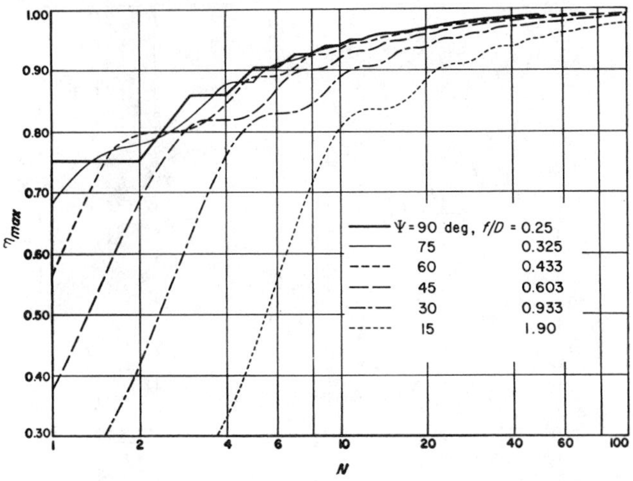

Fig. 5. Maximum aperture efficiency as a function of wave order.

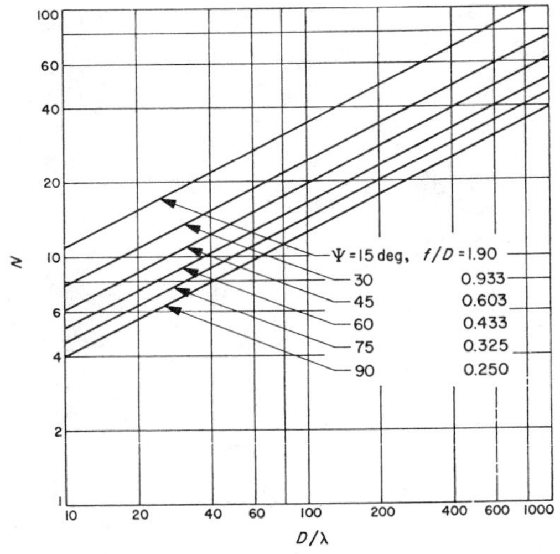

Fig. 6. Wave order as a function of aperture size.

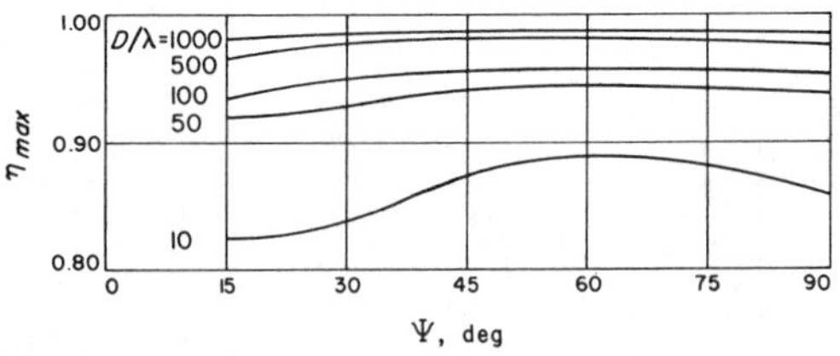

Fig. 7. Maximum aperture efficiency versus aperture angular size.

B. Figure of Merit

If the antenna system is to be used in a low-noise receiving configuration, it is appropriate to define an antenna figure-of-merit (FM), which is the quotient of aperture efficiency and system noise temperature

$$FM \triangleq \frac{\eta T_0}{T_0 + T_s} \tag{34}$$

where

$T_s \triangleq 240°$ Kelvin × (fractional rear hemisphere spill-over) (35)

$T_0 \triangleq$ (total system noise temperature) $- T_s$. (36)

Equation (35) is not rigorous; however, it is mathematically convenient and sufficiently accurate for FM optimization. For this optimization, the paraboloid edge angle ψ is varied in the region of the pattern cutoff angle ψ_0 such as to maximize (34). A parameter $\Delta\psi_{\text{opt}}$ may be defined such that (34) is optimized

$$\Delta\psi_{\text{opt}} \triangleq \psi_{\text{opt}} - \psi_0. \tag{37}$$

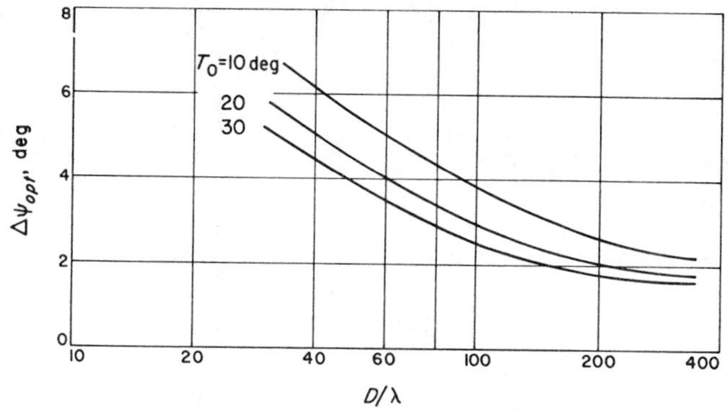

Fig. 8. $\Delta\psi_{\text{opt}}$ as a function of aperture size.

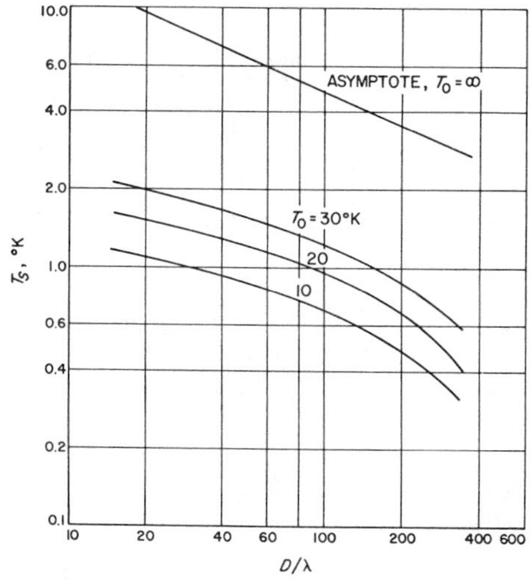

Fig. 9. Optimized spillover noise versus aperture size.

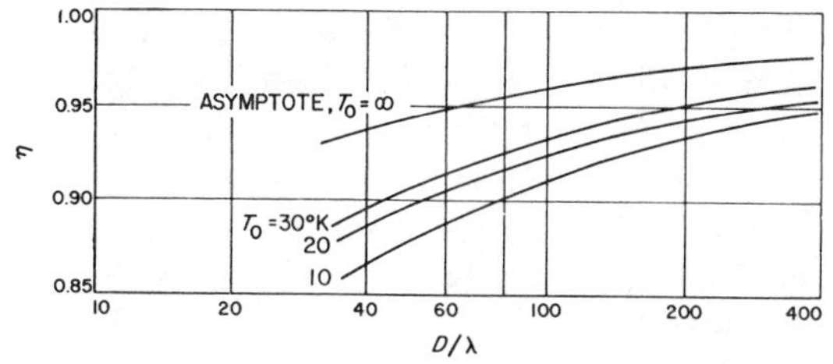

Fig. 10. Optimized aperture efficiency versus aperture size.

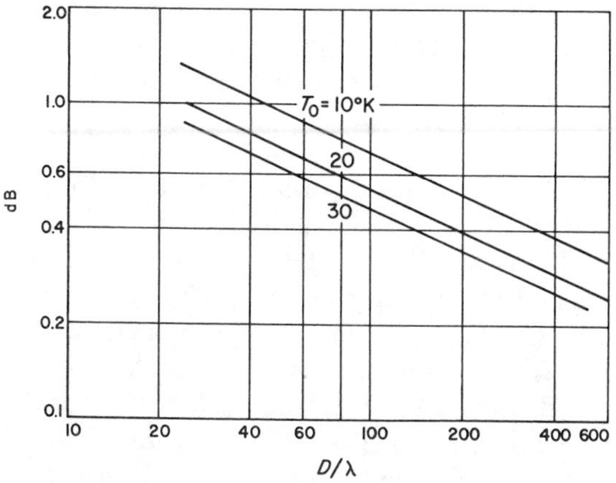

Fig. 11. Figure-of-merit versus aperture size.

Fig. 8 shows this parameter for the case of $\psi_0=60°$. It should be noted that $\Delta\psi_{\text{opt}}$ will always be positive and approaches zero for large N and high system noise temperature. Figs. 9 and 10 show, respectively, the corresponding spillover noise contribution and aperture efficiency achieved by use of $\Delta\psi_{\text{opt}}$. Finally, Fig. 11 shows FM as a function of aperture size. The following section considers a synthesis technique for realizing the performance levels shown in Figs. 9 to 11.

IV. Spherical Wave Synthesis of Nonoptical Subreflectors

Fig. 12 shows the assumed geometry of the Cassegrain-type paraboloidal antenna feed system. In this section, a general equation for the desired surface is obtained; for the limiting case of a spherical wave from the feed horn and vanishing wavelength, the surface equation reduces to that of a hyperboloid.

The total vector field $\boldsymbol{E}_T$ in the vicinity of the subreflector is the sum of the incident field $\boldsymbol{E}_H$ from the feed horn and the scattered field $\boldsymbol{E}_s$.

$$\boldsymbol{E}_T = \boldsymbol{E}_H + \boldsymbol{E}_s. \tag{38}$$

A general horn pattern $\boldsymbol{E}_H$ is assumed as follows:

$$\boldsymbol{E}_H \overset{\Delta}{=} -\frac{[F_{H\gamma}(\gamma)\sin\xi\boldsymbol{a}_\gamma + F_{H\xi}(\gamma)\cos\xi\boldsymbol{a}_\xi]}{kr} \cdot e^{-jk}\left[r - 2a - \frac{\delta_H(\gamma)}{k}\right] \tag{39}$$

where

$F_{H\gamma}(\gamma)$ = feed horn polar radiation pattern
$F_{H\xi}(\gamma)$ = feed horn azimuthal radiation pattern
a = a geometrical constant
$\delta_H(\gamma)$ = feed horn phase pattern, assumed to be azimuthally symmetric.

The scattered field $\boldsymbol{E}_s$ is now set equal to the sum of spherical wave functions

$$\boldsymbol{E}_s \overset{\Delta}{=} F(\psi, \rho)(\sin\xi\boldsymbol{a}_\psi + \cos\xi\boldsymbol{a}_\xi) \tag{40}$$

where

$$F(\psi, \rho) = jZ_0 \sum_{n=1}^{n=N} a_{\text{TE}_n} h_n(k\rho)\left[\frac{P_n{}^1(\psi)}{\sin\psi} + \frac{dP_n{}^1(\psi)}{d\psi}\right]. \tag{41}$$

Equation (41) may be expressed as having a quasi-spherical wavefront, as follows:

$$F(\psi, \rho) \overset{\Delta}{=} \frac{F_s(,\psi\ \rho)}{\rho} e^{-j[k\rho - \delta(\rho,\psi)]}. \tag{42}$$

The restriction is now imposed that the subreflector be a surface of revolution. This restriction considerably simplifies the problem and results in a polarization independent solu-

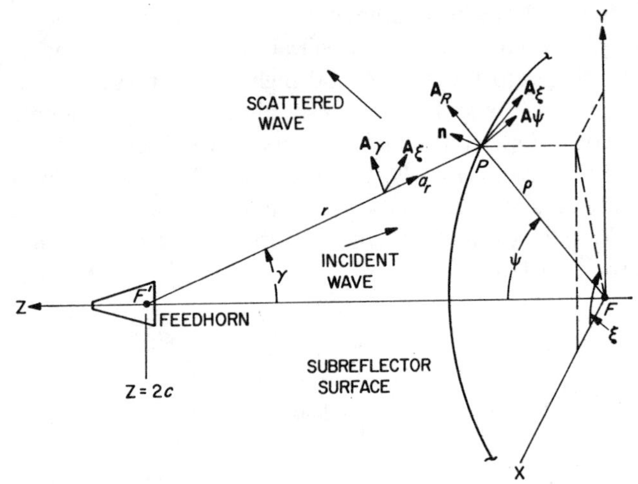

Fig. 12. Feed system geometry.

tion. From (38) to (42), two equations result:

$$F_{H\xi}(\gamma) = \frac{r}{\rho} F_s(\psi, \rho) = \frac{\sin \psi}{\sin \gamma} F_s(\psi, \rho) \tag{43}$$

$$\rho = \frac{c^2 - \left[a - \dfrac{\lambda\delta(\rho, \psi) - \lambda\delta_H(\gamma)}{4\pi} \right]^2}{c \cos \psi + a - \dfrac{\lambda\delta(\rho, \psi) - \lambda\delta_H(\gamma)}{4\pi}}, \tag{44}$$

and from the electric field boundary condition

$$\frac{F_{H\xi}(\gamma)}{F_{H\gamma}(\gamma)} = \cos (\psi + \gamma) - \frac{1}{\rho} \sin (\psi + \gamma) \frac{d\rho}{d\psi} \tag{45}$$

where $d\rho/d\psi$ is evaluated on the surface. Of interest is

$$\lim_{\lambda \to 0} \rho = \frac{c^2 - a^2}{c \cos \psi + a}, \tag{46}$$

which is the polar equation of the classical optically derived hyperboloid.

A. Comparison with Scattering Theory

Equation (44) defines an infinite subreflector surface which, when used with the feed horn patterns given by (43) and (45), will result in the high performance levels predicted in Section III. A computer program[28] has been written for the IBM 7094 to generate the required surface and feed horn patterns, given a few simple inputs such as N, λ, a, and c.

In any physical system, the subreflector must be truncated; it is intuitively suspected that truncation will be a small effect, as long as the truncation occurs in the region of low feed horn illumination. The effect of truncation has been quantitatively investigated by means of the subreflector scattering program.[8],[28] The latter uses the current integration method to numerically evaluate the scattered fields from a subreflector of revolution with arbitrary contour and illumination.

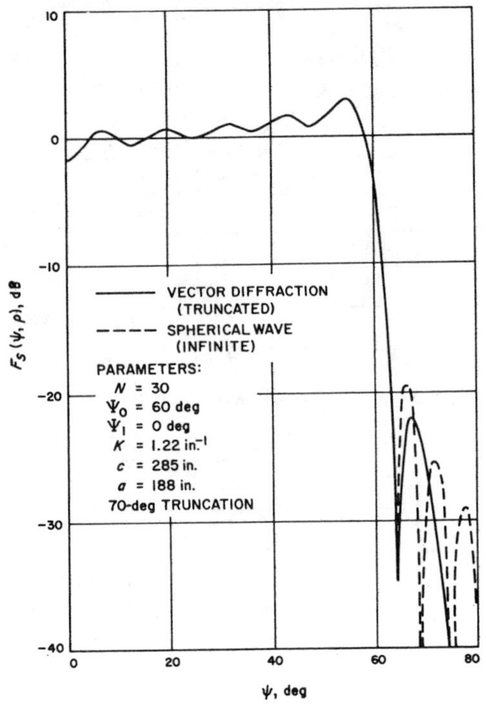

Fig. 13. Comparison of scattered fields by spherical wave and vector diffraction methods.

Fig. 13 shows a comparison of the scattered fields for the spherical wave formulation with those predicted by the current integration technique. The excellent agreement shown in Fig. 13 serves not only as a check on the spherical wave formulation but also as an unusual cross check between two completely different branches of electromagnetic theory. Of more practical significance is the demonstration that the ideal infinite subreflector may be truncated to a physically reasonable configuration without significant performance degradation. From Fig. 13 it can be seen that rear hemisphere spillover is essentially unchanged by a reasonable truncation. Also, it has already been shown[11] that the effect of forward spillover can be made negligibly small with a truncated subreflector.

B. Characteristics of the Synthesized Feed System

A typical required feed horn pattern is shown in Fig. 14, together with a four-waveguide-mode conical horn pattern as proposed by Ludwig.[16] As pointed out by Ludwig,[16] this four-mode pattern corresponds to $N \approx 25$ for a spherical wave synthesis of the horn pattern.

The nature of the synthesized subreflector surface is shown in Fig. 15 for the same case depicted in Fig. 14. Also shown is the experimentally derived nonoptical subreflector described in Potter.[10] The latter is seen to be an excellent fit to the spherical wave synthesis result. The case of $\alpha = 18.5°$ corresponds exactly to the shaped subreflector used in Potter.[11] The general result of a positive deviation beyond the optical edge at $\Psi = \psi$ appears to follow naturally from the requirement to use the feed horn pattern edge energy effectively; a similar result has been obtained with optical synthesis techniques.[14]

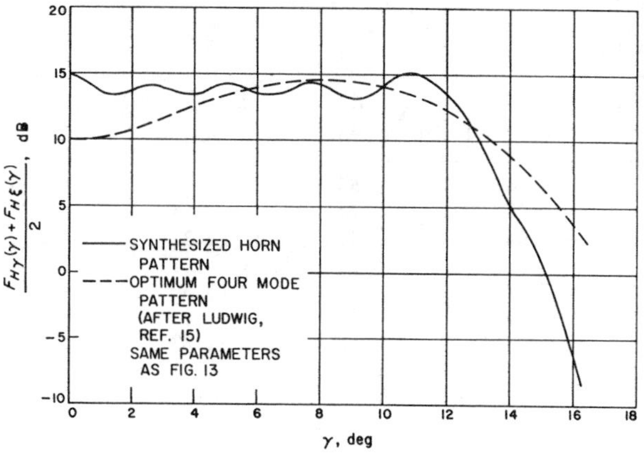

Fig. 14. Synthesized horn radiation patterns.

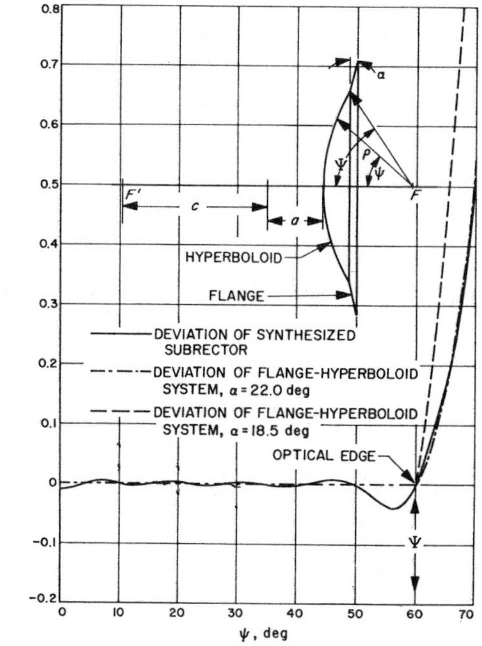

Fig. 15. Subreflector deviation from the optical case.

C. Application to Subreflector Matching

One of the most interesting applications of the spherical wave synthesis technique is that of subreflector matching. It is a simple matter to design such matching devices for prevention of subreflector scattering in the feed horn direction. A more difficult matter, however, is to design the vertex plate such that the redirected energy is used constructively in the antenna system. This latter problem is not amenable to solution by optical techniques because of the small wavelength sizes and radii of curvature involved. The boundary-value nature of the spherical wave technique is thus ideally suited to the problem.

Since the normalized wave functions $f_n(\Psi)$ are all unity for $\psi=0$ (see Fig. 2), the scattered axial far field is given simply by the sum of the coefficients A_n [see (26)]. Fig. 16 shows a comparison of a synthesized spherical wave vertex

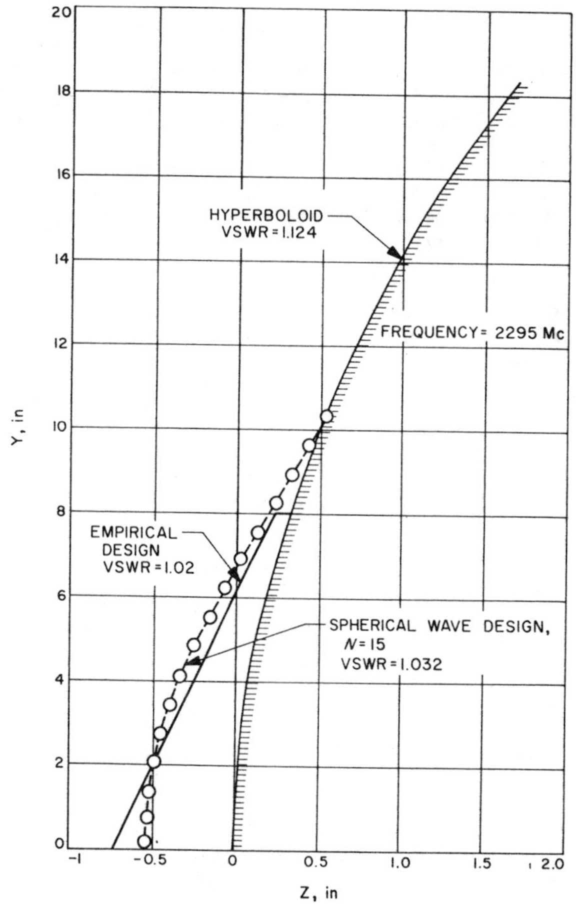

Fig. 16. Vertex plate designs.

plate with one which was previously empirically designed for good match and minimal effect upon the aperture efficiency. The VSWR values for the unmatched hyperboloid and the synthesized surface were calculated using standard current-integration techniques[29] and assuming a conventional horn pattern rather than the synthesized pattern.

V. Conclusions

Spherical wave theory has been applied in detail to the Cassegrainian-fed paraboloid design. Several applications of the method have been noted, including improved aperture illumination, reduced spillover and improved subreflector impedance matching. Perhaps the most important contribution of this analysis is to provide physical understanding to an important diffraction problem which is beyond the capabilities of geometric optics and which may be only crudely treated by geometric diffraction methods.

Appendix A

Spherical Wave Phase Error

A useful asymptotic expansion for the spherical Hankel functions has been developed by Stratton[17]

$$h_n^{(2)}(k\rho) = \frac{j^{(n+1)}}{k\rho} e^{-jk\rho}[P_{n+\frac{1}{2}}(k\rho) - jQ_{n+\frac{1}{2}}(k\rho)] \quad (47)$$

TABLE I

PHASE ERROR VERSUS τ

τ	$N=5$ (exact)	$N=5$ (approx.)	$N=10$ (exact)	$N=10$ (approx.)	$N=20$ (exact)	$N=20$ (approx.)	$N=50$ (exact)	$N=50$ (approx.)
0.250	−143.92°	−125.12°	−127.87°	−122.02°	−120.74°	−119.63°	−116.95°	−117.88°
0.500	−69.64	−74.71	−63.25	−69.74	−60.21	−66.74	−58.45	−64.79
1.000	−34.49	−35.80	−31.54	−32.86	−30.09	−31.31	−29.22	−30.38
2.000	−17.20	−17.39	−15.76	−15.94	−15.04	−15.20	−14.61	−14.76
4.000	−8.60	−8.62	−7.88	−7.90	−7.52	−7.54	−7.30	−7.32
8.000	−4.30	−4.30	−3.94	−3.94	−3.76	−3.76	−3.65	−3.65
16.000	−2.15	−2.15	−1.97	−1.97	−1.88	−1.88	−1.82	−1.82
32.000	−1.07	−1.07	−0.98	−0.98	−0.94	−0.94	−0.91	−0.91

where

$$P_{n+\frac{1}{2}}(k\rho) = 1 + \sum_{\substack{q=2 \\ q \text{ even}}}^{q=n} (-1)^{q/2} \cdot \frac{n(n+q) \prod_{i=1}^{i=q/2} [n^2 - (q-i)^2]}{2^q q!(k\rho)^q} \tag{48a}$$

$$Q_{n+\frac{1}{2}}(k\rho) = \sum_{\substack{q=1 \\ q \text{ odd}}}^{q=n} (-1)^{(q-1)/2} \cdot \frac{n(n+q) \prod_{i=1}^{i=(q+1)/2} [n^2 - (q-i)^2]}{2^q q!(k\rho)^q} . \tag{48b}$$

From these two expressions, the ratio of the terms T_q in the series is found to be

$$\frac{T_q}{T_{q-2}} = \frac{n^4\left[1-\left(\frac{q-1}{n}\right)^2\right]\left[1-\left(\frac{q-2}{n}\right)^2\right]}{2q(q-1)(k\rho)^2} \cdot \frac{n+q}{n+q-2} . \tag{49}$$

The terms of the series are thus monotonically decreasing and alternating. Convergence is rapid for $k\rho > n^2$. Combining (47) and (48) and using only the first terms of the series yields

$$\delta_n(k\rho) \approx \tan^{-1}\left[\frac{\dfrac{-n(n+1)}{2k\rho}}{1 - \dfrac{n(n^2-1)(n+2)}{8(k\rho)^2}}\right], \quad \text{radians} \quad k\rho \geq n^2 \tag{50}$$

where

$\delta_n(k\rho)$ = deviation from spherical wave.

It is convenient to define a far-field parameter τ, where

$$\tau = \frac{k\rho}{n^2} . \tag{51}$$

Table I gives the phase error δ_n versus τ for both the approximate method (50) and by exact calculation.

APPENDIX B

ELIMINATION OF FEED SYSTEM BACKLOBE

Using the asymptotic relationship for small ψ,[19]

$$P_n{}^1(\cos\psi) \approx \frac{n(n+1)}{2} \sin\psi, \quad (1-\cos\psi) \ll 1 \tag{52}$$

and the symmetry relationship

$$P_n{}^1[\cos(\pi-\psi)] = (-1)^{n-1} P_n{}'(\cos\psi) \tag{53}$$

one obtains

$$\lim_{\psi\to\pi} \frac{P_n{}^1(\cos\psi)}{\sin\psi} = \frac{(-1)^{n-1} n(n+1)}{2} \tag{54a}$$

$$\lim_{\psi\to\pi} \frac{dP_n{}^2(\cos)}{d\psi} = \frac{-(-1)^{n-1} n(n+1)}{2} . \tag{54b}$$

From (23) and (54), then,

$$f_n(\psi = \pi) = 0. \tag{55}$$

ACKNOWLEDGMENT

The author is indebted to A. Ludwig of the Jet Propulsion Laboratory for several interesting discussions on spherical waves and for stimulating the author's interest in performing the attendant computer programming.

REFERENCES

[1] P. W. Hannan, "Microwave antennas derived from the Cassegrain telescope," *IRE Trans. Antennas and Propagation*, vol. AP-9, pp. 140–153, March 1961.

[2] P. A. Jensen, "Designing Cassegrain antennas," *Microwaves*, vol. 1, December 1962.

[3] W. D. White *et al.*, "Scanning characteristics of two-reflector antenna systems," *1962 Internat'l Conv. Rec.*, pt. 1, p. 44–70.

[4] P. D. Potter, "Aperture illumination and gain of a Cassegrainian system," *IEEE Trans. Antennas and Propagation (Communications)*, vol. AP-11, pp. 373–375, May 1963.

[5] P. Foldes, "The capabilities of Cassegrain microwave optics systems for low noise antennas," in *Proc. 5th Agard Avionics Panel Conf.* (Oslo, Norway), vol. 4. New York: Pergamon, 1962, pp. 319–352.

[6] W. V. T. Rusch, "Scattering from a hyperboloidal reflector in a

Cassegrainian feed system," *IEEE Trans. Antennas and Propagation*, vol. AP-11, pp. 414–421, July 1963.

[7] ——, "A comparison of diffraction in Cassegrainian and Gregorian radio telescopes," *Proc. IEEE* (*Correspondence*), vol. 51, pp. 630–631, April 1963.

[8] ——, "Phase error and associated cross-polarization effects in Cassegrainian-fed microwave antennas," *IEEE Trans. Antennas and Propagation*, vol. AP-14, pp. 266–275, May 1966.

[9] P. Foldes and S. G. Kamlos, "Theoretical and experimental study of wideband paraboloid antenna with central reflector feed," *RCA Rev.*, vol. 21, pp. 94–116, March 1960.

[10] P. D. Potter, "Unique feed system improves space antennas," *Electronics*, vol. 35, pp. 36–40, June 22, 1962.

[11] ——, "The design of a very high power, very low noise Cassegrain feed system for a planetrary radar," in *Radar Techniques for Detection Tracking and Navigation, AGARDograph 100.* New York: Gordon and Breach Science Publishers.

[12] K. A. Green, "Modified Cassegrain antenna for arbitrary aperture illumination," *IEEE Trans. Antennas and Propagation* (*Communications*), vol. AP-11, pp. 589–590, September 1963.

[13] V. Galindo, "Design of dual reflector antennas with arbitrary phase and amplitude distributions," *IEEE Trans. Antennas and Propagation*, vol. AP-12, pp. 403–408, July 1964.

[14] W. F. Williams, "High efficiency antenna reflector," *Microwave J.*, vol. 8, pp. 79–82, July 1965.

[15] S. P. Morgan, "Some examples of generalized Cassegrainian and Gregorian antennas," *IEEE Trans. Antennas and Propagation*, vol. AP-12, pp. 685–691, November 1964.

[16] A. C. Ludwig, "Radiation pattern synthesis for circular aperture horn antennas," *IEEE Trans. Antennas and Propagation*, vol. AP-14, pp. 434–440, July 1966.

[17] J. A. Stratton, *Electromagnetic Theory.* New York: McGraw-Hill, 1941, ch. 7.

[18] C. J. Bouwkamp and H. B. G. Casimir, "On multipole expansions in the theory of electromagnetic radiation," *Physica*, vol. 20, pp. 539–554, 1954.

[19] E. Jahnke and F. Ende, *Tables of Functions.* New York: Dover, 1945.

[20] National Bureau of Standards, *Tables of Spherical Bessel Functions*, vols. 1 and 2. New York: Columbia University Press, 1947.

[21] S. L. Belousov, "Tables of normalized associated Legendre polynomials," in *Mathematical Tables Series*, vol. 18. New York: Macmillan, 1962.

[22] R. O. Gumprecht *et al.*, "Tables of the functions of first and second partial derivatives of Legendre polynomials," Engineering Research Institute, University of Michigan, Ann Arbor, 1951.

[23] T. T. Taylor, "A discussion of the maximum directivity of an antenna," *Proc. IRE* (*Correspondence*), vol. 36, p. 1135, September 1948.

[24] R. E. Collin and S. Rothschild, "Evaluation of antenna Q," *IEEE Trans. Antennas and Propagation*, vol. AP-12, pp. 23–27, January 1964.

[25] R. Joerg Irmer, "Spherical antennas with high gain," *IEEE Trans. Antennas and Propagation* (*Communications*), vol. AP-13, pp. 827–828, September 1965.

[26] L. J. Chu, "Physical limitations of omni-directional antennas," *J. Appl. Phys.*, vol. 19, pp. 1163–1175, December 1948.

[27] Hughes Aircraft Company, Fullerton, Calif., "Final report for low noise, high efficiency Cassegrain antenna studies," Contract NAS5-3282 (prepared for Goddard Space Flight Center, December 1963).

[28] A. C. Ludwig, Ed., "Computer programs for antenna feed system design and analysis," Jet Propulsion Lab., California Institute of Technology, Pasadena, Calif. Tech. Rept. 32-979 (to be published).

[29] S. Silver, Ed., "Microwave antenna theory and design," in *M.I.T. Radiation Laboratory Series.* New York: McGraw-Hill, 1949, p. 439.

On the Equivalent Parabola Technique to Predict the Performance Characteristics of a Cassegrainian System with an Offset Feed

WILLIAM C. WONG

Abstract—**The so-called equivalent parabola technique has been used extensively in the design of Cassegrainian systems. Hannan substantiated the validity of the technique by showing the excellent agreement between his measured data on a Cassegrainian system and the calculated results obtained using the equivalent parabola technique. However, his demonstration was restricted to an on-axis feed Cassegrainian. This paper derives an analytical expression involving only a 2-dimensional vector integral for the secondary vector field from the Cassegrain system with an offset feed by integrating over the surface currents on both the main dish and subdish with the ϕ integrations performed in closed form, and proceeds to compare its results with those obtained from the equivalent parabola technique, for different f/D of the main dish, eccentricities of the subdish, illumination patterns, and lateral displacements of the primary radiator. The agreement in general is good for Cassegrainians with an on-axis feed. For Cassegrainians with an offset feed, the error in using the equivalent parabola technique remains negligible for a beam scan of as much as four beamwidths.**

I. Introduction

THE CASSEGRAINIAN system has been treated analytically by many authors. Some applied the physical optics approximation [2]–[4] to both reflectors, under the assumption that the primary radiator is not laterally defocused, a constraint imposed to enable the ϕ integration to be carried out in closed form, thus reducing the 4-dimensional vector diffraction integral to a much more tractable 2-dimensional one. Others [1] applied the geometrical optics approximation to the subdish, but the physical optics approximation to the main dish. The latter technique is commonly known as the equivalent parabola technique in which the entire dual reflector system is replaced by a single equivalent parabola. The surface of the equivalent parabola is the contour traced out by the intersection of the rays parallel to the axis of the Cassegrain with the extension of the corresponding rays converging toward, or diverging from, the real focal point of the subdish. Since the contour converts a plane wave into a spherical wave or vice versa, it is paraboloidal. Although the dual reflector technique of calculating the secondary diffraction field is more exact, the equivalent parabola technique is several orders of magnitude simpler. The question therefore arises as to what extent the equivalent parabola technique is good, particularly when the primary radiator is laterally defocused. This paper therefore addresses itself to the problems of deriving an expression for the vector fields of a Cassegrainian system with an offset feed and of verifying the validity of the much simplified equivalent parabola technique.

Manuscript received October 5, 1972; revised November 24, 1972.
The author is with the Radio Frequency Laboratory, Systems Group, TRW, Inc., Redondo Beach, Calif. 90278.

II. Analytical Development

In describing a field, it is convenient to use a coordinate system whose origin is in the vicinity of the source, so that by invoking the parallel ray approximation any slight displacement of the source can be approximated by the linear term in the binominal expansion of the distance function in the phase. When the phase change can be approximated by a linear term, i.e., $|\bar{r} - \Delta\bar{r}| \approx r - \overline{\Delta r}\cdot\hat{r}$, the azimuthal integration can be carried out in closed form [6]. Thus the real focal point of a reflector, whether hyperbolic or parabolic, will be taken as the origin of the coordinate system for the incident field. Referring to the coordinate systems shown in Fig. 1, 0 and 0′ are therefore taken as the origins of the coordinate systems describing the fields incident on the subreflector and the main reflector, respectively. In general, the secondary complex vector field scattered from any reflector can be obtained by operating on the reflector surface currents $\bar{J}$ with the free-space tensor Green's function, $\bar{\bar{\Gamma}}$, as follows:

$$\bar{E}(\bar{r}_2) = \int \bar{\bar{\Gamma}}(\bar{r}_2,\bar{r}_2{}')\cdot\bar{J}(\bar{r}_2{}')\,d^2\bar{r}_2{}'. \tag{1}$$

The approximations usually used in the evaluation of this integral are the physical optics approximations to obtain the surface currents, parallel ray approximation to account for the lateral feed displacement, and the far-field approximation to represent the incident field in the vicinity of a reflector as a spherical wave. We shall incorporate these and further assume that the reflector is azimuthally symmetrical. The field from the primary radiator incident upon the subdish is completely arbitrary and it can be represented by a Fourier series

Reprinted from *IEEE Trans. Antennas Propagat.*, vol. AP-21, pp. 335–339, May 1973.

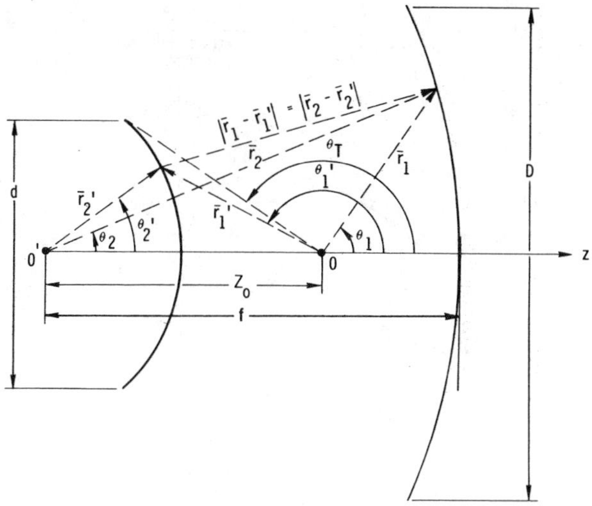

Fig. 1. Geometry of problem.

of the form

$$\bar{E}_i(\bar{r}_1') = \frac{\exp(-jkr_1')}{r_1'} \cdot \Big[\hat{a}_{\theta_1'} \sum_{m=0}^{M_1} (a_m(\theta_1') \sin m\phi_1' + b_m(\theta_1') \cos m\phi_1') + \hat{a}_{\phi_1'} \sum_{m=0}^{M_1} (c_m(\theta_1') \sin m\phi_1' + d_m(\theta_1') \cos m\phi_1')\Big]. \tag{2}$$

The radiation patterns of the open-ended circular waveguide and the conical horn excited in the dominant TE_{11} mode, for example, have a Fourier series representation given by (2), with $M_1 = 1$ and all the expansion coefficients, except b_1 and c_1, being zero. After some manipulations, the scattered vector far field from the subdish can be shown to be given by

$$\bar{E}(\bar{r}_2) = \frac{\exp(-jkr_2)}{r_2} \cdot \left[\hat{a}_{\theta_2} \int_{\theta_i}^{\theta_f} \sum_{m=0}^{M_1} (a_m' \sin m\psi + b_m' \cos m\psi)\, d\theta_{10}' + \hat{a}_{\phi_2} \cdot \int_{\theta_i}^{\theta_f} \sum_{m=0}^{M_1} (c_m' \sin m\psi + d_m' \cos m\psi)\, d\theta_{10}'\right] \tag{3}$$

where

$$a_m' = \gamma_m\{\cos\theta_2 \cos(\phi_2 - \psi) \cdot [a_m\alpha_1(J_{m-1} - J_{m+1}) - d_m g_1(J_{m-1} + J_{m+1})] - \cos\theta_2 \sin(\phi_2 - \psi) \cdot [b_m\alpha_1(J_{m-1} + J_{m+1}) + c_m g_1(J_{m-1} - J_{m+1})] - 2ja_m \sin\theta_2\beta_1 J_m\} \tag{4}$$

$$b_m' = \gamma_m\{\cos\theta_2 \sin(\phi_2 - \psi) \cdot [a_m\alpha_1(J_{m-1} + J_{m+1}) - d_m g_1(J_{m-1} - J_{m+1})] + \cos\theta_2 \cos(\phi_2 - \psi) \cdot [b_m\alpha_1(J_{m-1} - J_{m+1}) + c_m g_1(J_{m-1} + J_{m+1})] - 2jb_m \sin\theta_2\beta_1 J_m\} \tag{5}$$

$$c_m' = \gamma_m\{\sin(\phi_2 - \psi) \cdot [a_m\alpha_1(J_{m+1} - J_{m-1}) + d_m g_1(J_{m+1} + J_{m-1})] - \cos(\phi_2 - \psi) \cdot [b_m\alpha_1(J_{m+1} + J_{m-1}) + c_m g_1(J_{m-1} - J_{m+1})]\} \tag{6}$$

$$d_m' = \gamma_m\{\cos(\phi_2 - \psi) \cdot [a_m\alpha_1(J_{m+1} + J_{m-1}) - d_m g_1(J_{m-1} - J_{m+1})] - \sin(\phi_2 - \psi) \cdot [b_m\alpha_1(J_{m-1} - J_{m+1}) + c_m g_1(J_{m-1} + J_{m+1})]\} \tag{7}$$

$$\gamma_m = -j^m \frac{k}{2} r_{10}'^2 \sin\theta_1' \exp\{-jk[r_1'(1 - \cos\theta_1' \cos\theta_2) - (z_0 + \Delta z)\cos\theta_2]\} \tag{8}$$

$$\alpha_1 = g_1' \sin\theta_{10}' - g_1 \cos\theta_{10}' \tag{9}$$

$$\beta_1 = g_1' \cos\theta_{10}' + g_1 \sin\theta_{10}' \tag{10}$$

$$g_1 = \frac{-1}{r_{10}'(\theta_{10}')} \tag{11}$$

$$\psi = \tan^{-1} \frac{\Delta y + r_1' \sin\theta_2 \sin\phi_2}{\Delta x + r_1' \sin\theta_2 \cos\phi_2} \tag{12}$$

$$A = \frac{[(\Delta y + r_1' \sin\theta_2 \sin\phi_2)^2 + (\Delta x + r_1' \sin\theta_2 \cos\phi_2)^2]^{1/2}}{r_1'} \tag{13}$$

$$r_1' = \left[\frac{1}{g_1^2(\theta_{10}')} + \Delta z^2 + \frac{2\Delta z \cos\theta_{10}'}{g_1(\theta_{10}')}\right]^{1/2} \tag{14}$$

Δx and Δy are lateral feed displacements in the x and y directions, and Δz is the axial feed displacement. The phase term $\exp(+jkz_0 \cos\theta_2)$ in (8) accounts for the additional phase shift when the scattered field is referred to $0'$ rather than 0, the prime on g_1 denotes differentiation with respect to θ_1', r_{10}' is the distance of any point on the reflector from its focal point, and the argument of the Bessel functions is $kr_1'A \sin\theta_1'$.

Since the only restriction that has been placed on the shape of the reflecting surface is that it be azimuthally symmetrical, the scattered field given by (3) is good for both the subreflector and the main reflector, provided that the incident field is cast in the form shown in (2). Thus, if (3) is used to calculate the field scattered from the main reflector, its incident field, hence the field scattered from the subreflector must have a Fourier series representation shown in (2). Inspection of (2) and (3) shows that in the presence of lateral defocusing, the two Fourier series are not of the same form because ψ becomes dependent on both θ_2 and ϕ_2 as shown in (12). Whereas the series given by (2) is separable in θ_1' and ϕ_1', i.e., the Fourier coefficients are functions of θ_1' only and the expansion functions are functions of ϕ_1' only, the series given by (3) is not separable in θ_2 and ϕ_2. In the absence of lateral defocusing, ψ reduces to ϕ_2, all terms involving $\sin(\psi - \phi_2)$ in (4)–(7) vanish. The remaining series becomes separable in θ_2 and ϕ_2 and it also agrees with the results derived by Rusch [5] for an on-axis feed.

To be able to use (3) to calculate the vector field from the main reflector, the field scattered from the subreflector must be cast in the form given by (2), as follows:

$$\bar{E}(\theta_2,\phi_2) = \hat{a}_{\theta_2} \sum_{m=0}^{M_2} (f_m(\theta_2) \sin m\phi_2 + g_m(\theta_2) \cos m\phi_2) + \hat{a}_{\phi_2} \sum_{m=0}^{M_2} (h_m(\theta_2) \sin m\phi_2 + k_m(\theta_2) \cos m\phi_2). \quad (15)$$

In the event of no lateral defocusing, the upper limit of the series is equal that of the series given by (2), i.e., $M_2 = M_1$, but in the presence of lateral defocusing, M_2 depends on the magnitude of the lateral displacement. In general, $M_2 = 16$ has been found to be satisfactory to ensure the Fourier coefficients to be functions of θ_2 only. For a given value of θ_2, the expansion coefficients are given by

$$f_m(\theta_2) = \frac{\langle \bar{E}_2 \cdot \hat{a}_{\theta_2}, \sin m\phi_2 \rangle}{\pi} \quad (16)$$

$$g_m(\theta_2) = \frac{\langle \bar{E}_2 \cdot \hat{a}_{\theta_2}, \cos m\phi_2 \rangle}{\pi(1 + \delta_{m,0})} \quad (17)$$

$$h_m(\theta_2) = \frac{\langle \bar{E}_2 \cdot \hat{a}_{\phi_2}, \sin m\phi_2 \rangle}{\pi} \quad (18)$$

$$k_m(\theta_2) = \frac{\langle \bar{E}_2 \cdot \hat{a}_{\phi_2}, \cos m\phi_2 \rangle}{\pi(1 + \delta_{m,0})}. \quad (19)$$

The angular brackets in (16)–(19) denote integration with respect to ϕ_2 over the range of 2π. The integrals can be approximated by finite series, and the integrand E_2 in the integrals that is given by (3), is sampled at N_2 different values of ϕ_2. It should be noted in passing that if (3) is used to calculate the field scattered from a paraboloid, z_0 in (8) should be set to zero, since for a paraboloid, the incident and the scattered fields in general are referred to coordinates with the same origin.

III. Numerical Results

In order to describe a Cassegrainian system consisting of two reflectors and a cosine to the power n-type feed, no less than six parameters are required. A complete parametric study involving all the six parameters in this limited space is obviously not possible. In what follows, we therefore impose the restrictions that the phase center of the primary radiator is located at the vertex of the main dish and the diameter of the subdish is geometrically determined, i.e., for a given main dish, the subdish is chosen so that the tip angles at the virtual focal point of the subdish subtended by the subdish and the main dish, respectively, are equal.

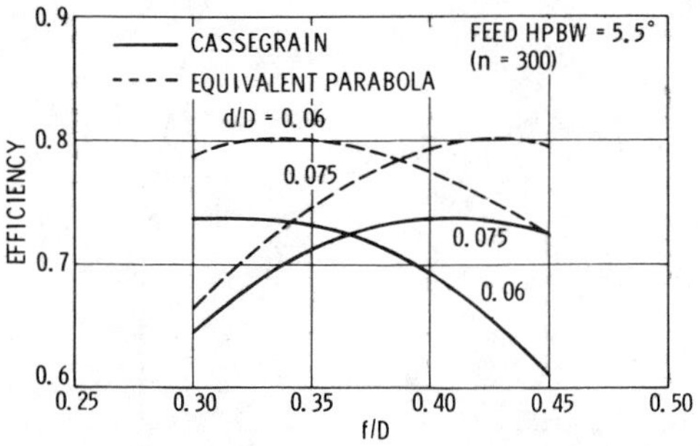

Fig. 2. Effects of f/D and d/D on efficiency for half-power beamwidth = 5.5°.

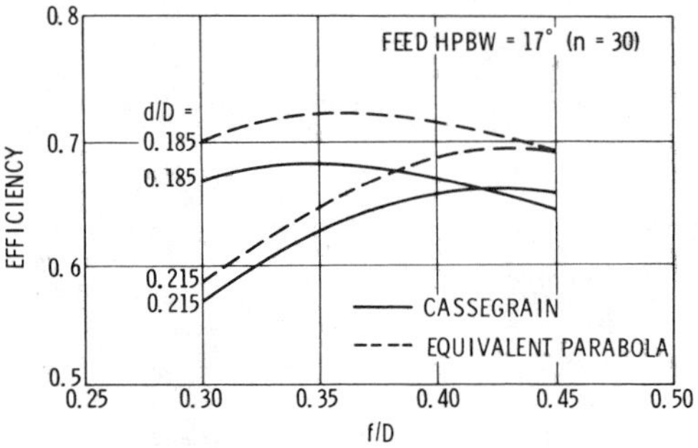

Fig. 3. Effects of f/D and d/D on efficiency for half-power beamwidth = 17°.

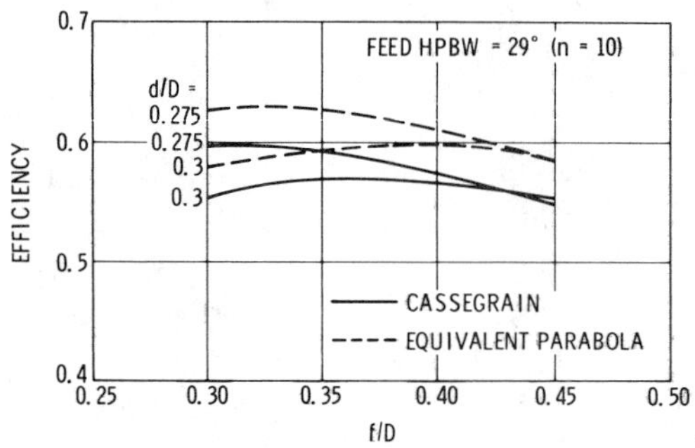

Fig. 4. Effects of f/D and d/D on efficiency for half-power beamwidth = 29°.

Figs. 2–5 were obtained for an on-axis feed Cassegrainian (solid curves) and its equivalent paraboloid (broken curves). The equivalent paraboloid has the same diameter as that of the Cassegrainian main reflector and its equivalent f/D is 0.25 tan $(\theta_T/2)$, where θ_T is the tip angle of the subreflector as defined in Fig. 1. The diameter of the main reflector is 240λ, and the primary excitation field is assumed to have a cosine to the power n type

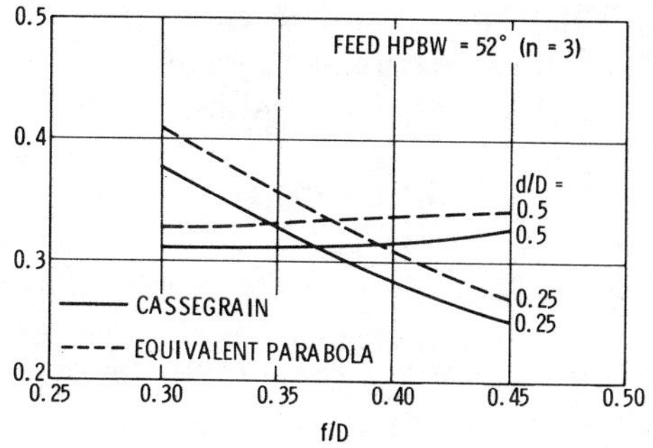

Fig. 5. Effects of f/D and d/D on efficiency for half-power beamwidth = 52°.

pattern so that all Fourier coefficients in (2) are zero except for b_1 and c_1, which are $\cos^n(\pi - \theta_1')$. The efficiency of the Cassegrainian and that of the equivalent paraboloid, referred to a uniformly illuminated paraboloid of the same diameter, are plotted against the f/D of the main reflector with the aperture blockage (d/D) and the illumination pattern beamwidth as parameters.

Since the subreflector merely plays the role of shifting the phase center of a spherical wave from the real focal point to the virtual focal point, reducing the edge taper of the beam incident on the subreflector also reduces the taper of the scattered beam. Thus for a given aperture blockage (d/D), increasing the f/D raises the illumination efficiency, increases the energy spillover at both reflectors, and decreases the effects of the blockage presented by the subreflector. For small d/D, the blockage effects are negligible, and the illumination efficiency and spillover are the only opposing forces affecting the antenna efficiency. Fig. 2 shows that for every d/D, there is a peak efficiency occurring at some optimum f/D. Below the optimum f/D, the illumination efficiency dominates and above the optimum f/D, the spillover dominates. That the blockage effects are negligible is evidenced by the fact that the peak efficiency is relatively constant for different d/D. At larger d/D, blockage effects become comparable to those of illumination and spillover. With three factors affecting the antenna efficiency, there are two optimum efficiencies, generally of different magnitudes, as evidenced by Fig. 5. The lowering of the peak efficiency with increasing d/D is clearly due to increased blockage.

Inspection of Figs. 2–5 shows that in all the cases considered, the difference between the efficiencies, obtained using the dual reflector Cassegrainian and its equivalent parabola, is less than 10 percent. The effect of lateral defocusing upon the secondary patterns, using, respectively, the complete Cassegrainian formulation and the equivalent parabola technique, is shown in Figs. 6 and 7. The Cassegrainian under investigation has a $D/\lambda = 253.5$, $d/D = 0.10414$, $f/D = 0.333$, eccentricity = 1.403, $\theta_T = 165.663°$, $\cos^3(\pi - \theta_1')$ illumination. The equivalent parabola has an f/D of 1.987. The primary radiator is laterally displaced with $\Delta x = 2\lambda$, 4λ, and 8λ.

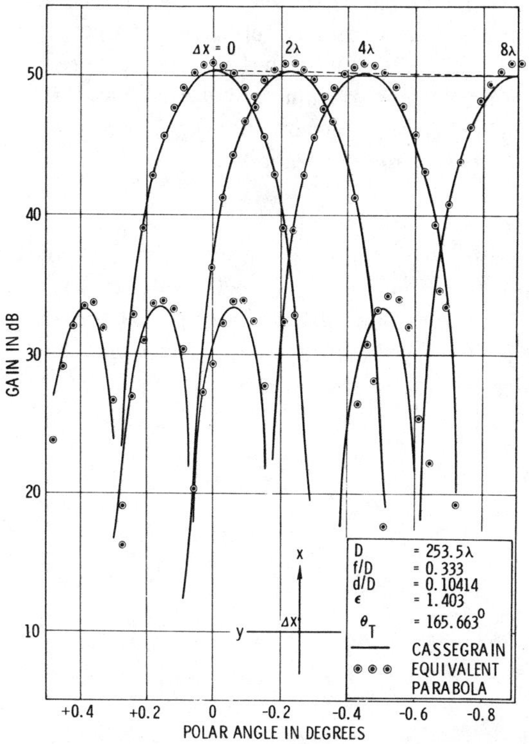

Fig. 6. Effects of lateral defocusing on secondary pattern.

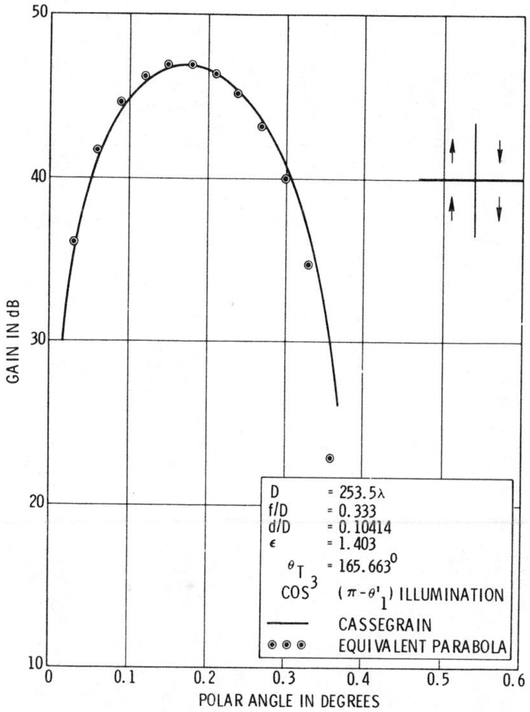

Fig. 7. Difference patterns in H plane.

An E-plane cut (Fig. 6) was obtained for both the Cassegrainian (solid curve) and its equivalent paraboloid (broken curves). The gain discrepancy remains small after the beam has been scanned by about four beamwidths. When four primary radiators with proper phasing and with $\Delta x = 2\lambda$ for each feed are clustered to form an amplitude comparison monopulse system, the resultant difference patterns are shown in Fig. 7, which shows that the gain discrepancy is negligible.

IV. Conclusion

A useful expression has been derived to obtain the vector diffraction field of a reflector antenna with an offset feed. Both the sum and difference patterns of a well designed Cassegrainian system can be predicted with a fair degree of accuracy using the equivalent parabola technique. The error in gain resulting from the use of the simplified technique remains small when the beam is squinted by about four beamwidths.

References

[1] P. W. Hannan, "Microwave antennas derived from the Cassegrain telescope," *IRE Trans. Antennas Propagat.*, pp. 140–153, Mar. 1961.

[2] W. V. T. Rusch, "Scattering from a hyperboloidal reflector in a Cassegrainian feed system," *IEEE Trans. Antennas Propagat.*, vol. AP-11, pp. 414–421, July 1963.

[3] A. Ludwig, "Calculation of scattered patterns from asymmetrical reflectors," Jet Propulsion Lab., Tech. Rep. 32–1430, Feb. 1970.

[4] H. Zucker and W. H. Ierley, "Computer-aided analysis of Cassegrain antennas," *Bell Syst. Tech. J.*, vol. 47, July–Aug. 1968.

[5] W. V. T. Rusch, *Analysis of Reflector Antennas*. New York: Academic Press, 1970, pp. 133–135.

[6] P. G. Ingerson and W. C. Wong, "The analysis of deployable umbrella parabolic reflectors," *IEEE Trans. Antennas Propagat.*, vol. AP-20, pp. 409–414, July 1972.

DESIGN OF CASSEGRAIN ANTENNAS EMPLOYING DIELECTRIC CONE FEEDS

Indexing terms: Reflector antennas, Antenna feeders, Microwave antennas, Antenna radiation patterns

A design procedure is described for a Cassegrain antenna employing a dielectric cone feed. Modal methods are used to determine the cone geometry, while a geometric-optics method is used to determine the subreflector profile. Good agreement is observed between measured and predicted radiation patterns for a 1·2 m-diameter parabolic antenna utilising the feed and operating at a frequency near 11 GHz.

The use of a low-permittivity dielectric cone to reduce spill-over in a Cassegrain antenna as shown in Fig. 1*a* was first proposed by Bartlett and Mosely[1] in 1966. However, as far as we are aware, no design theory has been presented previously. Two of the present authors have recently analysed the propagation and radiation characteristics of the dielectric cone feed,[2-4] and, in this letter, we describe our analysis of the Cassegrain antenna. Satisfactory agreement with experimental measurements has been obtained for an antenna operating near 11 GHz, which inspires confidence in the design procedure. Table 1 summarises the procedure.

The computation of the transverse-field distribution of the HE_{11} mode at the subdish is based on a theory which we have described previously.[2] Provided that the cone semi-angle θ_1 is small, we regard the cone as a cylinder of radius r_1, an assumption we have justified. To determine the subreflector profile, we use a geometric-optics method (see Fig. 1*d*). We assume that rays emerge from the cone apex, suffer reflection at the subdish and refraction at the cone–air interface. We then impose the condition that they should emerge from the cone as if they came from the paraboloid focus. Thus we obtain eqns. 1–3, which may be solved numerically to yield the subreflector polar co-ordinates R, θ as β is varied:

$$\frac{dR}{d\beta} = \frac{d\theta}{d\beta} R \tan \tfrac{1}{2}(\theta + \beta + \delta) \qquad (1)$$

$$(\theta + \beta + \delta) = \sin^{-1}\left[\frac{S}{R}\left\{\frac{\tan\beta}{\sec(\beta+\delta)}\right\}\left\{\frac{\tan(\beta+\delta)+\tan\theta_1}{\tan\beta+\tan\theta_1}\right\}\right] \qquad (2)$$

$$(\theta_1 + \beta + \delta) = \cos^{-1}\left\{\frac{\cos(\theta_1+\beta)}{(\bar{\varepsilon})^{\frac{1}{2}}}\right\} \qquad (3)$$

We emphasise that the aperture distribution $A(r)$, which can be obtained using an additional power-conservation equation in the above geometric-optics method, depends on the initial choice of parameters. We therefore examine $A(r)$, and, if the distribution is unsatisfactory, a modification to certain parameters must be made. For a small change in distribution, it is convenient to alter θ_1. However, if a large change is required, a new value of subreflector diameter D_s must be selected and the procedures of Table 1 repeated.

Following these procedures, we have designed a Cassegrain antenna with $D = 1{\cdot}2$ m for use at frequencies near

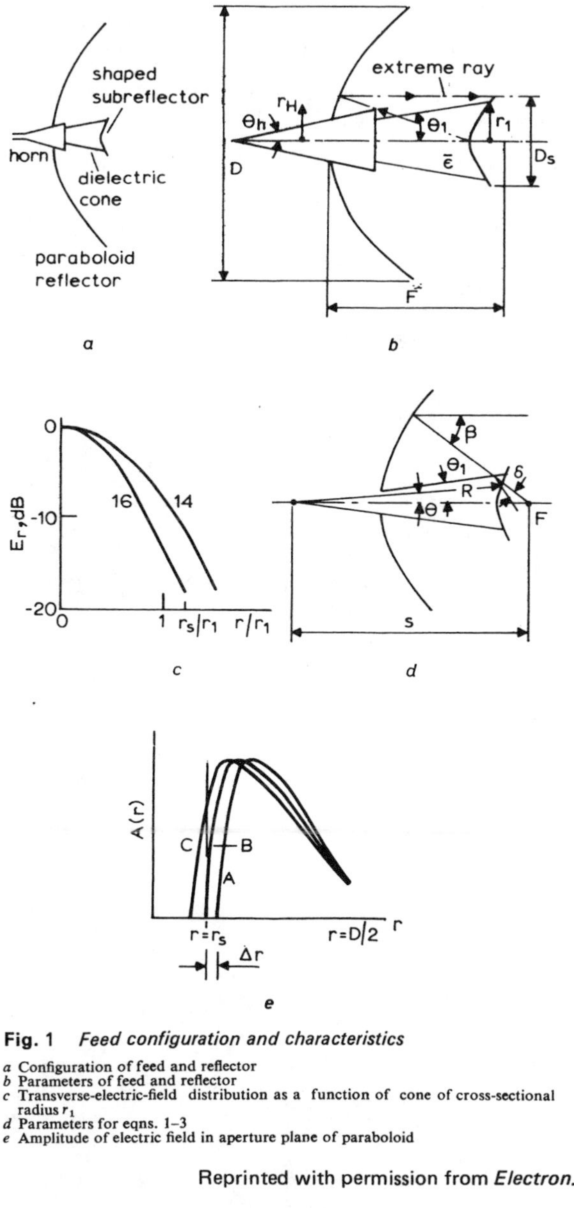

Fig. 1 *Feed configuration and characteristics*

a Configuration of feed and reflector
b Parameters of feed and reflector
c Transverse-electric-field distribution as a function of cone of cross-sectional radius r_1
d Parameters for eqns. 1–3
e Amplitude of electric field in aperture plane of paraboloid

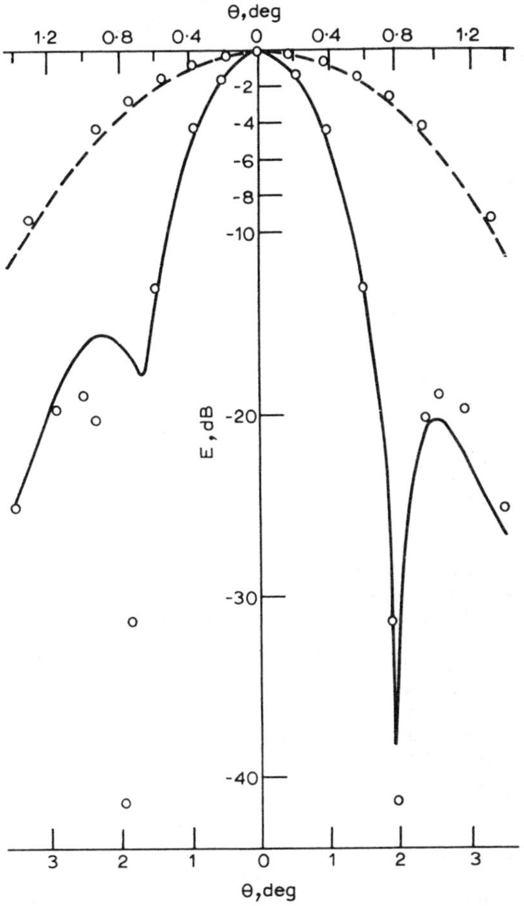

Fig. 2 *Measured and predicted E and H plane radiation patterns of Cassegrain antenna*

Principal parameters: $F/D = 0{\cdot}25$, $D = 1{\cdot}2$ m, $f = 10{\cdot}5$ GHz, $\bar{\varepsilon} = 1{\cdot}1$
The *E* plane pattern is on the left-hand side
The *H* plane pattern is on the right-hand side
—— measured
○○○ predicted
– – – refers to expanded scale

Table 1 DESIGN PROCEDURE FOR CASSEGRAIN ANTENNAS WITH DIELECTRIC CONE FEEDS

Procedure	Figure	Comment
(*a*) Select λ, D and F	1*b*	
(*b*) Select D_s	1*b*	D_s/D should be small for low aperture blockage; however, for the geometric-optics method of procedure 7 to apply, D_s/λ should exceed about 6.
(*c*) Select edge taper E		For a large value of the ratio of forward to backward radiation G_{fb}, E should be large. Following Reference 5, $G_{fb} = 20 \log_{10}(D/4F) - E + G$, where the terms E and G are expressed in decibels, G is the gain of the antenna on boresight and E is the subreflector edge taper. We have observed for a wide range of values of F/D that the subreflector-edge illumination is almost exactly identical to the main reflector-edge illumination, including the influence of space attenuation.
(*d*) Select ε, the permittivity of the cone		If $\varepsilon - \varepsilon_0$ is too small, excessive spillover will occur, and we suggest a lower limit for $\varepsilon/\varepsilon_0 = \bar{\varepsilon}$ of 1·04. An upper limit is imposed of about 1·4 in order to reduce the effects of higher-order modes.
(*e*) Determine the maximum radius r_1 of the cone cross-section	1*c*	From computed field distributions, r_1 is selected to ensure the required edge taper E at $r = r_1$
(*f*) Select θ_1, the cone semiangle	1*b*	θ_1 must be less than the critical angle for total internal reflection θ_c. The minimum value of θ_1 is then subject to a constraint on the location of the cone apex. The maximum value of $\theta_1 < \theta_c$ is imposed to prevent extreme rays from intersecting the feed-horn aperture after reflection at the subdish, since, for θ_1 large, the horn and subreflector are close together.
(*g*) Determine the subreflector profile	1*d*	The profile is obtained from a solution of eqns. 1–3. With the correct profile, a constant phase is maintained over the aperture of the paraboloid. The aperture distribution $A(r)$ is also obtained.
(*h*) Examine $A(r)$ from the standpoint of blockage	1*e*	Fig. 1*e* shows three possible distributions. In A, the extreme ray of Fig. 1*b* clears the subreflector, and, if Δr is too large, there is poor utilisation of the reflector and high sidelobes. In C, the extreme ray intersects the subdish; in this case, the input v.s.w.r. could be high. Distribution B represents a near optimum result.
(*i*) Select horn radius at aperture r_h for maximum efficiency		On the assumption of a cylindrical dielectric, the excitation efficiency has been computed using modal-matching techniques. For given ε and r_1, r_h is selected for maximum excitation efficiency of the HE_{11} mode assuming an H_{11} mode in the horn.
(*j*) Select horn semiangle θ_h		Ideally, the horn and cone apex should coincide for maximum excitation efficiency.

11 GHz. The subreflector was made from glass fibre and was glued to the dielectric cone.

Fig. 2 shows computed and measured E and H plane radiation patterns at 10·5 GHz. The predicted gain of 40·73 dB and efficiency of 65·5% agrees closely with the average gain figure of 40·71 dB and efficiency of 65·2% based on measurements.

The radiation patterns agree very closely over the main-lobe region, although there is some unexplained pattern asymmetry in the sidelobe regime.

The radiation patterns were measured on a high-performance antenna range, while gain was obtained using pattern-integration methods. The gain we have observed is about 1 dB higher than that normally obtained with a front-fed paraboloid of comparable F/D and D/λ. Also, because of the low F/D ratio (0·25) and high edge taper ($E = -20$ dB), the front/back ratio is high, typically 62 dB, while the level of wide-angle sidelobes is low. Also, typical coupling between two such antennas mounted adjacently has been observed to be -100 dB, a figure made possible by the high edge taper.

In conclusion, we have shown that, by combining modal and geometric-optics methods of design, a Cassegrain antenna, using a dielectric cone to support the subreflector, can be constructed having a performance close to that predicted theoretically.

We gratefully acknowledge W. C. Morgan's help in making available paraboloid reflectors, the range facilities of Andrew Antenna Systems, Lochgelly, Scotland, and K. R. Slinn for his assistance with the measurements. We also wish to thank our colleagues A. D. Olver, K. B. Chan and G. Poulton for many helpful suggestions, and J. Honour and J. Morris for their care in constructing the dielectric cones and subreflectors. C. E. R. C. Salema is indebted to the Gulbenkian Foundation and the Instituto de Alta Cultura, Portugal, for a research fellowship.

P. J. B. CLARRICOATS
C. E. R. C. SALEMA
S. H. LIM

3rd July 1972

Department of Electrical & Electronic Engineering
Queen Mary College
Mile End Road
London E1 4NS, England

References

1 BARTLETT, H. E., and MOSELEY, R. E.: 'Dielguides—highly efficient low noise antenna feeds', *Microwave J.*, 1966, **9**, pp. 53–58
2 CLARRICOATS, P. J. B., and SALEMA, C. E. R. C.: 'Propagation characteristics of low-permittivity cones', *Electron. Lett.*, 1971, **7**, pp. 483–485
3 CLARRICOATS, P. J. B., and SALEMA, C. E. R. C.: 'Design of dielectric cone feeds for microwave antennas'. Proceedings of the 1971 European microwave conference, paper B5/4
4 CLARRICOATS, P. J. B., and SALEMA, C. E. R. C.: 'Influence of launching horn on radiation characteristics of a dielectric cone feed', *Electron. Lett.*, 1972, **8**, pp. 200–202
5 KRITIKOS, H. N.: 'The extended aperture method for the determination of the shadow region radiation of parabolic reflectors', *IEEE Trans.*, 1963, **AP–11**, pp. 700–704

Design of Dual-Reflector Antennas with Arbitrary Phase and Amplitude Distributions

VICTOR GALINDO, STUDENT, IEEE

Summary—A synthesis method based on geometrical optics for designing a dual-reflector antenna system with an arbitrary phase and amplitude distribution in the aperture of the second reflector is presented. The first reflector may be illuminated by a pattern with an arbitrarily curved phase front. A pair of first-order ordinary nonlinear differential equations of the form $dy/dx = f(x, y)$ are developed for the system. Questions concerning uniqueness, existence and bounds for the solutions can be answered. Calculations and numerical results for the design of a uniform amplitude and phase dual-reflector system are presented.

Manuscript received March 14, 1963; revised November 14, 1963. The research reported in this paper was made possible in part through support received from the Departments of Army, Navy and Air Force Office of Scientific Research under grant AF-AFOSR-62-340.

The author is with the Department of Electrical Engineering, University of California, Berkeley, Calif.

I. Introduction

CONVENTIONAL dual-reflector antenna systems have been based largely on the Cassegrain parabola-hyperbola design or the Gregorian parabola-ellipse design.[1] Some highly specialized exceptions

[1] P. W. Hannon, "Microwave antennas derived from the Cassegrain telescope," IRE Trans. on Antennas and Propagation, vol. AP-9, pp. 140–153; March, 1961.

Reprinted from *IEEE Trans. Antennas Propagat.*, vol. AP-12, pp. 403–408, July 1964.

have also been reported.[2] The designs are all based on the principles of geometrical optics and are limited accordingly. That is, the reflectors must be large and have a large radius of curvature compared to the wavelength in addition to other restrictions.

A generalization of the design techniques for a dual-reflector antenna that is also based on geometrical optics and is similarly limited in application is presented here. It has been found that an arbitrary phase and amplitude distribution can be developed in the aperture of the larger reflector of a dual-reflector system with an arbitrarily curved phase front illuminating the smaller reflector. The design procedure is sufficiently general so that many useful variations of the design objectives are clearly possible. For example, the large reflector of the system may be specified to have a circular cross section and arbitrary phase distribution (uniform, for example) and the appropriate subreflector is then found for a given primary radiation pattern.

The synthesis method utilizes the analytical expression of the geometric optics principles together with a geometry for the reflectors, such as those illustrated in Fig. 1, to develop a pair of first-order nonlinear ordinary differential equations of the form

$$\frac{dy}{dx} = f(x, y) \tag{1}$$

which lead to the cross sections of each reflector when subject to boundary conditions such as

$$y(x = x_{\max}) = 0. \tag{2}$$

The above differential equation can, in general, be solved readily by high-speed machine computations. A pair of such solutions for a uniform phase and amplitude design is presented.

A desirable feature of the form of the differential solution (1) is that frequently much information can be predicted about the final solution before a machine computation is attempted. The questions of existence and upper bounds on the size of the reflector ($|y|_{max}$) can be answered by considering little more than the boundary conditions of the problem. If $f(x, y)$ is found in a suitable form, then lower bounds for the radius of curvature of the reflectors can also be found. The mathematical details of the above procedure for finding the bounds, existence and uniqueness proofs, etc., are left to any one of many discussions on the subject in the literature.[3] An alternative formulation of the problem is found in Galindo.[4]

[2] S. Silver, "Microwave Antenna Theory and Design," McGraw-Hill Book Co., Inc., New York, N. Y., sec. 13.9; 1949.

[3] H. T. Davis, "Introduction to Non-Linear Differential and Integral Equations," U. S. Atomic Energy Commission, Washington, D. C., pp. 88–93; September, 1960.

[4] V. Galindo, "Design of dual reflector antennas with arbitrary phase and amplitude distributions," *PTGAP International Symposium Proceedings*, pp. 91–95; July, 1963.

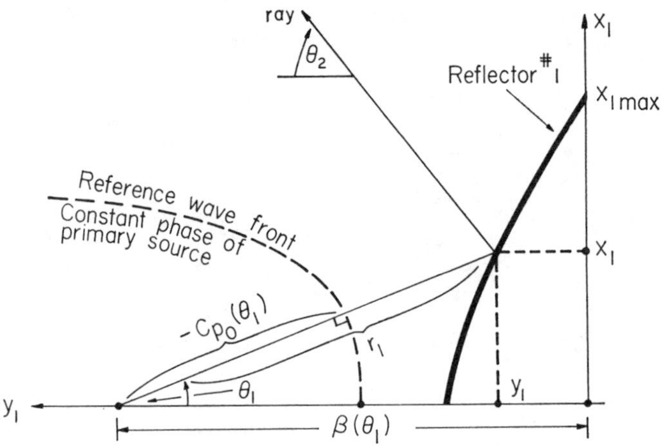

Fig. 1—Constant phase front reference of $I_1(\theta_1)$.

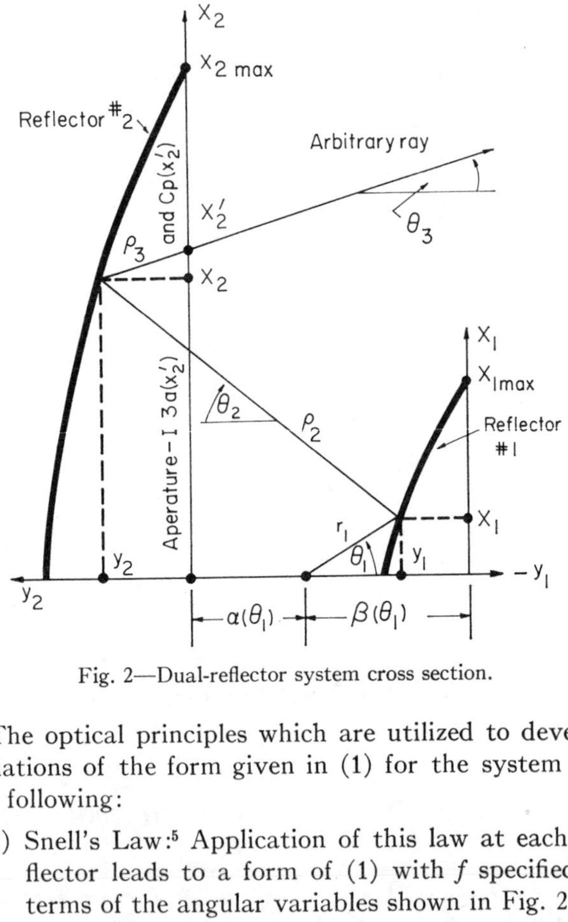

Fig. 2—Dual-reflector system cross section.

The optical principles which are utilized to develop equations of the form given in (1) for the system are the following:

1) Snell's Law:[5] Application of this law at each reflector leads to a form of (1) with f specified in terms of the angular variables shown in Fig. 2.
2) Conservation of energy flow[6] along the ray trajectories: The requirement that the energy flow be conserved in any solid angle bounded by ray trajectories leads to one equation of constraint for the system.
3) The surfaces of constant phase form normal surfaces[7] to the ray trajectories, and this normal congruence is maintained after any number of reflec-

[5] Silver, *op. cit.*, sec. 4.8.

[6] *Ibid.*, sec. 4.4.

[7] *Ibid.*, sec. 4.2.

tions[8] (theorem of Malus). This principle leads to another constraint for the system.

It can be shown that these principles are not independent.[2,5–8] The equations describing these conditions separately will, however, be used to obtain a solution. The optical principles together with the geometrical bounds of the system and the primary pattern, and the phase and amplitude distributions in the large reflector define a unique optical system (sometimes equivalent alternative systems exist, as with the Cassegrain and Gregorian systems).

The design method is described for surfaces of revolution in particular, although the antenna design method can readily be extended to include doubly curved surfaces where the surface coordinates are separable, and possibly also under other special conditions.[9] When the curvature in one plane is specified for the dual-reflector system, the design procedure for the curvature in the perpendicular plane can be the same as reported in this paper with an extension similar to that made by Dunbar[10] for a single doubly curved reflector.

II. Analytic Expressions of Optical Principles and Formal Solution

The principles stated earlier can be expressed as the following basic equations from which the solution is developed.

Snell's law at reflector number 1 leads to

$$\frac{dy_1}{dx_1} = \tan\left[\frac{\theta_1 - \theta_2(x_1, y_1, x_2, y_2)}{2}\right] \tag{3}$$

where

$$\theta_2 = \arc\tan\left(\frac{x_2 - x_1}{\alpha + \beta + y_2 - y_1}\right). \tag{4}$$

The quantities $\alpha(\theta_1)$ and $\beta(\theta_1)$ are prescribed such that $\alpha(\theta_1) + \beta(\theta_1) =$ given constant (see Fig. 2). The intersection of a primary field ray with the y_1 axis determines $\alpha(\theta_1)$ and $\beta(\theta_1)$. The rays are found from the known or given form of the primary field which arises from an arbitrary, not necessarily a point, source.

Snell's law at reflector number 2 leads to

$$\frac{dy_2}{dx_2} = -\tan\left[\frac{\theta_2 - \theta_3}{2}\right] \tag{5}$$

where

$$\theta_3 = \arc\tan\left(\frac{x_2' - x_2}{y_2}\right). \tag{6}$$

The conservation of energy principle is expressed differentially as

$$I_1(\theta_1)\sin\theta_1\left(\frac{d\theta_1}{dx_2'}\right) = I_{3a}(x_2')x_2'. \tag{7}$$

The quantity $I_1(\theta_1)$ is the power density of the primary illumination. The quantity $I_{3a}(x_2')$ is the power density flow normal to the aperture of reflector number 2 ($y_2 = 0$). It is chosen arbitrarily except for the required normalization

$$\int_{\theta_1 \min}^{\theta_1 \max} I_1(\theta_1)\sin\theta_1 d\theta_1 = \int_{x_2' \min}^{x_2' \max} I_{3a}(x_2')x_2' dx_2'. \tag{8}$$

The phase at the aperture can be assigned arbitrarily as $C_p(x_2')$ (we assume $k = 2\pi/\lambda = 1$). The path length from the primary illumination is given by

$$C_p(x_2') = r_1(\theta_1, y_1) + \rho_2(\theta_1, y_1, x_2, y_2) + \rho_3(x_2, y_2; x_2') + C_{p_0}(\theta_1). \tag{9}$$

The quantity $C_{p_0}(\theta_1)$ is defined by the primary illumination as illustrated in Fig. 1. An arbitrary primary illumination $[I_1(\theta_1)]$ phase front is chosen as a reference. The aperture phase $C_p(x_2')$ must be defined with respect to this reference. The remaining quantities in (9) are easily found as

$$r_1 = [\beta(\theta_1) - y_1]\sec\theta_1 \tag{10}$$

$$\rho_2 = \sqrt{(x_2 - x_1)^2 + (\alpha + \beta + y_2 - y_1)^2} \tag{11}$$

and

$$\rho_3 = \sqrt{(x_2' - x_2)^2 + y_2^2} = \frac{y_2}{\sqrt{1 - \left(\frac{dC_p}{dx_2'}\right)^2}}. \tag{12}$$

The right-hand side of (12) is found by expressing the theorem of Malus from the diagram of Fig. 3 as

$$\frac{dC_p}{dx_2'} = \sin\theta_3. \tag{13}$$

Since $C_p(x_2')$ is assumed given, then $\theta_3(x_2')$ is thus determined directly.

We will now choose as a single independent variable the quantity x_2'. The remaining dependent variables will be considered functions of x_2'. We will then derive the two differential equations

$$\frac{d\theta_1}{dx_2'} = f_{\theta_1}(\theta_1; x_2') \tag{14}$$

and

$$\frac{dy_1}{dx_2'} = f_{y_1}(\theta_1, y_1; x_2'). \tag{15}$$

[8] *Ibid.*, sec. 4.9.
[9] B. Ye Kimber, "On two reflector antennas," *Radio Eng. Electron. Phys.*, vol. 6, pp. 914–921; June, 1962.
[10] A. Dunbar, "Calculation of doubly curved reflectors for shaped beams," Proc. IRE, vol. 36, pp. 1289–1296; October, 1948.

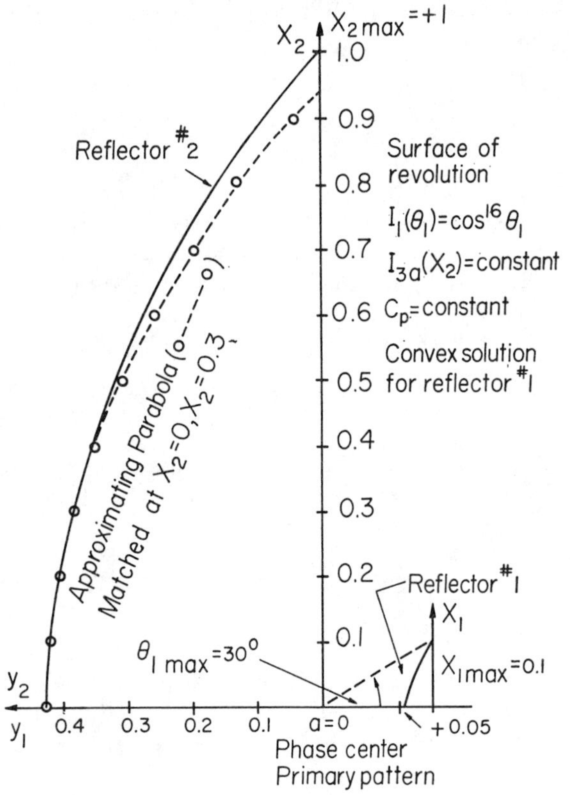

Fig. 3—Dual-reflector uniform phase-amplitude convex solution.

Upon solving these equations all the remaining dependent variables can be immediately found as functions of x_2'. Although the choice of x_2' as the independent variable and θ_1 and y_1 for the differential equations is somewhat arbitrary, this choice does allow all the following expressions (f_{θ_1}, f_y, etc.) to be found in explicit form. Furthermore, the same formulation is used for an extension of the synthesis which accounts to some extent for the diffraction effects of the small reflector.[11] An equivalent formulation in terms of $dy_1/dx_1 = f_1(x_1, y_1)$ and $dy_2/dx_2 = f_2(x_2; y_2)$ can also be found.[4]

The quantity $d\theta_1/dx_2'$ is found directly from (7) as

$$\frac{d\theta_1}{dx_2'} = f_{\theta_1}(\theta_1; x_2') = \frac{I_{3a}(x_2')x_2'}{I_1(\theta_1)\sin\theta_1}. \tag{16}$$

This equation can be solved immediately (by numerical methods) subject to the appropriate boundary conditions (see Fig. 1 and Davis,[3] for example). If $I_1(\theta_1)\sin\theta_1$ and $I_{3a}(x_2')x_2'$ are integrable, then we can solve (16) very easily as

$$\int_{\theta_{1\,\min}}^{\theta_1} I_1(\theta_1)\sin\theta_1 d\theta_1 = P_1(\theta_1) = \int_{x_{2\,\min}}^{x_2'} I_{3a}(x_2')x_2' dx_2' = P_{3a}(x_2'). \tag{17}$$

This integral solution will be of especial value if we can find $\theta_1(x_2')$ explicitly from (17). Otherwise it appears just as well to solve (16) since the numerical solution can be used simultaneously to solve (15). That is, the solutions of (14) and (15) can be set up on the same computer program.

Before finding dy_1/dx_2' explicitly we will find dx_1/dx_2' as a function of $(\theta_1, y_1, dy_1/dx_2'; x_2')$, $x_1(\theta_1, y_1)$, $x_2(\theta_1, y_1; x_2')$, and $y_2(\theta_1, y_1; x_2')$ explicitly. Upon finding dy_1/dx_2' we will then have obtained a complete solution for the system.

From the diagrams of reflector number 1 it is seen directly that

$$x_1 = x_1(\theta_1, y_1) = [\beta(\theta_1) - y_1]\tan\theta_1. \tag{18}$$

To find $dx_1/dx_2'(\theta_1, y_1, dy_1/dx_2'; x_2')$ we use

$$\frac{dx_1}{dx_2'} = \frac{\partial x_1}{\partial\theta_1}\frac{d\theta_1}{dx_2'} + \frac{\partial x_1}{\partial y_1}\frac{dy_1}{dx_2'} \tag{19}$$

with

$$\frac{\partial x_1}{\partial\theta_1} = \beta_{\theta_1}\tan\theta_1 + (\beta - y_1)\sec^2\theta_1 \tag{20}$$

where $\beta_{\theta_1} = d\beta/d\theta_1$ and

$$\frac{\partial x_1}{\partial y_1} = -\tan\theta_1. \tag{21}$$

We can find $y_2(\theta_1, y_1; x_2')$ and $x_2(\theta_1, y_1; x_2')$ explicitly by solving (9) and (13) simultaneously. This gives

$$y_2 = y_2(\theta_1, y_1; x_2') = -\frac{[(x_1 - x_2')^2 + (\alpha + \beta - y_1)^2 - B^2]}{2\left[(\alpha + \beta - y_1) + \dfrac{\dfrac{dC_p}{dx_2'}(x_1 - x_2') + B}{\sqrt{1 - \left(\dfrac{dC_p}{dx_2'}\right)^2}}\right]} \tag{22}$$

and

$$x_2 = x_2(\theta_1, y_1; x_2') = x_2' + \frac{[(x_1 - x_2')^2 + (\alpha + \beta - y_1)^2 - B^2]}{2\left[(x_1 - x_2') + \dfrac{(\alpha + \beta - y_1)\sqrt{1 - \left(\dfrac{dC_p}{dx_2'}\right)^2} + B}{\left(\dfrac{dC_p}{dx_2'}\right)}\right]} \tag{23}$$

[11] V. Galindo and W. J. Welch, Electronics Research Laboratory, University of California, Berkeley, Consolidated Quarterly Progress Rept. No. 11; November 15, 1963.

with

$$B = C_p - (\beta - y_1) \sec \theta_1 - C_{p0}(\theta_1). \qquad (24)$$

Now from (3) and (4) we obtain

$$\frac{dy_1}{dx_2'} = f_1(\theta_1, y_1; x_2') \frac{dx_1}{dx_2'} \qquad (25)$$

where

$$\theta_1, y_1; x_2') = \tan \left[\frac{\theta_1 - \arctan\left(\dfrac{x_2 - x_1}{\alpha + \beta + y_2 - y_1}\right)}{2} \right]. \qquad (26)$$

Substituting (18)–(24) into (25) and (26) we obtain explicitly

$$\frac{dy_1}{dx_2'} = f_{y_1}(\theta_1, y_1; x_2') = \frac{[\beta_{\theta_1} \tan \theta_1 + (\beta - y_1) \sec^2 \theta_1] f_{\theta_1} f_1}{(1 + \tan \theta_1 f_1)}. \qquad (27)$$

Upon solving (27) for $y_1(x_2')$ and either (16) or (17) for $\theta_1(x_2')$, all the remaining dependent variables, θ_2, x_1, y_2 and x_2, can be found also as functions of x_2' by applications of (4), (18), (22) and (23), respectively.

III. Note on Existence, Bounds, and Uniqueness

For the purpose of determining existence, uniqueness, and bounds (for y_1 and y_2) of the solution, it is convenient to use the form of the equations given in (3) and (5).[4] For given boundary conditions of the problem, $x_{1\,\max}$, $x_{1\,\min}$, $\theta_{1\,\max}$, $\theta_{1\,\min}$, $\theta_{2\,\max}$, $\theta_{2\,\min}$, $x_{2\,\max}$, $x_{2\,\min}$; $\alpha(\theta_{1\,\max})$, $\beta(\theta_{1\,\max})$, etc. (these conditions are not independent as is easily seen from Fig. 1); it is frequently possible to estimate bounds for the solutions $y_1(x_1)$ and $y_2(x_2)$ in addition to predicting uniqueness merely by using the forms given in (3) and (5). The methods for employing this useful property are easily seen from a study of the theory of these equations such as given in Davis.[3]

IV. Uniform Phase and Amplitude Examples

The solution for the dual-reflector system which will produce a uniform phase and amplitude distribution in the aperture of reflector Number 2 is found for a primary illumination $I_1(\theta_1) = \cos^n \theta_1$ on a surface of revolution. It is assumed that the primary phase fronts are circular with center location given by the *constants* α and β. In this case we will take the phase center of primary radiation as the phase reference for the system so that $C_{p0}(\theta_1) = 0$. The aperture amplitude distribution will be given by $I_{3a}(x_2') =$ constant to be found from (8). The aperture phase distribution $C_p(x_2')$ is also given as a constant which is determined from (9) and the boundary conditions. We may note immediately that in this case

$$[\theta_3 = 0 \text{ and } x_2' = x_2]. \qquad (28)$$

The boundary conditions will be chosen so that

$$\theta_{1\,\min} = x'_{2\,\min} = x_{2\,\min} = 0.$$

Four more independent boundary conditions are available. We will choose

$$x_{2\,\max},\ x_{1\,\max},\ \theta_{1\,\max} \text{ and } \alpha.$$

The remaining conditions are found from Fig. 1 as

$$\left\{ \begin{aligned} \beta &= \frac{x_{1\,\max}}{\tan \theta_{1\,\max}} \\ \theta_{2\,\max} &= \arctan \left[\frac{x_{2\,\max} - x_{1\,\max}}{\alpha + \beta} \right] \\ C_p(x_2') &= \left(\frac{\beta}{\cos \theta_{1\,\max}}\right) + \left(\frac{\alpha + \beta}{\cos \theta_{2\,\max}}\right) \end{aligned} \right\} \qquad (29)$$

Since the over-all scale of the geometry is arbitrary, the value of $x_{2\,\max}$ will be chosen as either

1) $x_{2\,\max} = +1$, which leads to a convex type of solution for reflector number 1 $[y_1(x_1)]$, or
2) $x_{2\,\max} = -1$, which leads to a concave type of solution for reflector number 1.

In this example we find that (17) is integrable and that $\theta_1(x_2') = \theta_1(x_2)$ can be found explicitly as

$$\theta_1 = \arccos \left(\sqrt[n+1]{1 - M x_2^2}\right) \qquad (30)$$

with

$$M = 1 - \cos^{n+1} \theta_{1\,\max}. \qquad (31)$$

From (22), with $dC_p/dx_2' = \sin \theta_3 = 0$ and $C_{p0}(\theta_1) = 0$, we obtain

$$y_2 = y_2(\theta_1, y_1; x_2') = \frac{(C_p - r_1)^2 - (x_2' - x_1)^2 - (\alpha + \beta - y_1)^2}{2(\alpha + \beta - y_1 - C_p - r_1)} \qquad (32)$$

where r_1 is given by (10).

It is now possible to construct the function [(27)]

$$\frac{dy_1}{dx_2'} = f_{y_1}(\theta_1, y_1; x_2') = f_{y_1}(y_1; x_2') \qquad (33)$$

explicitly and solve for y_1, x_1, y_2 and x_2 as functions of x_2'.

It should be mentioned that the alternative solution discussed previously in terms of $dy_1/dx_1 = f_1(x_1, y_1)$ and $dy_2/dx_2 = f_2(x_2, y_2)$ leads in the present example to explicit equations for $f_1(x_1, y_1)$ and $f_2(x_2, y_2)$.[4]

Numerical computations were made for the following special case:

$$n = 16,\ x_{1\,\max} = 0.1,\ \alpha = 0 \quad \text{and } \theta_{1\,\max} = 30°.$$

The values chosen are reasonably practical since

$$\cos^{16} \theta_{1\,\max} \approx \frac{1}{10}\ (-10 \text{ db})$$

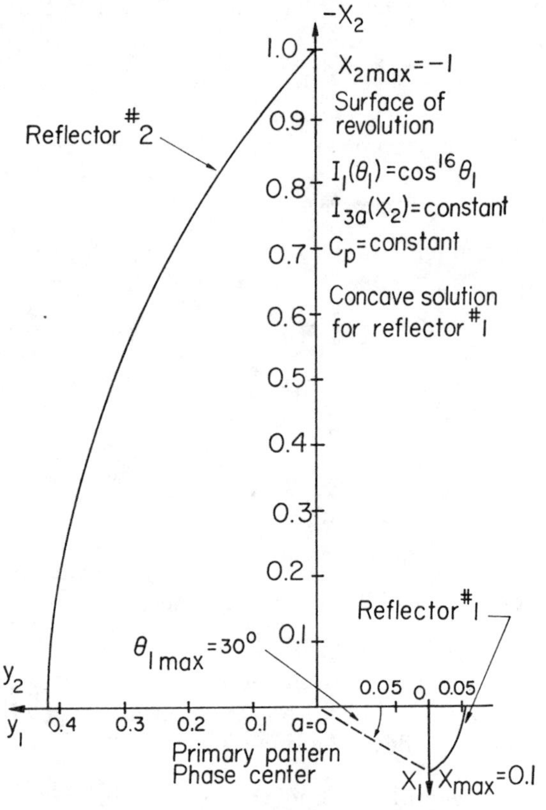

Fig. 4—Dual-reflector uniform phase-amplitude concave solution.

and

$$\left| \frac{x_{1\,\max}}{x_{2\,\max}} \right|^2 = 0.01$$

or 1 per cent optical blockage by reflector Number 1.

The results for the convex (Cassegrain type) solution are depicted in Fig. 3. A parabola, matched at two points, is drawn for comparison with reflector Number 2. The results for the concave (Gregorian type) solution are depicted in Fig. 4. In both cases a numerical and graphical check of the results indicates that the solutions conform to the optical constraints required. The smoothness of the contours (large radii of curvature) indicates that the optical design will be a useful one in this case.

V. Acknowledgment

The author wishes to thank particularly Prof. W. Welch of the University of California, Berkeley, for his helpful assistance and guidance. Much appreciation is also due F. Hennessey of Dalmo-Victor Co., Belmont, Calif., for his suggestion of the problem, Dr. Joseph Carter of the Stanford University Computation Center, Calif., for his work on the numerical results, and the reports group at this university for their preparation of the manuscript.

High Efficiency Antenna Reflector

WILLIAM F. WILLIAMS
WDL DIVISION, PHILCO CORPORATION

INTRODUCTION

With the introduction of the Cassegrain system into the antenna field, it was logical to use a parabolic main reflector and a hyperbolic subreflector. The effectiveness, or efficiency, of these reflectors as antennas is determined primarily by: (1) the ability of the energy source to illuminate only the reflectors while minimizing the energy that radiates elsewhere and (2) by the ability of this source to illuminate the parabola evenly, making maximum use of the entire reflector surface. The first item above is termed "spillover efficiency" and the second "illumination efficiency." The illumination efficiency is 100 per cent when the incident energy density on the reflector is a constant over the entire surface.

In both the prime focus feed system and the true Cassegrain system, the burden of maximizing (or optimizing) these efficiencies is placed upon the antenna feed itself. The ideal feed would have a radiation pattern which would be perfectly uniform within the angle subtended by the reflector to be illuminated, and which would fall abruptly to zero for all angles outside this region. Such a feed is, of course, not even remotely practicable. Horn (feed) patterns always gradually taper from their central maxima to nulls. If all this energy is intercepted by the reflector (for maximum spillover efficiency), then the illumination is far from uniform, and the illumination efficiency becomes very poor. Consequently, any attempt to obtain nearly uniform illumination will result in a great loss of energy in spillover.

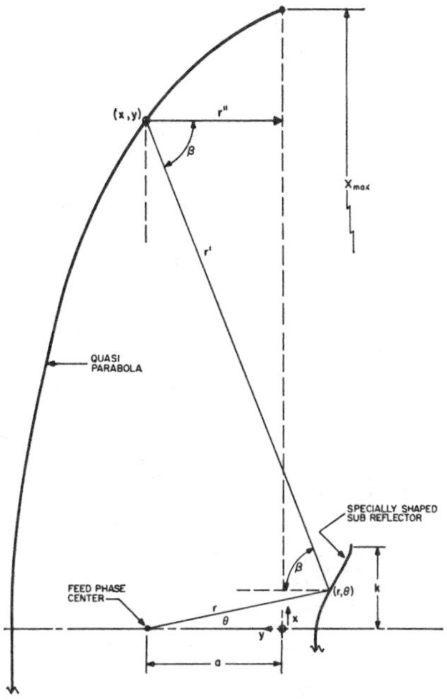

Figure 1 — Dual reflector system.

For example, a 3 dB taper at the reflector edge would achieve a spillover efficiency of only approximately 50 per cent; thus, a compromise must be made between these two antithetic effects. A common choice for both the prime focus system and the Cassegrain system is a 10 dB taper of the illumination pattern at the parabola edge. This selection results in a combination of spillover and illumination efficiency of from 75 - 80 per cent, about the best possible with existing parabola systems.

Now, if one starts with a given Cassegrain-type feed pattern and phase center (instead of starting with a parabola-hyperbola reflector system), what reflector system can be developed that will maximize illumination and spillover efficiency? The constraint to be applied will be that which gives uniform illumination and the resulting 100 per cent illumination efficiency. With the feed pattern given, a choice will be made for very small spillover, and the equations will be developed for surfaces which yield this uniform illumination.

THE THEORETICAL SOLUTION

The dual reflector system to be discussed is shown in Figure 1. A feed phase center is located as shown; it has a power radiation pattern $F(\theta)$. The requirements to be placed on this configuration are that the path length $r+r'+r''$ remain a constant for all θ and that the final illumination across the aperture, $I(x)$, remain constant, i.e., be uniform. This will give two equations of a required four to obtain a relationship $G(x, y, r, \theta)$. The other two equations are obtained by applying Snell's law at the two reflecting surfaces. The general problem and solution were presented in a paper by V. Galindo[1] of the University of California.

Reprinted with permission from *Microwave J.*, vol. 8, pp. 79-82, July 1965.

THE PLANAR PHASE FRONT

The equation for equal path lengths resulting in the plane phase front is obtained from trigonometry:

$$r+y+\frac{x-r\sin\theta}{\sin\beta}=C\ (\text{constant}) \qquad (1)$$

where (x, y) and (r, θ) are the coordinates of points on the main reflector and the subreflector, respectively.

UNIFORM ILLUMINATION

Remembering that the illustrated system is symmetrical about the y axis, the total power within the increment $d\theta$ of the pattern $F(\theta)$ will be $F(\theta)\ 2\pi \sin\theta\ d\theta$. The total radiated power from $\theta=0$ to any angle θ will then be

$$\int_0^{\theta} F(\theta)\ 2\pi \sin\theta\ d\theta$$

Similarly, the total power within the increment dx of the main antenna aperture is $I(x)\ 2\pi x\,dx$, where I(x) is the illumination function of the antenna aperture. Again, the total power within the region x is

$$\int_0^{x} I(x)\ 2\pi\ x\ dx$$

(The effects of aperture blocking by the subreflector are neglected.)

The object of the surface development is to make the function I(x) equal a constant. Then, if the power within the angle θ is related to that within x, the following is written:

$$C\int_0^{x} x\ dx = \int_0^{\theta} F(\theta)\ \sin\theta\ d\theta$$

where C is the illumination function, a constant. Now if this equation is normalized by dividing by the total power, there results:

$$\frac{\int_0^{x} x\ dx}{\int_0^{x_{max}} x\ dx} \qquad \frac{\int_0^{\theta} F(\theta)\ \sin\theta\ d\theta}{\int_0^{\theta_{max}} F(\theta)\ \sin\theta\ d\theta}$$

$$x^2 = x^2_{max}\ \frac{\int_0^{\theta} F(\theta)\ \sin\theta\ d\theta}{\int_0^{\theta_{max}} F(\theta)\ \sin\theta\ d\theta} \qquad (2)$$

Snell's Law for Equal Angles of Incidence and Reflection

The application of Snell's law to these two surfaces defines a relationship between the angles shown and the first derivatives (slopes) of the surfaces. These are:

$$\frac{1}{r}\frac{dr}{d\theta} = \tan\frac{\theta+\beta}{2} \qquad (3)$$

$$-\frac{dy}{dx} = \tan\frac{\beta}{2} \qquad (4)$$

Solutions of the Equations

These four equations now have the five dependent variables x, y, r, θ and β. These equations — one algebraic, two differential and one integral — are subject to exact solution only by the use of a computer. However, a consideration of requirements reveals the general results expected.

In the parabola-hyperbola case, the illumination taper on the parabola is approximated by the taper of the feed upon the hyperbola. We note that the excess of energy in the central portion of the horn beam must be spread to the outer regions of the main reflector. Therefore, we would expect the new subreflector surface to have a smaller radius of curvature in the central region (compared to the hyperbola) to reflect these central rays to the outer region.

In Figure 2, the ray is shown incident on the hyperbola and reflected at the angle β_1 back to the parabola or main reflector. The same ray is reflected at the angle β_2 for the special surface which is helping to spread the energy evenly across the dish. In fact, to a first order approximation, one might consider the new subreflector as just the device to obtain uniform illumination without regard to path length or maintaining the planar phase front.

In this case, a parabola could be assumed and the solution for a new subreflector surface giving uniform illumination could be obtained. Then, the path length constraint could be applied which would yield the small changes in parabola surface necessary to recover the uniform phase front. This can be done without the aid of a computer, and it was done for one example. The maximum difference between the parabola and new surface is about 0.06 of the subreflector diameter; this means the parabola should also be corrected by this amount.

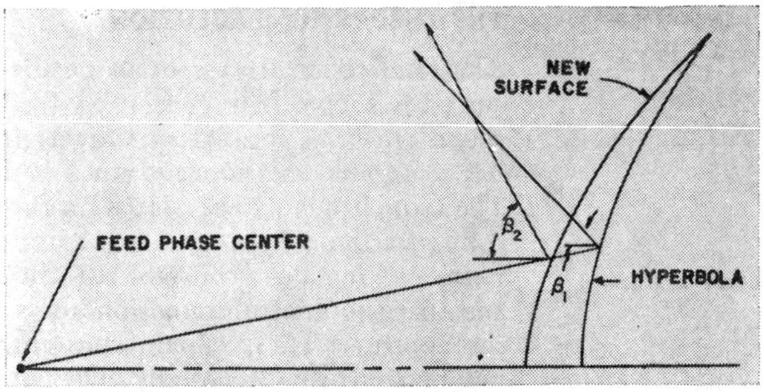

Figure 2 — Ray paths for hyperbola and new surface.

A computer program has been developed to solve the above equations. The following sample problem was solved to establish the technique:

20 dB illumination taper on the subreflector

$F(\theta) = (\cos\theta)^{84.5}$

$k = 0.5$ feet

$x_{max} = 5.0$ feet

$\theta_{max} = 18.73°$

Figure 3 shows the result of this computer solution. The new surfaces are now 0.0107D from the equivalent F/D=0.25 parabola. The variation becomes extreme as we demand more energy spreading from the new surface. If we place an entirely new parabola through the ends of this special surface, the new F/D is 0.239, and its maximum variation from the

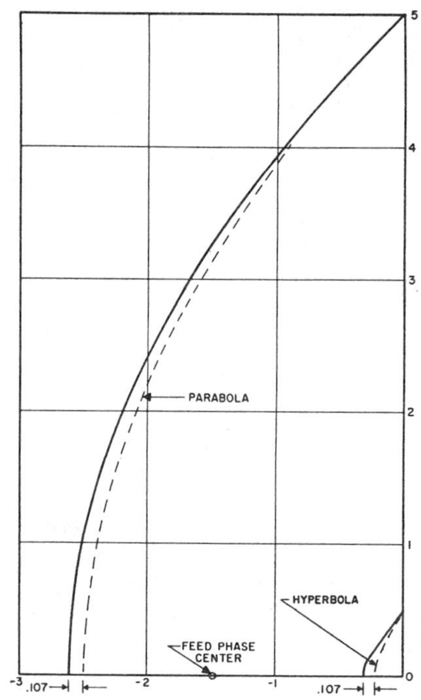

Figure 3 — Surface solution using $F(\theta) = (\cos\,\theta)^{84.5}$.

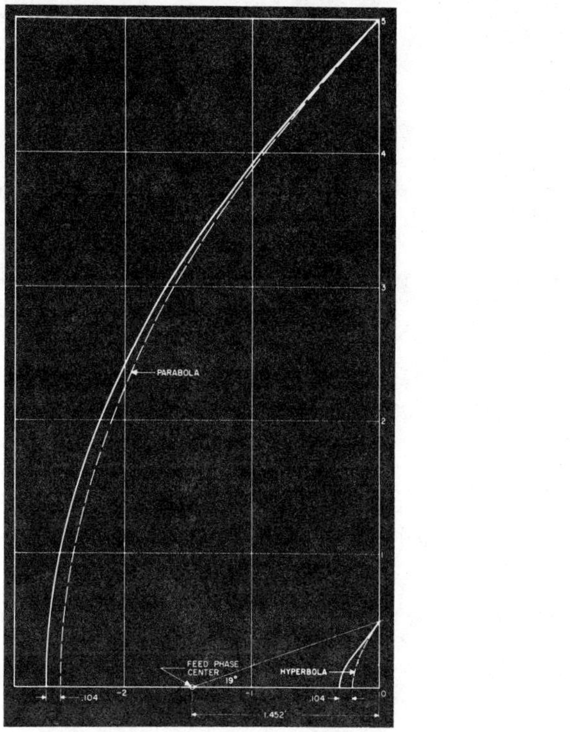

Figure 4 — Solution using equivalent F/D=0.25, 22 dB taper, F(θ) from measured patterns.

special surface is again about 0.001 of the main reflector diameter.

THE FEED HORN PATTERN

The conical feed horn has lower sidelobes than square or rectangular apertures because the circular aperture shape effectively provides a taper in the E plane, in addition to the usual H-plane cosine taper. Hence, spillover efficiency is usually higher. Integration of conical horn patterns shows that they contain 2.7 per cent energy in all of the sidelobes and 2.7 per cent energy in all of the cross-polarization components; thus, at least 5.4 per cent of the feed horn energy is lost.

The shape of the high efficiency antenna reflector is a function of, among other things, the shape of the beam power pattern of the feed horn. It would be highly desirable from a construction standpoint if the main and subreflectors had circular symmetry around the boresight axis. This can only be obtained if the patterns in all planes of the illuminating feed horn have equal beamwidths at all levels down to the illumination at the edge of the subreflector.

A conical feed horn having these properties was invented by the Jet Propulsion Labs in a research phase of a contract with NASA, NAS 7-100.[2,3,4] Also, a summary of this work is contained in the **microwave journal**.[5]

A model of this horn was constructed and tested for this study. The resulting E- and H-plane patterns were nearly identical. 98.7 per cent of the total energy was contained in a symmetrical main beam with 0.6 per cent lost in sidelobes and 0.7 per cent lost in cross polarization.

It should be pointed out that the broadening of the dual mode horn pattern results in a beam that falls off more slowly than the normal horn pattern, when both patterns are normalized with respect to beamwidth between the same power levels. This has the effect of reducing the aperture efficiency about as much as the spillover efficiency is improved, unless one uses the special reflector methods described in this report for providing constant illumination across the aperture. The reduced spillover losses, however, do lower the antenna noise temperature in cases where the spillover around the subreflector could intercept the earth.

FINAL DESIGN COMPUTATIONS

The pattern of this feed horn design was used to solve for reflector surfaces having maximum efficiency. This was done for two cases: (1) an equivalent F/D=0.25 and (2) an equivalent F/D=0.4.

Figure 4 is the representation of the solution for one case. The choice of 19° for θ_{max} gives a taper of just under 22 dB for the subreflector and a spillover efficiency of almost 98 per cent. Since the theoretical illumination efficiency is 100 per cent, a total overall efficiency of 98 per cent will be achieved, compared to about 78 per cent for the most ideal regular Cassegrain. Therefore, in this system as defined, a full 1 dB increase in gain can be expected. The result is very similar to the solution of 20 dB taper for the illumination of $(\cos\,\theta)^{84.5}$.

Figure 5 shows the solution for the equivalent F/D=0.4 case. Again, the measured $F(\theta)$ is used with θ_{max}= 19°, the only difference being in the dimension a (see Figure 1) which now becomes a negative 0.9475 feet. In this case, the F/D=0.4 parabola has a difference from the new surface

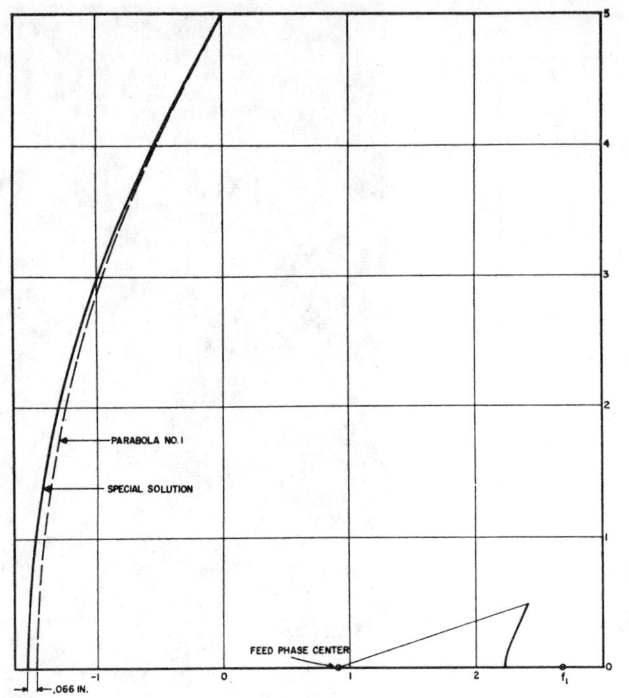

Figure 5 — Solution using equivalent F/D=0.4, 22 dB taper, F(θ) from measured pattern.

of about 0.0066 parts of a diameter. This agrees very well, being only slightly larger than the value 0.006 that was obtained using the approximation. Also, a different parabola can be solved which contains the end points of this new surface. When this is done, a parabola of F/D=0.399 results; and the maximum difference between it and the special surface is less than 0.0006 parts of a diameter. It appears that as the equivalent F/D ratios for these new surfaces become larger, the surface more closely approximates a paraboloid.

CONCLUSIONS

A technique has been described for improving the illumination efficiency of Cassegrain-type antennas to a theoretical 100 per cent while, at the same time, decreasing the spillover and noise temperature. A gain improvement on the order of 1 dB is indicated. This means, for example, that a 30 foot antenna will give the performance of a 33.5 foot parabola; a 40 foot unit the performance of a 45 foot parabola; and a 60 foot antenna would be equivalent to a 67.5 foot parabola.

The results of this program indicate a very promising antenna design. Although noise temperature change has not yet been evaluated, it is apparent from the very small spillover around the subreflector edge that some improvement can be expected to accompany the 1 dB gain increase.

The unknown effect of diffraction losses are not introduced in this treatment since the complete analysis used only geometrical optics.

ADDENDUM

Since this paper was originally written, an experiment was prepared to verify the predictions of increased antenna efficiency. Two reflectors of 2 foot diameter were built — one a parabola with F/D=0.4 and the other a special high efficiency surface. Two identical multimode horns were constructed for use with the systems — one illuminating a hyperbola with an 11 dB taper, the other used with the special high efficiency subreflector with a 20 dB taper. The two antennas were mounted adjacent to each other, and patterns were measured at the design frequency of 70 Gc.

Preliminary results are almost exactly as predicted. The new surface has a narrower beamwidth by about 10 per cent and has first sidelobes of about −17 dB, compared to −23 dB lobes from the parabola. Repeated measurements of relative gain indicate that the new surface has a greater gain by about 1 dB.

REFERENCES

1. Galindo, V., "Design of Dual Reflector Antennas with Arbitrary Phase and Amplitude Distributions," presented at PTGAP International Symposium, Boulder, Colo., July 1963.
2. Rusch, W. T. and A. C. Ludwig, "Analysis of Low Noise Antennas," *Technical Report No. 37-19,* Jet Propulsion Labs, p. 189.
3. Ludwig, A. C., P. Potter and W. T. Rusch, "Antennas for Space Communications," *Technical Report No. 37-20,* Jet Propulsion Labs, p. 126.
4. Ludwig, A. C. and W. T. Rusch, "Antennas for Space Communications," *Technical Report No. 37-21,* Jet Propulsion Labs, p. 187.
5. Potter, P. D., "A New Horn Antenna with Suppressed Sidelobes and Equal Beamwidths," the *microwave journal,* June 1963, pp. 71-78.

Shaping of Subreflectors in Cassegrainian Antennas for Maximum Aperture Efficiency

G. W. COLLINS

Abstract—This paper describes a systematic technique for obtaining high aperture efficiency in a Cassegrainian feed system by special shaping of the subreflector. Theoretical expressions are derived that determine the subreflector shape, which gives a uniform illumination of a paraboloid for a given primary feed pattern. Experimental data are then presented, which demonstrate the enhanced performance of a Cassegrainian feed system using a subreflector shaped according to the theoretical expressions.

Statement of Problem

IN RECENT years, a number of papers have appeared describing techniques for obtaining high aperture efficiency with reflector antennas. These techniques utilize the basic Cassegrainian geometry, and each provides for some type of special shaping of the reflector surface(s). The Dielguide [1] feed makes use of a dielectric guiding structure between the horn and the subreflector to simultaneously reduce forward spillover and produce a uniform subreflector illumination. A specially shaped subreflector then scatters the energy toward the main reflector. The dual shaped reflector [2]–[5] provides uniform aperture amplitude and phase illumination by special shaping of both the subreflector and the main reflector.

In the technique proposed in this paper, only the much smaller subreflector is of a special design, and no dielectric materials are used. It was shown by Williams [5], that when the dual shaped reflector system is used, the specially shaped main reflector may be approximated closely by an equivalent paraboloid with slightly reduced focal length. This idea was extended by Bruning [6], who showed that for small main reflector diameters, a best fit paraboloid may be used in place of the specially shaped surface. Since axial defocusing in a paraboloid produces only quadratic phase error, it is evident that the special shaping of the subreflector introduces primarily quadratic phase errors that are corrected for by special shaping of the main reflector. Thus, all required shaping may be done at the subreflector surface, provided the subreflector focal point is defined to be simply that point about which the scattered energy most nearly appears to be a spherical wave. The technique then becomes one of determining the subreflector shape required to provide a uniform amplitude distribution, and then experimentally determining the focal point of the resultant surface.

Manuscript received August 7, 1972; revised November 14, 1972.
The author is with Radiation, Inc., Division of Harris-Intertype Corporation, Melbourne, Fla. 32901.

Analytical Approach

The Cassegrainian feed system under consideration is shown in Fig. 1. The antenna is symmetrical about the major axis. A feed phase center is located as shown and has a power radiation pattern $F(\theta)$. The requirement to be placed on this configuration is that the scatter pattern function $I(\beta)$ (relative power) be constant or uniform within the angular region $\beta_1 \leq \beta \leq \beta_m$ and zero elsewhere. This will give one equation of two required to obtain the subreflector surface $G(r,\theta)$. The other equation is obtained by applying Snell's law at the reflecting surface of the subreflector.

The incident power is proportioned to $2\pi F(\theta) \sin \theta \, d\theta$. Similarly, the reflected power is proportional to $2\pi I(\beta) \sin \beta \, d\beta$. Equating the power within the angle θ to that within the angle β gives,

$$\int_{\beta_1}^{\beta_m} I(\beta) \sin \beta \, d\beta = \int_0^{\theta} F(\theta) \sin \theta \, d\theta. \qquad (1)$$

By making the lower limit β_1, the effects of blockage by the primary feed are taken into account.

Normalizing by dividing by the total power, and performing the integration with respect to β,

$$\cos \beta_1 - \cos \beta = (\cos \beta_1 - \cos \beta_m) \frac{\int_0^{\theta} F(\theta) \sin \theta \, d\theta}{\int_0^{\theta_m} F(\theta) \sin \theta \, d\theta}. \qquad (2)$$

Application of Snell's law to the "perfectly conducting" subreflector surface defines a relationship between the angles shown and the first derivative of the surface. It is

$$\frac{1}{r}\frac{dr}{d\theta} = \frac{\tan(\theta + \beta)}{2}. \qquad (3)$$

Equations (2), (3) are sufficient to specify completely the subreflector contour for a given subreflector diameter

Reprinted from *IEEE Trans. Antennas Propagat.*, vol. AP-21, pp. 309–313, May 1973.

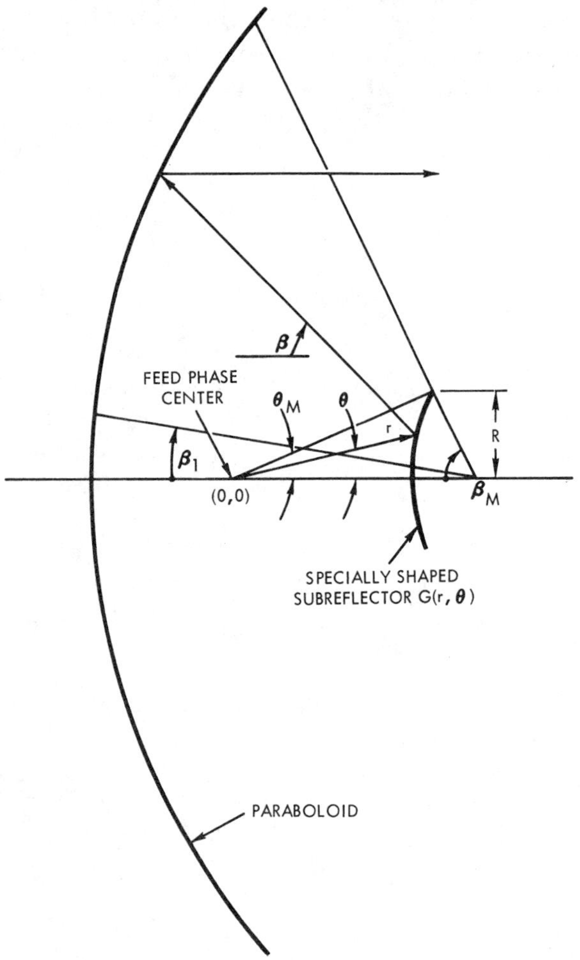

Fig. 1. Geometry of Cassegrainian feed system with specially shaped subreflector.

2R. These equations may be readily programmed for numerical solution on a digital computer. Thus, for any given primary feed pattern and desired reflector configuration (i.e., β_1,β_m,θ_m), the required subreflector shape may be determined.

The design procedure for a feed system of this type is readily apparent. First, it is necessary to design a horn providing an appropriate radiation pattern $F(\theta)$. After an adequate pattern $F(\theta)$ has been obtained, the subreflector may be designed using values of β_1,β_m,θ_m, the subreflector diameter dictated by the overall antenna design, and (2) and (3). The remaining step is to experimentally determine the subreflector focal point by measuring phase patterns. This point is then made coincident with the main reflector prime focus when the feed is installed. These steps properly completed should provide a highly efficient antenna system with near optimum gain.

Experimental Investigation

Equations (2) and (3) were used to design two subreflectors for subsequent pattern measurement. The first has 19 in diameter, designed to be tested in an $8\frac{1}{2}$-ft diameter paraboloid. The second has 28 in diameter. Scatter patterns were measured on each of the subreflectors, and these patterns were evaluated to determine illumination efficiency. The smaller subreflector was tested in the paraboloid to confirm the calculated results. These feed designs and the measured and calculated results are described in the following sections.

Description of Feed Designs

A horn pattern suitable for feeding a shaped subreflector should meet the following requirements: a) correct beamwidth, b) axial symmetry, c) stationary phase center, and d) constant beamshape over the operating band. These requirements are met with the corrugated conical horn [7]. The inside surface is corrugated with chokes that have the effect of broadening the E-plane pattern to the width of the H-plane pattern. Proper choice of the horn flare angle assures a minimum beamwidth variation with frequency and provides a stationary phase center. The horn provides near perfect axial symmetry and no sidelobes. In addition, the 20-dB beamwidth varies only 12 percent over a 25 percent frequency range. Because of its excellent characteristics, this horn was used as a primary feed for the subreflectors.

The 19-in diameter subreflector was designed using the primary feed pattern at a frequency of 7.25 GHz as $F(\theta)$. The feed was designed for use in a paraboloid with a focal length to diameter ratio of 0.417; hence, β_m was 62°. In order to minimize forward spillover, the 20 dB point of the horn pattern was chosen as the subreflector intercept; thus, θ_m was 36°. In addition, the horn subtended a half-angle of 14° as viewed from the subreflector focus. Therefore, this value was chosen for β_1. The subreflector shape computed on the basis of this input data is shown in Fig. 2.

The 28-in diameter subreflector was designed utilizing the feed pattern at 8.0 GHz as $F(\theta)$. As before, the 20 dB point was chosen as the intercept making $\theta_m = 35°$. Other input data were: $\beta_m = 61°$ and $\beta_1 = 9°$. The computed subreflector shape is also shown in Fig. 2.

Feed Pattern Measurements

The subreflectors described in the preceding section were tested utilizing standard amplitude and phase pattern measurement equipment. Each horn/subreflector combination was rotated about the subreflector focal point, and the amplitude and phase patterns were recorded as a function of the angle β.

The scatter patterns measured on the 19-in subreflector at 7.25 GHz are shown in Fig. 3. A high percentage of the energy is confined to the region ±62° from center with spillover lobes in all cases 12 dB below the peak. The void near $\beta = 0°$ is due to blockage introduced by the primary feed horn.

A typical pattern measured with the 28-in diameter subreflector is shown in Fig. 4. The general features are similar to those measured on the smaller model; however, in general the percentage of energy between $\pm\beta_m$ is higher, and the spillover levels are generally lower. The

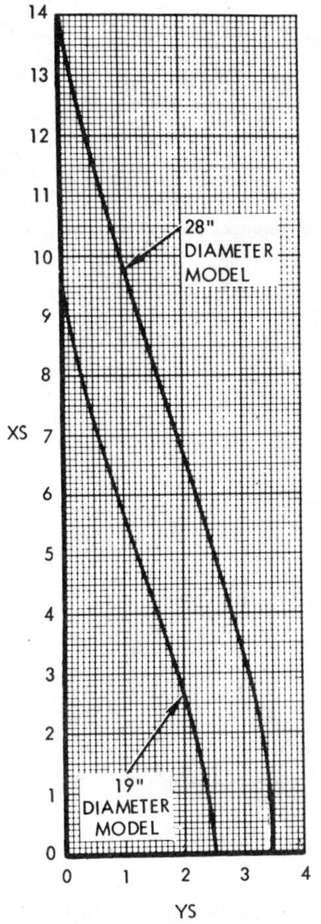

Fig. 2. Cross section of experimental subreflectors.

pattern shape does not vary significantly over the 6.8–8.4 GHz frequency band. The major variation is in the wide angle spillover, which is higher at the lower frequencies. This feature is due to the broadening of the primary feed pattern, which results in more energy radiated past the edge of the subreflector into the region near 135°.

The phase patterns while not perfectly uniform, indicate that a definite feed phase center does exist and is constant in position in both pattern planes. As will be shown subsequently, the degree of phase nonuniformity is not particularly severe in relation to its effect on antenna gain.

Pattern Analysis

In order to establish a quantitative measure of the scatter pattern performance, a figure of merit η_f is defined as follows:

$$\eta_f = \eta_{sp}\eta_{at}\eta_\phi \tag{9}$$

where η_{sp} is the percentage of the scattered energy contained in the sector $\pm\beta_m$, η_{at} a measure of the degree of amplitude uniformity [8] and η_ϕ is a measure of the degree of phase uniformity.

The computed values of η_f are displayed in Table I for the measured scatter patterns from both the 19- and the 28-in subreflectors. Several observations are worthy of note. Firstly, in spite of the deviations from uniform amplitude and phase and the presence of spillover, η_f is in all cases greater than 80 percent. Secondly, the performance of the 28-in subreflector is approximately 5 percent higher than the smaller model. This is as expected, since the geometric optics approximation is more nearly simulated on the larger model. Finally, it should be noted that maximum performance of the 28-in subreflector occurs at the design frequency, as expected. However, the degradation at other frequencies is not particularly severe.

Secondary Patterns and Gain Measurements

In order to confirm the computed performance of the shaped subreflector feed, the horn and 19-in diameter subreflector were mounted in a paraboloid for secondary pattern and gain measurements. The results of these tests are presented in the following paragraphs.

Description of Antenna

The horn and subreflector were mounted in an $8\frac{1}{2}$-ft diameter parabolic reflector. The horn was mounted in a metal support tube which contained the input waveguide section and attached to the paraboloid at the vertex. A Styrofoam cone was attached to the horn aperture and provided a low dielectric constant ($\epsilon \approx 1.02$) support for the subreflector. In this manner, the entire feed was supported by the metal tube, negating any necessity for spares to support the subreflector.

The reflector f/D ratio was 0.417, so that the feed design β_m and the reflector intercept were the same (62°). The reflector surface tolerance was 0.012-in rms, more than adequate for use at 7.25 GHz.

Analysis of Antenna Performance

The overall antenna performance is established primarily by the feed figure of merit discussed previously. However, two other factors related to the reflector system are important. These are: a) blockage and b) gain loss due to reflector surface tolerance.

The projection of the subreflector on the antenna aperture represents a block region and gives rise to a reduction in antenna efficiency. An equation to account for this effect has been derived [9] and is given as

$$\eta_b = \left[1 - \left(\frac{d_s}{d_r}\right)^2\right]^2 \tag{10}$$

where d_s is the subreflector diameter, and d_r the paraboloid diameter. For a 19-in diameter subreflector in a 102-in diameter paraboloid, the blockage efficiency is 0.931.

The gain loss due to random surface errors has been investigated [10] and is given by,

$$\eta_{\phi s} = e - \frac{(4\pi\delta)^2}{\lambda} \tag{11}$$

where δ is the rms surface accuracy. For a surface tolerance of 0.012 in, the gain reduction due to these errors at 7.25 GHz is approximately 0.990.

The total aperture efficiency and gain may now be computed utilizing the feed figure of merit. The total

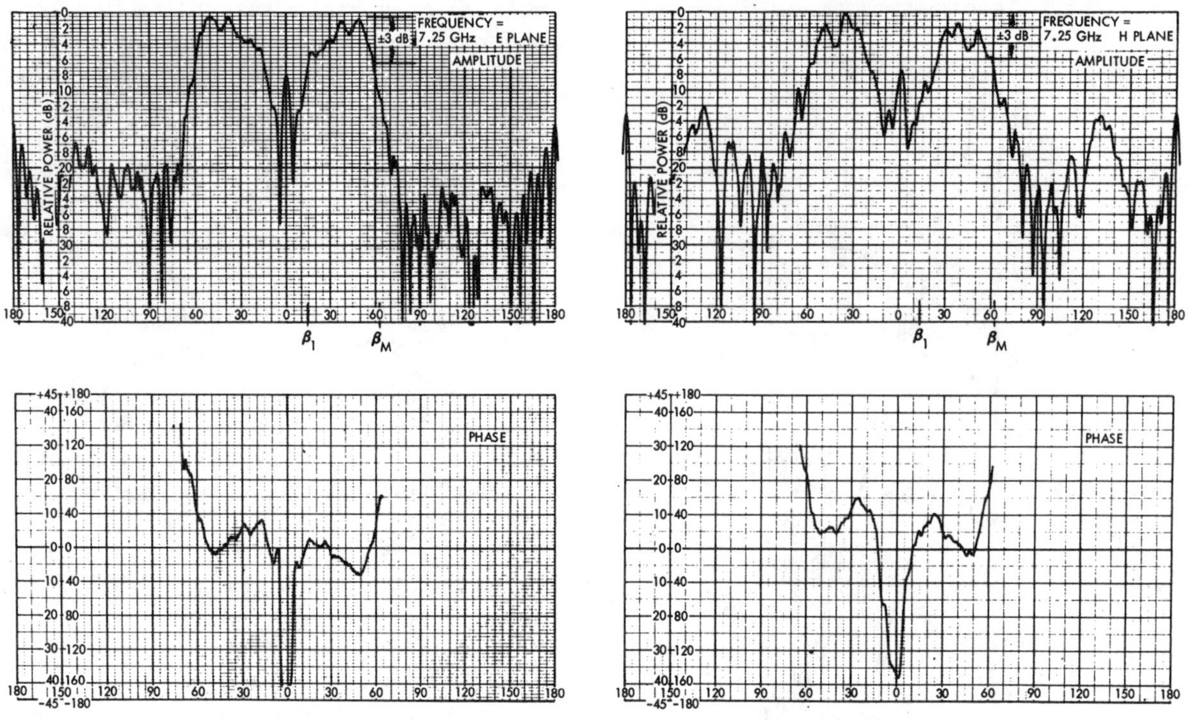

Fig. 3. Scatter patterns of 19-in (11.6λ) diameter subreflector.

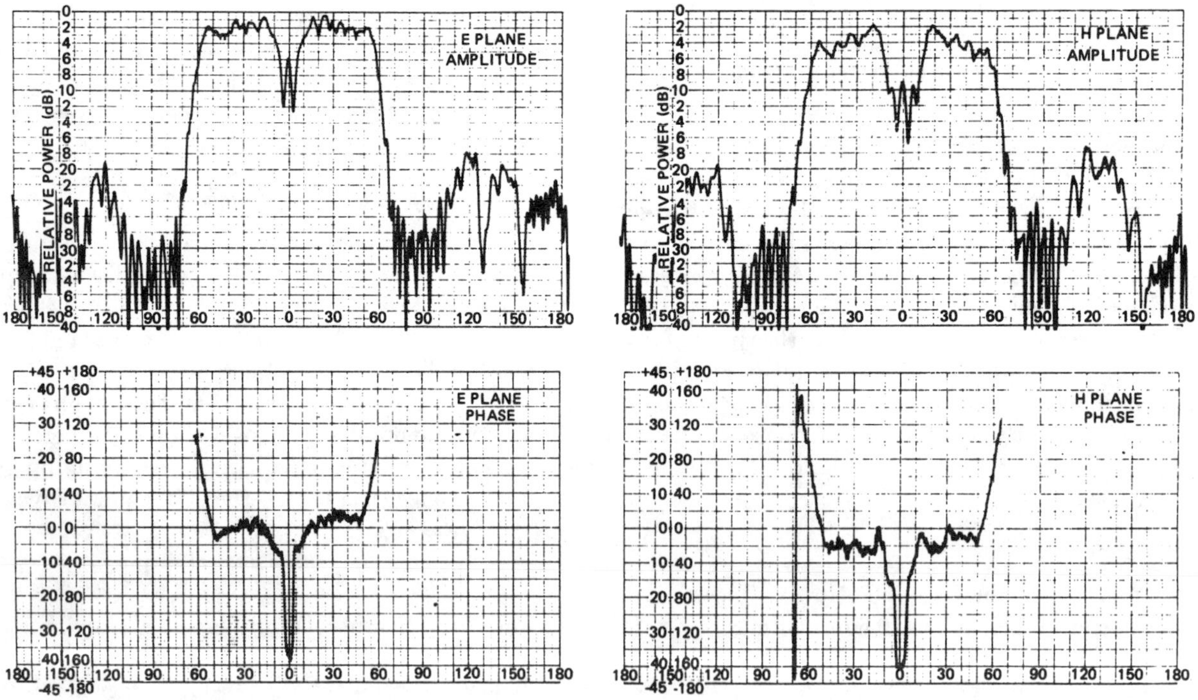

Fig. 4. Scatter patterns of 28-in (18.9λ) diameter subreflector (8.0 GHz).

efficiency is,

$$\eta = \eta_f \eta_b \eta_\phi \tag{12}$$

and the antenna gain is

$$g = \eta \frac{4\pi A}{\lambda 2} \tag{13}$$

where A is the projected area of the paraboloid. From Table I, $\eta_f = 0.801$ at 7.25 GHz so that the total aperture efficiency has the value of 0.738. Using (13), the gain is 44.7 dB.

Measured Antenna Performance

Patterns were measured on the antenna and are shown in Fig. 5. The patterns are approximately symmetrical,

TABLE I

FEED FIGURE OF MERIT

	19" Diameter			28" Diameter		
Freq. (GHz)	$\eta_{sp}\ \eta_{at}$	η_ϕ	η_f	$\eta_{sp}\ \eta_{at}$	η_ϕ	η_f
6.80				0.891	0.953	0.849
7.25	0.839	0.955	0.801			
7.40				0.844	0.970	0.849
8.00				0.899	0.960	0.862
8.40				0.896	0.935	0.835

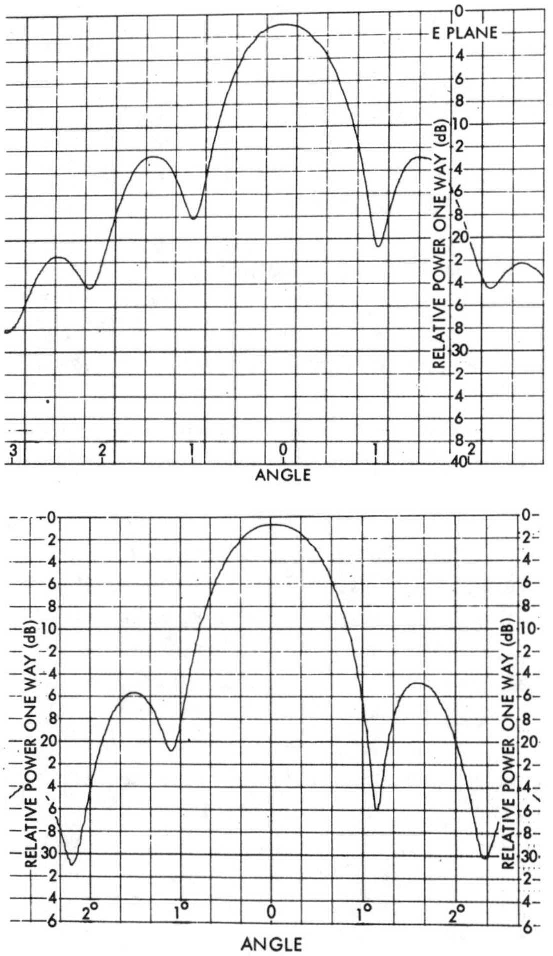

Fig. 5. Measured secondary patterns (7.25 GHz).

indicating proper vertical and horizontal alignment. The nulls between the main beam and first sidelobe are not as deep as usually experienced. This can be attributed to the lack of flat phase scatter patterns. The antenna gain was measured via the substitution method, and was 44.6 dB. The calculated and measured antenna gain agree very closely.

SUMMARY

The technique described provides a means for designing a feed system with total aperture efficiency in the 70–80 percent range by shaping only the subreflector surface. It was shown that by shaping the subreflector to provide a uniform illumination (ray-optics approximation), the resulting feed system provides a nearly constant amplitude distribution, high slopes at the main reflector edge, and low spillover. In addition, it was shown by phase measurements that the resulting wave front is reasonably spherical and a definite subreflector phase center exists.

Analysis of the feed scatter patterns revealed that for a subreflector 11.6λ in diameter, a figure of merit of 83.9 percent was obtained and that for a subreflector 18.9λ in diameter, a figure of merit of 86.2 percent was obtained. The total aperture efficiency of the former feed design in a paraboloid 63λ in diameter was 73.8 percent. This performance was confirmed by gain measurement.

By way of comparison with previous work, it should be noted that the phase efficiency shown in Table I agrees substantially with the gain loss calculated due to phase errors in [6]. It was shown that the best fit paraboloid introduces a gain loss of approximately 0.2 dB in the uniform illumination case. This corresponds to a value of η_ϕ of 95.5 percent, which is typical of the values shown in Table I.

ACKNOWLEDGMENT

The contributions to the work by Dr. A. W. Revay are gratefully acknowledged, and the assistance of L. Sheppard in programming the subreflector equations is due a special note of thanks.

REFERENCES

[1] R. E. Moseley and H. E. Bartlett, "Dielguides-highly efficient low noise antenna feeds," *Microwave J.*, p. 53, Dec. 1966.
[2] B. Y. Kimber, "On two-reflector antennas," *Radio Eng. Electron Phys.* (USSR), vol. 6, pp. 914–921, June 1962.
[3] K. A. Green, "Modified Cassegrain antenna for arbitrary aperture illumination," *IRE Trans. Antennas Propagat.* (Commun.), vol. AP-11, pp. 589–590, Sept. 1963.
[4] V. Galindo, "Design of dual reflector antenna with arbitrary phase and amplitude distributions," presented at PTGAP Int. Symp., Boulder, Colo., July 1963.
[5] W. F. Williams, "High efficiency antenna reflector," *Microwave J.*, p. 79, July 1965.
[6] J. H. Bruning, "A best fit paraboloid solution to the shaped dual reflector antenna," presented at Symp. U. S. Air Force Antenna Research and Development Program, Nov. 15, 1967.
[7] G. W. Collins, "Multimode horns using corrugated surfaces," presented at IEEE G-AP Symp., Los Angeles, Calif., Sept. 20–23, 1971.
[8] P. D. Potter, "The aperture efficiency of large paraboloidal antennas as a function of their feed system radiation characteristics," Jet Propulsion Lab., Pasadena, Calif., Tech. Rep. 32-149, Sept. 25, 1961.
[9] J. Ruze, "Feed support blockage loss in parabolic antennas," *Microwave J.* p. 76, Dec. 1968.
[10] ——, "Effect of aperture distribution errors on the radiation pattern," Antenna Lab Memo, Air Force Cambridge Res. Cent., Jan. 22, 1952.

Bibliography for Part IV

[1] H. Gniss and G. Ries, "Remarks on the concept of equivalent parabolas for Cassegrain antennas," *Electron. Lett.*, vol. 6, pp. 737-739, Nov. 12, 1970.

[2] H. E. Bartlett and R. E. Moseley, "Dielguides—Highly efficient low noise antenna feeds," *Microwave. J.*, vol. 9, pp. 53-58, Dec. 1966.

[3] P. J. B. Clarricoats and C. E. R. C. Salema, "Antennas employing conical dielectric horns," *Proc. Inst. Elec. Eng.*, vol. 120, parts I and II, pp. 741-749 and 750-756, July 1973.

[4] K. A. Green, "Modified Cassegrain antenna for arbitrary aperture illumination," *IEEE Trans. Antennas Propagat.*, vol. AP-11, pp. 589-590, Sept. 1963.

[5] B. Ye Kinber, "On two-reflector antennas," *Radio Eng. Electron. Phys.*, vol. 7, pp. 914-921, June 1962.

[6] P. Foldes and S. G. Komlos, "Theoretical and experimental study of wideband paraboloid antenna with central reflector feed," *RCA Rev.*, vol. 21, pp. 94-116, Mar. 1960.

[7] E. J. Wilkinson and A. J. Applebaum, "Cassegrain systems," *IRE Trans. Antennas Propagat.*, vol. AP-9, pp. 119-120, Jan. 1961.

[8] W. D. White and L. K. De Size, "Focal length of a Cassegrain reflector," *IRE Trans. Antennas Propagat.*, vol. AP-9, p. 412, July 1961.

[9] M. Viggh, "Designing for desired aperture illuminations in Cassegrain antennas," *IEEE Trans. Antennas Propagat.*, vol. AP-11, pp. 198-199, Mar. 1963.

[10] W. V. T. Rusch, "A comparison of diffraction in Cassegrainian and Gregorian radio telescopes," *Proc. IEEE*, vol. 51, pp. 630-631, Apr. 1963.

[11] P. D. Potter, "Aperture illumination and gain of a Cassegrainian system," *IEEE Trans. Antennas Propagat.*, vol. AP-11, pp. 373-375, May 1963.

[12] P. A. Jensen, "A low-noise multimode Cassegrain monopulse feed with polarization diversity," in *IEEE Northeast Electron. Res. and Eng. Meeting Dig.*, Nov. 1963, pp. 94-95.

[13] S. P. Morgan, "Some examples of generalized Cassegrainian and Gregorian antennas," *IEEE Trans. Antennas Propagat.*, vol. AP-12, pp. 685-691, Nov. 1964.

[14] D. C. Hogg and R. A. Semplak, "An experimental study of near-field Cassegrainian antennas," *Bell Syst. Tech. J.*, vol. 43, pp. 2677-2704, Nov. 1964.

[15] R. J. Gunderman, H. F. Mathis, and L. A. Zurcher, "A two-reflector, nonshadowing antenna," *IEEE Trans. Antennas Propagat.*, vol. AP-14, pp. 474-475, May 1965.

[16] J. S. Cook, E. M. Elam, and H. Zucker, "The open Cassegrain antenna: Part I: Electromagnetic design and analysis," *Bell Syst. Tech. J.*, vol. 44, pp. 1255-1300, Sept. 1965.

[17] B. Claydon, "A study of the performance of Cassegrain aerials," *Marconi Rev.*, vol. 30, pp. 98-115, 2nd Quarter 1967.

[18] H. Zucker and W. H. Ierley, "Computer-aided analysis of Cassegrain antennas," *Bell Syst. Tech. J.*, vol. 47, pp. 897-932, July-Aug. 1968.

[19] J. Dijk and E. J. Maanders, "Optimising the blocking efficiency in shaped Cassegrain systems," *Electron Lett.*, vol. 4, pp. 372-373, Sept. 6, 1968.

[20] S. D. Slobin and W. V. T. Rusch, "Case study of phase-center relationships in an asymmetric Cassegrainian feed system," *IEEE Trans. Antennas Propagat.*, vol. AP-17, pp. 698-703, Nov. 1969.

[21] B. L. J. Rao and S. N. C. Chen, "Illumination efficiency of a shaped Cassegrain system," *IEEE Trans. Antennas Propagat.*, vol. AP-18, pp. 411-412, May 1970.

[22] P. J. Wood, "Field correlation theorem with application to reflector aerial diffraction problems," *Electron. Lett.*, vol. 6, pp. 326-327, May 28, 1970.

[23] S. B. Cohn, "Flare-angle changes in a horn as a means of pattern control," *Microwave J.*, vol. 13, pp. 41-46, Oct. 1970.

[24] P. J. Wood, "Field correlation diffraction theory of the symmetrical Cassegrainian antenna," *IEEE Trans. Antennas Propagat.*, vol. AP-19, pp. 191-197, Mar. 1971.

[25] J. Dijk, E. J. Maanders, and J. P. F. Sniekers, "On the efficiency and radiation patterns of mismatched shaped Cassegrainian antenna systems," *IEEE Trans. Antennas Propagat.*, vol. AP-20, pp. 653-655, Sept. 1972.

[26] W. D. Fitzgerald, "The efficiency of near-field Cassegrainian antennas," *IEEE Trans. Antennas Propagat.*, vol. AP-20, pp. 648-650, Sept. 1972.

[27] G. S. Brown, "A simplification in the analysis of four- and five-horn fed Cassegrainian reflectors when the horns have nearly symmetric patterns," *IEEE Trans. Antennas Propagat.*, vol. AP-21, pp. 382-384, May 1973.

[28] M. Misuzawa and T. Kitsuregawa, "A beam-waveguide feed having a symmetric beam for Cassegrain antennas," *IEEE Trans. Antennas Propagat.*, vol. AP-21, pp. 884-886, Sept. 1973.

[29] B. L. J. Rao, "Bifocal dual reflector antennas," *IEEE Trans. Antennas Propagat.*, vol. AP-22, pp. 711-714, Sept. 1974.

[30] S. von Hoerner, "The design of correcting secondary reflectors," *IEEE Trans. Antennas Propagat.*, vol. AP-24, pp. 336-340, May 1976.

[31] R. J. Langley and E. A. Parker, "Filtering secondary mirror for a dual-band Cassegrain reflector antenna," *Electron. Lett.*, vol. 12, pp. 366-367, July 22, 1976.

Part V
Polarization Effects

This part leads off with a useful paper by Hanfling that complements the discussions on polarization of the aperture field given in the papers by Cutler, Jones, and Koffman in Part I. Considerable insight into the fact that offset (i.e., unsymmetrical) paraboloids possess cross-polarized lobes in the plane orthogonal to the plane of symmetry, even when the feed is an ideal Huygens source, can be gained by comparing Figs. 1 and 4 in Hanfling's paper. For the offset reflector in Fig. 4, the feed has been tilted so that its axis no longer coincides with the optical axis of the reflector. If the feed were not tilted, then the situation would be exactly the same as that in Fig. 1 for the symmetrical paraboloid. In that case, a Huygens source feed would produce a linearly polarized aperture field. The feed axis must be tilted, however, in order to avoid excessive spillover in the offset case. Further light is thrown on this subject in an illuminating paper by Jacobsen in Part VI of this book, devoted to offset reflector systems.

The whole subject of cross polarization in reflector antennas had been somewhat confused for many years and had given rise to certain misconceptions and some controversy. Thus, a discussion by Watson and Ghobrial in 1973 [1] prompted a response in which Wood [2] disputed certain conclusions reached by the former. This was quickly followed by an exchange between authors [3], [4] in which they attempted to reconcile their disagreements. The difficulty arose, at least in part, because of the lack of a precise and unambiguous definition of cross polarization. Ludwig points out this surprising deficiency in the second paper of this part and attempts to correct it. He discusses three different definitions that are commonly used, shows where they are applicable and how they differ, and he recommends the third definition as the proper one for describing antenna radiation patterns. Although not dealing specifically with reflector antennas, Ludwig's paper is included here because of its importance in connection with the cross-polar characteristics of reflectors. His third definition now seems to have gained general acceptance among antenna engineers when discussing those characteristics. One misconception that is disposed of by Ludwig is that reflector curvature causes cross-polarized aperture fields in a reflector fed by a short dipole at its focus, so that shallow reflectors have better cross-polar patterns than deep ones. Under his third definition, it is evident that cross polarization exists in the dipole pattern itself and that it increases with increasing radiation angle. A deep reflector simply intercepts, and reflects, more of this cross-polar component than a shallow one would.

The third and fourth papers are by Thomas and by Minnett and Thomas, and they deal with axially symmetric reflectors. Thomas, in his succinct but very informative paper, shows that minimum cross polarization in the secondary pattern occurs when the field of the feed is polarization-pure, as is the case for a corrugated horn radiating the balanced hybrid (HE_{1n}) mode. The cross-polarized component becomes vanishingly small, however, only when the reflector diameter-to-wavelength ratio becomes very large, i.e., in the geometric optics limit. Thus, there really are two contributors, in general, to the cross-polarized field radiated by reflector antennas. The first, and usually the dominant one, is the feed, while the second is the reflector itself. Even when the feed is pure a low level of cross-polarized radiation will occur due to depolarization by the reflector. This arises for two reasons. The first has to do with the presence of the factor cos θ (and its departure from unity) in the expressions for the fields scattered by the reflector. The second reason is that these fields (except on the axis, where $\theta = 0$) contain contributions from the axial, or z-directed, components of current that flow on the reflector surface. Although small, these currents are not negligible and do give rise to a low level of cross polarization in the intercardinal planes. Thus, there really is an effect due to reflector curvature after all, but it is small (vanishingly so in very large reflectors), and it is masked by the cross polarization of the feed except in the region of near exact ($\gamma = 1$) balanced hybrid mode conditions. It might help in diminishing the confusion that evidently exists in this area if the term "depolarization" were reserved to describe these characteristics that belong to reflectors themselves, and to use the term "cross polarization" to mean the net effect of impure feed polarization plus reflector depolarization. Thomas shows that dual reflector systems are inferior to prime focus-fed reflectors when the feed polarization is pure because of the greater depolarization introduced by the dual systems.

It is theoretically possible to achieve zero cross polarization, as shown in the fourth paper by Minnett and Thomas. Reflector depolarization can be eliminated if the surface is anisotropic, for example, if it is circumferentially grooved, in the manner of a corrugated horn. It is interesting that the feed, in this case, is required to radiate the balanced EH_{1n} mode ($\gamma = -1$) which is converted to the desired HE_{1n} field after reflection at the anisotropic surface.

The final paper, by Safak and Delogne, treats cross polarization in both front-fed and Cassegrainian paraboloids. Their principal conclusion is that the cross-polarization performance in either case is highly dependent on feed design and on main reflector size, improving as the aperture diameter increases.

Aperture Fields of Paraboloidal Reflectors by Stereographic Mapping of Feed Polarization

Abstract—**For any paraboloidal reflector operating in a focused condition, the aperture-field orientation may be obtained by first stereographically mapping the feed-polarization pattern and then overlaying the paraboloid aperture on the projection. Examples shown are for electric and magnetic dipoles, and Huygens source feeds pointed at central and offset sections of the paraboloidal reflector.**

Introduction

The radiation properties of a focused paraboloidal reflector excited by a short electric dipole, a small magnetic dipole, and a Huygens source have been described [1]. Conventional techniques were used to obtain the orientation of the aperture fields from the currents on the surface of the paraboloid. The aperture fields for these radiators are shown in Fig. 1. The analysis showed that the Huygens source produced a vertically oriented electric-field loci (family of parallel straight lines) in the aperture of the paraboloid; this, in turn, resulted in low-amplitude cross-polarized radiation in the far field of the reflector.

A method will be described herein in which the orientation of the aperture fields can be obtained directly from the stereographic map of the feed far-field polarization. An estimate of the antenna's cross-polarized far-field radiation can then be made without detailed pattern computations. The presentation demonstrates the usefulness of stereographic mapping, not only for determining the orientation of the fields in the aperture of paraboloidal reflectors, but also in general for calculating, visualizing, and presenting antenna polarization.

Stereographic Mapping of Feed Polarization

The far-field polarization of a feed is determined by the relative excitation and relative orientation of its component radiators. Each of these radiators can be resolved into combinations of the two fundamental radiating elements, the electric dipole and magnetic dipole. The far electric field is shown in Table I for x, y, and z

Manuscript received April 9, 1969; revised September 23, 1969

Reprinted from *IEEE Trans. Antennas Propagat.*, vol. AP-18, pp. 392–396, May 1970.

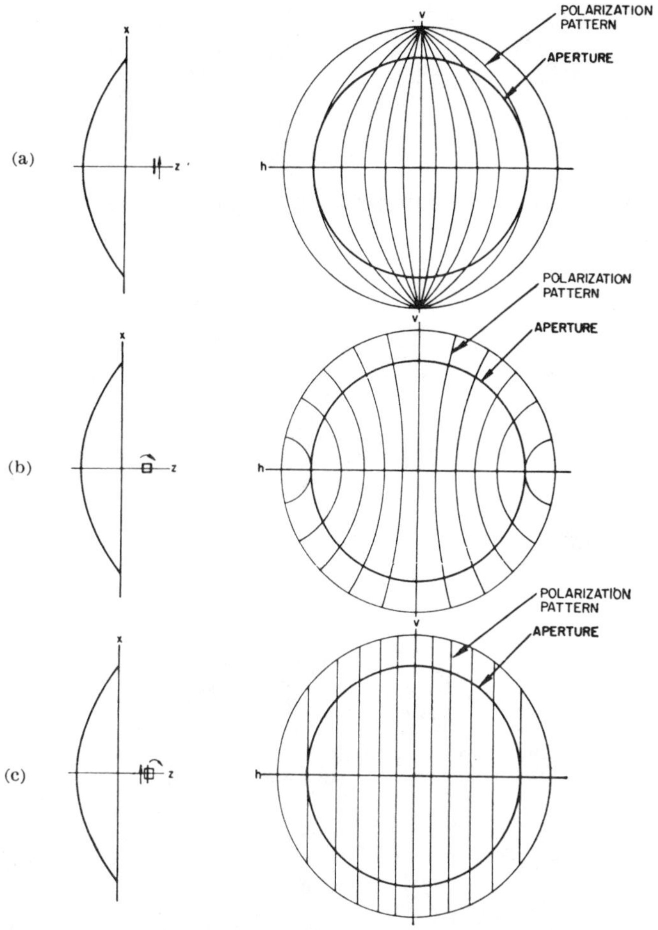

Fig. 1. Fields in aperture of paraboloidal reflector from stereographic projection. (a) Electric dipole feed. (b) Magnetic dipole feed. (c) Huygens source feed.

orientations of a short current element of length l and a small current loop of side a.

By the stereographic projection, the spherically distributed fields radiated from a feed can be conformally mapped onto a plane. This projection is illustrated in Fig. 2 where the ray FA emanates from the feed to the radiation sphere and the line PAH projects the radiated fields to the plane v,h. The projection transforms circles on the sphere to circles on the plane; in particular, circles on the sphere that pass through the pole P become straight lines on the plane [2].

As a general procedure, the projected E_θ- and E_ϕ-field components are resolved into vertical and horizontal components on the plane

$$E_v = E_\theta \cos\phi - E_\phi \sin\phi \tag{1}$$

$$E_h = E_\theta \sin\phi + E_\phi \cos\phi. \tag{2}$$

From the complex ratio $E_h/E_v = P$, the state of polarization (axial ratio r, tilt angle τ, and handedness) referenced to the vertical direction may be determined [3]. A pictorial presentation of the states of polarization at selected points or regions on the plane (here within a unit circle corresponding to the forward hemisphere) defines the polarization pattern of the feed.

For example, the Huygens source can be represented by a crossed pair of dipoles, one electric and one magnetic, with electric currents I_e and I_m in time quadrature and related by $-jI_el = I_m 2\pi a^2/\lambda$ to equalize their strength. For the x-directed electric dipole and y-directed magnetic dipole, the far electric fields are given in Table I. The resultant far electric field (radiation null in negative z direction),

$$E = K(1 + \cos\theta)[(\cos\phi)i_\theta - (\sin\phi)i_\phi] \tag{3}$$

has components in the θ and ϕ directions. By stereographic mapping and using (1) and (2),

$$E_v = -j(1 + \cos\theta)$$
$$E_h = 0. \tag{4}$$

It is evident from (4) that the polarization pattern is a family of straight vertical lines. The polarization patterns of the electric and magnetic dipoles are families of circular arcs (Fig. 1).

Certain composite radiators, such as crossed Huygens sources when time phased in quadrature, and collinear electric and magnetic dipoles give circular polarization for the entire plane, while others such as crossed electric or magnetic dipoles with currents in phase quadrature, contain in various directions, many possible polarization states, Fig. 3 [5]. The amplitude pattern is represented by the size of the circles and ellipses.

Calculation of Aperture Fields

The electric fields in the aperture of a paraboloidal reflector are determined from the currents on the surface of the reflector, which are excited by the far fields of the feed. The aperture fields are given by

$$E_a = J\eta/2 \cos \tfrac{1}{2}\theta \tag{5}$$

and the surface current J on the reflector by

$$J = (2/\eta)[\rho(n\cdot E) - E(n\cdot\rho)] \tag{6}$$

where n is the normal of the reflector, ρ is the radius vector from the feed, $E = E_\theta + E_\phi$, and η is the free-space impedance.

$$n = -[i_x \sin\tfrac{1}{2}\theta \cos\phi + i_y \sin\tfrac{1}{2}\theta \sin\phi + i_z \cos\tfrac{1}{2}\theta] \tag{7}$$

$$\rho = i_x \sin\theta\cos\phi + i_y \sin\theta \sin\phi + i_z \cos\theta. \tag{8}$$

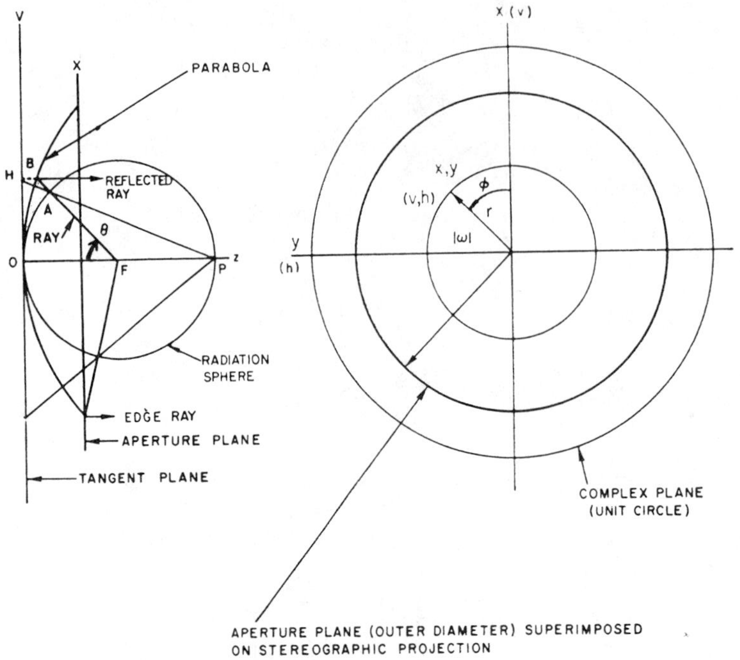

Fig. 2. Construction to obtain aperture fields from feed-polarization pattern.

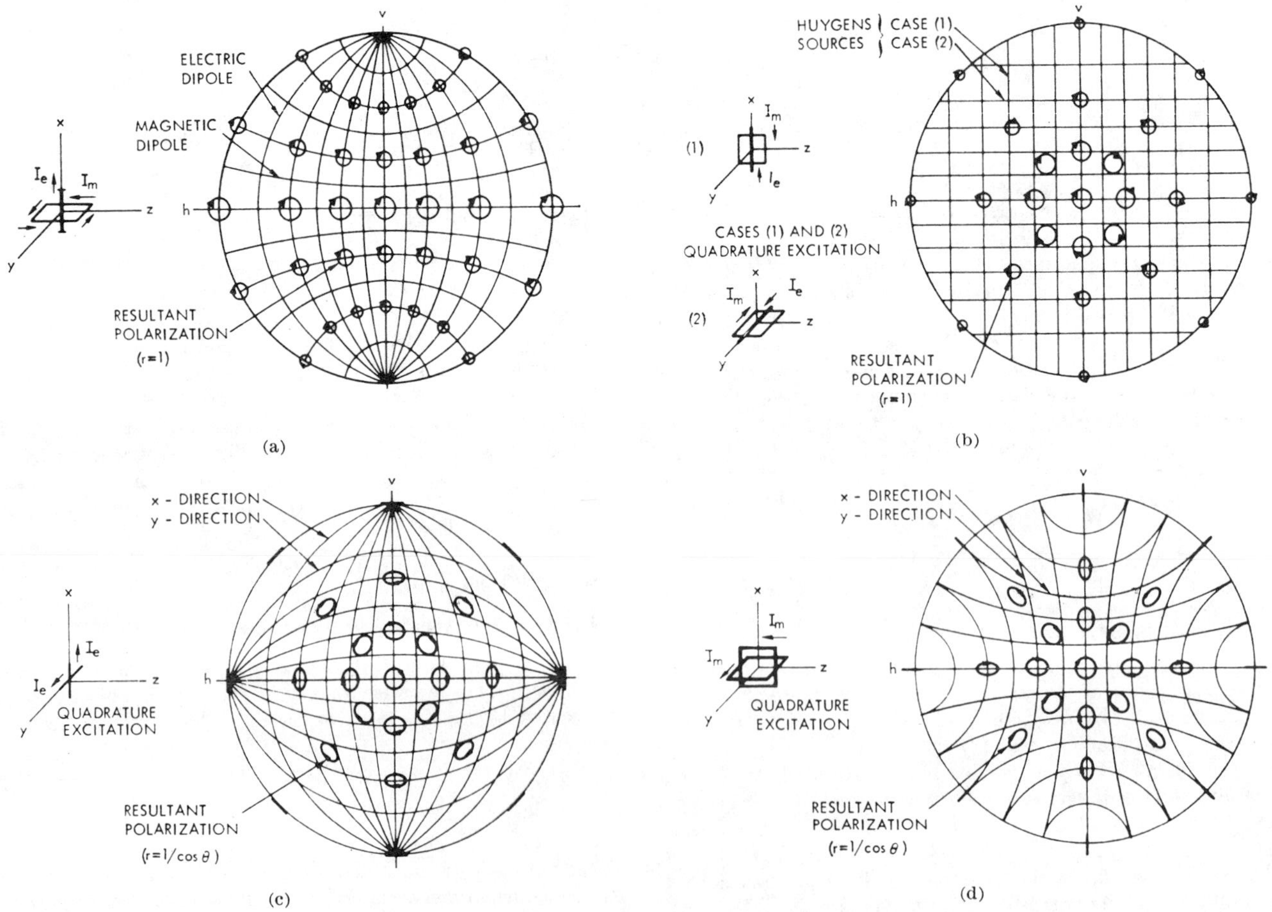

Fig. 3. Polarization patterns of composite radiators. (a) Polarization pattern of collinear electric and magnetic dipoles. (b) Polarization pattern of crossed Huygens sources. (c) Polarization pattern of crossed electric dipoles. (d) Polarization pattern of crossed magnetic dipoles.

TABLE I
MATHEMATICAL DESCRIPTION OF FAR ELECTRIC FIELDS OF DIPOLE ELEMENTS

$\uparrow$	Electric Dipole	Magnetic Dipole
z	$-K(\sin\theta)\boldsymbol{i}_\theta$	$K'(\sin\theta)\boldsymbol{i}_\phi$
x	$K[(\cos\theta\cos\phi)\boldsymbol{i}_\theta - (\sin\phi)\boldsymbol{i}_\phi]$	$-K'[(\sin\phi)\boldsymbol{i}_\theta + (\cos\theta\cos\phi)\boldsymbol{i}_\phi]$
y	$K[(\cos\theta\sin\phi)\boldsymbol{i}_\theta + (\cos\phi)\boldsymbol{i}_\phi]$	$K'[(\cos\phi)\boldsymbol{i}_\theta - (\cos\theta\sin\phi)\boldsymbol{i}_\phi]$
	$K = -j\omega\mu I_e l/4\pi R$	$K' = \omega\mu I_m a^2 k/4\pi R$

Arrow indicates orientation of electric dipole current I_e and axis of loop of electric current I_m.

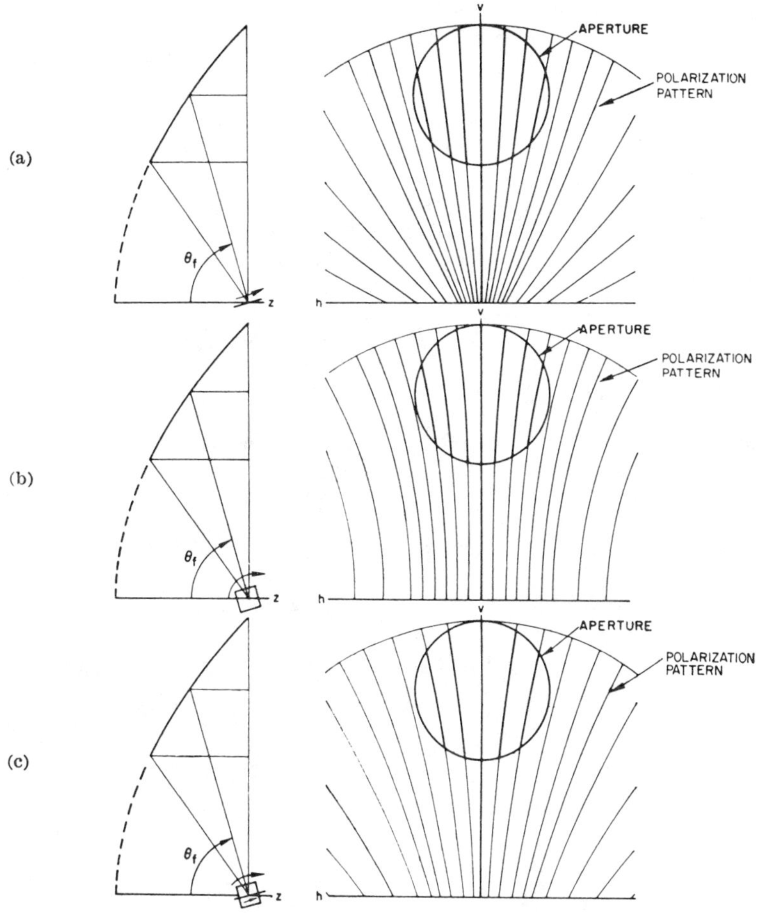

Fig. 4. Fields in aperture of offset paraboloidal reflector from stereographic projection of basic feed-polarization patterns. (a) Electric dipole. (b) Magnetic dipole. (c) Huygens source.

By substituting in (6), the components of current can be found

$$J_x = (2/\eta)\cos\tfrac{1}{2}\theta(E_\theta\cos\phi - E_\phi\sin\phi) \quad (9)$$

$$J_y = (2/\eta)\cos\tfrac{1}{2}\theta(E_\theta\sin\phi + E_\phi\cos\phi). \quad (10)$$

It is assumed that the z component of current does not contribute to the forward radiated field (generally neglected in radiation from paraboloidal reflectors). The aperture fields can be found from (5)

$$E_x = -(E_\theta\cos\phi - E_\phi\sin\phi) \quad (11)$$

$$E_y = -(E_\theta\sin\phi + E_\phi\cos\phi). \quad (12)$$

Equations (11) and (12) have the same form as that obtained by a stereographic map of the spherically distributed feed polarization, (1) and (2).

The stereographic projection of the feed polarization not only provides the same form of electric-field pattern, but can be constructed to give the actual fields in the aperture of the paraboloid. This construction is illustrated in Fig. 2. The tangent plane to the radiation sphere about the feed is also taken tangent to the vertex O of the parabola. For any point on the radiation sphere A, the ray from the feed FB intersects the aperture plane in the same Cartesian coordinates as the stereographic projector PH from the pole intersects the tangent plane. The points P,A,H are on a straight line if HB produced is the reflected ray. Also, the equality between the v,h Cartesian coordinates of space and the x,y Cartesian coordinates of the aperture plane can be seen from the parabola geometry with a radiation sphere of diameter $2F$

$$r = 2F\tan\tfrac{1}{2}\theta = |w|. \quad (13)$$

EXAMPLES

The use of this construction is illustrated in Fig. 1 for center-fed paraboloid antennas and in Fig. 4 for offset-fed paraboloid antennas both of $0.33F/D$ ratio and with the basic feeds. In Fig. 1 the feed is pointed on-axis and in Fig. 4 at $\theta = \theta_f = 74°$ to an offset section of the reflector. The polarization pattern of the feed is projected from the same pole P for both feed orientations; then the projected aperture circle of the paraboloid is drawn on the polarization pattern. Special use has been made of the conformal properties of stereographic mapping to draw the polarization locus for the feed oriented at θ_f. The polarization loci are families of circular arcs.

Conclusion

The above treatment can be carried over to more complicated feeds, and doubly offset paraboloid sections. Also, the stereographic mapping technique may be used to compute, visualize, and present the far-field polarization of antennas.

Jerome D. Hanfling
Missile Systems Div.
Raytheon Company
Bedford, Mass.

References

[1] E. M. T. Jones, "Paraboloid reflector and hyperboloid lens antennas," *IRE Trans. Antennas and Propagation*, vol AP-2, pp. 119–127, July 1954.
[2 J. D. Hanfling, "Mapping of the far field polarization of antennas by the stereographic proj ction," M.E.E. thesis. Polytechnic Ins\ of Brooklyn, Brooklyn, N. Y., June 1960.
——, Wheeler Lab., Smithtown, N. Y., Rept. 892P, March 1966.
[3] H. G. Booker, V. H. Rumsey, G. A. Deschamps, M. L. Kales, and J. I. Bohnert, "Techniques for handling elliptically polarized waves," *Proc. IRE*, vol. 39, pp. 533–556, May 1951.
[4] C. C. Cutler, "Parabolic-antenna design for microwaves," *Proc. IRE*, vol. 35, pp. 1284–1294, November 1947.
[5] J. F. Ramsey, J. P. Thompson, and W. D. White, "Polarization tracking of antennas," *1962 IRE Conv. Rec.*, vol. 10, pt. 1, pp. 13–42.

The Definition of Cross Polarization

ARTHUR C. LUDWIG

Abstract—**There are at least three different definitions of cross polarization used in the literature. The alternative definitions are discussed with respect to several applications, and the definition which corresponds to one standard measurement practice is proposed as the best choice.**

The use of orthogonal polarization to provide two communications channels for each frequency band has led to interest in the polarization purity of antenna patterns. It is a surprising fact that there is no universally accepted definition of "cross polarization" at the present, and at least three different definitions have been used either explicitly or implicitly in the literature. The *IEEE Standard* [1] definition is "The polarization orthogonal to a reference polarization." For circular polarization this is adequate, but for linear or elliptical polarization the direction of the reference polarization must still be defined.

We will first briefly present the definitions known to the author. Only the case of nominally linear polarization will be considered since the extension to elliptical polarization is straightforward. The three alternative definitions are: 1) in a rectangular coordinate system, one unit vector is taken as the direction of the reference polarization, and another as the direction of cross polarization [2]; 2) in a spherical coordinate system the same thing is done using the unit vectors tangent to a spherical surface [3], [4]; and 3) reference and cross polarization are defined to be what one measures when antenna patterns are taken in the usual manner [2, pp.

Manuscript received May 30, 1972; revised August 3, 1972. This work was supported by the European Space Research and Technology Centre.

The author was with the Laboratory of Electromagnetic Theory, Technical University of Denmark. He is now with the Jet Propulsion Laboratory, California Institute of Technology, Pasadena, Calif. 91103.

Reprinted from *IEEE Trans. Antennas Propagat.*, vol. AP-21, pp. 116–119, Jan. 1973.

557–564], [5]. These cases, which will be defined more precisely, are illustrated in Fig. 1.

Two different cases in which cross polarization is of interest may be distinguished: 1) describing the secondary radiation pattern of a complete antenna system, and 2) describing the source or primary field distribution. In the first case, it is desirable to have a definition which applies for all pattern angles, and which is easily related to channel interference, or other requirements such as ensuring that cross polarized fields do not exceed a given sidelobe specification. In the second case, it is desirable to have a simple relationship between source cross polarization and secondary pattern polarization. A second common application is the calculation of antenna feed or aperture illumination efficiency, where cross polarization must be included as a gain loss factor [6]. In this case, "cross polarization" really means fields which are antisymmetric in the aperture and therefore do not contribute to radiation on-axis. So it is desirable that the definition be consistent with this usage also.

We will now express the three definitions precisely and in terms of the same antenna pattern coordinate system as illustrated in Fig. 1. This will be done by deriving unit vectors $\hat{i}_{\text{ref}}$ and $\hat{i}_{\text{cross}}$ such that

$$\begin{aligned} \boldsymbol{E}\cdot\hat{i}_{\text{ref}} &\equiv \text{the reference polarization component of } \boldsymbol{E} \\ \boldsymbol{E}\cdot\hat{i}_{\text{cross}} &\equiv \text{the cross polarization component of } \boldsymbol{E}. \end{aligned} \tag{1}$$

Definition 1 is a trivial case with

$$\begin{aligned} \hat{i}_{\text{ref}}^{(1)} &\equiv \hat{i}_y = \sin\theta\sin\phi\,\hat{i}_r + \cos\theta\sin\phi\,\hat{i}_\theta + \cos\phi\,\hat{i}_\phi \\ \hat{i}_{\text{cross}}^{(1)} &\equiv \hat{i}_x = \sin\theta\cos\phi\,\hat{i}_r + \cos\theta\cos\phi\,\hat{i}_\theta - \sin\phi\,\hat{i}_\phi. \end{aligned} \tag{2}$$

For the second case, the polarization unit vectors are defined in a system of rectangular and spherical coordinates related to the system shown in Fig. 1 by

$$\tilde{x} = x \qquad \tilde{y} = z \qquad \tilde{z} = -y. \tag{3}$$

Then, by definition 2, we have

$$\hat{i}_{\text{ref}}^{(2)} \equiv \hat{i}_{\tilde{\theta}}$$

$$= \frac{-\sin^2\theta\sin\phi\cos\phi\,\hat{i}_x + (1-\sin^2\theta\sin^2\phi)\hat{i}_y - \sin\theta\cos\theta\sin\phi\,\hat{i}_z}{\{1-\sin^2\theta\sin^2\phi\}^{1/2}}$$

$$= \frac{\sin\phi\cos\theta\,\hat{i}_\theta + \cos\phi\,\hat{i}_\phi}{\{1-\sin^2\theta\sin^2\phi\}^{1/2}} \tag{4a}$$

$$\hat{i}_{\text{cross}}^{(2)} \equiv -\hat{i}_{\tilde{\phi}}$$

$$= \frac{\cos\theta\,\hat{i}_x - \sin\theta\cos\phi\,\hat{i}_z}{\{1-\sin^2\theta\sin^2\phi\}^{1/2}}$$

$$= \frac{\cos\phi\,\hat{i}_\theta - \cos\theta\sin\phi\,\hat{i}_\phi}{\{1-\sin^2\theta\sin^2\phi\}^{1/2}}. \tag{4b}$$

It should be noted that these equations depend on the choice of the relative orientation of the pattern and polarization coordinate systems. There are really two cases of definition 2 obtained by interchanging the subscripts ref and cross in (4). If this interchange is made in definitions 1 or 3, one obtains a result equivalent to rotating the coordinate system by 90° about the z axis (neglecting unimportant sign changes), but this is *not* true for definition 2. It is straightforward to show that

$$\begin{aligned} &\hat{i}_{\text{ref}}^{(2)}(\theta,\phi)\cdot\hat{i}_{\text{ref}}^{(2)}(\theta,\phi+90°) \\ &= \hat{i}_{\text{cross}}^{(2)}(\theta,\phi)\cdot\hat{i}_{\text{cross}}^{(2)}(\theta,\phi+90°) \\ &= \frac{\sin^2\theta\sin\phi\cos\phi}{\{(1-\sin^2\theta\sin^2\phi)(1-\sin^2\theta\cos^2\phi)\}^{1/2}}. \end{aligned} \tag{5}$$

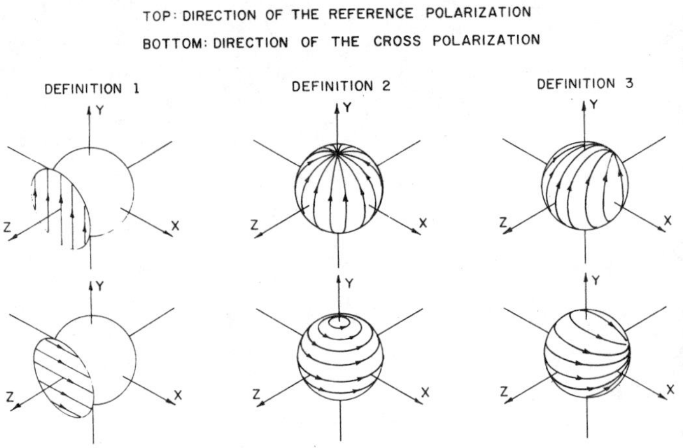

Fig. 1. Alternate polarization definition.

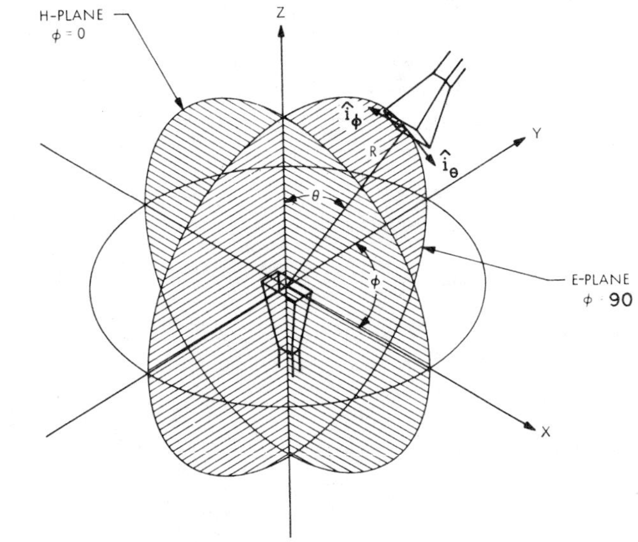

Fig. 2. Antenna pattern measurement system.

The third definition is simple in practice, but tricky to formulate precisely. The pattern measurement method that will be described is probably the one most commonly used by antenna engineers. It is sometimes presented as a standard method for testing feeds or small aperture antennas, [2, pp. 557–564], [7], but it can also be used for large antennas. In the terminology used by Scientific-Atlanta, this method may be implemented by mounting the antenna being tested, shown at the origin of the system as illustrated in Fig. 2, on a model tower or an elevation-over-azimuth positioner with an auxiliary polarization axis [8]. The elevation angle is always zero (z axis horizontal) so this axis is actually not required. Each pattern cut begins at $\theta = 0$, where the polarization axis is used to set the pattern cut angle ϕ by rotating the antenna being tested about the z axis. The probe is rotated about its axis by a second polarization positioner to align the probe polarization parallel to the polarization of the antenna being tested for a principal polarization pattern, or orthogonal to the polarization of the antenna being tested for a cross polarization pattern. Note that the orientation of the polarization of the antenna being tested, at the point $\theta = 0$, is the basic polarization reference direction by definition. Then a pattern is taken by varying θ by rotating in azimuth. This is equivalent to the probe traversing a great circle as illustrated in Fig. 2. The probe remains fixed about its axis so it retains the same relative orientation with respect to the unit vectors $\hat{i}_\theta$ and $\hat{i}_\phi$. Therefore, the measured pattern is given by

$$M(\theta) = \boldsymbol{E}(\theta,\phi)\cdot\{\sin\beta\,\hat{i}_\theta + \cos\beta\,\hat{i}_\phi\} \tag{6}$$

TABLE I
SOURCE CURRENT CONTRIBUTIONS TO RADIATED PATTERN CROSS POLARIZATION

Source Current Polarization	Contribution to Secondary Pattern Cross Polarization		
	Definition 1	Definition 2	Definition 3
$\hat{i}_x$	$1 - \sin^2\theta\cos^2\phi$	$\dfrac{\cos\theta}{(1-\sin^2\theta\sin^2\phi)^{1/2}}$	$1 - \cos^2\phi(1-\cos\theta)$
$\hat{i}_y$	$-\sin^2\theta\sin\phi\cos\phi$	0	$-(1-\cos\theta)\sin\phi\cos\phi$
$\hat{i}_z$	$-\sin\theta\cos\theta\cos\phi$	$\dfrac{-\sin\theta\cos\phi}{(1-\sin^2\theta\sin^2\phi)^{1/2}}$	$-\sin\theta\cos\phi$

where $E(\theta,\phi)$ is the field of the transmitting antenna, and the pattern cut angle ϕ and probe polarization angle β are fixed for a given pattern. For a transmitted field polarized in the $\hat{i}_y$ direction at $\theta = 0$, this alignment procedure leads to $\beta = \phi$ for a reference polarization pattern $R(\theta,\phi)$ and $\beta = \phi \pm 90°$ for a cross polarized pattern $C(\theta,\phi)$. Therefore, ignoring unimportant sign differences,

$$R(\theta,\phi) = E(\theta,\phi)\cdot\{\sin\phi\,\hat{i}_\theta + \cos\phi\,\hat{i}_\phi\}$$

$$C(\theta,\phi) = E(\theta,\phi)\cdot\{\cos\phi\,\hat{i}_\theta - \sin\phi\,\hat{i}_\phi\}. \quad (7)$$

This is equivalent to

$$\hat{i}_{\text{ref}}^{(3)} \equiv \sin\phi\,\hat{i}_\theta + \cos\phi\,\hat{i}_\phi$$

$$= -(1-\cos\theta)\sin\phi\cos\phi\,\hat{i}_x + \{1-\sin^2\phi(1-\cos\theta)\}\hat{i}_y - \sin\theta\sin\phi\,\hat{i}_z \quad (8a)$$

$$\hat{i}_{\text{cross}}^{(3)} \equiv \cos\phi\,\hat{i}_\theta - \sin\phi\,\hat{i}_\phi$$

$$= \{1-\cos^2\phi(1-\cos\theta)\}\hat{i}_x - (1-\cos\theta)\sin\phi\cos\phi\,\hat{i}_y - \sin\theta\cos\phi\,\hat{i}_z. \quad (8b)$$

The procedure defined in the preceding makes it easy to avoid probe polarization misalignment which is a severe pitfall in cross polarization measurements. To show this, suppose that $\beta = (\phi + 90°) + \epsilon$. Then from (6) and (7)

$$M(\theta) = C(\theta,\phi)\cos\epsilon - R(\theta,\phi)\sin\epsilon. \quad (9)$$

To illustrate what this means in practice, suppose that the true cross polarization is negligible, but a misalignment of $\epsilon = 1.5°$ is present. Then one would measure a "cross polarized" pattern which is actually the reference polarization pattern suppressed by 31.6 dB (i.e., sin 1.5°). Cross polarization patterns which do not have a null on axis are a symptom of this error.

Before discussing the relative merits of these definitions, it is necessary to relate the polarization of source currents J to the polarization of the radiated pattern $E(\theta,\phi)$, which is given by [9]

$$E(\theta,\phi) = F(\theta,\phi) - \{F(\theta,\phi)\cdot\hat{i}_r\}\hat{i}_r \quad (10a)$$

where

$$F(\theta,\phi) = -\frac{j\omega\mu\exp(-jkr)}{4\pi r}\int J\exp(-jk\varrho\cdot\hat{i}_r)\,dS \quad (10b)$$

and the undefined terms are unimportant for present purposes. In order to directly relate the components of J at any point to the components of F at any pattern angle, it is necessary to use unit vectors which do not vary as a function of either the pattern coordinates or the integration coordinates; the only apparent choice is rectangular unit vectors. Then the x, y, and z components of J are uniquely related to the x, y, and z components of F, respectively. However, the components of F and E do not have such a simple relation. From (10a) it can be found that they are coupled by a polarization matrix.

$$\begin{bmatrix}E_x\\E_y\\E_z\end{bmatrix} = \begin{bmatrix}1-\sin^2\theta\cos^2\phi & -\sin^2\theta\sin\phi\cos\phi & -\sin\theta\cos\theta\cos\phi\\ -\sin^2\theta\sin\phi\cos\phi & 1-\sin^2\theta\sin^2\phi & -\sin\theta\cos\theta\sin\phi\\ -\sin\theta\cos\theta\cos\phi & -\sin\theta\cos\theta\sin\phi & 1-\cos^2\theta\end{bmatrix}\cdot\begin{bmatrix}F_x\\F_y\\F_z\end{bmatrix}. \quad (11)$$

Similar factors relating the source current polarization to the radiated pattern cross polarization, for all three definitions, are given in Table I. It is seen that in all cases the dominant cause of cross polarization is the i_x source currents. The $\hat{i}_y$ source current contribution to cross polarization is suppressed by a factor which is in excess of 52 dB for $\theta < 4°$. The $\hat{i}_z$ contribution is suppressed by a factor which is in excess of 23 dB for $\theta < 4°$. Therefore, it is reasonable to define the $\hat{i}_x$ component as the cross polarized source currents, and to use the common terminology of longitudinal currents for the $\hat{i}_z$ component to distinguish the fact that it is a far less serious source of secondary pattern cross polarization. Of course, the $\hat{i}_y$ currents are the reference polarization currents.

Now we will compare the three definitions. Since the far-field fields of any antenna are tangent to a spherical surface, it is immediately apparent that definition 1 is fundamentally inappropriate for these applications. However, as noted above, it is the only apparent definition which applies to the case of source currents. Definitions 2 and 3 involve unit vectors tangent to a sphere so they are appropriate for the case of primary or secondary fields. For evaluating secondary patterns for the application of orthogonal channels, we postulate the following ideal case: the transmitting antenna has two ports, which radiate two patterns that are orthogonal at every pattern angle in the coverage region. Clearly, it is then possible to receive the two channels without any interference anywhere in the coverage region (in fact, this is far easier than the transmit problem since the receiving antenna must be free of cross polarization only very close to its axis). Now, since any field is everywhere orthogonal to some other field, this still leaves the "perfect" pattern undefined. A logical choice is a pattern which is orthogonal to itself after a 90° rotation about the z axis. This is also a realistic choice since it corresponds to adding an orthomode transducer to an otherwise circularly symmetric antenna. A pattern with no cross polarization by definitions 1 or 3 satisfy this requirement; this is not true for definition 2 as shown by (5), and as previously pointed out by Kreutel and Di Fonzo [4].

For relating source current distributions to secondary patterns, it is logical that a perfect source distribution radiate a perfect pattern; only definition 2 is compatible with this requirement, as shown in Table I.

For evaluating primary feed patterns for paraboloidal reflectors a logical requirement is that a perfect feed cause a perfect surface current distribution. By definition 2 an infinitesmial electric dipole pattern contains no cross polarization (if the subscripts ref and cross are reversed in definition 2, an infinitesimal magnetic dipole is perfect). However, it is well known that this feed causes substantial cross polarization [10]. This question has also been treated in an excellent paper by Koffman [11], where it is shown that a necessary and sufficient condition for zero cross polarized surface currents is

$$E_\theta \cos \phi = E_\phi \sin \phi. \tag{12}$$

From (8) it can be seen that this is identically equivalent to definition 3. Koffman gives an example of a Huygens source as satisfying (12). It is also possible to show that a physically circular feed with equal E- and H-plane amplitude *and phase* patterns is also perfect by this definition [6]. From (2) it may be seen that a perfect feed by definition 1 does not satisfy (12) either, so only definition 3 satisfies this requirement.

Finally, it has been previously shown by the author [6] that a feed with no cross polarization by definition 3 is optimum from the viewpoint of antenna aperture efficiency. Again, this is not true for either definition 1 or 2.

From the preceding discussion, it is clear that definition 1 is the proper choice for describing source current polarizations. It is the author's opinion that definition 3 is the best choice for describing antenna patterns. The only disadvantage of definition 3 is its imperfect relationship with the source current definition. However, this point is muddled in any case by the existence of longitudinal source currents. Definition 2 has the disadvantage of two perfect secondary patterns rotated 90° with respect to each other not being orthogonal, the serious point that a perfect primary pattern can produce a very poor secondary pattern, and its incompatibility with feed efficiency usage.

As an illustration of how definition 2 can be misleading, it may be noted that the cross polarized currents on a paraboloid illuminated by an infinitesimal electric dipole are frequently attributed to the reflector curvature. As a result, it is widely accepted that increasing the reflector f/D ratio substantially reduces cross polarization [12]. By definition 3, an electric dipole has substantial cross polarization which increases rapidly with increasing pattern angle (i.e., as the E- and H-plane edge illuminations diverge). A paraboloid with lower f/D subtends a larger pattern angle, and this is the reason that the cross polarization in the secondary pattern becomes worse. If the E- and H-plane edge illumination is held constant it is not difficult to show that the cross polarized currents are actually *independent* of the f/D ratio. Also, for certain practical feeds it has been pointed out that cross polarization may actually increase with increasing f/D [4]. The f/D ratio does effect the longitudinal currents, and also can have an effect in the case of defocused feeds, but with a proper definition it is seen that secondary pattern cross polarization is far less dependent on the f/D ratio than it seems from definition 2.

Acknowledgment

The author would like to acknowledge several interesting discussions on this topic with D. C. Patel of the European Space Research and Technology Centre.

References

[1] "IEEE standard definitions of terms for antennas," *IEEE Trans. Antennas Propagat.*, vol. AP-17, pp. 262–269, May 1969.
[2] S. Silver, *Microwave Antenna Theory and Design*. New York: McGraw-Hill, 1949, pp. 423, and 557–564.
[3] V. P. Narbut and N. S. Khmel'nitskaya, "Polarization structure of radiation from axisymmetric reflector antennas," *Radio Eng. Electron. Phys.*, vol. 15, pp. 1786–1796, 1970.
[4] D. F. Di Fonzo and R. W. Kreutel, "Communications satellite antennas for frequency reuse," presented at G-AP Int. Symp., paper 12-1, Sept. 1971.
[5] J. D. Kraus, *Antennas*. New York: McGraw-Hill, 1950, ch. 15.
[6] A. C. Ludwig, "Antenna feed efficiency," Jet Prop. Lab., Calif. Inst. Technol., Pasadena, Space Programs Summary 37–26, vol. IV, pp. 200–208, 1965.
[7] "Test procedures for antennas," IEEE Publ. No. 149, Jan. 1965.
[8] J. S. Hollis, T. J. Lyon, and L. Clayton, "Microwave antenna measurements," Scientific-Atlanta, Inc., Atlanta, Ga., Tech. Rep., 1970.
[9] W. V. T. Rusch and P. D. Potter, *Analysis of Reflector Antennas*. New York: Academic Press, 1970, p. 44.
[10] E. M. T. Jones, "Paraboloid reflector and hyperboloid lens antennas," *IRE Trans. Antennas Propagat.*, vol. AP-2, pp. 119–127, July 1954.
[11] I. Koffman, "Feed polarization for parallel currents in reflectors generated by conic sections," *IEEE Trans. Antennas Propagat.*, vol. AP-14, pp. 37–40, Jan. 1966.
[12] R. E. Collin and F. J. Zucker, *Antenna Theory*. New York: McGraw-Hill, 1969, part 2, p. 44.

CROSSPOLARISATION CHARACTERISTICS OF AXIALLY SYMMETRIC REFLECTORS

Indexing terms: Antenna feeders, Reflector antennas

The crosspolarisation levels of symmetric front-fed paraboloids and dual-reflector antenna systems have been determined for a wide range of reflector and feed parameters. It is shown that minimum crosspolarisation is achieved when the feed radiates a polarisation-pure field, and, in this case, the front-fed paraboloid is superior.

There is considerable interest concerning the crosspolarisation performance of axially symmetric front-fed and dual-reflector systems. In addition, some controversy exists[1-3] regarding their relative performance and the optimum type of feed required. In this letter, these problems are examined by both the geometric optics and vector diffraction methods to provide quantitative answers.

Geometric-optics solution: Let the orthogonal components of the field scattered by a symmetric reflector (diameter D) in a direction specified by polar angles (θ, ϕ) be given by $E_\theta(\theta)\cos\phi$ and $E_\phi(\theta)\sin\phi$. If the corresponding components of the incident field radiated by the feed in a direction (ξ, η) are $E_\xi(\xi)\cos\eta$ and $E_\eta(\xi)\sin\eta$, then, in the geometric-optics limit $(\lambda/D \rightarrow 0)$, the two fields[4, 5] are related by $E_\theta(\theta)/E_\phi(\theta) = E_\xi(\xi)/E_\eta(\xi)$, with $\phi = \eta$. Consequently, if the feed is 'balanced', so that the incident field satisfies $E_\xi(\xi) = -E_\eta(\xi)$, both the incident and output fields are 'pure' in the sense that they are free of crosspolarisation.

We may note first that input fields which induce reflector currents to flow in parallel paths do not, in general, yield pure output fields. Koffman[6] showed that such currents would be induced in a reflector of eccentricity e by a feed consisting of crossed electric and magnetic dipoles adjusted so that

$$E_\xi(\xi) = 1 - e\cos\xi \qquad E_\eta(\xi) = e - \cos\xi \quad . \quad . \quad (1)$$

For a paraboloid $(e = -1)$, eqn. 1 reduces to the field of a balanced Huygens source, giving $E_\xi(\xi) = -E_\eta(\xi) = 1 + \cos\xi$. In this particular case, zero crosspolarisation of the output field exists, at least in the limit $\lambda/D \rightarrow 0$.

For any other reflector shape, however, the reflected fields will contain a crosspolarised component, because eqn. 1 does not permit pure input fields. It can be shown that in the $\phi = \pm 45$ planes the ratio of the crosspolarised to the copolarised field of the reflected wave is $(1/m)\tan^2\theta/2$, where m is the reflector magnification factor $[m = (e+1)/(e-1)]$. For a hyperboloidal reflector with input and output semiangles ξ_0 and θ_0 of 15 and 60, respectively $(m = 4{\cdot}38)$, this ratio is $-22{\cdot}4$ dB at θ_0. This is actually much worse than the crosspolarised ratio for a dipole feed, which is given by $(1/m^2)\tan^2\theta/2$ and, for the reflector above, is only -35 dB.

In general, the geometric-optics solution gives the smoothed levels for the copolarised and crosspolarised components of the reflected fields, provided that the polarisation of the input field is not too close to the pure condition. When diffraction effects are ignored, these levels will be independent of λ/D. As the input field approaches purity, the crosspolarised level predicted by this method tends to zero. However, reflector depolarisation then becomes dominant and must be calculated by the vector-diffraction method.

Vector-diffraction solution: The field scattered by an annular zone of the reflector will be calculated from the distribution of current induced on the surface.[7] Diffraction effects caused by feed supports and subreflector blockage will be ignored. The factors $E_\theta(\theta)$ and $E_\phi(\theta)$, expressed in terms of the circumferential and radial currents on the reflector, $J_\eta(\xi)$ and $J_t(\xi)$, respectively, are given by

$$\left.\begin{aligned} E_\theta(\theta) &= -Z_0\,A(\theta,\xi)\,J_\eta(\xi) + Z_0\,C(\theta,\xi)\,J_t(\xi) \\ E_\phi(\theta) &= Z_0\,B(\theta,\xi)\,J_\eta(\xi) - Z_0\,D(\theta,\xi)\,J_t(\xi) \end{aligned}\right\} \quad . \quad (2)$$

where

$$\left.\begin{aligned} A(\theta,\xi) &= \cos\theta\,J_1(w)/w \\ B(\theta,\xi) &= J_1'(w) \\ C(\theta,\xi) &= \cos\theta\,g(\xi)\,J_1'(w) + j\sin\theta\,h(\xi)\,J_1(w) \\ D(\theta,\xi) &= g(\xi)\,J_1(w)/w \\ \text{and} \quad w &= k\rho\sin\theta\sin\xi \end{aligned}\right\} \quad (3)$$

The functions $g(\xi)$, $h(\xi)$ depend only on the shape of the reflector profile and result from the resolution of J_t into components, respectively, parallel and perpendicular to the aperture plane. From eqn. 2, pure output polarisation requires

$$A(\theta,\xi)/D(\theta,\xi) = B(\theta,\xi)/C(\theta,\xi) = -J_t(\xi)/J_\eta(\xi) \quad (4)$$

From the boundary condition and the relation $\boldsymbol{E}^i = Z_0\boldsymbol{H}^i$, the current $\boldsymbol{J}$ in terms of $\boldsymbol{E}^i$ is given by

$$J_t(\xi) = 2E_\xi(\xi)/Z_0 \qquad J_\eta(\xi) = 2E_\eta(\xi)/Z_0\,p(\xi) \quad . \quad (5)$$

where $p(\xi)$ equals the secant of the angle of incidence, Inspection of eqns. 3, 4 and 5 shows that pure polarisation can be approached only if θ is small $(\cos\theta \rightarrow 1, \sin\theta \rightarrow 0)$ and if the incident field satisfies

$$\frac{E_\eta(\xi)}{E_\xi(\xi)} = -g(\xi)\,p(\xi) = \frac{e - \cos\xi}{1 - e\cos\xi} \quad . \quad . \quad . \quad . \quad (6)$$

This is the field of an unbalanced Huygens source adjusted so that the current flows in parallel paths (eqn. 1). Eqn. 6 can also be closely approximated over a limited range of ξ by using the appropriate value of mode-content factor[8] γ for the HE_{11}-mode corrugated waveguide feed. For $e > 1$, a value of γ greater than unity is required, and, for $e < 1$, $\gamma < 1$. In general, the crosspolarised level, when $\boldsymbol{E}^i$ satisfies eqn. 6, is less than that provided by a balanced source only within a small angular region around the axis which contracts as $\lambda/D \rightarrow 0$. For larger values of θ, eqn. 4 cannot be satisfied, because of asymmetry introduced when the radiation from the axial component of $J_t(\xi)$ becomes significant [second term of $C(\theta, \xi)$] and when the $\cos\theta$ factors in $A(\theta, \xi)$ and $C(\theta, \xi)$ depart from unity. Over a practical range of θ, it is found that the best overall result is achieved with a balanced source. This is in agreement with the geometric-optics solution.

Fig. 1*a* shows the level of the first crosspolarised sidelobe of a paraboloidal reflector with $\theta_0 = 60°$ as a function of γ. The normalised radius ka of the feed is chosen to give maximum reflector gain for each value of γ; when the feed is balanced $(\gamma = 1)$, $ka = 4{\cdot}0$. Also shown is the corresponding level for the feed. The values for the often used TE_{11}-mode feed are given by $1/\gamma = 0$. Fig. 1*a* shows that minimum crosspolarisation occurs when $\gamma = 1$, except for small D/λ, where the optimum value of γ is slightly greater than unity. As D/λ decreases, θ is no longer small across the main beam and the crosspolarised level increases because the reflector introduces depolarisation.

As γ departs from unity, the crosspolarisation of the feed increases very rapidly, and, for relatively large D/λ, masks the depolarisation caused by the reflector. The crosspolarisation of the output fields then tends to be independent of D/λ. Curves for other values of θ_0 are similar to those shown in Fig. 1*a*, except near $\gamma = 1$, where the crosspolarised level for a given value of D/λ increases as θ_0 increases, as shown in Fig. 2.

For other reflector shapes, such as hyperboloidal and ellipsoidal, large values for θ in eqns. 3 are usually involved, so that depolarisation is greater than for the paraboloidal reflector. Fig. 1*b* shows the crosspolarised performance of a Cassegrain system with $\xi_0 = 15°$ and $\theta_0 = 60°$ (feed radius

Reprinted with permission from *Electron. Lett.*, vol. 12, pp. 218–219, Apr. 29, 1976.

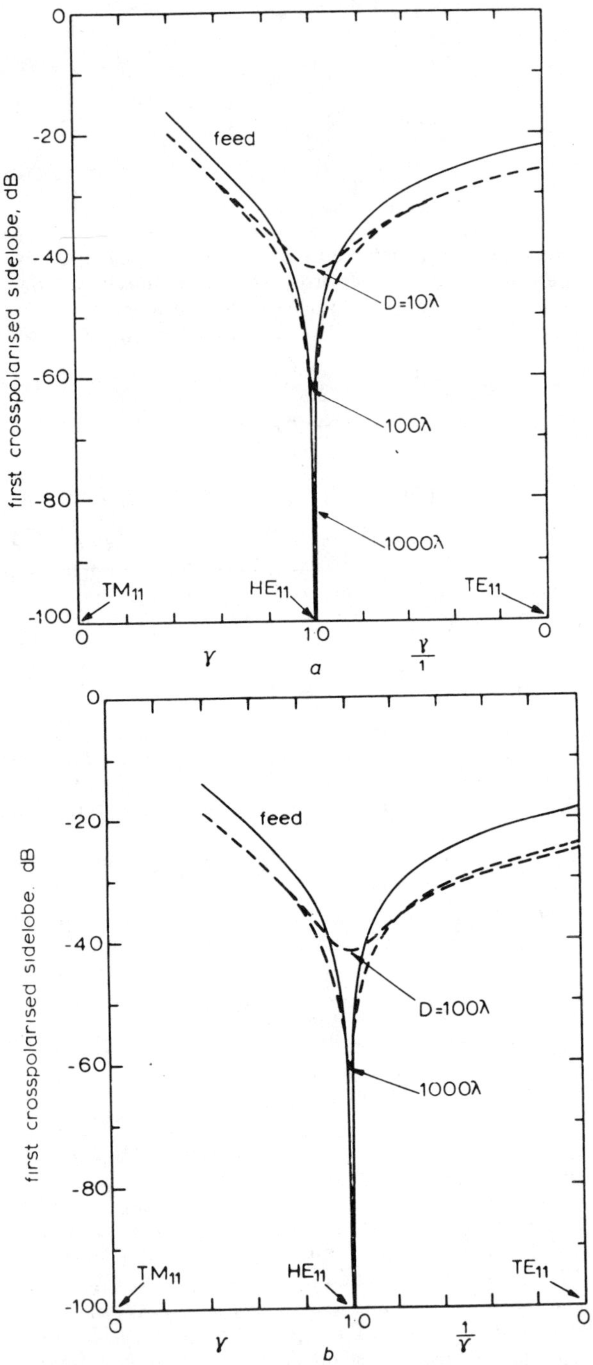

Fig. 1 *Level of first crosspolarised sidelobe as function of the mode-content factor γ of cylindrical HE_{11} feed*

D = main reflector diameter, d = subreflector diameter
a Paraboloid, $\theta_0 = 60°$
b Cassegrain, $\xi_0 = 15°$, $\theta_0 = 60°$, $D/d = 10$
——— feed
— — — — main reflector

$ka = 15$ at $\gamma = 1$). The minimum crosspolarised level for a given value of D/λ again occurs when $\gamma = 1$.

Fig. 2 summarises the performance of front-fed Gregorian and Cassegrain systems for a range of paraboloidal input angles and diameters. In each case, a balanced HE_{11} feed is assumed. The curves for the dual-reflector systems are virtually independent of the input angle ξ_0, and, for a given value of D, the crosspolarised level increases as the subreflector diameter d decreases. The curves for a TE_{11} feed are also

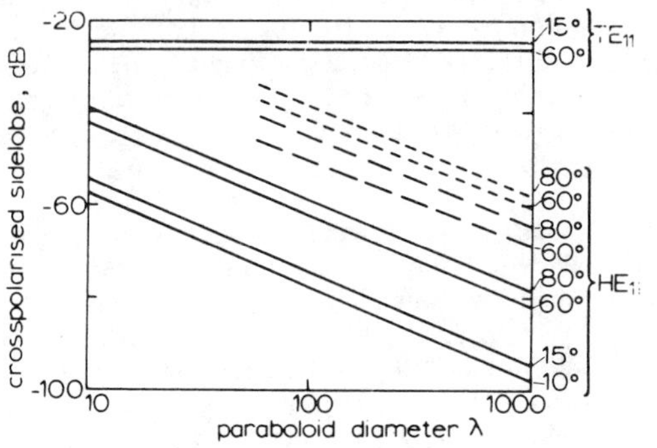

Fig. 2 *Main-reflector crosspolarisation characteristics using balanced ($\gamma = 1$) HE_{11} feed and TE_{11} feed*

The parameter is θ_0, the semiangle of the main reflector
——— Paraboloid
– – – – Gregorian, $D/d = 10$
- - - - - Cassegrain, $D/d = 10$

shown for comparison. They are computed for a front-fed paraboloid, but apply closely for dual-reflector systems.

Fig. 2 shows that the crosspolarised performance of dual-reflector systems illuminated by a balanced feed is inherently inferior to that of a front-fed reflector, in agreement with Wood.[1, 2] For both reflector systems, the low levels of cross-polarisation predicted for large D/λ when $\gamma = 1$ may be difficult to achieve in practice, because of possible deviations from the predicted circularly symmetric amplitude and phase patterns of the feed, scattering from the feed or subreflector supports and reflector surface irregularities. Some of these effects are being investigated.

B. MACA. THOMAS *26th March 1976*

Division of Radiophysics, CSIRO
PO Box 76
Epping, NSW 2121, Australia

References

1 WOOD, P. J.: 'Depolarisation with Cassegrainian and front-fed reflectors', *Electron. Lett.*, 1973, **9**, pp. 181–183
2 WOOD, P. J.: 'Crosspolarisation with Cassegrainian and front-fed reflectors', *ibid.*, 1973, **9**, pp. 597–598
3 WATSON, P. A., and GHOBRIAL, S. I.: 'Crosspolarisation in Cassegrain and front-fed antennas', *ibid.*, 1973, **9**, pp. 297–298
4 POTTER, P. D.: 'Aperture illumination and gain of a Cassegrainian system', *IEEE Trans.*, 1963, **AP-11**, pp. 373–375
5 RUSCH, W. V. T., and POTTER, P. D.: 'Analysis of reflector antennas' (Academic Press, 1970), section 2.3.
6 KOFFMAN, I.: 'Feed polarization for parallel currents in reflectors generated by conic sections', *IEEE Trans.*, 1966, **AP-14**, pp. 37–40
7 SILVER, S.: 'Microwave antenna theory and design' (McGraw-Hill, 1949)
8 THOMAS, B. MACA.: 'Theoretical performance of prime-focus paraboloids using cylindrical hybrid-mode feeds', *Proc. IEE*, 1971, **118**, (11), pp. 1539–1549

SYMMETRIC REFLECTORS WITH ZERO DEPOLARISATION

Indexing terms: Antenna feeders, Polarisation, Reflector antennas

The polarisation performance of a symmetric reflector antenna illuminated by an ideal feed is limited by depolarisation introduced by the reflector, even if other sources of cross-polarisation are negligible. It is shown that depolarisation can theoretically be eliminated by the use of anisotropic reflectors.

In symmetric reflector antennas used in radioastronomy and satellite communication, low crosspolarisation levels are often of great importance. Feeds are known which, theoretically, are capable of radiating pure polarised fields. However, even if the feed is ideal, the purity of its radiated field is degraded at each reflection in a multiple-reflector antenna. This sets a minimum theoretical level for the crosspolarised output attainable with conducting reflector surfaces.[1] It is shown below that, in principle, this limitation can be removed if the surfaces are reactive and anisotropic.

We are concerned here with axially symmetric reflector systems, and the fields of interest are those in which the $\boldsymbol{E}$ and $\boldsymbol{H}$ distributions are identical, but displaced by 90° about the axis of symmetry. As a result, the flow of energy is symmetric about the axis.[2,3] Fields of this type exist throughout the interior and exterior regions of cylindrical and conical anisotropic feeds adjusted to the balanced condition. It can also be shown that they are excited throughout the near- and far-field regions of a balanced Huygens source consisting of orthogonal electric and magnetic current elements ($\boldsymbol{J}$ and $\boldsymbol{K}$) adjusted so that $\boldsymbol{K} = \pm Z_0 \boldsymbol{J}$. All fields in this general class consist of equal TE and TM components, with either the same or opposite signs, and we shall therefore refer to them as balanced fields.

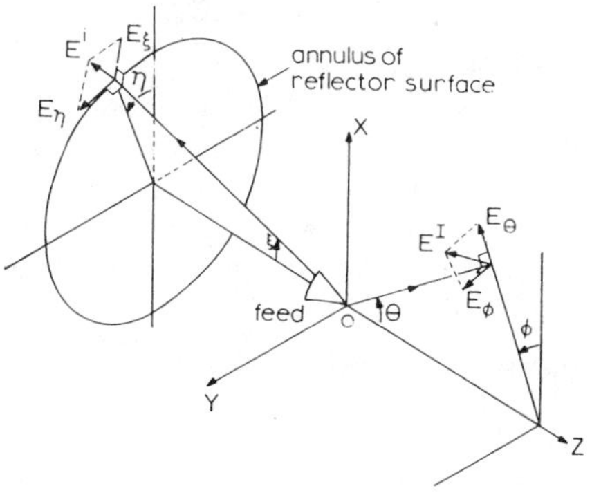

Fig. 1 *Co-ordinate systems for field $\boldsymbol{E}^i$ incident on an annulus of a symmetric reflector and for the far-zone reflected field $\boldsymbol{E}^r$*

A few diameters beyond the feed aperture (Fig. 1) balanced fields have the form

$$\left.\begin{aligned} \boldsymbol{E}^i &= \boldsymbol{i}_\xi E_\xi(\xi)\cos\eta + \boldsymbol{i}_\eta E_\eta(\xi)\sin\eta \\ &\text{with } E_\xi(\xi) = \pm E_\eta(\xi) \end{aligned}\right\} \quad (1)$$

where the dependence on time and distance has been suppressed for convenience. The field corresponding to the lower sign in eqn. 1 has zero crosspolarisation (corresponding to Ludwig's third definition[4]) and in this sense may be called a 'pure' balanced field. For brevity we shall designate it as HE_1, using hybrid wave terminology. In the second balanced field (upper sign), which will be designated EH_1, the polarisation direction varies with η and the field is zero along the axis. Because of these apparently undesirable characteristics, there has been no reported development of EH_1 feeds.

Consider a symmetric, perfectly conducting reflector (Fig. 1) illuminated by a field $\boldsymbol{E}^i$ given by eqn. 1. The reflected field $\boldsymbol{E}^r$ is balanced if it satisfies

$$\left.\begin{aligned} \boldsymbol{E}^r &= \boldsymbol{i}_\theta E_\theta(\theta)\cos\phi + \boldsymbol{i}_\phi E_\phi(\theta)\sin\phi \\ &\text{with } E_\theta(\theta) = \pm E_\phi(\theta) \end{aligned}\right\} \quad (2)$$

where the phase factor is omitted, as before. The lower sign is necessary for pure polarisation. The expressions for $E_\theta(\theta)$ and $E_\phi(\theta)$ produced by induced currents $J_\eta = J_\eta(\xi)\sin\eta$ and $J_t = J_t(\xi)\cos\eta$ on an annulus of reflector surface are found to be

$$\left.\begin{aligned} E_\theta(\theta) &= -Z_0\,A(\theta,\xi)J_\eta(\xi) + Z_0\,C(\theta,\xi)J_t(\xi) \\ E_\phi(\theta) &= Z_0\,B(\theta,\xi)J_\eta(\xi) - Z_0\,D(\theta,\xi)J_t(\xi) \end{aligned}\right\} \quad (3)$$

where the functions $A(\theta,\xi)$, $B(\theta,\xi)$, $C(\theta,\xi)$, $D(\theta,\xi)$ depend only on the reflector profile. Purity of the reflected field therefore requires

$$A(\theta,\xi)/D(\theta,\xi) = B(\theta,\xi)/C(\theta,\xi) = -J_t(\xi)/J_\eta(\xi) \quad (4)$$

and detailed analysis shows[1] that this is possible only within a small cone of directions centred on the axis ($\theta = 0$). Thus, even when the incident field is pure, the purity of the output field is degraded by reflector depolarisation.

Suppose now that we do not specify the reflector surface and simply require balanced fields (either HE_1 or EH_1) before and after reflection. Balanced fields require anisotropic boundary surfaces to ensure that $\boldsymbol{E}$ and $\boldsymbol{H}$ are treated equally. It follows that all surfaces encountered by the fields in an antenna system must satisfy this requirement if field balance is to be preserved throughout. For example, we have found that our cylindrical HE_1 feeds[2,5] perform better when the interior corrugations extend to part of the exterior ground plane.* It is reasonable therefore to suppose that reflector depolarisation might be eliminated with anisotropic reflector surfaces.

A 1st-order proof of this hypothesis will be given in outline. Fields incident on an anisotropic reflector induce both electric currents $\boldsymbol{J}$ and magnetic currents $\boldsymbol{K}$ (with components $K_\eta = K_\eta(\xi)\cos\eta$, $K_t = K_t(\xi)\sin\eta$). Eqn. 3, generalised to include the contribution from $\boldsymbol{K}$, becomes

$$\left.\begin{aligned} E_\theta(\theta) &= -Z_0\,AJ_\eta(\xi) + BK_\eta(\xi) + Z_0\,CJ_t(\xi) + DK_t(\xi) \\ E_\phi(\theta) &= -AK_\eta(\xi) + Z_0\,BJ_\eta(\xi) - CK_t(\xi) - Z_0\,DJ_t(\xi) \end{aligned}\right\} \quad (5)$$

where the functional dependence of A, B, C and D on ξ and θ has been dropped for brevity. The reflector surface may be characterised by two orthogonal surface reactances, normalised to the intrinsic impedance Z_0 of free space:

$$X_\eta = \frac{-jK_t}{Z_0\,J_\eta} \qquad X_t = \frac{jK_\eta}{Z_0\,J_t} \quad (6)$$

Eliminating $\boldsymbol{K}$ from eqn. 5 gives

$$\left.\begin{aligned} E_\theta(\theta) &= Z_0\,J_\eta(\xi)(-A + jX_\eta D) + Z_0\,J_t(\xi)(C - jX_t B) \\ E_\phi(\theta) &= Z_0\,J_\eta(\xi)(B - jX_\eta C) + Z_0\,J_t(\xi)(-D + jX_t A) \end{aligned}\right\} \quad (7)$$

The induced electric currents are related to the total tangential magnetic field $\boldsymbol{H}^T = \boldsymbol{H}^i + \boldsymbol{H}^r$ by the boundary conditions $J_\eta = -H_t^T$ and $J_t = -H_\eta^T$. The reflected field $\boldsymbol{H}^r$ may be calculated from the mismatch between the surface reactance and the wave impedance to obtain

$$\frac{H_t^T}{H_t^i} = \frac{2}{1 + j(X_\eta/p)} \qquad \frac{H_\eta^T}{H_\eta^i} = \frac{2}{1 + jpX_t}$$

where p equals the secant of the angle of incidence and is a function of ξ. Finally, with the relation $\boldsymbol{E}^i = Z_0\boldsymbol{H}^i$, the induced currents are given by

$$J_\eta = \frac{2E_\eta}{pZ_0(1 + j(X_\eta/p))} \qquad J_t = \frac{2E_\xi}{Z_0(1 + jpX_t)} \quad (8)$$

Substituting eqn. 8 in eqn. 7 gives the reflected fields in terms of the incident fields:

$$\left.\begin{aligned} E_\theta(\theta) &= -2\left\{\frac{E_\eta(\xi)(A - jX_\eta D)}{p + jX_\eta} + \frac{E_\xi(\xi)(B - (C/jX_t))}{p + (1/jX_t)}\right\} \\ E_\phi(\theta) &= 2\left\{\frac{E_\eta(\xi)(B - jX_\eta C)}{p + jX_\eta} + \frac{E_\xi(\xi)(A - (D/jX_t))}{p + (1/jX_t)}\right\} \end{aligned}\right\} \quad (9)$$

By inspection,

$$E_\theta(\theta) = \pm E_\phi(\theta) \quad (10)$$

* The importance of this principle, especially for small apertures, appears to have been overlooked.[6] The unbalancing effect of a smooth ground plane is evident in the theoretical and experimental results published recently by Parini *et al.*[7]

Reprinted with permission from *Electron. Lett.*, vol. 12, pp. 291-293, May 27, 1976.

when

$$E_\xi(\xi) = \mp E_\eta(\xi) \quad . \quad . \quad . \quad . \quad . \quad . \quad . \quad . \quad . \quad . \quad (11)$$

and

$$X_\eta X_t = -1 \quad . \quad . \quad . \quad . \quad . \quad . \quad . \quad . \quad . \quad . \quad (12)$$

Thus a surface satisfying eqn. 12 reflects balanced fields for all θ, when the incident fields are balanced. This condition is identical with that previously established for cylindrical anisotropic waveguides propagating balanced hybrid modes.[2] The relationship of the signs in eqns. 10 and 11 shows that field conversion takes place on reflection; thus an EH_1 incident field converts to an HE_1 reflected field, and vice versa. When all the surfaces of a multiple-reflector system satisfy the condition of eqn. 12, field balance is preserved throughout the system and EH_1 and HE_1 fields alternate in successive regions between reflectors. It is essential that an HE_1 field should emerge from the exit aperture for the output polarisation of the antenna system to be pure. The number of reflectors therefore determines whether the feed should be an HE_1 or EH_1 type.

It is interesting to note that a linearly polarised plane wave, i.e. an HE_1 field, incident normally on a smooth reflector, generates axial HE_1 waves along the axis which are exactly balanced only at the focal plane.[8] With a balanced anisotropic reflector, however, it is found that the reflected fields represent axial EH_1 waves which are balanced everywhere, as expected from the above theory.

Two particular surface structures satisfying eqn. 12 are of interest. If the surface permits only circumferential currents, as in Fig. 2*a*, the surface reactances are $X_t = \infty$ and $X_\eta = 0$. This corresponds to a circumferentially slotted surface, analogous to the usual corrugated waveguide, and has the advantage that the edge of the reflector does not perturb the current lines. Another structure, which is theoretically possible, constrains the currents to flow radially, as in Fig. 2*b*, by means of radial slots adjusted to make $X_t = 0$ and $X_\eta = \infty$. This is analogous to a longitudinally slotted waveguide.

The polarisation performance of a large front-fed smooth paraboloid ($D/\lambda \geqslant 100$) is likely to be limited in practice by feed deficiencies and support scattering, rather than reflector depolarisation. However, in a Cassegrain system the subreflector produces about 20 dB more crosspolarisation than the paraboloid[1] and, especially in small antennas ($D/\lambda \simeq 100$), may not do justice to the best-balanced feeds which can be

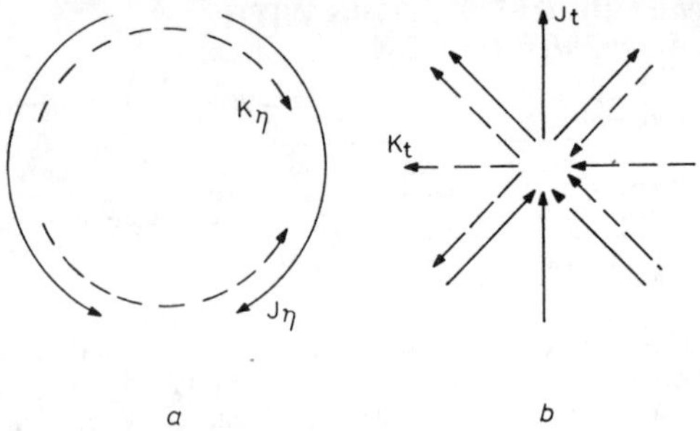

Fig. 2 *Projection on aperture plane of surface currents induced by an EH_1 field incident on symmetric anisotropic reflector*

For an HE_1 incident field the relative directions of J and K are reversed
a Currents constrained circumferentially
b Currents constrained radially
—— J
- - - - K

made. In this case there appears to be scope for removing this limitation by using an anisotropic subreflector with a smooth main reflector. We are therefore investigating EH_1 balanced feeds for this application.

H. C. MINNETT
B.MACA. THOMAS

3rd May 1976

Division of Radiophysics
CSIRO, Sydney, Australia

References

1 THOMAS, B. MACA.: 'Crosspolarisation characteristics of axially symmetric reflectors', *Electron. Lett.*, 1976, **12**, pp. 218–219
2 MINNETT, H. C., and THOMAS, B. MACA.: 'A method of synthesizing radiation patterns with axial symmetry', *IEEE Trans*, 1966, **AP-14**, pp. 654–656
3 RUMSEY, V. H.: 'Horn antennas with uniform power patterns around their axes', *ibid.*, 1966, **AP-14**, pp. 656–658
4 LUDWIG, A. C.: 'The definition of cross-polarisation', *ibid.*, 1973, **AP-21**, pp. 116–119
5 THOMAS, B. MACA.: 'Prime-focus one- and two-hybrid-mode feeds', *Electron. Lett.*, 1970, **6**, pp. 460–461
6 KNAPP, C. M., and WIESENFARTH, H. J.: 'On the radiation from an open-ended corrugated pipe carrying the HE_{11} mode', *IEEE Trans.*, 1972, **AP-20**, pp. 644–648
7 PARINI, C. G., CLARRICOATS, P. J. B., and OLVER, A. D.: 'Crosspolar radiation from open-ended corrugated wave uides', *Electron. Lett.*, 1975, **11**, pp. 567–568
8 MINNETT, H. C., and THOMAS, B. MACA.: 'Fields in the image space of symmetrical focusing reflectors', *Proc. IEE*, 1968, **115**, (10), pp. 1419–1430

Cross Polarization in Cassegrainian and Front-Fed Paraboloidal Antennas

MEHMET SAFAK AND PAUL P. DELOGNE

Abstract—**This paper presents the formulas relating feed polarization to aperture plane polarization, shows that the cross polarization in reflector antennas is closely related to the feed design, gives the ideal feed polarizations yielding zero aperture cross-polarized fields, and also puts an end to the controversy recently arisen about the relative susceptibility of Cassegrainian and front-fed antennas to depolarization effects.**

Introduction and Definitions

Frequency reuse by means of cross-polarized channels is intended for future satellite communication systems. One of the most important problems with respect to this application is the design of reflector antennas with a very high polarization purity, at least over the main lobe.

First of all, the polarization purity of an antenna must be clearly defined, even for a single polarization antenna. For doing this, it is necessary to define in each radial direction a nominal polarization. Considering the spherical coordinate system, the nominal polarization in a given direction (θ,ϕ) is defined by a unit complex vector in the base subtended by the unit vectors $(\bar{u}_\theta,\bar{u}_\phi)$, say

$$\bar{e}_n(\theta,\phi) = e_{n\theta}\bar{u}_\theta + e_{n\phi}\bar{u}_\phi \tag{1}$$

where $e_{n\theta}$ and $e_{n\phi}$ are given functions of (θ,ϕ), with

$$|\bar{e}_n|^2 = |e_{n\theta}|^2 + |e_{n\phi}|^2 = 1. \tag{2}$$

Of course, it is immediately possible to define in the base $(\bar{u}_\theta,\bar{u}_\phi)$, a unit vector

$$\bar{e}_\perp(\theta,\phi) = e_{\perp\theta}\bar{u}_\theta + e_{\perp\phi}\bar{u}_\phi \tag{3}$$

which is orthogonal to $\bar{e}_n$ for all radial directions, i.e.,

$$\bar{e}_\perp \cdot \bar{e}_n^* = 0. \tag{4}$$

More specifically, defining a complex angle ψ by

$$\tan\psi = \frac{e_{n\phi}}{e_{n\theta}} \tag{5}$$

and writing

$$\bar{e}_n = e_{n\theta}(\bar{u}_\theta + \tan\psi\bar{u}_\phi) \tag{6}$$

Manuscript received July 13, 1974; revised January 13, 1976.
M. Safak was with the Telecommunications and Microwaves Laboratory, Louvain University, Louvain-la-Neuve, Belgium. He is now with the Electrical Engineering Department, Technological University, Eindhoven, The Netherlands.
P. P. Delogne is with the Telecommunications and Microwaves Laboratory, Louvain University, Louvain-la-Neuve, Belgium.

one has, apart from an arbitrary phase,

$$\bar{e}_\perp = e_{n\theta}\tan\psi(\bar{u}_\theta - \cot^*\psi\bar{u}_\phi). \tag{7}$$

Now, let the fields radiated by the antenna be

$$\bar{E}(R\bar{u}_R) = \frac{e^{-jkR}}{R}\bar{F}(\theta,\phi) = \frac{e^{-jkR}}{R}(F_n\bar{e}_n + F_\perp\bar{e}_\perp) \tag{8}$$

the cross polarization of the antenna is defined as

$$C(\theta,\phi) = -20\log\left|\frac{F_\perp}{F_n}\right|. \tag{9}$$

This definition also applies to double polarization antennas; the two nominal polarizations should then obviously be taken orthogonal, i.e., $\bar{e}_n$ and $\bar{e}_\perp$.

Choice of the Nominal Polarization

In a specific application to a very directive antenna, the choice of the nominal polarization depends, among other factors, on the main type of antenna misalignment or on the variation of arrival angle that can occur. Right- or left-handed circular polarization in all directions is of course ideal with respect to roll (rotation about broadside), pitch or yaw misalignment, but can be difficult to achieve. If roll is not important, linear polarization in broadside direction is useful, but what should be the nominal polarization in other directions?

Considering the z axis as broadside and, for that particular direction a nominal polarization parallel to the x axis, the following definitions have been proposed by Ludwig [1].

1) A nominal polarization identical to that of a Hertz-dipole parallel to the x axis, that is,

$$\bar{e}_n = \frac{\cos\theta\cos\phi\bar{u}_\theta - \sin\phi\bar{u}_\phi}{\sqrt{1-\sin^2\theta\cos^2\phi}}. \tag{10}$$

In this case,

$$\bar{e}_\perp = \frac{\sin\phi\bar{u}_\theta + \cos\theta\cos\phi\bar{u}_\phi}{\sqrt{1-\sin^2\theta\cos^2\phi}}. \tag{11}$$

2) A nominal polarization identical to that of a Huygens source with electric field parallel to the x axis, that is,

$$\bar{e}_n = \cos\phi\bar{u}_\theta - \sin\phi\bar{u}_\phi. \tag{12}$$

In this case,

$$\bar{e}_\perp = \sin\phi\bar{u}_\theta + \cos\phi\bar{u}_\phi. \tag{13}$$

Reprinted from *IEEE Trans. Antennas Propagat.*, vol. AP-24, pp. 497–501, July 1976.

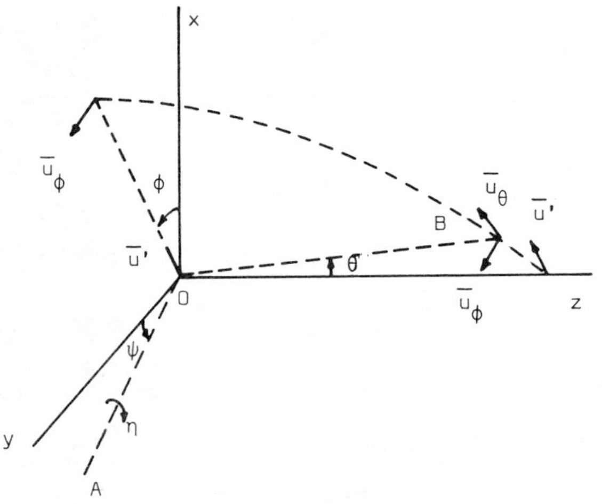

Fig. 1. Effect of antenna misalignment.

Some comments on Huygens sources will be done in the next paragraph. Let us now only note that definitions (10) and (12) are nearly equivalent for small angles θ; indeed, $\cos\theta > 0.99$ for $\theta < 8°$, so that the cross polarization of (10) with respect to (12) is less than 40 dB for these angles. So, both definitions are practically equivalent for the main lobe of all earth stations and also of all geostationary satellites. But they are of course not if larger angles are involved as in the case of the antenna feed or of the feed-primary reflector system of a Cassegrainian structure.

It is useful to see what is the performance of the nominal polarization (12) in case of antenna misalignment. If it is supposed that no roll (rotation about the z axis) exists, every misalignment is equivalent to a rotation of the antenna through some angle η about some axis OA located in the xy plane and making an angle ψ with Oy. As Fig. 1 shows, the direction OB transforming into the z-direction has an azimuth $\phi = \psi$ and a colatitude $\theta = \eta$; this rotation transforms $\bar{u}_\theta$ into $\bar{u}' = \cos\phi\bar{u}_x + \sin\phi\bar{u}_y$ while $\bar{u}_\phi = -\sin\phi\bar{u}_x + \cos\phi\bar{u}_y$ remains unchanged. It results that $\bar{e}_n$ given by (12) transforms into $\bar{u}_x$; a receiving antenna located on the original broadside direction will not suffer depolarization due to that type of antenna misalignment and it can be said that the nominal linear polarization (12) is the ideal one.

Huygens Source

The nominal polarization (12) is currently referred to as that of a Huygens source. In fact, a Huygens source is defined as a plane aperture (in the xy plane) illuminated by fields whose tangential components are related by

$$\bar{H}_t = \frac{1}{\eta}\bar{u}_z \times \bar{E}_t \tag{14}$$

where η is the instrinsic impedance of free space. This relation is also that existing between the fields of a uniform plane wave propagating in the z-direction. However, it is well known from the uniqueness theorem that $\bar{E}_t$ or $\bar{H}_t$ defines completely the fields for $z > 0$, so that one can only choose arbitrarily one of these components. It will be shown that (14) is possible only for an infinitely large aperture with uniform illumination.

Indeed, considering the two-dimensional Fourier transform of $\bar{E}_t$

$$\bar{f}(k_x,k_y) = \iint_{-\infty}^{\infty} \bar{E}_t(x,y)e^{j(xk_x+yk_y)}\,dx\,dy \tag{15}$$

and defining a vector $\bar{k} = k_x\bar{u}_x + k_y\bar{u}_y + k_z\bar{u}_z$, with k_z such that $\bar{k}\cdot\bar{k} = k_0{}^2$, the Fourier transform of $\bar{H}_t$ is found equal to [2]

$$\bar{g}(k_x,k_y) = \frac{1}{k_0\eta}\bar{k} \times \bar{f} \tag{16}$$

while (14) implies

$$\bar{g} = \frac{1}{\eta}\bar{u}_z \times \bar{f}. \tag{17}$$

As a result, (14) is only possible if $\bar{k} = k_0\bar{u}_z$, that is if $k_x = k_y = 0$. However, a Fourier transform which is nonzero for these values only is a Dirac impulse corresponding to $\bar{E}_t$ constant on the whole plane.

In fact, (14) can only be a good approximation in the case of large apertures with illuminations which vary slowly compared to the wavelength. This is the case for the aperture formed by a fairly large paraboloid with a smooth illumination. The problem of how well a horn or waveguide aperture conforms to (14) will be examined later.

Now, as explained in Collin and Zucker [2], there exist three methods of calculating the fields radiated by a planar aperture, by using $\bar{E}_t$, $\bar{H}t$, or both, respectively. They, all three, yield the same result provided exact values of the aperture fields are used. But if one uses values of $\bar{H}_t$ related to $\bar{E}_t$ by the approximate expression (14), the three results are, respectively,

$$E_\theta = \frac{j}{\lambda}\frac{e^{-jkR}}{R}(f_x\cos\phi + f_y\sin\phi)$$

$$E_\phi = \frac{j}{\lambda}\frac{e^{-jkR}}{R}\cos\theta(f_y\cos\phi - f_x\sin\phi) \tag{18}$$

$$E_\theta = \frac{j}{\lambda}\frac{e^{-jkR}}{R}\cos\theta(f_x\cos\phi + f_y\sin\phi)$$

$$E_\phi = \frac{j}{\lambda}\frac{e^{-jkR}}{R}(f_y\cos\phi - f_x\sin\phi) \tag{19}$$

and

$$E_\theta = \frac{j}{\lambda}\frac{e^{-jkR}}{R}\frac{1+\cos\theta}{2}(f_x\cos\phi + f_y\sin\phi)$$

$$E_\phi = \frac{j}{\lambda}\frac{e^{-jkR}}{R}\frac{1+\cos\theta}{2}(f_y\cos\phi - f_x\sin\phi) \tag{20}$$

where f_x and f_y are evaluated in

$$k_x = k_0\cos\phi\sin\theta$$

$$k_y = k_0\sin\phi\sin\theta. \tag{21}$$

It is seen that the results (18) to (20) are different. This is due to the fact that (14) is approximate. Namely, it is seen that if the illumination $\bar{E}_t$ has only an x-component, (19) has the nominal polarization (10), (20) has the nominal polarization (12), while (18) has the polarization of a magnetic dipole parallel to the y axis. However, as was noted earlier, all three polarizations are equivalent with an error less than 40 dB for angles θ smaller than 8°. Consequently, *the ideal polarization of the illumination on the aperture plane of a large reflector antenna is any way with an electric field having a constant direction.* The next step is to see how to realize this polarization in a reflector antenna.

Aperture Plane Cross Polarization

Consider a Cassegrainian system as shown in Fig. 2. Let $\bar{e}_i$ be the polarization vector of the incident electric field in the plane perpendicular to $\bar{s}_i$, $\bar{e}_0$ in the plane perpendicular to $\bar{s}_0$, and $\bar{e}_r$

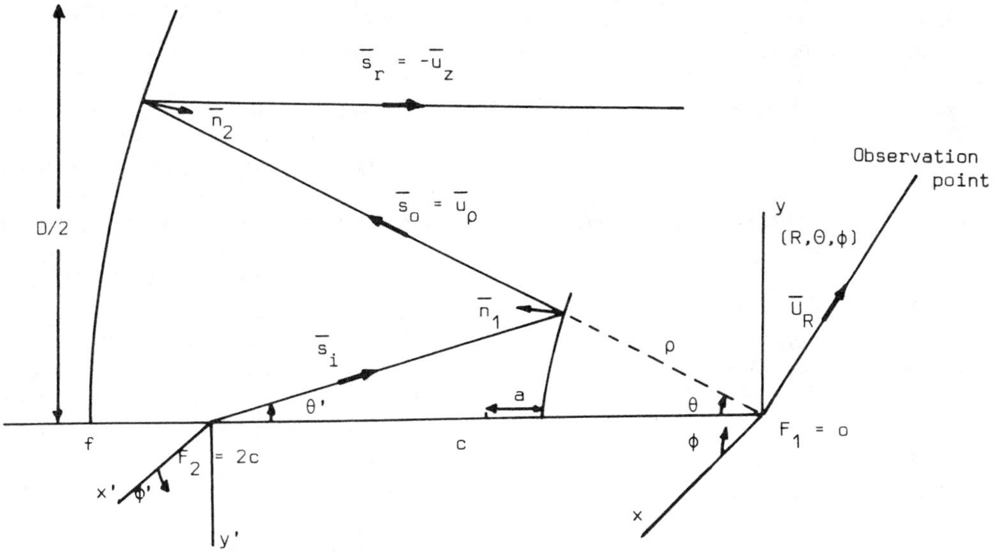

Fig. 2. Geometry of Cassegrainian system.

TABLE I
APERTURE PLANE POLARIZATION IN FRONT-FED ANTENNAS

Aperture Plane Polarization	Contribution to Aperture-Plane Polarization		
	x-directed dipole	y-directed dipole	z-directed dipole
e_{rx}	$-1 + (1 - \cos\theta)\cos^2\phi$	$(1 - \cos\theta)\sin\phi\cos\phi$	$\sin\theta\cos\phi$
e_{ry}	$(1 - \cos\theta)\sin\phi\cos\phi$	$-1 + (1 - \cos\theta)\sin^2\phi$	$\sin\theta\sin\phi$

TABLE II
APERTURE PLANE POLARIZATION IN CASSEGRAIN ANTENNAS

Aperture Plane Polarization	Contribution to Aperture-Plane Polarization		
	x-directed dipole	y-directed dipole	z-directed dipole
e_{rx}	$1 - \dfrac{(\varepsilon - 1)^2(1 - \cos\theta)\cos^2\phi}{q^2}$	$-\dfrac{(\varepsilon - 1)^2(1 - \cos\theta)\sin\phi\cos\phi}{q^2}$	$\dfrac{(\varepsilon^2 - 1)\sin\theta\cos\phi}{q^2}$
e_{ry}	$-\dfrac{(\varepsilon - 1)^2(1 - \cos\theta)\sin\phi\cos\phi}{q^2}$	$1 - \dfrac{(\varepsilon - 1)^2(1 - \cos\theta)\sin^2\phi}{q^2}$	$\dfrac{(\varepsilon^2 - 1)\sin\theta\sin\phi}{q^2}$

in the plane orthogonal to $\bar{s}_r$, i.e., xy plane. Let $\bar{n}_1$ and $\bar{n}_2$ denote the unit normal vectors on hyperboloidal and paraboloidal surfaces, respectively. From the electromagnetic boundary conditions on a perfect conductor one has

$$\bar{e}_0 = -\bar{e}_i + 2(\bar{n}_1 \cdot \bar{e}_i)\bar{n}_1 \tag{22}$$

$$\bar{e}_r = -\bar{e}_0 + 2(\bar{n}_2 \cdot \bar{e}_0)\bar{n}_2. \tag{23}$$

From (22) and (23) one easily obtains

$$\begin{bmatrix} e_{0\theta} \\ e_{0\phi} \end{bmatrix} = \begin{bmatrix} -1 & 0 \\ 0 & 1 \end{bmatrix} \begin{bmatrix} e_{i\theta}' \\ e_{i\phi}' \end{bmatrix} \tag{24}$$

and

$$\begin{bmatrix} e_{rx} \\ e_{ry} \end{bmatrix} = \begin{bmatrix} -\cos\phi & \sin\phi \\ -\sin\phi & -\cos\phi \end{bmatrix} \begin{bmatrix} e_{0\theta} \\ e_{0\phi} \end{bmatrix} \tag{25}$$

which are the reflecting matrices of hyberboloid and paraboloid, respectively. Combining the last two equations yields the total reflecting matrix of the Cassegrainian system

$$\begin{bmatrix} e_{rx} \\ e_{ry} \end{bmatrix} = \begin{bmatrix} \cos\phi & \sin\phi \\ \sin\phi & -\cos\phi \end{bmatrix} \begin{bmatrix} e_{i\theta}' \\ e_{i\phi}' \end{bmatrix} = \begin{bmatrix} \cos\phi' & -\sin\phi' \\ -\sin\phi' & -\cos\phi' \end{bmatrix} \begin{bmatrix} e_{i\theta}' \\ e_{i\phi}' \end{bmatrix}. \tag{26}$$

From (25) and (26), it is not difficult to see that *sources with the nominal polarization* (12) *or* (13) *located at the foci of front-fed or Cassegrainian antennas yield zero cross polarization at the aperture plane.* Then the whole problem reduces to the design of a feed having the nominal polarization given by (12) or (13) in all directions subtended by the subreflector in a Cassegrainian system or by the main reflector in a front-fed system Any other feed will yield cross polarized currents on the aperture plane.

For the sake of completeness, the contributions of x-, y-, and z-directed dipoles to aperture cross polarization of front-fed and Cassegrainian antennas are given in Tables I and II, respectively, where $\varepsilon = c/a$ and $q = (1 + 2\varepsilon\cos\theta + \varepsilon^2)^{1/2}$. Table II clearly shows that ε plays a very important role in the cross polarization characteristics of Cassegrainian antennas for dipole feeding; as it approaches unity, aperture cross polarization vanishes. On the other hand, comparison of these tables shows that aperture cross polarization in Cassegrainian systems is $(\varepsilon - 1)^2/q^2$ times less than that in front-fed antennas. For small values of θ, this factor is approximately equal to $1/M^2$ where $M = (\varepsilon + 1)/(\varepsilon - 1)$ is the magnification factor. This difference can readily be observed in [3, Fig. 11].

Comments on Waveguide Apertures and Horns

It is well known [4] that the tangential components of the fields of a waveguide mode propagating in the positive z-direction are related by

$$\text{TM mode} \qquad \bar{H}_t = \frac{k_0}{k_g \eta} \bar{u}_z \times \bar{E}_t \tag{27}$$

$$\text{TE mode} \qquad \bar{H}_t = \frac{k_g}{k_0 \eta} \bar{u}_z \times \bar{E}_t \tag{28}$$

where $k_g = 2\pi/\lambda_g$ is the phase constant of the mode. For the same mode propagating in the negative z-direction, $\bar{u}_z$ should be replaced by $(-\bar{u}_z)$. Consequently, insofar as it can be said that the waveguide aperture illumination contains only waveguide modes, two conditions are necessary in order to simulate a Huygens source defined by (14): k_g must be nearly equal to k_0 and the mouth reflection must be negligible. Both conditions are better fulfilled if the waveguide dimensions are large compared to the free space wavelength, but again, supposing they are, the radiated fields can only be calculated with a good accuracy for angles θ smaller than 8°. For larger angles, different formulas yield different polarizations and nothing can be said about the correctness of the Huygens source characteristics. Of course, the larger the waveguide mouth, the smaller its beamwidth.

A result of this discussion is that large feeds with constant polarization and, consequently, small feed beamwidth should be used. This influences the reflector antenna structure; for instance, a front-fed antenna with a small f/D ratio should be avoided. Also in a Cassegrain structure the distance between the two foci should be large enough.

In the case of a horn feed, the flare angle should be small to avoid a spherical wavefront incompatible with (14). Another solution is to extend the horn mouth by a cylindrical section.

Of course, the waveguide should have an electric field with a constant direction. In this respect, a rectangular or square (in case of double polarization) waveguide seems to be preferable to a circular one. Calculations have confirmed this for fairly large apertures ($a/\lambda > 1$). For small waveguide apertures, however, the situation is inverted [5]. In fact, the cross-polar performance of rectangular horns over the main lobe improves with the aperture size, while that of the circular one degrades. This conclusion is supported by several results spread in the literature and is perfectly consistent with the arguments developed above.

Diffraction Effects

A very important question is to know how far diffraction effects can destroy the cross-polar performance of a well designed feed-reflector system. There is probably no general answer to this question but it is the authors' opinion that some elementary precautions can probably solve the problem at least for the main lobe of high gain antennas. A first source of diffraction is the blockage of a part of the aperture by the feed or subreflector and by its struts, but the power captured by these structures is scattered with a much smaller directivity than that of the antenna and it is not believed that it will destroy the cross-polar performance over a large part of the main lobe. Nevertheless, care should be taken to avoid currents with large components in the y-direction when the aperture is illuminated with the electric field parallel to the x axis and inversely; consequently, struts should lie in the xz and yz planes rather than in the 45° planes.

Offset antenna structures have no similar diffraction effects, but they have per se very bad cross-polar characteristics if the feed or feed-subreflector system have a polarization approaching (12) in which $\theta = 0$ corresponds to the axis of this system, as is the case for most feeds. The ideal solution would be to use an offset structure with a source having the polarization (12) in which $\theta = 0$ is the paraboloid axis—not the source axis—but such a source does not seem to have been studied.

Diffraction by the main reflector edge can be responsible for some cross-polar radiation off boresight, but this effect can be minimized by a suitable taper of the illumination.

Far-Field Cross Polarization

The far-field pattern of a reflector antenna which is given by [6]

$$\bar{E}(\theta,\phi) = \bar{F}(\theta,\phi) - \{\bar{F}(\theta,\phi) \cdot \bar{U}_R\}\bar{U}_R \tag{29}$$

where

$$\bar{F}(\theta,\phi) = -\frac{jk\eta}{4\pi}\frac{e^{-jkR}}{R}\iint \bar{J}_s \exp(jk\rho\bar{u}_\rho \cdot \bar{u}R)\, dS \tag{29a}$$

can directly be related to aperture plane polarization. Surface current distribution on the paraboloidal surface one expressed in terms of $\bar{e}_r$ as [6]

$$\begin{aligned}\bar{J}_s &= K\sqrt{G_f(\theta,\phi)}\,\frac{e^{-jk\rho}}{\rho}\,[\bar{n}_2 \times (-\bar{u}_z \times \bar{e}_r)] \\ &= -K\sqrt{G_f(\theta,\phi)}\,\frac{e^{-jk\rho}}{\rho}\Big[\cos\frac{\theta}{2}\, e_{rx}\bar{u}_x + \cos\frac{\theta}{2}\, e_{ry}\bar{u}_y \\ &\quad + \sin\frac{\theta}{2}(\cos\phi e_{rx} + \sin\phi e_{ry})\bar{u}_z\Big]\end{aligned} \tag{30}$$

where $K = [(2P_t/\pi\eta)]^{1/2}$, P_t is the total radiated power and $G_f(\theta,\phi)$ is the gain function of the feed. By taking x-polarized aperture plane currents as reference, far-field copolar $R(\theta,\phi)$ and crosspolar $C(\theta,\phi)$ patterns are found from (29), (12), and (13) as

$$\begin{bmatrix} R(\Theta,\Phi) \\ C(\Theta,\Phi) \end{bmatrix} = \begin{bmatrix} 1 - (1 - \cos\Theta)\cos^2\Phi & (1 - \cos\Theta)\sin\Phi\cos\Phi & -\sin\Theta\cos\Phi \\ (1 - \cos\Theta)\sin\Phi\cos\Phi & 1 - (1 - \cos\Theta)\sin^2\Phi & -\sin\Theta\sin\Phi \end{bmatrix} \begin{bmatrix} F_x \\ F_y \\ F_z \end{bmatrix} \tag{31}$$

where

$$F_x = C_1 \iint \sqrt{G_f(\theta,\phi)}\cos\frac{\theta}{2}\frac{e_{rx}}{\rho}\exp[-jk\rho(1 - \bar{u}\rho \cdot \bar{u}R)]\, dS \tag{31a}$$

$$F_y = C_1 \iint \sqrt{G_f(\theta,\phi)}\cos\frac{\theta}{2}\frac{e_{ry}}{\rho}\exp[-jk\rho(1 - \bar{u}\rho \cdot \bar{u}R)]\, dS \tag{31b}$$

$$F_z = C_1 \iint \sqrt{G_f(\theta,\phi)} \sin\frac{\theta}{2} \frac{(\cos\phi e_{rx} + \sin\phi e_{ry})}{\rho} \cdot \exp[-jk\rho(1 - \bar{u}\rho \cdot \bar{u}R)]\, dS \quad (31c)$$

in which $C_1 = (jk\eta/4)C(e^{-jkR}/R)$ and the integration is to be performed on the paraboloidal surface with respect to (ρ,θ,ϕ).

It is interesting to note that (31) and Table I are the same except for a sign factor. This shows that the radiation from the aperture plane is that of a dipole instead a Huygens source. This comes from the fact that surface current distribution (30) is calculated from $\bar{H}_t$ only. So the far-field calculations correspond to (19). In fact, there is a constant proportionality between f_x, F_x, and f_y, F_y from (19) and (31). As D/λ increases, the angle Θ, which sees the main lobe and first few sidelobes, gets smaller so the term $(1 - \cos\Theta)$ approaches zero yielding zero far-field cross polarization provided polarization purity is achieved at the aperture plane.

Conclusions

All these considerations lead us to the conclusion that the cross polarization performance of reflector antenna system largely depends on the feed system and λ/D ratio. For this reason, superior performance of Cassegrainian or front-fed paraboloids with respect to polarization purity is closely related to the feed design. Thus this paper puts an end to the controversy recently arisen ([7]–[9]) about the immunity of these two types of antenna systems against depolarization effects. A detailed treatment may be required to see to what extent other factors may influence the cross polarization performance of reflector antennas.

References

[1] A. C. Ludwig, "The definition of cross-polarisation," *IEEE Trans. Antennas Propagat.*, vol. AP-21, pp. 116–119, January 1973.

[2] Collin and Zucker, *Antenna Theory.* New York: McGraw-Hill, 1969, Part 1, pp. 62–74.

[3] Watson and Ghobrial, "Off-axis polarisation characteristics of Cassegrainian and front-fed paraboloidal antennas," *IEEE Trans. Antennas Propagat.*, vol. AP-20, pp. 691–698, 1972.

[4] S. Silver, *Microwave Antenna Theory and Design.* M.I.T. Radiation Laboratory, 1964, p. 335.

[5] J. H. Cowan, "Dual-band reflector-feed element for frequency-reuse applications," *Electronics Letters*, vol. 9, no. 25, pp. 596–597, 13th December 1973.

[6] Collin and Zucker, *Antenna Theory.* New York: McGraw-Hill, 1969, Part 2, pp. 41–48.

[7] P. J. Wood, "Depolarisation with Cassegrainian and front-fed reflectors," *Electronics Letters*, vol. 3, pp. 181–183, 3rd May 1973.

[8] Watson and Ghobrial, "Cross-polarisation in Cassegrain and front-fed antennas," *Electronics Letters*, vol. 9, no. 14, pp. 297–298, 12th July 1973.

[9] P. J. Wood, "Cross-polarisation with Cassegrainian and front-fed reflectors," *Electronics Letters*, vol. 3, no. 25, pp. 597–598, 13th December 1973.

Bibliography for Part V

[1] P. A. Watson and S. I. Ghobrial, "Off-axis polarization characteristics of Cassegrainian and front-fed paraboloidal antennas," *IEEE Trans. Antennas Propagat.*, vol. AP-20, pp. 691–698, Nov. 1972.

[2] P. J. Wood, "Depolarisation with Cassegrainian and front-fed reflectors," *Electron. Lett.*, vol. 9, pp. 181–183, May 3, 1973.

[3] P. A. Watson and S. I. Ghobrial, "Cross-polarisation in Cassegrain and front-fed reflectors," *Electron. Lett.*, vol. 9, pp. 297–298, July 12, 1973.

[4] P. J. Wood, "Cross polarisation with Cassegrainian and front-fed reflectors," *Electron. Lett.*, vol. 9, pp. 597–598, Dec. 13, 1973.

[5] L. E. Raburn, "The calculation of reflector antenna polarized radiation," *IRE Trans. Antennas Propagat.*, vol. AP-8, pp. 43–49, Jan. 1960.

[6] W. V. T. Rusch, "Phase error and associated cross polarization effects in Cassegrainian fed microwave antennas," *IEEE Trans. Antennas Propagat.*, vol. AP-14, pp. 266–275, May 1966.

[7] T. S. Chu, "Restoring the orthogonality of two polarizations in radio communication systems, I," *Bell Syst. Tech. J.*, vol. 50, pp. 3063–3069, Nov. 1971.

[8] ——, "Restoring the orthogonality of two polarizations in radio communication systems, II," *Bell Syst. Tech. J.*, vol. 52, pp. 319–327, Mar. 1973.

[9] P. A. Watson and S. I. Ghobrial, "Cross polarizing effects of a water film on a parabolic reflector at microwave frequencies," *IEEE Trans. Antennas Propagat.*, vol. AP-20, pp. 668–671, Sept. 1972.

[10] S. I. Ghobrial, "Some data for the design of low cross polarisation feeds," *Electron. Lett.*, vol. 9, pp. 465–466, Oct. 4, 1973.

[11] M. M. A. El Futuh and S. I. Ghobrial, "The effect of deviation of feed polarization characteristics from that of a Huygens source on cross polarization in reflector antennas," *Radio Electron. Eng.*, vol. 44, pp. 269–272, May 1974.

[12] T. S. Chu, M. J. Gans, and W. E. Legg, "Quasi-optical polarization diplexing of microwaves," *Bell Syst. Tech. J.*, vol. 54, pp. 1665–1680, Dec. 1975.

[13] D. J. Brain, "Parametric study of the cross polarisation efficiency of parabolic reflectors," *Electron. Lett.*, vol. 12, pp. 245–246, May 13, 1976.

[14] A. B. Harris and B. Claydon, "Cross polarisation in beam-waveguide feeds," *Electron. Lett.*, vol. 12, pp. 529–531, Sept. 30, 1976.

[15] M. J. Gans, "Cross polarization in reflector-type beam waveguides and antennas," *Bell Syst. Tech. J.*, vol. 55, pp. 289–316, Mar. 1976.

[16] S. I. Ghobrial, "Loss in gain and boresight cross polarisation in reflector antennas with surface errors," *Electron. Lett.*, vol. 13, pp. 623–624, Sept. 29, 1977.

Part VI
Offset or Unsymmetrical Reflectors

There is an ambiguity associated with the word "offset" as it is currently used in antenna engineering. In this part, an offset paraboloid is taken to mean one which is not symmetrical about its axis of revolution; the portion of the reflector surface lying on one side of that axis is discarded altogether. Since the feed must still be located on or very close to the axis, this arrangement removes the feed from the region of highest aperture field intensity and reduces, or even eliminates, blockage. Of course, the axis of the feed must be tilted to throw its cone of illumination onto that part of the reflector surface that does remain; otherwise, spillover would be excessive. The other meaning associated with "offset" is the one used in the next part, in which the phase center of the feed is displaced (i.e., offset) laterally from the optical axis in order to squint the beam.

The first paper, by Chu and Turrin, takes up the important question of cross polarization in offset reflectors, They show, both theoretically and experimentally, that an offset section of a paraboloid will give rise to cross-polarized radiation in the plane of asymmetry, even when a linearly polarized Huygens source feed is used. This was alluded to in the introductory comments to Part V. They also point out the interesting fact that when a feed designed for circularly polarized excitation is used, the secondary beam remains circularly polarized, i.e., there is no cross-polarized component. The beam, however, is squinted away from the optical axis of the system, the displacement being in the plane of asymmetry and changing sign with the sense of the polarization. Dragone and Hogg follow with a paper that discusses offset dual reflector systems, with particular emphasis on the gain, sidelobe level, and reflection coefficient as compared to their more conventional symmetrical counterparts.

A useful contribution is made in the third paper, by Ingerson and Wong. They define an offset axis as the line through the focus that bisects the angle subtended at the focal point by the parabolic arc of the reflector in its plane of symmetry. It follows that the offset focal plane is the plane through the focus and perpendicular to the offset axis. With these definitions, the focal region characteristics of offset reflectors are quite analogous to those of symmetrical reflectors. A beam deviation factor is given for transverse displacements of the feed in the focal plane, and a gain loss curve is shown for displacements along the offset axis. This short paper has great relevance for multiple beam forming reflectors in which the feed cluster does not create blocking. For this reason, it could well have been included in Part VII of this book.

On the other hand, the fourth paper, by Dijk, Maanders, and Thurlings, might very well have appeared in Part V, since it deals almost exclusively with the polarization properties of reflectors. The emphasis, however, is on offset systems, and for that reason it has been placed in this part. In their introduction, the authors present a brief, interesting review of many of the topics that are covered in both this part and in Part V. To summarize their findings, offset reflectors can compete favorably with symmetric paraboloids only when simple electric dipole-type feeds are used in each case. If Huygens source feeds are used, then the comparison is clearly unfavorable to the offset reflector since the symmetric reflector has only its own small depolarization in this case. For most offset reflectors, they find little differences in the cross polarization between electric dipole and Huygens source feeds.

In the fifth paper of this part, Rudge considers offset reflectors having feeds that are offset from the axis. He therefore makes use of both meanings of the term "offset," as they were defined above. Two theoretical models are described for far-field pattern prediction of both the co-polar and cross-polar fields, and there is no inherent restriction to angles near the boresight axis. The models also can predict the effects of moderate lateral displacements of the feed (i.e., feed offsets) and hence are of value in connection with the design of multiple beam antennas, which form part of the subject matter for Part VII. This paper is followed by one by Rudge and Adatia in which they show that the cross polarization introduced by offset reflectors can be cancelled by proper feed design. In a cylindrical waveguide feed, the introduction of a particular higher mode (the TE_{21} in smooth wall guide or the HE_{21} in corrugated guide) will accomplish this result. By the same token, this kind of feed will eliminate beam squint in circularly polarized offset reflectors.

Jacobsen, in the final paper of the part, reiterates the view that the offset reflector itself does not contribute to the cross-polarized radiation that is found in the plane of asymmetry, but that this cross-polar radiation is merely that of the primary illuminating field after collimation by the reflector. Parenthetically, the effort involved in arranging for stereoscopic viewing of some of the figures in this paper is well repaid in terms of a better understanding of this sometimes vexing question. The author's discussion of why a polarization-pure Huygens source feed gives rise to cross polarization in an offset reflector is particularly illuminating. It is simply a consequence of the tilting of the feed, so that its axis no longer coincides with the optical axis of the reflector. Like Rudge, Jacobsen proposes to reduce or eliminate the undesirable cross polarization by means of improved feed design. He suggests that one way is to introduce a small amount of TE_{20}-mode radiation, in addition to the dominant TE_{01}, in a rectangular feed horn. Another way is to use a linear array of small-aperture round waveguide feeds with their axes parallel to the optical axis of the offset reflector. The primary illumination is then tilted to illuminate the reflector by appropriate phasing of the array elements. Jacobsen gives an example using only two such elements.

Depolarization Properties of Offset Reflector Antennas

TA-SHING CHU AND R. H. TURRIN

Abstract—The cross polarized radiation for linearly polarized excitation and the beam displacement for circularly polarized excitation have been investigated for offset reflector antennas. Numerical calculations are given to illustrate the dependence upon the angle θ_0 between the feed axis and the reflector axis as well as upon the half-angle θ_c subtended at the focus by the reflector. In the case $\theta_0 = \theta_c = 45°$, measured results have been obtained for both linearly and circularly polarized excitations with a dual mode feed illuminating an offset paraboloid.

The cross polarized radiation of horn reflector and open Cassegrainian antennas rises sharply to rather high values off the beam axis; however, in general, the maximum cross polarized radiation of offset reflector antennas can be made small by using a small angle between the feed and reflector axes. The cross polarization caused by offset is compared with that caused by an unbalanced feed pattern. The effect of the longitudinal current distribution and of departure of the surface from a paraboloid on cross polarization are also examined. The clarification of these cross polarization properties is found to be valuable in the design of reflector antennas.

I. Introduction

IN ORDER to increase the communication capacity of a transmission system by using orthogonal polarizations, it is essential to maintain the orthogonality, thereby preventing crosstalk. This requirement is easily fulfilled when the antenna is transmitting or receiving on the beam maximum. However poor pointing accuracy can rapidly increase depolarization in the vicinity of the beam maximum. In some satellite communication systems the ground stations are distributed over the beamwidth of the satellite antenna; in that case any polarization degradation within the satellite antenna beamwidth will give rise to cross polarization coupling.

The condition [1] that the directions of stationary polarization[1] and maximum gain coincide is satisfied by all center-fed paraboloids where the aperture distribution of the polarization vectors are symmetrical with respect to the center of the aperture. Since it is difficult to predict the effect of aperture blocking on the sidelobe level and cross polarization properties, various versions of offset antennas such as the open Cassegrainian antenna [2] have been proposed. The purpose here is to discuss the depolarization properties of offset reflector antennas which can be conveniently characterized by two angles. Section II will calculate the cross polarization for linearly polarized excitation and the beam displacement for

Manuscript received October 30, 1972; revised November 24, 1972.
The authors are with Bell Laboratories, Holmdel, N. J. 07733.

[1] Stationary polarization here means that the slope of the cross polarized radiation pattern is zero in the direction of the main beam maximum.

Reprinted from *IEEE Trans. Antennas Propagat.*, vol. AP-21, pp. 339-345, May 1973.

circularly polarized excitation. Section III will describe the measured results of an experimental example. Section IV will discuss the comparison of cross polarization in center-fed and offset paraboloids, the depolarization in offset Cassegrainian antennas, and other effects contributing to depolarization.

Jones [3] showed that there will be no cross polarization in the aperture field of a center-fed paraboloid if the feed is a plane wave source, i.e., a combination of electric and magnetic dipoles at right angles to each other with ratio equal to the plane wave impedance. The deficiency in simulating this plane wave feed by a horn may be attributed to the questionable validity of the Kirchhoff approximation for a small horn. Recently, corrugated horns [4], [5] and dual-mode horns [6], [7] with circularly symmetric radiation patterns have been recognized as capable of eliminating cross polarization in a center-fed paraboloidal aperture. Theoretical prediction based on these feed patterns can be better realized in practice. We will use balanced feed radiation (defined by (5) in Section II) with circularly symmetric patterns to calculate the cross polarization properties of offset reflectors, in which case the cross polarization is only a consequence of the offset; it diminishes with a decrease in the angle between the feed axis and reflector axis.

II. Cross Polarization and Beam Shift

Let an offset reflector be illuminated by a feed as shown in Fig. 1 where θ_0 is the angle between the feed axis and the reflector axis, and θ_c is the half-angle subtended by the reflector at the focus. In this section we will calculate the cross polarization properties of the reflector as a function of these two angles. The far-field expression of the feed radiation can be written

$$\bar{E}_f = E_\theta'\hat{\theta}' + E_\phi'\hat{\phi}'. \tag{1}$$

The reflected field from the paraboloid is

$$\bar{E}_r = -\bar{E}_f + 2\hat{n}(\bar{E}_f\cdot\hat{n}) \tag{2}$$

where $\hat{n}$ is a unit vector normal to the reflector surface. Substituting (1) into (2) and using the transformations $(\hat{\rho}',\hat{\theta}',\hat{\phi}') \rightarrow (\hat{x}',\hat{y}',\hat{z}') \rightarrow (\hat{x},\hat{y},\hat{z})$ (see Appendix) and the normal vector $\hat{n} = -(\hat{\rho}+\hat{z})/(2t)^{1/2}$ for a paraboloidal surface, one obtains the following form:

$$\begin{aligned}\bar{E}_r = \frac{\hat{x}}{t}\{&[\sin\theta'\sin\theta_0 - \cos\phi'(1+\cos\theta'\cos\theta_0)]E_\theta'\\ &+\sin\phi'(\cos\theta'+\cos\theta_0)E_\phi'\}\\ +\frac{\hat{y}}{t}\{&-\sin\phi'(\cos\theta'+\cos\theta_0)E_\theta'\\ &+[\sin\theta'\sin\theta_0 - \cos\phi'(1+\cos\theta'\cos\theta_0)]E_\phi'\}\end{aligned} \tag{3}$$

where

$$t = 1 + \cos\theta'\cos\theta_0 - \sin\theta'\sin\theta_0\cos\phi'. \tag{4}$$

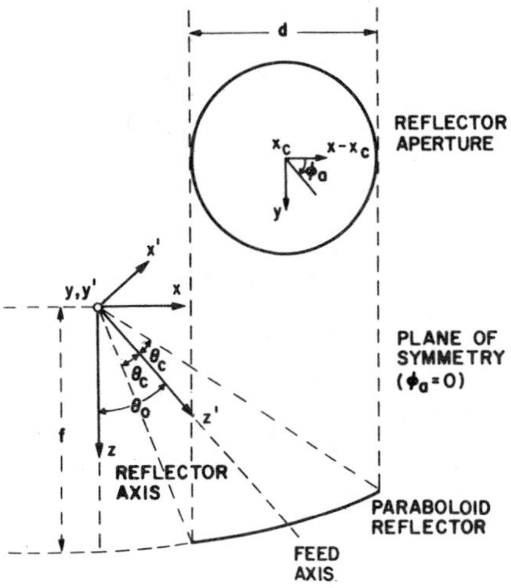

Fig. 1. Geometry of offset reflector.

The $\hat{z}$ component is absent in the aperture because of the unique focusing property of paraboloid [8]. However z component currents flow on the reflector surface.

Let us consider a balanced feed radiation which can be written in the following form:

$$\bar{E}_f = F(\theta',\phi')\begin{bmatrix}\cos\phi' & & \sin\phi' & \\ & \hat{\theta}' \mp & & \hat{\phi}' \\ \sin\phi' & & \cos\phi' & \end{bmatrix}\frac{\exp(-jk\rho)}{\rho} \tag{5}$$

corresponding to two principal linear polarizations along the x and y axes, respectively. An important special case of (5) is a circularly symmetric radiation pattern where $F(\theta',\phi')$ is independent of ϕ'. The principal polarization component of the reflected field becomes

$$\begin{aligned}M = \bar{E}_r\cdot\begin{matrix}\hat{x}\\ \hat{y}\end{matrix} = \frac{F(\theta',\phi')}{t\rho}[&\sin\theta'\sin\theta_0\cos\phi' - \sin^2\phi'\\ &\cdot(\cos\theta_0+\cos\theta') - \cos^2\phi'(1+\cos\theta_0\cos\theta')]\end{aligned} \tag{6}$$

while the cross polarization component is

$$\begin{aligned}N = \bar{E}_r\cdot\begin{matrix}\hat{y}\\ \hat{x}\end{matrix} = \mp\frac{F(\theta',\phi')}{t\rho}[&\sin\theta'\sin\theta_0\sin\phi'\\ &-\sin\phi'\cos\phi'(1-\cos\theta')(1-\cos\theta_0)]\end{aligned} \tag{7}$$

where $M^2 + N^2 = F^2/\rho^2$ and N vanishes when $\theta_0 = 0$.

One notes that the rotation of the polarization vector due to offset in a paraboloidal aperture has the same magnitude and is in the same sense for any orientation of the incident linear polarization. It follows that the reflected field of a circularly polarized wave will remain circularly polarized but in opposite rotating sense, and with a phase shift $\tan^{-1}(N/M)$. If the feed radiation is circularly polarized everywhere, no cross polarization will appear in the radiation from the reflector; however, a small beam displacement will occur because of variation in the phase shift across the reflector.

The projection of the intersection of a circular cone (with vertex at the feed) and the offset paraboloid onto the xy plane is a circle [2] with center

$$x_c = \frac{2f \sin \theta_0}{\cos \theta_c + \cos \theta_0} \tag{8}$$

and the diameter

$$d = \frac{4f \sin \theta_c}{\cos \theta_c + \cos \theta_0} \tag{9}$$

where f is the focal length of the paraboloid. The relationships between points on the paraboloid and their projections onto the xy plane are

$$x = \rho(\cos \theta_0 \sin \theta' \cos \phi' + \sin \theta_0 \cos \theta') \tag{10}$$

$$y = \rho \sin \theta' \sin \phi' \tag{11}$$

$$\rho = \frac{2f}{1 + \cos \theta} = \frac{2f}{1 + \cos \theta' \cos \theta_0 - \sin \theta' \sin \theta_0 \cos \phi'}. \tag{12}$$

Then the far-field pattern of the antenna is given for small angles by

$$A = \int_{-\pi}^{\pi} \int_0^{\theta_c} \frac{LF}{\rho} \exp \left\{ j \frac{2\pi v}{d} [(x - x_c) \cos \phi_a + y \sin \phi_a] \right\} \cdot \rho^2 \sin \theta' \, d\theta' \, d\phi' \tag{13}$$

where $v = \sin \theta_a / (\lambda/d)$ and L is a factor for the polarization component being calculated. The aperture surface element can be reduced to $\rho^2 \sin \theta' \, d\theta' \, d\phi'$ [2]. Assuming the circularly symmetric feed pattern to be the H-plane pattern of an open-end circular waveguide excited by TE_{11} mode

$$F(\theta') = \left\{ \left[1 - \left(\frac{u_{11}}{u} \right)^2 \right]^{1/2} + \cos \theta' \right\} \frac{J_1'(u \sin \theta')}{1 - \left(\dfrac{u \sin \theta'}{u_{11}} \right)^2} \tag{14}$$

where u is the circumference of the waveguide in wavelengths, and u_{11} is the first root of $J_1'(u) = 0$. This is a good representation of a small aperture dual mode feed [6]. Substituting (6) for L in (13) yields, for the principal polarization component,

$$P = 2 \int_0^{\pi} \int_0^{\theta_0} M \exp \left[j \frac{2\pi v}{d} (x - x_c) \cos \phi_a \right] \cdot \cos \left[\frac{2\pi v}{d} y \sin \phi_a \right] \rho \sin \theta' \, d\theta' \, d\phi'. \tag{15}$$

For the cross polarization component, substituting (7) for L in (13) gives

$$C = 2j \int_0^{\pi} \int_0^{\theta_0} N \exp \left[j \frac{2\pi v}{d} (x - x_c) \cos \phi_a \right] \cdot \sin \left(\frac{2\pi v}{d} y \sin \phi_a \right) \rho \sin \theta' \, d\theta' \, d\phi'. \tag{16}$$

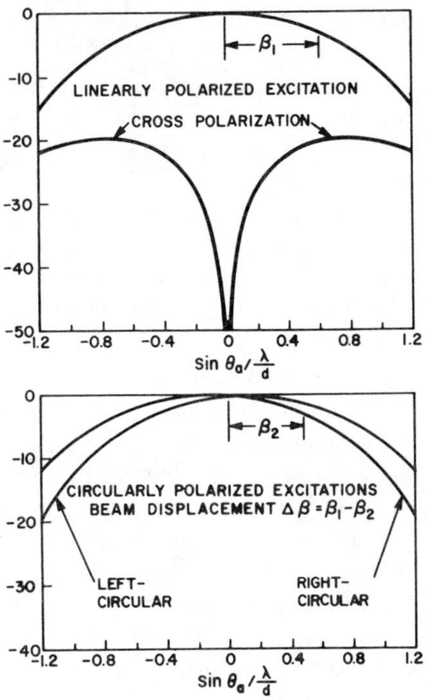

Fig. 2. Radiation patterns of offset paraboloid (f/D — 0.25).

Putting $L = \exp [j \tan^{-1}(N/M)]$, (13) will give the displaced radiation pattern for circular polarization as

$$K = 2 \int_0^{\pi} \int_0^{\theta_c} \exp \left[j \frac{2\pi v}{d} (x - x_c) \cos \phi_a \right] \cdot \cos \left[\tan^{-1} \frac{N}{M} + \frac{2\pi v}{d} y \sin \phi_a \right] F\rho \sin \theta' \, d\theta' \, d\phi'. \tag{17}$$

Owing to cancellation by symmetry, the cross polarization will vanish when $\phi_a = 0°$. The numerical calculation will be made for $\phi_a = 90°$ where maximum cross polarization is expected. The parameter u in (14) is selected to give a 10 dB taper for the feed pattern. For example, when $\theta_c = \theta_0 = 45°$, numerical integration for (15)–(17) results in the plots in Fig. 2. The cross polarization has a maximum located beyond the -3 dB point of the main beam; however, at the -3 dB point it is only slightly below the maximum. In the far field of the main beam, the cross polarized and principally polarized components are in phase quadrature whereas they are in phase at the aperture of the reflector. The offset reflector is also characterized by a lack of polarization stationarity at the beam maximum, consequently the cross polarization rises sharply off axis.

The beam displacement $\Delta\beta$, as defined in Fig. 2, is the shift of the circularly polarized beam with respect to the physical plane of symmetry. The direction of the shift is towards the right for left-handed circular polarization and towards the left for right-handed circular polarization.[2] Since the half-power beamwidth $2\beta_1$ of the linearly

[2] Circular polarization sense by IEEE definition: Wave receding from observer having clockwise rotation of the electric field is right circularly polarized.

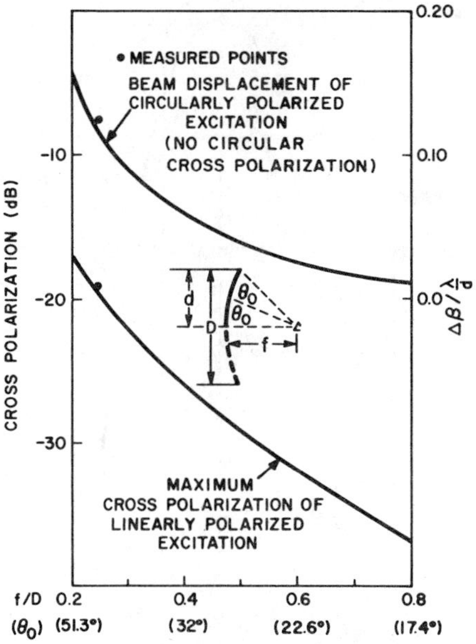

Fig. 3. Cross polarization and beam displacement versus f/D ratio ($\theta_0 = \theta_c$).

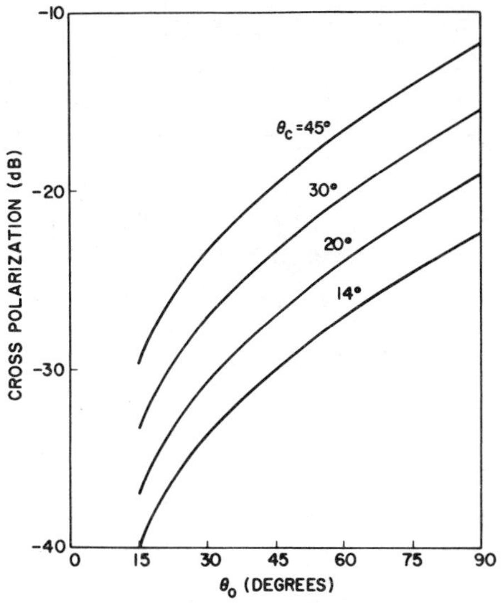

Fig. 4. Maximum cross polarization of linearly polarized excitation.

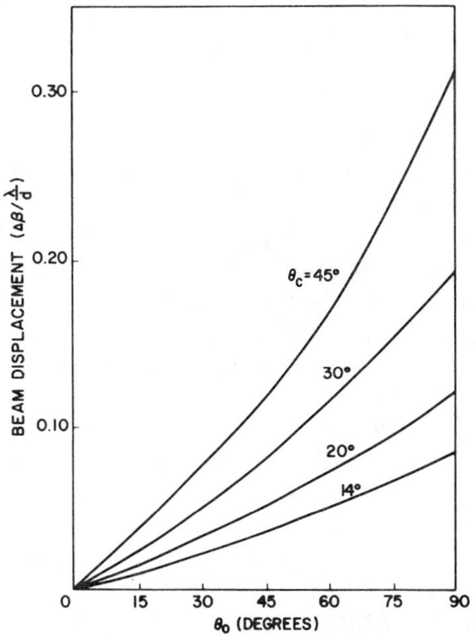

Fig. 5. Beam displacement of circularly polarized excitation (no circular cross polarization).

polarized pattern equals that of the circularly polarized pattern, $\Delta\beta = \beta_1 - \beta_2$ where β_2 is the angle between the axis and the half-power direction of the circularly polarized pattern. The beam displacement for circularly polarized excitations results in some sacrifice of power if two opposite circular polarizations are used simultaneously within a specified angular region. The cross polarization of linearly polarized excitation for an offset reflector can be regarded as the consequence of the differential gain of its two opposite circularly polarized components. This interpretation is analogous to the depolarization of circularly polarized excitation for a pyramidal horn, a consequence of the difference between the patterns of its two linearly polarized components.

Numerical data on the maximum cross polarization for linearly polarized excitation and the beam displacement for circularly polarized excitations are summarized in three figures. Fig. 3 shows the variation with respect to f/D ratio (or the angle θ_0) for the case of $\theta_0 = \theta_c$. It is seen that offset reflectors with large f/D ratios of the order of unity will introduce little cross polarization. Using θ_0 as abscissa and θ_c as parameter, their effects on the maximum cross polarization and the beam displacement are presented, respectively, in Figs. 4 and 5. A feed pattern of 10 dB taper has been used in these calculations. Computation also shows that increasing the taper to 20 dB reduces the maximum cross polarization only about 1 dB.

The previously computed cross polarization of -23 dB for an open Cassegrainian antenna [2] with $\theta_0 = 47.5°$ and $\theta_c = 30.5°$ checks very well with our results. It is also interesting to compare these cross polarization data with those of the pyramidal and conical horn reflectors [9], [10], where $\theta_0 = 90°$. One recalls that cross polarizations are present for both linearly and circularly polarized excitations of the horn reflector because this antenna is not only offset but also has different aperture distributions for longitudinal and transverse polarizations. The aperture distribution of the horn reflector is a result of single-mode excitation of the horn.

III. Measured Results

This section describes an 18.5 GHz experimental test of the cross polarization properties of an offset reflector antenna. The reflector consists of a paraboloidal section which was molded with Ultracal (a nonshrinking molding plaster) on a standard spun aluminum paraboloid with an f/D ratio of 0.25. This partial paraboloid corresponds to the case of $\theta_0 = \theta_c = 45°$, and its projection onto the plane normal to the axis of the paraboloid is a circular aperture of 12 in diameter.

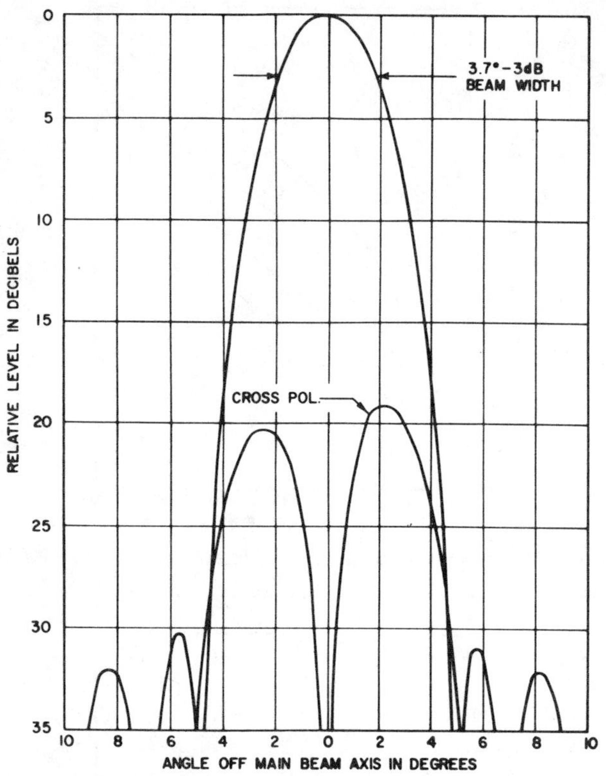

Fig. 6. Measured radiation pattern of offset fed paraboloid antenna linearly polarized at 18.5 GHz. Aperture diameter is 12 in, $f/D = 0.25$.

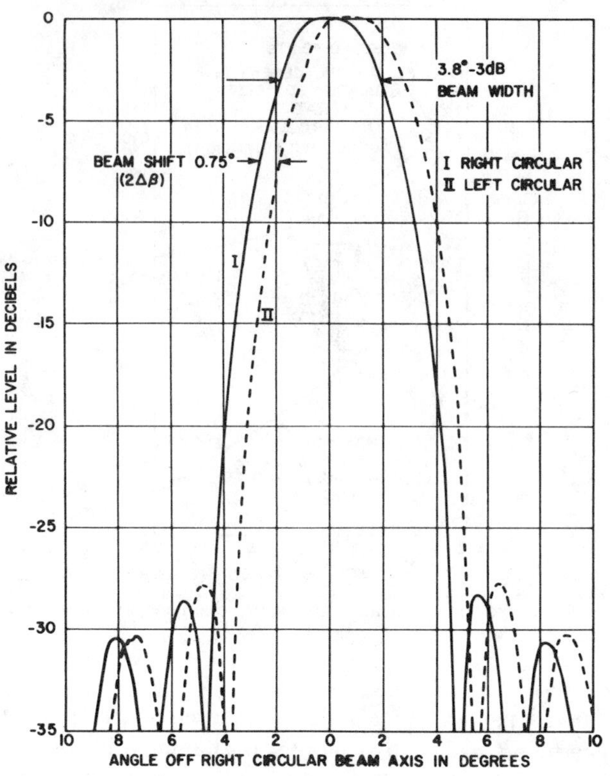

Fig. 7. Measured radiation patterns of offset fed paraboloid antenna with circular polarization at 18.5 GHz. Aperture diameter is 12 in, $f/D = 0.25$.

A dual mode feed of circularly symmetric radiation provides a pattern taper of -10 dB at the periphery of the offset paraboloidal section. Because of the geometrical difference in path length from focal point to the paraboloidal section there exists an amplitude asymmetry of 6 dB in the aperture distribution for the case $f/D = 0.25$. These conditions of illumination are the same as those of the theoretical calculations.

Radiation patterns were recorded for various polarization combinations in a plane transverse to the plane of symmetry of the offset paraboloid because this plane is expected to show maximum cross polarization for linearly polarized excitation and maximum beam displacement for circular polarization. Since the cross polarization patterns in the symmetry plane $\phi_a = 0°$ are theoretically null patterns, no attempt was made to measure them. Fig. 6 shows the in-line and cross polarized patterns with the principal linear polarization oriented in the plane of symmetry. The patterns were virtually identical with polarization either in the plane of symmetry or orthogonal to it.

Fig. 7 shows a composite of two measured patterns for the two senses of circular polarization, respectively. The angular displacement between the two patterns of opposite circular polarizations is twice the displacement of each circular polarization beam relative to the physical plane of symmetry with the direction as expected from the calculation in the preceding section. The maximum undesired circular polarization, which ideally should be zero, has been measured to be less than -38 dB relative to the beam maximum. This residual depolarization can be attributed to the effect of the longitudinal current distribution (see Section IV-C), the experimental environment, and the imperfection of the feed.

To compare with the calculated results, the measured maximum cross polarization and beam displacement for the case of $\theta_0 = \theta_c = 45°$ have been plotted in Fig. 3 where agreement is within about 5 percent. No elaborate efforts have been made to improve the imperfect symmetry of the cross polarization lobes in Fig. 6. The low sidelobe levels in both Figs. 6 and 7 can be explained by the nonorthogonality between the plane of the reflector edge and the plane of the measured patterns.

IV. Discussion

A. Comparison of Cross Polarization in Center-Fed and Offset Paraboloids

The cross polarization discussed in the preceding section originates entirely from the offset between the feed axis and the reflector axis. This kind of cross polarization vanishes when the angle of offset approaches zero. It is well known that cross polarization can also occur without offset when the feed radiation is unbalanced and does not satisfy (5). The characteristics of these two cross polarization mechanisms are enumerated in Table I.

Fig. 8 illustrates the cross polarization characteristics in the paraboloidal aperture for offset and unbalanced feed, respectively. In the offset case, the two principal linear polarizations rotate in the same direction at a given

TABLE I
COMPARISON BETWEEN TWO CROSS POLARIZATION MECHANISMS

Offset Paraboloid with Balanced Feed	Unbalanced Feed [(5) invalid]
1. Unstationary (cusp) polarization at beam maximum.	1. Stationary (saddle point) polarization at beam maximum.
2. The polarization vectors rotate in the same direction for two principal linear polarizations in the paraboloidal aperture [Fig. 8(a)].	2. The polarization vectors rotate in opposite directions for two principal linear polarizations in the paraboloidal aperture [Fig. 8(b)].
3. Cross polarized radiation in phase quadrature with principal polarization.	3. Cross polarized radiation in phase[a] with principal polarization.
4. Cancellation of cross polarization in one principal plane only.	4. Cancellation of cross polarization in both principal planes.
5. Maximum cross polarization in the other principal plane.	5. Maximum cross polarization in the 45° planes.
6. No cross polarization but beam displacement for circularly polarized excitation.	6. Equal cross polarization lobes in all planes for circularly polarized excitation.

[a] If the feed phase centers of the E-plane and the H-plane do not coincide, there will be a phase difference between cross and principal polarizations.

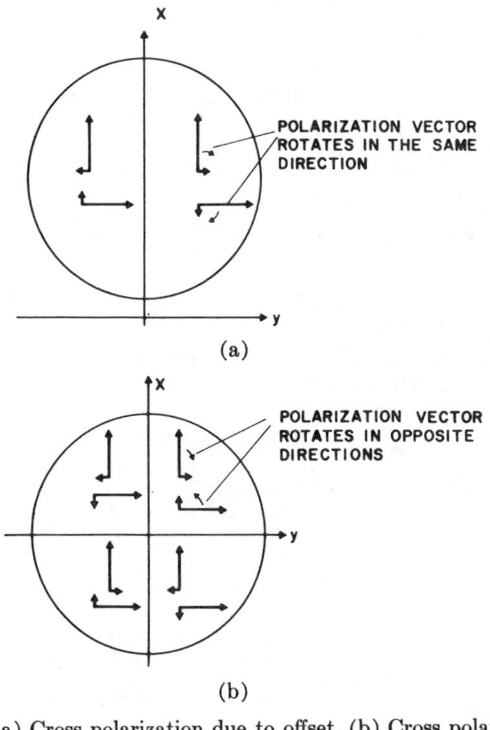
X
POLARIZATION VECTOR ROTATES IN THE SAME DIRECTION
y
(a)
X
POLARIZATION VECTOR ROTATES IN OPPOSITE DIRECTIONS
y
(b)

Fig. 8. (a) Cross polarization due to offset. (b) Cross polarization due to unbalanced feed.

point of the aperture and the cross polarizations are antisymmetrical with respect to only one principal axis. On the other hand, for the unbalanced feed, the two linear polarizations rotate in opposite directions, and the cross polarizations are antisymmetrical with respect to both principal axes. These differences in the aperture distributions lead to the different properties of the cross polarized radiation.

The consequences of both mechanisms of cross polarization remain the same if the shape of the aperture on the paraboloid is made arbitrary but retains symmetry about the principal axes. For example, illuminating an elliptically shaped section of the offset paraboloid by a circularly symmetric feed pattern will result in an elliptically shaped far-field radiation pattern with the same cross polarization properties as an offset circular aperture. However some sacrifice in gain will occur due to spillover.

B. Depolarization in Offset Cassegrainian Antennas

A Cassegrainian antenna has an equivalent focal length much longer than that of its main reflector [11]. Thus an offset Cassegrainian antenna may have a large effective f/D ratio and hence a small angle between the feed axis and the main reflector axis. The limiting case is that of two partial paraboloids, confocal and coaxial, which form a near field Cassegrainian or Gregorian antenna without aperture blocking. Bisected near-field Cassegrainian telescopes have been used in infrared propagation experiments [12]. In that case the feed axis is parallel to the main reflector axis, and there will be no cross polarization in the aperture of the main reflector provided no cross polarization illuminates the subreflector [13].[3] In other words, the combination of a plane wave feed and a confocal-paraboloid subreflector gives an effective balanced feed radiation which satisfies (5) in spite of the asymmetry of the geometrical configuration. This example illustrates that balanced feed radiation is not necessarily circularly symmetric.

Some versions of offset Cassegrainian antennas are expected to have poor off-axis polarization discrimination (of the order of 20 dB) because of the relatively large angle between the feed and reflector axes at certain orientations of the antenna. This poor polarization performance is the price paid for the mechanical advantage of the feed location in the open Cassegrainian antenna [2]. It is tempting to introduce additional subreflectors to bring the feed axis parallel to the reflector axis for open Cassegrainian antennas.

C. Other Effects Contributing to Depolarization

1) *Longitudinal Current Distribution:* The cross polarization contributed by the longitudinal current distribution on a reflector antenna is expected to be of the order of the isotropic level. Therefore neglecting the longitudinal current distribution, as done in this paper, is only valid for a large reflector with high gain. In particular, for a center fed paraboloid uniformly illuminated by a feed radiation pattern which satisfies (5), the reflected field (obtained by geometrical optics ray tracing) contains no cross polarization in the paraboloidal aperture. However, it can be deduced from the radiation integral [14] of the current distribution on the reflector surface that the voltage ratio of the maximum cross polarization to the isotropic level is approximately $D/8f$.

2) *Nonparaboloidal Surface Shape:* In the design of a multibeam or scanning antenna, one often likes to use spherical or torus reflectors which are slight perturbations of true paraboloids. It is of interest to investigate the

[3] The cross polarizations caused by the subreflector and the main reflector cancel each other.

contribution of such changes in the surface shape to the cross polarization.

The reflected field $\bar{E}_r$ from a perfectly conducting surface can be expressed in terms of the incident field $\bar{E}_i$ and the unit vector normal $\hat{n}$ to the surface by geometrical optics as in (2) which is rewritten as

$$\bar{E}_r = -\bar{E}_i + 2\hat{n}(\hat{n}\cdot\bar{E}_i). \quad (18)$$

If the surface deviates slightly from a paraboloid, i.e., $\hat{n}' = \hat{n} + \bar{\Delta}$, then the reflected field becomes

$$\bar{E}_{r'} = -\bar{E}_i + 2\hat{n}'(\hat{n}'\cdot\bar{E}_i). \quad (19)$$

Subtracting (18) from (19) and taking its component along the direction of cross polarization, one obtains

$$(\bar{E}_{r'} - \bar{E}_r)\cdot\hat{X} = 2(\hat{n}\cdot\hat{X})(\bar{\Delta}\cdot\bar{E}_i) + 2(\bar{\Delta}\cdot\hat{X})(\hat{n}\cdot\bar{E}_i) + 2(\bar{\Delta}\cdot\hat{X})(\bar{\Delta}\cdot\bar{E}_i). \quad (20)$$

It is obvious from (20) that the cross polarization will be reduced to a second-order effect if the $\hat{n}$ is nearly perpendicular to both $\hat{X}$ and $\bar{E}_i$. This latter condition is certainly satisfied by a reflector with large f/D ratio. In the case of a multibeam reflector antenna a large f/D ratio is often required in any case for reducing the phase error.

It is also evident from (20) that the cross polarization vanishes when both $\hat{n}$ and $\bar{\Delta}$ lie in, or parallel to, a constant plane of incidence. This observation is simply a rediscovery of the fact that no depolarization is produced by a cylindrical reflector.

IV. Conclusion

The cross polarization of linearly polarized excitation and the beam displacement of circularly polarized excitation have been presented in Figs. 4 and 5 versus two characteristic angles for an offset reflector with a circularly symmetric feed pattern. An experimental example has verified the theoretical calculations. The nonstationary polarization at the beam maximum of an offset reflector implies stringent pointing tolerance. A larger effective f/D ratio always reduces the cross polarization. Some asymmetrical configurations of a multireflector antenna such as the combination of two partial confocal paraboloids may keep the feed axis parallel to the reflector axis and hence avoid cross polarization due to offset.

Appendix

With reference to Fig. 1, the transformations for the unit vectors $(\hat{\rho}',\hat{\theta}',\hat{\phi}') \rightarrow (\hat{x}',\hat{y}',\hat{z}') \rightarrow (\hat{x},\hat{y},\hat{z})$ are the standard relations

$$\hat{\rho}' = \sin\theta'\cos\phi'\hat{x}' + \sin\theta'\sin\phi'\hat{y}' + \cos\theta'\hat{z}'$$

$$\hat{\theta}' = \cos\theta'\cos\phi'\hat{x}' + \cos\theta'\sin\phi'\hat{y}' - \sin\theta'\hat{z}'$$

$$\hat{\phi}' = -\sin\phi'\hat{x}' + \cos\phi'\hat{y}'$$

and

$$\hat{x}' = \hat{x}\cos\theta_0 - \hat{z}\sin\theta_0$$

$$\hat{y}' = \hat{y}$$

$$\hat{z}' = \hat{x}\sin\theta_0 + \hat{z}\cos\theta_0.$$

The position vectors in (x,y,z) and (x',y',z') are identical to each other; i.e., $\rho \equiv \rho'$ and $\hat{\rho} \equiv \hat{\rho}'$. Representing both sides of the last equation in terms of $(\hat{x},\hat{y},\hat{z})$ yields

$$\sin\theta\cos\phi = \sin\theta'\cos\phi'\cos\theta_0 + \sin\theta_0\cos\theta'$$

$$\sin\theta\sin\phi = \sin\theta'\sin\phi'$$

$$\cos\theta = -\sin\theta'\cos\phi'\sin\theta_0 + \cos\theta_0\cos\theta'.$$

Using the preceding transformations, one may obtain (4) by lengthy but straightforward algebra.

Acknowledgment

The authors wish to thank Ms. W. L. Mammel for assistance with the computation.

References

[1] T. S. Chu and R. G. Kouyoumjian, "An analysis of polarization variation and its application to circularly-polarized radiators," *IRE Trans. Antennas Propagat.*, vol. AP-10, pp. 188–192, Mar. 1962.

[2] J. S. Cook, E. M. Elam, and H. Zucker, "The open Cassegrain antenna: Part 1—Electromagnetic design and analysis," *Bell Syst. Tech. J.*, vol. 44, pp. 1255–1300, Sept. 1965.

[3] E. M. T. Jones, "Paraboloid reflector and hyperboloid lens antennas," *IRE Trans., Antennas Propagat.*, vol. AP-2, pp. 119–127, July 1954.

[4] H. C. Minnett and B. MacA. Thomas, "A method of synthesizing radiation patterns with axial symmetry," *IRE Trans. Antennas Propagat.*, (Commun.), vol. AP-14, pp. 654–656, Sept. 1966.

[5] V. H. Rumsey, "Horn antennas with uniform power patterns around their axes," *IRE Trans. Antennas Propagat.* (Commun.), vol. AP-14, pp. 656–658, Sept. 1966.

[6] R. H. Turrin, "Dual mode small-aperture antennas," *IRE Trans. Antennas Propagat.* (Commun.), vol. AP-15, pp. 307–308, Mar. 1967.

[7] P. D. Potter, "A new horn antenna with suppressed side lobes and equal beamwidths," *Microwave J.*, vol. 6, pp. 71–78, June 1963.

[8] S. Silver, *Microwave Antenna Theory and Design*, (Radiation Lab. Series vol. 12). New York: Dover, 1965, p. 148.

[9] A. B. Crawford, D. C. Hogg, and L. E. Hunt, "A horn-reflector antenna for space communication," *Bell Syst. Tech. J.*, vol. 40, pp. 1095–1116, July 1961.

[10] J. N. Hines, T. Li, and R. H. Turrin, "The electrical characteristics of conical horn-reflector antenna," *Bell Syst. Tech. J.*, vol. 42, pp. 1187–1211, July 1963.

[11] P. W. Hannan, "Microwave antennas derived from the Cassegrain telescope," *IRE Trans. Antennas Propagat.*, pp. 140–153, Mar. 1961.

[12] T. S. Chu and D. C. Hogg, "Effects of precipitation on propagation at 0.63, 3.5, and 10.6 microns," *Bell Syst. Tech. J.*, vol. 47, pp. 723–759, May–June 1968.

[13] A. Saleh, private communication.

[14] M. S. Afifi, "Scattered radiation from microwave antennas and the design of a paraboloidal-plane reflector antenna," Ph.D. dissertation, Tech. Univ., Delft, The Netherlands, June 1967.

The Radiation Pattern and Impedance of Offset and Symmetrical Near-Field Cassegrainian and Gregorian Antennas

C. DRAGONE AND D. C. HOGG, FELLOW, IEEE

Abstract—**Most Cassegrainian and Gregorian antennas have axial symmetry, in which case the subreflector and associated supporting members partially block the aperture. Consequently, relatively high sidelobes appear in the radiation pattern, and a reflection is produced in the transmission line of the feed. These undesirable effects can be largely eliminated using asymmetrical configurations. Here we compare axisymmetrical and offset near-field Cassegrainians and Gregorians; expressions for the reflection coefficient and increase in sidelobe level are given. The offset designs are found to have superior performance in both respects.**

Manuscript received June 1, 1973; revised November 12, 1973.
The authors are with the Crawford Hill Laboratory, Bell Laboratories, Holmdel, N. J. 07733.
[1] To obtain low cross polarization, Chu [3] has pointed out that the axis of the radiation pattern of the feed must be parallel to the axis of the main reflector.

I. INTRODUCTION

Axisymmetric antennas of the near-field Cassegrain [1] (or Gregorian) type consist of two confocal paraboloids (Fig. 1) and a feed which illuminates the subreflector through a centrally located aperture. The field produced by the feed, approximately a plane wave, is transformed by the subreflector into a spherical wave which, neglecting diffraction by the supports of the subreflector (not shown in Fig. 1), is transformed by the main reflector into a uniform phase front. However, a portion of the spherical wave from the subreflector is intercepted by the feed which partly absorbs and partly scatters the radiation; thus some of this energy represents impedance mismatch. Also, the interference forthcoming from high sidelobes generated by blockage imposes limitations in sharing of frequency bands by terrestrial and satellite communication systems [2].

Blockage can be minimized at the expense of symmetry, as in the offset antennas of Fig. 2. In Figs. 2(a) and (b), blockage is eliminated completely; in Fig. 2(c) there is some blockage, but the effect is small since the intensity over the blocked area is very low. In all of the designs, the field distribution over the aperture of the main reflector is essentially a replica of the field over the aperture of the feed.

The polarization properties of offset antennas are discussed in [3] where it is shown that antennas discussed here will have very low cross polarization,[1] provided the feed has perfect polarization properties and a large effective focal length, which is characteristic of the near-field design.

Reprinted from *IEEE Trans. Antennas Propagat.*, vol. AP-22, pp. 472–475, May 1974.

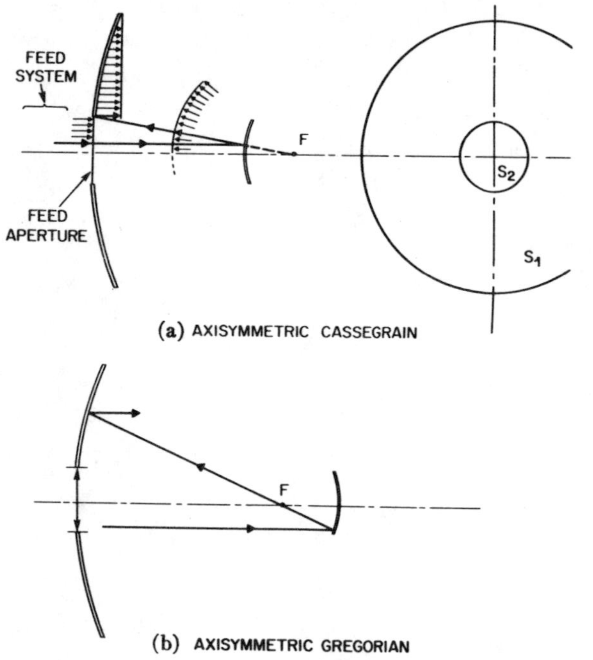

Fig. 1. Conventional axially symmetrical antennas of Cassegrainian and Gregorian type.

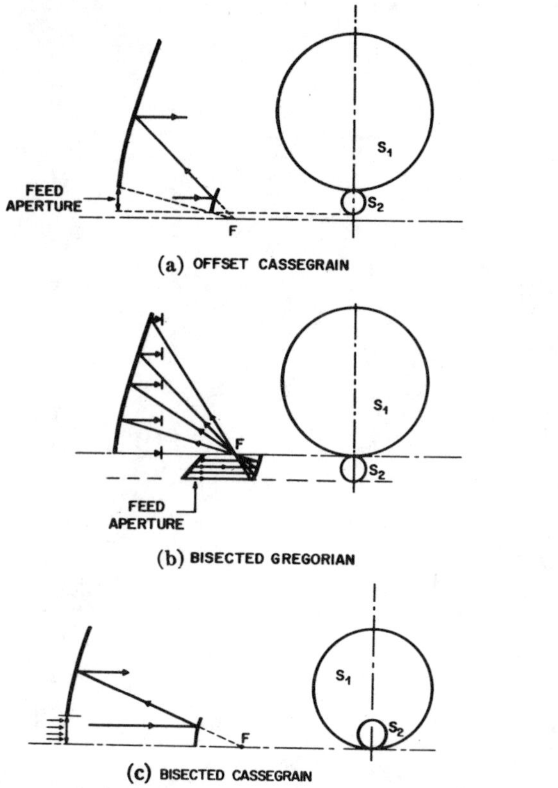

Fig. 2. (a) Cassegrainian configuration without blockage. (b) Gregorian configuration without blockage. (c) Bisected Cassegrainian configuration with some blockage.

II. REFLECTION COEFFICIENT[2]

In Fig. 1(a), the plane tangent to the vertex of the subreflector is illuminated by a linearly polarized field E_2. The field E_2' produced by the wave reflected from the subreflector is polarized in the direction of E_2. At a point in S_2,

$$E_2' = E_2 \exp(-j\psi) \tag{1}$$

where

$$\psi = k(R - f_2) \tag{2}$$

R being the distance of the point from the focus and f_2 the focal length of the subreflector. If the vertex of the subreflector is the origin of a coordinate system with the xy plane tangent at the vertex

$$\psi \simeq \frac{k}{f_2}\frac{x^2 + y^2}{2}. \tag{3}$$

The reflection coefficient on the feed transmission line (produced by the reflected wave) is given by

$$\begin{aligned} |\Gamma| &= \left|\int_{S_2} E_2E_2'\,dx\,dy\right| \Big/ \left|\int_{S_2} |E_2|^2\,dx\,dy\right| \\ &= \left|\int_{S_2} E_2^2 \exp(-j\psi)\,dx\,dy\right| \Big/ \left|\int_{S_2} |E_2|^2\,dx\,dy\right|. \end{aligned} \tag{4}$$

The principle of stationary phase can be applied to obtain an estimat of $|\Gamma|$ for large k, the major contribution arising from the area in the vicinity of the vertex. If E_{20} denotes the values of E_2 at the vertex, then

$$\int_{S_2} |E_2|^2 \exp(-j\psi)\,dx\,dy \simeq |E_{20}|^2 \int_0^\infty \int_0^\infty \exp\left(-j\frac{k}{f_2}\frac{x^2+y^2}{2}\right) dx\,dy \tag{5}$$

convergence being assured if k has a small imaginary component. Then, for large k,

$$\int_{S_2} |E_2|^2 \exp(-j\psi)\,dx\,dy \simeq -j\lambda f_2 |E_{20}|^2 \tag{6}$$

valid provided the vertex is within S_2. If the vertex is on the boundary of S_2, as in Fig. 2(c),

$$\int_{S_2} |E_2|^2 \exp(-j\psi)\,dx\,dy = \tfrac{1}{2}(-j\lambda f_2 |E_{20}|^2). \tag{7}$$

For Figs. 1 and 2(c), (4) becomes

$$|\Gamma| = \begin{cases} \dfrac{2\beta}{\alpha}, & \text{(symmetrical)} \\[2ex] \dfrac{\beta}{\alpha}, & \text{(offset)} \end{cases} \tag{8}$$

where $\alpha = ka_2^2/f_2$, a_2 being the subreflector radius. The quantity

$$\beta = \frac{\pi a_2^2 |E_{20}|^2}{\displaystyle\int_{S_2} |E_2|^2\,dx\,dy} \tag{10}$$

is the ratio between the power density at the vertex of S_2 and the average power density over S_2.

Consider a circularly symmetric field distribution with quadratic dependence on the distance ρ_2 from the center of S_2, i.e.,

$$E_2 = 1 - A\left(\frac{\rho_2}{a_2}\right)^2. \tag{11}$$

Let $A = 0.7763$, which corresponds to a feed with a -13-dB illumination. Then in the cases of Figs. 1 and 2(c), (10) results in $\beta = 2.355$ and 0.1177, and, from (8) and (9),

$$|\Gamma| = \begin{cases} \dfrac{4.71}{\alpha}, & \text{(symmetrical)} \\[2ex] \dfrac{0.1177}{\alpha}, & \text{(offset).} \end{cases}$$

[2] The good match obtainable with a bisected Cassegrainian (Fig. 2(c)) was recognized when used as an infrared telescope [4] feeding high-gain amplifiers.

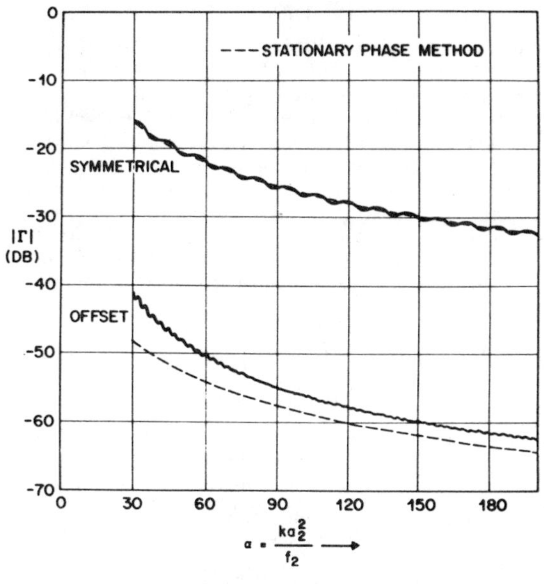

Fig. 3. Reflection coefficients for cases of Figs. 1 and 2(c), assuming an illumination taper of 13 dB at edge of subreflector.

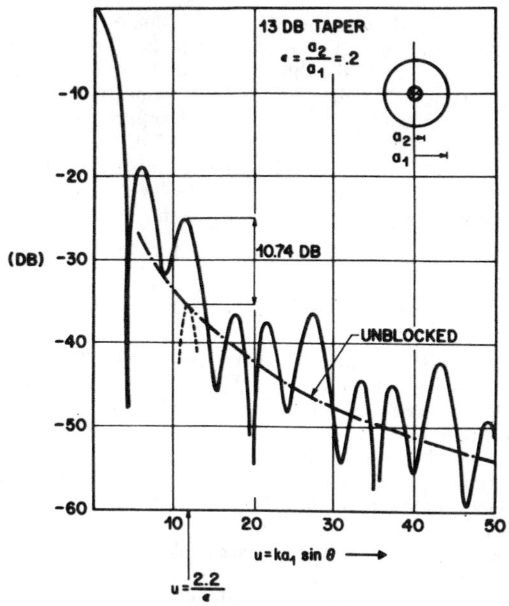

Fig. 4. Radiation pattern of blocked aperture of Fig. 1 ($a_2/a_1 = 0.2$).

Fig. 3 shows curves of $|\Gamma|$ versus α calculated from (3), (4), and (11), as well as curves given by the principle of stationary phase. In the offset case the reflection coefficient is <-40 dB, negligible for most practical purposes. The asymptotic relation (8) gives accurate results for α of practical interest.[3] In an experiment [1] at 6 GHz, on an axisymmetric 5-m near-field Cassegrainian with a 1.8-m focal length using a subreflector with $a_2 = 0.381$ m, $f_2 = 0.287$ m, and $\beta \approx 2.2$ (corresponding to an average illumination of -11 dB at the edge of the subreflector), the reflection coefficient was -22 dB; the calculated value, from (8), is -21.5 dB.

Although ray tracing in Figs. 2(b) and (c) shows that the latter involves some energy captured by the feed and the former does not, (9) formally applies to both cases. In the case of Fig. 2(a), $|\Gamma|$ is near zero, since the difference in phase between the two waves E_2 and E_2' has no stationary point over the aperture of the subreflector.

It is interesting that the reflection coefficient accounts for only a small fraction of the total power intercepted by the feed aperture. Using the aperture distribution of (11), one finds that the ratio between the power intercepted by the feed and the total power in the spherical wave reflected by the subreflector, is, for configurations of Figs. 1 and 2(c), $U = 0.09, 0.024$, (-10.46 dB, -16.2 dB), respectively, using a -13 dB taper and $a_2/a_1 = 0.2$. The difference between U and $|\Gamma|^2$ is power not accepted by the feed which is radiated[4] thereby degrading the radiation pattern; this depends on the particular scattering properties of the feed and will not be considered further here.

III. RADIATION PATTERNS

The field distribution over the aperture of the main reflector in Figs. 1 and 2(c) is assumed zero within the blocked region S_2. By superposition the overall antenna radiation pattern is approximated by the combination [5] of the fields from the unblocked and blocking apertures.

The envelope of the sidelobes of an unblocked aperture with a circularly symmetric field distribution typical of illuminations produced by dual mode [6] and hybridmode feed systems is shown in Figs. 4 and 5 by the dash-dot lines. By expressing the aperture field as a polynomial it can be shown that the *largest increase* in sidelobe level caused by the blockage is given by the simple expression

$$10 \log_{10} [(1 + 1.032\epsilon^{1/2}\gamma)^2\eta] \qquad (12)$$

where $\epsilon = a_2/a_1$. γ is the ratio between the field amplitude at the

[3] For Fig. 2(c), convergence $|\Gamma|$ to the asymptotic relation (9) is much slower; this is because the stationary point is not a point of maximum illumination, as in the symmetrical case.

[4] Some of this power is again radiated toward the subreflector, resulting in multiple bounces between feed and subreflector; this gives rise to a poor transmission characteristic for broad-band signals.

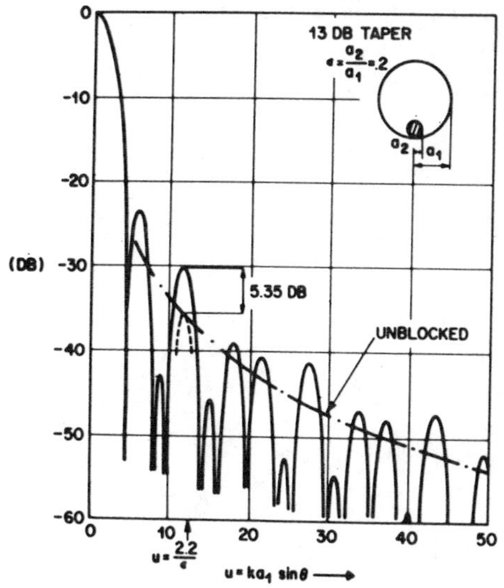

(a)

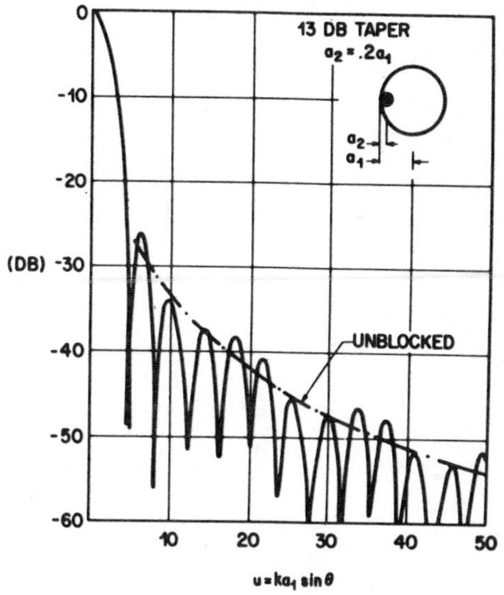

(b)

Fig. 5. Radiation pattern for marginally blocked aperture of Fig. 2(c) ($a_2/a_1 = 0.2$). (a) Horizontal plane. (b) Vertical plane.

center of S_2 and that at the edge of S_1 (thus $\gamma = E(0)/E(1)$ for central blockage). η is the ratio between the (power) gain of the aperture without and with blockage (for the cases of Figs. 1 and 2(c), $10 \log_{10} \eta = 0.68$ dB and 0.25 dB, respectively). As seen in Fig. 4, both the exact computation and the approximate (12) result in "largest increase" in sidelobe level of 10.7 dB.

Now consider the marginally blocked aperture of Fig. 2(c). According to (12) the increase in sidelobe level is much lower since γ is now given by

$$\gamma = \frac{E(1-\epsilon)}{E(1)} \qquad \epsilon = a_2/a_1. \tag{13}$$

For the assumed illumination, (13) gives $\gamma = 1.679$, which according to expression (12) with $\epsilon = 0.2$ and $10 \log_{10} \eta = 0.25$ dB corresponds to an increase in sidelobe level of 5.23 dB. Fig. 5(a) shows the exact radiation pattern in the horizontal plane; the increase in level is 5.35 dB, close to the value given by (12). The pattern in the orthogonal plane is shown in Fig. 5(b). The increase in level, for all sidelobes, is less than that given by (12), a consequence of the asymmetry.

The patterns of Figs. 4 and 5 were calculated for -13-dB illumination at the edge of S_1; for a -16-dB taper the maximum increase in sidelobe level has also found to be given accurately by (12).

The offset Cassegrainian and bisected Gregorian (Figs. 2(a) and (b)) are unblocked antennas, thus the envelopes of the patterns are given by the dash–dot lines in Fig. 5.

IV. DISCUSSION

Comparison of the properties of asymmetrical with axisymmetric antennas where an illumination taper of -13 dB and a blockage ratio $\epsilon = a_2/a_1 = 0.2$ are assumed, shows: the antennas of Figs. 2(a) and (b) have gain higher by about 0.7 dB, the level of near sidelobes is lower by about 10 dB, and their reflection coefficient is lower by 25 to 30 dB. The configuration of Fig. 2(c) has similar advantages: the gain is higher by about 0.4 dB, sidelobes lower by 5 dB, and reflection coefficient down by 25 to 30 dB when compared with the axisymmetric case. An important advantage of the various offset antennas is that they can be constructed without use of spars (or struts) which produce aperture blockage in the symmetrical case.

ACKNOWLEDGMENT

We thank R. Kompfner for his interest in the bisected Gregorian design.

REFERENCES

[1] D. C. Hogg and R. A. Semplak, "An experimental study of near-field Cassegrainian antennas," *Bell Syst. Tech. J.*, vol. 43, pp. 2677–2704, Nov. 1964.
[2] L. C. Tillotson, "A model of a domestic satellite communications system," *Bell Syst. Tech. J.*, no. 10, pp. 2111–2137, Dec. 1968.
[3] T. S. Chu and R. H. Turrin, "Depolarization properties of offset reflector antennas," *IEEE Trans. Antennas Propagat.*, vol. AP-21, pp. 339–345, May 1973.
[4] T. S. Chu and D. C. Hogg, "Effects of precipitation on propagation at 0.63, 3.5 and 10.6 microns," *Bell Syst. Tech. J.*, vol. 47, no. 5, pp. 723–759, May–June 1968.
[5] C. C. Cutler, "Parabolic-antenna design for microwaves," *Proc. IRE*, vol. 35, pp. 1284–1294, Nov. 1947.
[6] R. H. Turrin, "Dual mode small-aperture antennas," *IEEE Trans. Antennas Propagat.*, vol. AP-15, pp. 307–308, Mar. 1967.

FOCAL REGION CHARACTERISTICS OF OFFSET FED REFLECTORS

P. G. Ingerson and W. C. Wong
Radio Frequency Laboratory
TRW Systems Group
Redondo Beach, California 90278

Abstract

The focal region characteristics of offset fed reflector antennas have been investigated both analytically and experimentally. In order for offset reflectors to have analogous focal region properties to those of front fed parabolic reflectors, an offset axis and an offset focal plane are defined. The definitions of the offset axis and offset focal plane are essential for satisfactory operations of offset reflectors illuminated by a cluster of feeds. Numerical results using physical optics approximation are presented to show the dependence of the beam deviation factor of offset reflectors upon the edge angle (Θ_E) and the orientation of the feed axis (Θ_o).

Introduction

The offset fed reflector antenna has become increasingly attractive for satellite communications where low sidelobe levels are essential to achieve good isolation between adjacent high gain beams operated over the same frequency band. To meet these requirements, the use of offset reflectors is becoming desirable since the severe limiting effects of aperture blockage on front fed paraboloid reflectors can be entirely removed. The focal region behavior of offset reflectors then becomes of major interest in dealing with scanned beams and multiple beam antennas (or a shaped beam antenna) involving a cluster of feeds.

The focal region properties of front fed paraboloidal reflectors have been extensively investigated [1-4] in recent years. Although a few papers on offset reflectors have appeared in the literature [5,6,7,8], the general focal region behavior has never been discussed. A few authors have presented data on laterally scanned beams, but the analysis was restricted to the $\phi = \pm 90°$ plane (see Figure 1). It will be shown that feed movements along the axis of the parent paraboloid (the paraboloid from which the offset section is described) causes the beam to scan vertically in the $\phi = 0°\text{-}180°$ plane. Similarly, lateral feed motion in the focal plane of the parent paraboloid, except in the $\phi = \pm 90°$ plane will not produce the familiar results:

- Beam scan proportional to feed displacement (i.e., a beam deviation factor).
- Equi-phase plane of secondary beam peaks.
- Maximum* gain plane for lateral scanned beams.

It has been found, however, that by redefining a new focal plane and axis, the analogous properties are restored.

Offset Axis and Offset Focal Plane

For a front fed reflector, the axial defocusing curve is obtained by displacing the feed along the reflector axis. The secondary beam remains symmetrically disposed to the reflector axis, independent of the axial feed displacement. For an offset fed section of the parent reflector, however, the direction of the axis of the secondary beam is a function of axial defocusing as shown in Figure 2. This is because axial defocusing by ΔZ in an offset reflector antenna produces in the X-Z plane a non-uniform phase that changes monotonically from $\Delta Z \cos\Theta_L$ to $\Delta Z \cos\Theta_U$, where Θ_U and Θ_L are the angles subtended at the focal point by the edges of the reflector in X-Z plane. To insure that the direction of the axis of the secondary beam is always along the reflector axis, the feed must be moved along a direction Θ_o so that the phase variation across the aperture in the X-Z plane is symmetrical with respect to that direction. It can easily be shown that the path length, from the feed phase center to the focal plane, of a typical ray is $2f - \Delta Z' \cos(\Theta - \Theta_o)$ where f is the focal length of the reflector, $\Delta Z'$ is the distance through which the feed is displaced along the direction Θ_o, and Θ is the acute angle between the typical ray and the reflector axis. Clearly, the typical ray is an even function of Θ about Θ_o for $\Theta_L \le \Theta \le \Theta_U$ if Θ_o is given by:

$$\Theta_o = \frac{1}{2} (\Theta_U + \Theta_L) \qquad (1)$$

Defining the offset axis as the direction Θ_o, and $\Delta Z'$ as the defocusing along the offset axis, a calculated offset axial defocusing curve along with some measured data are plotted in Figure 3. In both the measured and calculated results, the primary feed is oriented so that the feed axis coincides with Θ_o. The close agreement between the measured and calculated results is evident.

If the offset axial defocusing is produced by feed displacement along the offset axis, then the equi-phase surface (offset focal plane), using the parallel ray approximation, is the plane containing the focal point and perpendicular to the offset axis.* Calculations show that this is indeed the case if the reference phase center is chosen as the projection on the parent focal plane of the intersection of the offset axis with the reflector surface.

*For small angles of scan, Reference [4].

*Dr. J. W. Duncan, TRW Systems Group, Private Communication.

Reprinted from *1974 Int. IEEE/AP-S Symp. Program & Digest*, June 10–12, 1974, pp. 121–123.

Beam Deviation Factor

Figure 4 shows the measured and calculated beam squints in the principal planes due to lateral feed displacements in the offset focal plane. Good agreement between the measured and calculated results is apparent. It is of interest to note that the beam squints in the principal planes corresponding to equal lateral displacements in the two planes are practically the same, indicating that the beam deviation factor (BDF) is independent of the azimuthal position of the feed. Figure 5 shows the calculated BDF as a function of the edge angle Θ_E (measured with respect to offset axis Θ_o) for various Θ_o. The primary feed is assumed to have a $\cos^n$ type voltage pattern with n adjusted to give a -10 dB edge taper (exclusive of space loss). We see that for a given Θ_o (or Θ_E), the BDF decreases with increasing Θ_E (or Θ_o). This is to be expected, as either increasing θ_E for a given θ_o or increasing θ_o for a given θ_E decreases the f/D, which is given by

$$f/D = \frac{\cos\theta_E + \cos\theta_o}{4 \sin\theta_E} , \qquad (2)$$

thus decreasing the BDF.

It has been found that for a given set of θ_E and θ_o the ratio of the BDF (from Figure 5) to the f/D (from Equation 2) is constant for a given θ_E. This means that the BDF of an offset reflector can be calculated from that of a front fed reflector with the same edge angle using the following equation:

$$(BDF)_{offset} \; (f/D)_{front\ fed} = (BDF)_{front\ fed} \; (f/D)_{offset} \qquad (3)$$

Conclusion

An offset axis and an offset focal plane have been defined for offset reflectors so that the general focal region properties are analogous to those of front fed paraboloids. The theoretically predicted focal region behaviors of a particular offset reflector are also experimentally verified. The dependence of the BDF upon Θ_o and Θ_E is presented for offset reflectors illuminated by a $\cos^n$ voltage illumination function with n adjusted to give a -10 dB edge taper.

References

[1] Y. T. Lo, "On the Beam Deviation Factor of a Parabolic Reflector," IRE Trans. Antennas and Propagation (Commun.), vol. AP-8, pp. 347-349, May 1960.

[2] J. Ruze, "Lateral-Feed Displacement in a Paraboloid," IEEE Trans. Antennas and Propagation, vol. AP-13, pp. 660-665, September 1965.

[3] P. G. Ingerson and W.V.T. Rusch, "Radiation from a Paraboloid with an Axially Defocused Feed," IEEE Trans. Antennas and Propagation, vol. 21, No. 1, January 1973.

[4] W.V.T. Rusch and A. C. Ludwig, "Determination of the Maximum Scan-Gain Contours of a Beam-Scanning Paraboloid and Their Relation to the Petzval Surface," IEEE Trans. Antennas and Propagation, vol. AP-21, No. 2, March 1973.

[5] D. F. DiFonzo, "Offset and Symmetrical Reflector Antennas," M.S. Thesis, San Fernando Valley State College, January 1972.

[6] T. S. Chu and R. H. Turrin, "Depolarization Properties of Offset Reflector Antennas," IEEE Trans. Antennas and Propagation, vol. AP-21, May 1973.

[7] A. W. Rudge, "Offset-Reflector Antennas with Offset Feeds," Electronic Letters, November 17, 1973, pp. 611-613.

[8] J. A. Janken, W. J. English, and D. F. DiFonzo, "Radiation from Multimode Reflector Antennas," 1973 G-AP Symposium Digest, pp. 306-309.

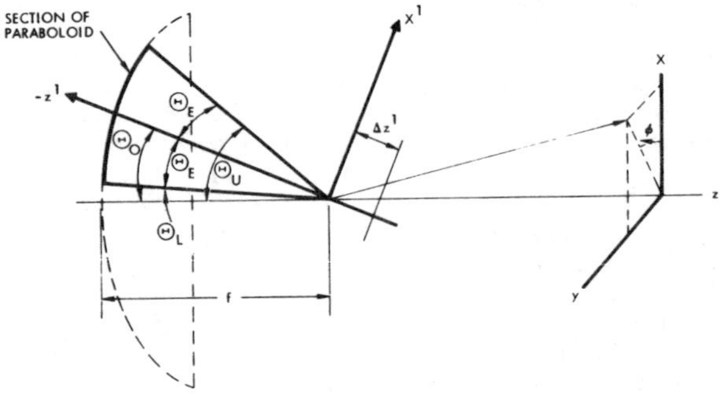

Figure 1. Offset paraboloidal reflector antenna.

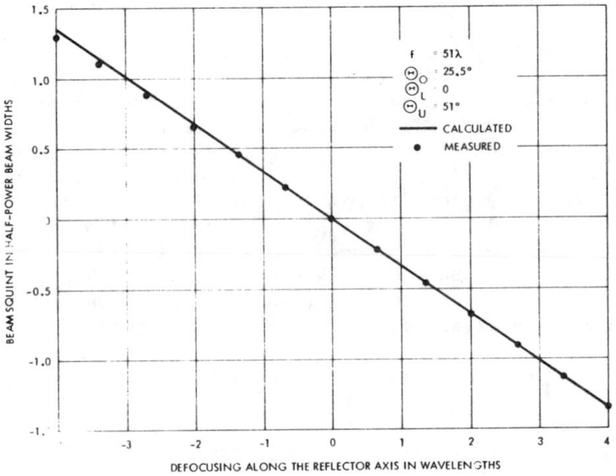

Figure 2. Beam squint characteristics.

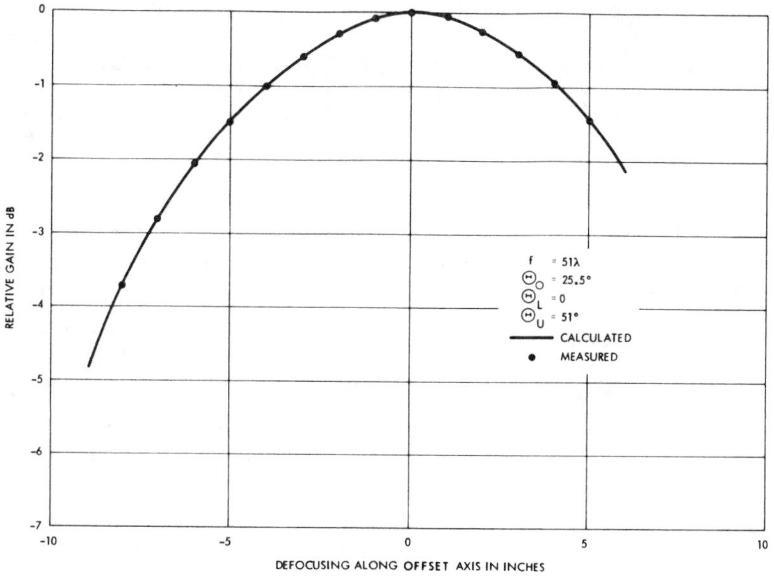

Figure 3. Defocusing curve.

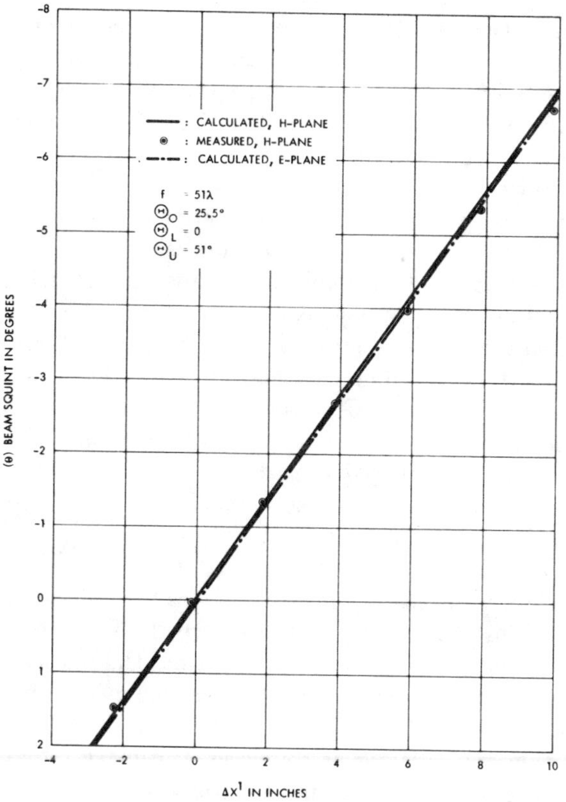

Figure 4. Beam deviation characteristics.

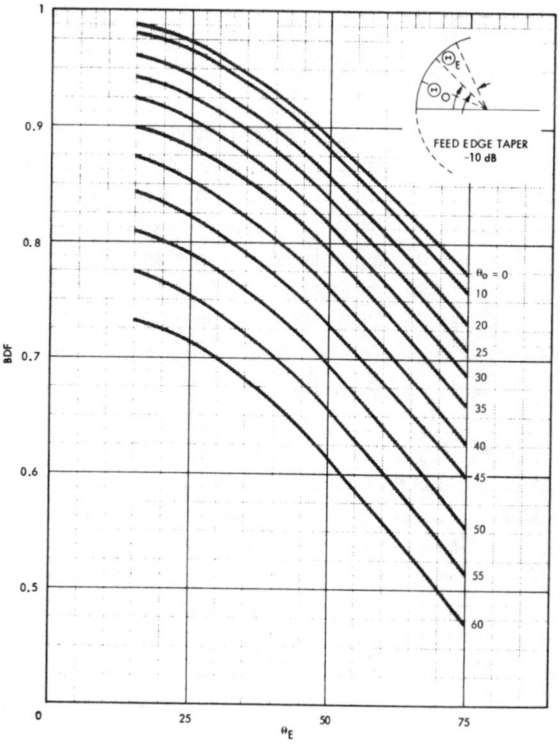

Figure 5. Beam deviation factor vs. the edge angle Θ_E for different offset angle Θ_O.

The Polarization Losses of Offset Paraboloid Antennas

JACOB DIJK, CHRISTIAAN T. W. VAN DIEPENBEEK, EDUARD J. MAANDERS, SENIOR MEMBER, IEEE, AND LAMBERT F. G. THURLINGS

Abstract—**In this paper the electric field in the aperture of offset front-fed paraboloid antennas and open Cassegrainian antennas, excited by an electric dipole or Huygens source in the focus, is compared with the fields of front-fed circularly symmetrical paraboloid reflector antennas and classical Cassegrainian antennas. The aperture field forms the basis of expressions to calculate the polarization efficiency of all four types of antenna. Computed results are given, showing that offset antennas can compete with front-fed paraboloids if they are excited by an electric dipole; the classical Cassegrainian antenna, however, shows better results. If offset antennas are excited by a Huygens source, the result is very unfavorable compared with the symmetrical antennas which show no cross polarization.**

I. INTRODUCTION

IT HAS BEEN known for several years that, if a paraboloid reflector antenna is fed by a linearly polarized electrical dipole, the antenna system will radiate energy not only in the main polarization, but also a fair amount in an unwanted polarization called cross-polarization or depolarization.

Condon [1] was one of the first to give a detailed analysis of this phenomenon. It appears that cross-polarized lobes, also called Condon lobes, are formed having a maximum in planes at 45° to the principal plane. Silver [2, p. 423] also mentioned this cross polarization, mainly as an abstract of Condon's work.

Cutler [3] gives a physical explanation as to the relation between aperture electric field lines and the polarization of the dipole feed, and explains the very unfavorable situation that occurs if the focus of the paraboloid falls between the aperture and apex of the paraboloid. This work has been continued by Jones [4], who investigated the radiation characteristics of paraboloid reflector antennas excited in their foci by a short electrical dipole feed, a short magnetic dipole feed, and a plane wave source, being a combination of an electric and a magnetic dipole. If this dipole pair is represented by dipole fields of equal intensity, commonly known as a Huygens source, it has been proved that the cross-polarized component of the aperture illumination could be made to disappear [5].

Koffman [5] has extended this work by considering other conical sections of revolutions as well as the paraboloid. The cross-polarized pattern of the reflector excited by any arbitrary feed system may be calculated, using the methods of Afifi [6], while Potter [7] has found an analytical expression for the polarization loss or polarization efficiency. It is the latter expression that will also be reviewed in this paper. Potter [8] has also found a similar expression for Cassegrainian antennas, which will be included in the present study.

Watson and Ghobrial [9] have investigated the cross-polarization isolation at off-axis incidence for classical Cassegrainian antennas and front-fed paraboloidal reflectors. It was shown that the Cassegrainian antenna was much superior to the equivalent front-fed antenna.

Not much is known so far about offset paraboloids and open Cassegrainian [10] antennas. Hanfling [11] has shown a stereographic mapping method that contains the aperture field lines of an offset antenna excited by several field sources, but without further details. Graham [12], [13] describes the polarization of offset antennas and states that an offset Cassegrainian antenna can be designed to have low cross-polarization losses, which was experimentally discovered by letting the axis of the main and subreflector differ only a few degrees. No calculations have been mentioned.

Since plans exist for frequency reuse, above 10 GHz, by polarization diversity, the interest in cross-polarization problems has recently increased considerably.

Ludwig [14] has published a paper on the definition of cross-polarization, and Kinber and Tischenko [15] calculated the current distribution of various reflector antennas with different illumination. Unfortunately no numerical results are given. Chu and Turrin [16] have discussed the beamshift of offset antennas with circular polarization and have calculated the level of cross-polarization sidelobes. The poor polarization performance of the open Cassegrainian antenna has been predicted.

It is the purpose of this paper to obtain a more detailed insight into the cross-polarization losses of offset antennas. For this purpose we shall compare the front-fed paraboloid, the true Cassegrainian antenna, the offset front-fed paraboloid, and the open Cassegrainian antenna. In all the cases we shall use a short electrical linearly polarized dipole and a Huygens source as a primary radiator. We will compare the aperture electric fields, define the polarization efficiency, and calculate this for different configurations. Finally, we will show a practical example.

II. APERTURE FIELDS OF REFLECTOR ANTENNAS ILLUMINATED BY AN ELECTRIC DIPOLE

Let us consider a short electric dipole of length l [2, p. 92], lying along the x axis of a Cartesian coordinate

Manuscript received October 3, 1972; revised January 23, 1974.
J. Dijk, E. J. Maanders, and L. F. G. Thurlings are with the Eindhoven University of Technology, Eindhoven, the Netherlands.
C. T. W. van Diepenbeek is with the Max Planck—Institut für Radioastronomie, Bonn, Germany.

Reprinted from *IEEE Trans. Antennas Propagat.*, vol. AP-22, pp. 513–520, July 1974.

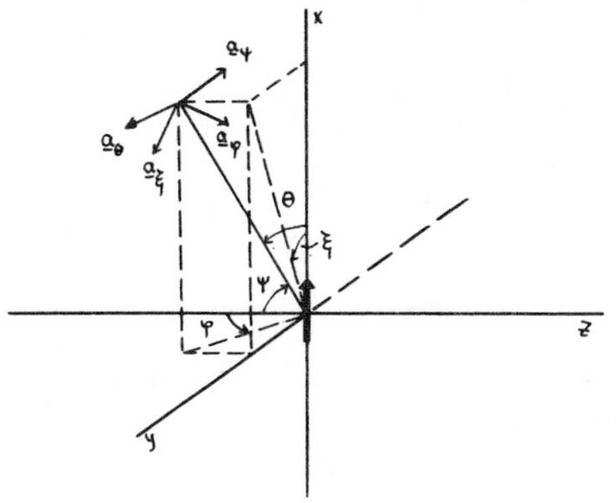

Fig. 1. Electric dipole oriented along positive x axis of Cartesian coordinate system.

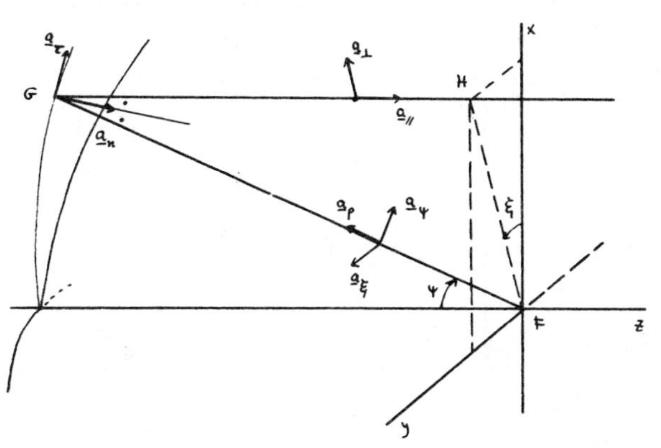

Fig. 2. Geometry of parabolic reflector with incident and reflected rays and vectors.

system (Fig. 1), with a current I flowing in the direction of the positive x axis. Expressed in ρ,ψ,ξ coordinates, the far zone components of the complex electric field are

$$\bar{E} = E_\psi \bar{a}_\psi + E_\xi \bar{a}_\xi$$

or

$$\bar{E} = \frac{j\eta Il \exp(-jk\rho)}{2\lambda\rho}(-\bar{a}_\psi \cos\psi \cos\xi + \bar{a}_\xi \sin\xi) \quad (1)$$

where $\eta = 120\,\pi\,\Omega$, $\bar{a}_\psi$ and $\bar{a}_\xi$ are unit vectors along the ψ and ξ axes, respectively, and k is the wavenumber.

In x,y,z coordinates (1) becomes

$$\bar{E} = -E_0[\bar{x}(\cos^2\psi \cos^2\xi + \sin^2\xi) - \tfrac{1}{2}\bar{y}\sin^2\psi \sin 2\xi + \tfrac{1}{2}\bar{z}\sin\psi\cos\xi] \quad (2)$$

where

$$E_0 = \frac{j\eta Il \exp(-jk\rho)}{2\lambda\rho}.$$

If the dipole is oriented along the positive y axis, it is readily seen that the electric field becomes

$$\bar{E} = (-\bar{a}_\psi \cos\psi \sin\xi - \bar{a}_\xi \cos\xi) \quad (3)$$

or

$$\bar{E} = -E_0[-\tfrac{1}{2}\bar{x}\sin^2\psi \sin 2\xi + \bar{y}(\cos^2\psi \sin^2\xi + \cos^2\xi) + \tfrac{1}{2}\bar{z}\sin 2\psi \sin\xi]. \quad (4)$$

These fields will induce surface currents in any arbitrary reflector using geometrical optical techniques.

Using the method employed by Jones [4], the aperture field may now be found by calculating the surface-current density of the reflector $\bar{K} = 2(\bar{n} \times \bar{H}_i)$, $\bar{H}_i$ being the initial field and $\bar{n}$ the unit vector normal to the surface at the point of incidence and projecting $\bar{K}$ on the aperture.

A simpler way to find the aperture field may be followed by investigating what happens with the fields $E_\psi \bar{a}_\psi$ and $E_\xi \bar{a}_\xi$ at the point of incidence. From Fig. 2 it is readily seen that the vector $E_\xi \bar{a}_\xi$ is perpendicular to the plane comprising the z axis, radius ρ from focus to the surface of the reflector, the reflected ray, and the vector $\bar{n}$ at the point of incidence (plane FGH). After reflection this vector remains perpendicular to the surface, but its direction reverses. Therefore,

$$\bar{E}_\xi^r = -E_\xi \bar{a}_\xi \quad (5)$$

the index r indicating reflection.

The vector $E_\psi \bar{a}_\psi$ lies in plane FGH and is perpendicular to the radius. To find out what happens with $E_\psi \bar{a}_\psi$ we will use Fig. 2 and define the indices n and τ as the directions normal and tangential to the paraboloid surface at the point of incidence. We now resolve E_ψ in $E_{\psi,n}$ and $E_{\psi,\tau}$ resulting in

$$E_{\psi,n} = E_\psi \sin\tfrac{1}{2}\psi$$
$$E_{\psi,\tau} = E_\psi \cos\tfrac{1}{2}\psi. \quad (6)$$

After reflection, $E_{\psi,n}$ is continuous and $E_{\psi,\tau}$ retains its sign. Therefore,

$$E_{\psi,n}{}^r = E_\psi \sin\tfrac{1}{2}\psi$$
$$E_{\psi,\tau}{}^r = -E_\psi \cos\tfrac{1}{2}\psi. \quad (7)$$

By means of the vectors $\bar{a}_\perp$ and $\bar{a}_{||}$ (Fig. 2) and by resolving $E_{\psi,n}$ and $E_{\psi,\tau}$ along these vectors it is readily found, using (7) that

$$E_\perp{}^r = -E_{\psi,n}{}^r \sin\tfrac{1}{2}\psi + E_{\psi,\tau}{}^r \cos\tfrac{1}{2}\psi = -E_\psi, \quad (8)$$

$$E_{||}{}^r = E_{\psi,n}{}^r \cos\tfrac{1}{2}\psi + E_{\psi,\tau}{}^r \sin\tfrac{1}{2}\psi = 0. \quad (9)$$

The unit vector $\bar{a}_\perp$ may be written

$$\bar{a}_\perp = \bar{x}\cos\xi + \bar{y}\sin\xi.$$

If we use an electric dipole oriented along the positive x axis, the reflected field $\bar{E}_\psi{}^r$ follows from (1), (8), and (10) resulting in

$$\bar{E}_\psi{}^r = E_0 \cos\psi \cos\xi(\cos\xi, \sin\xi, 0) \quad (11)$$

and (5) becomes $\bar{E}_\xi{}^r = -E_0 \sin\xi \bar{a}_\xi$.

By means of (1), (5), and (11) the aperture field E_A yields

$$\bar{E}_A = E^r(x,y,z) = E_0 \cos\psi \cos\xi(\cos\xi, \sin\xi, 0) - E_0 \sin\xi(-\sin\xi, \cos\xi, 0)$$

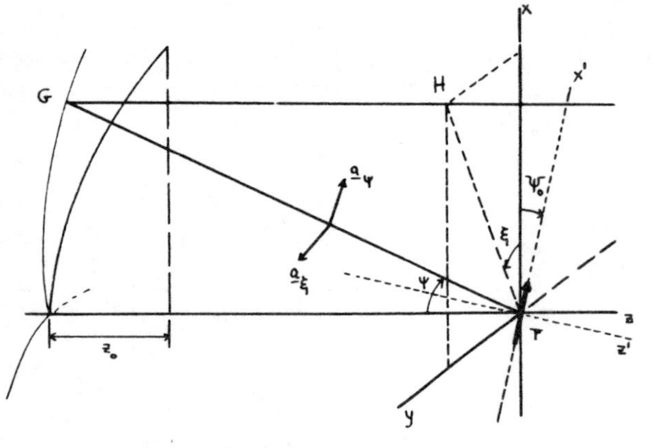

Fig. 3. Offset paraboloidal reflector.

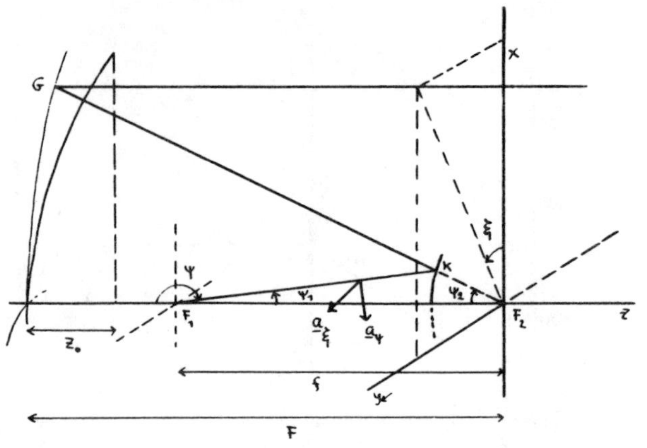

Fig. 4. Geometry of classical Cassegrainian antenna.

or

$$\bar{E}_A = E_0[\{1 - \cos^2 \xi(1 - \cos \psi)\}\bar{x} - \tfrac{1}{2} \sin 2\xi(1 - \cos \psi)\bar{y}] \quad (12)$$

where

$$E_0 = \frac{j\eta Il \exp[-jk(F + z_0)]}{2\lambda\rho}.$$

Using the same technique we readily find the aperture field, if the dipole is oriented along the positive y or z axis [17].

The offset paraboloid is illustrated in Fig. 3. If the electric dipole is located in the focus of the paraboloid oriented along the positive x' axis of an x',y',z' coordinate system, the aperture field appears to be

$$\begin{aligned}\bar{E}_A = {} & [E_0 \cos \Psi_0\{1 - \cos^2 \xi(1 - \cos \psi)\} \\ & + E_0 \sin \Psi_0 \sin \psi \cos \xi]\bar{x} \\ & + [E_0 \cos \Psi_0\{-\tfrac{1}{2} \sin 2\xi(1 - \cos \psi)\} \\ & + E_0 \sin \Psi_0 \sin \psi \sin \xi]\bar{y}. \end{aligned} \quad (13)$$

The results for the case when the dipole is oriented along the y' or z' axis are found elsewhere [17].

The same technique used for the front-fed paraboloid may be employed to calculate the fields in the aperture of a Cassegrainian antenna. However, there are some fundamental differences because the dipole field is reflected twice before it arrives at the main reflector aperture. Therefore, the components $\bar{E}_\xi$ and $\bar{E}_\psi$ have to be known after this double reflection in order to calculate this aperture field. If the electric dipole is located in focus F_1 and oriented along the positive x axis (Fig. 4), the aperture field is

$$\bar{E}_A = -E_0'[\{1 - \cos^2 \xi(1 - \cos \psi_1)\}\bar{x} - \tfrac{1}{2} \sin 2\xi(1 - \cos \psi_1)\bar{y}]. \quad (14)$$

In (14) $E_0' = j\eta Il \exp(-jkr_a)/2\lambda\rho'$, where ρ' is the distance between the primary focus and the surface of the main reflector. It is readily found that $r_a = f/e + F + Z_0$ and $\rho' = 2F/(1 + \cos \psi) + f/e$, where f is the distance between the two hyperboloid foci, e the hyperboloid eccentricity, and Z_0 the depth of the paraboloid. If the dipole is oriented along the positive y or z axis similar equations may be found [17].

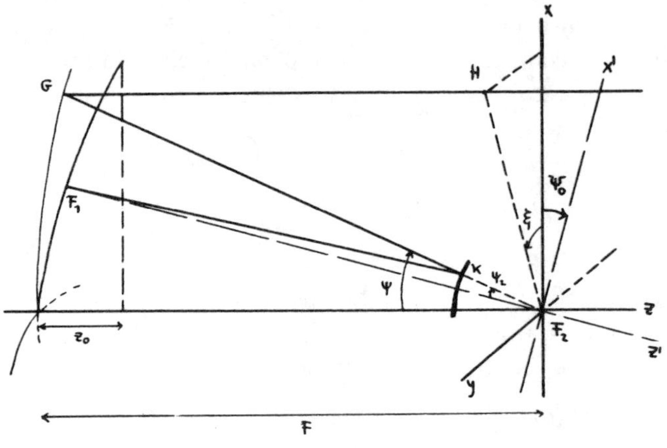

Fig. 5. Geometry of the open Cassegrainian antenna.

The calculation of the aperture field of an open Cassegrainian antenna is much more complicated than the previous ones. The geometry is presented in Fig. 5. In general, the planes KGH (with the z axis) and F_1KG (with the z' axis) will not coincide. Therefore, the ray from the primary focus F_1 to the subreflector and the ray reflected from the paraboloid (GH) will generally not be located in the same plane. The calculation of the aperture field leads to long algebraic equations. The reader is referred to a report recently issued [17], where a detailed description of these equations is given.

III. THE APERTURE FIELDS OF REFLECTOR ANTENNAS ILLUMINATED BY A HUYGENS SOURCE

A combination of an electric dipole and a magnetic dipole of equal intensity and crossly oriented is often called a Huygens source [4]. If this source is located in the focus of a paraboloid antenna in such a way that the electric dipole orients along the positive x axis and the magnetic dipole along the positive y axis, the aperture fields are readily found by superposition of the aperture fields caused by illumination with electric and magnetic

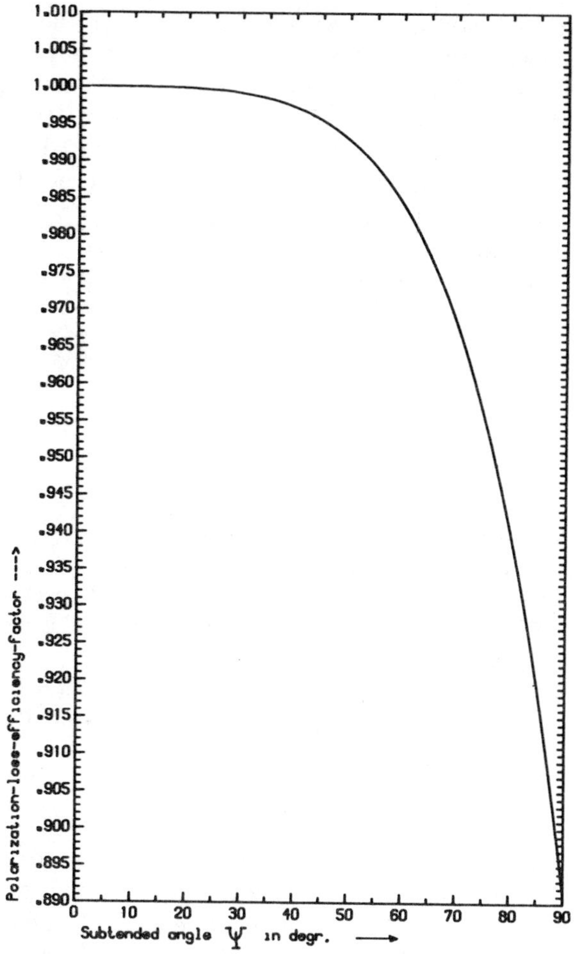

Fig. 6. Polarization loss efficiency factor of parabolic reflector.

dipoles. The reader is referred to a report [17] recently issued for the aperture fields of reflector antennas fed by a magnetic dipole.

In accordance with Jones [4], we find

$$\bar{E}_A = E_0(1 + \cos\psi)\bar{x}. \tag{15}$$

$$\eta_p = \frac{\left| \int_0^{2\pi} \int_0^{\Psi} \frac{E_{mp}(\psi,\xi)}{E_0} \tan \tfrac{1}{2}\psi \, d\psi \, d\xi \right|^2}{\left| \int_0^{2\pi} \int_0^{\Psi} \left[\left(\frac{E_{mp}(\psi,\xi)}{E_0} \right)^2 + \left(\frac{E_{cp}(\psi,\xi)}{E_0} \right)^2 \right]^{1/2} \tan \tfrac{1}{2}\psi \, d\psi \, d\xi \right|^2}. \tag{18}$$

In the same way if the electric dipole is oriented along the $+y$ axis and the magnetic dipole along the $-x$ axis

$$\bar{E}_A = E_0(1 + \cos\psi)\bar{y}. \tag{16}$$

It appears that in both cases the cross-polarization component disappears. A classical Cassegrainian antenna shows similar results and the cross-polarization component disappears as well.

The aperture field of an offset antenna illuminated by a Huygens source may be found by combining the aperture fields originated by an electric dipole oriented along the positive x' axis and a magnetic dipole oriented along the positive y' axis, or by an electric dipole oriented along the positive y' axis and the magnetic dipole along the negative x' axis. As will be noticed, the cross-polarization component does not disappear.

If we try to find the aperture fields of an open Cassegrainian antenna, it appears also that no simplification takes place. Therefore, it is of little value to rewrite the equations found before. As will be noticed, the cross-polarization component in the aperture does not disappear either.

IV. THE POLARIZATION EFFICIENCY

In accordance with Potter [7], the polarization efficiency of an antenna is defined by the ratio of antenna gain including the effects of cross polarization, to antenna gain if the cross-polarized energy were zero everywhere. Thus

$$\eta_p = \frac{\left| \int_0^{2\pi} \int_0^{\Psi} E_{mp}(\psi,\xi) \rho^2 \sin\psi \, d\psi \, d\xi \right|^2}{\left| \int_0^{2\pi} \int_0^{\Psi} [E_{mp}{}^2(\psi,\xi) + E_{cp}{}^2(\psi,\xi)]^{1/2} \rho^2 \sin\psi \, d\psi \, d\xi \right|^2} \tag{17}$$

where $E_{mp}(\psi,\xi)$ represents the electric field in the aperture with principal polarization and $E_{cp}(\psi,\xi)$ that of the cross-polarization. By means of (17) and the equations for the electric field in the aperture found in the previous paragraphs it is now possible to calculate the polarization efficiency.

In the case that a front-fed paraboloid is investigated, the distance ρ between paraboloid and focus is $\rho = 2F/(1 + \cos\psi)$, and because all the fields involved are proportional to $\exp(-jk(F + z_0)/\rho$, (17) may be replaced by (18).

If the paraboloid is illuminated by an electric dipole oriented along the $+x$ axis, the aperture fields to be used are

$$E_{mp} = E_0[1 - \cos^2\xi(1 - \cos\psi)] \tag{19}$$

$$E_{cp} = -\tfrac{1}{2}E_0 \sin 2\xi(1 - \cos\psi) \tag{20}$$

where

$$E_0 = \frac{j\eta Il \exp[-jk(F + z_0)]}{2\lambda\rho}.$$

It is possible to simplify (18) by substituting (19) and (20), but this does not increase the insight into the problem. An approximation of this equation as carried out by Potter [7], has the drawback that it gives only reliable results for very shallow paraboloid reflectors with subtending angles of less than 60°. The results of (18), computed without any approximation, applied to front-fed paraboloid reflector antennas are presented in Fig. 6.

In the case of a classical Cassegrainian antenna (Fig. 4), the integration is carried out over the angles ξ and ψ_2. We can now replace (17) by

$$\eta_p = \frac{\left| \int_0^{2\pi} \int_0^{\Psi} \frac{E_{mp}(\xi,\psi_2)}{E_0} \cdot \tan\left(\tfrac{1}{2}\psi_2\right) d\xi\, d\psi_2 \right|^2}{\left| \int_0^{2\pi} \int_0^{\Psi} \left[\left(\frac{E_{mp}(\xi,\psi_2)}{E_0}\right) + \left(\frac{E_{cp}(\xi,\psi_2)}{E_0}\right) \right]^{1/2} \tan\left(\tfrac{1}{2}\psi_2\right) d\xi\, d\psi_2 \right|^2}. \quad (21)$$

If the subreflector is illuminated by an electric dipole, oriented along the positive x axis, the aperture fields to be used are

$$\frac{E_{mp}}{E_0'} = -1 + \cos^2 \xi \left[1 - \cos\xi \cdot 2 \arctan \left\{ \frac{e-1}{e+1} \tan\left(\tfrac{1}{2}\psi_2\right) \right\} \right] \quad (22)$$

$$\frac{E_{cp}}{E_0'} = \tfrac{1}{2} \sin 2\xi \left[1 - \cos \xi \cdot 2 \arctan \left\{ \frac{e-1}{e+1} \tan\left(\tfrac{1}{2}\psi_2\right) \right\} \right]. \quad (23)$$

Fig. 7 shows the computed results, where the polarization efficiency is given in relation to the subtended angle of the main reflector with the magnification ratio $M = e + 1/e - 1$ as a parameter.

When an offset paraboloid antenna is investigated (18) may still be used, however, the integration limits will differ. As explained before [17], ψ will have to be integrated between $\Psi_0 - \Psi$ and $\Psi_0 + \Psi$. The integration limits of ξ, ξ_L, and ξ_R are

$$\xi_L = -\arccos \left(\frac{\cos \Psi - \cos \Psi_0 \cos \psi}{\sin \Psi_0 \sin \psi} \right) \quad (24)$$

$$\xi_R = -\arccos \left(\frac{\cos \Psi - \cos \Psi_0 \cos \psi}{\sin \Psi_0 \sin \psi} \right) \quad (25)$$

where Ψ_0 is the offset angle and Ψ the angular aperture of the mainreflector. (In the open Cassegrainian antenna Ψ is called Ψ_2). Equation 18 is then written as

$$\eta_p = \frac{\left| \int_{\Psi_0-\Psi}^{\Psi_0+\Psi} \int_{\xi_L}^{\xi_R} \frac{E_{mp}(\xi,\psi)}{E_0} \tan \tfrac{1}{2}\psi\, d\psi\, d\xi \right|^2}{\left| \int_{\Psi_0-\Psi}^{\Psi_0+\Psi} \int_{\xi_L}^{\xi_R} \frac{(E_{mp}^2 + E_{cp}^2)^{1/2}}{E_0} \tan \tfrac{1}{2}\psi\, d\psi\, d\xi \right|^2} \quad (26)$$

where the main and cross polarized fields for the offset paraboloid and open Cassegrainian antenna have been discussed in the previous sections for various illuminations.

In the case of an open Cassegrainian antenna the efficiency factor becomes a little more complicated. It is readily shown that the factor E_0 in the aperture fields is equal to that of the classical Cassegrainian antenna and that the integration limits are the same as for the offset antenna. Fig. 8 shows the polarization efficiency of an offset paraboloid illuminated by an electric dipole oriented along the positive x' axis and positive y' axis, respectively, as well as illuminated by a Huygens source. The results obtained with an open Cassegrainian antenna are given in Fig. 9. The eccentricity of the hyperboloid subreflector was 1.5. For both offset and open Cassegrainian antennas the offset angle served as a parameter.

V. A PRACTICAL EXAMPLE

In the previous section a Huygens source was presented with equal intensities of a magnetic and an electric dipole. However, many feed patterns may be divided in electric and magnetic dipoles with unequal intensities. In this section we work out a practical example.

A popular feed system used to illuminate a reflector surface is the open waveguide excited with the TE_{10} mode described by Silver [2, p. 343] and Jones [4]. The field components of a rectangular waveguide excited in the TE_{10} mode and the electric field vector oriented along the x axis is, in accordance with Silver, represented by

$$\bar{E}_\psi(\psi,\xi) = C \frac{\cos \xi}{\rho} \left[1 + \frac{\beta_{10}}{k} \cos \psi \right] F(\psi,\xi) \exp(-jk\rho) \bar{a}_\psi$$

$$\bar{E}_\xi(\psi,\xi) = -C \frac{\sin \xi}{\rho} \left[\frac{\beta_{10}}{k} + \cos \psi \right] F(\psi,\xi) \exp(-jk\rho) \bar{a}_\xi \quad (27)$$

where

$$F(\psi,\xi) = \frac{\cos[(\pi a/\lambda) \sin \psi \cos \xi]}{[(\pi a/\lambda) \sin \psi \cos \xi]^2 - (\pi/2)^2} \cdot \frac{\sin[(\pi b/\lambda) \sin \psi \sin \xi]}{(\pi b/\lambda) \sin \psi \sin \xi}.$$

In this equation it is assumed that the reflection coefficient at the opening of the waveguide is zero. The symbols a and b are waveguide dimensions, and C is a coefficient depending upon the wavelength and dimensions [2, p. 343]. Further, β_{10} stands for the phase constant for the

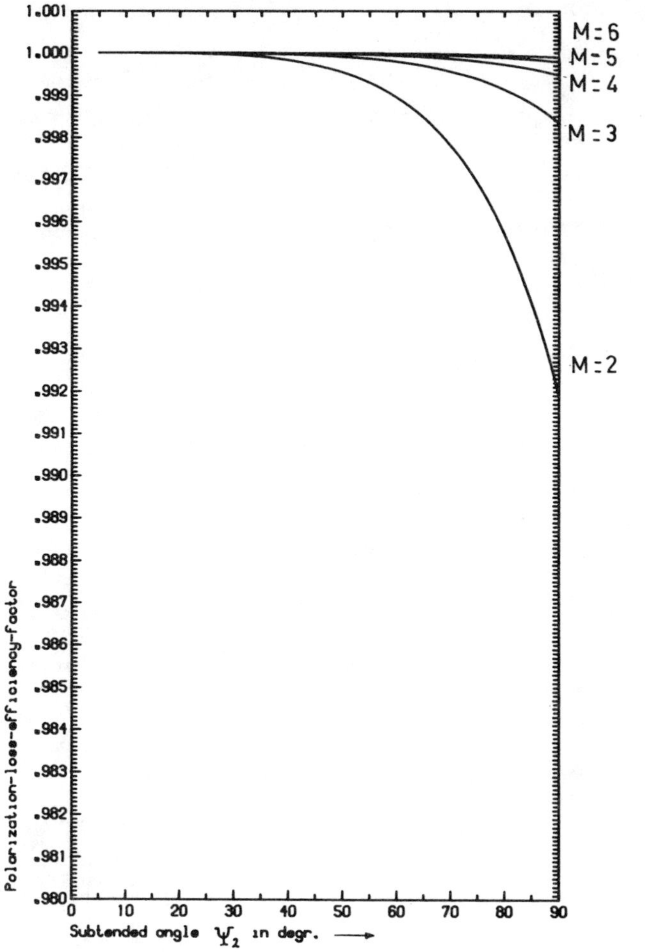

Fig. 7. Polarization loss efficiency factor of classical Cassegrainian antenna.

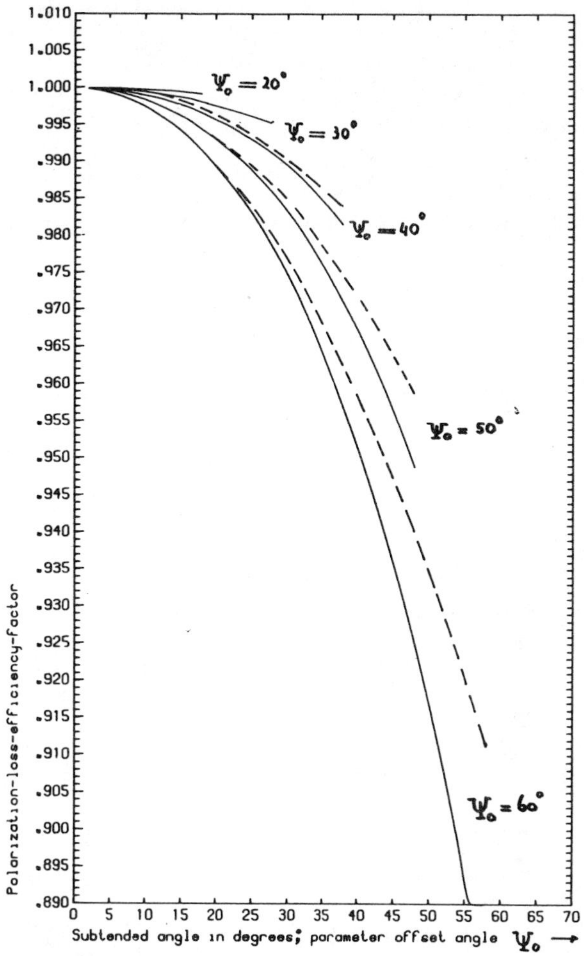

Fig. 8. Polarization loss efficiency factor of offset paraboloid reflector, offset angle Ψ_0 being a parameter, illuminated by —— E dipole oriented along x' axis and Huygens source and - - - E dipole oriented along y' axis.

TE_{10} mode, and k the propagation constant, equal to $2\pi/\lambda$.

The polarization vector is

$$\bar{a}_i = \frac{\cos\xi}{\rho}\left(1 + \frac{\beta_{10}}{k}\cos\psi\right)\bar{a}_\psi - \frac{\sin\xi}{\rho}\left(\cos\psi + \frac{\beta_{10}}{k}\right)\bar{a}_\xi. \tag{28}$$

If the dimensions of the waveguide are such that $\beta_{10}/k = 1$ the polarization vector reduces to

$$\bar{a}_i = \cos\xi\bar{a}_\psi - \sin\xi\bar{a}_\xi \tag{29}$$

which is equal to that of a Huygens source. However, in practice this cannot be realized as normally

$$\frac{\beta_{10}}{k} = \frac{\lambda}{\lambda_{g10}} \tag{30}$$

where λ_{g10} is the wavelength in the guide [2, p. 205]

$$\lambda_{g10} = \frac{\lambda}{[1 - (\lambda/2a)^2]^{1/2}} \tag{31}$$

for the TE_{10} mode. Therefore, $\beta_{10} = k$ only for $\lambda \ll a$.

Nevertheless, this polarization vector is very popular and is used by several authors such as Afifi [6], Carter [18], and Tartakovski [19], as it simplifies the complicated mathematical work considerably. If we want to study the cross-polarization properties of antennas illuminated by this feed, we must know the waveguide dimensions, frequency range, and cutoff frequency.

If we study a rectangular waveguide in the X band (8200–12 400 MHz) the dimensions a and b are 0.900×0.400 in and the cutoff frequency is 6560 MHz. The proportions of the lowest and highest frequencies to the cutoff frequency are 1.25 and 1.90, a relationship that is also found for waveguides in other frequency bands. Let λ_1 (be the longer wavelength) = 3.66 cm and λ_2 (the shorter) = 2.42 cm. The wavelength in the waveguide for λ_1 is then

$$\lambda_{g10} = \frac{3.66}{[1 - (3.66/4.57)^2]^{1/2}} = 6.11 \quad \text{cm}$$

and for λ_2

$$\lambda_{g10} = \frac{2.42}{[1 - (2.42/4.57)^2]^{1/2}} = 2.85 \quad \text{cm}.$$

From (30) we then obtain for

$$\beta_{10/k} = \lambda_1/\lambda_{g10} = 3.66/6.11 = 0.60$$

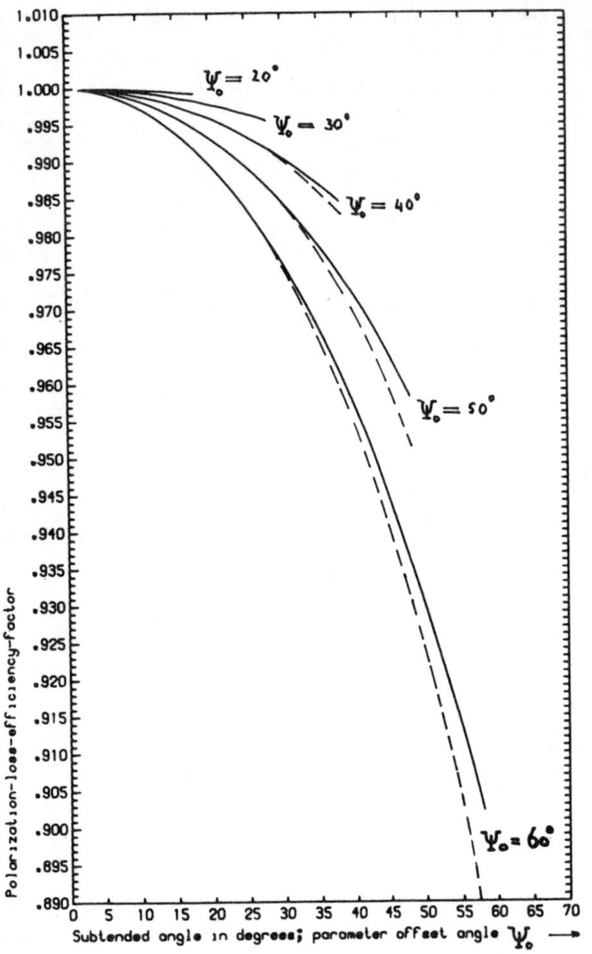

Fig. 9. Polarization loss efficiency factor of open Cassegrainian antenna, offset angle Ψ_0 being a parameter, illuminated by —— E dipole oriented along x' and y' axis and - - - Huygens source.

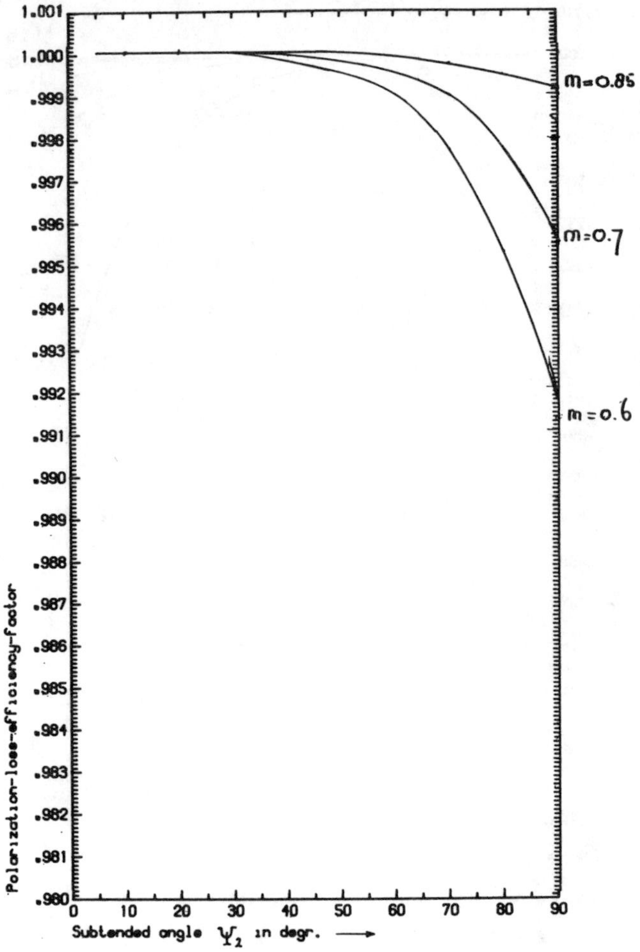

Fig. 10. Polarization loss efficiency factor of a circularly symmetrical paraboloid antenna illuminated by open waveguide excited with the TE_{10} mode.

and

$$\beta_{10/k} = \lambda_2/\lambda_{g10} = 2.42/2.85 = 0.85$$

The polarization vector for λ_1 is now

$$\bar{a}_i = \frac{\cos\xi}{\rho}(1 + 0.60\cos\psi)\bar{a}_\psi - \frac{\sin\xi}{\rho}(0.60 + \cos\psi)\bar{a}_\xi \tag{32}$$

and for λ_2

$$\bar{a}_i = \frac{\cos\xi}{\rho}(1 + 0.85\cos\psi)\bar{a}_\psi - \frac{\sin\xi}{\rho}(0.85 + \cos\psi)\bar{a}_\xi. \tag{33}$$

The polarization properties apparently depend on the frequency. If such a feed is used to illuminate a front-fed paraboloid antenna, it is readily found by means of the theory developed in Section II that the aperture field

$$\bar{E}_A = E_0 \frac{\cos\xi}{1+\cos\psi}(1 + m\cos\psi)(\cos\xi, \sin\xi, 0)x$$

$$- E_0 \frac{\sin\xi}{1+\cos\psi}(m + \cos\psi)(-\sin\xi, \cos\xi, 0)$$

or

$$E_A = \frac{E_0}{1+\cos\psi}[\{(1 + m\cos\psi)\cos^2\xi + (m+\cos\psi)\sin^2\xi\}\bar{x} + \{(1+m\cos\psi)\cos\xi\sin\xi - (m+\cos\psi)\cos\xi\sin\xi\}\bar{y}] \tag{34}$$

where m is any value between 0.60 and 0.85. E_0 is the amplitude factor of the feed system and is in accordance with Silver [2, p. 343]

$$E_0 = \frac{\cos[(\pi a/\lambda)\sin\psi\cos\xi]}{[(\pi a/\lambda)\sin\psi\cos\xi)^2 - 1/2\pi]} \times \frac{\sin[(\pi b/\lambda)\sin\psi\sin\xi]}{(\pi b/\lambda)\sin\psi\sin\xi}. \tag{35}$$

The results for three different values of m are given in Fig. 10.

VI. CONCLUSIONS

It has been demonstrated that by calculating the aperture electric fields of antennas with a paraboloid (main) reflector, expressions may be derived for polarization efficiency or polarization loss. These expressions are found not only for front-fed paraboloids, but also for classical Cassegrainian antennas, front-fed offset paraboloids, and open Cassegrainian antennas. Both electric dipole excitation and excitation by a Huygens source are investigated as they give a good insight into the problems and facilitate comparative studies. Moreover, there are a number of realistic feeds, such as a rectangular horn excited in the TE_{01} mode, having polarization properties close to the Huygens source. An example of this kind has been worked out, showing that the polarization losses decrease considerably if the polarization vector approaches that of a Huygens source. If investigations are required for feeds with polarization properties different from those discussed here, the same techniques may be used.

After the electric aperture field has become known, an expression may be found for the polarization efficiency η_p. Carrying out the computation, it is readily seen that the front-fed paraboloid has very bad polarization properties, becoming worse for deep paraboloids. In the case in which the focus falls within the aperture plane ($\Psi_2 = 90°$), the polarization efficiency falls to 89 percent (Fig. 6). On the other hand, the true Cassegrainian antenna has much better properties, which not only depend upon the subtended angle by the main reflector, but also on the magnification ratio $M = (e+1)/(e-1)$, which has been introduced as a parameter (Fig. 7). The result becomes worse for low M values and deep main paraboloids; however, for $M = 2$ and $\Psi_2 = 90°$, the true Cassegrainian antenna still retains a polarization efficiency of 99 percent, which is considerably more than in the case of front-fed paraboloids with equal Ψ_2. Offset paraboloid antennas show an increase in the losses at increasing subtended angle and increasing offset angle. If we compare the front-fed paraboloid with the offset paraboloid, it appears that the former shows better results for equal subtended angles than the offset antenna with an electric dipole polarized along the x' axis; e.g., a front-fed paraboloid with a subtended angle of 60° has a polarization efficiency of 98.5 percent, while an offset paraboloid with subtended and offset angles of 60° shows an efficiency of only 91 percent (Fig. 8). If the dipole is polarized along the y' axis, the efficiency even drops to 89 percent.

If we study the results obtained with an open Cassegrainian antenna illuminated by an electric dipole, it appears that not much difference is noticed if the dipole is oriented along the x'' axis or y'' axis. At offset angles and subtended angles of about 60 degrees it appears that the efficiency drops to 90 percent, which is of the same order as for offset front-fed paraboloids (Fig. 9). The results obtained by illumination by a Huygens source, for both offset antennas and open Cassegrainian antennas, are similar to those obtained by illumination by an electric dipole. The results clearly depend on the offset and subtended angles rather more than on the polarization of the feed. At offset angles and main reflector subtended angles of about 60°, an efficiency of about 90 percent is noticed again.

We also investigated the losses of open Cassegrainian antennas in relation to the eccentricity of the hyperboloid subreflector. Using eccentricities of 2.0 and 2.5, the results are very similar to those with eccentricities of 1.5.

Compared with the symmetrical front-fed paraboloid antenna and the classical Cassegrainian antenna, offset antennas are very unfavorable when illuminated by a Huygens source. The Huygens source gives zero polarization losses for symmetrical paraboloid reflector antennas, but the losses of offset antennas are of the same order as those calculated for offset antennas illuminated by an electric dipole. This conclusion is supported by the fact that for eccentricities differing from $e = 1.5$ similar results are obtained.

More study is required to find out whether feeds may be designed having polarization properties that may improve the polarization losses of offset antennas. However, the present study makes the use of offset antennas for purposes where a polarization discrimination of more than 30 dB is required, very questionable.

REFERENCES

[1] E. U. Condon, "Theory of radiation from paraboloid reflectors," Westinghouse Rep. no. 15, Sept. 24, 1941.

[2] S. Silver, *Microwave Antenna Theory and Design*. New York: McGraw-Hill, 1949.

[3] C. C. Cutler, "Parabolic antenna design for microwaves," *Proc. IRE*, vol. 35, pp. 1284–1294, Nov. 1947.

[4] E. M. T. Jones, "Paraboloid reflector and hyperboloid lens antennas," *IRE Trans. Antennas Propagat.*, vol. 2, pp. 119–127, July 1954.

[5] I. Koffman, "Feed polarization for parallel currents in reflectors generated by conic sections," *IEEE Trans. Antennas Propagat.*, vol. AP-14, pp. 37–40, Jan. 1966.

[6] M. Afifi, "Scattered radiation from microwave antennas and the design of a paraboloid plane reflector antenna," Ph.D dissertation, Delft Univ. of Technology, the Netherlands, 1967.

[7] P. D. Potter, "The aperture efficiency of large paraboloidal antennas as a function of their feed system radiation characteristics," Jet Propulsion Lab., Pasadena, Calif. Tech. Rep., no. 32-149, Sept. 25, 1961.

[8] P. D. Potter, "Aperture illumination and gain of a Cassegrainian system," *IEEE Trans. Antennas Propagat.*, vol. AP-11, pp. 373–375, May 1963.

[9] P. A. Watson and S. I. Ghobrial, "Off-axis polarization characteristics of Cassegrainian and front-fed paraboloidal antennas," *IEEE Trans. Antennas Propagat.*, vol. AP-20, pp. 691–699, Nov. 1972.

[10] J. S. Cook, E. M. Elam, and H. Zucker, "The open Cassegrain antenna," Bell Syst. Tech. J., vol. 44, pp. 1255–1299, Sept. 1965.

[11] J. D. Hanfling, "Aperture fields of paraboloidal reflectors by stereographic mapping of feed polarization," *IEEE Trans. Antennas Propagat.*, vol. AP-18, no. 3, pp. 392–396, May 1970.

[12] R. Graham, "The polarization characteristics of offset Cassegrain aerials," in European Microwave Conf., London Sept. 8–12, 1969, p. 352.

[13] ——, "The polarization characteristics of offset Cassegrain aerials," presented at the Int. Conf. Radar and Future, IEE, London, Oct. 23–25, 1973.

[14] A. C. Ludwig, "The definition of cross polarization," *IEEE Trans. Antennas Propagat.*, vol. AP-21, pp. 116–119, Jan. 1973.

[15] B. E. Kinber and V. A. Tischenko, "Polarization of radiation of axisymmetric reflector antennas," *Radio Eng. Electron. Phys.*, vol. 17, pp. 528–534, Apr. 1972, (published January 1973).

[16] T. S. Chu and R. H. Turrin, "Depolarization properties of offset reflector antennas," *IEEE Trans. Antennas Propagat.*, vol. AP-21, pp. 339–345, May 1973.

[17] J. Dijk, C. T. W. van Diepenbeek, E. J. Maanders, and L. F. G. Thurlings, "The polarization losses of offset antennas," Eindhoven Univ. Tech., the Netherlands, TH Rep. 73-E-39, June 1973.

[18] D. Carter, "Wide angle radiation in pencil beam antennas," J. Appl. Phys., vol. 26, Nr. 6, pp. 645–652, June 1955.

[19] L. B. Tartakovski, "Side radiation from ideal paraboloid with circular aperture," *Radio Eng. Electron. Phys.*, vol. 4, no. 4, no. 6, pp. 14–28, 1959.

Multiple-Beam Antennas: Offset Reflectors with Offset Feeds

ALAN W. RUDGE, MEMBER, IEEE

Abstract—**Two simple mathematical models are described for the prediction of the vector radiation fields from offset parabolic reflector antennas with offset feeds. Experimental support for the predictions obtained from the models has been obtained by comparisons with measured data from antenna systems operating at 30 GHz. The principal radiation characteristics of the offset-reflector with offset-feed configuration are discussed.**

I. Introduction

IN THE DESIGN of antennas with a frequency reuse capability, an offset portion of a parabolic reflecting surface used in conjunction with either a single or multiple-element primary-feed offers a number of advantages. Compared to its full-paraboloidal counterpart, the offset-reflector avoids aperture-blocking effects, reduces the reflector reaction upon the primary-feed, and offers a reduction in astigmatism for off-axis feed locations [1], [2]. For practicable designs the offset structure leads to the use of larger effective focal-length to diameter ratios, with higher gain primary-feeds and a subsequent reduction in mutual coupling between adjacent feed-elements. Although these advantages are significant, the asymmetry of the offset configuration raises certain questions regarding the vector-radiation characteristics of the overall antenna. In a multiple-beam application, which may include polarization diversity, the asymmetry is compounded by the off-axis location of the primary-feed elements. In such cases the beam cross-over levels and the levels of coma-lobes and cross-polarized sidelobes are of particular interest.

The vector-radiation fields from offset parabolic reflector antennas have been previously considered by Cook *et al.* [3] and more recently by Chu and Turrin [4] and Rudge *et al.* [5], [6]. The electric field distribution in the focal region of offset-reflectors under conditions of a normally incident plane-wave has been studied by Bem [7] and polarization losses for offset reflectors with dipole and Huygen-source primary-feeds have been calculated by Dijk *et al.* [8].

In the work described here two simple mathematical models are described that provide predictions of both the principally polarized (copolar) and cross-polarized (cross-polar) radiation from offset parabolic reflector antennas with either linearly or circularly polarized primary-feed elements. The theoretical approach adopted here differs from that employed in [3] and [4] in that the predictions are not inherently limited to a small angular range about the antenna boresight. In addition, the theoretical models developed can accommodate small offsets in the location of the primary-feed relative to the reflector geometric focus and can thus be employed in the study and design of multiple-beam antennas. The models have been developed with the prime objective of producing sufficiently accurate, yet comparatively simple, expressions that when programmed are suitable for repeated use in a design-optimization mode.

Experimental support for the mathematical models has been obtained by comparing their predictions with measured data obtained from antenna systems operating at a frequency of 30 GHz. The models have been applied to examine the radiation characteristics of the offset-reflector with offset-feed configuration and some of the principal features are reported.

II. General Approach

The geometry of the offset reflector is shown in Fig. 1. The basic parameters of the reflector are shown as the focal length F of the parent paraboloid, the offset angle θ_0, and the half angle θ^* subtended at the focus by any point on the reflector rim. The physical contour of the reflector is elliptical but its projection into the $x'y$ plane produces a true circle.

The far-field radiation arising from a known tangential electric field distribution in an infinite plane can be determined exactly, as shown for example in the text by Collin and Zucker [2]. In dealing with the offset reflector, the infinite surface is chosen as the $x'y$ plane and the electric fields outside of the projected aperture region are assumed negligible. An electromagnetic field distribution is introduced around the boundary of the projected aperture to satisfy the continuity criterion, and thus the predicted radiation fields satisfy the radiation conditions in the forward hemisphere. The neglect of the electric fields outside of the projected-aperture region is acceptable providing that the dimensions of the aperture are large relative to the electrical wavelength (λ). The tangential electric field distribution within the projected-aperture region is determined by use of the physical optics approximation. That is, the electric field $\boldsymbol{E}_r$ reflected from the offset reflector is obtained from [1], [2]

$$\boldsymbol{E}_r = 2(\boldsymbol{a}_n \cdot \boldsymbol{E}_i)\boldsymbol{a}_n - \boldsymbol{E}_i \tag{1}$$

where $\boldsymbol{E}_i$ is the incident electric field at the reflector and $\boldsymbol{a}_n$ is the surface unit normal. The incident field $\boldsymbol{E}_i$ is taken as the radiation field of the primary-feed.

Manuscript received July 24, 1974; revised October 30, 1974. This work was largely carried out under contracts with the European Space Research Organization and the Royal Radar Establishment, Malvern, England.
The author was with the Department of Electronic and Electrical Engineering, University of Birmingham, Birmingham, England. He is now with the ERA-IITRI RF Technology Centre, Leatherhead, Surrey KT22 7SA, England.

Reprinted from *IEEE Trans. Antennas Propagat.*, vol. AP-23, pp. 317–322, May 1975.

The physical-optics aperture-field method was selected in preference to the surface-current method after comparisons between the two techniques indicated that there were no significant differences between the far-field predictions obtained from the methods over a cone of angles subtending a half-angle of at least 30° about the antenna boresight. Since the aperture-field method leads to the more simple mathematical expressions, it offers a significant reduction in the required computational effort.

The distant radiation field from a linearly polarized antenna can be completely specified in terms of two spatially orthogonal vector components. The definition of these vectors in terms of a copolarized and cross-polarized component is, to some extent, an arbitrary one and at least three different definitions are commonly used in the literature. The definition employed here has the particular advantage that the predicted field components at any point in space, correspond directly to the components measured using standard antenna-range techniques [1]. This definition has been discussed in a useful paper by Ludwig [9] and has found favor with a number of authors in recent publications [10], [11]. Employing this definition, with the antenna having its principal electric vector along the y axis, the copolar (E_p) and cross-polar (E_q) "measured-field components" can be related to the field components E_ψ, E_Φ in a classical spherical coordinate system r,ψ,Φ, by the matrix expression

$$\begin{bmatrix} E_p \\ E_q \end{bmatrix} = \begin{bmatrix} \sin\Phi & \cos\Phi \\ \cos\Phi & -\sin\Phi \end{bmatrix} \begin{bmatrix} E_\psi \\ E_\Phi \end{bmatrix}. \qquad (2)$$

III. The Mathematical Model for Linearly-Polarized Antennas

The model employs a conventional spherical coordinate system (r,ψ,Φ) with origin at the center of the reflector projected-aperture, and makes use of the spatial Fourier transform formulation for the radiation from an infinite surface [2]. Using (2), the normalized radiation patterns of the offset antenna may be expressed directly in terms of a linear copolar field component (E_{pn}) and a cross-polar component (E_{qn}) as

$$\begin{bmatrix} E_{pn} \\ E_{qn} \end{bmatrix} = \frac{1+\cos\psi}{2F_p(0,0)} \begin{bmatrix} 1 - t^2\cos 2\Phi & t^2 \sin 2\Phi \\ t^2 \sin 2\Phi & 1 + t^2 \cos 2\Phi \end{bmatrix} \cdot \begin{bmatrix} F_p(\psi,\Phi) \\ F_q(\psi,\Phi) \end{bmatrix} \qquad (3)$$

where the functions F_p and F_q are the spatial Fourier transforms of the copolar and cross-polar components of the tangential electric field in the projected aperture plane and $t = \tan\psi/2$.

The Fourier transforms are the transverse Cartesian components of the vector

$$F(\psi,\Phi) = \int_{x'}\int_{y} \varepsilon(x',y) \exp\left[jkR'(x',y,\psi,\Phi)\right] dx'\, dy \qquad (4)$$

where ε is the tangential electric field distribution in the projected-aperture plane and R' is the distance from a general point in the projected-aperture plane to a far-field point. The transform functions F_p, F_q can be conveniently expressed in terms of an offset primary spherical coordinate system (ρ,θ,ϕ) with origin at the reflector focus, i.e., with θ measured to the inclined z axis, and ϕ to the x axis, of Fig. 1. Ignoring multiplying constants the transform integrals can be written

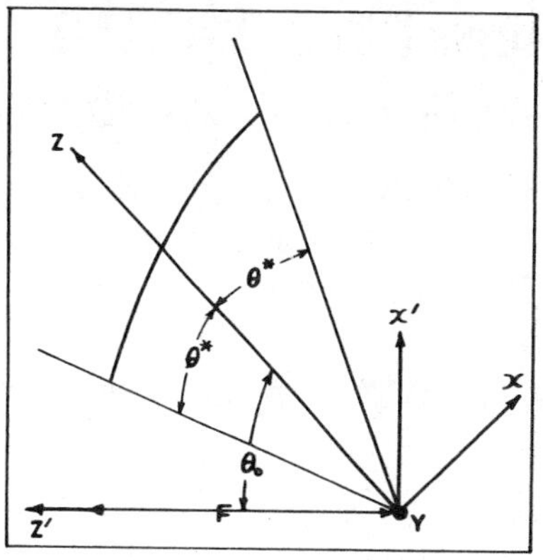

Fig. 1. Offset reflector geometry.

$$F_i(\psi,\Phi) = \int_0^{2\pi}\int_0^{\theta^*} \varepsilon_i \exp\left[-jkR\sin\psi\right] p^2 \sin\theta\, d\theta\, d\phi \qquad (5)$$

where i is either p or q, $k = 2\pi/\lambda$

$$R = p[(\sin\theta\cos\theta_0\cos\phi + \sin\theta_0\cos\theta)\cos\Phi - \sin\theta\sin\phi\sin\Phi] \qquad (6)$$

and the distance between a point on the reflector surface and the reflector geometric focus (p) is given by

$$p = \frac{2F}{(1+\cos\theta\cos\theta_0 - \sin\theta\sin\theta_0\cos\phi)}. \qquad (7)$$

The geometry involved in arriving at (5)–(7) from (4) is straightforward but somewhat lengthy. Relevant details of the geometry of offset parabolic surfaces can be found in [3]–[5] and will not be repeated here.

When the phase center of the primary-feed is located in the general vicinity of the reflector geometric focus, the incident fields at the reflector can be expressed in the form,

$$\boldsymbol{E}_i = \frac{[A_\theta(\theta,\phi)\boldsymbol{a}_\theta + A_\phi(\theta,\phi)\boldsymbol{a}_\phi]\exp[jk(R_1 - p)]}{p} \qquad (8)$$

where A_θ and A_ϕ are normalized functions describing the radiation pattern characteristics of the primary-feed radiation, $\boldsymbol{a}_\theta$ and $\boldsymbol{a}_\phi$ are the associated unit vectors, and the function R_1 is a phase-compensation term that accounts for small offsets in the primary-feed location, from the geometric focus. In the approach adopted here, the effect on the overall antenna radiation pattern of the amplitude variations in the field at the reflector, due to a small feed offset, have been assumed to be negligible. Thus if Δ_t is a small transverse offset, Δ_z a small axial offset, and the angle ϕ_0 (measured to the x axis in the xy plane) denotes the plane of the offset, then providing Δ_t and Δ_z are much less than F, it can be shown that [5], [12]

$$R_1 \approx \Delta_t \sin\theta \cos(\phi - \phi_0) + \Delta_z \cos\theta. \quad (9)$$

The unit normal to the reflector surface ($\boldsymbol{a}_n$) can be expressed in terms of the offset primary coordinate system (with unit vectors $\boldsymbol{a}_x, \boldsymbol{a}_y, \boldsymbol{a}_z$) as

$$\boldsymbol{a}_n = -\left\{\left(\frac{p}{4F}\right)^{1/2} [(\sin\theta\cos\phi - \sin\theta_0)\boldsymbol{a}_x + \sin\theta\sin\phi\boldsymbol{a}_y + (\cos\theta + \cos\theta_0)\boldsymbol{a}_z]\right\} \quad (10)$$

Hence making use of (1), (8), and (10), and assuming that the primary-feed has its principal electric vector in the $\boldsymbol{a}_y$ direction, then the Cartesian components of the tangential electric field in the projected-aperture plane can be resolved. The $\boldsymbol{a}_y$ directed, or copolar component of the aperture field (ε_p) and the $\boldsymbol{a}_x$ directed, or cross-polar component (ε_q) can be related directly to the radiation characteristics of the primary-feed by the matrix expression

$$\begin{bmatrix} \varepsilon_p \\ \varepsilon_q \end{bmatrix} = K \begin{bmatrix} s_1 & -c_1 \\ c_1 & s_1 \end{bmatrix} \begin{bmatrix} A_\theta \\ A_\phi \end{bmatrix} \quad (11a)$$

where

$$K = \frac{-\exp[jk(R_1 - 2F)]}{2F} \quad (11b)$$

$$s_1 = (\cos\theta_0 + \cos\theta)\sin\phi \quad (11c)$$

$$c_1 = \sin\theta\sin\theta_0 - \cos\phi(1 + \cos\theta\cos\theta_0) \quad (11d)$$

IV. Computation of the Model

The mathematical model for the offset reflector with a linearly polarized feed is effectively given by (3), (5), (9), and (11). To predict the overall antenna radiation it is necessary to specify the primary-feed directivity characteristics (A_θ, A_ϕ) and to compute the two-dimensional integrals given by (5). Expressions for a wide variety of primary-feed types are available in the literature [1], [2], [4], [5], although the cross polarization characteristics of the feed models are not always adequate and must be examined carefully [13]. Having calculated F_p and F_q, the computation of the radiation fields by means of (3) is trivial. It is apparent that the evaluation of the two-dimensional integrals represents the crux of the computational problem.

The computational problem can be alleviated by imposing a minor constraint upon the primary-feed functions A_θ, A_ϕ. Providing A_θ is an odd function and A_ϕ an even function, with respect to the variable phi (ϕ), which is certainly true for many practical primary-feed types, then (5) can be readily reduced to a half-range form with respect to phi with a subsequent saving in computation time. The two-dimensional integral can be accomplished numerically employing a variety of techniques including fast Fourier transforms. A Romberg integration technique was adopted for the models described here [5], [14]. This method involves the use of trapezoidal-rule integration with successive interval-halving and an extrapolation technique to remove increasing orders of error.

V. Circularly Polarized Antennas

The right- and left-hand components of the far-field radiation from a circularly polarized antenna can be simply defined in terms of the linearly polarized components as

$$\begin{bmatrix} E_R \\ E_L \end{bmatrix} = \frac{1}{2} \begin{bmatrix} 1 & j \\ 1 & -j \end{bmatrix} \begin{bmatrix} E_q \\ E_p \end{bmatrix}. \quad (12)$$

However, it will be both informative and useful with regard to minimizing computational effort, to express the normalized radiation patterns in the form

$$\begin{bmatrix} E_{Rn} \\ E_{Ln} \end{bmatrix} = \frac{1 + \cos\psi}{2F_0} \begin{bmatrix} 1 & t^2 \exp[j2\Phi] \\ t^2 \exp[-j2\Phi] & 1 \end{bmatrix} \cdot \begin{bmatrix} F_R \\ F_L \end{bmatrix} \quad (13)$$

where the subscript n denotes the normalization, F_R, F_L are the Fourier transformations of the right- and left-hand components of the reflector projected-aperture tangential electric field ($\varepsilon_R, \varepsilon_L$), and F_0 is the normalizing constant. In this form the circularly polarized aperture-field components can be related directly to the radiation characteristics of the primary-feed (A_θ, A_ϕ) by

$$\begin{bmatrix} \varepsilon_R \\ \varepsilon_L \end{bmatrix} = \frac{2FK}{p}(1 + \cos\theta) \begin{bmatrix} \exp[j\Omega] & -j\exp[j\Omega] \\ \exp[-j\Omega] & j\exp[-j\Omega] \end{bmatrix} \cdot \begin{bmatrix} A_\theta \\ A_\phi \end{bmatrix} \quad (14)$$

where

$$\Omega(\theta,\phi) = \arctan(s_1/c_1). \quad (15)$$

The numerical integration time of the Fourier transform functions can again be reduced by imposing a constraint upon the primary-feed radiation. For example, (5) can again be reduced to a half-range integration with respect to phi if the normalized circularly polarized primary-feed radiation ($\boldsymbol{E}_n$) takes the form

$$\boldsymbol{E}_n = (A_1(\theta)\boldsymbol{a}_\theta - jA_2(\theta)\boldsymbol{a}_\phi)\exp[-j\phi] \quad (16)$$

where the functions A_1, A_2 are independent of the phi variable. This expression is satisfied by the radiation fields from many practical types of circularly polarized feed including fundamental-mode conical-horns, dual-mode horns [15], and corrugated horns [16]. In these cases the reflector aperture-plane fields may be obtained from (14) as

$$\varepsilon_M = \frac{2FK}{p}(1 + \cos\theta)(A_1(\theta) \pm A_2(\theta))\exp -j[\phi \pm \Omega] \quad (17)$$

where M is either R or L, and L takes the upper sign. It is apparent that when $A_1(\theta) = A_2(\theta)$, which, for a circularly symmetric feed, is the condition for zero cross polarized primary-feed radiation [9], then the reflector aperture-plane field will be purely copolarized with a "beam-squinting" phase distribution given by $\Omega + \phi$.

VI. Experimental Verification

The measurements reported here were made over a period of many months at the University of Birmingham. The antenna range was investigated using field-probing

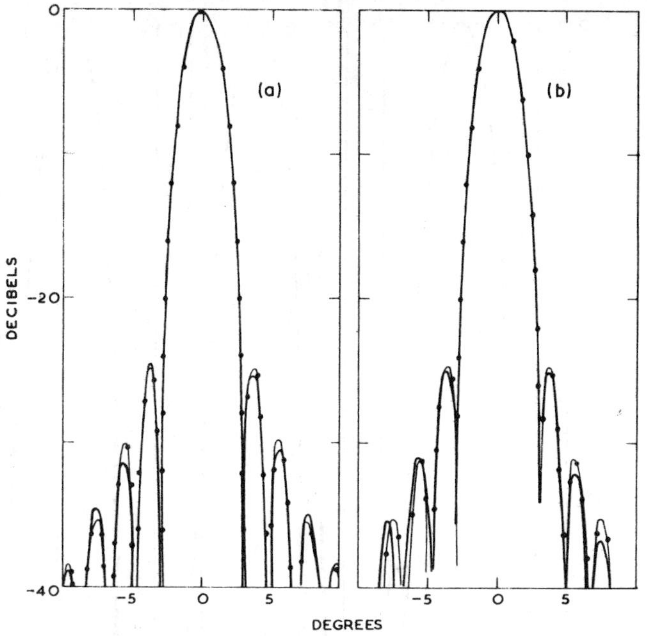

Fig. 2. Copolar radiation fields from offset reflector antenna ($F = 22.7\lambda$, $\theta_0 = 44°$, $\theta^* = 30°$) fed by linearly polarized rectangular horn with aperture dimensions $1.57\lambda \times 2.14\lambda$, —— measured; —•— predicted. (a) Plane of asymmetry, $\Phi = 90°$. (b) Plane of symmetry, $\Phi = 0°$.

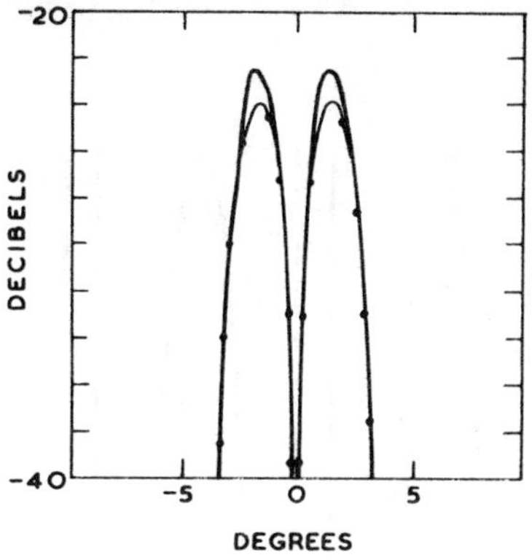

Fig. 3. Cross-polar radiation field in plane of asymmetry ($\Phi = 90°$) for offset antenna of Fig. 2; —— measured; —•— predicted.

techniques and the perturbation level for linearly polarized waves determined as −40 dB. The isolation between co-polarized and cross-polarized signals incident at the test antenna was measured as better than 37 dB employing a pyramidal-horn probe.

In Fig. 2 the copolar radiation patterns are shown for an offset reflector fed by a linearly polarized TE_{10} mode rectangular horn with aperture dimensions of $1.57\lambda \times 2.14\lambda$, which produces a −12-dB illumination taper at ±30°. The reflector surface was precision machined to a tolerance of better than 0.1 mm and has parameters $F = 22.7\lambda$, $\theta_0 = 44°$, $\theta^* = 30°$. The predicted radiation patterns in the principal-planes are shown superimposed upon experimental data obtained with a 30-GHz antenna system. The correlation can be seen to be excellent although the cross-polarized results shown in Fig. 3 are slightly inferior. The cross-polarized data is shown for the plane of asymmetry (yz' plane), which contains the peak levels of the reflector generated cross-polarized radiation. In the plane of symmetry ($x'z'$ plane) the reflector generated cross polarized radiation is zero.

In Fig. 4 the 30-GHz copolar and cross-polar radiation patterns are shown for a second offset reflector with $F = 30.4\lambda$, $\theta_0 = 35°$, and $\theta^* = 30°$. The reflector was again fed with the linearly polarized rectangular horn and the cases shown are for transverse primary-feed offsets of 0, 0.83λ, and 2.5λ, respectively. The feed offsets are along the y axis and correspond to an E-plane array of 4 beams with cross-over levels of −5 to −6 dB, a maximum coma-lobe level (on the outer beams) of approximately −21 dB and a peak cross-polar level of −26 dB. Reducing the primary-feed spacing (and hence the feed aperture dimensions) to provide a −3 dB beam cross-over level, results in an outer-beam coma-lobe level of approximately −18 dB with a −25 dB peak cross-polar level in the plane of asymmetry.

Fig. 5 shows the copolar radiation field of the first reflector fed by a circularly polarized corrugated scalar-horn feed with a diameter of 2.8λ. Employing circular polarization the offset reflector does not generate a cross polarized component and, since the scalar-feed produces a very low level of cross-polarized radiation, only the copolar result is of concern in this case. The scalar horn produces a −17 dB illumination taper at ±30° and thus the sidelobe levels are significantly lower than previously. In Fig. 6 the radiation pattern is shown in the plane of asymmetry with the horn moved transversely along the y axis by distances of 0, 1.4λ, and 2.8λ, respectively. For this case the coma-lobe levels have been reduced to better than −27 dB but the larger primary-feed horn implies a minimum beam cross-over level of approximately −19 dB.

VII. Features of the Offset Configuration

Having gained some confidence in the quality of the predictions obtained, the theoretical models were employed to examine the principal features of the offset-reflector configuration. A few of the many points of interest that emerged during the course of this work will be briefly mentioned.

For focused primary-feed locations the cross-polarized radiation generated by the reflector with linearly polarized illumination is primarily a function of the parameters θ_0 and θ^*. The peak level of the cross polarized radiation increases with increasing values of θ_0 and θ^* and is comparatively insensitive to the amplitude taper of the primary-feed illumination. The peak values of the reflector-generated cross polarization occur in the plane of asymmetry (yz' plane) at the angles corresponding approximately to the −6-dB levels of the main copolar beam. In the plane of symmetry ($x'z'$ plane) the reflector-generated cross polarization is zero. Using reflector parameters of the order of $\theta_0 = 35°$ and $\theta^* = 30°$ the peak values of the copolar and cross-polar sidelobes can be reduced below −25 dB using simple primary-feed types, which provide an illumination taper of at least −13 dB. To reduce the peak cross polarized levels to below −30 dB, the offset angle must be reduced to 30° or less. When a circularly polarized primary-feed illumination is employed, offset reflectors of this type

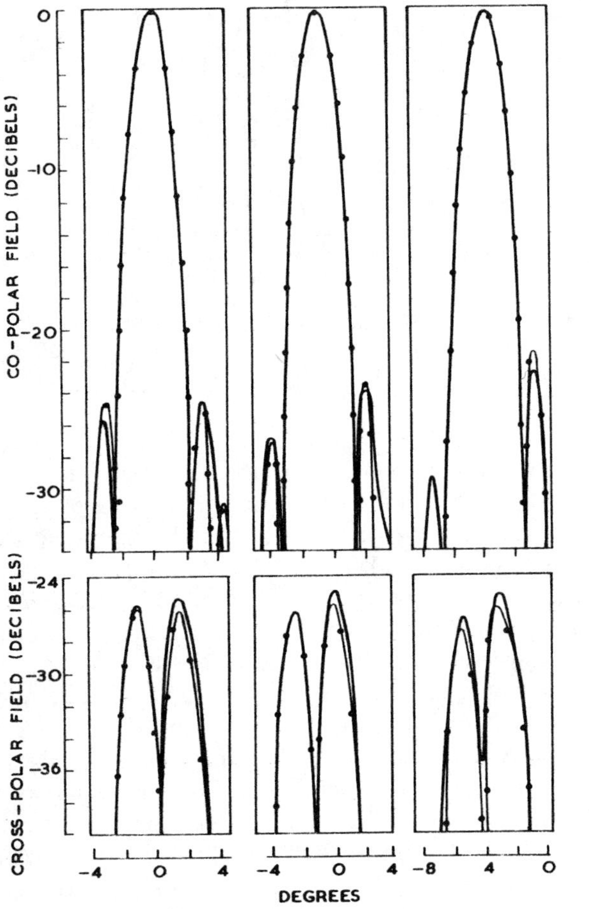

Fig. 4. Radiation fields from offset reflector ($F = 30.4\lambda$, $\theta_0 = 35°$, $\theta^* = 30°$) with primary-feed offset transversely by 0, 0.83λ, and 2.5λ, respectively; —— measured; —•— predicted.

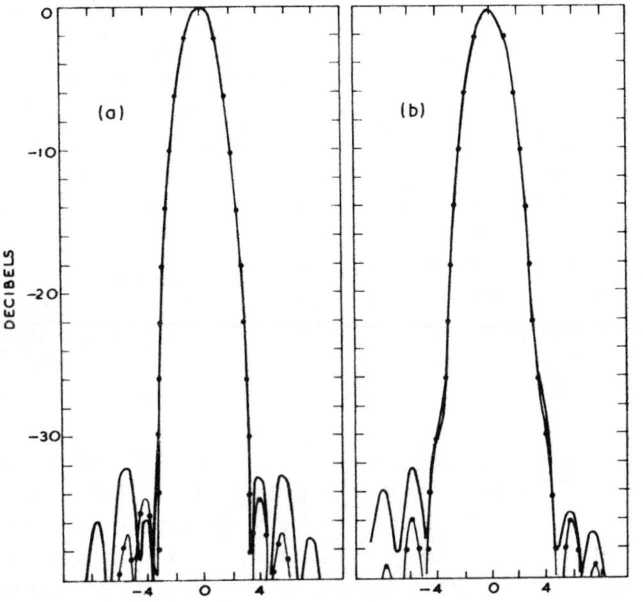

Fig. 5. Copolar radiation field from an offset reflector antenna ($F = 22.7\lambda$, $\theta_0 = 44°$, $\theta^* = 30°$) fed by circularly polarized scalar horn with aperture-diameter 2.8λ; (a) Plane of asymmetry, $\Phi = 90°$. (b) Plane of symmetry, $\Phi = 0°$.

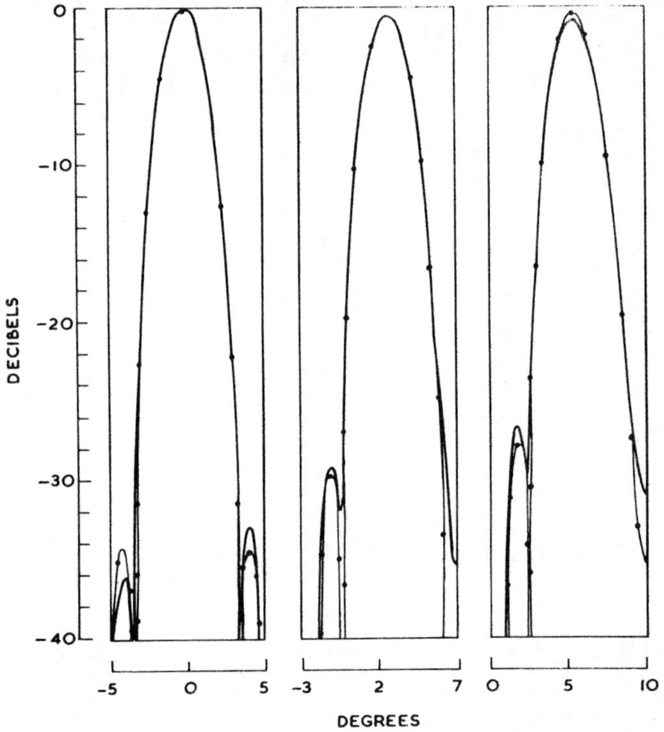

Fig. 6. Copolar radiation field from circularly polarized offset reflector antenna ($F = 22.7\lambda$, $\theta_0 = 44°$, $\theta^* = 30°$) with scalar horn feed offset transversely by 0, 1.4λ, and 2.8λ, respectively; —— measured; —•— predicted.

do not generate a cross-polarized component and the only significant cross-polarized radiation that occurs is due directly to the primary-feed. However, the beam formed by the antenna exhibits a small squint, the angle and direction of which is dependent upon the reflector offset-angle and the hand of polarization. When a duplex dual-polarized configuration is utilized, some loss of gain will occur in the boresight direction due to the squint-angle difference of the two beams. The results obtained for this focused configuration are in general agreement with those obtained by Chu and Turrin [4], which are valid for small angles about the antenna boresight.

For a linearly polarized multiple-beam application, a small transverse offset of the primary-feed results in the expected spatial shift of both the copolar and cross-polar distributions. However, it is significant that the peak levels of the cross-polarized lobes remain essentially unchanged and that these lobes are comparatively insensitive to phase-errors of the type introduced by small offsets in the primary-feed location. For offset feed locations, the formation of the copolar coma-lobe remains the most evident source of pattern deterioration. The results described earlier provide an indication of the tradeoff between beam cross-over levels and copolar and cross-polar sidelobes, which is available using an *E*-plane array of rectangular feed elements.

With circularly polarized multiple-beam antennas, the lack of a reflector-generated cross-polarized component is a useful attribute, although the beam-squint problem may be significant if polarization diversity is required. For the circularly polarized reflector described here, the squint of each beam from boresight is of the order of 0.07 of the antenna -3 dB beamwidth and the resultant loss of gain in the boresight direction is approximately 0.03 dB per beam. These figures could be improved to 0.05 and 0.02 dB, respectively, by reducing the offset angle from 44° to 35°. Although the reflector does not contribute to the cross polarized radiation the contribution from the circularly polarized primary-feed may not be negligible, particularly for multiple-beam applications. Fundamental mode conical horns, for example, which are a possible choice of feed-array element, radiate a very significant cross-polar component.

VIII. Conclusions

Two theoretical models have been described for the prediction of the far-field radiation from offset reflector antennas. The models can accommodate small offsets in the primary-feed location with respect to the reflector geometric focus and can thus be usefully employed in the study or design of multiple-beam antennas. The models are based upon the physical-optics approximation and employ a known aperture-field approach, the validity of which is not inherently restricted to angles close to the antenna boresight. Experimental verification of the models, and comparisons with the surface-current technique, have indicated that reliable predictions of the significant co-polarized and cross-polarized radiation can be obtained by use of these methods over a moderate range of angles about the antenna boresight.

The models have been employed to examine the radiation characteristics of the offset-reflector with offset-feed configuration and some of the principal features have been briefly discussed. A particularly interesting result is that the maximum levels of the reflector-generated cross polarized radiation are comparatively insensitive to small offsets in the primary-feed location from the geometric focus.

Acknowledgment

The author wishes to thank Dr. M. Shirazi for the computation of the mathematical models.

References

[1] S. Silver, *Microwave Antenna Theory and Design.* New York: McGraw-Hill, 1949.
[2] R. E. Collin and F. J. Zucker, *Antenna Theory,* Parts I and II. New York: McGraw-Hill, 1969.
[3] J. S. Cook, E. M. Elam, and H. Zucker, "The open Cassegrain antenna: Part I—Electromagnetic design and analysis," *Bell Syst. Tech. J.,* vol. 44, pp. 1255–1300, Sept. 1965.
[4] T. S. Chu and R. H. Turrin, "Depolarization properties of offset reflector antennas," *IEEE Trans. Antennas Propagat.,* vol. AP-21, pp. 339–345, May 1973.
[5] A. W. Rudge and M. Shirazi, "Multiple-beam antennas: Offset reflectors with offset feeds," University of Birmingham, U.K., Final Rep. ESRO/ESTEC Contract 1725/72PP, July 1973.
[6] A. W. Rudge, "Offset reflectors with offset feeds," *IEE Electron. Lett.,* vol. 9, pp. 611–613, Dec. 1973.
[7] D. J. Bem, "Electric-field distribution in the focal region of an offset paraboloid," *Proc. Inst. Elec. Eng.,* vol. 116, no. 5, pp. 579–684, 1969.
[8] J. Dijk *et al.,* "The polarization losses of offset paraboloid antennas," *IEEE Trans. Antennas Propagat.,* vol. AP-22, pp. 513–520, July 1974.
[9] A. C. Ludwig, "The definition of cross polarization," *IEEE Trans. Antennas Propagat.* (Commun.), vol. AP-21, pp. 116–119, Jan. 1973.
[10] P. J. Wood, "Depolarization with Cassegrain and front-fed reflectors," *IEE Electron. Lett.,* vol. 9, pp. 181–183, May 1973.
[11] P. A. Watson and S. I. Ghobrial, "Crosspolarization in Cassegrain and front-fed antennas," *IEE Electron. Lett.,* vol. 9, pp. 297–298, June 1973.
[12] J. Ruze, "Lateral-feed displacement in a paraboloid," *IEEE Trans. Antennas Propagat.,* AP-13, pp. 660–665, Sept. 1965.
[13] A. W. Rudge, T. Pratt, and A. Fer, "Cross-polarised radiation from satellite reflector antennas," in *Proc. AGARD Conf. on Antennas for Avionics,* Munich, Nov. 1973, pp. 16.1–16.8.
[14] W. V. T. Rusch and P. D. Potter, *Analysis of Reflector Antennas.* New York: Academic, 1970.
[15] P. D. Potter, "A new horn antenna with suppressed side lobes and equal beamwidths," *Microwave J.,* vol. 6, pp. 71–78, June 1963.
[16] P. J. B. Clarricoats and P. K. Saha, "Propagation and radiation behavior of corrugated feeds," *Proc. Inst. Elec. Eng.,* vol. 118, pp. 1177–1186, Sept. 1971.

NEW CLASS OF PRIMARY-FEED ANTENNAS FOR USE WITH OFFSET PARABOLIC-REFLECTOR ANTENNAS

Indexing terms: Antenna feeders, Reflector antennas

In many practical applications the performance of an offset parabolic-reflector antenna is limited by the offset reflectors' depolarising and beam-squinting properties. The letter describes a new class of primary-feed antennas which overcome these limitations. The new primary-feed types offer a significant improvement in the crosspolar and beam-squinting properties of the offset reflector antenna, without adding significantly to the complexity or mass of the primary-feed system.

In many antenna systems where high gain and very low sidelobe radiation is required, the offset parabolic reflector offers a number of significant advantages over its axisymmetric counterparts. In particular, since the primary feed and its supporting structure need not protrude into the optical path of the incident or reflected wavefronts, spurious scattering, and the associated gain loss due to blockage effects, can be avoided. The avoidance of blockage is a major factor in achieving high overall efficiencies and low sidelobe performance.[1,2]

For many practical applications the offset reflector configuration has a serious disadvantage, however, in that the reflector has undesirable depolarising properties when illuminated efficiently by a perfectly linearly polarised primary-feed radiation. The overall radiation field from such an antenna typically exhibits a pair of crosspolarised lobes, with peaks close to the -6 dB contour of the antenna main beam, in one of the principal planes of the reflector. In the other principal plane the crosspolarised field is zero.[3,4] The plane containing the peak crosspolarised lobes will be referred to here as the plane of asymmetry. When similarly illuminated by a purely circularly polarised primary feed, the offset reflector does not introduce a crosspolarised field, but the main beam of the antenna is squinted from the antenna boresight.[3-5] These depolarising and squinting phenomena are often undesirable in high-performance antennas, especially when frequency reuse is required. In many cases these effects constitute a major limitation on the application of the offset reflector, regardless of its other desirable properties.

To overcome these limitations in offset reflector antennas, without adding excessively to the complexity or mass of the

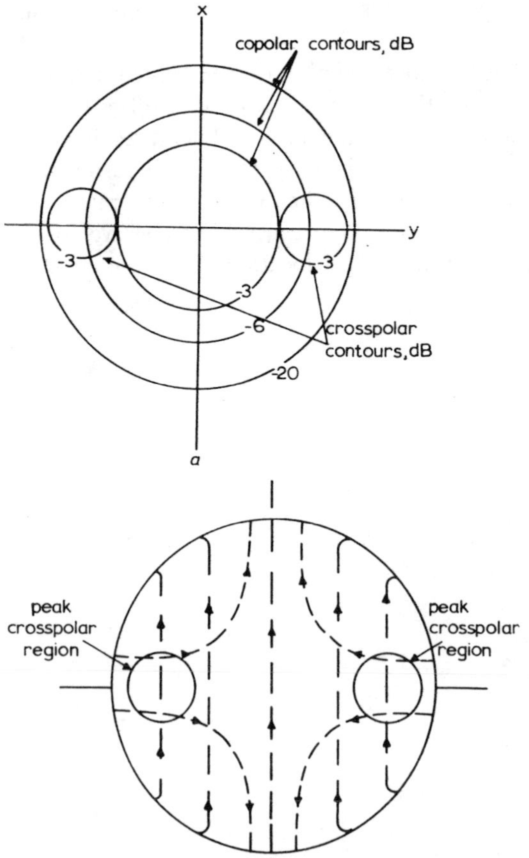

Fig. 1
a Contour plot of typical focal-plane field characteristics for an offset parabolic reflector. Crosspolarised field peak level typically -20 to -30 dB below peak copolarised field
b Aperture-field distribution of overmoded waveguide horn

Reprinted with permission from *Electron. Lett.*, vol. 11, pp. 597-599, Nov. 27, 1975.

feed radiator, a set of novel primary-feed devices have been devised and are under development at the RF Technology Centre.[6] This letter briefly describes the principle behind these feeds and shows predicted and preliminary experimental data for a prototype of one of the new feed types.

The general concept behind the improved feeds can be understood in a simple fashion by a consideration of the nature of the focal-plane electric-field distribution of an offset reflector operating in its reception mode. Fig. 1*a* shows an approximate contour map of the electric-field distribution with a distant source located on the antenna boresight. To provide a conjugate match to this incident field distribution, the primary-feed aperture-plane fields must exhibit similar polarisation properties. In Fig. 1*b* this general aperture-field characteristic is approximated by cylindrical waveguide modes in which a particular higher-order mode provides the necessary crosspolarisation characteristics. This waveguide mode is the TE_{21} in a cylindrical smooth-walled guide or the HE_{21} hybrid mode in a cylindrical corrugated guide. A similar effect can be achieved in a rectangular waveguide structure, and, in this case, the crosspolarisation characteristics are matched by the addition of the TE_{11} mode to the fundamental TE_{10} mode. For the corrugated cylindrical structures and the rectangular structures, the use of the fundamental mode plus the one higher-order mode is sufficient. For the smooth-walled cylindrical structure, a third mode, the TM_{11}, can be employed to improve the axisymmetry of the feed copolar radiation pattern, and to remove crosspolarised fields which otherwise radiate into the diagonal planes of the feed far-field pattern. This technique is well known in axisymmetric reflector design[7]. Although all three feed configurations are of interest, in view of space constraints, only the trimode cylindrical configuration will be discussed further here. The general structure for this feed is illustrated in Fig. 2.

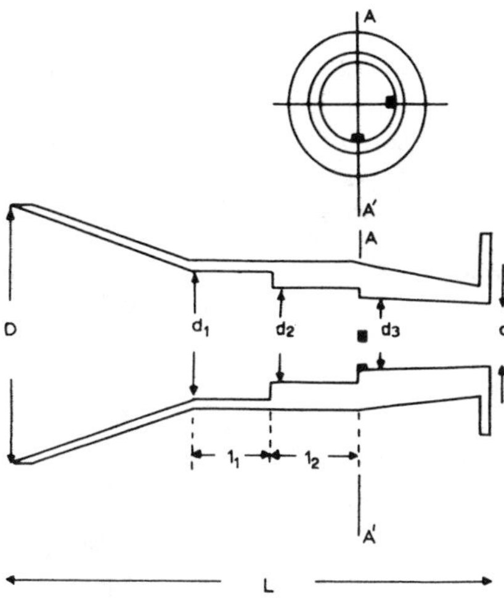

Fig. 2 *Experimental trimode offset reflector conical-horn feed*
Asymmetric discontinuity introduced by two small posts at right angles

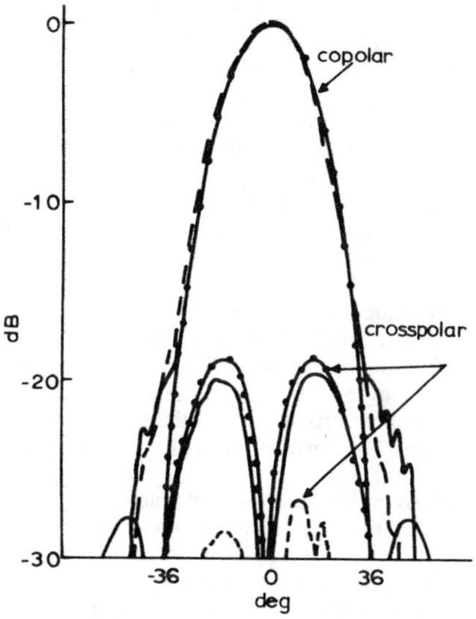

Fig. 3 *Predicted and measured copolar and crosspolar characteristics for 30 GHz trimode feed*
Predicted (*H*-plane): —●—
Measured (*H*-plane): ——
(*E*-plane): - - - -

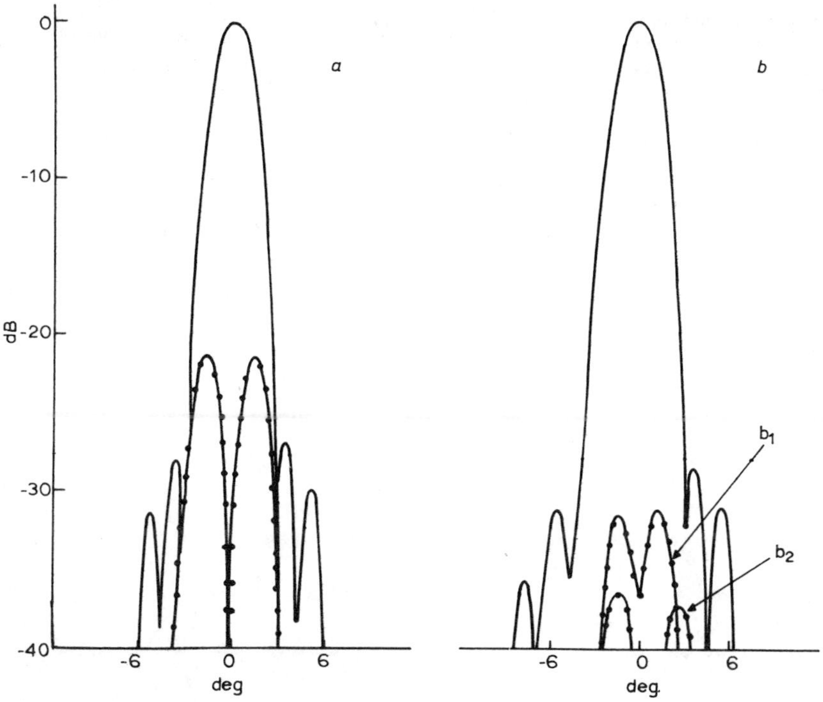

Fig. 4 *Measured radiation patterns in plane of asymmetry for precision offset reflector*
$F = 22 \cdot 7\lambda$, $\theta_0 = 44°$, $\theta^* = 30°$ fed by (*a*) a conventional axisymmetric Potter horn with diameter $2 \cdot 8\lambda$, (*b*) the trimode feed with diameter $2 \cdot 8\lambda$. Curves b_1 and b_2 refer to the measured crosspolar levels before and after temporary experimental modifications to the amplitude of the TE_{21} mode

Briefly, the trimode primary feed is essentially a small-flare-angle conical horn with two discontinuities or steps. The first step (d_3/d_2) is asymmetric and generates the TE_{21} mode. The diameter d_2 is chosen to cut off all higher modes above the TE_{21}, which is the first propagating mode above the fundametal. The second step (d_2/d_1) is axisymmetric and the guide dimensions cut off all modes above the TM_{11} mode. The symmetry of this discontinuity avoids the further generation of the TE_{21} mode. The amplitudes of the modes generated are governed by the ratios d_3/d_2 and d_2/d_1, and the relative phases of the modes are adjusted by the constant-diameter phasing section which follows each discontinuity. The mode amplitudes required are a function of the parameters of the offset reflector; in particular, the offset angle θ_0 and the semi-angle θ^* subtended by the reflector rim at the geometric focus. Typically, the mode amplitudes are of the order of -20 dB below the fundamental mode. The diameter of the primary-feed aperture is selected in the usual way to satisfy the illumination requirements of the reflector. The overall length of the feed is between 0·25 and 1·0 wavelengths greater than a conventional axisymmetric dual-mode feed of the Potter-horn variety.[7] The new dual-mode corrugated and rectangular feed types referred to above can be even more compact.

The predicted radiation pattern of a trimode feed with $D = 2{\cdot}8\lambda$ is shown in Fig. 3. The TE_{21} mode amplitude for this case is -20 dB below the fundamental TE_{11} mode. The TM_{11} mode amplitude is as for a standard Potter dual-mode horn. This feed would be well suited to feed an offset reflector with $\theta_0 = 35°$ and $\theta^* = 30°$. The required pair of cross-polarised lobes are generated in one principal plane, and the crosspolarised field is zero in the other principal plane. In Fig. 3, measured data are superimposed for a prototype trimode feed operating at 30 GHz. Some spurious cross-polarised radiation is evident around the -30 dB level in the E-plane of the horn, but the correlation is, in general, very satisfactory.

In Fig. 4*a* the 30 GHz overall radiation characteristics are shown, in the plane of asymmetry, for a precision offset reflector fed by a conventional low-crosspolar feed with diameter $2{\cdot}8\lambda$. The offset reflector has parameters $F = 22{\cdot}7\lambda$, $\theta_0 = 44°$, $\theta^* = 30°$, and the large crosspolar peaks are predictable for a reflector with this relatively large offset angle. In Fig. 4*b* the radiation characteristics in this plane are shown when the feed is replaced by the trimode feed. The reduction in the crosspolarised radiation (to the curve b_1) is very significant, if not entirely complete. The imperfect cancellation is predictable, and arises from the fact that the prototype feed is not optimised for the parameters of this particular offset reflector. By artificially modifying the amplitude of the TE_{21} mode, the measured performance is improved to that shown in Fig. 4*b*, curve b_2. This result demonstrates the level of performance which can be expected when the mode amplitude is correctly chosen to match the reflector θ_0, θ^* parameters.

The smooth-walled and corrugated structures referred to here can be designed for use in circularly polarised offset reflector systems. In such cases, the feed can be employed either to remove or to enhance the beam-squinting effects normally incurred in these configurations. Further development of all three feed configurations is proceeding.[6]

The principal advantages of the new class of primary-feed antennas described here can be briefly summarised as follows:

(*a*) The feeds are essentially optimised for operation with offset parabolic-reflector antennas. Used in conjunction with such reflectors, highly efficient antennas with low copolarised and crosspolarised sidelobe radiation can be constructed.

(*b*) The feeds provide a very significant improvement in the crosspolar performance of offset reflector antennas by cancelling the depolarising properties of the offset reflector.

(*c*) The corrugated dual-mode and conical trimode feeds will also remove beam-squinting effects when used for circularly polarised applications.

(*d*) The feeds are relatively simple to construct and do not involve any significant increase in complexity or mass over existing feeds designed for axisymmetric systems.

Acknowledgment: The authors gratefully acknowledge the assistance of N. Williams during the preliminary measurements described here, and the contribution of P. R. Foster during various discussions.

A. W. RUDGE
N. A. ADATIA

31st October 1975

RF Technology Centre
ERA-IITRI
Cleeve Road
Leatherhead
Surrey KT22 7SA, England

References

1 DRAGONE, C., and HOGG, D. C.: 'The radiation pattern and impedance of offset and symmetrical near-field Cassegrainian and Gregorian antennas', *IEEE Trans.*, 1974, **AP-22**, pp 472–475
2 RUDGE, A. W., FOSTER, P. R., *et al.*: 'Study of the performance and limitations of multiple-beam antennas'. ESTEC Contract 2277/74 HP, ERA-IITRI RF Technology Centre, 1975
3 CHU, T. S. and TURRIN, R. H.: 'Depolarisation properties of offset reflector antennas', *IEEE Trans.*, 1973, **AP-21**, pp. 339–345
4 RUDGE, A. W.: 'Multiple-beam antennas: offset reflectors with offset feeds', *ibid.*, 1975 **AP-23**, pp. 317–322
5 ADATIA, N. A., and RUDGE, A. W.: 'Beam-squint in circularly polarised offset-reflector antennas', *Electron. Lett.*, 1975, **11**, pp. 513–515
6 RUDGE, A. W., and ADATIA, N.: 'Improved primary-feed antennas for offset parabolic-reflector antennas'. British Patent Application 44505/75, Oct. 1975
7 POTTER, P. D.: 'A new horn antenna with suppressed sidelobes and equal beamwidths', *Microwave J.*, 1963, **6**, pp. 71–78

On the Cross Polarization of Asymmetric Reflector Antennas for Satellite Applications

JOHANNES JACOBSEN, SENIOR MEMBER, IEEE

Abstract—It is well known that focussed, axial symmetrical reflector antennas collimate the co- and cross-polar components of the primary field separately, i.e., the reflector does not create a contribution to the cross polarization of the far-field. By a simple extension of a classical physical argument it is demonstrated that this separability does not depend on the symmetry of the antenna, and that it, therefore, holds even for off-set fed reflectors. A new mathematical formulation of the collimation is derived in which this is shown. Yet the separability does depend on how the co- and cross-polar fields are defined, and the cross polarization of feeds for asymmetric reflectors is discussed in detail in the light of this. It is further suggested how to design low cross polarization feeds for off-set fed antennas. As a consequence of the separate collimation such feeds will lead to low cross-polarization of the secondary fields. Two simple examples are treated. The only limitations of the results are those due to the application of the aperture field version of the physical optics approximation.

Manuscript received December 28, 1975; revised April 11, 1976.

The author is with The European Space Technology Centre, Antennas and Propagation Section, Noordwijk, Holland.

I. INTRODUCTION

In order to save frequency bandwidth it is desirable to reuse the frequency by operating in two orthogonal polarizations either within the same beam or as a means of providing isolation between neighboring beams operating in the same frequency band. In such systems the isolation between channels depends on the suppression of cross polarization. In a typical system such suppression is required to be in the order of 30 dB within the complete coverage zone. This translates into an isolation due to the antenna hardware of 40 dB as atmospheric depolarization and other sources outside the onboard antenna will also contribute to the interchannel coupling.

Reprinted from *IEEE Trans. Antennas Propagat.*, vol. AP-25, pp. 276–283, Mar. 1977.

Off-set reflector antenna configurations offer several advantages compared to center-fed systems. The blockage by the feed horn (or subreflector in Cassegranian systems) and the supporting struts which decrease the gain in front-fed systems is avoided; this is particularly important in multibeam antennas, because the feed system is relatively large. The field scattered by the struts also has a large content of depolarized signal and, therefore, causes a limitation to which polarization purity can be achieved in front-fed systems [1]; this limitation is particularly severe in relatively small reflectors such as is typical for on-board X-band antennas. Furthermore, in a multibeam antenna, the coupling between the feeds via the reflector is considerably lower in an off-set system since there is no direct reflection back into the feed. Finally, there are mechanical advantages in that more freedom is allowed in the choice of supporting structures and in the means of thermal control.

Extensive literature has appeared in recent years dealing with analysis of cross polarization and other characteristics of off-set reflector antennas, and references have been given by several authors [2]–[6]. It has been observed that such antennas usually have a high level of cross polarization at least when operated in linear polarization, whereas for circular polarization the two (mutually orthogonal) linear cross-polar components may add up to a co-polar circular component causing a beam squint as shown by Chu and Turrin [3].

The purpose of the present paper is to demonstrate that the cross polarization is not a basic and unsurmountable characteristic of off-set reflectors. By using a definition of cross polarization, in which the reference is a Huygens source oriented orthogonal to the axis of the reflector system, it is shown that the reflector does not contribute to the cross polarization in the far-field as is sometimes stated, neither is it a necessary consequence of the asymmetry of the configuration. The cross polarization of the far-field is shown to be merely the cross polarization of the primary field after collimation by the reflector. It is suggested, therefore, that the cross polarization can be significantly decreased by improvements in the feed, and there seems to be no physical limitation to the extent to which this can be acheived independently of the type of symmetry involved in the antenna system. Two simple feeds for off-set illumination of elliptically contoured reflectors with low cross polarization are treated as examples of how the techniques work.

II. ON THE DEFINITION OF CROSS POLARIZATION

Co- and cross polarization is usually defined by comparing the source under consideration with a reference source [7]. The co-polar field of the given source is then taken to be the component of the field which is parallel to the field of the reference source and the cross-polar field is the orthogonal component. This means that the co-polar field of a given source is defined by

$$\bar{E}_{\text{co}} = \frac{\bar{E} \cdot \bar{E}_{\text{ref}}}{\bar{E}_{\text{ref}}{}^2} \bar{E}_{\text{ref}}, \tag{1}$$

where $\bar{E} = E(\xi, \psi)$ is the electric vector field of the given source, $\bar{E}_{\text{ref}} = \bar{E}_{\text{ref}}(\xi, \psi)$ is the electric vector field of the reference source ((1) is to be properly mended in directions where $\bar{E}_{\text{ref}} = 0$), and the cross-polar field is obtained by (1) by replacing $\bar{E}_{\text{ref}}$ with the orthogonal field. It should be clear that the definition of cross polarization then depends not only on which source has been chosen as the reference, but also on how it is oriented.

Various definitions have been discussed by Ludwig [8] who named them 1st, 2nd, and 3rd definition according to the reference field being 1) a plane wave, 2) the radiated E-field from a short electric dipole, and 3) the E-field radiated by a Huygens source, and recommended the 3rd definition be used in connection with feed systems because of the following essential merits.

a) Interchanging the co- and cross-polar fields as measured in any direction corresponds to a 90° rotation of the reference source.

b) The definition is logically associated with normally used primary feed measurement setup as described by Silver [9] and Hollis *et al.*, [10]. It should be noted that the reference Huygens source is understood to be oriented perpendicular to the $\xi = 0$ axis of the measurement coordinate system.

c) When using this definition rigorously the collimation performed by a focused reflector system treats the co- and cross-polar components of the primary field separately. Thus no coupling is introduced by the reflector. This last charateristic will be dealt with in detail in the following sections.

The recommendation of Definition 3 has been widely accepted in recent publications; yet, as will be pointed out in the following, the statement in item c) above holds only if the definition is used in a strict sense. Fig. 1, which corresponds to the relevant part of Fig. 1 of Ludwig's paper [8], shows the electrical field lines on a sphere surrounding a Huygens source which is orthogonal to the z axis and which has the electric dipole oriented towards the y and x axis, respectively. For convenience this will be referred to as p- and q-polarization throughout this paper; $\hat{p}$ and $\hat{q}$ are supposed to be unit vector fields with directions defined by the $\bar{E}$-fields from the two Huygens sources, so that the two polarization components, according to (1) are

$$E_p = \bar{E} \cdot \hat{p}$$

$$E_q = \bar{E} \cdot \hat{q} \tag{2}$$

where $\hat{p}$ and $\hat{q}$ are related to (ξ, ψ) coordinates by

$$\begin{Bmatrix} \hat{p} \\ \hat{q} \end{Bmatrix} = \begin{Bmatrix} \sin\psi & \cos\psi \\ \cos\psi & -\sin\psi \end{Bmatrix} \begin{Bmatrix} \hat{\xi} \\ \hat{\psi} \end{Bmatrix}. \tag{3}$$

III. RADIATION FROM FOCUSED REFLECTORS

A. Currents on a Focused Paraboloidal Reflector Surface

Let a parabolic reflector be illuminated by a Huygens source which is located in its focal point and oriented with the electric dipole in the x direction and the magnetic dipole in the y direction (i.e., the source is q-polarized). The associated coordinate system is defined in Fig. 2. On the basis of the paraboloidal geometry and by the use of the physical optics approximation by which the surface current density $\bar{J}$ on the reflector is given by $\bar{J} = 2(\hat{n} \times \bar{H})$, where $\bar{H}$ is the incident magnetic field and $\hat{n}$ is the outgoing unit normal to the surface, it has then been shown by Jones [11] and Koffmann

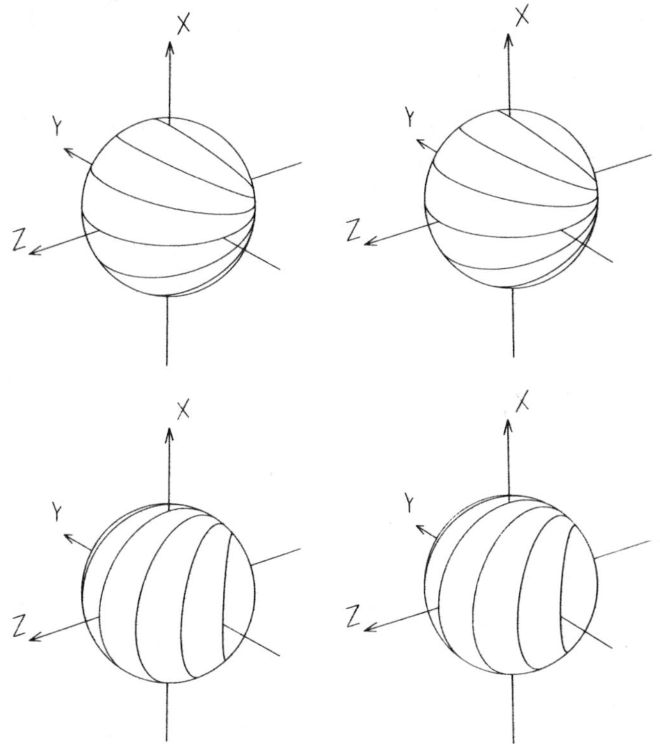

Fig. 1. $\hat{p}$ and $\hat{q}$ vector fields on unit sphere. Fields are defined by $\bar{E}$-field from Huygens source placed at origin and oriented orthogonal to z axis with electric dipole paralled to (a) y axis and (b) x-axis, respectively. This figure, and some of following, are drawn for stereoscopic viewing.

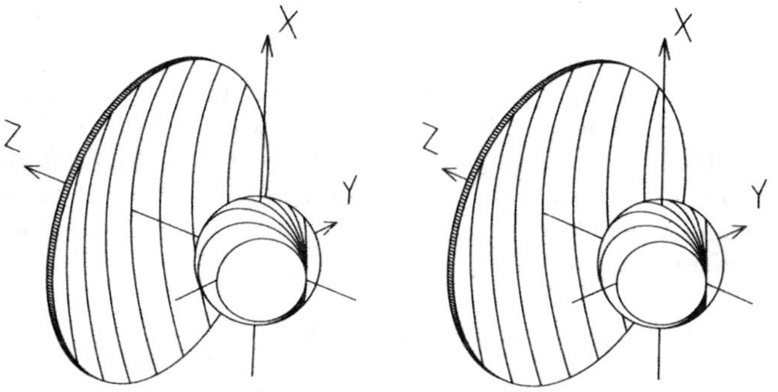

Fig. 2. Parabolic reflector illuminated by q-oriented linearly polarized source. Currents on reflector and q field lines on sphere surrounding feed are indicated.

[12] that the currents satisfy the condition $J_y = 0$ at all points on the reflector. Briefly speaking, this means that the far-field radiated by the currents on the reflector will be linearly polarised in the x direction, and the Huygens source is, therefore, an "ideal source" in this particular sense. Obviously, any other q-polarized feed will excite currents in the same direction even if the amplitude distribution is different (e.g., asymmetric), and the same conclusion for the far-field therefore holds. In Fig. 2 the currents on the reflector are shown, and a unit sphere with the q-field lines are shown also in order to illustrate the polarization performance of the feed.

It is worthy of note that the condition $J_y = 0$ is satisfied all over the surface. This means that the polarization purity of the far-field does not depend on a concellation of cross polarization contributions from various parts of the reflector. The secondary field therefore remains linearly polarized if some part of the reflector is removed or new parts are added. It may thus be deduced in general, that the polarization of the secondary field is independent of how the contour of the reflector is shaped. As an example an off-set-fed reflector antenna is shown in Fig. 3. Again, the vertical currents and the unit sphere showing the feed polarization are shown. The conclusion drawn here is in contrast to the frequently appearing statement that cross polarization is created by the reflector or by the asymmetry of the antenna configuration; a statement which seems to have impeded straight forward design of off-set reflector antennas for polarization diversity. A more rigorous treatment of the problem is given in the following.

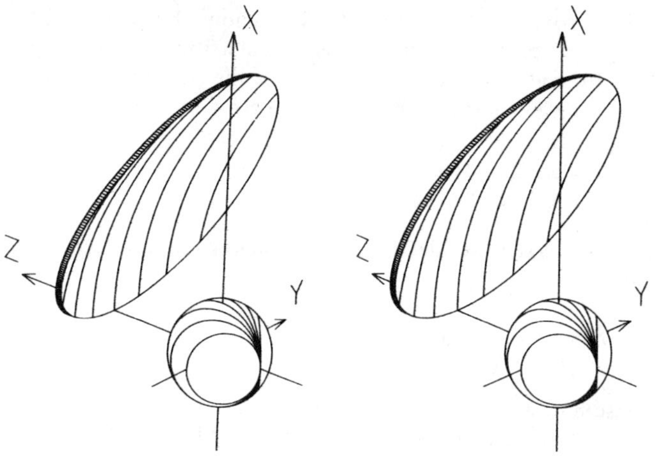

Fig. 3. Off-set parabolic reflector illuminated by q-oriented linearly polarized source. Currents on reflector and q field lines on sphere surrounding feed are indicated.

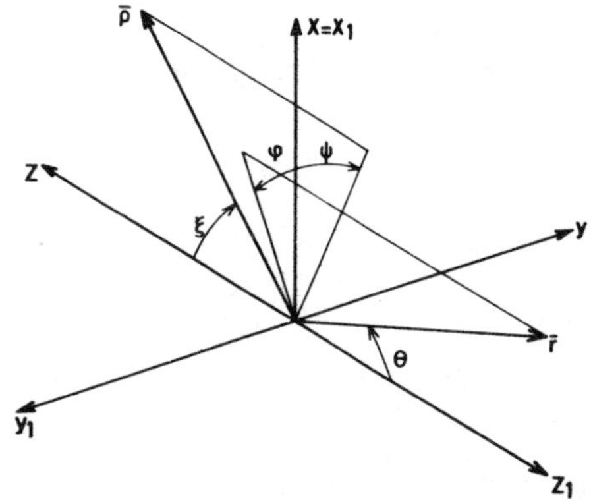

Fig. 4. Coordinate system for focused paraboloid and primary and secondary field.

B. Mathematical Formulation of the Reflector Transformation

In order to obtain the mathematical relationship between the primary and the secondary field for a focal-fed paraboloidal reflector the aperture technique will be used because of its simplicity. This technique is well established, and is used here as described by Collin and Zucker [13]. It is worthwhile to note that the technique has been shown by Rudge [14] to work well for multibeam antennas as well.

The coordinate system for the paraboloid and the feed is shown in Fig. 4. The feed radiation is described by

$$\bar{E}^i(\rho,\xi,\psi) = \bar{E}^i(\xi,\psi)\cdot\frac{e^{ik\rho}}{\rho}, \tag{4}$$

where

$$\bar{E}^i(\xi,\psi) = E_p{}^i(\xi,\psi)\hat{p} + E_q{}^i(\xi,\psi)\hat{q}. \tag{5}$$

Here $\hat{p}$ and $\hat{q}$ are the polarization vectors defined in Section II, and in the following $E_p{}^i$ will be referred to as the "co-polar" component essentially oriented towards the y axis, and $E_q{}^i$ as the "cross-polar" component. Throughout this paper the time dependence factor exp $(-i\omega t)$ is understood to multiply all field quantities, and $k = \omega\sqrt{\mu\epsilon}$ to be the free space wavenumber. The geometry of the reflector is given by

$$\rho = \frac{2f}{1 + \cos\xi} = \frac{f}{\cos^2(\xi/2)}, \tag{6}$$

where f is the focal length, and its outgoing normal unit vector by

$$\hat{n} = -\cos\left(\frac{\xi}{2}\right)\hat{\rho} + \sin\left(\frac{\xi}{2}\right)\hat{\xi}. \tag{7}$$

Assuming a physical optics reflection and neglecting a constant phase term, the aperture field is expressed by

$$\bar{E}^a = 2(\hat{n} \cdot \bar{E}^i)\hat{n} - \bar{E}^i. \tag{8}$$

The aperture field will now be evaluated in terms of the co-polar and cross-polar components of the primary field. First, $\bar{E}^i$ as given by (5) is expressed in spherical vectors, by using (3); carrying out the multiplication in (8) in spherical coordinates then gives

$$\begin{aligned}\bar{E}^a = &-\sin\xi\,(\sin\psi\, E_p{}^i + \cos\psi\, E_q{}^i)\,\hat{\rho}\\ &-\cos\xi\,(\sin\psi\, E_p{}^i + \cos\psi\, E_q{}^i)\,\hat{\xi}\\ &-(\cos\psi\, E_p{}^i - \sin\psi\, E_q{}^i)\,\hat{\psi}.\end{aligned} \tag{9}$$

By further substituting the following coordinate transformation

$$\begin{aligned}\hat{\rho} &= \sin\xi\cos\psi\,\hat{x} + \sin\xi\sin\psi\,\hat{y} + \cos\xi\,\hat{z}\\ \hat{\xi} &= \cos\xi\cos\psi\,\hat{x} + \cos\xi\sin\psi\,\hat{y} - \sin\xi\,\hat{z}\\ \hat{\psi} &= -\sin\psi\,\hat{x} + \cos\psi\,\hat{y},\end{aligned} \tag{10}$$

in (9). Taking the ρ-dependent term of (4) into account again, the following expression for $\bar{E}^a$ in Cartesian coordinates is then found:

$$\bar{E}^a = \frac{-1}{\rho}\{E_q{}^i(\xi,\psi)\hat{x} + E_p{}^i(\xi,\psi)\hat{y}\}, \tag{11}$$

neglecting a constant phase term. Here ρ is given by (6).

The aperture field is assumed (for physical reasons) to be a TEM wave and, hence, the radiation from the aperture is given in the coordinate system of Fig. 4 by [13].

$$\begin{aligned}E_\theta &= \frac{-ike^{ikr}}{2\pi r}(f_x\cos\phi + f_y\sin\phi)\\ E_\phi &= \frac{-ik\cos\theta\, e^{ikr}}{2\pi r}(f_y\cos\phi - f_x\sin\phi),\end{aligned} \tag{12}$$

where

$$\begin{aligned}\bar{f}(k_x,k_y) &= f_x\hat{x}_1 + f_y\hat{y}_1\\ &= \iint_{\text{aperture}} \bar{E}^a(x_1,y_1)e^{-i\bar{k}\cdot\bar{r}}\,dx_1\,dy_1\end{aligned} \tag{13}$$

$$\bar{k} = k_x\hat{x} + k_y\hat{y}$$

$$k_x = k\sin\theta\cos\phi$$

$$k_y = k\sin\theta\sin\phi$$

and

$$r = x_1\hat{x}_1 + y_1\hat{y}_1.$$

In the aperture the coordinates x_1 and y_1 are related to the primary feed coordinates by

$$\begin{aligned}x_1 &= \bar{\rho}\cdot\hat{x} = \rho\hat{\rho}\cdot\hat{x}\\ &= \frac{2f}{1+\cos\xi}\sin\xi\cos\psi\end{aligned} \tag{14a}$$

$$\begin{aligned}y_1 &= -\bar{\rho}\cdot\hat{y} = -\rho\hat{\rho}\cdot\hat{y}\\ &= \frac{-2f}{1+\cos\xi}\sin\xi\sin\psi.\end{aligned} \tag{14b}$$

Substituting the expression (11) for the aperture field in (13) gives

$$\bar{f}(k_x,k_y) = \iint_{\text{aperture}} \frac{1}{\rho}\{-E_q{}^i(\xi,\psi)\hat{x}_1 + E_p{}^i(\xi,\psi)\hat{y}_1\}e^{-i\bar{k}\cdot\bar{r}}\,dx_1\,dy_1, \tag{15a}$$

where the change of sign is caused by the change of coordinate system. By change of parameters (which is a quite lengthy but straightforward operation) this expression may be rewritten as

$$\bar{f}(k_x,k_y) = \iint r\{-E_q{}^i(\xi,\psi)\hat{x}_1 + E_p{}^i(\xi,\psi)\hat{y}_1\}e^{-i\bar{k}\cdot\bar{r}}\,d\xi\,d\psi \tag{15b}$$

where the integration is carried out over the range of (ξ,ψ) values subtended by the aperture. The expression of $\bar{f}(k_x,k_y)$, in (15b) is now used in (12) in order to find the far-field. Using also the transformation between $(\hat{p},\hat{q})$ and $(\hat{\theta},\hat{\phi})$ equivalent to (3), the far-field in terms of co- and cross-polar components turns out to be

$$\begin{aligned}E(\theta,\phi) &= E_p(k_x,k_y)\hat{p} + E_q(k_x,k_y)\hat{q}\\ &= \frac{-ike^{ikr}}{2\pi r}(f_y\hat{p} + f_x\hat{q}),\end{aligned} \tag{16}$$

which gives

$$E_p(k_x,k_y) = \frac{-ike^{ikr}}{2\pi r}\iint rE_p{}^i(\xi,\psi)e^{-i\bar{k}\cdot\bar{r}}\,d\xi\,d\psi \tag{17a}$$

$$E_q(k_x,k_y) = \frac{ike^{ikr}}{2\pi r}\iint rE_q{}^i(\xi,\psi)e^{-i\bar{k}\cdot\bar{r}}\,d\xi\,d\psi. \tag{17b}$$

This is the "reflector transformation" which expresses the co- and cross-polar components of the radiated field. It shows that when Ludwig's 3rd definition of polarization is used (in the present strict sense) then the co- and cross-polar components are transformed separately for symmetric as well as unsymmetric antennas. This is the mathematical proof of the more heuristic conclusion of the previous section.

For simplicity, the present analysis has been limited to parabolic antennas, but the conclusion will yet hold even for compound focused reflector systems. It may be shown [15] that the currents which flow on any reflector surface generated by a closed conic section (i.e., ellipsoid or hyperboloid) when it is illuminated by a Huygens source type of feed placed in one focal point will have the same direction at each point as the current produced by the same type of feed placed in the opposite focal point. Therefore, the secondary field from such a reflector will have the same polarization characteristics as the primary feed, and, thus when it is acting as a primary field for a new reflection, will lead to the same conclusion, although the actual transformation is not simple to carry out. Again, this conclusion is valid within the limits of the physical optics approximation. It is anticipated that the analysis can be extended to cover the case where the feed is moved off the focus in order to steer the beam off axis [2] or in order to improve the polarization characteristics of a Cassegranian antenna [16].

IV. CROSS POLARIZATION IN FEED SYSTEMS FOR ASYMMETRIC REFLECTOR ANTENNAS

A. Definitions

While the previous section covered focused reflector antennas irrespective of whether the configurations and feed patterns were symmetric or not, it is worthwhile to look at the feeds for off-set antennas in more detail, because for these it is less obvious how to deal with the definition of cross polarization than in the symmetric case. Consider, again, a linearly polarized primary source with the q-polarization as shown in Fig. 1(b). If this source is tilted, around the y axis, the polarization of course remains polarization clean assuming that the reference Huygens source is tilted together with the horn. However, such a tilted polarization reference makes less sense in relation to the reflector system because the condition for separate transformation of co- and cross-polarized fields is that the Huygens sources defining the p- and q-vector fields are oriented orthogonal to the axis of the reflector system. Therefore, it is useful to find the field components of the tilted feed horn in the p-, q-system defined in Fig. 1. In Fig. 5, the electric field lines of the tilted horn are shown together with the q-vector field from Fig. 1(b), and it is observed that the field lines cross. This means that the tilted feed does have a cross polarized component in the reflector coordinate system although it was polarization clean in the coordinate system oriented towards its own axis.

A simple example will demonstrate the effect of a reorientation of a feed horn. Consider an idealized circular horn situated in a set of coordinate systems as shown in Fig. 5. The associated (ξ,ψ,n) and (ξ',ψ',n') coordinates are defined in the usual manner. Let the radiation pattern of the horn by given by

$$E_{\xi'} = \sin\psi' \frac{J_1(ka\sin\xi')}{ka\sin\xi'}$$

$$E_{\psi'} = \cos\psi' \frac{J_1(ka\sin\xi')}{ka\sin\xi'}$$

where a is the radius of the horn aperture, and J_1 is the Bessel function of first order and first kind. The co- and cross-polarized field components in the same coordinate system are then

$$E_{p'} = \frac{J_1(ka\sin\xi')}{ka\sin\xi'}$$

$$E_{q'} = 0,$$

where p' and q' refer to Huygens sources with the electric di-

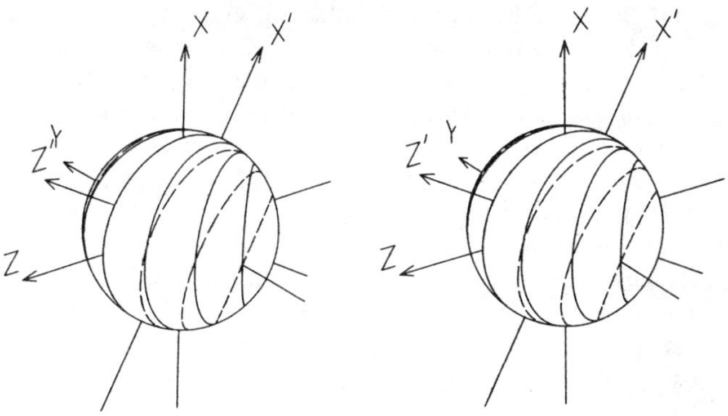

Fig. 5. E-field lines for tilted Huygens source compared to q-field lines.

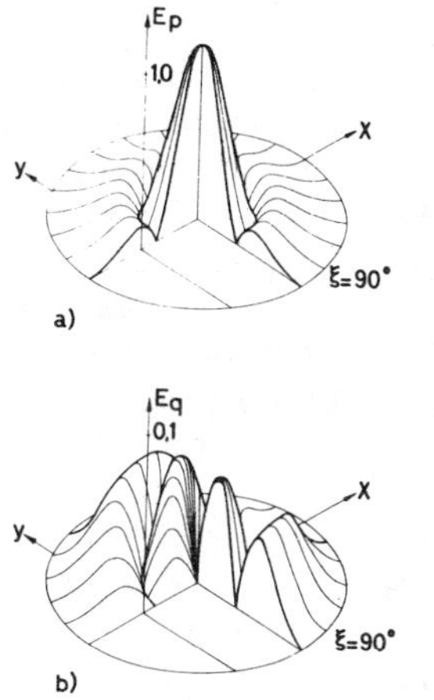

Fig. 6. Patterns for p- and q-polarized fields radiated by ideal, circular feed, which is tilted compared to reference coordinate system.

pole along the y' and the x' axis, respectively. The co- and cross-polarised field components, E_p and E_q, in the (x,y,z) coordinates may then be found by carrying out the appropriate coordinate transformations. The mathematical details of this are straightforward, and the resulting E_p and E_q patterns are shown in Fig. 6. With the parameters chosen (radius of the horn: $a = 1.15\ \lambda$, tiltangle: $\theta_0 = 45°$) the co-polar (E_p) radiation pattern is almost unchanged in the new coordinates, whereas, there is a significant cross-polar component (E_q); the pattern of this is antisymmetric around the $\phi = 0$ axis and has a maximum value of approximately 23 dB relative to the beam peak of E_p.

If the tilted horn is used as feed in an off-set reflector system, the cross polarized field component found above transforms into a cross polarized component of the secondary field. Such a cross polarized far-field component in off-set reflector antennas has been discussed by several authors based on studies of specific examples [2]–[4], [6], and the conclusion has been drawn, logically, but not entirely correct, that a long focal length is advantageous because the tilt angle θ_0 becomes smaller, and the cross polarization thus decreases. Some authors [6] have even extended the conclusion to state that the use of off-set antennas for purposes where a polarization discrimination of more than 30 dB is required is very questionable. In view of the separability of the collimation of co- and cross-polar fields it is now clear that this conclusion is not correct.

Bearing in mind that the cross-polarized far-field is merely the near-field cross polarization transformed according to (17b), the author [17] suggested two other obvious ways to suppress it. One is a mode matching technique similar to that used by Potter [18]. In the so-called Potter horn, the sidelobes and cross polarization radiated by the TE_{11} mode of a circular aperture are partially cancelled by adding in opposite phase a small portion of the TM_{11} mode which has (almost) identical sidelobe and cross-polar patterns. Obviously this technique can also be used to cancel the cross polarized radiation pattern in Fig. 6 by adding a suitable mode. This method will be referred to as the mode-matching technique, and an example in which a rectangular horn is treated will be given in Section V.

The second method is based on the observation that a necessary and sufficient condition for a plane aperture field to radiate a far-field with zero cross polarization is that the tangential E- and H-field components in the aperture satisfy the plane wave impedance relation. This can be seen from the expressions for the far-field radiated by an aperture as given by Collin and Zucker [13]. The requirement that E_ξ and E_ψ are to satisfy the relation for zero cross polarization, $E_\xi \cos\psi = E_\psi \sin\psi$, is easily seen to be equivalent to the requirement that $f_y = -\xi_0 g_x$ where ξ_0 is the free space impedance for plane waves. This implies that the equivalent electric and magnetic currents should be related as in a plane wave; hence, the equivalent surface source radiating a far-field with zero cross polarization is a Huygens layer. An open ended circular waveguide operated near cutoff radiates a field with very low cross polarization, and a Huygens source layer may, therefore, be approximately realized as an array of open ended circular waveguides. This method will be called the Huygens source technique, and a very simple example with an asymmetric beam will be given also in Section V.

B. The "Wedge Setup"

The equivalence between the Huygens source definition and the standard antenna test setup discussed in Section II exists only under the assumption that the Huygens reference source is oriented orthogonal to the $\xi = 0$ axis. This means that if one wished to use measured results for a primary field directly in the reflector transformation (17) in order to achieve the secondary field, then the primary field must be measured in a test-setup in which it is tilted the same amount relative to the $\xi = 0$ axis as it will eventually be tilted relative to the axis of the reflector system in which it is to be used as the primary source. This is illustrated in Fig. 7 where a wedge is shown to provide the appropriate alignment.

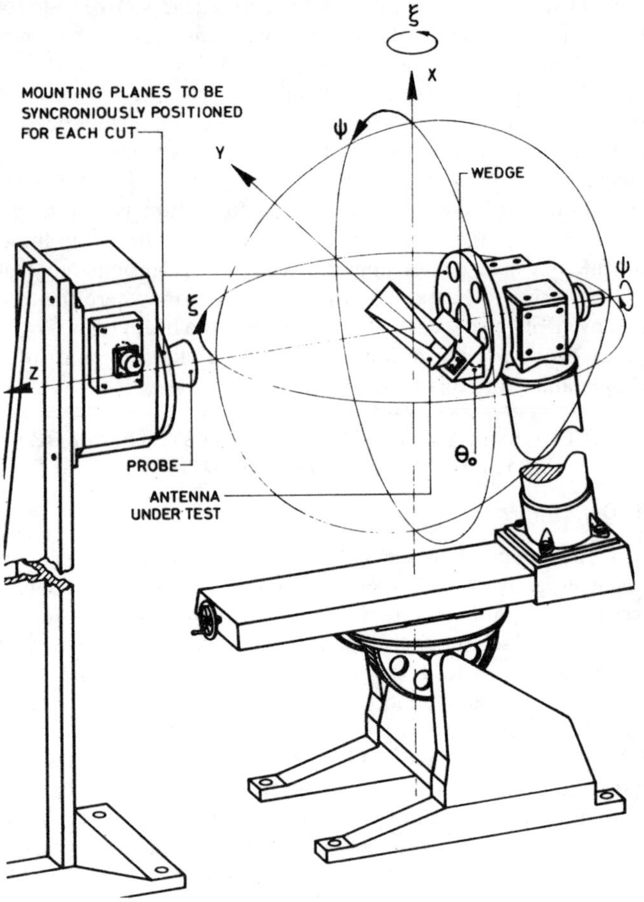

Fig. 7. "Wedge setup" for measurement of feed systems for off-set reflector antennas.

V. TWO EXAMPLES

A. The Mode Matching Technique

To demonstrate this technique a rectangular horn with a 2.5 by 2.5 λ aperture has been chosen. The aperture is situated in the (x',y',z') coordinate system shown in Fig. 5 with its center in the origin and its sides parallel to the x' and y' axis and radiates in the direction of the positive z' axis. The aperture is excited by the TE_{01}-mode where the first index refers to y' axis and the second to the x' axis; so, the horn is essentially p-polarized. The radiated field is found from a model in which the electrical aperture field is assumed to be the TE_{01} mode-field and the magnetic field is assumed to be zero. This model is similar to that used by Silver [9], and it is known to give a realistic prediction of the cross-polar as well as the copolar field [14] whereas a straight use of Silver's model leads to too optimistic values for the cross-polar field.

Because of the relatively large size of the aperture, a very low cross polarization (below 45 dB) is found when the pattern is calculated in the (x',y',z') coordinate system, but when the p and q components are evaluated a peak level for the q component of about 25 dB is found, as in the example treated in Section IV, whereas the p component is almost the same as the original co-polar component. The p and q components are shown in Figs. 8(a) and 8(b). The p and q component of the radiation from a TE_{20} mode excited in the same aperture are shown in Figs. 8(c) and (d). (The same theoretical model has been used.)

A striking similarity between the q-polar components of the TE_{01} and the TE_{20}-modes is evident, which means that a proper combination of the two modes can lead to a (partial) cancellation of the q component. This is shown in Figs. 8(e) and (f). In the $\psi' = 90°$ plane, the q component is below -41 dB inside the -10 dB points of the p pattern, i.e., a 15-dB improvement compared to the pattern of the TE_{01}-mode has been achieved in this plane. The cancellation is less perfect but still good in other planes, and a ratio of about -40 dB can be anticipated for a reflector antenna illuminated by the present feed.

It is observed from Fig. 8 that the q pattern of the TE_{20}-mode is considerably broader than the q pattern of the TE_{01}-mode. A better cancellation of the cross polarization could, therefore, be achieved, by equalizing the two patterns. This could be done, for instance, by adding a finned section to each side of the horn in the E-plane of the TE_{01}-mode with the fins in the y' direction. This would not effect the aperture distribution of the TE_{01}-mode but would give a broader aperture for the TE_{20}-mode and, therefore, serve to narrow the beam and to match it to the TE_{01}-mode.

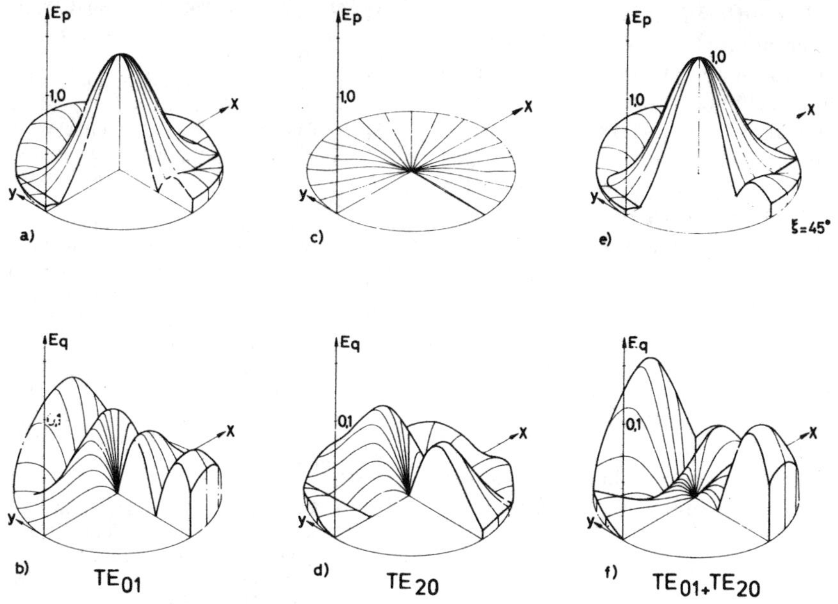

Fig. 8. Radiation patterns for rectangular horn excited by (a) and (b) TE_{01} mode; (c) and (d) TE_{20} mode; and (e) and (f) combined TE_{01} and TE_{20} mode.

B. The Huygens Source Technique

An open ended circular waveguide operated near cutoff is known to radiate a field with very low cross polarization thus approximating a Huygens source. A feed with two open ended circular waveguides has been discribed by Gruner and English [19]. The waveguides were fed in phase in order to produce an elliptical beam for illumination of an elliptical reflector. As the coupling between small circular apertures is very low [14], no reason is seen to abstain from phasing the two waveguides relative to each other in order to squint the beam so that it can illuminate an off-set, elliptical reflector.

As an example of this, the calculated p and q patterns for an array consisting of two open ended waveguides are shown in Figs. 9(a) and (b). The following parameters have been used: radius of the waveguide $a = 0.321\ \lambda$, distance between the centres $d = 0.70\ \lambda$, and phase difference $dp = -2.2$ rad. The beam maximum occurs at about $\xi = 25^\circ$, $\psi = 0^\circ$, and because the axis of the waveguides remains oriented towards the z axis (i.e., $\xi = 0^\circ$) the cross-polar component remains low. The element pattern has been obtained by using Chu's model as described by Silver [9] for the open ended waveguide. This model is known from experiments to give a realistic prediction of the co-polar pattern. For the cross-polar pattern, experiments have shown that the predicted levels are far too pessimistic when the waveguides are operated near cutoff [14], and in the present work 6 dB has been subtracted from the cross-polar values predicted by the Chu model. This is a rather arbitrary value, but is is considered to be conservative.

The results show a cross-polar peak level better than −32 dB relative to the co-polar beam peak inside the −6 dB contour of the co-polar beam. Thus secondary patterns can be expected to give cross-polar peaks in the range of 35–40 dB below the co-polar peak. By adding more elements to the feed array a more refined beam shape could be modelled.

The two examples shown here do not pretend to be optimized feeds. Rather, they have been chosen because of their simplicity in order to demonstrate the priniciples outlined in the previous sections. Recently, an elegant feed design based on mode matching technique has been presented by Rudge [20] who used a Potter horn as basic configuration and added the TE_{21}-mode to cancel the cross polarization.

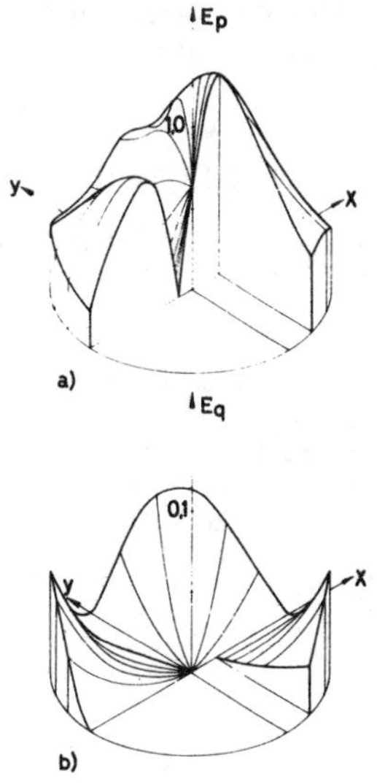

Fig. 9. Radiation patterns for Huygens source array of two elements.

VI. CONCLUSIONS

It is shown that co- and cross-polar components of the radiation from a feed system are collimated separately by focused paraboloidal reflectors, even for offset and for elliptical beam antennas. However, the separability depends on a rigorous use of the Huygens source definition (Ludwig's 3rd definition) of cross polarization by which it is assumed that the Huygens source with which the feed system under consideration is compared shall be properly aligned with the axis of the reflector system in which the feed is supposed to be used. A test setup which conforms with this definition is shown.

The consequence of the above separability is that cross polarization of a reflector antenna system may be removed by using a clean feed system. Two different methods for design of clean feed systems have been pointed out, one based on mode matching technique, one based on array technique. A theoretical example on each of these techniques have been studied and a considerable improvement of cross polarization performance has been observed. An experimental verification of the designs has been beyond the scope of the present investigation, but it is believed that the very simple principles behind the designs will prove useful in many practical antenna systems.

The proof of the separation of co- and cross-polar fields has only been carried out in detail for the parabolic reflector, but it has been pointed out that it holds also for compound, focused systems. The physical optics approximation has been used in its aperture field version. This is a well established technique, and the limitations its use impose on the results are mostly well known. However, it should be mentioned that edge effects, not treated by the theory, may have an unknown depolarization effect, in particular for nonsymmetric configurations.

ACKNOWLEDGMENT

The author wishes to express appreciation to F. Jensen, TICRA A/S, who produced the stereoscopic drawings and to J. Aasted, ESTEC, who encouraged me to publish this work and gave many helpful comments.

REFERENCES

[1] W. V. T. Rusch and O. Sorensen, "Aperture blocking of a focused paraboloid," Techn. Univ. of Denmark, Electromagnetics Institute, R. 126, July 1974.

[2] A. W. Rudge, "Multiple-beam antennas: Offset reflectors with offset feeds," *IEEE Trans. Antennas Propagat.*, vol. AP-23, pp. 317–322, May 1975.

[3] T. S. Chu and R. H. Turrin, "Depolarisation properties of offset reflector antennas," *IEEE Trans. Antennas Propagat.*, vol. AP-21, pp. 339–345, May 1973.

[4] M. J. Gans and R. A. Semplak, "Some far-field studies of an offset launcher," *Bell System Tech. Jour.*, vol. 54, pp. 1319–1340, Sept. 1975.

[5] P. J. Wood, "Depolarisation with cassegrain and front-fed reflectors," *IEE Electron. Lett.*, vol. 9, pp. 181–183, May 1973.

[6] J. Dijk, C. T. W. v. Diepenbeek, E. J. Maanders, and L. F. G. Thurlings, "The polarization losses of offset paraboloid antennas," *IEEE Trans. Antennas Propagat.*, vol. AP-22, pp. 513–520, July 1974.

[7] "IEEE Standard Definitions of Terms for Antennas," *IEEE Trans. Antennas Propagat.*, vol. AP-17, pp. 262–269, May 1969.

[8] A. C. Ludwig, "The definition of cross polarisation," *IEEE Trans. Antennas Propagat.*, vol. AP-21, pp. 116–119, Jan. 1973.

[9] S. Silver, *Microwave Antenna Theory and Design.* New York: McGraw-Hill, 1949.

[10] J. S. Hollis, T. J. Lyon, and L. Clayton, "Microwave antenna measurements," Scientific-Atlanta, Inc. Atlanta, Georgia, USA, July 1970.

[11] E. M. T. Jones, "Paraboloid reflector and hyperboloid lens antennas," *IRE Trans. Antennas Propagat.*, vol. AP-2, pp. 119–127, July 1954.

[12] I. Koffman, "Feed polarization for parallel currents in reflectors generated by conic sections," *IEEE Trans. Antennas Propagat.*, vol. AP-14, pp. 37–40, Jan. 1966.

[13] R. E. Collin and F. J. Zucker, *Antenna Theory.* New York: McGraw-Hill, 1969.

[14] A. W. Rudge, P. R. Foster, N. Williams, and N. Adatia, "Study of the performance and limitations of multiple-beam antennas, final report on ESTEC contract no.: 2277/74 HP," ERA, RF Technology Centre, England, Sept. 1975.

[15] G. T. Poulton, private communication.

[16] R. Graham, "The polarization characterisitcs of off-set cassegrain aerials," *International Conference on Radar and Future,* IEE, London, Oct. 23–25, 1973.

[17] J. Jacobsen, "Study of limitations of RF sensing systems due to distortions of large spacecraft antennas," Technical Specifications to ESTEC Contract No.: 2330/74 AK, Aug. 1974.

[18] P. D. Potter, "A new horn antenna with suppressed sidelobes and equal beamwidths," *Microwave J.*, vol. 6, pp. 71–78, June 1963.

[19] R. W. Gruner and W. J. English, "Antenna design studies for a U.S. domestic satellite," *COMSAT Techn. Rev.*, vol. 4, pp. 413–447, Fall 1974.

[20] A. W. Rudge and N. A. Adatia, "A new primary feed for offset reflector antennas," *IEE Electron. Lett.*, vol. 11, pp. 597–599, Nov. 1975.

Bibliography for Part VI

[1] M. J. Pagones, "Gain factor of an offset-fed paraboloidal reflector," *IEEE Trans. Antennas Propagat.*, vol. AP-16, pp. 536-541, Sept. 1968.

[2] A. W. Rudge, "Offset reflector antennas with offset feeds," *Electron. Lett.*, vol. 9, pp. 611-613, Dec. 27, 1973.

[3] M. J. Gans and R. A. Semplak, "Some far field studies of an offset launcher," *Bell Syst. Tech. J.*, vol. 54, pp. 1319-1340, Sept. 1975.

[4] H. Tanaka and M. Mizusawa, "Elimination of cross polarization in offset dual reflector antennas," *Electron and Commun. in Japan*, vol. 58-B, pp. 71-78, 1975.

[5] J. F. Kauffman and W. F. Croswell, "Off focus characteristics of the offset fed parabola," in *IEEE AP-S Symp. Dig.*, June 1975, pp. 358-361.

[6] N. A. Adatia and A. W. Rudge, "Beam squint in circularly polarized offset reflector antennas," *Electron. Lett.*, vol. 11, pp. 513-515, Oct. 16, 1975.

[7] H. P. Coleman, R. M. Brown, and B. D. Wright, "Paraboloidal reflector offset fed with a corrugated conical horn," *IEEE Trans. Antennas Propagat.*, vol. AP-23, pp. 817-819, Nov. 1975.

[8] A. G. P. Boswell and R. W. Ashton, "Beam squint in a linearly polarised offset reflector antenna," *Electron. Lett.*, vol. 12, pp. 596-597, Oct. 28, 1976.

[9] A. R. Valentino and P. P. Toulios, "Fields in the focal region of offset parabolic antennas," *IEEE Trans. Antennas Propagat.*, vol. AP-24, pp. 859-865, Nov. 1976.

[10] T. S. Chu, "Cancellation of polarization rotation in an offset paraboloid by a polarization grid," *Bell Syst. Tech. J.*, vol. 56, pp. 977-986, July-Aug. 1977.

Part VII
Lateral Feed Displacement, Scanning, and Multiple Beam Formation

When the feed in a reflector antenna is moved away from the focus in a direction transverse to the axis, the beam is displaced in the opposite direction and is said to be squinted, or scanned. Because such a displacement of the feed produces higher odd order, as well as linear phase terms in the aperture, the angle through which the beam is squinted is less than the angle (measured at the paraboloid vertex) through which the feed is displaced. The ratio of the beam angle to the feed angle is called the beam deviation factor, abbreviated BDF. The calculation of this factor is carried out in the first paper of this part, by Lo, in which analytical and experimental results are given as a function of f/D ratio.

Accompanying the beam squint due to feed displacement are beam broadening, loss in gain, and the incidence of coma lobes. These phenomena are analyzed by Ruze, in the second paper, and the range of validity is given for the approximations that he finds necessary to use. Figs. 3–8 in this paper are useful graphical presentations of the analytical results that enable the designer to estimate the magnitudes of the various effects as a function of feed displacement or of the number of beamwidths scanned. In general, scanning is limited to a very few beamwidths before the beam degradation becomes intolerable. In the third paper, Rudge and Withers present a method which has experimentally been shown to permit scanning through ±15 beamwidths with little pattern degradation and minimal gain loss. Their technique is based on the use of a number of feeds arrayed in the plane of scan and arranged to move on an appropriate locus in that plane. Basically, the feed array carries out a spatial Fourier transform of the distorted focal region fields which result when an off-axis plane wave is incident on the reflector.

The correct locus on which a feed should be moved and the orientation of the feed for optimum scanning of the beam are the subjects of investigation in paper four, by Rusch and Ludwig. The analysis is based on physical optics and results are given in relation to what is called the Petzval surface in optics. One interesting result that is not evident *a priori* is that a higher scan gain is obtained when the feed axis remains parallel to the reflector axis rather than being pointed to the vertex. This is true unless the f/D ratio is very large, and it holds up until the point where spillover loss begins to dominate.

Imbriale, Ingerson, and Wong have used a vector formulation that is more accurate than the scalar approximation used by Ruze to investigate the effects of large lateral displacement of the feed in a paraboloid. Their results are given in the fifth paper, and quite reasonable agreement with experimental results is shown for displacements of up to 16 wavelengths, corresponding to a beam scan of about 29 half power beamwidths. At this point, the pattern degradation is severe and the loss in gain amounts to about 14 dB. One significant conclusion is that the approximate scalar analysis does succeed in predicting the scan angle quite accurately, even for the largest displacements.

Although all of the above papers, and at least two in Part VI (those by Ingerson and Wong, and by Rudge), have great relevance for multiple beam formation in paraboloidal reflectors, they are incomplete in the sense that they are concerned only with the effects of a single feed when it is displaced from the axis. The formation of multiple, simultaneous beams requires the use of multiple, contiguous feeds in the focal plane, only one of which can be on axis and at the geometric focus. Thus, a new set of problems is introduced and new questions are raised. Is it physically possible to stack feed horns side by side to give adjacent beams with reasonable crossover levels; how severe will be the cross talk due to the sidelobes in adjacent beams; what is the nature of the matrix that is used to access the beams? Some of these questions are very briefly considered in the provocative short paper by Shelton. Many, however, remain unanswered at the present time, although there is a great deal of effort being expended in these areas because of the undoubted future importance of multiple beam antennas in satellite communications systems.

The seventh and final paper, by Ohm, represents a good, practical approach toward solving some of the problems peculiar to multiple beam forming antenna design. An offset paraboloid is suggested for the practical reason that aperture blocking by the feed cluster can thereby be avoided. These notions have been successfully put to practical test in an offset Cassegrainian system at 100 GHz. The results are reported in a paper by Semplak which appeared after this volume was in preparation [1].

Provided that spherical aberration can be either corrected or minimized, then spherical reflectors have great potential for the formation of multiple beams in space. These kinds of reflectors comprise the subject matter for Part IX in this volume and will not be further mentioned here. If multiple beams are required to be generated only in one plane, then toroidal reflectors have advantages, but again, the spherical aberration in that plane must be reckoned with. This approach has been adopted for an unattended earth terminal antenna and is described in detail in a paper by Hyde, Kreutel, and Smith [2].

On the Beam Deviation Factor of a Parabolic Reflector*

Y. T. LO†

IN a parabolic reflector antenna, the scanning can be achieved for a range by displacing the feed in the focal plane without resorting to a costly steerable mechanism for the whole antenna system. It is well-known that the range of scan by this method is limited by the increasing coma and astigmatism. However, it is still widely used in some specific applications. It is not the intention of this paper to investigate the aberrations of such a system, which may be referred to elsewhere, but rather to derive a formula for another characteristic of interest, namely the beam deviation factor. Such a formula does not seem to appear in literature, although there are available some experimental results for specific cases.

The beam deviation factor has been defined as the ratio of the beam deflection angle Θ_b to the angular displacement of the feed Θ_f, both measured from the axis of the reflector with the vertex as origin. Let the feed be at F' at a distance d from the focus F and let d be much smaller than the focal length f; then with a certain plausible approximation, the field at $P(R, \Theta\Phi)$, in Fig. 1 is given by

$$I = K \iint_A \frac{f(\phi, \theta)}{r} \cos [\beta\rho \sin \Theta \sin \Phi \sin \phi] \times \cos [\beta\rho(\sin \Theta \cos \Phi - d/r) \cos \phi]\rho d\phi d\rho, \quad (1)$$

where (ρ, ϕ, z) or (r, θ, ϕ) are the coordinates of a typical point q of integration.

K is a proportional constant, including the inverse R factor and the phase delay function due to r and R; A is the aperture of the reflector, *i.e.*, for $\rho=0$ to $D/2$ (D=diameter), and $\phi=0$ to 2π,

$r = f \sec^2 \theta/2$, and

$f(\phi, \theta)$ is the primary pattern of the feed.

If the plane containing F' is defined as one for which $\Phi=0$ and π, the maximum of I must appear in the same plane, since in practice $f(\theta, \phi)$ is real and positive for q in A. Thus if the maximum of I, I_m, occurs at Θ_m,

$$I_m = K \iint_A \frac{f(\phi, \theta)}{r} \cos [\beta\rho(\sin \Theta_m - d/r) \cos \phi]\rho d\phi d\rho, \quad (2)$$

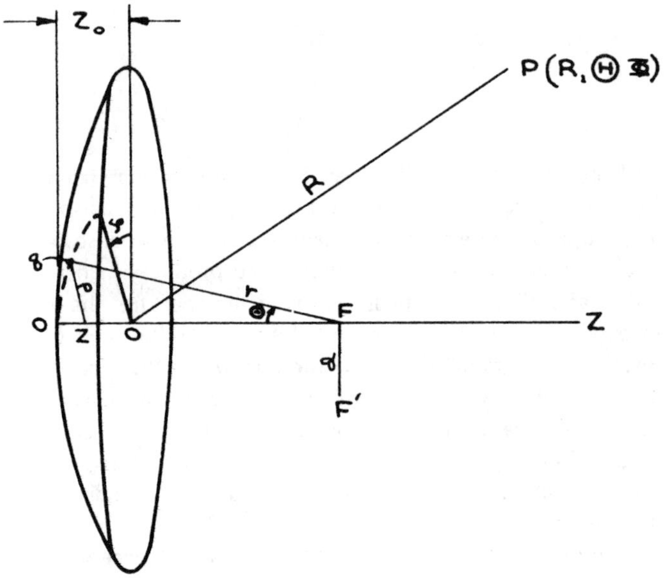

Fig. 1—Geometry of the reflector.

and Θ_m satisfies the following equation:

$$\iint_A \frac{f(\phi, \theta)}{r} \sin [\beta\rho(\sin \Theta_m - d/r) \cos \phi]\beta\rho^2 \cdot \cos \phi d\phi d\rho = 0. \quad (3)$$

Since for (ρ, ϕ) in A, $f(\phi, \theta)>0$, and

$$\sin [\beta\rho (\sin \Theta_m - d/r) \cos \phi] \cos \phi$$

is a continuous function of p and ϕ. This equation will be satisfied

$$\sin [\beta\rho (\sin \Theta_m - d/r) \cos \phi] \equiv 0$$

for some (ρ, ϕ) in A, by invoking mean value theorem for integrals. As $d=0$, the first condition can be met with $\Theta_m=0$, as expected. Were r a constant the condition could still be satisfied with $\Theta_m=\sin^{-1} d/r$. Now r is a variable with a range from f to $f+Z_0$ where Z_0=the depth of the dish. Therefore, it will be expected that in general there are many solutions of Θ_m to satisfy this second condition. For the maximum of the main beam, as shown in Fig. 2, the following inequality must be satisfied:

$$(1 + Z_0/f)^{-1}d/f < \sin \Theta_b < d/f.$$

It is easily seen that for a shallow dish Θ_b can be determined accurately without going to the complicated

* Manuscript received by the PGAP, February 8, 1960. This work was supported in part by Westinghouse Electric Corp., Baltimore, Md., during 1957.
† University of Illinois, Urbana, Illinois.

Reprinted from *IRE Trans. Antennas Propagat.*, vol. AP-8, pp. 347–349, May 1960.

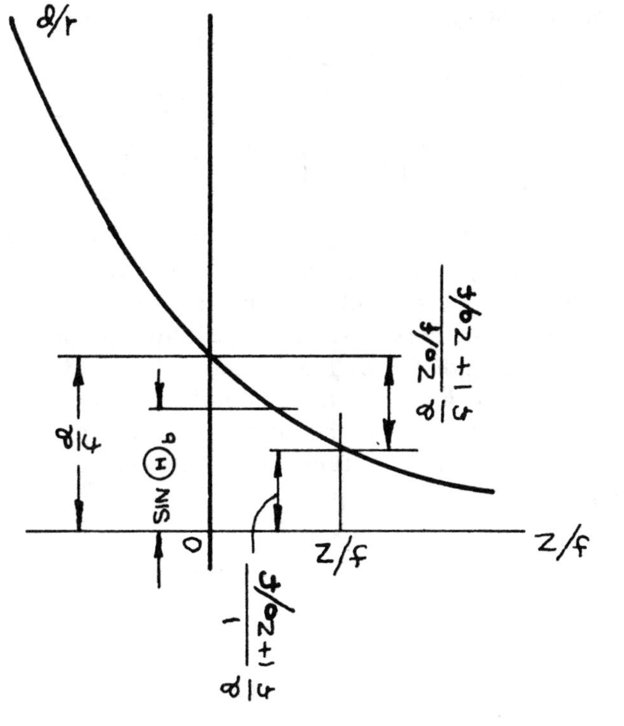

Fig. 2—The range of sin Θ_b which is a portion of the hyperbola d/r.

integration in (3). Now we can write

$$\sin \Theta_b = \frac{d}{f}\left(\frac{1}{1+Z_0/f} + k\,\frac{Z_0/f}{1+Z_0/f}\right)$$

$$= \frac{d}{f}\,\frac{1+kZ_0/f}{1+Z_0/f} = \frac{d}{f}\,\frac{1+k(D/4f)^2}{1+(D/4f)^2}, \tag{4}$$

where $k<1$. Or

$$\frac{\sin\Theta_b}{\tan\Theta_f} = \frac{1+k(D/4f)^2}{1+(D/4f)^2}\,. \tag{5}$$

If the usual definition of the beam deviation factor (BDF) is used,

$$\mathrm{BDF} = \frac{\Theta_b}{\tan^{-1} d/f} = \frac{\sin^{-1}\left[\dfrac{d}{f}\,\dfrac{1+k(D/4f)^2}{1+(D/4f)^2}\right]}{\tan^{-1} d/f}$$

$$\approx (\mathrm{BDF})_0\left\{1 + \tfrac{1}{3}(d/f)^2\left[\tfrac{1}{2}(\mathrm{BDF})_0^2 + 1\right] + O(d/f)^4\right\},$$

where

$$(\mathrm{BDF})_0 = \frac{1+k(D/4f)^2}{1+(D/4f)^2}\,. \tag{6}$$

It is seen that by the usual definition the BDF, up to the approximations assumed so far, depends also upon (d/f).[1]

Although k is a function of f, D, $f(\theta, \phi)$, its value is not critical, especially for large f/D as seen from (5) and (6). From one experimental point (or computation) it should be possible to predict the rest as a function of f/D with good accuracy. In Fig. 3, there shows the

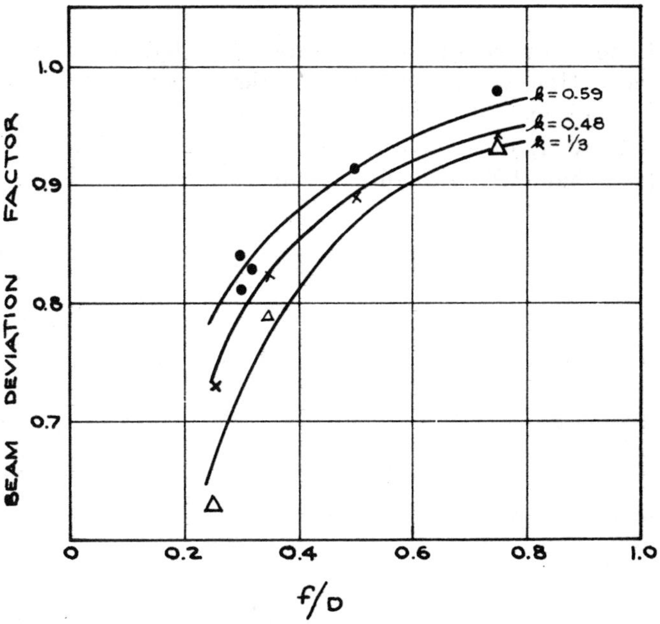

Fig. 3—Beam deviation factor as a function of f/D. Experimental data: Silver and Pao: ●; Keller and Coleman: x20-db taper, Δ 10-db taper; computed curve: ———.

computed curves and the experimental results by Silver and Pao,[2,3] and by Kelleher and Coleman.[1] For the former, there is no information available for $f(\theta, \phi)$ function, and $k=0.59$ is established by its point where $f/D=0.5$. For the latter, one case has a 10-db tapering, the other 20 db with $k=\frac{1}{3}$ and 0.48 respectively. The extraction of the experimental data from Kelleher and Coleman's report is rather difficult because they measured for various values of Θ_b, those indicated here being for $\Theta_b = 2\times$beamwidth. It seems that one can conclude that the values of k are from about 0.3 to 0.7, and that k becomes larger with higher tapering, as expected.

A second method to determine Θ_b is by assuming that the system is circularly symmetrical and $f(\theta, \phi) = \cos^n \theta$.

[1] K. S. Kelleher and H. P. Coleman, "Off-Axis Characteristics of the Paraboloidal Reflector," NRL Rept. No. 4088, Washington, D. C.; December, 1952.

[2] S. Silver and C. S. Pao, "Paraboloid Antenna Characteristics as a Function of Feed Tilt," Rad. Lab. Rept. No. 479, Cambridge, Mass.; 1944.

[3] S. Silver, "Microwave Antenna Theory and Design," Rad. Lab. Ser., McGraw-Hill Book Co., Inc., New York, N. Y., vol. 12, p. 487; 1949.

In this case (3) becomes

$$\int_0^{D/2} \frac{\cos^n \theta}{1 + (\rho/2f)^2} J_1\left[\beta\rho\left(\sin\Theta_b - \frac{d/f}{1 + (\rho/2f)^2}\right)\right] \cdot \rho^2 d\rho = 0. \quad (7)$$

Since

$$\rho' \equiv \rho/2f = \tan\theta/2,$$

$$\cos^n\theta = (1 - \rho'^2)^n(1 + \rho'^2)^{-n}.$$

It does not seem that (7) can be evaluated in a closed form. However, by using the series-expansion for the Bessel Function it can be integrated term by term since the integrand involves algebraic functions only. Suppose that D is so small that only the first term is significant; then

$$\Theta_b = \sin^{-1}\frac{d}{f}\left[1 - \frac{2}{3}\frac{Z_0}{f} + k'\left(\frac{Z_0}{f}\right)^2\right]$$
$$= \sin^{-1}\frac{d}{f}\left[1 - \frac{2}{3}\left(\frac{D}{4f}\right)^2 + k'\left(\frac{D}{4f}\right)^4\right], \quad (8)$$

where $k' = 1/2$, $13/18$, $15/18$ for $n = 0, 2, 3$, respectively. For small d/f, the first term corresponds to the reflection by a flat sheet, and the remaining terms appear as a result of the curvature of the reflector. If the result in (4) is expanded in Z_0/f and compared with (8), it will be found that $k = \frac{1}{3}$, a value we obtained before. It may also be mentioned that the same method can be carried out if the illumination function $f(\theta)$ be any polynomial in ρ, such as $(1-\rho^2)^n$ type.

Lateral-Feed Displacement in a Paraboloid

JOHN RUZE, FELLOW, IEEE

Abstract—**The beam shift and degradation of a paraboloidal reflector with an offset feed is analyzed by the scalar plane wave theory. Higher order coma terms are included with the feed at its optimum axial position. The beam characteristics for a tapered circularly symmetric illumination are presented. The range of validity of the approximate analysis is indicated.**

I. Introduction

THE BEAM degradation due to a lateral feed displacement in a paraboloidal mirror is part of the classical study of aberrations of optical systems. These have been extensively investigated by Nijboer and Nienhuis [1], Kingslake [2], and others. The text by Born and Wolf [3] provides an excellent presentation of the geometric and wave theory of optical aberrations.

However, these optical papers are not immediately applicable to antenna technology. The reasons for this may be summarized as:

1) In many optical instruments, we are primarily interested in the spatial distortion (ray aberration) of an image point from its desired or Gaussian focus; in an antenna, however, angular beam distortions, and factors such as gain, beam width, and side lobe level are of primary interest.

2) Optical systems usually have a higher "f" number (f/D ratio) than antenna reflectors, where values as low as 0.25 are not uncommon. This lower ratio increases the importance of the higher order aberrations.

3) The optical analysis is generally concerned with uniform illumination of the exit pupil, whereas antenna apertures are invariably tapered or apodized.

4) Optical systems further are many orders of magnitude greater in aperture to wavelength ratio. They follow much more closely geometric optics' behavior than antenna systems, where diffraction theory is mandatory.

5) Some optical systems are limited by other factors than the phase aberrations, so that much larger wavelength distortions may be tolerated.

6) The expansion of the aberration function, in terms of Zernike polynomials [1], is useful when dealing with a single aberration, but becomes unmanageable when higher order effects and apodization are included. With modern computing machines, it is preferable to deal with the original diffraction integral than with its general evaluation in series form [4].

Manuscript received April 2, 1964; revised March 11, 1965.
The author is with the M.I.T., Lincoln Laboratory, Lexington, Mass. (Operated with support from the U. S. Air Force.)

Reprinted from *IEEE Trans. Antennas Propagat.*, vol. AP-13, pp. 660–665, Sept. 1965.

Unfortunately, the antenna literature is not very extensive. We have the original experimental work of Silver and Pao [5]; the theoretical derivation, limited to primary coma, and experimental work of Kelleher and Coleman [6]; a paper on the beam deviation factor by Lo [7]; and a more extensive analysis by Sandler [8].

It is the purpose of this paper to present the properties of the offset fed paraboloid in the form of graphs of the significant characteristics. These are derived by means of the scalar plane wave theory, and series expansion of the phase aberration function. The results have the same small-angle limitation as the on-axis patterns normally used. In addition, the range of validity of the expansion is examined.

II. Analysis

Let us consider a paraboloid of diameter $D=2a$, with a spherical coordinate system centered at the focus (Fig. 1). The field at a far-field point with a focal feed is

$$E(\theta, \phi) = \int_0^{2\pi} \int_0^a f(r, \phi') e^{jk(p - \bar{p} \cdot \bar{R}_0)} r\,dr\,d\phi', \tag{1}$$

where $f(r, \phi')$ is an effective aperture distribution, and where we have suppressed constant factors. The bars represent vector quantities, and the subscript "0" their unit values.

With feed displacement to the lateral point "ϵ", we have

$$E(\theta, \phi) = \int_0^{2\pi} \int_0^a f(r, \phi') e^{jk(p' - \bar{p}' \cdot \bar{R}_0)} r\,dr\,d\phi', \tag{2}$$

where we have assumed that the magnitude of the effective aperture distribution has remained unchanged.[1]

We have the geometric relations:

$$p = \frac{2f}{1 - \cos\theta'} \tag{3a}$$

$$\bar{p} = p[\cos\phi' \sin\theta' \bar{x} + \sin\phi' \sin\theta' \bar{y} + \cos\theta' \bar{z}] \tag{3b}$$

$$\bar{R}_0 = [\cos\phi \sin\theta \bar{x} + \sin\phi \sin\theta \bar{y} + \cos\theta \bar{z}] \tag{3c}$$

[1] This implies that the vertex ray is the principal ray, and that the optical stop is at the mirror [11]. The feed should, therefore, be pointed at the vertex.

$$\bar{p}' = \bar{p} + \epsilon_x \bar{x} + \epsilon_z \bar{z} \tag{3d}$$

$$p' = p\left\{1 + \frac{2\epsilon_x}{p} \cos\phi' \sin\theta' + \frac{2\epsilon_z}{p} \cos\theta' + \frac{\epsilon^2}{p^2}\right\}^{1/2}. \tag{3e}$$

For feed displacements small compared to the focal length (small-field angle),

$$\frac{\epsilon_x}{p} < \frac{\epsilon_x}{f} \ll 1, \tag{4}$$

we can write the phase factor (2), neglecting terms higher than the square of this parameter, as

$$\begin{aligned} p' - \bar{p}' \cdot \bar{R}_0 = 2f &- \epsilon_x \cos\phi \sin\theta - \epsilon_z \cos\theta \\ &- p \sin\theta' \sin\theta \cos(\phi' - \phi) + \epsilon_x \cos\phi' \sin\theta' \\ &+ \epsilon_z \cos\theta' + \frac{\epsilon_x^2}{2p} + p \cos\theta'(1 - \cos\theta) \\ &- \frac{\epsilon_x^2}{2p} \cos^2\phi' \sin^2\theta'. \end{aligned} \tag{5}$$

The first three terms are independent of the integration coordinates and may be taken out of the integral; they represent a phase pattern of the far field. Recalling that $r = p \sin\theta'$, the next term is the normal phase factor due to an in-phase aperture. The fifth term represents the beam shift and comatic aberrations. The next three terms are field curvature (terms proportional to r^2), and higher terms of even power. The last term is the astigmatism.

The field curvature may be eliminated by axially refocusing the feed. This condition gives the Petzval surface in optics [9] which is defined as that locus which contains a sharp image when the other aberrations are absent. It may be obtained from (5), by setting all the field curvature terms to zero (including those hidden in the last two terms), as

$$\epsilon_z = \frac{\epsilon_x^2}{2f}. \tag{6}$$

For small aberrations, (6) defines the feed locus for sharpest nulls, as another paraboloid of focal length "$f/2$" tangent to the focal plane. The vertex feed distance must, therefore, be slightly increased as we scan off axis.

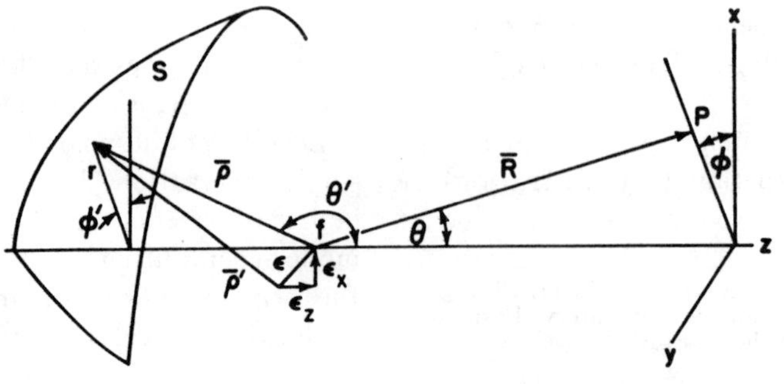

Fig. 1. Coordinate system.

For this optimum feed position, the magnitude of the far field can be written, from (2), as

$$|E(\theta,\phi)| = \int_0^{2\pi}\int_0^a f(r,\phi') \cdot e^{-jk[r\sin\theta\cos(\phi'-\phi)-\epsilon_x\sin\theta'\cos\phi']}rdrd\phi', \quad (2a)$$

where, for the present, we have neglected the astigmatism.

The second term of the exponential causes the beam shift and the beam degradation as it represents the phase departure from the nonscanned in-phase aperture.

Since

$$\sin\theta' = \frac{r/f}{1+(r/2f)^2} = \frac{r}{f}\left[1-\left(\frac{r}{2f}\right)^2+\left(\frac{r}{2f}\right)^4-\cdots\right], \quad (7a)$$

this phase departure may be written

$$\delta = \frac{2\pi}{\lambda}u_s r\cos\phi'\left[1-\left(\frac{r}{2f}\right)^2+\left(\frac{r}{2f}\right)^4-\cdots\right], \quad (7b)$$

where

$$u_s = \frac{\epsilon_x}{f} = \tan\theta_s$$

is a measure of the feed squint.

The first term of (7b) is a phase shift linear with "x" across the aperture and causes an undistorted beam shift, equal to the feed squint. The second term, proportional to "$u_s r^3\cos\phi'$", is what is known as primary coma, and creates beam degradation and a beam shift in the opposite direction. The remaining terms are higher order coma terms normally neglected in high "f/D" systems but retained here.

It is of interest to examine the ratio of the total coma aberration to the neglected astigmatism. We have for the ratio of the relative edge errors

$$\frac{\text{ASTIGMATISM}}{\text{Total Coma}} = \frac{2u_s(f/D)}{[1+(D/4f)^2]^2}. \quad (8)$$

We see that for normal parabolic antennas, the astigmatism is a small quantity. However, for astronomical mirrors and Cassegrainian systems of high magnification, the astigmatism becomes the limiting factor on the field of view. Dimitroff and Baker [10] give the comparative image errors for an $f/3$ and $f/10$ parabolic telescope. When the astigmatism becomes significant, the Petzval surface loses its utility and the feed focus will be poorly defined.

With the notation

$$u = \sin\theta; \qquad M(r) = 1+(r/2f)^2,$$

the exponential in the integral of (2a) may be written

$$kr\left[u\cos(\phi'-\phi) - \frac{u_s\cos\phi'}{M(r)}\right] = Akr\cos(\phi'-\alpha), \quad (9)$$

where "A" and "α" may be determined as

$$A^2 = u^2 - \frac{2uu_s}{M(r)}\cos\phi + \frac{u_s^2}{M^2(r)} \quad (9a)$$

$$\tan\alpha = \frac{u\sin\phi}{u\cos\phi - u_s/M(r)}. \quad (9b)$$

The magnitude of the far field is:

$$|E(\theta,\phi)| = \int_0^{2\pi}\int_0^a f(r,\phi')e^{jkrA\cos(\phi'-\alpha)}rdrd\phi'. \quad (2b)$$

For circularly symmetric illumination functions, the "ϕ'" integration can be performed with the result

$$E(\theta,\phi) = 2\pi\int_0^a f(r)J_0(krA)rdr. \quad (2c)$$

By means of computing machines, this integral can be evaluated for a specified illumination. We note that the pattern is symmetric about the plane of scan ($\phi=0$); for zero feed squint ($A=u$), it reduces to the normal circular diffraction pattern; and with off-axis feeds, it contains all the comatic aberrations (first order in ϵ_x/f).

Let us now consider the pattern in the plane of scan ($\phi=0$) where the major pattern distortions occur, and where

$$A = u - \frac{u_s}{M(r)}; \qquad \alpha = 0. \quad (9c)$$

For small feed displacements, the position of the beam maximum, "u_m," may be found as that value of "u" that minimizes the illumination-weighted squared phase error, or

$$\frac{\partial}{\partial u}\int_0^{2\pi}\int_0^a f(r)[krA\cos\phi']^2 rdrd\phi' \equiv 0.$$

Performing this operation, we have for the beam deviation factor,

$$BDF = \frac{u_m}{u_s} = \frac{\int_0^a \frac{f(r)r^3}{M(r)}dr}{\int_0^a f(r)r^3dr}. \quad (10)$$

This may be evaluated in closed form for various illumination functions, as has been done by Lo [7]. The field reduction, at the beam peak (u_m), can also be obtained from (2c).

III. Computed Results

For computational purposes, (2c) was put in a normalized form and evaluated on an IBM 7090.

$$E(w,0) = 2(p+1)\int_0^1 (1-r^2)^p J_0\left[\left(w - \frac{w_s}{M(r)}\right)r\right] r\,dr, \qquad (2d)$$

where

$$w_s = \frac{2\pi a}{\lambda}\tan\theta_s; \qquad w = \frac{2\pi a}{\lambda}\sin\theta.$$

Computations were made for the illumination function $(1-r^2)^p$, and for the 10 dB tapered illumination $f(r)=0.3+0.7(1-r^2)$, with $f/D=0.25$, 0.33, 0.4, 1.0, and 2.0.

Figure 2 shows typical scan plane patterns. We note that the gain drops with scan, the beam broadens, the beam scan is less than the feed squint, the sidelobe on the axis side (coma lobe) increases; whereas, the first sidelobe on the other side decreases, changes sign, and merges with the main beam and second sidelobe causing additional beam broadening. Complete patterns or image plane isophots may be found in Born and Wolf [3], Figs. 9.6 to 9.8. These are for primary coma only, and represent parabolic reflectors of large "f/D" ratio, with no astigmatism. They are expressed in maximum-edge coma error which, in our notation, is

$$\delta_c(a) = \frac{w_s}{M(a)}\left(\frac{D}{4f}\right)^2. \qquad (12)$$

The problem remained of expressing the large amount of computed data in terms of useful curves. It was determined that if the pattern characteristics were plotted against the quantity

$$X = \frac{\frac{w_m}{2w_0}(D/f)^2}{1+0.02(D/f)^2}, \qquad (13)$$

then over the region of interest the data was essentially independent of the f/D ratio. The factor "$w_m/2w_0$" is the number of half-power beamwidths scanned, a convenient variable. Figures 3 to 6 show these results.

The curves permit expressing the scanning limit in terms of a simple formula. If we choose 1 dB loss of gain (the Rayleigh limit) as our criteria, then for the 10 dB taper, $X=22$, and the number of beamwidths scanned is

$$\frac{w_m}{2w_0} = 0.44 + 22(f/D)^2. \qquad (14)$$

This criteria may not be adequate for many applications as the coma lobe has increased to -10.5 dB at the scan limit.

Figure 7 shows the beam deviation factor. To a first approximation (10), the beam deviation factor is independent of the number of beamwidths scanned. The machine computation shows its dependence on X. We have indicated in the figure the small increase, by means of vertical bars, as our parameter X approaches 50.

The question naturally arises as to the range of validity of the approximation made in going from the phase function of (2) to the truncated series expansion of (2b). As the two expressions are known, their maximum difference, or the phase error made by this analysis, can be calculated. This difference depends on the number of beamwidths scanned, the diameter in wavelengths, and the f/D ratio.

In Fig. 8, we show the loci of quarter wavelength phase errors. Also the 1 dB scanning limit from (14) is shown. The use of this figure requires an example. Consider an f/D of 0.4; from (14) or Fig. 6 or 8 we can scan 4 beamwidths for a 1 dB loss of gain. From Fig. 8 we have, that for a half-power beamwidth of 1.9 degrees or less, our approximate analysis is within a maximum phase error of a quarter wavelength. For greater beamwidths, larger errors will be incurred, and more precise and laborious methods, based on (2), may be necessary [4].

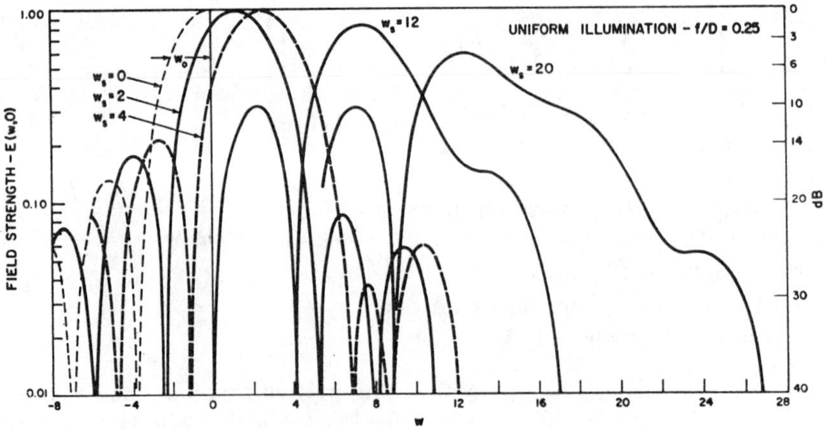

Fig. 2. Typical scanned patterns.

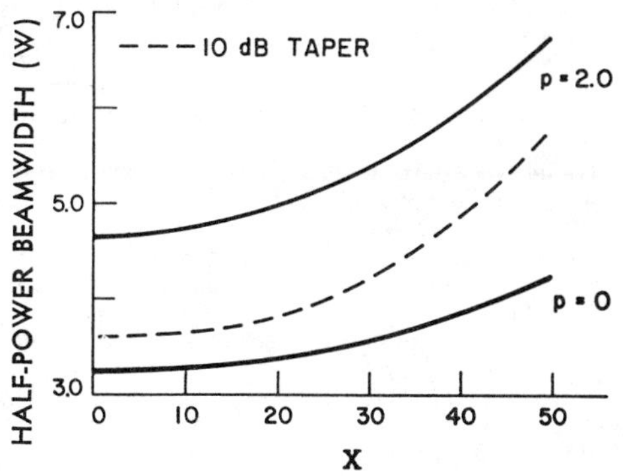

Fig. 3.   Half-power beamwidth.

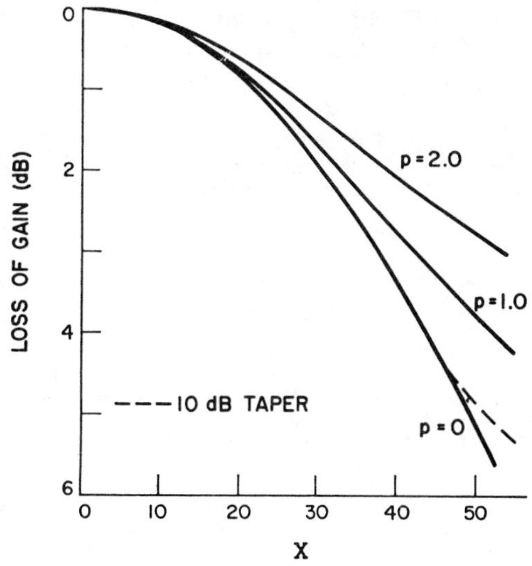

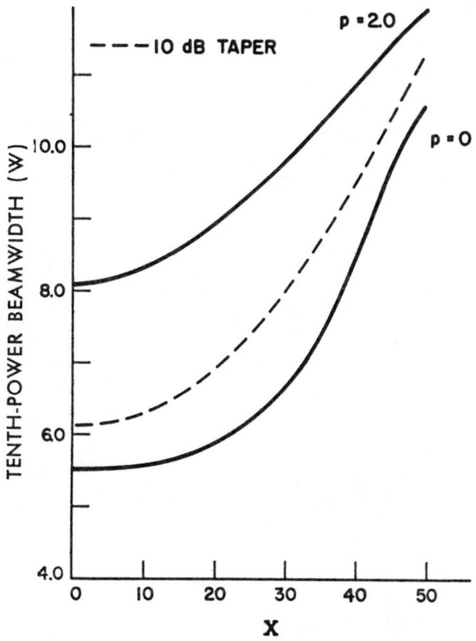

Fig. 4.   Tenth-power beamwidth.

Fig. 5.   Coma lobe.

Fig. 6.   Loss of gain.

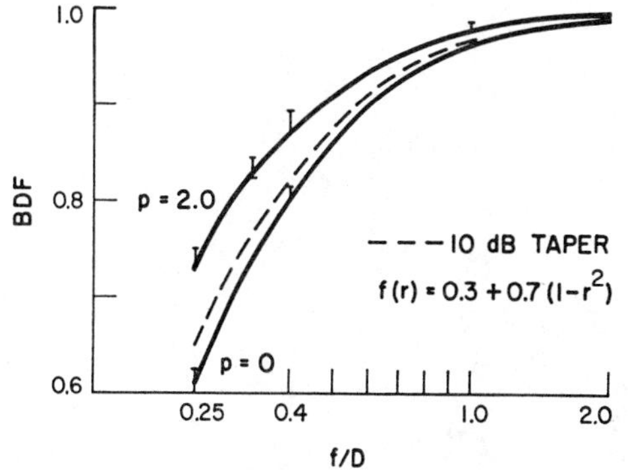

Fig. 7.   Beam deviation factor.

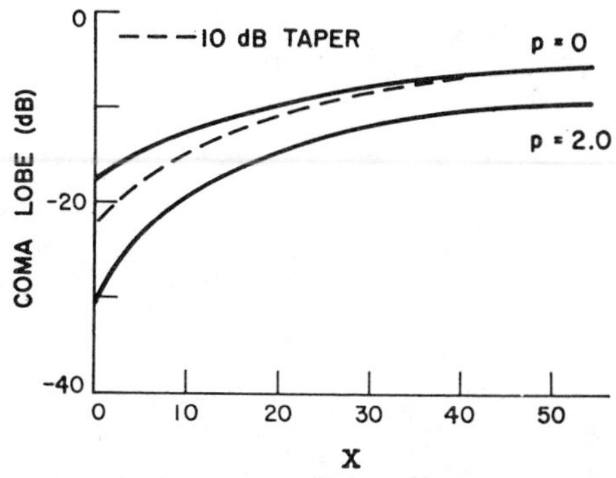

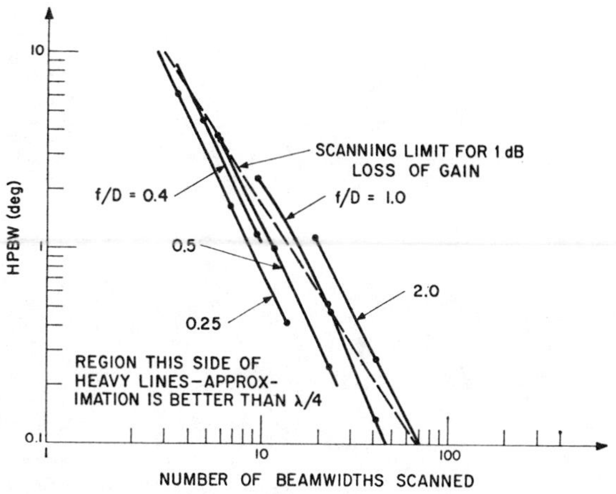

Fig. 8.   Region of validity.

ACKNOWLEDGMENT

Acknowledgment is made to Mrs. Margaret Gardner of the Lincoln Laboratory Computer Facility under the direction of Mrs. Virginia J. Mason. Mrs. Gardner programmed (2d), and made available the data for Section III of this paper.

The author also is indebted to D. S. Grey for many helpful discussions on the optical problem, to K. J. Keeping, and T. Hagfors for detailed reading and comments.

REFERENCES

[1] B. R. A. Nijboer, "The diffraction theory of aberrations," pts. 1 and 2, *Physica X*, pp. 679–692, October 1943; *Physica XIII*, pp. 605–620, December 1947; with K. Nienhuis, pt. 3, *Physica XIV*, pp. 590–608, January 1949.

[2] R. Kingslake, "The diffraction structure of the elementary coma image," *Proc. Phys. Soc. (London)*, vol. 6, pp. 147–158, 1948.

[3] M. Born and E. Wolf, *Principles of Optics*. New York: Pergamon, chs. 5 and 9, 1959.

[4] C. C. Allen, "Numerical integration methods for antenna pattern calculations," *IRE Trans. on Antennas and Propagation (Special Supplement)*, vol. AP-7, pp. S387–S401, December 1959.

[5] S. Silver and C. S. Pao, "Paraboloidal antenna characteristics as a function of feed tilt," Radiation Lab., Mass. Inst. Tech., Cambridge, Rept. 479, February 1944.

[6] K. S. Kelleher and H. P. Coleman, "Off-axis characteristics of paraboloidal reflectors," Naval Research Lab., Washington, D. C., Rept. 4088, December 1952.

[7] Y. T. Lo, "On the beam deviation factor of a parabolic reflector," *IRE Trans. on Antennas and Propagation*, vol. AP-8, pp. 347–349, May 1960.

[8] S. S. Sandler, "Paraboloidal reflector patterns for off-axis feed," *ibid.*, pp. 368–379, July 1960.

[9] L. C. Martin, *Technical Optics*, vol. 2. London: Sir Isaac Pitman & Sons, p. 74, 1954.

[10] G. Z. Dimitroff and J. G. Baker, *Telescopes and Accessories*. London: Churchill, p. 86, 1946.

[11] A. S. Filler, "Primary aberrations of mirrors," *Amer. J. Phys.*, vol. 29, pp. 687–694, October 1961.

New technique for beam steering with fixed parabolic reflectors

A. W. Rudge, Ph.D., Mem.I.E.E.E., and Prof. M. J. Withers, M.Sc., C.Eng., M.I.E.E.

Indexing term: Reflector antennas

Abstract

A technique is described which offers the potential of achieving wide-angle beam steering with fixed parabolic reflectors. The technique involves a primary-feed device with an aperture field distribution which can be adapted to match the distorted field distributions resulting from a parabolic reflector when an off-axis plane wave is incident. To provide an adaptation without deterioration of the system signal/noise ratio, which requires only a movement of the primary feed along a given locus and an adjustment of noninteracting phase shifters, the primary feed carries out a spatial Fourier transformation of the intercepted fields. The technique has been implemented in an experimental X band antenna and beam steering of ± 15 beamwidths achieved with negligible distortion of the directional pattern and less than 0·5 dB loss in gain.

List of principal symbols

f = focal length of parabolic reflector
d = diameter of reflector
θ = halfangle subtended at the focus by a point on the reflector
θ^* = maximum value of θ
$u = \sin\theta$
$\hat{u} = \sin\theta^*$
ϕ = rotational angle in reflector-aperture plane
t, ϕ' = focal-plane polar co-ordinates
x, y = focal-plane rectangular co-ordinates
$p = \sin\theta\cos\phi$
$q = \sin\theta\sin\phi$
$\hat{p} = \sin\theta^*\cos\phi$
$\hat{q} = \sin\theta^*\sin\phi$
ψ = beam-steering angle, from antenna boresight
β_n = component of phase error referred to reflector-aperture plane
$k = 2\pi/\lambda$
λ = wavelength

1 Introduction

In the majority of present-day applications of large parabolic-reflector antennas, it is required that the antennas' directional patterns should be steerable over large angles. This is normally accomplished by mechanically steering the reflector structure. In a large antenna, the reflector and associated backing structure may have a deadweight of the order of hundreds of tons, and thus the steering requirement constitutes a major part of the antenna-design problem, particularly with respect to the economics of the construction. The steering problem becomes further aggravated when the pattern-steering requirements demand rapid movement coupled with highly accurate positioning and maintenance of pointing under adverse weather conditions.

To achieve a relaxation in the specification pertaining to the mechanical steering requirements, it would be advantageous if a moderate degree of beam steering could be effected by means of a suitable primary-feed design. While this may involve movement of a complex feed assembly, the problem compares favourably with that involved in moving the massive reflector structure.

Parabolic-antenna radiation-pattern steering can be achieved over a small range of angles by displacing a conventional primary feed radially about the reflector vertex.[1,2] However, unless the reflector-aperture illumination is severely tapered, with a resultant reduction in aperture efficiency, the introduction of phase-error effects restricts the range of scanning to within a few beamwidths of the boresight before the radiation-pattern deterioration becomes excessive.

A number of studies have been reported describing methods leading to the reduction of distortion of the antenna radiation pattern at small to moderate scan angles. For example, Takeshima[3] has described a defocusing technique to achieve balancing of two or more aberrations, a compensatory phase-error technique involving tilting of the subreflector has been developed by Hannan[4] for Cassegrainian systems, and both Loux and Martin[5] and Assaly and Ricardi[6] have described focal-plane-array techniques which carry out weighting, phasing and summing of the intercepted energy.

The technique described here appears similar to the focal-plane-array techniques mentioned above. However, here the location of the feed is not restricted to the focal plane, and the signal processing employed is that of a spatial Fourier transformation of the intercepted electric fields. The advantages of this approach lie in the fact that, with a movement of the feed array along a defined locus, only an adjustment of phase shifters is required to achieve aberration-free scanning. The technique described here[7,8] is equally applicable to either transmission or reception, it provides a greater angle of scan than comparative systems, and requires no amplitude weighting of the intercepted energy to maintain optimum signal/noise performance of the antenna system.

2 Reflector electric-field distributions

2.1 Normally incident waves

It has been shown that the principal component of the electric-field distribution E in the focal region of a circular parabolic reflector can be related to the electric field F in the reflector-aperture plane by a scalar equation of the form[9,10]

$$E(t,\phi') = j\frac{k}{2\pi}\int_0^{2\pi}\int_0^{\hat{u}}\frac{F(u,\phi)}{\sqrt{(1-u^2)}}\exp\{jktu\cos(\phi-\phi')\}\,u\,du\,d\phi \quad (1)$$

where the geometry is given in Fig. 1, θ^* is the maximum halfangle subtended by the reflector from the geometric focus, $u = \sin\theta$, $\hat{u} = \sin\theta^*$ and $k = 2\pi/\lambda$, where λ is the operating wavelength.

While this equation is an approximation, it has been found to provide reasonable solutions for the principal components of the focal-plane electric-field distribution for paraboloids of any focal-length/diameter (f/d) ratio.[10]

For parabolic reflectors with a rectangular aperture, which will be pertinent to the following discussion, eqn. 1 may be expressed in rectangular co-ordinates as[9]

$$E(x,y) = jk\int_{-\infty}^{\infty}\int_{-\infty}^{\infty} G(p,q)\exp\{jk(xp+yq)\}\,dp\,dq \quad (2)$$

Paper 6444 E, first received 3rd September 1970 and in revised form 19th April 1971
Dr. Rudge and Prof. Withers were formerly with the Department of Electrical Engineering, University of Birmingham, Birmingham, England. Dr. Rudge is now with IIT Research Institute, 10 West 35 Street, Chicago, Ill. 60616, USA, and Prof. Withers is a visiting professor of telecommunications at the Instituto Technologico de Aeronáutica, São José dos Campos, São Paulo, Brazil

Reprinted with permission from *Proc. IEE*, vol. 118, pp. 857–863, July 1971.

where $p = u \cos \phi$
$q = u \sin \phi$
$x = t \cos \phi'$
$y = t \sin \phi'$

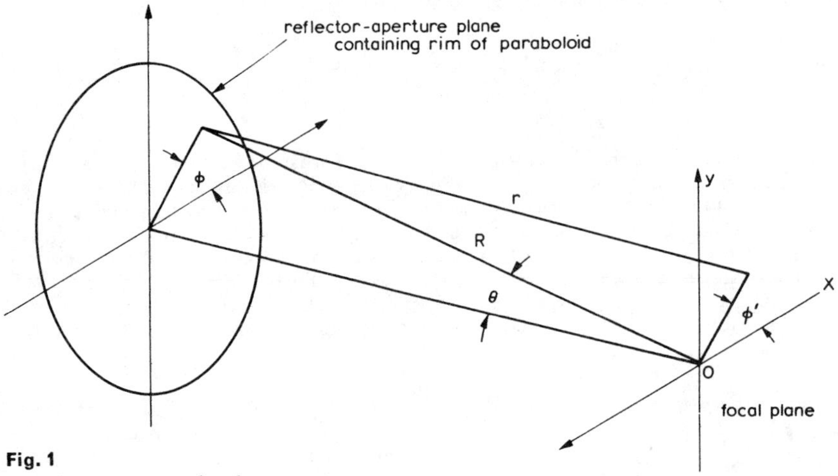

Fig. 1
Co-ordinate system employed

and the function G is defined as

$$G = \begin{cases} (1 - p^2 - q^2)^{-1/2} F(p, q) & |p| \leqslant \hat{p} \quad |q| \leqslant \hat{q} \\ 0 & |p| > \hat{p} \quad |q| > \hat{q} \end{cases} \quad \ldots \quad (3)$$

where $\hat{p} = \hat{u} \cos \phi$ and $\hat{q} = \hat{u} \sin \phi$

The function G has been termed the 'modified' aperture-field distribution since it can be considered as an amplitude-weighted version of the true aperture distribution.[9]

The existence of a spatial 2-dimensional Fourier-transform relationship between the modified-aperture-plane field distribution G and the focal-plane field E is indicated by eqn. 2.

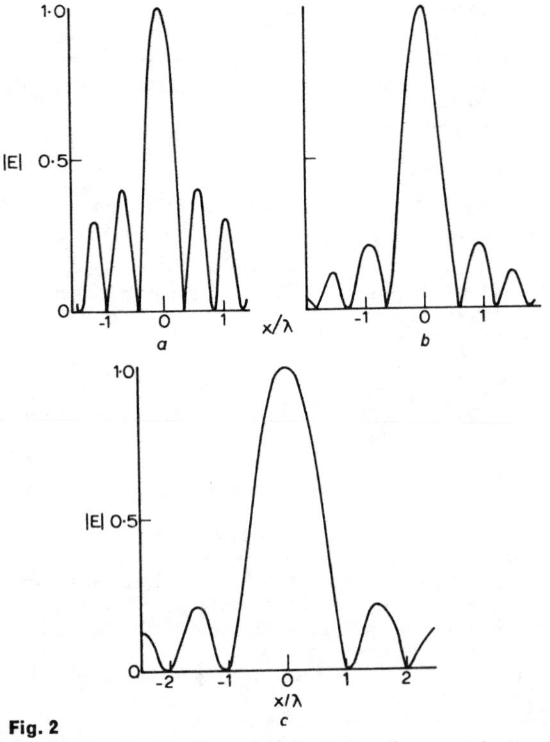

Fig. 2
Predicted focal-plane electric-field distributions from eqns. 4
a f/d = 0·25
b f/d = 0·5
c f/d = 1·0

For reflectors having f/d ratios greater than about 0·5, the transform relationship can be considered to exist directly between the reflector-aperture-plane field F and that of the focal plane.

When a linearly polarised plane wave is normally incident on a rectangular parabolic-reflector aperture, the aperture illumination observed from the reflector focal point has a uniform amplitude and phase distribution. The electric-field distribution along the principal axes of the focal plane may then be obtained from eqn. 2, putting the complex function F equal to a constant. The solutions take the form

$$E(x) \propto \begin{cases} \dfrac{2\hat{p} \sin kx\hat{p}}{kx\hat{p}} & \hat{p} < 0{\cdot}5 \\ \pi J_0(kx) & \hat{p} = 1{\cdot}0 \end{cases} \quad \ldots \quad (4)$$

and similarly for $E(y)$, replacing $\hat{p}$ with $\hat{q}$ by x and y. These functions are illustrated in Fig. 2.

2.2 Inclined incident waves

Eqns. 1–3 are not restricted to the case of a normally incident plane wave at the reflector aperture and may be usefully applied to incident waves inclined to the antenna boresight. However, for inclined incident waves, at a given angle of incidence, eqns. 1 and 2 express a relationship between the reflector-aperture plane and one of a set of transform planes which are normally inclined to the centre-line of the angular cone subtended by the perimeter of the reflector aperture (i.e. the angle subtended at the new 'focal point' must remain constant and equal to $2\theta^*$). This relationship is valid providing the original assumptions hold;[9, 10] in particular, that the spread of the energy in the transform plane is small compared with the distance separating this plane from the reflector-aperture plane, and that most of the energy is concentrated within the region about the new focal point. The locus of the transform planes is shown in Fig. 3.

The mathematical description of the function G, for a given angle of arrival of the incident wave, has still to be determined. We will first discuss the general effect on the reflector fields of an inclined incident wave.

Observed from the reflector geometric focus, the inclined incident wave effectively produces a reflector-aperture electric-field distribution which is nonuniform in both phase and amplitude. The aperture phase distribution β can be expressed as a series of the form[11]

$$\beta(a) = \beta_1 a + \beta_2 a^2 + \beta_3 a^3 + \beta_4 a^4 \quad \ldots \quad (5)$$

where $a = u/\hat{u}$ and β_n is the value of the respective phase term at the edge of the aperture. β_n will be a function of the angle of inclination of the incident wave and the reflector f/d ratio. The first four terms of the infinite series predominate and are commonly termed linear, focus, coma and spherical aberrations. Detailed analyses of these aberrations appear in the literature,[11-13] and it will be sufficient for our purpose to merely outline the effects of these phase deviations.

A linear phase distribution is not, in effect, an aberration, since it does not distort the focal-plane field distribution, but merely shifts it along the transverse axis. At points along this axis, however, the system geometry introduces both focus and

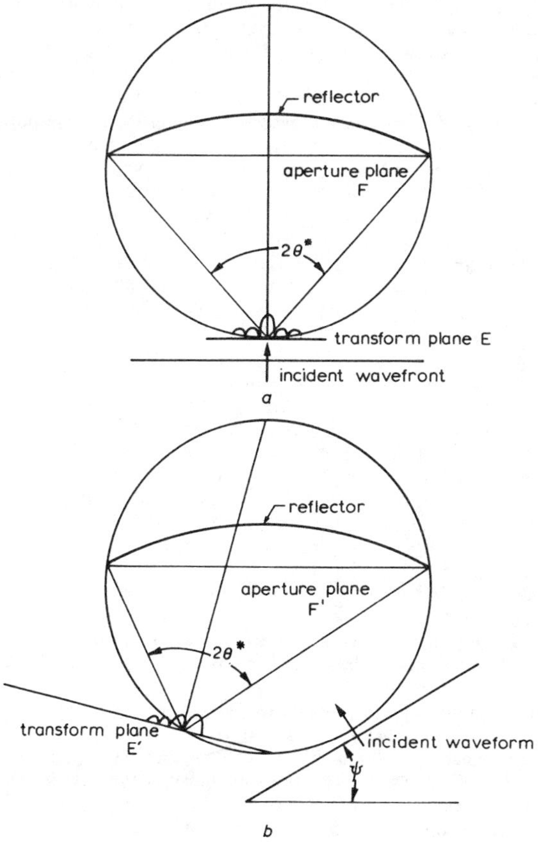

Fig. 3

Locus of transform planes with associated (typical) electric-field distributions

Circle diameter $= f\{1 + (d/2f)^2 \sec\theta^*$

coma aberrations. The focus error tends to defocus the field distribution in a symmetrical fashion, while the coma error introduces an asymmetrical distortion, which results in a rapid increase in the level of the minor lobes of the focal field distribution on one side of the main lobe.

To achieve even a small degree of beam steering in a high-gain fixed-reflector system, it is necessary to move the phase centre of the primary feed from its boresight location to obtain the required linear tilt in the phase distribution at the reflector aperture. However, once the primary feed is so moved, the effects described above are incurred, and, in addition to the desired linear term in the reflector-aperture phase distribution, there appear additional terms owing to focus, coma and spherical aberrations.

The focus and spherical terms are symmetrical errors and can be minimised by reducing the spacing between the reflector and the primary feed, thereby introducing compensatory phase errors by defocusing the aperture illumination. In addition, a primary-feed location can be determined which provides a relatively uniform aperture-amplitude distribution. Consequently, it is the coma phase error which initially constitutes the major limitation to achieving wide-angle beam steering.

To illustrate the effect of the coma phase error, consider a parabolic reflector curved in one dimension, with an f/d ratio not less than 0·5. We will consider a wave incident on the reflector aperture at an angle ψ_1 to the boresight and assume a transform plane with a phase centre which compensates for all but the coma aberration. Along the transverse axis of the transform plane x', the electric-field distribution will be given by[9, 10]

$$E(x') = K\int_{-\infty}^{\infty} G'(p) \exp(kx'p)dp \quad . \quad . \quad . \quad . \quad . \quad (6)$$

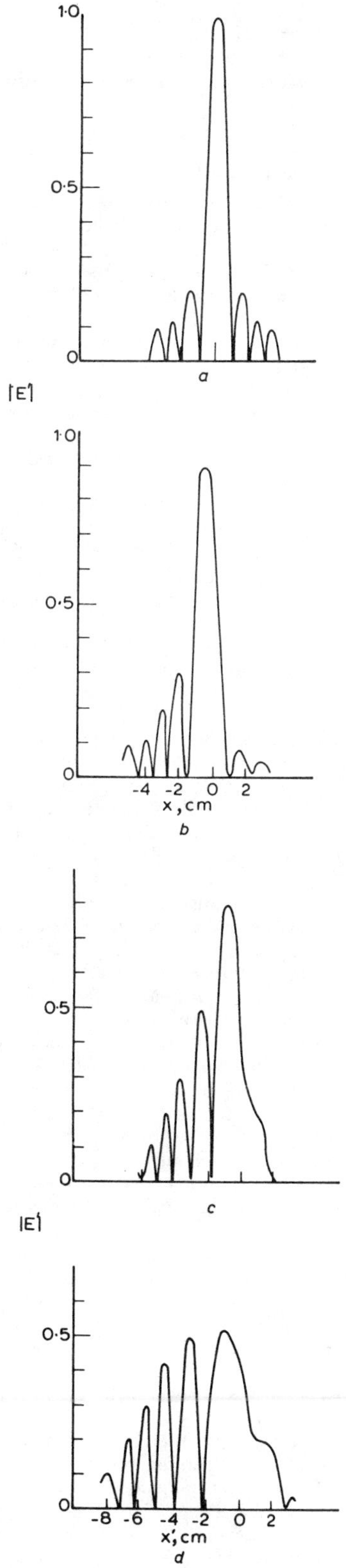

Fig. 4

Predicted electric-field distributions in transform planes, from eqn. 6

$\lambda = 3\cdot 2$cm, $f/d = 0\cdot 5$

a $\beta_3 = 0$
b $\beta_3 = 0\cdot 64\pi$
c $\beta_3 = 1\cdot 28\pi$
d $\beta_3 = 2\cdot 55\pi$

where K is a constant, and G' comprises a uniform amplitude distribution with a coma phase-error distribution of $\beta_3(\psi_1)$ at the edge of the aperture

$$G'(p) = \begin{Bmatrix} \exp\{j\beta_3(p/\hat{p})^3\} & |p| \leq \hat{p} \\ 0 & |p| > \hat{p} \end{Bmatrix} \quad . \quad . \quad . \quad (7)$$

Eqn. 6 has the form of an Airy function[14] which has been plotted elsewhere for another application.[15] The function is shown in Fig. 4 with $\lambda = 3\cdot2$cm and $\hat{p} = 0\cdot8$ for several values of $\beta_3(\psi_1)$.

3 Beam-steering technique

To design an efficient primary feed for a parabolic-reflector antenna, it is necessary to achieve a best-match condition between the electric-field distribution across the aperture of the primary feed on transmission and that field produced at the same surface by the reflector when it is illuminated by a distant source.[16] Applying this condition to the beam-steering problem, it is evident that, to achieve wide-angle beam steering, it is first necessary to move the primary feed to a region where it intercepts the energy from the reflector, and then effectively to adapt the primary-feed aperture field distribution, so that the required matching characteristics are obtained for the given angle of beam steering.

From Fig. 4, it can be seen that the variation in both the amplitude and phase distribution of the transform plane field at different β_3 (corresponding to different steered angles) is severe. Direct replacement of a conventional primary feed with a more extensive multielement array will thus demand an adaptive system with the capability of achieving combination of nonuniform signals. To avoid signal/noise-ratio degradation in the combining process, it is necessary to weight adaptively the gain of each array element with its own signal/noise ratio.[17]

To avoid these complications, an alternative approach has been proposed. The method employs the existence of the 2-dimensional spatial Fourier-transform relationship between the electric-field distributions in the modified aperture plane G and a set of transform planes. The technique consists of carrying out a second spatial Fourier transformation on the electric-field distribution in the relevant transform plane, so that, at the output of the transforming device, the amplitude of the field distribution always has the spatial form $|G|$ (*a* constant for a given reflector), while the phase distribution in this plane is an image of that in the reflector-aperture plane.

If the transforming device is followed by a suitable phase-shifting network and a combining matrix, compensation for aperture-plane phase errors can be made by means of phase-shifter adjustments at the primary feed, without adjustments to either the transforming-device or the combining-matrix components.[9] Beam steering can thus be achieved by a combination of movement of the primary feed along a defined locus with switching in preset values of phase in the phase-shifter network. The basic primary feed can be constructed as a passive device and can be used for either transmission or reception.

Employing an ideal Fourier-transforming device, i.e. one having infinite dimensions, the spatial distribution at the device output would be that given by eqn. 3, where p and q now correspond to linear distances across the output. Use of a finite device tends to impose an amplitude weighting on the output distribution, which is directly related to the reflector-aperture illumination taper produced by any finite primary feed. Although this weighting represents a decrease in reflector-aperture efficiency, it has practical applications in providing a reduction of sidelobe levels in the overall antenna radiation pattern. The effect of the amplitude weighting provides an additional bonus, in that the spatial output distribution of the finite transformer tends to be uniform and is thus in convenient form for combining under optimum signal/noise-ratio conditions.

A good approximation to the ideal Fourier transformation can be achieved provided that the collecting aperture of the finite device intercepts most of the energy in the transform plane. Examination of both Figs. 2 and 4 indicates that this can be achieved for both normal and inclined incident waves in the reflector aperture, with a primary-feed collecting aperture having dimensions of the order of wavelengths.

4 Operation of primary feed

For simplicity, the operation of the primary feed will be described in terms of a reflector curved in one dimension and having an f/d ratio of the order of $0\cdot5$. Consider Fig. 3*a*. The circle diagram illustrates a parabolic reflector with a normally incident plane wave. The array primary feed situated at the focus of the reflector 'sees' a uniform phase and amplitude distribution over the angle $2\theta^*$. The output of the primary-feed transforming device thus comprises a uniform phase and amplitude distribution, which may be summed directly to maximise the power output.

Now consider a plane wave incident at an angle to the reflector boresight. Viewed from the geometric focus, the reflector aperture has a nonuniform amplitude and phase distribution. The amplitude distribution is a consequence of the fact that, after reflection, the incident wavefront is no longer convergent in the immediate vicinity of the reflector geometric focus. If the primary feed is moved from the geometric focus along a locus which maintains the angle $2\theta^*$ constant, a point on this locus can be reached where the linear and focus phase errors and the nonuniform amplitude distribution introduced by the incident wave are largely compensated. Fig. 3*b* illustrates the configuration. However, while the observed aperture/amplitude distribution is now uniform over the angle $2\theta^*$, the phase distribution will still contain uncompensated phase-error terms resulting from the asymmetry of the configuration.

The electric-field distribution across the aperture of the primary feed will now be of the form shown in Fig. 4, while, at the output of the primary-feed transforming device, the field distribution will be that given by eqn. 7; i.e. a uniform amplitude distribution with a phase distribution containing a dominant coma term, which can be compensated by a phase-shifting network prior to summation. In this fashion, the array primary feed is capable of achieving beam steering over a range of angles, limited only by the constraint that the spread of energy in the relevant transform plane should not exceed the collecting aperture of the primary feed.

5 Experimental system

An experimental primary-feed system was constructed completely in X band waveguide for operation at 10GHz in conjunction with a 1-dimensionally curved parabolic reflector having a diameter of $1\cdot8$m and an f/d ratio of $0\cdot5$. The feed system, which has been described in an earlier publication,[9] is shown in block-schematic form in Fig. 5. The system comprises a linear array of eight waveguide feeds with an interelement spacing of $\lambda/2 \sin\theta^*$, followed by an 8-port Butler matrix, to provide a sampled spatial Fourier transformation.[9] Each of the matrix output ports is taken via an adjustable phase shifter to the combining matrix. The Butler matrix and the combining matrix are constructed individually as 1-piece units by a dip-brazing technique.[18] Overall, the experimental feed system introduces an insertion loss of the order of 1dB, this being largely attributable to the imperfect construction of the r.f. components forming the matrices.

The primary-feed subassembly was mounted on a movable carriage, to permit the aperture plane of the array to be moved around the locus shown in Fig. 3, and the combination of reflector and primary-feed assembly was rotatable, to allow measurements of the overall antenna radiation pattern.

For measurements of the electric-field distributions in the transform planes, the linear array was replaced by a single waveguide feed which was positionally coupled to an *xy* recorder. The single feed was then mechanically scanned across the relevant transform plane, and the measured electric-field spatial distribution recorded.

6 Measurements

The proposed beam-steering technique is based on the existence of the spatial Fourier-transform relationship between the electric-field distribution in the reflector modified-aperture plane and that of a defined transform plane. In addition to measurements of the system performance, it was considered desirable to obtain experimental verification of the theory on this point. This verification was achieved by means of additional field measurements in the transform planes. The measured field distributions in the transform planes were compared with the fields predicted in Fig. 4 and with the radiation-patterns of the overall antenna. An explanation of the radiation-pattern comparison follows.

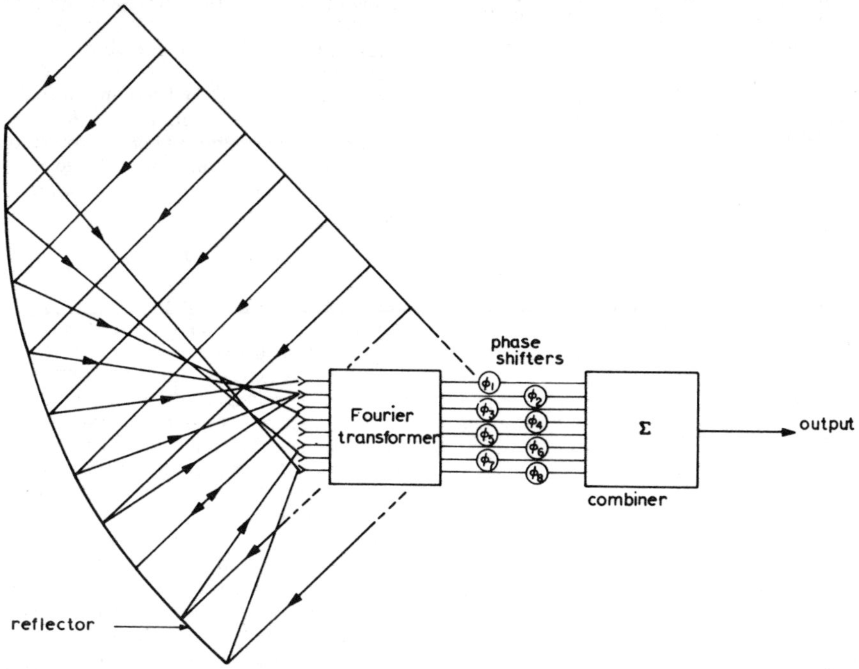

Fig. 5
Adaptive-primary-feed block schematic

It is a well established fact that the far-field radiation pattern of a parabolic-reflector antenna (in sin ψ measure) is related to the reflector-aperture field distribution by a spatial Fourier transformation.[1] Thus the antenna radiation pattern measured with a single-element primary feed illuminating the reflector from a point on the $2\theta^*$ locus (with the feed aperture plane aligned with the relevant transform place at that point) will constitute the Fourier transformation of the reflector-aperture field distribution including the coma phase errors arising from the geometrical asymmetry. The main beam of the antenna will appear at an angle ψ_1 to the boresight. If now, with the antenna aligned so that the target transmitter is situated at the same angle ψ_1 to the reflector boresight, a travelling feed is employed to measure the field distribution in the relevant transform plane (about the point at which the single feed had been located), for reflectors having f/d ratios of not less than about 0·5, this distribution should be an identity with the radiation pattern previously measured, the two distributions being related by a double Fourier transformation.

Measurements of the system performance comprised antenna-radiation-pattern measurements and power-output-level measurements using the array primary feed and steering the radiation pattern over angles of up to ± 15 beamwidths. The reflector system was positioned so that the target transmitter made an angle of ψ to the reflector boresight. The primary-array feed was moved along the $2\theta^*$ locus and locked initially at the point giving the best output power level. The array-feed phase shifters were then adjusted to maximise the output power. No significant interaction between the phase shifters was observed, as would be expected in view of the orthogonality of the beams formed by the matrix. The radiation pattern of the antenna was then recorded by rotating the complete antenna assembly. The array feed was then moved incrementally on the $2\theta^*$ locus, and the procedure repeated until the optimum radiation pattern with respect to gain and sidelobe levels was obtained. The power level at the system output was calibrated by comparison with the output from a reference horn antenna.

With the array feed removed, the field distribution in the relocated transform plane was recorded on the xy plotter by traversing the travelling feed in the plane defined by the aperture of the array. Finally, with the travelling feed at the centre of the relevant transform plane, the antenna radiation pattern was again recorded by rotating the assembly. Measurements were made at offset angles ψ of 0, 5, 10 and $\pm 15°$.

Some experimental results are shown in Fig. 6. The critical results for $\psi = 0$ and 15° are given, as these illustrate the range of the experiments. The results are shown for each ψ in the form

(*a*) antenna radiation pattern with a single-element primary feed
(*b*) relevant transform-plane field distribution
(*c*) antenna radiation pattern employing array primary feed.

7 Experimental results

The measure transform-plane field distributions were in good general agreement with these predicted by eqn. 6. In addition, the similarity of the relevant antenna radiation patterns employing the single-element feed and the associated transform-plane field distributions, for all ψ employed, provided further experimental support for the predicted double-Fourier-transform relationship (see Figs. 2, 4 and 6). For reflectors of very small f/d ratio, the antenna radiation pattern and the transform-plane field distribution will be related by the 'modified'-aperture distribution. In such cases, the transform-plane field will be modified and will no longer be directly similar to the antenna radiation pattern. However, for the f/d ratio of 0·5 employed in the experimental system, the curvature effect will be small, and, consequently, considerable similarity can be expected between the two distributions.

The results obtained from the beam-steering measurements were very encouraging. The antenna radiation pattern was steered over a range of $\pm 15°$ (or ± 15 beamwidths) with less than 0·5dB reduction in the system gain. Fig. 6 shows a redistribution of power in the sidelobe distribution at 15°, but the peak level of the sidelobe distribution has not changed significantly. The slightly higher sidelobe levels in the antenna radiation patterns in all cases involving use of the array feed are a result of the aperture-shadowing effect of the array hardware. This would be reduced to a negligible level in a practical configuration involving a large reflector, since the dimensions of the primary-feed aperture are governed by the f/d ratio, rather than the diameter of the reflector. Hence, to achieve ± 15 beamwidths of scanning at 10 GHz with a reflector having $f/d = 0 \cdot 5$ demands a primary-feed aperture of the order of 20 cm diameter, regardless of the diameter of the main reflector.

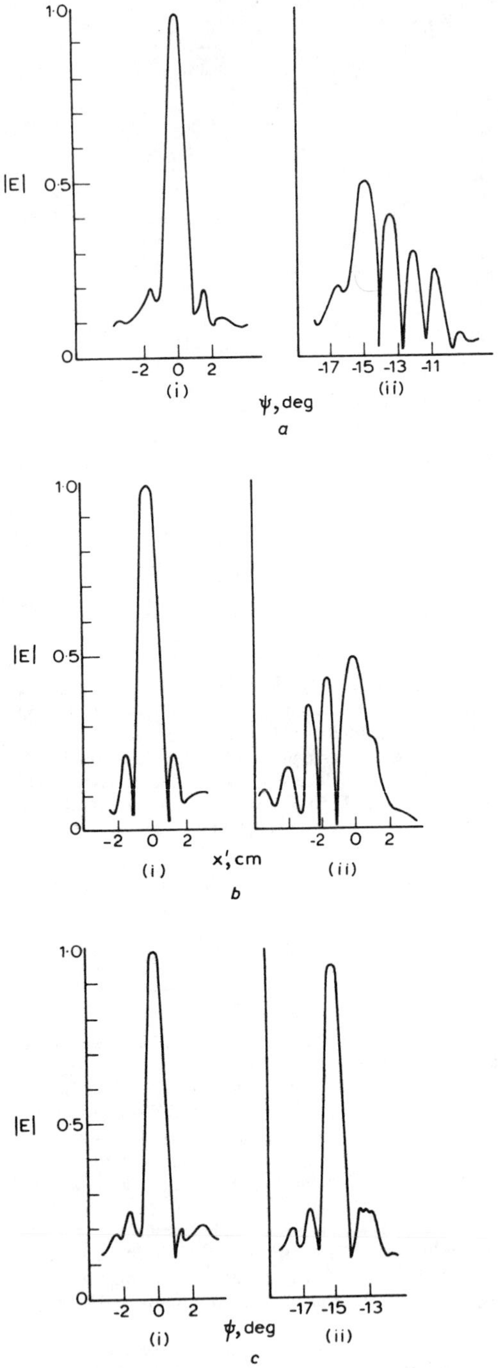

Fig. 6

Experimental results

a Steered antenna radiation patterns with conventional single-element primary feed
(i) Feed on axis
(ii) Feed offset
b Transform-plane field distributions
(i) Focal plane with target transmitter along antenna boresight
(ii) Relocated transform plane with target transmitter offset 15°
c Steered-antenna radiation patterns with adaptive primary feed

$\lambda = 3 \cdot 2$ cm, $f/d = 0 \cdot 5$

(i) Primary feed in focal plane
(ii) Primary feed offset

8 Conclusions

The beam-steering technique developed shows considerable promise with respect to achieving wide-angle beam steering of parabolic reflectors. While, owing to mechanical limitations, the experimental system did not exceed ± 15 beamwidths of beam steering, this does not constitute a fundamental limitation on the technique, nor does it necessarily represent the maximum range obtainable with the 8-element array feed. Further work is planned to determine the fundamental limitations and to determine an analytical relationship between the dimensions of the collecting aperture of the array feed and the maximum obtainable range of beam steering. From a superficial examination based on the geometry of Fig. 3, the overall limits of the beam-steering technique appear to be given by

$$\psi^* = 90 - \theta^* \quad . \quad . \quad . \quad . \quad . \quad . \quad . \quad . \quad . \quad . \quad (8)$$

where ψ^* is the maximum steered angle from the antenna boresite. Verification of this equation, however, awaits the outcome of further studies.

Any fixed-aperture antenna system will suffer from a reduction of gain proportion to $\cos^2 \psi_1$ when the antenna radiation pattern is steered to an angle ψ_1 from the boresight. At 15 beamwidths, with the experimental system, this would result in a $0 \cdot 3$ dB reduction in gain, which would, within the experimental tolerance, account for virtually all the $0 \cdot 5$ dB gain loss measured.

In many applications of large-reflector antennas (e.g. radar, radioastronomy and satellite communications), it is often desirable to have the facility of rapidly changing the direction of the antenna radiation pattern at electronic switching speeds. Early results of experiments aimed at the evaluation of purely electronic beam steering (i.e. by electronically controlling the adjustment of the phase shifters only) indicate that the range of such a technique is of the order of a few beamwidths. However, this would be sufficient to achieve conical scanning of the beam.

In addition to conventional beam-steering applications, the techniques described would be particularly suitable where an offset primary-feed configuration for a parabolic reflector is desired. Such a configuration would permit more than one independent radiation pattern to be formed by employing two or more primary-array feeds on the one reflector.

In the interest of clarity, the description of the beam-steering technique has been largely confined to reflectors curved in one dimension. Nevertheless, the basic technique may be applied similarly to the case of 2-dimensionally curved surfaces. Application of the spatial-transform technique to a 2-dimensionally curved reflector demands a 2-dimensional Fourier transformer. In this case, the use of alternative methods of achieving the spatial transformation, and, in particular, the use of a second reflector or a microwave lens, appears attractive in offering a reduction of the complexities involved in the construction of a large planar matrix.

9 Acknowledgments

The authors wish to thank D. E. N. Davies for valuable discussions and EMI Electronics Ltd., Feltham, England, for the loan of the Butler matrices. Thanks are also due to the University of Birmingham, England, IIT Research Institute, Chicago, Ill., USA, and the Instituto Tecnológico de Aeronáutica, São José dos Campos, Brazil, for facilities to undertake the work.

10 References

1 SILVER, S.: 'Microwave antenna theory and design' (McGraw-Hill, 1949)
2 JASIK, H.: 'Antenna engineering handbook' (McGraw-Hill, 1961)
3 TAKESHIMA, T.: 'Beam scanning of parabolic antenna by defocusing', *Electronic Engng.*, 1969, pp. 70–72
4 HANNAN, P. W.: 'Microwave antennas derived from Cassegrain telescopes', *IRE Trans.*, 1961, **AP-9**, pp. 140–153
5 LOUX, P. C., and MARTIN, R. W.: 'Efficient aberration correction with a transverse focal plane array technique', *IEEE Internat. Convention Record*, 1964, **12**, p. 125

6 ASSALY, R. N., and RICARDI, L. J.: 'A theoretical study of a multi-element scanning system for a parabolic cylinder', *IEEE Trans.*, 1966, **AP-14**, p. 601

7 RUDGE, A. W.: British Provisional Patent 23145/68, May 1968

8 RUDGE, A. W., and WITHERS, M. J.: 'Beam-scanning primary feed for parabolic reflectors', *Electron. Lett.*, 1969, **5**, pp. 39–41

9 RUDGE, A. W., and DAVIES, D. E. N.: 'Electronically controllable primary feed for profile-error compensation of large parabolic reflectors', *Proc. IEE*, 1970, **117**, (2), pp. 351–358

10 RUDGE, A. W.: 'Focal-plane field distribution of parabolic reflectors', *Electron. Lett.*, 1969, **5**, pp. 510–512

11 WELFORD, W. T.: 'Geometrical optics, optical instrumentation' (North Holland, 1962)

12 BORN, M., and WOLF, E.: 'Principles of optics' (Pergamon Press, 1964)

13 LUM, Y. F., and PAVLASEK, T. J. F.: 'The influence of aberrations and aperture inclinations on the phase and intensity structure in the image region of a lens', *IEEE Trans.*, 1964, **AP-12**, pp. 717–727

14 ABRAMOWITZ, M., and STEGUN, I.: 'Handbook of mathematical functions' (Dover, 1964), p. 446

15 KRAMER, S. A.: 'Doppler and acceleration tolerances of high-gain, wideband linear f.m. correlation sonars', *Proc. Inst. Elect. Electron. Engrs.*, 1967, **55**, pp. 627–636

16 RUDGE, A. W., and WITHERS, M. J.: 'Design of flared-horn primary feeds for parabolic reflector antennas', *Proc. IEE*, 1970, **117**, (9), pp. 1741–1749

17 BRENNAN, D. G.: 'On the maximum signal-to-noise ratio realizable for several noisy channels', *Proc. Inst. Radio Engrs.*, 1955, **43**, p. 1530

18 THRAVES, J.: 'An X-band Butler matrix array'. Proceedings of the 12th AGARD Symposium, Dusseldorf, 1966

Determination of the Maximum Scan-Gain Contours of a Beam-Scanning Paraboloid and Their Relation to the Petzval Surface

WILLARD V. T. RUSCH AND ARTHUR C. LUDWIG

***Abstract*—The scan-plane fields in the focal region of a beam-scanning paraboloid are determined from physical optics. Amplitude and phase contours are presented, and comparisons are made with the geometrical-optics results. Contours for maximum scan-gain are determined as a function of F/D and illumination taper and compared with the Petzval surface. Unless the F/D is very large or spillover is excessive, a higher scan gain is achieved when the axis of a directional feed is parallel to the axis of the reflector than when the feed is directed toward the vertex. The contour of maximum scan-gain is a function of both illumination taper and F/D. In general, larger F/D values tend to have a maximum-gain contour close to the focal plane, while the smaller F/D values tend to have a maximum-gain contour closer to the Petzval surface. Increasing the illumination taper moves the maximum-gain contour closer to the Petzval surface. Normalized maximum-gain contours are presented as a function of beamwidths of scan. The frequency dependence of these results is discussed.**

I. Introduction

PARABOLOIDAL antennas in radar, radio-astronomy, and microwave communication systems frequently employ lateral displacement of the feed to achieve a beam-scanning capability. Features of beam scanning achieved in this manner have been dealt with in [1]–[7]. However, the simple but important problem of determining the proper position and orientation of the feed to achieve maximum scan gain still remains relatively unresolved. This paper describes a numerical study of that problem.

In the receive mode, a linearly polarized plane wave, incident from some direction off axis, includes currents on the reflector which then give rise to a particular field distribution in the focal region. In the transmit mode, a linearly polarized directional point-source feed is conceptually placed at some point in the focal region in order to scan the far-zone transmitted beam off axis. These two situations are related by the reciprocity principle [8], [9], which is particularly simple when the transmitting feed is an infinitesimal dipole. For actual feeds the relationship is more complex, but it can be shown that the efficiency of an antenna is given by the correlation (over any closed surface) between the focal-region fields of the reflector receiving an incident plane wave and the fields of the transmitting feed [10], [11]. Thus the focal-region fields correspond to the optimum feed aperture distribution. This fact has been used with considerable success in the development of corrugated feeds, for example [10], [12]. Rudge has also applied this approach to the design of an array feed for a beam-scanning paraboloid, where the feed aperture fields are synthesized to match the focal region fields on a given surface [13]. The primary objective of this paper is to determine the optimum orientation and location of displaced feeds. This may be interpreted in terms of correlation as moving the feed aperture until the feed fields in the aperture most effectively match the focal region fields over that same surface.

II. Analysis by Geometrical Optics

Geometrical ray tracing has been used by several authors to determine surfaces of sharpest focus for scanned paraboloidal reflectors [14]–[19]. Fig. 1 indicates the reflected rays in the plane of scan[1] of a paraboloid receiving a plane wave from 16° off axis. If only single reflections are considered, two of these reflected rays will pass through a given point in the plane of scan. The envelope of these rays (AOB in Fig. 1) forms a caustic curve in the plane of scan [20]. This caustic will also define a forbidden region within which no rays will pass. Consequently, the geometrical-optics (GO) field is zero within this region.

A second caustic, COD (sometimes referred to as a "ridge line"), also forms a boundary beyond which no reflected rays from out of the plane of scan will cross the plane of scan. As a result, four rays will pass through every point in the plane of scan below COD in Fig. 1, two rays from points of reflection in the plane of scan, and two rays from points of reflection out of the plane

Manuscript received February 22, 1972; revised August 14, 1972. This work was supported by NASA under Contract NAS7-100.

W. V. T. Rusch is with the Department of Electrical Engineering, University of Southern California, Los Angeles, Calif. He is also a Consultant for the Jet Propulsion Laboratory, California Institute of Technology, Pasadena, Calif. 91103.

A. C. Ludwig is with the Jet Propulsion Laboratory, California Institute of Technology, Pasadena, Calif. 91103.

[1] The plane of scan contains the incident wave normal and the reflector axis.

Reprinted from *IEEE Trans. Antennas Propagat.*, vol. AP-21, pp. 141–147, Mar. 1973.

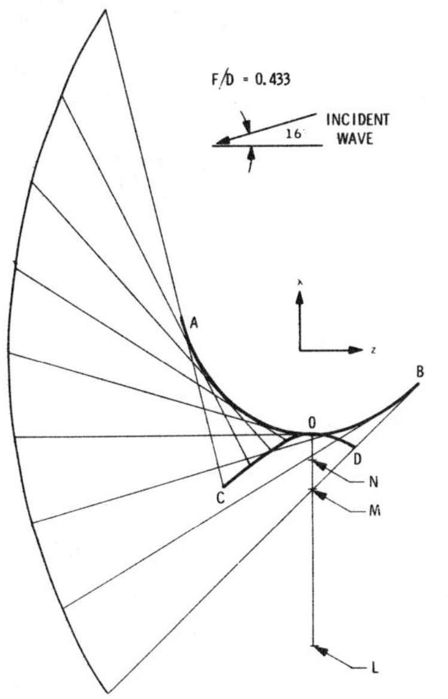

Fig. 1. Reflected rays in plane-of-scan of paraboloid receiving plane wave from 16° off axis.

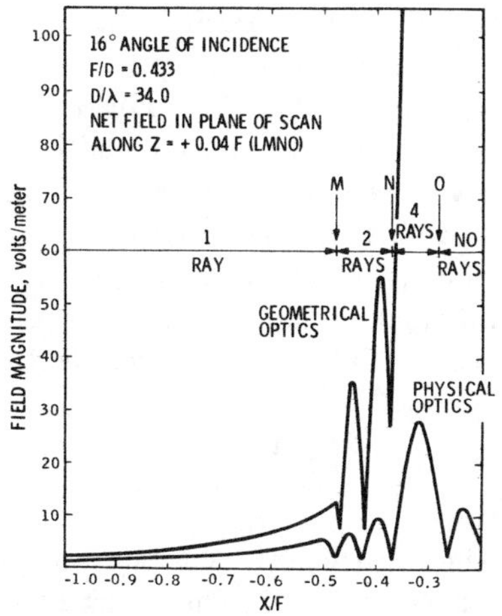

Fig. 2. Comparison of geometrical-optics and physical-optics fields.

of scan, one from above and one from symmetrically below. However, truncation of the reflector may limit this number to as few as one.

Superposition of the contributions of each ray passing through a point will give the total GO field at that point. The field intensity calculated from GO and physical optics (PO) is plotted in Fig. 2 for the line $LMNO$ of Fig. 1. In general, the GO field is significantly larger than the PO field. From L to M, only one geometric ray crosses the path and the field shows a slow, monotonic increase. From M to N two rays cross the path and two interference fringes are observed in both the GO and PO intensities. From N to the caustic COD four rays cross the path: the GO intensity becomes unrealistically large as the caustic is approached, while the PO field exhibits its primary maximum. Beyond O the GO field ceases to exist, while the PO field drops to smaller but nonzero values. Such results may be easily obtained using GO for a wide range of parameters which are useful in locating the general position of the most intense focal-region fields and for grossly characterizing these fields. However, the classical GO analysis is incapable of yielding useful intensity information on the focal-region fields.

III. Analysis by Physical Optics

The analytical PO technique has been used to produce the numerical results presented in this paper. PO assumes the surface currents on the reflector to be approximated by the GO currents; the free-space dyadic Green's function is then integrated over these currents to obtain the fields. Analytic justification for the use of this technique may be found in [21]. Only paraboloidal reflectors are considered, and such parasitic effects as aperture blocking, surface irregularities, etc., are not included in the analysis.

It has been pointed out in the literature that, except under special circumstances, the PO approximation fails to satisfy the reciprocity principle [22]. One set of special conditions for which the PO fields do satisfy the reciprocity principle is the focused paraboloid and a distant observation point on the reflector axis [23]. However, the fields considered primarily in this paper arise from defocused conditions. The reciprocal properties of these defocused PO fields were examined numerically in the following manner:

1) An infinitesimal unit-amplitude electric dipole was conceptually placed at point (x,y,z) in the focal region of the paraboloid. $E_t(x,y,z,\theta,\phi)$, the resulting PO field—exclusive of the $(\exp(-jkr)/r)$ factor—radiated to a distant field point in the direction (θ,ϕ), was numerically computed as described elsewhere [24].

2) An incoming unit-amplitude vector plane wave from the direction (θ,ϕ) then illuminated the reflector and the resulting field at (x,y,z), $E_r(x,y,z,\theta,\phi)$ was numerically computed.

The reciprocity principle requires that[2]

$$E_t(x,y,z,\theta,\phi) = E_r(x,y,z,\theta,\phi). \tag{1}$$

For all sets of parameters of the 34-wavelength paraboloid considered in this paper, the two fields agreed to within 0.4 percent or better, usually to within 0.2 percent. How much of this small difference is due to different numerical integration algorithms is unknown. It is felt that, on the basis of this evidence, the PO solution and the computer programs used were sufficient for the accuracies required for this study.

A. Scan-Plane Fields in Focal Region of Receiving Paraboloid

Phase and amplitude contours in the focal region plane of scan of a paraboloid receiving an incident plane wave are plotted for 3 different angles of incidence in Fig. 3. The electric vector is perpendicular to the plane of scan.

[2] A complete statement of reciprocity includes the currents on the reflector. However, the tangential E field is zero on the conductor, and (1) results.

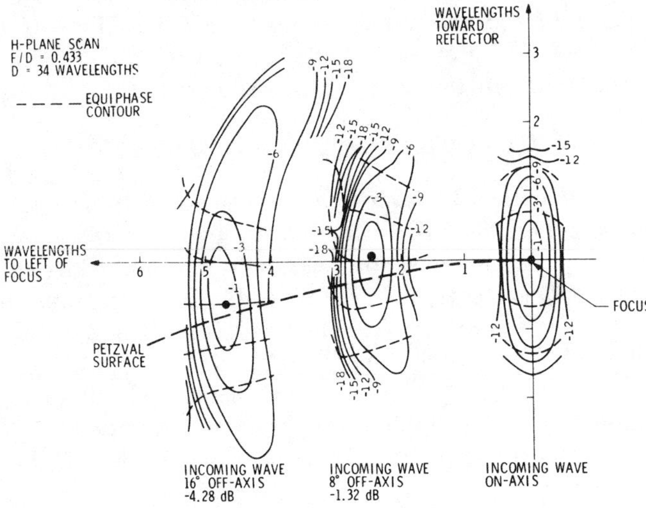

Fig. 3. Phase and amplitude contours in focal region plane-of-scan of receiving paraboloid.

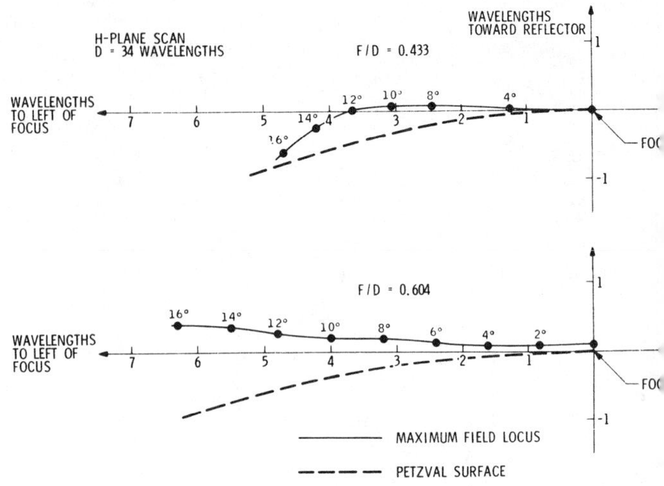

Fig. 4. Comparison of maximum field locus and Petzval surface for receiving paraboloid.

The origin is the reflector focus. The coordinate axes are the reflector axis marked in wavelengths toward the reflector (the vertex of which is 14.7 wavelengths from the origin) and a line at right angles to the reflector axis marked in wavelengths to the left of the focus. When the incident wave travels along the reflector axis, the set of contours surrounding the focus is generated. These contours are similar to results for the normal-incidence case [12], [25]. However, the phase is retained in its exact form [26], [27]. Consequently, the contours are not symmetrical about the focal plane, and closely agree with the physically measured values [28]. The maximum received field intensity occurs at the focus. The equiphase contours are separated by 180 electrical degrees. It is evident from the figure that, in terms of an equivalent wave travelling along the reflector axis, the effective wavelength is greater than the actual wavelength.

When the incident field arrives from a direction to the right of the reflector axis, the phase and amplitude contours shift to the left as shown. The maximum intensity points for 8° and 16° incidence are, respectively, 1.32 and 4.28 dB below the original maximum intensity for axial incidence. The amplitude and phase contours distort and disperse. At the maximum intensity point for each angle of incidence, the constant phase contour is approximately parallel to the aperture plane, i.e., perpendicular to the reflector axis.

Reciprocity permits the contours in Fig. 3 to be interpreted in terms of an infinitesimal electric dipole feed polarized perpendicular to the plane of scan, transmitting to a distant field point. For example, if the distant field point is eight degrees on the right side of the reflector axis, the set of amplitude and phase contours for eight degrees will yield the phase and magnitude of the field observed at that distant point. Placing the dipole at the maximum intensity point will transmit the maximum possible field to the distant point; this field, however, will be 1.32 dB below the field radiated to a distant axial field point when the dipole transmitting feed is at the focus.

The Petzval surface, a term from classical optics, is the surface of best focus in an optical system in the absence of astigmatism [29]–[31]. For a single mirror the radius of curvature of the Petzval surface is one half the radius of curvature of the mirror [31]. In the primary microwave reference to the subject, the Petzval surface of a paraboloidal mirror is derived by Ruze [7] to be another paraboloid of half the focal length, tangent to the focal plane at the focus, and described by the equation

$$\left(\frac{X}{\lambda}\right)^2 = 2\frac{F}{\lambda}\frac{Z}{\lambda} \tag{2}$$

where F is the focal length, λ is the wavelength, and X and Z are defined in Fig. 1. The Petzval surface is superimposed on Fig. 3. It comes close to, but does not pass through, the scan maxima.

Plots of the scan maxima, without the corresponding amplitude and phase contours, are shown in Fig. 4 for $F/D = 0.433$ and $F/D = 0.604$. The remaining parameters are the same as for Fig. 3. The appropriate Petzval surfaces for the two focal lengths are also plotted. As before, these curves can be interpreted in either the transmit or receive mode. The plot for $F/D = 0.433$ contains the same information as Fig. 3, except that additional scan maxima for 4°, 10°, 12°, and 14° are included. For $F/D = 0.604$ the maximum-field locus remains relatively far from the Petzval surface for scan angles as large as 16°, while the contour for $F/D = 0.433$ droops toward the Petzval surface more rapidly.

For axial incidence the maximum field intensity for $F/D = 0.604$ does not occur at the focus but a small fraction of a wavelength toward the reflector. This effect can best be understood in terms of an infinitesimal electric dipole transmitting feed illuminating the reflector. By moving the feed away from the focus, perfect phase synchronism is lost for a distant point on axis. However, by moving the feed closer to the reflector, more power is intercepted by the reflector and becomes available for secondary radiation. Thus the phase loss is initially overcome by a decrease in spillover, and the maximum occurs closer to the reflector. This effect, which is exclusively geometrical, becomes less pronounced for the deeper reflector ($F/D=0.433$). This example, however, illustrates the fact that the microwave problem is generally more complex than the scalar-optical problem. In addition to phase effects, the vector nature of the fields, the directionality of the sources, spillover, etc., must be taken into account.

For a paraboloidal mirror, the S (sagittal) focal surface from third-order Seidel aberration theory [31] is the focal plane, and the T (tangential) focal surface lies between the focal plane and the mirror. The T surface is three times as far from the Petzval surface as the S surface [31], [33]. The best compromise focus of an optical system is expected to lie between the S and T surfaces. In Fig. 4 the maximum field locus for $F/D=0.433$ out to 12° and the entire locus for $F/D=0.604$ satisfies this criterion. It will be shown later that at low illumination tapers and small scan angles the maximum field locus lies between the S and T surfaces. However, as the taper and/or the scan angle increases, the maximum gain locus crosses the focal plane and droops toward the Petzval surface defined in (2). This difference between the third-order optical and microwave vector diffraction theory results is attributed to: 1) the breakdown of third-order aberration theory at wide angles; 2) optical systems usually have a much larger f value (F/D ratio) than antenna reflectors, for which values may be as small as 0.25; and 3) the unique differences associated with the complete vector result not included in scalar optics, e.g., as illustrated in the previous paragraph.

B. Orientation of Directional Transmit Feed to Achieve Maximum Gain

The remainder of the paper is concerned with the problem of positioning a directional transmit feed to achieve maximum scan gain. The class of feed patterns to be considered has a spherical phase front and a linearly polarized, axially symmetric voltage pattern of the idealized form $\cos^n \theta$, where θ is the polar angle measured from the axis of the pattern maximum.

From a purely geometrical point of view, optimum gain for a given position of the feed (when scanned) is achieved by orienting the feed pattern to minimize spillover beyond the edge of the dish. Generally speaking, this is achieved by directing the feed maximum toward the reflector vertex. However, analysis has consistently revealed that unless spillover is excessively large (of the order of dB's), a *higher gain is achieved for moderate F/D values when the feed axis is parallel to the axis of the reflector than when the feed is pointed toward the vertex.* For example, if $F/D = 0.433$, $D/\lambda = 34.0$, and a $\cos^2 \theta$ feed (14.5 dB taper at the reflector edge), a maximum scanned gain at a scan angle of 20° (approximately 10 beamwidths of scan) is achieved if the feed is positioned 5.8 wavelengths laterally from the reflector axis (1/3 of the distance to the edge of the aperture). If the feed is directed parallel to the reflector axis, i.e., "straight ahead," the spillover loss is 0.19 dB greater than when the feed is directed toward the reflector vertex. However, the scan gain for the former orientation is 0.85 dB higher than for the latter orientation. For smaller scan angles the scan gain is also larger for the straight-ahead orientation, although the magnitude of the effect is proportionately reduced. Kelleher and Coleman [3] have pointed out that, based on geometrical ray tracing, the straight-ahead orientation yields an aperture field centered symmetrically in the aperture.

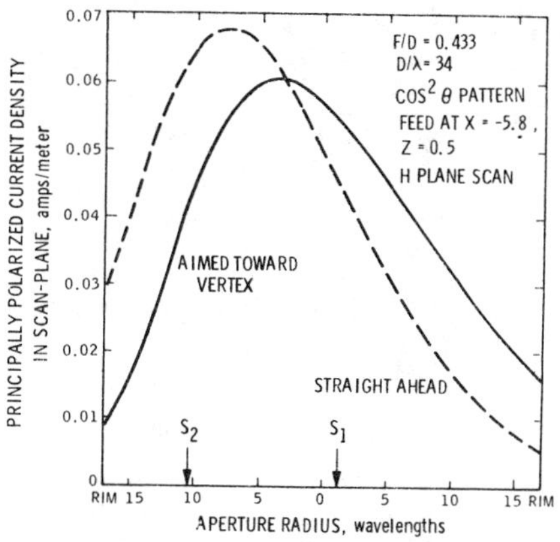

Fig. 5. Comparison of induced current density distributions for two orientations of feed.

An explanation of the straight-ahead orientation yielding maximum scan gain is based on the constant-phase contours in Fig. 3. The contours through the maximum scan intensity points are nearly perpendicular to the reflector axis. Consequently, by the reciprocity/correlation principle, the scan gain will be maximized when the phase front of the feed is parallel to this. A more quantitative derivation is contained in the Appendix.

It is also instructive to examine the components of reflector current for the two orientations of the feed. The principally polarized components of the induced current density in the scan plane are plotted in Fig. 5 for an H-plane scan with the two different orientations of the feed. The straight-ahead orientation yields a maximum current density that is 1.00 dB higher than the current density produced by the vertex-look orientation. (For an E-plane scan the difference is 0.95 dB). The two scan-plane stationary points (i.e., points of geometrical reflection) for that feed position-observation angle combination are indicated by S_1 and S_2, neither being particularly close to either current maximum. Since the relative phases of the contributions to the total field by various parts of the current distribution is a complicated function of position, the interpretation of these relative maxima is not completely clear. However, they probably account for the fact that the straight-ahead orientation produces 0.85 dB higher gain in spite of having 0.19 dB more spillover, e.g., compare $1.00 - 0.19 = 0.81$ dB versus

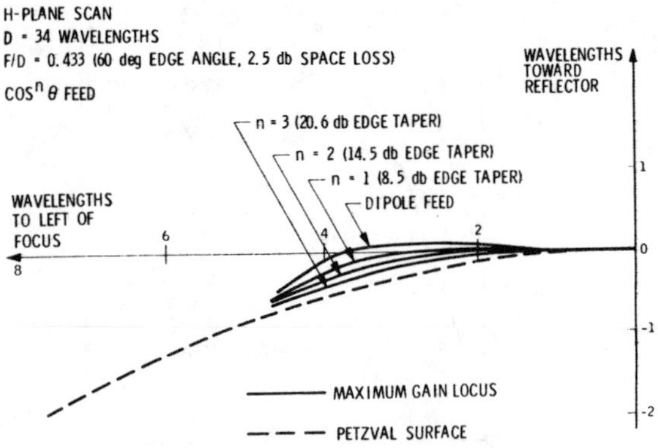

Fig. 6. Comparison of maximum transmit-gain contours and Petzval surface; $F/D = 0.433$.

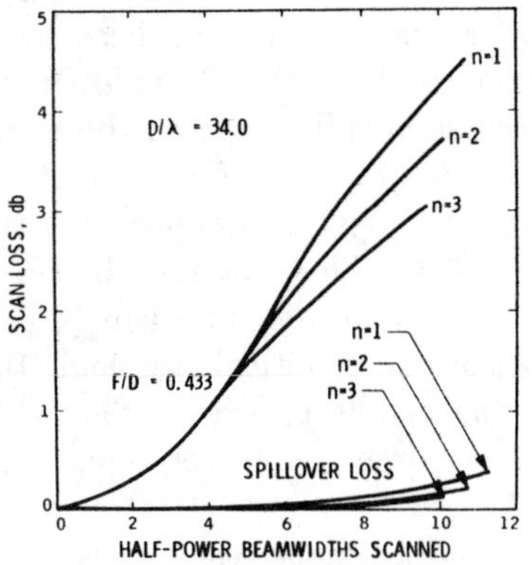

Fig. 7. Beam-scan losses; $F/D = 0.433$.

0.85 dB. Consequently, all maximum gain contours computed in this paper are for the feed axis directed parallel to the axis of the reflector.[3]

C. Location of Transmit Feed to Achieve Maximum Gain

Fig. 6 shows the maximum scan-gain contours for $F/D = 0.433$ and the four feed functions $n = 1,2,3$ and infinitesimal dipole. Fine structure of the order of 0.1 dB peak-to-peak has not been included. The dipole curve is identical to the maximum receive-field contour in Fig. 4. The Petzval surface is also plotted in the figure. It is evident that the maximum-gain contours are slightly on the focal-plane side of the Petzval surface, although the difference is a relatively small fraction of a wavelength. As the scan angle is increased, or the edge taper is increased, the maximum-gain contours approach the Petzval surface.

Fig. 7 presents the scan losses along the maximum-gain contours. The abscissa is half-power beamwidths (HPBWs) scanned for the 34-wavelength aperture. The scan-loss curves *do* include spillover loss, since spillover occurs as the feed is scanned laterally while pointing straight ahead. Also plotted in the figure, however, is the component of loss due exclusively to the decreasing fraction of the total power intercepted by the reflector as the feed is scanned. This spillover loss amounts to a fraction of a dB, while the remaining scan loss is due to phase defocusing and other effects.

Fig. 8 presents a similar set of maximum-gain contours for the same class of feeds but an F/D value of 0.604. Essentially, the same phenomena take place as for the deeper dish of Fig. 6, but there is a larger and more significant separation between the maximum-gain contours and the Petzval surface. For this value of F/D the contours lie relatively close to the focal plane before drooping toward the Petzval surface at larger scan angles. These results are consistent with the limited experimental data of [3]. Fig. 9 is a plot of scan loss versus beamwidths of scan for $F/D = 0.604$ and the 3 feeds under consideration. For a given edge taper this shallower reflector is subject

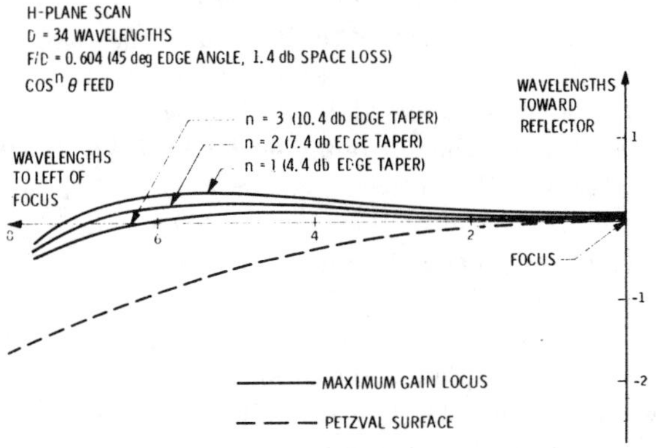

Fig. 8. Comparison of maximum transmit-gain contours and Petzval surface; $F/D = 0.604$.

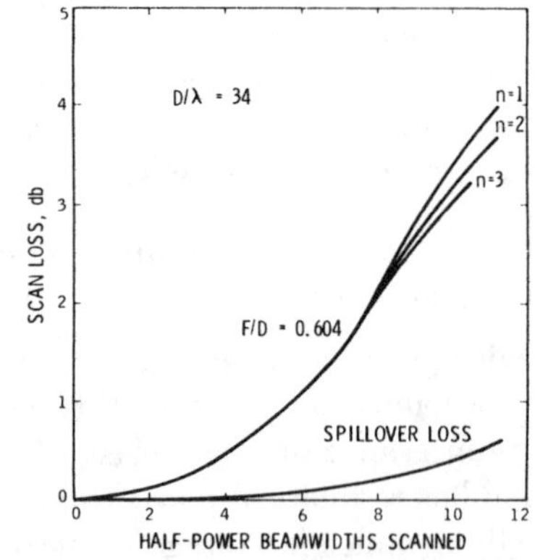

Fig. 9. Beam-scan losses; $F/D = 0.604$.

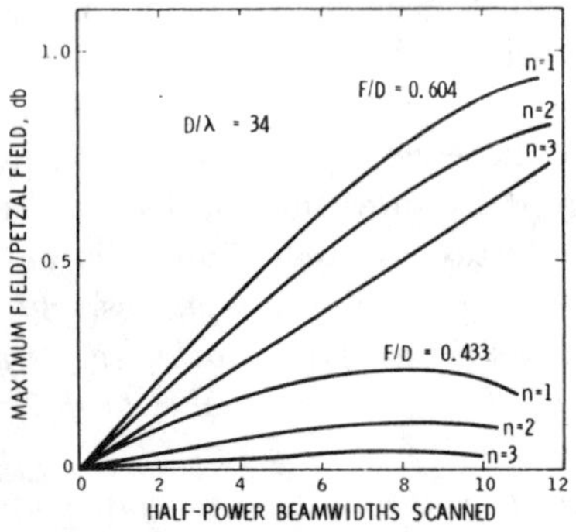

Fig. 10. Comparison of maximum-gain and Petzval fields.

[3] No attempt was made to find other orientations of the feed which would optimize the scan gain.

to larger spillover loss but less total scan loss with scan angle.

Fig. 10 presents the dB difference between the fields of the maximum-gain contour and the fields of the Petzval surface. For $F/D = 0.433$ the difference is only a tenth of a dB or so. However, for $F/D = 0.604$ the difference is a significant fraction of a dB. Since the reflectors with higher F/D values are generally used for beam scanning applications because of reduced scan loss, this effect is of practical interest in wide-angle scanning systems.

IV. Summary of Results and Conclusions

1) Unless the F/D is very large or spillover is excessive, a higher scan gain is achieved when the axis of a directional feed is parallel to the axis of the reflector than when the feed is directed toward the reflector vertex.

2) The contour of maximum scan-gain is a function of illumination taper and reflector F/D. In general, larger F/D values (greater than 0.5) tend to have a maximum-gain contour close to the focal plane, while the smaller F/D values tend to have a maximum-gain contour closer to the Petzval surface. Increasing the illumination taper moves the maximum-gain contour closer to the Petzval surface.

3) The maximum-gain contours in Figs. 6 and 8 are plotted on coordinate axes expressed in terms of wavelengths for two different F/D values. Similar contours are plotted in Fig. 11 for 10, 15, and 20 dB illumination edge taper for the same class of feed functions $\cos^n \theta$. However, the axial and lateral components of feed displacement are divided by the F/D value. Three solid maximum-gain curves corresponding to F/D values of 0.433 (60° edge angle), 0.604 (45° edge angle), and 0.687 (40° edge angle) are plotted for each edge taper.[4] Superimposed on each figure are dashed curves to indicate the scan angle in HPBWs. Normalizing the coordinate axes by the F/D values makes the maximum-gain contours considerably less sensitive to F/D value than the results exhibited in Figs. 6 and 8. In a sense, then, these curves are "universal" and can be used to select the feed position for maximum scan gain over a wide range of scan angle and reflector shape by simple interpolation.

4) Because of limitations imposed by available computer time and the requirements of the program study, all results to this point were computed for an aperture diameter of 34.0 wavelengths. Data for two additional maximum-gain contours were also generated for twice the frequency ($D = 68.0$ wavelengths). This limited study yielded the following results:

a) For a given illumination taper and F/D, the lateral component of scan (in wavelengths) for a particular scan angle (in HPBWs) was the same at both 34 and 68 wavelengths diameter. If this result can be extrapolated to other frequencies, it implies that the lateral component of the curves in Fig. 11 are also "universal."

b) For a given illumination taper and F/D, the axial component of scan for a particular scan angle was the same at the two frequencies for small scan angles. However, for large scan angles, the axial component of scan was smaller at 68 wavelengths than at 34 wavelengths.

c) For a given illumination taper F/D and scan angle (in HPBWs), the 68-wavelength gain underwent considerably less scan loss than the 34-wavelength gain.

[4] In the process of generating these curves, small variations of the order of 0.1 wavelength have been suppressed. However, these variations were not present for the 20-dB edge taper, indicating perhaps, the existence of "edge-diffracted" rays in the focal region for the less tapered illumination.

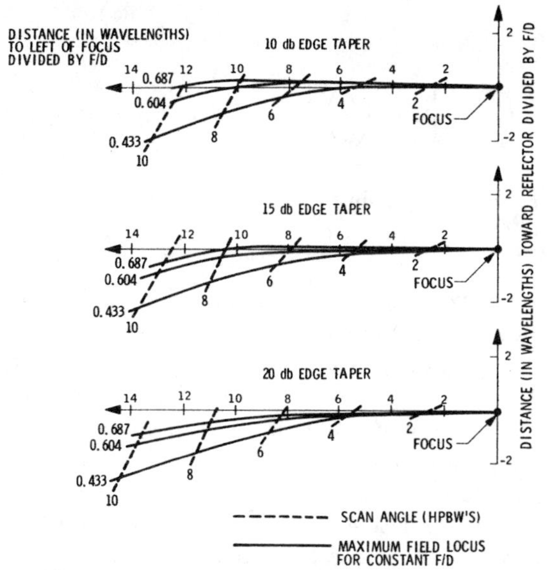

Fig. 11. Maximum transmit-gain contours versus beamwidths of scan; 10, 15, and 20 dB edge taper.

Appendix

The correlation between the focal region field $\bar{E}_1$ and the feed aperture field $\bar{E}_2$ is approximately[5]

$$\eta \cong \left| \int_S \bar{E}_1 \cdot \bar{E}_2 \, dS \right|^2 \tag{A-1}$$

where S is the feed aperture. Considering only the principle polarization,

$$E_1 \cong 1 - \left(\frac{r}{\alpha}\right)^2 \tag{A-2}$$

$$E_2 \cong \left[1 - \left(\frac{r}{\alpha}\right)^2\right] \exp\,(jk_{\text{eff}} r \cos \phi \sin \delta) \tag{A-3}$$

where

- r, ϕ circular coordinates of the feed aperture
- α the feed aperture radius, wavelengths
- δ the tilt of the feed aperture relative to the focal-region field phase front
- $k_{\text{eff}} = 2\pi/\lambda_{\text{eff}}$.

[5] The true correlation involves both $\bar{E}$ and $\bar{H}$ fields [10].

Then

$$\eta \cong \left| \int_0^{2\pi} \int_0^{\alpha} \left[1 - \left(\frac{r}{\alpha} \right)^2 \right]^2 \exp\,(jk_{\text{eff}} r \cos\phi \sin\delta) r\, dr\, d\phi \right|^2$$

$$= \left[4\pi\alpha^2 \frac{J_3(k_{\text{eff}}\alpha \sin\delta)}{(k_{\text{eff}}\alpha \sin\delta)^3} \right]. \qquad \text{(A-4)}$$

For the example considered in Section III-B, $\lambda_{\text{eff}} = 1.23\lambda$ (cf., Fig. 3). If $\alpha = 0.53\lambda$, the field of a circular aperture closely approximates the assumed $\cos^2\theta$ feed pattern. If $\alpha = 0.71\lambda$ the received field contours are closely matched. For the purpose of evaluating (A-4), an average value of $\alpha = 0.62\lambda$ was used. If the hypothetical feed is pointed toward the vertex of the paraboloid, it would be rotated by $\delta = 21.5°$ relative to the focal-region phase front, and equation (A-4) gives a predicted loss (relative to $\delta = 0°$) of 0.74 dB. This value is in reasonable agreement with the computed value of 0.85 dB.

References

[1] S. Silver and C. S. Pao, "Paraboloid antenna characteristics as a function of feed tilt," Radiation Lab., MIT, Cambridge, Mass., Rep. 479, 1944.

[2] F. B. Hildebrand, "The alternation in the radiated field of a paraboloid due to a shift in the position of the dipole feed," Radiation Lab. MIT, Cambridge, Mass., Rep. 1078, Feb. 20, 1946.

[3] K. S. Kelleher and H. P. Coleman, "Off-axis characteristics of the paraboloidal reflector," Naval Research Laboratory, Washington, D. C., NRL Rep. 4088, Dec. 31, 1952.

[4] Y. T. Lo, "On the beam deviation factor of a parabolic reflector," *IRE Trans. Antennas Propagat.* (Commun.), vol. AP-8, pp. 347–349, May, 1960.

[5] S. S. Sandler, "Paraboloidal reflector patterns for off-axis feed," *IRE Trans. Antennas Propagat.*, vol. AP-8, pp. 368–379, July 1960.

[6] R. C. Hansen, *Microwave Scanning Antennas.* New York: Academic Press, 1964, vol. I, ch. 2.

[7] J. Ruze, "Lateral-feed displacement in a paraboloid," *IEEE Trans. Antennas Propagat.*, vol. AP-13, pp. 660–665, Sept. 1965.

[8] J. Van Bladel, *Electromagnetic Fields.* New York: McGraw-Hill, 1964, p. 205.

[9] R. E. Collin and F. J. Zucker, *Antenna Theory.* New York: McGraw-Hill, 1969, pt. 1, pp. 93–98.

[10] B. MacA. Thomas, "Matching focal-region fields with hybrid modes," *IEEE Trans Antennas Propagat.* (Commun.), vol. AP-18, pp. 404–405, May 1970.

[11] A. W. Rudge and M. J. Withers, "Design of flared-horn primary feeds for parabolic reflector antennas," *Proc. Inst. Elec. Eng.*, vol. 117, no. 9, pp. 1741–1744, Sept. 1970.

[12] H. C. Minnett and B. MacA. Thomas, "Fields in the image space of symmetrical focusing reflectors," *Proc. Inst. Elec. Eng.*, vol. 115, no. 10, pp. 1419–1430, Oct. 1968.

[13] A. W. Rudge and M. J. Withers, "New technique for beam steering with fixed paraboloid reflectors," *Proc. Inst. Elec. Eng.*, vol. 118, no. 7, pp. 857–863, July 1971.

[14] "Calculation of the caustic (focal) surface when the reflecting surface is a paraboloid of revolution and the incoming rays are parallel," Parke Math. Lab., Concord, Mass., Study 3, Contract AF 19(122)-484 for AFCRL, May 1952.

[15] "Calculations of the caustic surface of a paraboloid of revolution for an incoming plane wave of twenty degrees incidence," Parke Math. Lab., Concord, Mass., Rep. 1, Contract AF19(604)-263 for AFCRL, May 1952.

[16] I. W. Kay and M. Goldberg, "Investigation of electromagnetic fields in the focal regions of a paraboloid receiving off-axis," Conductron, Ann Arbor, Mich., Rep. D5220-448-P410, Contract AF19(628)-5812, June 24, 1966.

[17] F. S. Holt, "Application of geometrical optics to the design and analysis of microwave antennas," AFCRL, Bedford, Mass., AFCRL-67-0501, Sept. 1967.

[18] R. E. Collin, and F. J. Zucker, *Antenna Theory.* New York: McGraw-Hill, 1969, pt. 2, pp. 16-103.

[19] M. S. Affifi, "Aberration and dispersion off the focus of a parabola," in *1971 G-AP Symp. Dig.*, p. 219.

[20] J. B. Scarborough, "The caustic curve of an off-axis parabola," *Appl. Opt.*, vol. 3, no. 12, pp. 1445–1446, Dec. 1964.

[21] W. H. Watson, "The field distribution in the focal plane of a paraboloidal reflector," *IEEE Trans. Antennas Propagat.*, vol. AP-12, pp. 561–569, Sept. 1964.

[22] R. G. Kouyoumjian, "Asymptotic high frequency methods," *Proc. IEEE*, vol. 53, pp. 864–876, Aug. 1965.

[23] W. V. T. Rusch and P. D. Potter, *Analysis of Reflector Antennas.* New York, Academic Press, 1970, 141–142.

[24] A. C. Ludwig, "Computation of Radiation patterns involving numerical double integration," *IEEE Trans. Antennas Propagat.* (Commun.), vol. AP-16, pp. 767–769, Nov. 1968.

[25] E. M. Kennaugh and R. H. Ott, "Fields in the focal region of a parabolic receiving antenna," Antenna Lab., the Ohio State Univ. Research Foundation, Columbus, Ohio, Rep. 1223-16, Contract AF33(616)-8039, Aug. 31, 1963.

[26] H. Gniss and G. Ries, "Feldbild um den Brennpunkt., von Parabolreflektoren mit kleinem *f/D*-Verhaeltnis," *Arch. Elek. Ubertragung*, vol. 23, no. 10, pp. 481–488, Oct. 1969.

[27] P. G. Ingerson and W. V. T. Rusch, "Studies of an axially defocused paraboloid," in *1969 G-AP Symp. Dig.*, pp. 62–68.

[28] M. Landry and Y. Chasse, "Measurement of electromagnetic field intensity in focal region of wide-angle paraboloid reflector," *IEEE Trans. Antennas Propagat.*, vol. AP-19, pp. 539–543, July 1971.

[29] A. E. Conrady, *Applied Optics and Optical Design.* New York: Dover, 1960, p. 290.

[30] F. A. Jenkins and H. E. White, *Fundamentals of Optics*, 3rd ed. New York: McGraw-Hill, p. 150.

[31] H. P. Brueggemann, *Conic Mirrors.* London, England: Focal Press, p. 30.

[32] K. G. Habell and A. Cox, *Engineering Optics.* London, England: Pitman & Sons, pp. 62–93.

[33] L. C. Martin, *Technical Optics*, volume II. New York: Pitman Publ., p. 75.

Large Lateral Feed Displacements in a Parabolic Reflector

WILLIAM A. IMBRIALE, MEMBER, IEEE, PAUL G. INGERSON, MEMBER, IEEE, AND WILLIAM C. WONG, MEMBER, IEEE

Abstract—**The radiation patterns of a parabolic reflector with large lateral-feed displacements are computed utilizing both the vector current method and scalar aperture theory, and compared to experimental results. The theory is general enough to include asymmetric primary pattern illumination. The scalar and vector solutions are derived from the same initial equation so that the approximations used in obtaining the scalar solution are clearly displayed. Results from the vector and scalar theories are compared and the range of validity of the approximate analysis is indicated.**

I. INTRODUCTION

As THE TREND toward multiple beams in spacecraft antennas continues, and in particular when multiple beam systems are obtained utilizing multiple feeds in a parabolic reflector, the evaluation of the performance of reflector systems when there is large lateral feed displacement becomes increasingly important. Previous analyses have treated the case of a few wavelengths of lateral displacement but there is little discussion of the limits of the range of scan, nor whether the concept of a beam deviation factor is valid for large (10 to 20 wavelengths) amounts of scan [1]–[5]. The purpose of this paper is to examine reflector performance under the condition of large feed displacements. In particular, the complete solution taking into account all of the phase errors resulting from lateral feed displacement is compared to the approximate formulation of Ruze [1]. The complete and approximate solutions are derived from the same initial equation so that the approximations used in obtaining the Ruze type of solution are clearly displayed. The approximate formulation is generalized to include asymmetric primary illumination. The calculated results are compared to experimental measurements obtained using a 9-ft reflector.

II. ANALYSIS

The far-field secondary pattern for the reflector system of Fig. 1 is given by

$$\bar{E}(P) = \frac{-j\omega\mu}{4\pi}\frac{\exp(-jkR)}{R}\int_{\text{surface}}[\bar{J}_s - (\bar{J}_s\cdot\bar{a}_R)\bar{a}_R]\cdot\exp(jk\bar{\rho}\cdot\bar{a}_R)\,dS \quad (1)$$

where $\bar{J}_s$ is the reflector surface current, $k = 2\pi/\lambda$, $\bar{a}_R$ is the unit position vector of the far-field point, dS is the incremental surface area, and the quantities R, ρ, and P are as defined in Fig. 1.

Manuscript received November 13, 1972; revised May 8, 1974.
The authors are with TRW Systems Group, Redondo Beach, Calif. 90278.

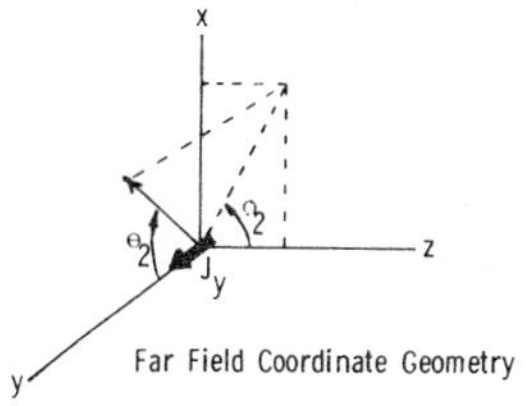

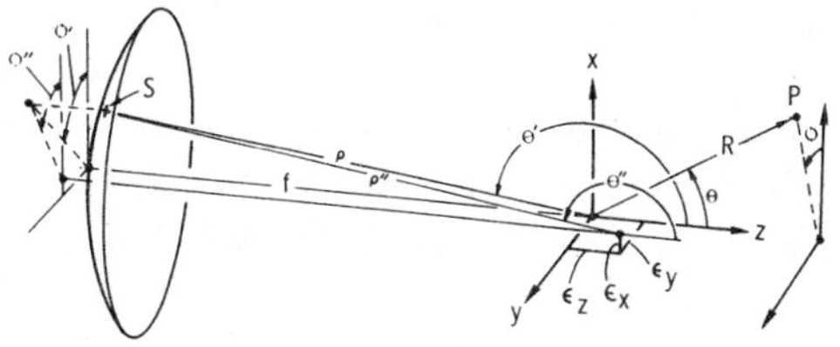

Fig. 1. Reflector coordinate system.

The solution will be determined in the following manner. 1) The incident field is formulated in the double primed coordinate system, hence the induced current $\bar{J}_s$ obtained using the physical optics approximation is expressed in the same coordinate system. 2) A change of variables converts the $\bar{J}_s$ to the reflector (prime) coordinate system. 3) For the complete solution a two-dimensional numerical integration of (1) is required and for the scalar solution, approximations can be made to reduce (1) to a one-dimensional integral. Before exploring the solution further, it will prove convenient to rewrite (1) in a slightly different form.

Letting

$$[\bar{J}_s - (\bar{J}_s\cdot\bar{a}_R)\bar{a}_R] = \bar{G}(\theta',\phi',\theta,\phi)\frac{\exp[-jk\rho''(\theta',\phi')]}{\rho''(\theta',\phi')} \quad (2)$$

and further utilizing the definitions,

$$r = \frac{\rho}{a}\sin\theta' \qquad u = \frac{\pi D}{\lambda}\sin\theta \qquad D = 2a$$

the two-dimensional vector integral for the far zone electric field given in (1) can be cast in the form

$$\bar{E}(P) = -j\frac{\exp(-jkR)}{R}\frac{k\eta_0 a^2}{2f}\int_0^1\left[\int_0^{2\pi}\bar{G}(\theta',\phi',\theta,\phi)\csc\frac{\theta'}{2}\cdot\exp(-jk(\rho''-\rho\cos\theta\cos\theta'))\left(\frac{\rho}{\rho''}\right)\cdot\exp(jur\cos(\phi-\phi'))\,d\phi'\right](1-\cos\theta')r\,dr. \quad (3)$$

Reprinted from *IEEE Trans. Antennas Propagat.*, vol. AP-22, pp. 742–745, Nov. 1974.

The only remaining step for the vector solution is to specify the vector quantity $\bar{G}$. To this end we cast the incident field in its most general form in the double primed coordinate system as follows:

$$\bar{E}_{\text{inc}} = \frac{\exp -jk\rho''}{\rho''} (F_1(\theta'',\phi'')\hat{a}_{\theta''} + F_2(\theta'',\phi'')\hat{a}_{\phi''}). \quad (4)$$

Recalling that the physical optics approximation implies that $\bar{J}_s = 2\hat{n} \times \bar{H}_{\text{inc}}$, where $\hat{n}$ is the outward normal from the surface and that

$$\bar{H}_{\text{inc}} = \frac{\hat{a}_{\rho''} \times \bar{E}_{\text{inc}}}{\eta_0}$$

we can calculate the current $\bar{J}_s$ induced on the reflector surface by the incident E-field from (4). A simple trigonometric manipulation is then required to transform the double primed angles (θ'',ϕ'') into the (θ',ϕ') coordinate system. This transformation is necessary so that the incident field given by (4) is stationary with respect to the feed. Knowing $\bar{J}_s$, the vector $\bar{G}$ given by (2) can easily be obtained by vector manipulation. Equation (3), along with the quantities F_1 and F_2 from the incident field allow one to calculate the effects of lateral scan from the complete vector description of the problem.

The scalar aperture result will now be developed from (3). The incident field is approximated in the following manner:

$$\bar{E}_{\text{inc}} = \frac{\exp(-jk\rho'')}{\rho''} (E_P(\theta') \sin\phi' \hat{a}_{\theta'} - H_P(\theta') \cdot \cos\phi' \hat{a}_{\phi'}). \quad (5)$$

The surface current is again obtained using the physical optics approximation where $E_P(\theta')$ is the E-plane radiation pattern and $H_P(\theta')$ is the H-plane radiation pattern.

There will in general be x, y, and z directed current components. For the scalar case, which assumes fields near the boresight, we ignore the z and x directed current components and utilize only the y directed currents so that the secondary radiation field has a principal polarization vector component ($\hat{a}_{\theta_2}$ directed) in the (θ_2,ϕ_2) coordinate system of Fig. 1. The vector $\bar{G}$ is therefore approximated by

$$\bar{G} = -(1 - \sin^2\theta \sin^2\phi)^{1/2} \frac{2}{\eta_0} \left(-\sin\frac{\theta'}{2} [\sin^2\phi' E_P(\theta') + \cos^2\phi' H_P(\theta')] \right) \hat{a}_{\theta_2}. \quad (6)$$

In the vector formulation when the feed is laterally defocused, the illumination function is assumed stationary with respect to the feed and the distance function ρ'' is obtained exactly using the law of cosines. In the scalar formulation, the illumination function is assumed stationary with respect to the reflector so that the transformation from double primed to single primed coordinates is not necessary, and the lateral feed displacement is accounted for by the distance function ρ'' in the phase only, ρ'' being approximated by the parallel ray approximation, as follows. Let $\bar{\rho}'' = \bar{\rho} - \bar{d}$, where the feed displacement $\bar{d} = \epsilon_x \bar{a}_x + \epsilon_y \bar{a}_y + \epsilon_z \bar{a}_z$ whence, if we assume that $(d/\rho) \ll 1$, we obtain

$$\rho'' \approx \rho - (\epsilon_x \sin\theta' \cos\phi' + \epsilon_y \sin\theta' \sin\phi' + \epsilon_z \cos\theta'). \quad (7)$$

Further, we make the assumption that the observation point is near the boresight so that $\cos\theta \approx 1$ and $\sin\theta \ll 1$, which implies that $\rho(1 - \cos\theta\cos\theta') \sim 2f$, and the square root of (6) can be replaced by 1. With these approximations, (3) can be shown to reduce to

$$E(P) = -j \exp(-j2f) \frac{\exp(-jkR)}{4\pi R} k \left(\frac{a^2}{f}\right) \cdot \int_0^1 \exp(jk\epsilon_z \cos\theta') I(r)(1 - \cos\theta') r\, dr \quad (8)$$

where

$$I(r) = \int_0^{2\pi} [\sin^2\phi' E_P(\theta') + \cos^2\phi' H_P(\theta')] \cdot \exp[jk(\epsilon_x \sin\theta' \cos\phi' + \epsilon_y \sin\theta' \sin\phi')] \cdot \exp[jur\cos(\phi - \phi')]\, d\phi'. \quad (9)$$

Note that this would be identical to the formulation given by Ruze if we used the assumed illumination function

$$f(r,\phi') = \sin^2\phi' E_P(\theta') + \cos^2\phi' H_P(\theta'). \quad (10)$$

Now, making use of some trigonometric manipulations similar to those employed by Ruze, incorporating the plane wave to cylindrical wave transformation, and using the following definitions

$$\omega = [(ur)^2 + 2k\epsilon_R \sin\theta' ur \cos(\phi - \xi) + (k\epsilon_R \sin\theta')^2]^{1/2}$$

$$\alpha = \tan^{-1} \frac{(ur\sin\phi + k\epsilon_R \sin\theta' \sin\xi)}{(ur\cos\phi + k\epsilon_R \sin\theta' \cos\xi)}$$

$$\epsilon_R = (\epsilon_x^2 + \epsilon_y^2)^{1/2}$$

$$\xi = \tan^{-1} \frac{\epsilon_y}{\epsilon_x} \quad (11)$$

we obtain the final result

$$E(P) = -j \exp(-2jf) \frac{\exp(-jkR)}{R} \left(\frac{\pi D}{\lambda}\right) \left(\frac{D}{4f}\right) \cdot \int_0^1 (1 - \cos\theta') \exp(jk\epsilon_z \cos\theta') \cdot \left\{ J_0(\omega) \frac{E_P + H_P}{2} + \frac{E_P - H_P}{2} J_2(\omega) \cos 2\alpha \right\} r\, dr. \quad (12)$$

Note that (12) reduces to Ruze's result for $\epsilon_z = 0$ and $E_P = H_P$.

Equation (12) can now be interpreted as the principal

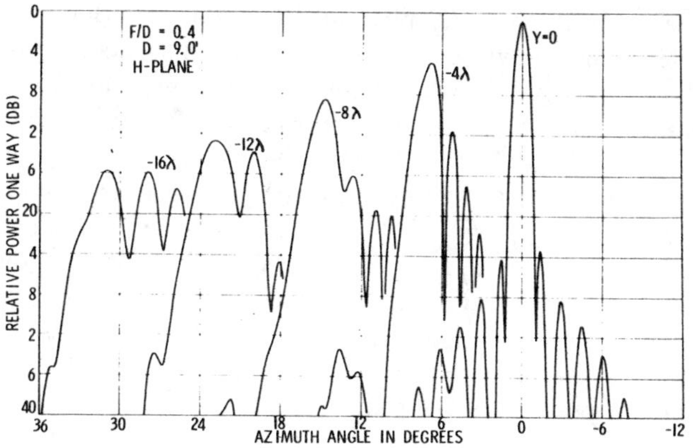

Fig. 2. Measured secondary patterns as function of lateral primary feed displacement.

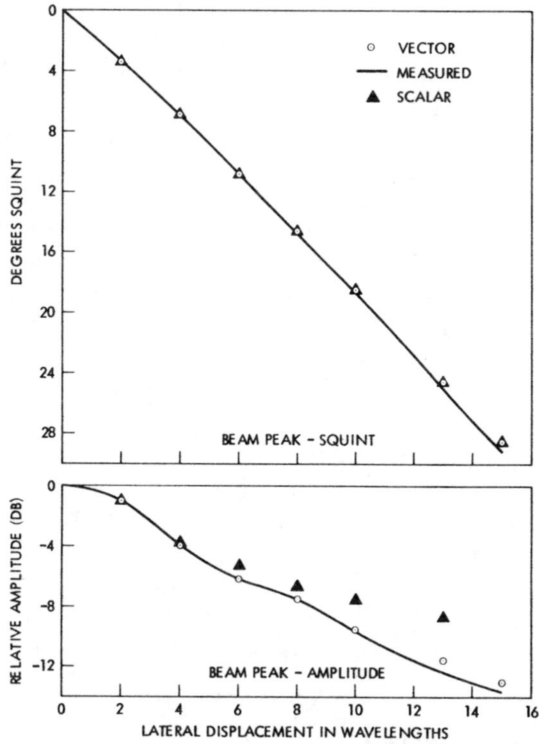

Fig. 3. Secondary beam squint and beam scan loss as function of lateral displacement of primary feed in focal plane.

polarization component in the (θ_2,ϕ_2) coordinate system utilizing approximations for the incident field ρ'' and some small angle assumptions.

III. DESCRIPTION OF EXPERIMENTAL MEASUREMENTS

Measurements were performed on a 1750-ft antenna test range. A 9-ft diameter high precision experimental reflector, with an $f/D = 0.4$ and a rms surface error of 0.004 in, was used with a circular 1-in inside diameter open ended waveguide primary feed. At the testing frequency of 7.9 GHz the primary feed had a 3-dB beamwidth of 64° in the E-plane and 70° in the H-plane, which gives an edge taper of approximately 9.5 dB in the E-plane and 8.7 dB in the H-plane. A standard Scientific Atlanta positioner was used to support the reflector but a special track was added to the base arm of the positioner. A carriage was mounted on this track, and a thin support mast was placed vertically up from the carriage and supported the feed in front of the reflector. This experimental arrangement allows minimum blockage and accurate movement of the feed axially and laterally. Further details of the experimental setup can be found in [6].

The primary feed's radiation patterns were recorded everywhere over the far-field sphere. In particular, the functions $F_1(\theta'',\phi'')$ and $F_2(\theta'',\phi'')$ of (4) and the functions $E_P(\theta')$ and $H_P(\theta')$ of (5) were extracted from the measured data, digitized, and used as input for the computer programs implementing (3) and (12).

A sample of the measured secondary patterns as a function of lateral primary feed displacement in the focal plane with the feed aligned parallel to the z axis is shown in Fig. 2. The axis of the feed was maintained parallel to the z axis because a higher scan gain was achieved than by directing the feed towards the apex [7]. The measured data is summarized and compared to the calculated data in the following figures. Fig. 3 shows the pattern peak gain and angular position as a function of lateral defocusing in the focal plane. The half-power beamwidth for the focused feed is approximately 1°, hence the beam has been scanned 29 half-power beamwidths at the extreme end. "Vector" refers to the results using the vector formulation of (3), and "scalar" refers to the results using (12). It is noted that for beam squint the agreement is very good for both calculations and, in fact, the slight discrepancy at the extreme angles was traced to the primary fixture sag at this extreme range. Observe that the scalar analysis yields higher gain than both the vector formulation and the measured data.

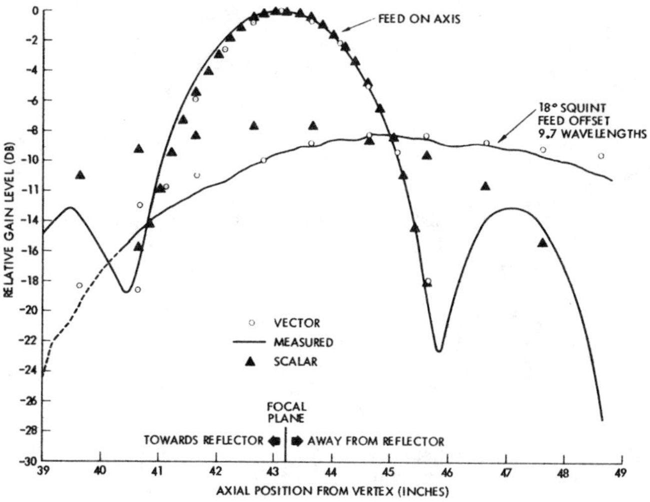

Fig. 4. Comparison of theoretical and measured gains as function of axial defocusing.

Fig. 4 explores the scan-plane fields for axial feed movements with no lateral displacement and lateral displaced 9.7 wavelengths in the H-plane (approximately 18 half-power beamwidths of scan). Several pertinent results are illustrated. The scalar formulation indicates that the highest gain occurs in the focal plane whereas the measured

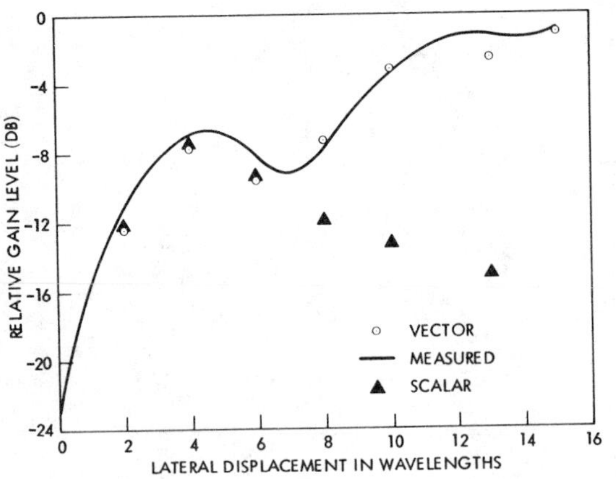

Fig. 5. Coma-lobe gain level relative to beam peak as function of lateral displacement of primary feed in focal plane.

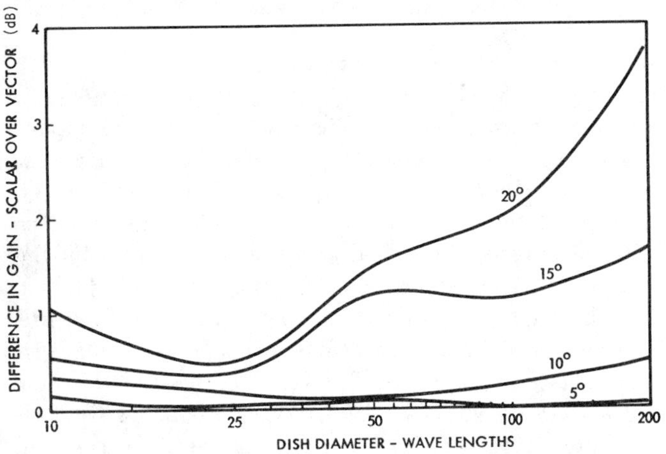

Fig. 6. Comparison of scalar and vector gain versus dish diameter.

data peaks off the focal plane. The difference in the measured and the scaled peak gain is small. Since the data in Fig. 3 was measured in the focal plane, it is to be expected that the scalar gain, when it differs from the vector gain, would yield values too high. The vector calculation correlates well with the measured data. Even though the antenna range on which the data was measured was approximately $2\frac{1}{2}D^2/\lambda$, there is a very slight near-field shift in the measured on-axis defocusing curve, but this shift introduces negligible error into the data in Figs. 2, 3, and 5.

Fig. 5 presents the coma-lobe gain level relative to the beam peak as a function of lateral displacement. Again, the approximate analysis can be misleading for scans greater than 10 half-power beamwidths.

For the measured data the reflector diameter was 72 wavelengths. A plot showing the difference between the peak gain in the focal plane of the scalar and the vector calculation for various scan angles and dish diameters is shown in Fig. 6. For a 10° scan there is very little peak gain error for any dish diameter from 10 to 200 wavelengths even though at 10 wavelengths the scan is 1.5 half-power beamwidths and at 200 wavelengths the scan is about 29 half-power beamwidths. This plot indicates that the peak gain difference is more dependent upon the actual angular scan as contrasted to the scan measured in half-power beamwidths. For the various cases calculated the scalar and vector formulations compared favorably for the angular position of the peak.

IV. CONCLUSIONS

Vector and scalar analyses were used to examine reflector performance under the condition of large feed displacement. Results from the two formulations were compared to experimental data obtained using a precision reflector with minimal blockage. Several pertinent observations can be made. The beam peak angle position is accurately predicted by both the vector and scalar theories. The peak gain is accurately predicted by the vector theory, but the scalar theory is several dB in error for large scan angles. The scalar theory can be greatly in error for predicting the coma-lobe peak. A comparison of the peak gain difference between the two theories indicates that the accuracy of the scalar theory is more dependent upon the actual amount of angular scan rather than upon the scan in terms of half-power beamwidths. The scalar analysis indicates that the highest gain level for a given lateral displacement occurs with the feed in the focal plane, whereas the vector formulation correctly predicts scan-plane fields.

REFERENCES

[1] J. Ruze, "Lateral-feed displacement in a paraboloid," *IEEE Trans. Antennas Propagat.*, vol. AP-13, pp. 660–665, Sept. 1965.

[2] Y.T. Lo, "On the beam deviation factor of a parabolic Reflector," *IRE Trans. Antennas Propagat.* (Commun.), vol. AP-8, pp. 347–349, May 1960.

[3] K. S. Kelleher and H. P. Coleman, "Off-axis characteristics of the paraboloidal reflector," National Research Laboratories, Washington, D.C., Rep. 4088, Dec. 1952.

[4] S. S. Sandler, "Paraboloidal reflector patterns for off-axis feed," *IRE Trans. Antennas Propagat.*, vol. AP-8, pp. 368–379, July 1960.

[5] S. Silver, *Microwave Antenna Theory and Design.* New York: McGraw-Hill, 1949.

[6] W. A. Imbriale, P. G. Ingerson, and W. C. Wong, "Experimental verification of the analysis of umbrella parabolic reflectors," *IEEE Trans. Antennas Propagat.*, vol. AP-21, pp. 705–708, Sept. 1973.

[7] W. V. T. Rusch and A. C. Ludwig, "Determination of the maximum scan-gain contours of a beam-scanning paraboloid and their relation to the Pitzval surface," *IEEE Trans. Antennas Propagat.*, vol. AP-21, pp. 141–147, Mar. 1973.

Multiple-Feed Systems for Objectives

Before the advent of large high-performance array-type antennas, the microwave optical focusing objective, such as the lens or reflector, was the basic tool. Multiple beams have occasionally been formed, as in stacked beam height finders, by arraying feeds in the neighborhood of the focus.

The problem of multiple feeds for a focusing objective is a long-standing one in the field of microwave antennas. The dilemma is that simple independent feed radiators cannot be placed close enough together to obtain good beam crossover level if reasonable aperture efficiency is specified.

Since, in one dimension, the feed system is just a linear array, one might expect that a solution can be obtained in terms of array factors and independent of element patterns, thereby eliminating the element pattern and spacing problem.

Based on the results for a linear-array multiple-beam system, one concludes that the secondary patterns, for maximum efficiency and minimum beam spacing, should be of the general form, $\sin \theta/\theta$. Further, the aperture illumination at the objective should be of uniform amplitude. Thus, the pattern of an individual feed should be as nearly rectangular as possible, and finally, the amplitude distribution at the feed radiator should be of the form $\sin x/x$. The preceding chain of conclusions is based on the approximation that an aperture distribution and its far-field radiation patterns are Fourier transforms of each other.

Perhaps the best place to start on the problem, based on the previously mentioned ground rules, is to determine the accuracy with which the rectangular feed pattern can be realized. The maximum slope that can be achieved at the edge of the rectangular pattern is proportional to the aperture of the feed system. If the multiple-element feed system is considered as an array, the approximation of the required primary beam can be described by Fig. 1. The entire feed array is used to generate a set of pencil beams covering the objective, and the directivity of these beams, and, therefore, the slope at the edge of the composite pattern is proportional to the total aperture of the array.

At this point, since it starts to look like the solution to the problem may be an array fed by a hybrid matrix, one might ask why the objective is used at all; why not just use an array as the primary radiator and save all of the intervening steps? The answer is related to the original choice between the array and the geometrical focusing objective. The important factors are gain and coverage. If one needs a versatile antenna of low or medium gain, the array is usually the best answer. Further, if both high gain and large-solid-angle coverage, such as five or ten thousand square degrees, are required, the array is probably still called for. For the case of a high-gain antenna with a relatively small number of simultaneous beams, such as ten or twenty, the array would represent a great waste of complexity, and a focusing objective is the best answer. Therefore, although the solution may seem complex, it must be remembered that once a multiple-feed system is designed, it can be used with any focusing objective of any size.

Referring again to Fig. 1, it is seen that the relative phase of the narrow beams determines the effective phase center of the fan beam that they comprise. If the narrow beams are fed in phase, the phase center of the fan beam is in the center of the array. If they are fed in progressive phase, the phase center is displaced from the center of the feed array. Thus, the ports of the hybrid matrix generating the pencil beams illuminating the objective can be fed by a second, smaller matrix to realize the optimally spaced phase centers corresponding to the desired aperture distributions.

Since no study or design program related to this technique has been carried out, a detailed theoretical example is presented. The radiating elements and feed matrices are shown schematically in Fig. 2. A 16-element array is fed by a 16-port matrix. The initial design choice is the number of beams to use, which is determined by the angle subtended by the objective. In the case shown, the outer two beams on each side are discarded, and the angular coverage in $\psi(=2\pi d \sin\theta)/\lambda$ is between $\pm(3\pi/4)$ and $\pm(13\pi/16)$, corresponding to the half-power points and first nulls of the outer beams. If the radiators are half-wave spaced, the actual range of angular coverage is $\theta=\pm 48.6°$ to $\theta=\pm 54.4°$. The maximum allowable element spacing in this case is approximately $19\lambda/32$, for which the real angle coverage falls in the range $\theta=\pm 39.3°$ to $\theta=\pm 44.3°$.

Once the number of output beams has been selected, the size of the second matrix is determined—12 ports in the example.

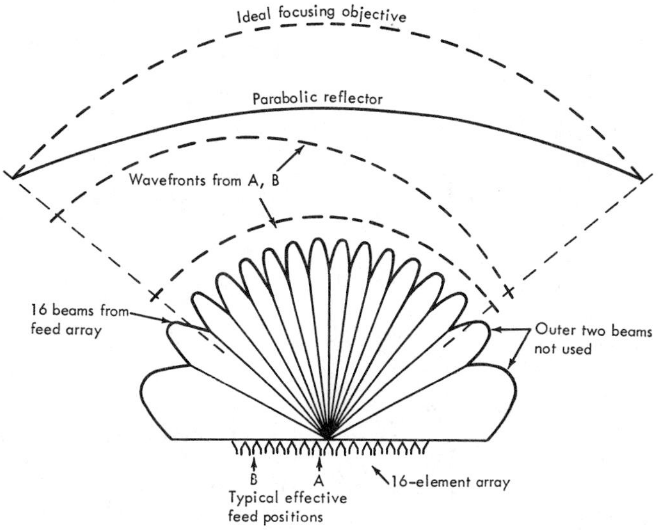

Fig. 1. Illumination of focusing objective with multiple beams from feed array.

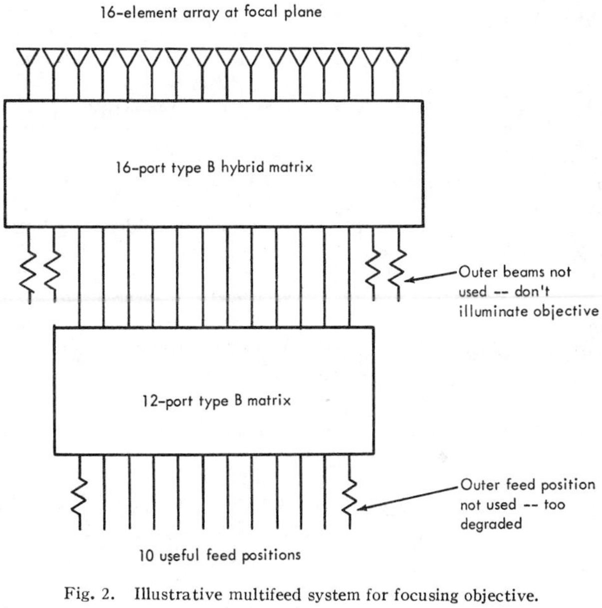

Fig. 2. Illustrative multifeed system for focusing objective.

Manuscript received July 2, 1965.

Reprinted from *IEEE Trans. Antennas Propagat.*, vol. AP-13, pp. 992–994, Nov. 1965.

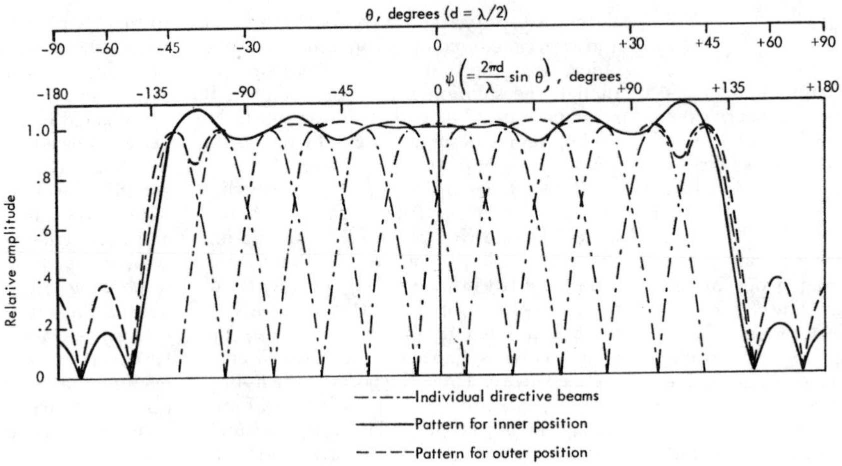

Fig. 3. Representative patterns from multifeed system.

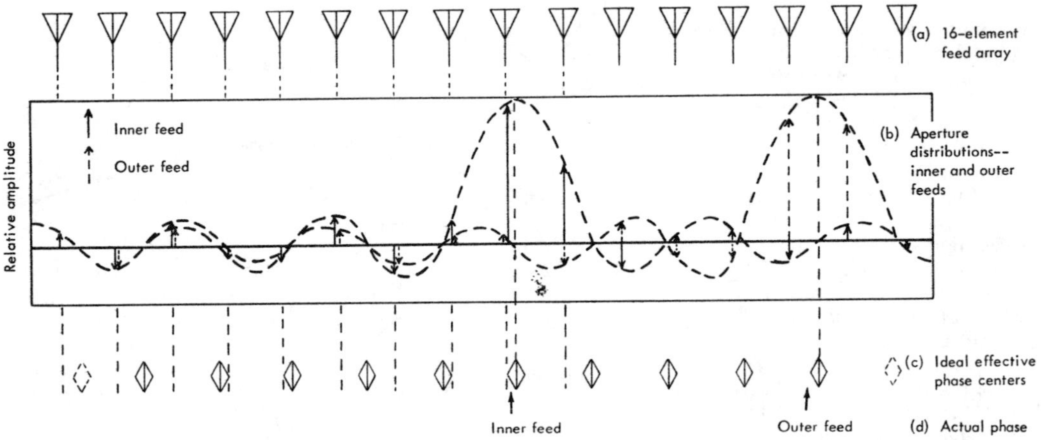

Fig. 4. Aperture distributions and phase centers for multifeed system.

There are 12 effective feed positions located along the 16-element array. The performance of the system for each feed position is measured by the quality of the radiation pattern and by its effective phase center. These parameters have been calculated for the first and fifth off-axis positions. The patterns are shown in Fig. 3, and the aperture distributions and phase centers are illustrated in Fig. 4. It is found that the characteristics degrade slightly for the fifth position off axis. The radiation pattern becomes more distorted and a poorer approximation to the desired rectangular shape. The phase center deviates further from the desired position. The degradation is not excessive, however. The inner feed position has an effective phase center that is displaced from the ideal by only 0.027 element spacing. The outer feed position is displaced by 0.219 element spacing. When it is considered that the effective feed positions are separated by 1.333 element spacings, these errors are within reason.

The outermost feed positions are discarded because the performance will be sharply degraded by the splitting of the "main lobe" of the aperture distribution. One of the strongly excited elements slips off one end of the array and reappears on the other end with resultant pattern and phase-center distortion. Thus, ten feed positions are obtained.

The relationship between secondary-pattern beamwidth and beam spacing is a function of the focal length and aperture of the objective and the primary pattern beamwidth. It is interesting to note, however, that optimum secondary-pattern characteristics are obtained if the Abbe sine condition is satisfied; that is, the aperture is given by $2f \sin \theta_0$, where f is the focal length and θ_0 the half angle subtended by the objective. The sine condition is the criterion for optimum off-axis focus, and examples of objectives which satisfy it are the Luneberg lens and the spherical zoned-plate reflector.

This design approach is not perfect, of course, because some energy is inevitably spilled over the edge of the focusing objective, and the fan-shaped feed patterns also tend to deteriorate as the phase center is displaced farther from the center of the feed system. However, the important point is that the type of system shown in Fig. 2 does approach perfection in the limit of very many elements in the feed array.

PAUL SHELTON
Inst. for Defense Analyses
Arlington, Va.

A Proposed Multiple-Beam Microwave Antenna for Earth Stations and Satellites

By E. A. OHM

(Manuscript received October 3, 1973)

An offset Cassegrainian antenna with essentially zero aperture blockage is expected to support closely spaced well-isolated beams suitable for earth stations and satellites. Each beam is fed with a separate small-flare-angle corrugated horn and has good area efficiency over a 1.75:1 bandwidth. Each beam also has good cross-polarization properties. The antenna is compact, and the design appears practical for a 4- and 6-GHz earth station, a 20- and 30-GHz earth station, and a 20- and 30-GHz satellite.

I. INTRODUCTION

Satellite communication systems with large capacities can be achieved if the satellites and earth stations are provided with multiple-narrow-beam antennas.[1] The capacity is proportional to the number of satellites, and thus it is important to use as many as practical in the limited orbital space. A moderate number of the resulting closely spaced satellites can be served by a single antenna at each earth-station site if the antenna is patterned after the offset Cassegrainian antenna shown in Fig. 1. This design allows an orderly expansion in communication capacity by the addition of feed horns. Since only one antenna is needed at each site, the design also permits a large saving in earth-station costs. Good multiple-beam performance can be achieved across all up/down pairs of satellite frequency bands, including those well below 10 GHz. At 20 and 30 GHz, a large earth-station antenna with acceptable thermal and wind distortion is hard to achieve. However, with the design outlined here, these problems can be largely overcome because the main reflector and subreflector can be fixed in position, thus allowing a stiffer structure. The steering of each beam is achieved by moving one of the feed horns, resulting in a steerable angle sufficient for tracking near-synchronous satellites.

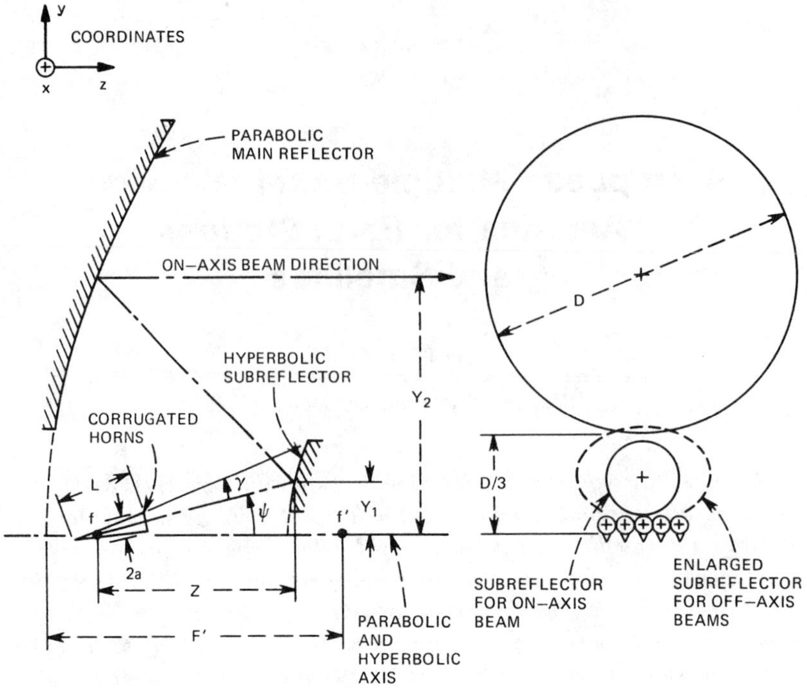

Fig. 1—Geometry of antenna and feed system. The feed horns are scaled for a 3m 20- and 30-GHz satellite antenna. For a 30m 4- and 6-GHz earth-station antenna, L and $2a$ are half as large as shown.

The offset Cassegrain is also appropriate for use aboard a satellite because all beams, including those moderately far off-axis, have high area efficiencies and low side-lobe levels. However, good results on a satellite are restricted to bands well above 10 GHz because the antenna size is limited by the launch vehicle.

It has been previously shown that a multiple-beam antenna can be achieved in a variety of ways,[2–5] where each approach has emphasized one feature desired in a practical antenna. By combining several of these with a corrugated feed horn[6] and an enlarged subreflector, it is possible to achieve a compact antenna with exceptionally good multiple-beam characteristics. In particular, in the offset Cassegrainian antenna shown in Fig. 1:

(*i*) An offset design essentially eliminates beam blockage, thus allowing a significant reduction in side-lobe level.[7] This, in turn, results in higher isolation between beams and a lower antenna noise temperature.

(*ii*) The Cassegrainian feed system is compact and has a large focal-length-to-diameter (F/D) ratio.[8] The large F/D ratio reduces aberrations to an acceptable level, even when a beam is moderately far off-axis.

(*iii*) A corrugated feed horn is essentially a Gaussian-beam launcher[9] and, as such, it can be used to achieve beams with low side-lobe levels. The corresponding feed-horn aperture[6] is small enough to allow the beams to be closely spaced.

(*iv*) An enlarged subreflector, as indicated by the dashed line in Fig. 1, allows the main reflector to be properly illuminated, even when a beam is moderately far off-axis.

These features can be achieved over a wide range of antenna parameters. Using the results developed here, the sample calculations summarized in Table I show that (*i*) the off-axis beam angles are practical, (*ii*) the coma aberration is small, (*iii*) the feed-horn dimensions are reasonable, and (*iv*) the isolations between beams are large.

II. OFF-AXIS DESIGN CRITERIA

Consider a parabolic reflector that is circularly symmetric and illuminated with a feed at its prime focus. If the aperture is large in wavelengths and the prime focal-length-to-diameter, F'/D, ratio is 2 or more, it is well-known that a beam can be scanned over tens of beamwidths by lateral displacement of the feed.[2] A Cassegrainian antenna normally has a secondary focal length F larger than F', and thus a larger F/D ratio.[8] Consequently, a scanned beam can also be obtained by displacing a feed at the secondary focus.[5] For the small off-axis angle reported in Ref. 5 (4 beamwidths $\equiv 0.9°$), the on-axis and off-axis beam characteristics are nearly identical, and the residual differences can be readily explained in terms of an equivalent parabola.[5,8] The equivalent parabola, in turn, has characteristics identical to those of a prime-focus parabola. Consequently, the prime-focus theory[2] can be used to predict the off-axis equivalent-parabola results, and thus the Cassegrainian results. This chain of reasoning assumes that the equivalent-parabola concept is valid for the antenna parameters (F'/D and F/D ratios and off-axis angles) considered here. In support of this assumption, it is of interest to note that the chief off-axis beam parameter of a prime-focus parabola, namely,[2]

$$X' = \frac{N\left(\frac{D}{F'}\right)^2}{1 + 0.02\left(\frac{D}{F'}\right)^2}, \tag{1}$$

where N is the off-axis angle in half-power beamwidths, has a value in Ref. 5 of about 30. Thus, the equivalent-parabola concept is valid for X' values at least through 30. Furthermore, the known results indicate that the region of validity can be extrapolated to X' values well beyond 30. In particular, Ref. 5 shows that the coma lobe, which is the first side lobe aimed toward the on-axis direction, increases very slowly as a function of off-axis beam angle. From Ref. 2, it is also known that an increase in coma-lobe level is a sensitive leading indicator of serious aberration problems, and that X' increases rapidly with coma-lobe level. It follows that X' in Ref. 5 can be much larger than 30 before a larger increase in coma-lobe level signals the onset of serious aberrations. The upper limit of X' should and can be calculated but, in the meantime, some of the results in Table I include an engineering judgment that the equivalent-parabola concept is valid for X' values through 45. Even if the upper limit turns out to be somewhat less, the offset Cassegrain can still support a respectable number of multiple beams, i.e., for $X' = 30$, the number of 1°-spaced beams from the earth-station antenna of Table I is 7 rather than 11.

An important parameter of an off-axis beam is the third-order phase error across the beam at the antenna aperture. This error, $\Delta\phi$, increases the level of the coma lobe.[2] For a symmetrical parabola illuminated with a feed displaced laterally from the prime focus, the peak value of $\Delta\phi$ at the edge of the aperture can be calculated from eq. (12) of Ref. 2. Similarly, when an offset parabola (as in Fig. 1) is illuminated with a feed displaced laterally from the prime focus (in the x direction in Fig. 1), the maximum third-order phase error, $\Delta\phi'$, which occurs at the side edge of the aperture, can be calculated from[10]

$$\Delta\phi' = \frac{2\pi}{32}\frac{F'}{\lambda}\frac{\sin\theta}{(F'/D)^3}\frac{1}{1+(Y_2/2F')^2}, \tag{2}$$

where F' is the prime focal length, θ is the off-axis angle of the beam, D is the diameter of the offset aperture, and Y_2 (see Fig. 1) is the offset height of the aperture. Equation (2) assumes that the feed is also displaced slightly in the longitudinal direction (the $-z$ direction in Fig. 1) to cancel field curvature.

Comparison of eqs. (12) and (13) of Ref. 2 shows that $\Delta\phi'$ is proportional to X'. Noting that $\Delta\phi'$ in (2) is defined in terms of the aperture diameter, D, independently of whether the aperture is centered or offset, it follows that D in eq. (1) should be interpreted in the same way, i.e., it is the diameter of the offset aperture, D, and not the diameter of the aperture of the full parabola ($8/3\ D$ in Fig. 1).

If the prime-focus feed illuminating the offset parabola is replaced with a Cassegrainian feed system, as in Fig. 1, and the equivalent-parabola concept is valid, F' in (1) and (2) can be replaced with the Cassegrainian focal length F. In Fig. 1, F is the distance Z times the ratio of centerline-ray heights where they intercept the main and subreflector heights, i.e., $F = Z(Y_2/Y_1)$. For the antenna parameters listed in Table I, the values of $\Delta\phi$ calculated from (2) are substantially less than 90°. For these values, the first side lobe, or coma lobe, is increased in amplitude, but the side lobes which are positioned further out, i.e., those that determine the minimum spacing of well-isolated beams, are virtually unchanged. Accordingly, in the remainder of this paper, it is assumed that $\Delta\phi$ is zero. The corresponding values of X, which are calculated from (1) after replacing F' by F, are found to be 4.5 or less. From the plots given in Ref. 2, the off-axis and on-axis beam characteristics are essentially identical for these values of X.

III. BEAM SPACING

Suppose the amplitude distribution across an unblocked aperture is that of a dominant-mode Gaussian beam, that the amplitude at the edge is truncated at the −15-dB point, and that the phase front is uniform. The envelope of the resulting radiation pattern is shown in Fig. 2. For the offset Cassegrain shown in Fig. 1, the above amplitude and phase distribution can be achieved by placing a corrugated feed horn[6] at the secondary focal point, f. Comparison of Dragone's results[9] with the standard Gaussian-beam equations[11] shows that the radius of the beam, ω, at the −8.686-dB (or $1/e$ amplitude) point, is related to the feed-aperture radius, a, by

$$\omega = 0.647\, a. \tag{3}$$

The comparison also shows that the phase-front radius is equal to the slant length of the feed-horn, L. Using Gaussian-beam equations,[11] the beam parameters in any other region in the feed system can be calculated. One result is that the required feed-horn length, L, can be found from the half-angle, γ, subtended at the focus f by the subreflector, and the illumination taper, T, in dB, at the edge of the subreflector.

$$L = 0.076 \frac{\lambda}{\gamma^2} T_{\mathrm{dB}}. \tag{4}$$

Equation (4) includes the feed-horn design criterion[6]

$$a^2/\lambda L = 1, \tag{5}$$

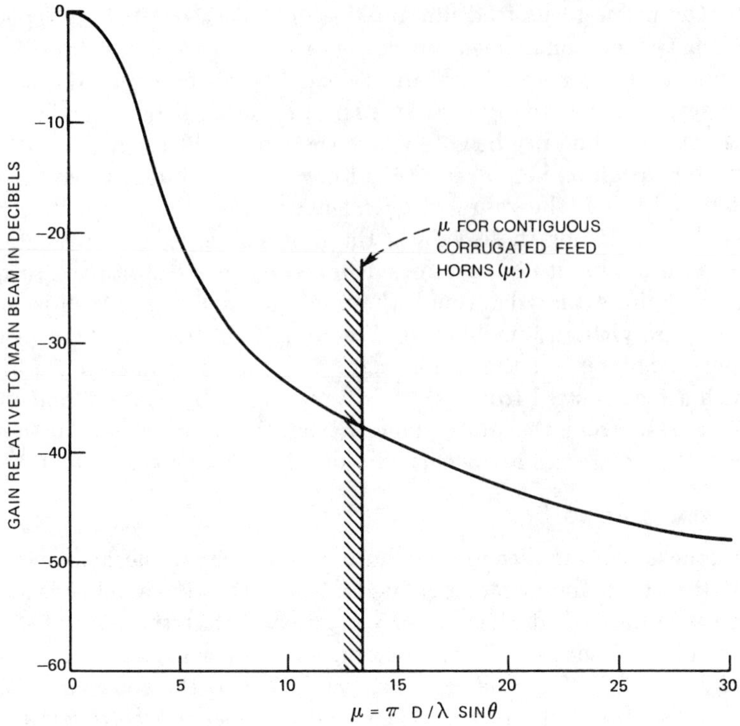

Fig. 2—Estimate of the side-lobe envelope resulting from a Gaussian illumination taper truncated at the −15 dB point, courtesy of T. S. Chu.

where, for a 1.75:1 bandwidth, λ is specified at the low end of the frequency range. Equation (4) is strictly valid only when $\gamma \gg \lambda/D_{\text{sub}}$, where D_{sub} is the diameter of the subreflector. For an equivalent parabola with focal length F,[8] it can be shown that the γ criterion is automatically satisfied when the F/D ratio is less than 5.

The corresponding feed-horn aperture radius, a, is found by solving $a^2/\lambda L = 1$ for a, and substituting L from (4):

$$a = 0.275 \frac{\lambda}{\gamma} \sqrt{T_{\text{dB}}}. \tag{6}$$

Suppose the antenna shown in Fig. 1 has a diameter-to-wavelength ratio, D/λ, in the hundreds, an equivalent focal length, F, and an F/D ratio larger than 2. Then if a second feed-horn is placed adjacent to the on-axis feed, the second beam will be aimed in an off-axis direction,

$\theta_1 = 2a/F$. Inserting a from Eq. (6) and noting[8] that $\gamma = D/2F$,

$$\theta_1 = 1.1 \frac{\lambda}{D} \sqrt{T_{\mathrm{dB}}}. \qquad (7)$$

Inserting (7) into the parameter on the abscissa of Fig. 2, the value of u for contiguous corrugated feed horns is

$$u_1 = 3.46\sqrt{T_{\mathrm{dB}}}. \qquad (8)$$

For $T = 15$ dB, $u_1 = 13.4$. From Fig. 2, the -3-dB beamwidth is 3.62; thus, u_1 corresponds to $13.4/3.62 = 3.7$ beamwidths. For $u_1 = 13.4$, Fig. 2 shows that the side-lobe envelope level is -37 dB; this is approximately equal to the isolation of two beams spaced θ_1 degrees apart. The isolations for typical beam spacings are included in Table I. In the earth-station example, the minimum beam spacing is 0.6°, but the corresponding isolations, 37 and 43 dB at 4 and 6 GHz, respectively, are too small for allowable adjacent-satellite interference.[12] These isolations can be increased to 45 and 49 dB, respectively, by increasing the beam (and satellite) spacing to 1°. The increased beam spacing also allows room between feed horns, so they can be moved individually to track small errors in satellite positioning.

IV. AREA EFFICIENCY

Suppose an off-axis plane wave is incident on the main-reflector aperture shown in Fig. 1. The rays intercepted and reflected by the main reflector are displaced laterally with respect to those from an on-axis beam. But if the subreflector surface is sufficiently broadened, each of these rays will be intercepted and focused to a new point that is displaced laterally with respect to focal point f. To accommodate off-axis beams in the horizontal plane, the subreflector width is increased; similarly, for beams in the vertical plane, the height is increased, as indicated by the dashed line in Fig. 1.

The lateral displacement of the focus, corresponding to an off-axis beam at an angle θ, is equal to θ times the equivalent focal length, F. It is assumed that a separate corrugated feed horn is optimally positioned about the focus of each off-axis beam, i.e., each feed is pointed such that the original on-axis amplitude distribution is maintained across the main-reflector aperture, and each feed is longitudinally positioned to minimize aberrations.

The phase center of a corrugated horn can be calculated as a function of frequency.[9] This in turn allows the longitudinal position of the feed to be optimized for broadband performance.

Assuming the foregoing precautions are observed, each beam of the antenna in Fig. 1 has a computed gain about 1 dB less than that obtainable from an aperture with a uniform amplitude distribution. The underlying reasons for the good area efficiency, 80 percent, are (*i*) the main reflector does not have to be enlarged to accommodate off-axis beams, and (*ii*) the F/D ratio of a Cassegrainian antenna is fairly large.

V. POLARIZATION CROSS-COUPLING

T. S. Chu and R. H. Turrin have shown that the cross-coupling of an offset reflector is a function of (*i*) the angle between the feed axis and the reflector axis and (*ii*) the half-angle subtended at the focus by the reflector.[13] In an offset Cassegrainian antenna with a moderate F/D ratio, these angles are fairly small; thus, the cross-coupling is very small. In particular, in Fig. 1, $\psi = 14°$ and $\gamma = 8.5°$. For linearly polarized excitation, the cross-polarized lobes have a peak value of -45 dB. It is anticipated that, in beams with small off-axis angles, as in Table I, the cross-coupling will be about the same.

VI. MULTIPLE-BEAM ANTENNA PARAMETERS

The off-axis beam parameters and corresponding feed-horn dimensions of an offset Cassegrainian antenna fed with corrugated horns can be calculated once the main-aperture diameter and operating

Table I — Multiple-beam antenna parameters

	Earth Station at 4/6 GHz	Satellite at 20/30 GHz
Aperture diameter, D	30 meters	3 meters
Wavelength, λ	7.5 cm/5 cm	1.5 cm/1 cm
Beamwidth, β	0.165°/0.11°	0.33°/0.22°
Primary focal length, F'	30 meters	3 meters
Off-axis beam angle, θ	5°	4°
No. of beamwidths, $N = \theta/\beta$	30/45	12/18
Off-axis parameter, X'	30/45	12/18
Main-reflector offset, Y_2	25 meters	2.5 meters
Subreflector offset, Y_1	5 meters	0.5 meter
Equivalent focal length, F	100 meters	10 meters
F/D ratio	3.33	3.33
Coma aberration, $\Delta\phi$	18°/27°	14°/21°
Off-axis parameter, X	3.0/4.5	1.2/1.8
Feed-horn length, L	3.8 meters	76 cm
Feed-horn diameter, $2a$	1.03 meters	20.5 cm
Beam spacing θ_1	0.6°	1.2°
Isolation at θ_1 spacing	37 dB/43 dB	37 dB/43 dB
Isolation at 1° spacing	45 dB/49 dB	—
No. of available beams	16 (in a row)	18 (within U. S.)

wavelengths are specified. Typical results for an earth-station antenna at 4 and 6 GHz and a satellite antenna at 20 and 30 GHz are given in Table I. Similar results for other diameters and wavelengths can be found by following the text and performing the calculations in the order listed in Table I.

VII. CONCLUSIONS

An offset Cassegrainian antenna fed with corrugated horns is expected to have well-isolated multiple beams that are broadband and dual-polarized. The antenna has good area efficiency and is relatively compact. This combination of properties makes the antenna well-suited for earth stations and satellites.

VIII. ACKNOWLEDGMENT

The author wishes to thank R. F. Trambarulo and T. S. Chu for stimulating discussions, and T. S. Chu for the envelope profile shown in Fig. 2. The author also wishes to thank M. J. Gans for his calculation of area efficiency and his off-axis radiation patterns, which show that the further-out side lobes are not affected by coma aberration.

REFERENCES

1. L. C. Tillotson, "A Model of a Domestic Satellite Communication System," B.S.T.J., *47*, No. 10 (December 1968), pp. 2111–2137.
2. John Ruze, "Lateral Feed Displacement in a Paraboloid," IEEE Trans. on Antennas and Propagation, September 1965, pp. 660–665.
3. T. S. Chu, "A Multibeam Spherical Reflector Antenna," IEEE Antennas and Propagation Int. Symp., Program and Digest, December 9, 1969, pp. 94–101.
4. Henry Zucker, "Offset Parabolic Reflector Antenna," U. S. Patent 3,696,435, filed Nov. 24, 1972.
5. William C. Wong, "On the Equivalent Parabola Technique to Predict the Performance Characteristics of a Cassegrain System with an Offset Feed," IEEE Trans. on Antennas and Propagation, *AP-21*, No. 3 (May 1973).
6. S. K. Buchmeyer, "Corrugations Lock Horns with Poor Beamshapes," Microwaves, January 1973, pp. 44–49.
7. C. Dragone and D. C. Hogg, "The Radiation Pattern and Impedance of Offset and Symmetrical Near-Field Cassegrainian and Gregorian Antennas," IEEE Trans. on Antennas and Propagation, *AP-22*, No. 3 (May 1974), pp. 472–475.
8. Peter W. Hannan, "Microwave Antennas Derived from the Cassegrain Telescope," IRE Trans. on Antennas and Propagation, March 1961, pp. 140–153.
9. C. Dragone, unpublished work, June 1972.
10. H. Zucker, unpublished work, December 1969.
11. H. Kogelnik and Tingye Li, "Laser Beams and Resonators," Appl. Opt., *5*, No. 10 (October 1966), pp. 1550–1567.
12. AT&T Application for a Domestic Communications Satellite System, before the FCC, March 3, 1971, Table IX.
13. Ta-Shing Chu and R. H. Turrin, "Depolarization Properties of Offset Reflector Antennas," IEEE Trans. on Antennas and Propagation, *AP-21*, No. 3 (May 1973), pp. 339–345.

Bibliography for Part VII

[1] R. A. Semplak, "100 GHz measurements on a multiple beam offset antenna," *Bell Syst. Tech. J.*, vol. 56, pp. 385-398, Mar. 1977.

[2] G. Hyde, R. W. Kreutel and L. V. Smith, "The unattended earth terminal multiple beam antenna," *COMSAT Tech. Rev.*, vol. 4, pp. 231-262, Fall 1974.

[3] S. S. Sandler, "Paraboloidal reflector patterns for off-axis feed," *IRE Trans. Antennas Propagat.*, vol. AP-8, pp. 368-379, July 1960.

[4] J. B. Scarborough, "The caustic curve of an off-axis parabola," *Appl. Opt.*, vol. 3, pp. 1445-1446, Dec. 1964.

[5] T. Takeshima, "Beam scanning of parabolic antenna by defocusing," *Electron. Eng.*, vol. 41, pp. 70-72, Jan. 1969.

[6] A. W. Rudge and M. J. Withers, "Beam-scanning primary feed for parabolic reflectors," *Electron. Lett.*, vol. 5, pp. 39-41, Feb. 6, 1969.

[7] M. S. Afiffi, "Aberration and dispersion off the focus of a parabola," in *IEEE G-AP Symp. Dig.*, 1971, p. 219.

[8] O. Sorensen and W. V. T. Rusch, "Application of the geometrical theory of diffraction to Cassegrain subreflectors with laterally defocused feeds," *IEEE Trans. Antennas Propagat.*, vol. AP-23, pp. 698-702, Sept. 1975.

[9] B. H. C. Liesenkötter, "Raising the crossover level of dual beam parabolic antennas," *Electron. Lett.*, vol. 12, pp. 559-560, Oct. 14, 1976.

Part VIII
Phase Errors and Tolerance Theory

The first paper in this section, by Cheng, presents a simple analysis whereby the maximum loss in gain can be estimated when phase errors are present in the aperture of an antenna. The exact nature of the phase and amplitude distributions need not be known. To estimate the maximum change in beamwidth, it turns out to be necessary to know either the amplitude distribution or else the slope of the secondary pattern near the half-power point.

The next two papers are concerned with the validity of the simulation technique that makes use of axial defocusing of the feed in a reflector antenna so that Fraunhofer patterns may be measured in the Fresnel region. An inconsistency in Cheng's treatment (second paper) is resolved by Chu in the third paper.

Perhaps the earliest statistical treatment of the effect of random phase errors in the aperture of an antenna was given by Ruze in 1952. He published this work in an Italian journal [1] and about ten years later produced the well-known review paper which appears here, as the fourth in this part. It ranks high among the most oft-quoted antenna papers of all time. It is followed by two papers in similar vein. The one by Vu is concerned only with the effect of phase errors on the boresight gain. In the sixth paper, Zarghamee notes that the assumption of a uniform error distribution almost invariably produces a pessimistic estimate of the gain, although the effect is not serious until the surface deviations that cause the phase errors become an appreciable fraction of a wavelength. The same assumption, he finds, may significantly influence the scatter pattern even when the "roughness" is small.

The seventh and eighth papers are of importance for large ground-based reflector antennas in which astigmatism induced by gravitational deformation is likely to occur. Cogdell and Davis present a brief but useful discussion of astigmatism in their paper, and describe a procedure for focusing an antenna and for diagnosing the existence of astigmatism. Von Hoerner and Wong introduce the important concept of homologous deformation whereby a large antenna is deliberately designed to deform under tilting, but in such a way as to remain a paraboloid of revolution with a different focal length. It is clear, then, that the effects of distortion can be rendered innocuous by commanding the feed to follow the focal point as it changes with tilt angle.

The determination of errors in the surface profile of large reflector antennas can be a difficult and time-consuming task, even with the use of sophisticated optical surveying techniques. Bennett *et al.* describe a holographic technique, in the ninth paper, that can be applied to reflectors at microwave frequencies. The method is simple, in principle, and requires only a fixed reference antenna, the antenna under test (with its scanning pedestal), and a source, which in the case of large antennas may be a radio star. Not only can surface profile errors be determined with this powerful technique, but other vital information is also obtained, for example, the *E* and *H* plane phase centers of the feed. Accurate prediction of both near and far field radiation patterns may also be made from the holographic data.

In the tenth and final paper of this part, Ingerson and Rusch return to a consideration of deterministic (as opposed to random) phase errors of the kind caused by axial displacement of the feed from the focal point. They show that defocusing effects are not symmetrical about the focus and that, when defocusing is deliberately used to achieve beam broadening, the feed should produce a well-tapered aperture distribution. For less tapered illumination, the main beam becomes bifurcated before appreciable broadening occurs.

Effect of Arbitrary Phase Errors on the Gain and Beamwidth Characteristics of Radiation Pattern*

D. K. CHENG†

Summary—**Simple expressions have been obtained for predicting the maximum loss in antenna gain when the peak value of the aperture phase deviation is known. It is not necessary to know the exact amplitude or phase distribution function as long as the phase errors are relatively small; and the same expressions may be used for both rectangular and circular aperture cases. Relations have also been established such that the maximum change in the main-lobe beamwidth can be predicted from the knowledge of the amplitude distribution function and the peak phase deviation.**

INTRODUCTION

IT IS KNOWN that for a given amplitude illumination function, a uniform phase distribution over the aperture plane of a microwave antenna reflector gives a maximum gain. A uniform phase distribution requires an exact parabolic surface in addition to a correct primary feed. Any deviation from the exact parabolic surface will introduce phase errors, which in turn will cause a reduction in gain. Unfortunately phase errors are often quite arbitrary and it is in general not possible to insert them under integral signs, weight them properly with the amplitude function, and perform integrations. Ruze[1] has investigated the effect of random phase errors on the radiation pattern as a statistical problem and obtained approximate formulas for the reduction in gain. However, as a statistical problem, only the average behavior of a large number or an ensemble of seemingly identical antennas and the probability distribution of the members of the ensemble about an average radiation pattern can be discussed; the individual patterns will differ from the system-average pattern. By a least-square analysis, Spencer[2] obtained an approximate expression for the fractional loss in gain due to small phase errors. In order to estimate the loss quantitatively it would be necessary to determine the plane least-square solution of the wavefront from a complete knowledge of the amplitude illumination function and the phase-error function over the aperture. The integration process involved is in general very difficult to carry out.

In practice, it is desirable to be able to predict the *maximum* effect on the gain, main-lobe beamwidth, etc., if the peak phase error is given for an *individual* antenna, even when the exact phase distribution is not known or too complicated for analysis. This paper presents a simplified approach with which the maximum loss in gain and the maximum change in beamwidth due to small arbitrary phase errors can be estimated.

PHASE-ERROR EFFECT ON ANTENNA GAIN

Consider the case of a rectangular aperture with separable, symmetrical field distribution. The maximum value of gain function, or simply gain, can be written as

$$G = \frac{2\pi A}{\lambda^2} \frac{\left| \int_{-1}^{1} f(x)dx \right|^2}{\int_{-1}^{1} |f(x)|^2 dx}, \tag{1}$$

in which all notations are conventional. The aperture-field distribution function $f(x)$ is in general

$$f(x) = F(x)\epsilon^{j\phi(x)}, \tag{2}$$

where $F(x)$ is the amplitude illumination function and $\phi(x)$ represents the phase function. It is implied in (1) that the phase error is small and that maximum radiation occurs along the axis of the reflector. $\phi(x)$ may vary in an unknown manner across the aperture but it is assumed that the maximum deviation from an average value is known or can be estimated:

$$|\Delta\phi(x)| = |\phi(x) - \overline{\phi(x)}| \leqq m, \qquad -1 \leqq x \leqq 1. \tag{3}$$

In (3), $\overline{\phi(x)}$ is the average value; it will have no effect on the gain since the term $\epsilon^{j\overline{\phi(x)}}$ can be taken out from under the integral sign. Only the phase deviation $\Delta\phi(x)$ from this average value is of importance. Substituting (2) in (1), one has

$$G = \frac{2\pi A}{\lambda^2} \frac{\left| \int_{-1}^{1} F(x)\epsilon^{j\Delta\phi(x)}dx \right|^2}{\int_{-1}^{1} |F(x)|^2 dx}. \tag{4}$$

When there is no phase error, the maximum gain is

$$G_0 = \frac{2\pi A}{\lambda^2} \frac{\left| \int_{-1}^{1} F(x)dx \right|^2}{\int_{-1}^{1} |F(x)|^2 dx}. \tag{5}$$

Hence, for $F(x) \geqq 0$,

$$\frac{G}{G_0} = \frac{\text{gain with phase error}}{\text{gain without phase error}} = \frac{\left| \int_{-1}^{1} F(x)\epsilon^{j\Delta\phi(x)}dx \right|^2}{\left| \int_{-1}^{1} F(x)dx \right|^2}. \tag{6}$$

* Original manuscript received by the PGAP, February 18, 1955; revised manuscript received, April 14, 1955.

† Electrical Engrg. Dept., Syracuse University, Syracuse, N. Y.

[1] J. Ruze, "Effect of Aperture Distribution Errors on the Radiation Pattern," Antenna Lab. Memo., AF Cambridge Res. Center; January 22, 1952.

[2] R. C. Spencer, "A Least Square Analysis of the Effect of Phase Errors on Antenna Gain," Rep. No. E5025, AF Cambridge Res. Center; January, 1949.

Reprinted from *IRE Trans. Antennas Propagat.*, vol. AP-3, pp. 145–147, July 1955.

Examining the numerator of (6) for small $\Delta\phi(x)$ values, i.e., when

$$\epsilon^{j\Delta\phi(x)} \cong 1 - \tfrac{1}{2}[\Delta\phi(x)]^2 + j\Delta\phi(x), \quad (7)$$

which holds for $(\Delta\phi)^3/3! \ll \Delta\phi$, or $\Delta\phi \ll \sqrt{6} = 2.45 = \pi/1.28$,

$$\left| \int_{-1}^{1} F(x)\epsilon^{j\Delta\phi(x)}dx \right|^2 \cong \left\{ \int_{-1}^{1} F(x)\left[1 - \frac{(\Delta\phi)^2}{2}\right]dx \right\}^2 + \left\{ \int_{-1}^{1} F(x)\Delta\phi dx \right\}^2. \quad (8)$$

The first term on the right-hand side of (8) is

$$\left\{ \int_{-1}^{1} F(x)[1 - (\Delta\phi)^2/2]dx \right\}^2 = \left\{ \int_{-1}^{1} F(x)dx - (1/2)[\Delta\phi(\xi)]^2 \int_{-1}^{1} F(x)dx \right\}^2 \geqq \left(1 - \frac{m^2}{2}\right)^2 \left| \int_{-1}^{1} F(x)dx \right|^2. \quad (9)$$

Eq. (9) is the result of the application of the mean-value theorem [$\Delta\phi(x)$ is continuous and $F(x)$ is positive within the range of integration; $-1 \leqq \xi \leqq +1$] and relation (3). The second term on the right-hand side of (8) is

$$\left\{ \int_{-1}^{1} F(x)\Delta\phi dx \right\}^2 \geqq 0. \quad (10)$$

Substituting (9) and (10) in (8), and then back into (6), one obtains

$$\frac{G}{G_0} \geqq \left(1 - \frac{m^2}{2}\right)^2. \quad (11)$$

Eq. (11) is a useful relation because it sets the lower bound for the gain when the maximum phase deviation m is known; it is independent of the amplitude illumination function $F(x)$ and the exact variation of $\Delta\phi(x)$. From (11), one also readily obtains the maximum fractional reduction in gain

$$\frac{\Delta G}{G_0} = 1 - \frac{G}{G_0} \leqq m^2\left(1 - \frac{m^2}{4}\right). \quad (12)$$

It can be shown that for reflectors with circular aperture and symmetrical field distribution, (6) will be changed to the following:

$$\frac{G}{G_0} = \frac{\left| \int_0^1 F(\rho)\epsilon^{j\Delta\phi(\rho)}\rho d\rho \right|^2}{\left| \int_0^1 F(\rho)\rho d\rho \right|^2}. \quad (13)$$

Provided (7) is satisfied for small phase deviations, the derivation procedure is entirely similar, and one also obtains (11) and (12) as the result.

As an example, if $m = \pi/16 = 0.1964$ (corresponding to $\lambda/32$ in terms of wavelength λ), the *minimum* G/G_0 is 0.9618 which corresponds to a *maximum* reduction in gain of 0.169 db. The *maximum* fractional reduction in gain from (12) is 0.0382 or 3.82 per cent.

When the phase distribution function is such that the maximum radiation does not occur along the axis of the aperture, for instance, if $\phi(x)$ is not an even function of x in the rectangular aperture case, then the gain formula as given by (6) does not give the maximum gain ratio, but the lower bound as given by (11) still holds and is on the safe side.

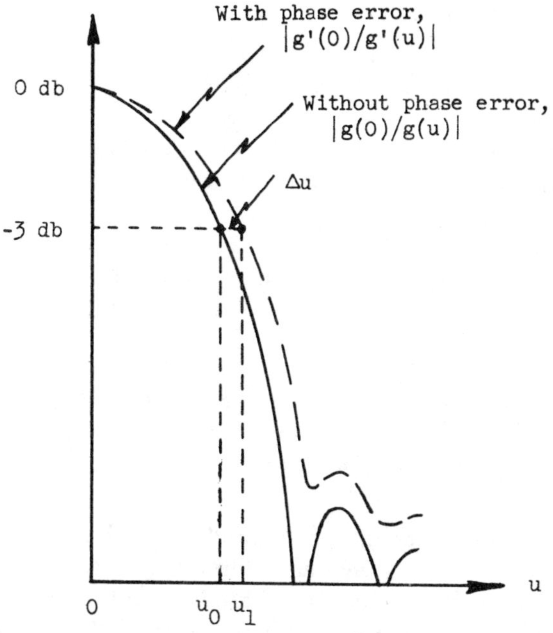

Fig. 1—Normalized radiation patterns. $u = (\pi D/\lambda)\sin\theta$

Phase-Error Effect on Beamwidth

When the phase error is small, an estimate on the maximum change in the main-lobe beamwidth of the radiation pattern can also be obtained. Consider this time the case of a circular aperture. It is necessary to assume here that both the amplitude and the phase distribution functions in the aperture plane are circularly symmetrical. In Fig. 1 are plotted two normalized radiation patterns, one without phase error $|g(0)/g(u)|$ and one with phase error $|g'(0)/g'(u)|$, where

$$g(u) = \int_0^1 F(\rho)J_0(u\rho)\rho d\rho \quad (14)$$

$$g'(u) = \int_0^1 F(\rho)J_0(u\rho)\rho\epsilon^{j\Delta\phi(\rho)}d\rho. \quad (15)$$

The condition for equal normalized radiation level is

$$\left| \frac{g'(0)}{g'(u_1)} \right| = \left| \frac{g(0)}{g(u_0)} \right|, \quad (16)$$

which would be equal to $10^{3/20}$ at 3 db down. It is noted that for $u \leqq 2.405 = \pi/1.3$, the integrand in (14) is always greater than or equal to zero and the following inequality holds:

$$|g'(u)| \leqq g(u). \tag{17}$$

When the small phase error condition as indicated in (7) is satisfied (with x changed to ρ in this case), one may go through the same reasoning that led to (11) and get the relationship

$$|g'(u)| \geqq \left(1 - \frac{m^2}{2}\right) g(u). \tag{18}$$

Combining (16), (17) and (18), one obtains

$$g(u_1) \geqq |g'(u_1)| \geqq \left(1 - \frac{m^2}{2}\right) g(u_0). \tag{19}$$

For small phase errors, the change in beamwidth is small and it is convenient to write

$$u_1 = u_0 + \Delta u, \qquad \Delta u \ll 1. \tag{20}$$

Substitution of (14), (15) and (20) in (19) yields the following result:

$$\Delta u \leqq \frac{m^2}{2} \frac{\int_0^1 F(\rho) J_0(u_0\rho)\rho d\rho}{\int_0^1 F(\rho) J_1(u_0\rho)\rho^2 d\rho} = \frac{m^2}{2}\left[\frac{g(u)}{-\frac{d}{du}g(u)}\right]_{u_0}. \tag{21}$$

Eq. (21) furnishes an upper bound in the change in half-beamwidth for a given maximum phase error; it is dependent upon the amplitude illumination function $F(\rho)$.

For the case of a rectangular aperture, the following expression is obtained by a similar procedure (for $u \leqq \pi/2$):

$$\Delta u \leqq \frac{m^2}{2} \frac{\int_0^1 F(x) \cos(u_0 x) dx}{\int_0^1 F(x)\, x \sin(u_0 x) dx} = \frac{m^2}{2}\left[\frac{g(u)}{-\frac{d}{du}g(u)}\right]_{u_0}. \tag{22}$$

It is seen that (22) is entirely similar to (21). Results listed in Table I are for the simplest case of uniform amplitude illumination function. The maximum change in beamwidth for other typical amplitude illumination functions can similarly be computed; there is no need to know the exact phase deviation curve. Eqs. (21) and (22) do not give useful information for very small u_0-values but are very helpful in estimating changes in 3-db beamwidth. Moreover, it can be proved that when u_0 is small compared with unity, Δu is always smaller than u_0.

TABLE I

MAXIMUM 3-DB BEAMWIDTH CHANGES FOR UNIFORM AMPLITUDE FUNCTION

		Rectangular aperture	Circular aperture
Amplitude function		$F(x)=1$	$F(\rho)=1$
Half 3-db beamwidth (no phase error), u_0		0.45π	0.51π
$m=0.1$	Δu	$\leqq 0.00918$	$\leqq 0.011$
	$\Delta u/u_0$	$\leqq 0.65$ per cent	$\leqq 0.68$ per cent
$m=0.2$	Δu	$\leqq 0.0367$	$\leqq 0.044$
	$\Delta u/u_0$	$\leqq 2.6$ per cent	$\leqq 2.8$ per cent

CONCLUSION

It has been shown that for small phase errors, simple expressions (11) or (12) can be used to compute the maximum possible loss in antenna gain when the peak values of the phase deviation is known. It is not necessary to know the exact amplitude or phase distribution function in the aperture; and the same expressions apply to both rectangular and circular aperture cases. Similarly, (21) and (22) can be used to compute the maximum possible change in half 3-db beamwidth, which is dependent upon the amplitude illumination and is different for the rectangular and circular aperture cases.

ACKNOWLEDGMENT

The author wishes to acknowledge the valuable assistance of Pranas Grusauskas in the final preparation of this paper.

On the Simulation of Fraunhofer Radiation Patterns in the Fresnel Region*

DAVID K. CHENG†

Summary—**Physical limitations on the size of obstacle-free test sites give rise to the need of making radiation-pattern measurements on high-gain antennas at a reduced distance. The general practice is to defocus the primary source along the principal axis of the antenna reflector by a small distance so that Fraunhofer patterns may be simulated in the Fresnel region. This note summarizes and compares three different approaches with which the proper amount of defocus may be determined.**

THE PERFORMANCE of an antenna is usually specified in terms of the characteristics of its radiation pattern in the Fraunhofer region. This is because of the fact that Fraunhofer radiation patterns do not change with the distance from the antenna as long as the far-zone approximations are satisfied. Radiation patterns in the Fresnel region tend to be very complex, and they change considerably with distance.

While there is no clear-cut boundary between the Fraunhofer and the Fresnel regions, a common and acceptable criterion is that $2D^2/\lambda$ represents a safe far-zone distance, where D is the maximum dimension of the antenna aperture and λ is the operating wavelength. At a distance of $2D^2/\lambda$ the maximum path-length difference between the contribution from the edge of the aperture and that from the center corresponds to $\lambda/16$ or $\pi/8$ radians. In practice, an unobstructed, open space with a dimension of $2D^2/\lambda$ is often not available for testing high-resolution antennas. For example, the $2D^2/\lambda$ distance for a 20-foot antenna at 3 cm would be about 1.53 miles. Higher gain requirements would demand even larger test sites. Unfortunately, calculation of Fraunhofer radiation patterns from measurements in the Fresnel region is not of practical value at the present time both because of the difficult and laborious process of extrapolation and because of the inherent difficulties in making accurate amplitude and phase measurements.[1] The need for the technique of testing microwave antennas at reduced ranges is, therefore, both real and urgent.

A commonly used practice is to displace the primary source of the antenna assembly slightly from the focal position in a direction away from the reflector. The

* Manuscript received by the PGAP, February 27, 1956; revised manuscript received February 4, 1957. This work was supported by the Rome Air Development Center, USAF, under Contract No. AF 30(602)-1360.

† Elec. Eng. Dept., Syracuse Univ., Syracuse, N. Y.

[1] A. F. Kay, "Far Field Data at Close Distances," Final Rep. for Contract No. AF 19(604)-1126, Tech. Res. Group, New York, N. Y.; October, 1954.

Reprinted from *IRE Trans. Antennas Propagat.*, vol. AP-5, pp. 399–402, Oct. 1957.

present note summarizes three different methods with which the required amount of defocus may be determined.[2] The results are plotted and compared.

The Geometrical Approach

It is a well-known fact that for an effective point source of excitation, the best radiation pattern will be obtained from a paraboloidal reflector at a field point in the far zone when the source is located at the focal point of the reflector. Geometrically, this may be explained by equal path length from the source to all points in an aperture plane by virtue of the inherent property of a focused paraboloid. If the field point is far enough away from the reflector, the path lengths from the aperture points to the field point will be again approximately equal, resulting in an optimum additive effect. When the field point lies in the quasi-near zone (Fresnel region), the path-length differences from the points in an aperture plane to the field point must be compensated in some way in order that the measured radiation pattern may approach the true far-zone (Fraunhofer) pattern. This is done by slightly defocusing the source along the reflector axis in the direction away from the reflector. Since the amount of on-axis defocus is the only adjustable variable here, one cannot expect to achieve equal path length for all points in the aperture plane. For simplicity, the conventional approach is to make the path length from the source to the field point by way of the apex of the paraboloid equal to that by way of the points on the edge of the reflector.

Refer to Fig. 1, which represents a cross section of a symmetrical paraboloidal reflector with focal length $\overline{OF}=f$ and defocused point source at F', the above requirement is equivalent to making

$$\overline{F'O} + \overline{OO'} = \overline{F'A} + \overline{AB} \tag{1}$$

Call $\overline{AA'}=D$ (aperture diameter), $\overline{O'P}=\overline{BP}=R$ (distance of measurement), $\overline{FF'}=\epsilon$ (defocus distance). Eq. (1) can be reduced to give

$$\epsilon = \frac{f^2}{R}\left[\left(\frac{R}{R-f}\right)+\left(\frac{D}{4f}\right)^2\right]. \tag{2}$$

When $(f/R)^2 \ll 1$, it is accurate enough to write (2) as

$$\epsilon = \frac{f^2}{R}\left[1+\frac{f}{R}+\left(\frac{D}{4f}\right)^2\right]. \tag{3}$$

Normalizing all quantities with respect to the focal length and introducing new notations $\epsilon'=\epsilon/f$, $R'=R/f$, and $D'=D/f$, one can rewrite (3) as

$$\epsilon' = \frac{1}{R'}\left[1+\frac{1}{R'}+\left(\frac{D'}{4}\right)^2\right]. \tag{4}$$

The normalized amount of defocus needed is seen to increase when R' decreases and when D' increases. As R' approaches infinity, ϵ' correctly goes to zero.

[2] D. K. Cheng, "Microwave aerial testing at reduced ranges," *Wireless Eng.* vol. 33, pp. 234–237; October, 1956.

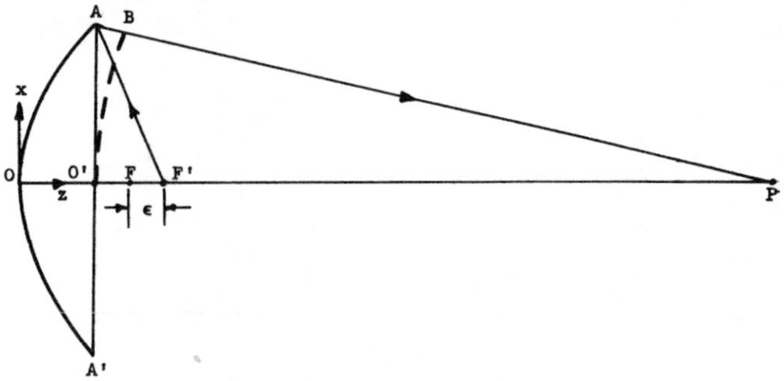

Fig. 1—Geometry of on-axis defocusing arrangement for paraboloidal reflector.

The Aperture-Phase Approach

The defocusing problem can also be approached from a consideration of the phase distribution in an aperture plane of the reflector together with the diffraction integral for the field at a point in space. When the point under consideration is in the quasi-near zone of a paraboloidal reflector, the normalized diffraction integral which gives the field pattern in a horizontal plane can be approximated as[3]

$$I(u) = \int_0^1 F(r)^{-jkD^2r^2/8R} rJ_0(ur)dr. \tag{5}$$

In (5), r is the radial dimension of the aperture plane normalized with respect to $D/2$; $u=(\pi D/\lambda)\sin\theta$, θ being the azimuth angle; $k=2\pi/\lambda$; and $F(r)$ is the circularly symmetrical amplitude illumination function over the aperture. The explicit exponential term is the Fresnel-region contribution; terms above the second order are neglected. When R is very large, (5) reduces to the far-zone pattern function

$$I_0(u) = \int_0^1 F(r)rJ_0(ur)dr. \tag{6}$$

When the primary source is displaced from the focus of a paraboloidal reflector along the reflector axis in the direction away from the reflector with a view to simulating far-zone patterns in the quasi-near zone, there will be a relative phase variation over the aperture. It has been found that this phase variation referred to the center point can be approximated satisfactorily by[4]

$$\delta \cong -2\epsilon\left[1-\frac{r^2}{\left(\frac{4f}{D}\right)^2+1}\right]. \tag{7}$$

Eq. (7) is exact for $r=0$ (center) and $r=1$ (edge of aperture). For other values of r, the $|\delta|$ given by (7) is slightly too large; the error decreases when the (f/D)

[3] S. Silver, "Microwave Antenna Theory and Design," M.I.T. Rad. Lab. Ser., McGraw-Hill Book Co., Inc., New York, N. Y., vol. 12, ch. 6; 1949.

[4] D. K. Cheng and S. T. Moseley, "On-axis defocus characteristics of the paraboloidal reflector," IRE Trans., vol. AP-3, pp. 214–216; October, 1955.

ratio of the reflector increases. The diffraction integral now becomes

$$I(u) = \int_0^1 F(r)\epsilon^{jk[\delta - D^2r^2/8R]} rJ_0(ur)dr. \tag{8}$$

In order to simulate Fraunhofer radiation patterns in the Fresnel region, the exponent under the integral sign in (8) should be made to vanish. This yields

$$\frac{\epsilon}{f} = \frac{f}{R}\left[1 + \left(\frac{D}{4f}\right)^2\right]$$

or

$$\epsilon' = \frac{1}{R'}\left[1 + \left(\frac{D'}{4}\right)^2\right] \tag{9}$$

which checks with (4) when $R' = R/f \gg 1$. If it is desirable to write

$$R' = nD'^2/\lambda' \tag{10}$$

with $\lambda' = \lambda/f$, n a numeric, then (9) reduces to

$$\frac{\epsilon'}{\lambda'} = \frac{\epsilon}{\lambda} = \frac{1}{n}\left[\left(\frac{1}{D'}\right)^2 + \left(\frac{1}{4}\right)^2\right]. \tag{11}$$

Eq. (11) shows that for a given value of D', (ϵ/λ) plotted vs n gives a hyperbola in linear scales, and a straight line in log-log scales.[4] It is noted that for $n=2$ ($R = 2D^2/\lambda$), appreciable defocus is still necessary.

The Ellipsoidal-Reflector Approach

The purpose of defocusing the primary source in the case of a paraboloidal reflector is to simulate far-zone radiation patterns at points in the quasi-near zone. In terms of geometrical optics, it is quite easy to see that this could be achieved by means of an ellipsoidal reflector. If the primary source is placed at one of the two foci of an ellipsoidal reflector, the reflected rays will converge at the other.

The equation in the xz plane of a cross section of an ellipsoidal reflector with focal lengths f_1 and f_2 is

$$z = \frac{f_1 + f_2}{2}\left[1 - \sqrt{1 - \frac{x^2}{f_1 f_2}}\right]. \tag{12}$$

Subject to the condition

$$\sqrt{1 - \frac{x^2}{f_1 f_2}} \cong 1 - \frac{x^2}{2f_1 f_2} \tag{13}$$

(13) can be approximated as

$$z = \frac{f_1 + f_2}{4f_1 f_2} x^2 \tag{14}$$

which is the equation for a parabola of focal length

$$f = \frac{f_1 f_2}{f_1 + f_2} \tag{15}$$

or

$$\frac{1}{f} = \frac{1}{f_1} + \frac{1}{f_2}\cdot \tag{16}$$

Hence, for reflected rays to converge at $R = f_2$, the primary source should be placed at $z = f_1$, and

$$\epsilon = f_1 - f = \frac{f^2}{R - f} \tag{17}$$

or, in normalized form,

$$\epsilon' = \frac{1}{R' - 1} = \frac{1}{R'}\left[1 + \frac{1}{R'} + \frac{1}{(R')^2} + \cdots\right]\cdot \tag{18}$$

Eq. (18) should be compared with both (4) and (9). ϵ' can also be expressed in terms of the focal lengths as

$$\epsilon' = \frac{f_1}{f_2} \tag{19}$$

which is extremely simple.

The basis of the ellipsoidal-reflector approach lies in the fact that an ellipsoidal reflector of focal lengths f_1 and f_2 approximates a paraboloidal reflector of focal length f as given by (15) or (16). An examination of (13) shows that it implies the condition $(1/8)(x^2/f_1f_2)^2 \ll 1$. Now the maximum value of x is $D/2 \leq 2f_1$. This reduces the condition to

$$\frac{f_2}{f_1} \gg \sqrt{2} \tag{20}$$

which is undoubtedly true in practice. An ellipsoidal reflector with focal lengths f_1 and f_2 has its semimajor and semiminor axes equal to $(f_1+f_2)/2$ (arithmetical mean) and $\sqrt{f_1 f_2}$ (geometrical mean), respectively; it approaches very closely a paraboloidal reflector when (20) is satisfied. As an example, with $f_2 = R = 50f_1$ the maximum error is less than 0.09 per cent.

Comparison of Defocusing Methods

Curves plotting ϵ' vs R' based upon (4), (9), and (19) from the three different approaches discussed above are shown in Fig. 2. It is seen that except for small values of R', the required ϵ' from the geometrical approach is nearly the same as that from the aperture-phase approach, both of which increase with increasing D'. The required ϵ' from the ellipsoidal-reflector approach is the smallest of the three methods and is independent of D'.

A review of the geometrical approach reveals that there is really no plausible justification in requiring equal path length from the source to the field point by way of the apex and by way of the points on the edge of the paraboloidal reflector only; the path lengths by way of the intermediate points on the reflector would then all be longer. Besides, there is no guarantee that the rays emanating from F' will be reflected to pass through the point P except the ray along the principal axis. The approximation (7) used in the aperture-phase approach is

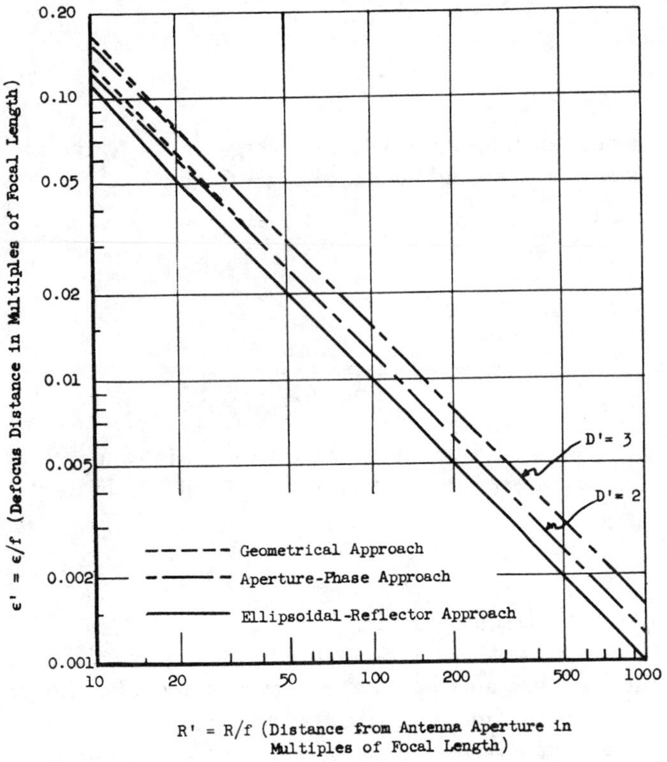

Fig. 2—Comparison of defocusing methods.

exact for $r=0$ and $r=1$ only; for $0<r<1$, δ, given by (7), is numerically too large resulting in an ϵ' which is also too large. It can be shown that the maximum error in δ introduced by (7) is

$$1 - \frac{1}{\sqrt{1+q}}\left(2 - \frac{1}{\sqrt{1+q}}\right) \tag{21}$$

where $q=(D/4f)^2$. For a reflector with $q=0.35$ or $D'=2.36$, the maximum error is about 5 per cent.

Although the geometrical approach and the aperture-phase approach yield approximately the same results, the aperture-phase approach makes it clear that this method would not be useful when R is too small because it would then be necessary to include terms higher than the second order in the exponent that appears in (5); the geometrical approach gives no indication of this resctriction. It is believed that ϵ' in (18) derived from the ellipsoidal-reflector approach gives the most nearly correct results because the approximation implied by (13) is very good; it does not restrict its correctness only to the edge of the reflector.

It should be noted that in all three methods the required amount of defocus is not a function of the operating wavelength and that diffraction phenomena are neglected.

A Note on Simulating Fraunhofer Radiation Patterns in the Fresnel Region

Abstract—**This communication resolves the inconsistency among the earlier results of Cheng [1]. The region of validity for this simulation technique is clarified by calculating the residual phase deviation.**

A test range for the measurement of antenna radiation patterns should be free of scattering obstacles and satisfy the far zone ($2D^2/\lambda$) criterion. The practical difficulty of providing such test sites for large antennas stimulates the interest in simulating Fraunhofer radiation patterns by measurements in the Fresnel region. The required defocusing for the feed of a paraboloidal reflector in order to achieve this simulation was discussed by Cheng [1]. However, he obtained inconsistent answers from different approaches. It is the purpose of this communication to resolve this ambiguity by calculating the residual phase deviation.

The focal plane phase front of a defocused paraboloid deviates from that of a focused paraboloid by $\delta = \epsilon(1 + \cos\theta)$ as shown in Fig. 1(a). Making use of the relation $\tan\theta/2 = rD/4f$ yields

$$\delta = \frac{-2\epsilon}{1 + (D/4f)^2 r^2} = -2\epsilon\left[1 - \frac{r^2}{(4f/D)^2 + r^2}\right] \tag{1}$$

where r is the normalized radius of the reflector aperture. Cheng proposed the following approximation for (1)

$$\delta = -2\epsilon\left[1 - \frac{r^2}{(4f/D)^2 + 1}\right]. \tag{2}$$

Within the preceding approximation, a spherical wavefront of quadratic approximation [exp $(-jkD^2r^2/8R)$] will be obtained from the defocused paraboloid if the defocus distance is

$$\epsilon = \frac{f^2}{R}\left[1 + \left(\frac{D}{4f}\right)^2\right]. \tag{3}$$

Equations (1) and (2) coincide at $r = 0$ and $r = 1$. The difference between (1) and (2) at intermediate points $(0 < r < 1)$ is

$$\frac{\Delta}{\lambda} = -\frac{1}{2}\frac{r^2(1 - r^2)(D/4f)^2}{1 + (D/4f)^2 r^2}\frac{(D/2)^2}{\lambda R}. \tag{4}$$

Cheng also compared an ellipsoidal reflector with a defocused paraboloid. The equation of an ellipsoidal reflector with focal lengths f_1 and R is

$$z = \frac{f_1 + R}{2}\left[1 - \left(1 - \frac{\rho^2}{f_1 R}\right)^{1/2}\right]. \tag{5}$$

Taking the first two terms of the binomial expansion for the square root, (5) becomes a paraboloid $z = \rho^2/4f$, where the focal length of the paraboloid is $f = f_1R/(f_1 + R)$. The required defocus from this approach is

$$\epsilon = f_1 - f = \frac{f^2}{R[1 - f/R]} \approx \frac{f^2}{R} \tag{6}$$

where $f/R \ll 1$. One notes that f/R is also small compared with $(D/4f)^2$ in the case of a large microwave paraboloidal antenna. The deviation between the wavefronts reflected from an ellipsoid and a defocused paraboloid is $\Delta z(1 + \cos\theta)$ as shown in Fig. 1(b). Estimating Δz by the third term in the square root expansion of (5), we have the phase deviation

$$\frac{\Delta}{\lambda} = \frac{1}{2}\frac{r^4(D/4f)^2}{1 + (D/4f)^2 r^2}\frac{(D/2)^2}{\lambda R}. \tag{7}$$

Manuscript received October 9, 1970; revised January 25, 1971.

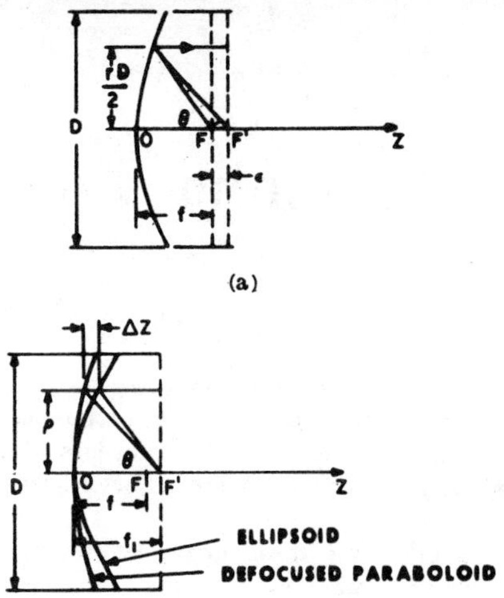

Fig. 1. (a) Defocusing of paraboloid. (b) Ellipsoidal-reflector approach.

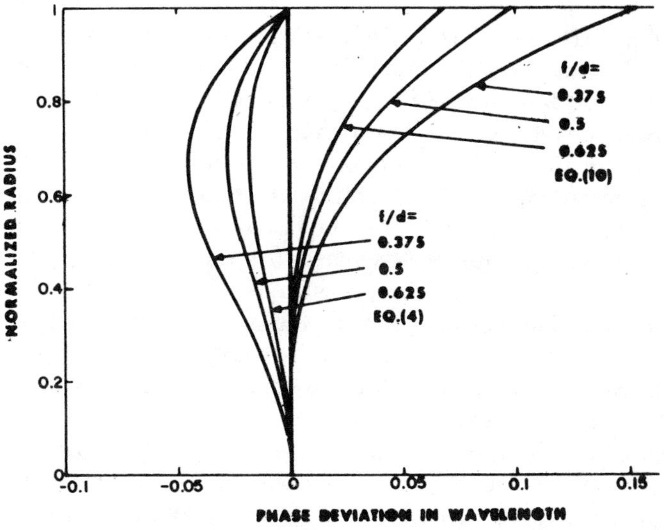

Fig. 2. Residual phase errors of defocused paraboloids with unity Fresnel number. $(D/2)^2/\lambda R = 1$.

Numerical comparison between (4) and (7) in the case of unity Fresnel member $(D/2)^2/\lambda R$ has been plotted in Fig. 2 for various f/D ratios. It is seen that the maximum total phase deviation with the defocus of (6) is more than three times that with the defocus of (3). Cheng [1] suggested (6) as the most nearly correct answer perhaps because he overlooked the effect of the large factor outside the bracket in (5). The ellipsoidal reflector approach will also give an answer identical to (3) if three terms of the square root expansion of (5) are taken and if coincidences at the center and the edge are imposed.

The phase deviation in (4) is similar to that of an optimally defocused spherical reflector for obtaining approximate plane phase front, while (7) is similar to that of feeding a spherical reflector at the half-radius point [2].

T. S. Chu
Crawford Hill Lab.
Bell Telephone Labs., Inc.
Holmdel, N. J. 07733

References

[1] D. K. Cheng, "On the simulation of Fraunhofer radiation patterns in the Fresnel region," *IRE Trans. Antennas Propagat.*, vol. AP-5, Oct. 1957, pp. 399–402.
[2] A. S. Dunbar, "Optics of microwave directive systems for wide-angle scanning," Naval Res. Lab., Rep. R-3312, Sept. 7, 1948.

Reprinted from *IEEE Trans. Antennas Propagat.*, vol. AP-19, pp. 691–692, Sept. 1971.

Antenna Tolerance Theory—A Review

JOHN RUZE, FELLOW, IEEE

Abstract—**The theoretical basis of antenna tolerance theory is reviewed. Formulas are presented for the axial loss of gain and the pattern degradation as a function of the reflector surface rms error and the surface spatial correlation.**

Methods of determining these quantities by astronomical or ground-based electrical measurements are described. Correlation between the theoretical predictions and the performance of actual large antenna structures is presented.

I. Introduction

THE REQUIREMENT of precise optics for good image quality is well known in optical technology, and methods of testing and contour shaping have been developed to obtain precisions in excess of one part in 10^7. Optical systems of very large D/λ (diameter to wavelength), ratio are therefore common. Large antennas, such as required for radio astronomy or interplanetary probes, are engineering structures subject to gravity, wind, and thermal strains. Contour measurement and adjustment to the accuracy desired is also extremely difficult. Normal civil engineering structures have a precision of about one part in a thousand. Significant progress has been made in recent large parabolic antennas both in precision of construction (one part in 30 000) and in the computer prediction of deformation under various loads. Nevertheless, the tolerance of the structure sets a limit on the highest frequency of operation and thereby on the D/λ ratio. It is desirable to review the theory of aperture errors and their effect on the antenna radiation pattern.

We begin with a simple approach and attempt to develop a tolerance theory in an heuristic manner. The axial gain of a circular aperture with an arbitrary phase error or aberration $\delta(r, \phi)$ may be written as

$$G(0) = \frac{4\pi}{\lambda^2} \frac{\left| \int_0^{2\pi} \int_0^a f(r, \phi) e^{j\delta(r,\phi)} r dr d\phi \right|^2}{\int_0^{2\pi} \int_0^a f^2(r, \phi) r dr d\phi}, \tag{1}$$

where $f(r, \phi)$ is the in-phase illumination function in terms of the aperture coordinates r, ϕ.

For small phase errors, the exponential may be expanded in a power series with the result that the ratio of the gain to the no-error gain G_0 is

$$\frac{G}{G_0} \approx 1 - \overline{\delta^2} + \bar{\delta}^2 \tag{2}$$

where

$$\overline{\delta^2} = \frac{\int_0^{2\pi} \int_0^a f(r, \phi) \delta^2(r, \phi) r dr d\phi}{\int_0^{2\pi} \int_0^a f(r, \phi) r dr d\phi}$$

$$\bar{\delta} = \frac{\int_0^{2\pi} \int_0^a f(r, \phi) \delta(r, \phi) r dr d\phi}{\int_0^{2\pi} \int_0^a f(r, \phi) r dr d\phi}.$$

In general, the phase reference plane can be chosen so that $\bar{\delta}$, the illumination weighted mean phase error, is zero. The loss of gain is then simply

$$\frac{G}{G_0} = 1 - \overline{\delta_0^2}, \tag{3}$$

where $\overline{\delta_0^2}$ is calculated from the mean phase plane.

This simple relation (3), that the fractional loss of gain is equal to the weighted mean-square phase error was probably first pointed out by Marechal [1] and, in antenna technology, by Spencer [2]. It is valid for any illumination and reflector deformation, provided the latter is small in wavelength measure. It indicates that for a one dB loss of gain the rms phase variation about the mean phase plane must be less than $\lambda/14$ or, for shallow reflectors, the surface error must be less than $\lambda/28$.

We next seek a more exhaustive analysis valid for large phase errors and one that would give information on the radiation pattern. The problem is common to a class of problems, illustrated in Fig. 1, where a plane wave is distorted into an error phase front. Alternately, we can say that to a narrow or "diffracted limited" direction of transmission is added a wider angular spectrum of scattered energy.

If we have detail knowledge of the phase front error, the radiation pattern or angular spectrum can be obtained by machine computation of the standard Kirchhoff integral [3]. Unfortunately, such detail knowledge is not available, and we must fall back on various statistical estimates of the character of the surface distortion and obtain a probable radiation pattern.

Manuscript received October 1, 1965; revised December 1, 1965. This paper was presented at the 1965 Symposium of the IEEE Group on Antennas and Propagation.

The author is with the Lincoln Laboratories, Massachusetts Institute of Technology, Lexington, Mass. (Operated with the support of the U. S. Air Force.)

Reprinted from *Proc. IEEE*, vol. 54, pp. 633–640, Apr. 1966.

We can begin the analysis by subdividing the aperture into N subregions, each with a phase error and with no relation or correlation with contiguous regions. This crude model is shown in Fig. 2, where the aperture phase front is represented by a number of hatboxes of random heights. The axial field is the sum of these individual vector contributions. With no phase error the power sum of N unit vectors is N^2 (see Fig. 3). If we now assume that the phase of each vector is randomly in error by an amount taken from a Gaussian population of standard deviation δ, in radians, then the expected or average power sum is

$$\bar{P} = N^2 e^{-\overline{\delta^2}} + N(1 - e^{-\overline{\delta^2}}). \tag{4}$$

The first term may be considered as the coherent power and the second as the incoherent. For small or large errors, we get the limiting forms for the coherent or incoherent addition of waves. The distribution of the sum is also of interest [4]. For small phase errors, the distribution is Gaussian in voltage with a standard deviation $\sqrt{N}\delta$, so that the distribution becomes relatively more peaked with a larger number of vectors and smaller phase errors. For large phase errors, we have the well-known Rayleigh distribution characterized by the mean power N.

The expected or average radiation function of the crude model shown in Fig. 2 can be derived. The procedure is briefly as follows [5]: the field at a general far-field point is expressed as a Kirchhoff surface integral. The power pattern is obtained by multiplying by the conjugate integral yielding a double surface integral of the two running surface vector variables. A correlation function is defined as a function of the vector difference of these variables. The average or expected value is then obtained. To perform the integration, assumptions must be made on the spatial nature of the correlation and on the frequency distribution of the phase errors. For the model chosen, these assumptions are that the phase values are completely correlated in a diameter "$2c$" and completely uncorrelated for larger distances. In addition, the various phases come from a Gaussian population of rms error "δ." As in all statistical problems, the number of components must be large so that

$$N \approx \left(\frac{D}{2c}\right)^2 \gg 1. \tag{5}$$

The result of this process is

$$G(\theta, \phi) = G_0(\theta, \phi) e^{-\overline{\delta^2}} + \left(\frac{2\pi c}{\lambda}\right)^2 (1 - e^{-\overline{\delta^2}}) \Lambda_1\left(\frac{2\pi c u}{\lambda}\right), \tag{6}$$

where

$G_0(\theta, \phi)$ is the no-error radiation diagram whose axial value is $\eta(\pi D/\lambda)^2$

η is the aperture efficiency

u is $\sin\theta$

$\Lambda_1(\ )$ is the Lambda function.

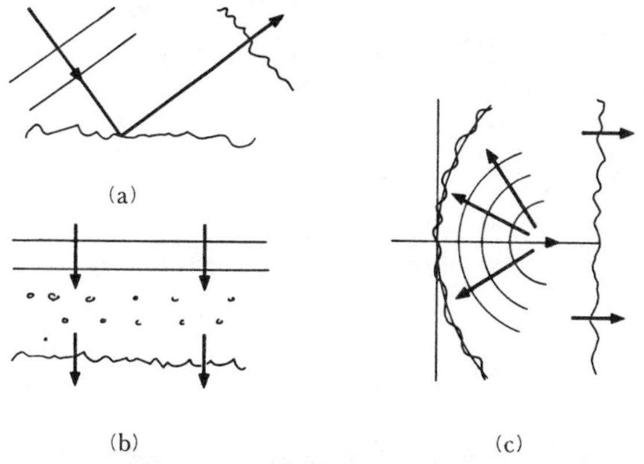

Fig. 1. A class of problems. (a) Reflection from a rough surface. (b) Transmission through a random medium. (c) Diffraction from an imperfect paraboloid.

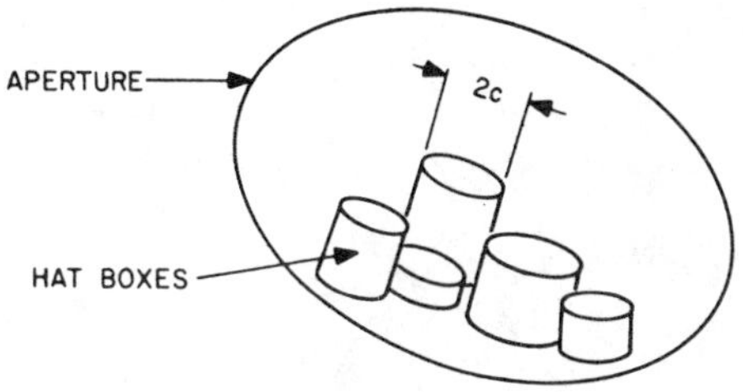

Fig. 2. Aperture subdivided into a number of hatboxes.

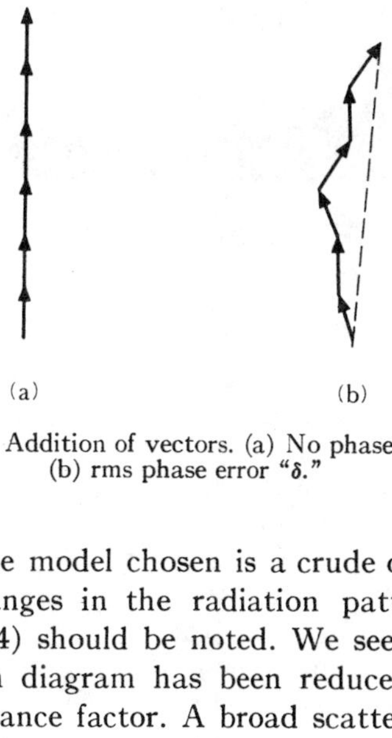

Fig. 3. Addition of vectors. (a) No phase error. (b) rms phase error "δ."

Although the model chosen is a crude one, (6) illustrates the changes in the radiation pattern and its similarity to (4) should be noted. We see that the no-error radiation diagram has been reduced by an exponential tolerance factor. A broad scattered field has been added whose "beamwidth" is inversely proportional to the size of the correlated region in wavelengths, so that smooth reflectors (large c) scatter more directively and rough reflectors (small c) more diffusely. For small phase errors, the relative magnitude of the axial scattered field is

$$\frac{1}{\eta}\left(\frac{2c}{D}\right)^2 \overline{\delta^2}. \tag{7}$$

The model chosen can be considerably improved by replacing the hatboxes with hats as shown in Fig. 4. If

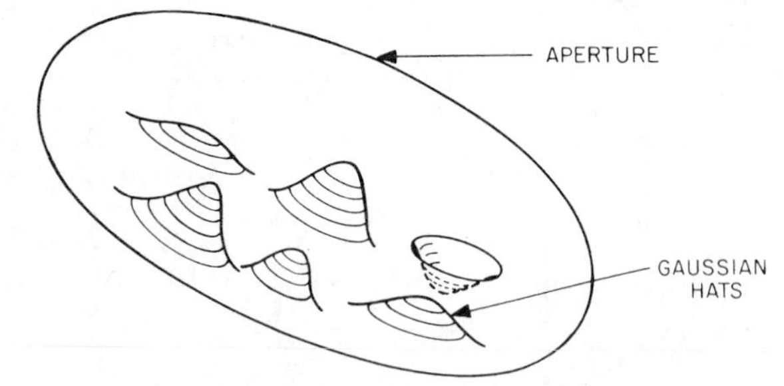

Fig. 4. Aperture subdivided into a number of hats.

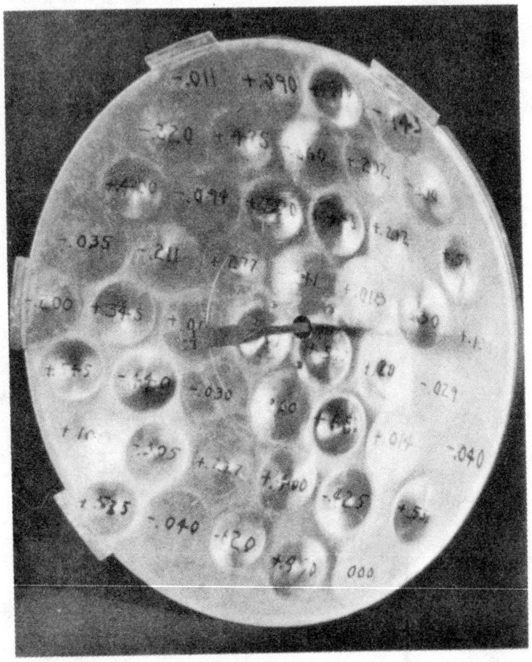

Fig. 5. Special model constructed to test theory.

the phase front distortions are assumed to be of Gaussian shape, the required integrations can again be performed [5] with the following result:

$$G(\theta, \phi) = G_0(\theta, \phi)e^{-\overline{\delta^2}} + \left(\frac{2\pi c}{\lambda}\right)^2 e^{-\overline{\delta^2}} \sum_{n=1}^{\infty} \frac{\overline{\delta^2}^2}{n \cdot n!} e^{-(\pi c u/\lambda)^2/n}. \tag{8}$$

Although (8) is more complex, the general effects are similar to those discussed previously.

We have considered a two-dimensional distribution of errors. It is of interest to present the one-dimensional case derived by Bramley [6] in our notation

$$G(\theta) = G_0(\theta)e^{-\overline{\delta^2}} + \frac{\sqrt{\pi} c}{\lambda} e^{-\overline{\delta^2}} \sum_{n=1}^{\infty} \frac{\overline{\delta^{2n}}}{\sqrt{n} \cdot n!} e^{-(\pi c u/\lambda)^2/n}. \tag{9}$$

The gain reduction and pattern degradation predicted by (8) was checked in the original reference [5] by the construction of a special model, Fig. 5, which fulfilled the statistical assumptions necessary for the theoretical development.

II. Discussion

From (8), we can write the reduction of axial gain as

$$\frac{G}{G_0} = e^{-\overline{\delta^2}} + \frac{1}{\eta}\left(\frac{2c}{D}\right)^2 e^{-\overline{\delta^2}} \sum_{n=1}^{\infty} \frac{\overline{\delta^2}^n}{n \cdot n!}. \tag{10}$$

In the region of interest, i.e., reasonable tolerance losses, and for correlation regions that are small compared to the antenna diameter, the second term may be neglected and we have for the gain

$$G = G_0 e^{-\overline{\delta^2}} = \eta \left(\frac{\pi D}{\lambda}\right)^2 e^{-(4\pi\epsilon/\lambda)^2}, \tag{11}$$

where we define "ϵ" as the effective reflector tolerance in the same units as λ; i.e., that rms surface error on a shallow reflector (large f/D), which will produce the phase front variance $\overline{\delta^2}$. In Fig. 6 we plot the loss of gain (11) as a function of the rms error and the peak surface error. The ratio used, 3:1, is one found experimentally for large structures and results, in part, from the truncation used in the manufacturing process (i.e., large errors are corrected).

It should be noted that for small errors (11) is identical with (3), with the exception that the former is independent of the illumination function and the latter is not. For the statistical analysis, it was necessary to assume a uniform distribution of errors, for which case the illumination dependence factors out in (3) and becomes identical to (11).

For deep (nonshallow) reflectors, the surface tolerance is not exactly equal to the effective tolerance "ϵ." In addition, structural people at times measure the reflector deformations normal to the surface and at times in the axial direction. The relation between these quantities is

$$\epsilon = \frac{\Delta z}{1 + (r/2f)^2} \tag{12a}$$

$$\epsilon = \frac{\Delta n}{\sqrt{1 + (r/2f)^2}}. \tag{12b}$$

The result is that the tolerance gain loss in dB, as computed from the reflector axial or normal mean square error, is too high by a factor A. This factor is given in Fig. 7. For shallow reflectors, this correction factor approaches unity.

Equation (11) indicates that if a given reflector is operated at increasing frequency, the gain, at first, increases as the square of the frequency until the tolerance effect take over and then a rapid gain deterioration occurs. Maximum gain is realized at the wavelength of

$$\lambda_m = 4\pi\epsilon, \tag{13}$$

where a tolerance loss of 4.3 dB is incurred. This maximum gain is

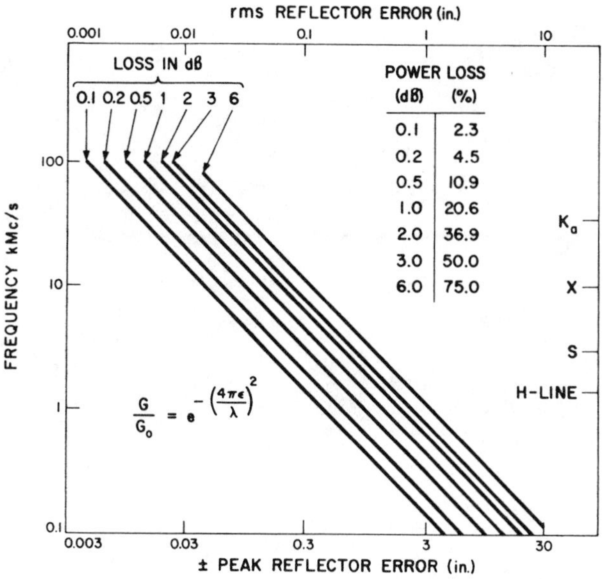

Fig. 6. Gain loss due to reflector tolerance.

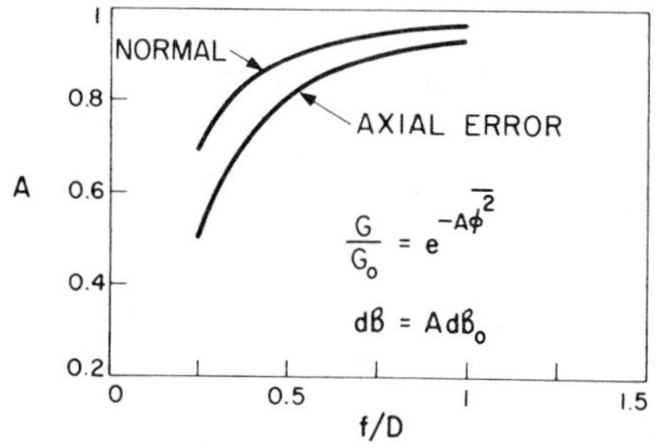

Fig. 7. Correction factor due to reflector curvature.

$$G_{\max} \approx \frac{\eta}{43}\left(\frac{D}{\epsilon}\right)^2 \tag{14}$$

and is proportional to the square of the precision of manufacture (D/ϵ).

This behavior is illustrated in Fig. 8, where we show some of the world's large antennas. The frequency region where the smaller and more precise structure is superior to the larger and coarser antenna, and the converse, is evident.

Next we consider the effect of surface errors on the radiation diagram. In Fig. 9, we show the pattern of a 12-dB tapered circular aperture with random phase errors and with $D=20c$. We plot from (8) the expected power diffraction and scatter patterns for mean-square phase errors of 0.2, 0.5, 1.0, 2.0, and 4.0 in radian squared measure. These correspond to tolerance gain losses of 0.87, 2.2 4.3, 8.6, and 16.6 dB, respectively. The complete radiation diagram is the *power sum* of the diffraction and scatter patterns. It should be noted that the diffraction pattern is reduced by the exponential tolerance factor and that the energy lost appears in the scattered pattern, which broadens as the surface error increases.

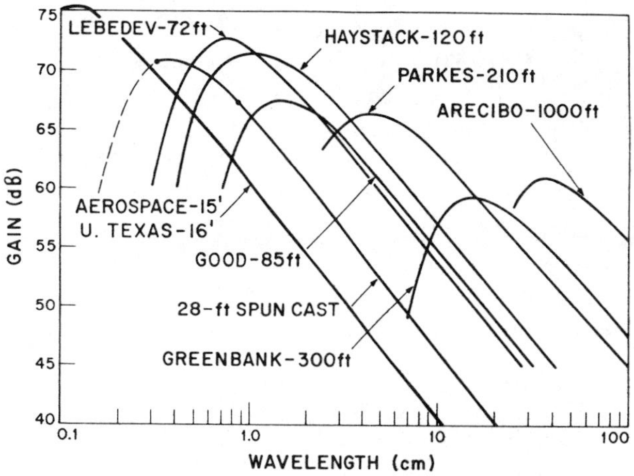

Fig. 8. Gain of large paraboloids (based on published estimates).

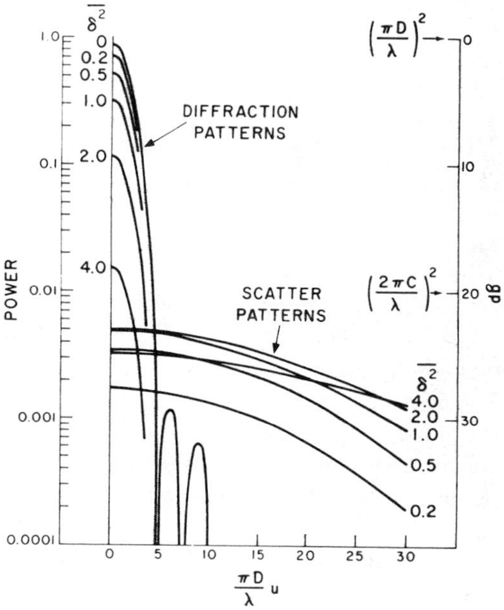

Fig. 9. Radiation patterns of phase distorted circular aperture, 12 dB illumination taper, $D=20c$.

With further increase in loss, the diffraction pattern is submerged in the scattered energy and disappears. Scheffler [7],[1] in a similar analysis, has pointed out that for large phase errors the scattered pattern approaches

$$G_s(\theta) = \left(\frac{2\pi c}{\lambda}\right)^2 \frac{\left[1 - e^{-\overline{\delta^2}}\right]}{\overline{\delta^2}} e^{-(\pi c u/\lambda)^2/\overline{\delta^2}}, \tag{15}$$

so that for extremely large phase errors the radiated energy is scattered over an angular region with the intensity equal to

$$G_s(\theta) = \left(\frac{c}{2\epsilon}\right)^2 e^{-(cu/4\epsilon)^2}. \tag{16}$$

[1] This reference was first brought to my attention by Dr. P. Mezger of the National Radio Astronomy Observatory.

We note that under these extreme conditions the beamwidth is defined by the average surface slopes and is wavelength independent, a result we would have expected from geometric optics.

Before leaving the theoretical discussion, it should be recalled that the distribution in the focal plane has the same shape as the radiated angular spectrum. Therefore, the same relation (8) can be used to determine the spot size due to surface imperfections or small scale atmospheric inhomogeneities.

III. Application to Antenna Structures

The experimental check of the theory afforded by the specially constructed model (Fig. 5) merely verifies the mathematical development. We turn now to practical structures and list those factors which deviate from the theoretical assumptions.

1) The surface errors are not random, but to a large part are due to calculable gravity, wind, and thermal strains. However, analysis of actual antenna photogrammetric measurements indicates that the reflector deviations, if not strictly random and Gaussian, are distributed in a bell-shaped curve [8], [9].
2) The actual reflector errors are not uniformly distributed over the aperture. Again, photogrammetric measurements and deformation calculations after structural compensation indicate that this condition is not grossly violated [8], [9].
3) The theory assumes a fixed, circular correlation region. As the contour adjustment points are normally spaced in a uniform grid, there is a tendency for this condition; however, various structural factors such as pie-panel segments would yield elliptical correlated regions of varying size.
4) The theory also requires that the number of uncorrelated regions in the aperture be large, that is $D \gg 2c$. It has been found that for compensated structures the number of regions is related to the panel size or spacing of the target points.
5) It was also assumed that the spatial phase correlation function had a particular shape, namely Gaussian. Another smooth deformation surface would have yielded slightly different functional forms in the shape of the scattered power.
6) Finally, we have developed a statistical theory and obtained the average power pattern of the ensemble of such antennas. We apply the theory to one sample.

Therefore, a check of the performance of actual antenna structures with the above theory is necessary. Correlation has been obtained between frequency-gain measurements and optical photogrammetric measurements [9]. We present here other confirmation.

In Fig. 10, we show a horn reflector antenna [10]. The gain of this antenna was precisely measured over 6:1 range of frequencies [11]. Equation (11) can be written as

Fig. 10. Horn reflector antenna.

$$10 \log G\lambda^2 = 10 \log \eta(\pi D)^2 - \left(\frac{4\pi\epsilon}{\lambda}\right)^2 10 \log e, \tag{17}$$

which is the straight line

$$y = a - bx$$

when $G\lambda^2$ in dB is plotted against reciprocal wavelength squared. The vertical intercept is a measure of the aperture efficiency and the reflector tolerance can be obtained from the slope.

The experimental data is shown in Fig. 11, where outside of a gain droop at low frequencies, due to diffraction effects, the data follow a straight line with a mean deviation of 0.166 dB. The indicated aperture efficiency also lies between the calculated efficiencies of 78.34 and 76.13 percent for the two polarizations used. The predicted surface tolerance is an effective value of 33 mils or 48 mils normal to the parabolic surface. The agreement of the measured data with the predicted straight line relationship is a confirmation of antenna tolerance theory. In addition, this procedure, combined with a linear regression analysis of the experimental data, to establish confidence limits, is probably the most convenient and accurate method of determining the surface precision [12].

We next consider the determination of the size of the correlation region by means of electrical measurements. The temperature measured on an extended astronomical source is equal to the product of the fractional enclosed power and the source brightness temperature. With no surface errors, practically all the radiated power is enclosed by the source if it is at least several beamwidths in extent. With reflector errors, some of the scattered energy is outside of the source and the measured temperature is decreased. This reduction depends on both

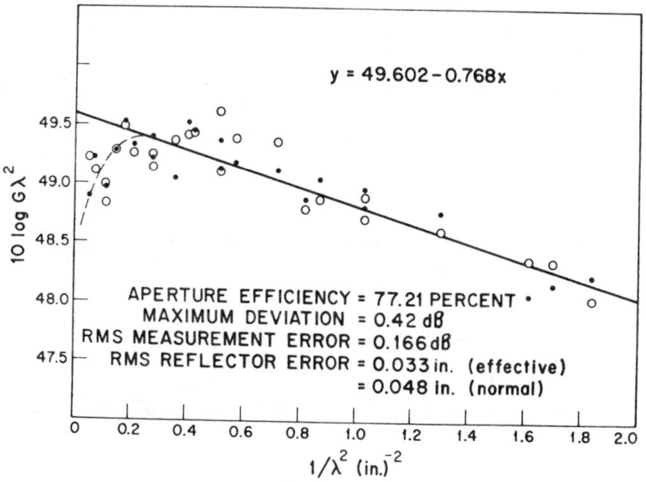

Fig. 11. Gain vs. frequency-horn reflector antenna.

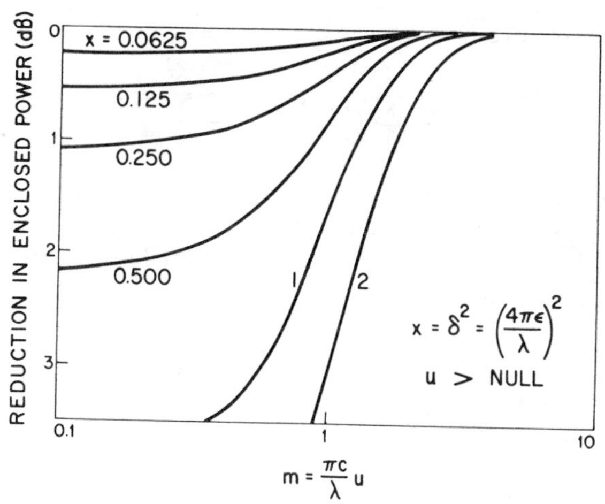

Fig. 12. Reduction in enclosed power.

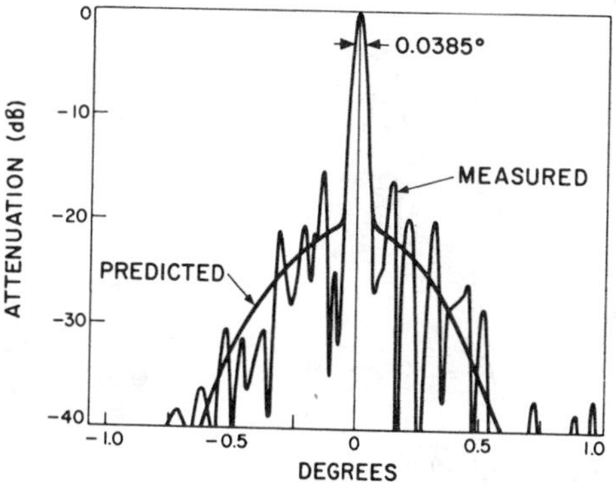

Fig. 13. Comparison of measured and predicted patterns, HAYSTACK (15.745 Gc/s).

the rms surface error and the size of the correlation region.

The fractional enclosed power in a cone angle $u_0 = \sin\theta_0$ (several beamwidths) can be obtained by integration of (8) with the result:

$$EP(\theta_0) = [1 - S]\left[1 - e^{-\overline{\delta^2}} \sum_{n=1}^{\infty} \frac{\overline{\delta^2}^{\,n}}{n!} e^{-(\pi c u_0/\lambda)^2/n}\right] \quad (18)$$

where S is the fractional energy very widely scattered by aperture blockage (feed supports, etc.). It should be noted that, either with no surface error or with large cone angle, the enclosed power is a constant.

Fig. 12 shows the reduction of enclosed power as a function of the rms error and the enclosed cone angle. If temperature measurements are now made of the same source at two different frequencies, the enclosed power is different. If we enter the temperature ratio (after correction for atmospheric effects and spectral index) as an ordinate into Fig. 12 and the frequency ratio as an abscissa, then we can obtain a set of values of tolerance error and correlation intervals which satisfy this condition (with a known source cone angle). If the reflector tolerance is known from point source gain measurements, the required correlation radius is determined.

This type of measurement was applied to the HAYSTACK radio telescope (120-foot diameter in a metal space frame radome) at the frequencies of 7750 and 15 500 Mc/s. The moon was used as an extended source and the planet Jupiter as a point source. By means of the procedure outlined, it was concluded that the rms surface error ϵ was 0.053 inch and that the correlation radius c was 4.4 feet.

A check of antenna tolerance theory is obtained by comparison of the predicted antenna pattern based on the astronomically determined values of (ϵ, c) and the experimentally measured pattern with a ground-based transmitter. Figure 13 shows this comparison at the frequency where the tolerance effect is significant. The agreement of the predicted pattern and the actual measured characteristic is excellent considering the statistical nature of the problem and that the sidelobe peaks should be 3 dB higher than the average intensity. A corresponding pattern, taken with a similar feed, at 7750 Mc/s where the tolerance effects are not significant showed sidelobe levels of about 25 dB down.

From the above measurements and those cited in the references, it may be inferred that the present status of antenna tolerance theory is such that the behavior of large antennas may be determined by the specification of two quantities: the rms surface error and the correlation interval. These quantities may be determined from electrical ground-based or astronomical data. Detail correlation of these electrically determined values and actual mechanical measurements is lacking. However, available photogrammetric or other structural estimates are not in variance with the theoretical predictions.

Appendix

A. Derivation of (4)

Consider the power sum of N unit vectors whose phases "δ" come from a normal distribution of zero mean and variance $\overline{\delta^2}$

$$P = \sum_i^N \sum_j^N e^{j(\delta_i - \delta_j)} = \sum_i^N \sum_j^N e^{jy_{ij}}. \tag{19}$$

y is another statistic, defined as the difference of two samples taken from the original distribution. It can be readily shown that it is normally distributed with zero mean and variance $2\overline{\delta^2}$ or

$$W(y) = \frac{1}{\sqrt{4\pi\overline{\delta^2}}} e^{-y^2/4\overline{\delta^2}}, \tag{20}$$

now

$$\overline{P} = \sum_i^N \sum_j^N \overline{\cos y} + i\,\overline{\sin y},$$

for $i \neq j$

$$\overline{\cos y} = \int_{-\infty}^{\infty} \cos y W(y) dy = e^{-\overline{\delta^2}} \tag{21}$$

$$\overline{\sin y} = \int_{-\infty}^{\infty} \sin y W(y) dy = 0. \tag{22}$$

Therefore

$$\overline{P} = N^2 e^{-\overline{\delta^2}} + N(1 - e^{-\overline{\delta^2}}). \tag{23}$$

B. Derivation of (8)

The gain function of an aperture is written:

$$G(\theta_1\phi) = \frac{4\pi}{\lambda^2} \frac{\left| \int f(\bar{r}) e^{j\bar{k}\cdot\bar{r}} e^{j\delta(r,\phi)} dS \right|^2}{\int f^2(\bar{r})\, dS}, \tag{24}$$

where

$\bar{r}$ is an aperture vector position variable,
$\bar{k} = (2\pi/\lambda)\bar{p}_0$ is a vector in the direction of observation,
$\delta(r, \phi)$ is the aperture phase perturbation function,
dS is an elemental aperture area,
$f(\bar{r})$ is the aperture illumination function.

The numerator may be written as:

$$\iint f(\bar{r}_1) f(\bar{r}_2) e^{j\bar{k}\cdot(\bar{r}_1 - \bar{r}_2)} e^{j(\delta_1 - \delta_2)}\, dS_1 dS_2,$$

denoting $\bar{\tau} = \bar{r}_1 - \bar{r}_2$ as the vector difference between the two aperture running variables and $y(\bar{\tau}) = \delta_1 - \delta_2$ as the phase difference of the two points. We have

$$\iint f(\bar{r}_1) f(\bar{r}_1 + \bar{\tau}) e^{j\bar{k}\cdot\bar{\tau}} e^{jy(\bar{\tau})}\, dS_1 dS_\tau$$

defining $\phi(\bar{\tau})$ as the illumination correlation function

$$\phi(\bar{\tau}) = \frac{\int f(\bar{r}_1) f(\bar{r}_1 + \bar{\tau})\, dS_1}{\int f^2(\bar{r}_1)\, dS_1}. \tag{25}$$

We now rewrite (24) as:

$$G(\theta_1 \phi) = \frac{4\pi}{\lambda^2} \int \phi(\bar{\tau}) e^{j\bar{k}\cdot\bar{\tau}} e^{jy(\bar{\tau})}\, dS_\tau$$

and the average or expected value

$$\overline{G(\theta_1 \phi)} = \frac{4\pi}{\lambda^2} \int \phi(\bar{\tau}) e^{j\bar{k}\cdot\bar{\tau}} [\overline{\cos y(\tau)} + i\,\overline{\sin y(\tau)}]\, dS_\tau \tag{26}$$

for τ large compared to "c," the phase correlation distance where the two phase samples are uncorrelated, $y(\tau)$ is normally distributed with zero mean and variance $2\overline{\delta^2}$. When τ approaches zero, $y(\tau)$ approaches zero with zero variance. Some convenient form must be assumed for the variance function. Taking

$$\overline{y^2(\tau)} = 2\overline{\delta^2}[1 - e^{-\tau^2/c^2}] \tag{27}$$

from (21) and (22), we have

$$\overline{\cos y(\tau)} = e^{-\overline{\delta^2}[1 - e^{-\tau^2/c^2}]}$$

$$\overline{\sin y(\tau)} = 0.$$

Equation (26) may be rewritten as

$$\overline{G(\theta, \phi)} = \frac{4\pi}{\lambda^2} e^{-\overline{\delta^2}} \int \phi(\bar{\tau}) e^{j\bar{k}\cdot\bar{\tau}} e^{\overline{\delta^2} e^{-\tau^2/c^2}}\, dS_\tau$$

$$= \frac{4\pi}{\lambda^2} e^{-\overline{\delta^2}} \sum_{n=0}^{\infty} \int \phi(\bar{\tau}) e^{j\bar{k}\cdot\bar{\tau}} \frac{\overline{\delta^2}^n}{n!} e^{-n\tau^2/c^2}\, dS_\tau.$$

The first term is the unperturbed pattern $G_0(\theta_1\phi)$

$$\overline{G(\theta_1\phi)} = G_0(\theta_1\phi) e^{-\overline{\delta^2}} + \frac{4\pi}{\lambda^2} e^{-\overline{\delta^2}} \sum_{n=1}^{\infty} \int \phi(\bar{\tau}) e^{j\bar{k}\cdot\bar{\tau}} \frac{\overline{\delta^2}^n}{n!} e^{-n\tau^2/c^2}\, dS_\tau.$$

Due to the exponential factor, the remaining terms have their principal contribution for $\tau < c$. As we have assumed that c is small compared to the aperture dimensions, the illumination correlation function (25) may be assumed as unity in evaluating these terms. The angular integration can be immediately performed with the result:

$$G(\theta_1\phi) = G_0(\theta_1\phi) e^{-\overline{\delta^2}} + \frac{8\pi^2}{\lambda^2} e^{-\overline{\delta^2}} \sum_{n=1}^{\infty} \frac{\overline{\delta^2}^n}{n!} \int J_0\left(\frac{2\pi}{\lambda} u\tau\right) e^{-n\tau^2/c^2} \tau\, d\tau.$$

The integral can be evaluated by extending the limits and recalling that

$$\int_0^\infty J_0\left(\frac{2\pi}{\lambda} u\tau\right) e^{-n\tau^2/c^2} \tau d\tau = \frac{c^2}{2n} e^{-(\pi cu/\lambda)^2/n},$$

with the final result

$$G(\theta_1\phi) = G_0(\theta_1\phi)e^{-\overline{\delta^2}} + \left(\frac{2\pi c}{\lambda}\right)^2 e^{-\overline{\delta^2}} \sum_{n=1}^{\infty} \frac{\overline{\delta^2}^n}{n \cdot n!} e^{-(\pi cu/\lambda)^2/n}. \quad (28)$$

In this derivation, we have assumed that we are dealing with highly directive antennas. The obliquity factor has, therefore, been suppressed and we have used the small angle formulation of the Kirchhoff integral.

C. The Function

$$S(m, x) = e^{-x} \sum_{n=1}^{\infty} \frac{x^u}{n \cdot n!} e^{-m^2/n}$$

x \ m	0.0	0.5	1.0	2.0	3.0	4.0
0.2	0.1723	0.1351	0.0655	0.0042	0.0001	—
0.5	0.3458	0.2739	0.1379	0.0120	0.0007	—
1.0	0.4848	0.3907	0.2093	0.0263	0.0026	0.0002
2.0	0.4986	0.4162	0.2505	0.0521	0.0087	0.0013
3.0	0.4111	0.3549	0.2367	0.0690	0.0159	0.0031
4.0	0.3242	0.2877	0.2091	0.0779	0.0227	0.0054

D. Data for Fig. 12

$$\text{Ratio } \frac{\text{Enclosed Power to Angle } u_0}{\text{Total Power}} \text{ dB}$$

x \ m	0.25	0.5	1.0	1.414	2.0	2.83
0.125	0.50	0.42	0.20	0.08	0.01	0.001
0.25	1.00	0.83	0.40	0.16	0.03	0.003
0.50	2.02	1.65	0.80	0.35	0.09	0.011
1.00	4.00	3.21	1.59	0.76	0.25	0.046
2.00	7.80	6.03	3.07	1.67	0.69	0.185

Acknowledgment

The author is indebted to Dr. J. L. Meeks and Dr. S. Weinreb for the radiometric measurements of the HAYSTACK antenna. Also, to Dr. J. W. Findlay and Dr. Mezger of the National Radio Astronomy Observatory for many helpful discussions.

References

[1] A. Marechal, "The diffraction theory of aberrations," *Rep. Progr. in Phys. (GB)*, vol. XIV, p. 106, 1951; for English summary see E. Wolf.

[2] R. C. Spencer, "A least square analysis of the effect of phase errors on antenna gain," Air Force Cambridge Research Center, Bedford, Mass., AFCRC Rept. E5025, January 1949.

[3] A. R. Dion, "Investigation of effects of surface deviations on HAYSTACK antenna radiation patterns," M.I.T. Lincoln Lab., Lexington, Mass., Rept. 324, July 1963.

[4] P. Beckman, "The probability distribution of the vector sum of *N* unit vectors with arbitrary phase distributions," *ACTA Tech. (Czechoslovakia)*, vol. 4, no. 4, pp. 323–334, 1959.

[5] J. Ruze, "The effect of aperture errors on the antenna radiation pattern," *Suppl. al Nuovo Cimento*, vol. 9, no. 3, pp. 364–380, 1952. This work used for the theoretical basis of antenna tolerance, was first prepared by the author as a Ph.D. dissertation under the direction of Prof. L. J. Chu at M.I.T. in 1952.

[6] E. N. Bramley, "Some aspects of the rapid directional fluctuations of short radio waves reflected from the ionosphere," *Proc. IEE (London)*, vol. 102B, pp. 533–540, 1955.

[7] H. Scheffler, "Uber die Genauigkeitsforderungen bei der Herstellung optischer Flachen fur astronomische Teleskope," *Z. Astrophys. (Germany)*, vol. 55, pp. 1–20, 1962.

[8] J. W. Findlay, "Operating experience at the National Radio Astronomy Observatory," *Ann. N. Y. Acad. Sci.*, vol. 116, pp. 25–40, June 1964.

[9] P. G. Mezger, "An experimental check of antenna tolerance theory using the NRAO 85-foot and 300-foot telescopes," *1964 Internat'l Symp. on Antennas and Propagation*, pp. 181–185.

[10] R. W. Friis and A. S. May, "A new broadband microwave antenna system," *Trans. AIEE (Communication and Electronics)*, vol. 77, pp. 97–100, March 1958.

[11] A. Sotiropoulos and J. Ruze, "HAYSTACK calibration antenna," M.I.T. Lincoln Lab., Lexington, Mass., Tech. Rept. 367, December 1964.

[12] J. Ruze, "Reflector tolerance determination by gain measurement," *1964 NEREM Conv. Rec.*, pp. 166–167.

[13] R. H. T. Bates, "Random errors in aperture distributions," *IRE Trans. on Antennas and Propagation*, vol. AP-7, pp. 369–372, October 1959.

[14] H. G. Booker, J. A. Ratcliffe, and D. H. Shinn, "Diffraction from an irregular screen with application to ionospheric problems," *Phil. Trans. (GB)*, vol. 242, ser. A, pp. 579–609, 1950.

[15] R. N. Bracewell, "Tolerance theory of large antennas," *IRE Trans. on Antennas and Propagation*, vol. AP-9, pp. 49–58, January 1961.

[16] B. V. Brande, N. A. Esepkina, N. L. Kaidanovskii, and S. E. Khaikin, "The effects of random errors on the electrical characteristics of high-directional antennae with variable-profile reflectors," *Radioteknika i Elektronika (USSR)*, vol. 5, no. 4, pp. 75–92, 1960.

[17] D. K. Cheng, "Effect of arbitrary phase errors on the gain and beam width characteristics of radiation pattern," *IRE Trans. on Antennas and Propagation*, vol. AP-3, pp. 145–147, July 1955.

[18] A. Consortini, L. Rouchi, A. M. Scheggi, and G. Toroldo DiFrancia, "Gain limit and tolerances of big reflector antennas," *Alta Frequenza (Italy)*, vol. 30, pp. 232–276, March 1961.

[19] C. Dragone and D. C. Hogg, "Wide angle radiation due to rough phase fronts," *Bell Sys. Tech. J.*, vol. 42, pp. 2285–2296, September 1963.

[20] J. Robieux, "Influence of the precision of manufacture on the performance of aerials," *Am. Radio Elect.*, vol. 11, pp. 29–56, January 1956.

[21] Ya. S. Shifrin, "The statistics of the field of a linear antenna," *Radio Engrg. Electronic Phys.*, vol. 8, pp. 351–358, March 1963.

[22] R. A. Shore, "Partially coherent diffraction by a circular aperture," in *Electromagnetic Theory and Antennas*, E. C. Jordan, Ed. New York: Pergamon, 1963, pp. 787–795.

The Effect of Phase Errors on the Forward Gain

The effect of random phase errors, which are caused by reflector surface irregularities, on the radiation characteristics of a reflector antenna had been treated by a number of authors in the past. By treating the case as a statistical problem, Ruze [1] had obtained an approximate expression for the average loss in gain in the forward direction in terms of the standard deviation of the error. Cheng [2], on the other hand, proposed a way to estimate the upper bound in the loss as a function of the peak phase error. Cheng's results are, however, not very useful in practice because estimates based on his formulas tend to be too conservative; in other words, they tend to underestimate the actual capability of the antenna system. Ruze's idea of finding the "average" performance of the antenna is basically sound, but to know the "average" radiation characteristics is not enough because the actual radiation pattern of a given antenna at any particular time is likely to be different from "average" pattern. One is therefore interested in finding the actual loss in the forward gain if the rms error at any particular instant is known. The problem becomes even more important when it had been learned that, with the Haystack antenna, the effects of snow and wind loads are completely eliminated. It is therefore possible to obtain the error distribution at any antenna pointing direction, and to optimize the feed position simply by feeding the bearing data of the antenna to an appropriate computer.

Since we are only interested in the gain in the forward direction, the antenna can be replaced by a circular aperture (Fig. 1). If the illumination is uniform across the aperture, the field strength in the forward direction in the Fraunhofer region is given by

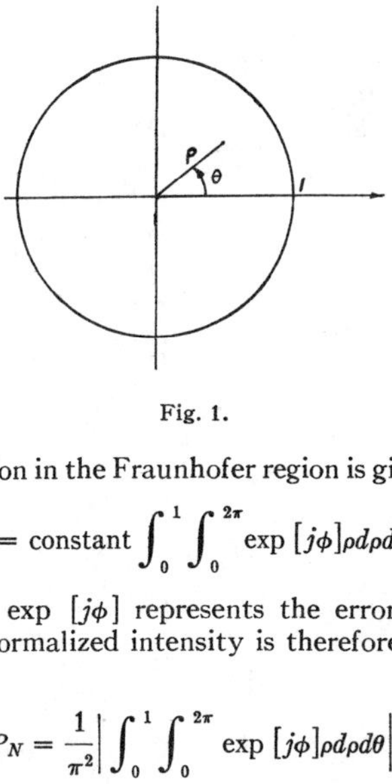

Fig. 1.

$$E_0 = \text{constant} \int_0^1 \int_0^{2\pi} \exp[j\phi]\rho d\rho d\theta \quad (1)$$

where $\exp[j\phi]$ represents the error term. The normalized intensity is therefore equal to

$$P_N = \frac{1}{\pi^2}\left| \int_0^1 \int_0^{2\pi} \exp[j\phi]\rho d\rho d\theta \right|^2 = \left| \frac{\int_0^1 \int_0^{2\pi} \exp[j\phi]\rho d\rho d\theta}{\int_0^1 \int_0^{2\pi} \rho d\rho d\theta} \right|^2 .$$

For small value of ϕ, we have

$$\exp[j\phi] \doteq \left(1 - \frac{\phi^2}{2}\right) + j\phi$$

$$P_N = \left| \frac{\int_0^1 \int_0^2 \left[\left(1 - \frac{\phi^2}{2}\right) + j\phi\right] \rho d\rho d\theta}{\int_0^1 \int_0^2 \rho d\rho d\theta} \right|^2$$

$$= \left| 1 - \frac{\overline{\phi^2}}{2} + j\bar{\phi} \right|^2$$

$$\doteq 1 - [\overline{\phi^2} - (\bar{\phi})^2]$$

where $\overline{\phi^2}$ and ϕ are the mean value of ϕ^2 and ϕ, respectively. But $\overline{\delta^2}$, the mean square phase deviation is given by

$$\overline{\delta^2} = \overline{(\phi - \phi)^2} = \frac{\int_0^1 \int_0^{2\pi} (\phi - \bar{\phi})^2 \rho d\rho d\theta}{\int_0^1 \int_0^{2\pi} \rho d\rho d\theta}$$

$$= \overline{\phi^2} + (\bar{\phi})^2 - 2\bar{\phi}\cdot\bar{\phi}$$

$$= \overline{\phi^2} - (\bar{\phi})^2$$

$$\therefore P_N = 1 - \overline{\delta^2}. \quad (2)$$

This result has been derived directly from the optical case. In practice, however, it is very unlikely to find that the antenna is uniformly illuminated, and (2) does not apply. One must therefore find a different way to solve a more general problem with the errors not necessarily small and the amplitude illumination function of any form.

Referring to Fig. 2, it can be shown [3] that, when errors are present, the field strength in the forward direction is given by

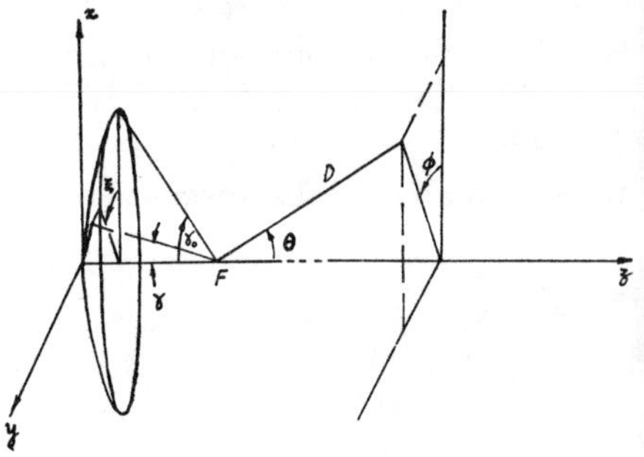

Fig. 2.

$$E = C \int_0^{2\pi} \int_0^{\gamma_0} [G(\xi, \gamma)]^{1/2} e^{j\phi} \tan\frac{\gamma}{2} d\gamma d\xi \quad (3)$$

Manuscript received June 14, 1965.

Reprinted from *IEEE Trans. Antennas Propagat.*, vol. AP-13, pp. 981-982, Nov. 1965.

where C=constant, $F(\xi, \gamma)$ represents the amplitude illumination function, and $\exp[j\phi]$ again represents the error term.

Imagine that we rotate the reflector about the axis of the antenna while keeping the feed horn fixed. We shall find that the field strength at any point in the far field region will change because the phase errors caused by surface irregularities of the reflector upset the symmetry of the radiation pattern. There is one exception however; that is, the field strength in the forward direction is not disturbed in any way by this rotation, as it is in the direction of the axis of rotation. This is true irrespective of the shape of the radiation pattern. Now, since we only perform the rotation in our mind, the error pattern is not changed by gravitational or any other effects. On the other hand the phase error corresponding to any other point (x, y, z) which is fixed relative to a fixed system of reference axes will change with the rotation. In other words, if we change the magnitude of the angle of rotation in a random manner, we therefore achieved a random variation of phase errors. The problem can therefore be treated as a statistical problem. It is important to note, that while the actual field strength in any direction is different from its "average" value, the average field strength in the forward direction is exactly equal to its actual value. The actual field strength in the forward direction in the presence of the error is therefore given by

$$E_{act} = \overline{E} = C\int_0^{2\pi}\int_0^{\gamma_0}[G(\xi, \gamma)]^{1/2} \cdot (\overline{\cos\phi + j\sin\phi}) \tan\frac{\gamma}{2}\, d\gamma d\xi. \quad (4)$$

Now

$$\phi = \frac{2\pi}{\lambda}\epsilon \cdot 2\cos\frac{\gamma}{2} \quad (5)$$

where $|\epsilon|$ is the magnitude of the surface irregularity, and λ is the wavelength of the operating frequency. Since it is quite justifiable to assure that ϵ is normally distributed with zero mean, the same thing can be said of ϕ, the standard deviation of which is given by

$$\phi_{s.d.} = \frac{4\pi}{\lambda}\sigma\cos\frac{\gamma}{2} \quad (6)$$

where σ is the standard deviation of ϵ. We also have

$$\overline{\sin\phi} = \frac{1}{\sqrt{2\phi_{s.d.}^2}}\int_{-\infty}^{+\infty}\sin\phi \cdot \exp\left[-\frac{\phi^2}{2\phi_{s.d.}^2}\right] d\phi = 0$$

$$\overline{\cos\phi} = \frac{1}{\sqrt{2\pi\phi_{s.d.}^2}}\int_{-\infty}^{+\infty}\cos\phi \cdot \exp\left[-\frac{\phi^2}{2\phi_{s.d.}^2}\right] d\phi = \exp\left[-\frac{\phi_{s.d.}^2}{2}\right]$$

or

$$\overline{\cos\phi} = \exp\left[-\frac{8\pi^2}{\lambda^2}\sigma^2\cos^2\frac{\gamma}{2}\right]. \quad (7)$$

Equation (4) can be rewritten as

$$E_{act} = \overline{E} = C\int_0^{2\pi}\int_0^{\gamma_0}[G(\xi, \phi)]^{1/2} \cdot \exp\left[-\frac{8\pi^2\sigma^2}{\lambda^2}\cos^2\frac{\gamma}{2}\right]\tan\frac{\gamma}{2}\, d\gamma d\xi.$$

The actual value of the field strength in the forward direction can, therefore, be estimated for any error pattern irrespective of the amplitude illumination function.

Acknowledgment

The author is indebted to Prof. Willoughby for valuable discussions.

The Bao Vu
Dept. of Elec. Engrg.
University of Adelaide
Adelaide, Australia

References

[1] J. Ruze, "The effect of aperture errors on the antenna radiation pattern," *Supplemento del Nuovo Cimento*, vol. 9, no. 3, pp. 364–380, 1952.
[2] D. K. Cheng, "Effect of arbitrary phase errors on the gain and beamwidth characteristics of radiation pattern," *I.R.E. Trans. on Antennas and Propagation*, vol. AP-3, pp. 145–147, July 1955.
[3] S. Silver, *Microwave Antenna Theory and Design.* New York: McGraw-Hill, 1949, ch. 12.

On Antenna Tolerance Theory

MEHDI S. ZARGHAMEE

Abstract—**To predict the loss of gain of antennas due to surface deviations which are not distributed uniformly over the aperture, an extension of Ruze's theory is presented. It is found that the assumption of uniform error distribution, in general, underestimates the axial gain of an antenna whose surface deviations have regional variations over the aperture. This effect becomes significant only when the surface deviations cannot be considered small as compared to the wavelength. Furthermore, it is found that the assumption of a uniform distribution of error may have a significant effect on the predicted scatter even when surface deviations are not large. Assuming that the deviations from uniform distribution are also random, a correction term to the theory is also presented.**

NOTATION

A = area of aperture
c = correlation radius
f = aperture illumination function
G = antenna gain
G_0 = no error gain, a function of direction of observation
$\bar{k} = 2\pi\bar{p}/\lambda$, where $\bar{p}$ is a unit vector in the direction of observation
$\bar{r}$ = aperture position vector (Fig. 1)
$u = \sin\theta$
δ = phase error, a function of position
ϵ = effective surface deviation, a function of position
ϵ_0 = rms of effective surface deviations [see (9)]
η_0 = fourth root of the second variance of surface deviations [see (10)]
(θ, ϕ) = angles defining the direction of observation (Fig. 1)
λ = wavelength
σ = standard deviation of phase-error distribution function, a function of position
σ_0 = averaged standard deviation of phase error over the aperture [see (7)].

INTRODUCTION

THE DEVIATIONS of an antenna reflector from its ideal shape cause, in general, loss of gain and pattern degradation. These deviations may result from manufacturing and rigging tolerances and from gravity, wind, and thermal effects. The effects on surface deviations of manufacturing and rigging tolerances (including nondeterministic errors in the measuring instruments employed) are usually random in nature and can be estimated through stochastic analyses.[1] Recently developed automated computation techniques permit structural engineers to predict accurately the deformations of the reflector surface caused by known wind, gravity, and temperature changes.[2]

The effects of surface deviations on the radiation pattern and gain may be predicted from the actual distribution of surface deviations over the aperture.[1],[3] A simpler, approximate method for computing these effects is presented by

Manuscript received April 10, 1967; revised June 2, 1967. This work was partially supported by M.I.T. Lincoln Laboratory, Lexington, Mass.
The author is with the Structural Mechanics Division, Simpson Gumpertz and Heger Inc., Cambridge, Mass.

Reprinted from *IEEE Trans. Antennas Propagat.*, vol. AP-15, pp. 777–781, Nov. 1967.

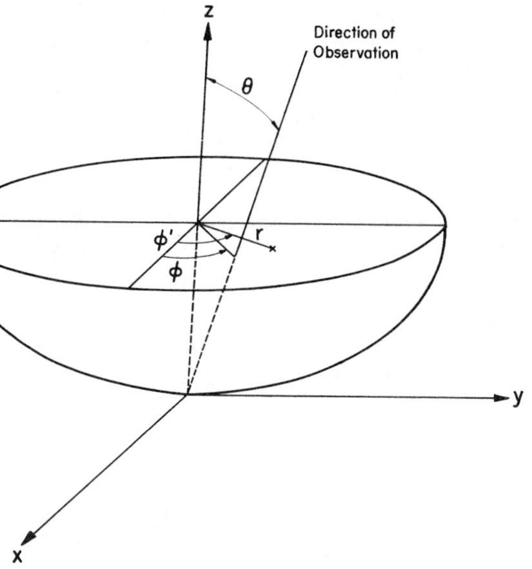

Fig. 1. Aperture position and direction of observation coordinate systems.

Ruze,[4],[5] in which only two quantities, namely the root mean square of the effective surface deviations and the correlation radius, are employed as the measure of the deviation of the reflector from its ideal shape. (Effective surface deviation is defined as one half the change in the RF path length. As shown in the Apppendix, it is equal to the axial component of the normal deviation from the ideal surface.) In the Ruze formulation, the assumption was made that the effective surface deviation at any point is a random sample from a *single* Gaussian distribution with zero mean and a standard deviation equal to the rms of the effective surface deviation of the reflector. An additional assumption was that the surface deviations are correlated in small regions. The gain was then expressed as follows:

$$G(\theta, \phi) = G_0(\theta, \phi) e^{-(4\pi\epsilon_0/\lambda)^2} + \left(\frac{2\pi c}{\lambda}\right)^2 e^{-(4\pi\epsilon_0/\lambda)^2} \sum_{n=1}^{\infty} \frac{1}{n n!} \left(\frac{4\pi\epsilon_0}{\lambda}\right)^{2n} e^{-(\pi c u/\lambda)^2/n}. \quad (1)$$

The actual effective surface deviations of a reflector may have a distribution which differs significantly from the aforementioned assumption of uniform error distribution over the aperture. For example, the random errors of many paraboloidal reflectors frequently increase with radial distance from the axis; the magnitude of the surface deformations also may exhibit nonrandom regional variations over the aperture.

It is of interest to determine the range of validity of (1) in predicting both the loss of axial gain and the half-power beamwidth for a reflector with nonuniform distribution of its surface deviations.

In this paper an equation which is a generalization of (1) to the cases where the error has nonuniform distribution over the aperture is presented. Also, for cases where the deviations from a uniform distribution can be considered random over the aperture, a correction term to the usual tolerance theory is proposed.

Theoretical Development

The gain of an antenna may be expressed by the following equation:

$$G(\theta, \phi) = \frac{4\pi}{\lambda^2} \frac{\left| \int_A f(\bar{r}) e^{j\bar{k}\cdot\bar{r}} e^{j\delta(\bar{r})} dS \right|^2}{\int_A f^2(\bar{r}) dS}. \quad (2)$$

Consider the phase-error function δ at point $\bar{r}$ as a random sample from a Gaussian distribution with zero mean and a standard deviation $\sigma(\bar{r})$, a function of position within the aperture. Let us further assume that the phase errors are so correlated that for the difference in the phase errors at points $\bar{r}_1$ and $\bar{r}_2$, we can write

$$\sigma^2(\bar{r}_1 - \bar{r}_2) = \left[\sigma^2(\bar{r}_1) + \sigma^2(\bar{r}_2)\right](1 - e^{-\tau^2/c^2}) \quad (3)$$

where τ is the distance between $\bar{r}_1$ and $\bar{r}_2$ and c is referred to as the radius of the correlation region. This assumption on the shape of the spatial phase-correlation function is not believed to affect the shape of the scattered power significantly if it is replaced with a similar smooth function.

For c sufficiently small as compared to the dimensions of the aperture, the expected gain of the antenna may be expressed as follows:

$$G(\theta, \phi) = \frac{4\pi}{\lambda^2} \frac{\left| \int_A f(\bar{r}) e^{-\sigma^2/2} e^{j\bar{k}\cdot\bar{r}} dS \right|^2}{\int_A f^2(\bar{r}) dS} + \left(\frac{2\pi c}{\lambda}\right)^2 \sum_{n=1}^{\infty} \frac{1}{n n!} e^{-(\pi c u/\lambda)^2/n} \frac{\int_A f^2(\bar{r}) e^{-\sigma^2} (\sigma^2)^n dS}{\int_A f^2(\bar{r}) dS} \quad (4)$$

where $u = \sin\theta$. Equation (4) reduces to (1) if $\sigma(\bar{r})$ is assumed to be constant over the aperture.

The complexity of (4) reduces its suitability for use in approximate design; therefore, certain simplifications will be made in this equation for this purpose. However, these simplifications limit the applicability of the theory to the prediction of the loss of gain alone. To predict the scatter, (4) must still be employed. If the corresponding simplifications are performed for (1), it reduces to the following well-accepted simpler form:

$$G = G_0 e^{-(4\pi\epsilon_0/\lambda)^2}. \quad (5)$$

Considering only the first term of (4), the axial gain of the antenna can be written in the following form:

$$G = \frac{4\pi}{\lambda^2} e^{-\sigma_0^2} \frac{\left| \int_A f(\bar{r}) e^{-\xi/2} dS \right|^2}{\int_A f^2(\bar{r}) dS} \quad (6)$$

where $\xi = \sigma^2(\bar{r}) - \sigma_0^2$, and σ_0^2 is the averaged variance of the phase error defined as

$$\sigma_0^2 = \frac{\int_A \sigma^2(\bar{r})f(\bar{r})dS}{\int_A f(\bar{r})dS}. \tag{7}$$

Note that the average ξ over the surface is zero. Then, if we assume that ξ at each point on the aperture is a random sample from a single Gaussian distribution with zero mean and standard deviation σ_ξ, the expected value of gain can be written as follows:

$$G = G_0 e^{-\sigma_0^2} e^{\sigma_\xi^2/4}. \tag{8}$$

To express (8) in terms of surface deviations, let us introduce the rms of the effective surface deviations, defined as follows:

$$\epsilon_0^2 = \text{rms}^2 = \frac{\int_A \epsilon^2 f(\bar{r})dS}{\int_A f(\bar{r})dS} \tag{9}$$

where the function ϵ is the effective surface deviation from the best-fit paraboloid for deterministic errors and it is the standard deviation of the effective surface deviations for random errors that have Gaussian distributions with zero means. Let us also define a quantity called the second variance of surface deviations as follows:

$$\eta_0^4 = \text{second variance of surface deviations}$$

$$= \frac{\int_A (\epsilon^2 - \epsilon_0^2)^2 f(\bar{r})dS}{\int_A f(\bar{r})dS}. \tag{10}$$

Then (8) may be written as follows:

$$G = G_0 e^{-(4\pi\epsilon_0/\lambda)^2} e^{\frac{1}{4}(4\pi\eta_0/\lambda)^4}. \tag{11}$$

Discussion

To show the accuracy of the usual tolerance theory [(1) and (5)] in predicting the loss of gain of antennas due to surface deviations, the axial loss of a uniformly-illuminated paraboloidal reflector with various radially-linear distributions of surface deviations is calculated and the results are compared with the corresponding values obtained assuming uniform error distribution. For the purposes of this comparison, (5) and (6) are employed instead of the more general (1) and (4). The effect of neglecting the second parts of these equations, which become important for large tolerance losses, is not expected to alter the conclusions reached from the results obtained herein.

The function σ is assumed to be varying linearly with radius, that is

$$\sigma(\bar{r}) = \sigma_0 + \nu(r - r_0) \tag{12}$$

where the constant σ_0 is the averaged standard deviation of the phase error in (7), and ν and r_0 are arbitrary constants. Let us consider a number of possible values for these constants and compare the values for the tolerance loss computed from (6) to the corresponding values obtained from (5). Four cases are considered for this purpose, as follows.

Case I: The function $\sigma(\bar{r})$ vanishes at the outer edge of the reflector.

Case II: The function $\sigma(\bar{r})$ vanishes at the center of the reflector.

Case III: An intermediate condition in which the rate of change of σ with radius is one half that for Case I.

Case IV: An intermediate condition in which the rate of change of σ with radius is one half that for Case II.

The results are shown in Fig. 2 and indicate that the assumption of uniform distribution of surface deviations over the aperture involves errors that become significant for high-tolerance losses.

For a 120-ft (36.58 m) uniformly-illuminated paraboloidal reflector with surface deviations that are distributed as in Cases I and II and an rms $\epsilon_0=0.1$ in (2.54 mm), the axial gain was computed from (5), (6), and (11), and the results are compared in Fig. 3. The agreement between the results of (6) and those obtained using the simpler theory (11) is quite good in most ranges of interest.

For the cases examined, a point of interest is that the usual antenna tolerance theory always underestimates the gain. The difference between the actual axial gain and that predicted by assuming uniform distribution of error is negligible for wavelengths which are large compared with surface errors and becomes increasingly more pronounced as the frequency increases. For σ_0 equal to unity, which corresponds to $\lambda/\epsilon \approx 12.5$ and to a loss of axial gain of 4.34 dB as computed by (5), the usual tolerance theory underestimates the axial gain by 8.6 percent for surface deviations assumed in Case II with a uniform illumination. This error increases with an increase in the taper of the illumination function until it reaches 12.5 percent for the case where the illumination function vanishes at the edge. (In these calculations the illumination was assumed to be radially parabolic.) If we employ (11) instead, the error in predicting the axial gain reduces by a factor of at least four.

In the general case of arbitrary distribution of error over the aperture, the tendency of (5) to underestimate the gain can also be shown by examining (8). This equation has been derived assuming that the variable ξ, the deviation from mean variance of phase errors, has a Gaussian distribution with zero mean. If the distribution is not Gaussian, then the expected value of exp $(-\xi/2$ is) is given by

$$E(e^{-\xi/2}) = \int_{-\infty}^{\infty} f_\xi(t)e^{-t/2}dt \tag{12}$$

where f_ξ is the frequency distribution function for ξ. If f_ξ is symmetric about its zero mean then, from the mean-value theorem,

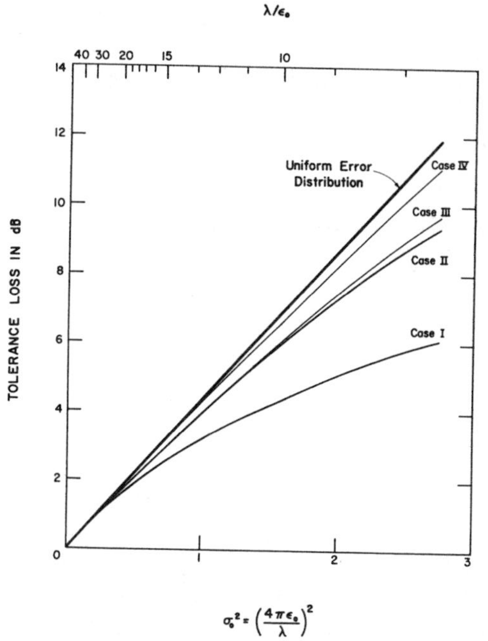

Fig. 2. Tolerance loss for various distributions of surface deviations.

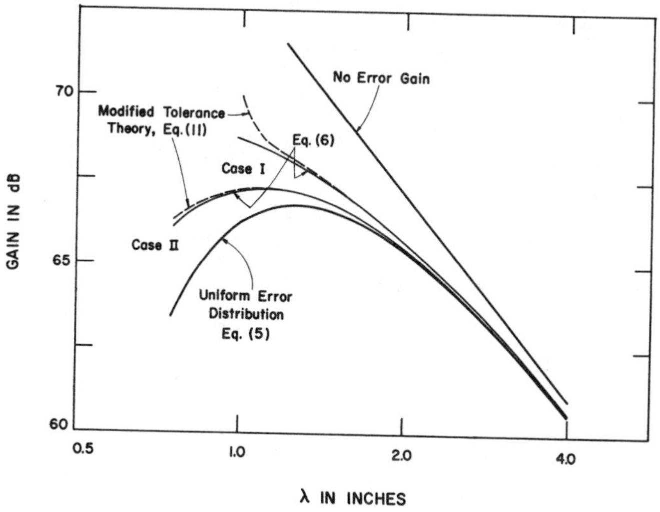

Fig. 3. Comparison of calculated gain of a 120-ft (36.58 m) antenna with $\epsilon_0=0.1$ in (2.54 mm) and with various distributions of surface deviations.

$$E(e^{-\xi/2}) = \cosh \frac{\bar{t}}{2} \int_{-\infty}^{\infty} f_\xi(t)dt = \cosh \frac{\bar{t}}{2} \qquad (13)$$

for some real $\bar{t}$. Obviously, $E(e^{-\xi/2})$ is always greater than or equal to unity, which shows that the usual tolerance theory always underestimates the gain. Only for f_ξ having extremely large skew can the expected value of exp $(-\xi/2)$ be less than unity.

The increase in scatter with an increase in the rms error may be predicted by employing (1). For this purpose a size must be assumed for the correlation radius c. For the two limiting cases of radially-linear variation of σ, namely for Cases I and II, the half-power beamwidth was also calculated as a function of the rms error of the reflector, employing (4). The computation was performed for a uniformly-illuminated circular aperture of 120-ft (36.58 m) radius and $c=5.6$ ft (1.71 m). The results are shown in Fig. 4.

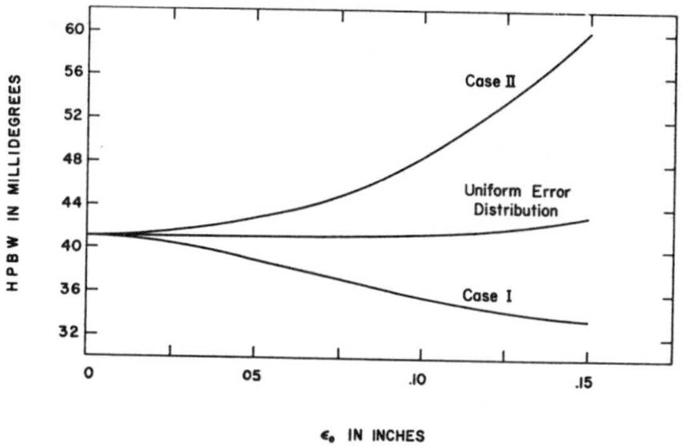

Fig. 4. HPBW of a 120-ft (36.58 m) uniformly illuminated reflector with various distributions of surface deviations; $\lambda=1$ in (25.4 mm); $c=5.6$ ft (1.7 m).

It can be observed that the half-power beamwidth is significantly affected by the distribution of the variance of phase error. For antennas having less error at the center the half-power beamwidth is greater than that predicted by (1). When an antenna has its greatest surface deviations at the center, the half-power beamwidth decreases with increasing rms. This effect of the error distribution on the half-power beamwidth may be explained by noting that when the antenna has its larger errors at its edge the effective taper of the illumination function is increased, which corresponds to an increase in the half-power beamwidth. Similarly, for an antenna with its larger errors at its center, we expect a reduction of the effective taper and thus a decrease in the half-power beamwidth; for large center errors, we have effectively an annular ring.

Conclusions

A generalization of Ruze's eq. (1) is derived including the effect of variation of phase error over the aperture and the accuracy of his equation is examined. It is found that the assumption of uniform error distribution, in general, underestimates the gain. This error is small for effective surface deviations less than about a twentieth of the wavelength, but becomes rapidly larger as the surface deviations increase. In radio astronomy applications, deviations in excess of a twentieth wavelength are not uncommon.

If the rms of the surface deviations is calculated[5] by substituting into (5) measurements of the gain at two different frequencies, the resulting rms value, in general, will be too small, since the error in (5) increases with frequency.

The distribution of the surface deviations is found to have an effect on beamwidth. In the absence of large feed-support surface deformations, the centrally supported antennas have their larger surface deviations near their edges. As is shown here the beamwidths of this class of antennas are larger than the values predicted using (1). This phenomenon has in fact been observed for many antennas.

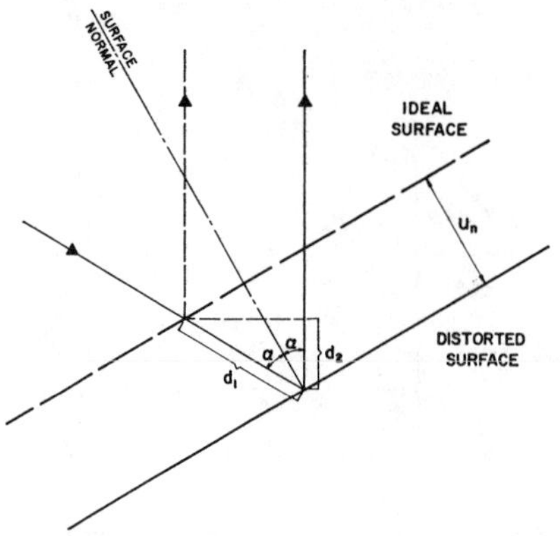

Fig. 5. Change in RF path length.

Appendix

The total change in the RF path length at a point on a reflector surface at which the normal surface deviation from the theoretical surface is equal to u_n (Fig. 5) is d_1+d_2. On the other hand, the axial component of the normal deviation may be expressed as follows:

$$\epsilon = u_n \cos \alpha = d_1 \cos^2 \alpha$$

$$= \tfrac{1}{2}(d_1 + d_1 \cos 2\alpha)$$

$$= \tfrac{1}{2}(d_1 + d_2).$$

It is thus shown that the effective surface deviation at a point is equal to the axial component of the normal deviation of the surface at that point.

Acknowledgment

The author wishes to express special appreciation to Dr. J. Ruze of Lincoln Laboratory for his invaluable guidance and helpful comments and criticisms and to Dr. H. Simpson for his suggestions during the preparation of this manuscript.

References

[1] M. S. Zarghamee and H. Simpson, "Feasibility study for rerigging the Haystack antenna," M.I.T. Lincoln Lab., Lexington, Mass., Tech. Rept. ESD-TR-67-235, February 1967.

[2] H. Simpson and J. Antebi, "Space frame analysis and applications to other types of structures," *Proc, SHARE Design Automation Workshop* (Atlantic City, N. J., June 1964), pp. 24–26.

[3] A. R. Dion, "Investigation of effect of surface deviations on haystack antenna radiation patterns," M.I.T. Lincoln Lab., Tech. Rept. 324, July 1963.

[4] J. Ruze, "The effect of aperture errors on the antenna radiation pattern," *Suppl. Nuove Cimento*, vol. 9, pp. 364–380, 1952.

[5] J. Ruze, "Antenna tolerance theory—A review," *Proc. IEEE*, vol. 54, pp. 633–640, April 1966.

Astigmatism in Reflector Antennas

JOHN R. COGDELL AND JOHN H. DAVIS

***Abstract*—The characteristics of the astigmatic phase error in large parabolic reflector antennas are described. A procedure for focusing an antenna and diagnosing the presence and degree of astigmatism is given.**

Astigmatism is a term used to describe one of several common aberrations (imperfections) in optical reflectors [1]. Astigmatism is common in microwave reflector antennas and is often the dominant error in degrading antenna performance. The astigmatic problem may be caused by feed displacement [2] or feed phase center problems [3, pp. 157–160, and other references cited therein], but the more serious problems are due to the shape of the reflector. While good discussions of common aberrations, including astigmatism, in microwave antennas exist [3, p. 139], workers with large reflector antennas seemingly fail to realize the importance of astigmatism or recognize its effects on antenna properties. The purpose of this communication is to describe the nature of the astigmatic phase error and its effects on the antenna properties and to outline a simple procedure for detecting astigmatism in reflector antennas. This discussion is placed within the context of the practical matter of locating the optimum focus of the reflector.

No reflector can be manufactured without errors. The errors in a reflector may be traced to various physical causes but naturally fall into two classes. Errors which decorrelate over a region small with respect to the antenna have been treated by Ruze [4] and others with probabilistic models. Errors which affect the entire antenna structure lead to the various aberrations, of which astigmatism is often the dominant form.

Let us investigate the nature of large scale errors in reflectors. A perfect parabola will transform a spherical wave originating from its true focus into a plane wave perpendicular to its axis. If the reflector is imperfect, the surface of constant phase will deviate from the aperture plane by an amount $z_r(x,y)$ which may be expressed as a power series

$$z_r(x,y) = \sum_{n=0}^{\infty}\sum_{n=0}^{\infty} a_{nm} \frac{x^n y^m}{(D/2)^{n+m}}$$

where x and y are Cartesian coordinates in the aperture plane and D is the antenna diameter. The error coefficients a_{nm} are normalized to be the maximum deviation of the (n,m)th term in the series at the edge of the antenna.

Another source of phase error is feed displacement from the true focus. If the spherical waves originate at a point displaced from the focus by x', y', and z', then an additional error $z_f(x,y)$ will result, given by

$$z_f(x,y) = F + \frac{1}{4F}(x^2+y^2) - \left[(x-x')^2 + (y-y')^2 + \left(F + z' - \frac{x^2+y^2}{4F}\right)^2\right]^{1/2}$$

where F is the focal length of the reflector. Adding the two results and converting to phase error in the aperture plane, we may expand the two expressions as

$$\varphi(x,y) = \frac{2\pi}{\lambda}(z_r + z_f) = \text{constant term (absolute phase)}$$

+ first-order terms (beam steering)
+ second-order terms (defocusing and astigmatism)
+ third- and higher order odd terms (coma and others)
+ fourth- and higher order even terms (spherical aberration and others).

The constant and first order terms are without physical significance. The third- and higher order odd terms lead to various effects, notably a large principal sidelobe (coma). Normally one would equalize the principal sidelobes through lateral adjustment of the equivalent feed position during the focusing procedure, thus reducing coma.

The second-order term takes the form:

$$\text{second-order term} = \left[\frac{\alpha(x^2-y^2) + \gamma(x^2+y^2)}{(D/2)^2}\right]\frac{2\pi}{\lambda}$$

where

$$\gamma = \left(\frac{z'}{8}\left(\frac{D}{F}\right)^2 + \frac{a_{20}+a_{02}}{2}\right)$$

$$\alpha = \frac{a_{20}-a_{02}}{2}, \text{ the astigmatism parameter.}$$

For the second-order term to take the simple form given in the preceding with no xy term, the x,y axes must be aligned with the direction of maximum astigmatism. We see two second-order terms: a radially symmetric error which is parabolic due to both reflector errors (a_{20} and a_{02}) and axial feed position (z'). Note that this term can be reduced to zero by setting the axial position of the feed properly. This feed position will maximize the peak gain of the antenna. With the feed in the maximum gain position, the remaining second-order phase error is the astigmatism term, proportional to α. Thus we reach the following conclusion: The proper feed location in the lateral plane is that which eliminates coma (equalizes sidelobes) and in the axial dimension is that which maximizes gain. With this optimum feed position, phase errors up to fourth order are eliminated except astigmatism, which is second order. It is precisely this which makes astigmatism so important to recognize, as it cannot be eliminated through focusing.

Physically, astigmatism can be thought of as an effect of squeezing the reflector at opposite edges. Opposite sides would move near the focus while the other two quadrants would move further away as the reflector acts like a shell. Thus the phase in the aperture plane leads in opposite quadrants and lags in the orthogonal quadrants. Gravitational sag tends to produce a deformation of this type, as will thermal expansion in certain cases. We would expect to see the effects of astigmatism in nearly all large movable antennas, which is indeed the case. In the following, we discuss the effects that astigmatism produces on the measurable characteristics of the antenna.

It is important in the evaluation of a large antenna to be aware of the effects of astigmatism on the gain and pattern of the antenna for several reasons. For one, moving the feed position cannot eliminate astigmatism and hence proper focusing of the antenna amounts to reducing the phase error to the astigmatic form. In other words, one is through focusing when one obtains the characteristics of pure astigmatism. Another benefit of recognizing the effects of astigmatism is that the engineer can weight the benefits of reworking

Manuscript received October 13, 1972; revised January 17, 1973. This work was supported in part by the NASA under Grant NGL 44-012-006.

The authors are with the Electrical Engineering Research Laboratory, Department of Electrical Engineering, University of Texas, Austin, Tex. 78712.

Reprinted from *IEEE Trans. Antennas Propagat.*, vol. AP-21, pp. 565–567, July 1973.

(in some way) the antenna surface or structure to reduce severe astigmatism. Finally, exploring the effects of astigmatism leads us to a simple procedure for detecting the presence and degree of astigmatism in an antenna.

The effects of astimgmatism are most easily seen in the patterns of the antenna. Let us consider the effects on the patterns with astigmatism present as the feed is moved axially. Recall the phase error is

$$\varphi(x,y) = \left[\frac{\alpha(x^2 - y^2) + \gamma(x^2 + y^2)}{(D/2)^2}\right]\frac{2\pi}{\lambda}.$$

As γ is varied by moving the feed axially, the amount of parabolic phase errors in each principal direction is varied. This will be evidenced in the patterns by broadened beamwidths, increased sidelobes, and an absence of nulls, with patterns in x and y behaving differently due to the astigmatism. For example, if $\gamma = -\alpha$, there is no phase error in the x direction but there is a 2α parabolic error in the y direction. This condition will produce a well formed pattern in the x direction, with a narrow beamwidth, small sidelobes, and deep nulls. The pattern in the y direction will be defocused and will exhibit a broadened main lobe, high principal sidelobes, and poor nulls. On the other hand, with $\gamma = +\alpha$, the opposite effects will be seen: the y pattern will be narrow and well formed and the x pattern will be broad and badly formed. If the feed is placed in the compromise position, both patterns will be equally defocused but the gain will be maximum.

Of the several effects of astigmatism which we have described above, the effect which gives the best quantitative evaluation of the astigmatism of a reflector is that of beam broadening. Half-power beamwidths can be conveniently and precisely measured on a pattern range or through astronomical measurements of point sources [5], or the solar limb [6]. The 3-dB beamwidths are generally a good diagnostic for the sharpness of the focus of an antenna. For example, if one varies the axial feed position for a perfect reflector and examines the beamwidths, one finds both beamwidths minimum at the best focus, i.e., the focus for maximum gain.

In the previous section we have described how the astigmatic antenna focuses differently in orthogonal planes. Thus we would expect the two beamwidths to minimize at two separate feed positions. This is shown in Fig. 1, which gives calculated beamwidths versus axial feed position for several values of the astigmatism parameter. These calculations are for $F/D = 0.5$ and a 10 dB illumination taper. Note that the beamwidths do minimize at different axial foci, and that the distance between the axial foci increases with the astigmatism parameter. Note that the beamwidths are equal at the compromise focus but are broadened with increasing astigmatism.

Comparison of experimental beamwidth data with theoretical curves like those in Fig. 1 allows one to estimate the astigmatism. The data plotted on Fig. 1 were taken on the University of Texas 16-ft antenna at 94.0 GHz in 1969. From the data we judged the astigmatism parameter to be 1.3, indicating a displacement of the reflector of 0.66 mm at the edge from a true parabola. This astigmatism has since been corrected by a reshimming of the antenna backup structure.

We might summarize the focusing and diagnostic procedures as follows:

1) Locate *a priori* focus. Mechanical or optical measurements can locate the focus approximately. Fine adjustments must be based on the patterns.

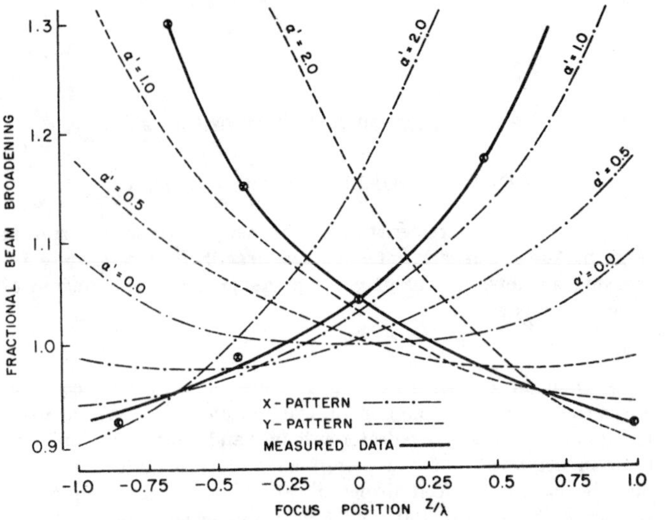

Fig. 1. Relative beamwidths versus axial defocusing for several values of astigmatism parameter $\alpha' = 2\pi\alpha/\lambda$, in radians of phase error at edge of reflector. This chart allows one to estimate degree of astigmatism from measured beamwidths. (Data plotted here are from the University of Texas 16-ft antenna in 1969.)

2) Remove coma. Lateral movements are made to remove coma. To correct coma, move the feed away from the prominent (coma) sidelobe until symmetry is obtained in both directions. Coma should be checked in more than two orthogonal planes in the pattern.

3) Locate the direction of maximum astigmatism. With the feed axially defocused from the maximum gain position, make a contour of the beam at the 3 or 10 dB level. If the beam is elliptical, astigmatism is possibly present. If defocusing on the opposite side of the maximum rotates the ellipse by 90°, then astigmatism is present. If not, unequal illumination tapering is indicated.

4) Measure degree of astigmatism. Measure beamwidths in major and minor axis directions versus axial feed position. Data can then be compared with theoretical calculations as in Fig. 1 to estimate astigmatism.

The preceding procedure is premised on using a pattern range. All of the preceding procedures, or equivalent, can be accomplished through radio astronomical methods in the centimeter- and millimeter-wavelength range, although not as conveniently as on a pattern range. In the event that the antenna astigmatism depends on some independent variable such as ambient temperature or antenna elevation angle, the evaluation procedure is complicated though not confused. Modifications would need to be developed to fit the specific case.

References

[1] M. Born and E. Wolf, *Principles of Optics*, 4th ed. Oxford, England: Pergamon Press, Ch. IX, 1970.
[2] J. Ruze, "Lateral-feed displacement in a paraboloid," *IEEE Trans. Antennas Propagat.*, vol. AP-13, pp. 660–665, Sept. 1965.
[3] R. C. Hansen, Ed., *Microwave Scanning Antennas*. New York: Academic Press, 1964, vol. I, pp. 139, 157–160 and other references cited therein.
[4] J. Ruze, "Antenna tolerance theory—a review," *Proc. IEEE*, vol. 54, pp. 633–640, Apr. 1966.
[5] J. R. Cogdell *et al.*, "High resolution millimeter reflector antennas," *IEEE Trans. Antennas Propagat.*, vol. AP-18, pp. 515–529, July 1970
[6] E. Jocobs and H. King, "2.8-minute beamwidths, millimeter wave antenna measurements and evaluation," in *1965 IEEE Int. Conv. Rec.*, pt. 5, pp. 92–100.

Gravitational Deformation and Astigmatism of Tiltable Radio Telescopes

SEBASTIAN VON HOERNER AND WOON-YIN WONG

Abstract—**An ideal telescope structure would deform, when tilted, in a homologous way, from one paraboloid of revolution to another one. Conventionally designed telescopes approach this condition already to some degree, special designs to a very high degree, and a measure for the degree of this approach is suggested. An equation is presented for the deviations from homology if an antenna is adjusted at zenith angle θ and observes at angle ϕ; it contains only two structural parameters for alt–azimuth mounts (three for polar mounts). The choice of the best adjustment angle is discussed; this is basically different for both types of mount, and examples are given. Supported at two elevation bearings, conventional telescopes will generally show a strong gravitational astigmatism which may be corrected at the secondary mirror, thus improving the astronomical performance considerably. Two possible correction methods are suggested. Several of the equations presented are checked with the example of the 140-ft telescope at Green Bank, W. Va. Good agreement is obtained between detailed computer analysis and equations, and between analysis and astronomically obtained data. Suggested improvements for the 140-ft (applicable also to other telescopes) would diminish the deviations from homology by a factor of 2.5, and the remaining residuals then are brought below the internal inaccuracy of the surface panels. The astigmatic correction thus is advisable as well as sufficient.**

I. Introduction

If a radio telescope is adjusted to a perfect paraboloid of revolution in zenith position (or any other zenith angle θ), and

Manuscript received January 13, 1975; revised March 18, 1975. The National Radio Astronomy Observatory is operated by Associated Universities, Inc., under contract with the National Science Foundation.

The authors are with the National Radio Astronomy Observatory, Green Bank, W. Va.

then is tilted in elevation, its structure must deform under the influence of the changing direction of gravity, which usually will degrade its efficiency. The ideal case would be a structure which deforms in a "homologous" way, from one paraboloid to another one, thus yielding a perfect mirror for any angle of tilt. A mathematical method was developed since 1964 for the structural synthesis of telescopes with homologous deformations [1], [2] and good results were obtained in the design of a 65-m telescope for $\lambda = 3.5$ mm wavelength [3].

For a convenient description, imagine a surface is perfect in the absence of gravity. Then let gravity be switched on in zenith position, and call H_z the rms deviation of the deformed surface from its best-fit paraboloid. Change gravity to horizon position and call H_h the deviations of the surface from a new best-fit paraboloid. As will be shown, the rms of both,

$$H_0 = \sqrt{\tfrac{1}{2}(H_z^2 + H_h^2)}, \tag{1}$$

which we call the "standard deviation from homology," is the rms deviation during a full turn of 360° in elevation. The quantities H_z and H_h can be obtained for a given telescope from a structural computer analysis, as well as from astronomical observations of the efficiency at various zenith angles ϕ, although with less accuracy.

The definitions of the parameters H_z and H_h can be modified in various ways, depending on the details and the weight function used when getting the best-fit paraboloid and the rms deviations from it. The most unrefined definition is an unweighted use of the gravitational deformations of the surface parallel to telescope axis. The more realistic modification uses the difference in optical path-length, weighted with the illumination pattern of

Reprinted from *IEEE Trans. Antennas Propagat.*, vol. AP-23, pp. 689–695, Sept. 1975.

the feed. This latter case will give smaller values for the parameters than the first one.

Normal conventional telescopes, not especially designed for homology, approach homologous deformations already to some degree. Let Δz_0 = rms (Δz_s) be the rms value of the actual structural deformations Δz_s due to gravity of all surface points, in both zenith and horizon position. For all comparable structures, Δz_0 is proportional to the square of the diameter. As a measure for this approach, we define a "homology ratio"

$$q = \Delta z_0 / H_0.$$

The conventionally designed 140-ft telescope at Green Bank, for example, has $q = 2.0$; whereas a new 25-m design of NRAO, especially designed for homologous deformations, yields $q = 80.3$ (see Table I).

For given values of H_z and H_h, we shall derive formulas for the telescope performance at any observational zenith angle ϕ, and for any adjustment angle θ. The question of the "optimum θ" is discussed and examples are given. Defining the ratio

$$g = H_h / H_z \tag{3}$$

the best adjustment angle $\theta(g)$ can be derived as a function of g, at least for alt–azimuth mounts in a unique way, while polar mounts need a compromise with personal judgment.

Tiltable telescopes, supported at two elevation (or declination) bearings, and constrained tangentially at the elevation wheel, have two planes of symmetry: both planes include the telescope's focal axis (z), with the elevation axis (y) being in one plane and perpendicular to the other plane. Describing the gravitational deformations of such a structure in gradually increasing detail, the deformations are in first order homologous: rigid-body translation and rotation, and a change of focal length. In second order, this type of symmetry will show a (nonhomologous) astigmatism, meaning two different focal lengths for the xz plane and the yz plane. In third and higher order, one could develop the deformations into a series of cylinder functions; but we shall stop here and regard all deviations from homology and astigmatism as "residuals." In practice, this means finding the best-fit paraboloid, calling Δz the deviations from it, and obtaining a least squares value for the amplitude a of the astigmatism, by minimizing the residuals.

If it turns out that the remaining rms residual is small as compared to rms (Δz), then a considerable improvement of an existing telescope is possible with relatively easy means: let a secondary Cassegrain mirror be mechanically squashed, by an elevation-dependent amount, such that its astigmatism just corrects that of the main mirror.[1]

This type of improvement is suggested for the 140-ft telescope at Green Bank. We will show that its deviations from homology then can be decreased by a factor of 2.5. The same astigmatic correction could also be obtained by placing two cylindrical dielectric lenses in front of the secondary (both lenses rotatable about the focal axis). This solution looks easier mechanically, but may give limits in wavelength and some absorption. Most conventional telescopes will show a strong astigmatism which then could be corrected, improving the performance considerably.

The formulas of this paper will be derived for general use. But since fairly elaborate computer analyses and observational data of the 140-ft telescope are now available, we will check our formulas with the example of the 140-ft data. Although the 140-ft has a polar mount which would need the inclusion of a third parameter H_a, for gravity in hour angle direction (y) which actually is included in our computer programs, we will limit most of our discussions to the (more reasonable and mostly prevailing) case of alt–azimuth mounts; which means movement along the meridian for polar mounts. Finally, resulting from the present investigation, several suggestions are presented for future improvement of the 140-ft telescope.

[1] If $a(\phi)$ has always the same sign, then a specially-designed secondary, with one diagonal cable pulled over a motor-driven axis, is sufficient. If $a(\phi)$ changes sign for varying angle ϕ, then a perpendicular spring-loaded diagonal cable would be needed in addition, or a second cable with its motor.

II. Some Formulas and their Derivation

A. Structures in General

Let a structure be tilted about a horizontal axis. How will its gravitational deformations depend on ϕ, the angle of tilt? Let δ be defined as the displacement vector of the points of the structure, K the stiffness matrix, and P the force vector in the direction of gravity, then the displacement of the structure is

$$[\delta] = [K]^{-1}[P]. \tag{4}$$

In a co-rotating system fixed in the structure, the displacement of joints at a given tilt angle is

$$\delta(\phi) = a \cos \phi + b \sin \phi \tag{5}$$

where a and b are defined as displacement vectors at two orthogonal positions with the corresponding gravitational force vectors P_a and P_b, and the corresponding stiffness matrix K_a and K_b, in which a and b are defined as

$$[a] = [K_a]^{-1}[P_a], \qquad [b] = [K_b]^{-1}[P_b]. \tag{6}$$

Equation (5) can be written as

$$\delta(\phi) = c \cos (\phi - \Omega) \tag{7}$$

with $c^2 = a^2 + b^2$, and $\Omega = \arctan (b/a)$. Equation (7) means: if we rotate the structure through a full 360°, each of its deformations describes a simple cosine wave, with some amplitude c and phase shift Ω. Conditions for the validity of (7) are: small deformations (Hooke's law), no partly loose joints, and the determinant of matrix A must not be zero or close to it (no joint with almost coplanar members, no structural parts with extreme cantilever, no "oilcanning"). In short, every reasonable structure, with reasonable loads, must follow (7). If actual measurements do not agree, there is something wrong, either with the measurements or with the structure. An example will be given later.

B. Deviation from Homology

We apply this result to a telescope structure, adjusted to a perfect paraboloid in the absence of gravity. How will the deviations H_ϕ from a best-fit paraboloid (a new fit for each ϕ) depend on the zenith distance ϕ? The best-fit programs actually used minimize the weighted path-length difference to be discussed in Section II-E. However, for the simplicity of equations, and the transparency of concepts, the discussions of this and the two following sections will regard only the unweighted z-deformation Δz which is parallel to the optical axis of the antenna. Call Δz_b the value of Δz for the best-fit paraboloid. Then, for a number s of surface points,

$$H_\phi^2 = \frac{1}{s} \sum_{k=1}^{s} (\Delta z_k - \Delta z_{bk})^2. \tag{8}$$

During a least squares fit, the parameters of the best-fitting paraboloid are obtained by the inverse of a matrix, and the resulting values Δz_b of this paraboloid then can again be expressed as a matrix operation on the actual deformations Δz:

$$\Delta z_{bk} = \sum_{j=1}^{s} B_{kj}\,\Delta z_j. \tag{9}$$

The same reasoning which yielded (7) will now yield the form

$$\Delta z_{bk}(\phi) = d_k \cos(\phi - \alpha_k) \tag{10}$$

and applying (7) to the surface deformations Δz gives $\Delta z_k = c_k \cos(\phi - \Omega_k)$. The difference between both terms can be written as

$$\Delta z_k(\phi) - \Delta z_{ak}(\phi) = e_k \cos(\phi - \beta_k) \tag{11}$$

which yields, for the sum in (8),

$$\begin{aligned}\sum_{k=1}^{s} (\Delta z_k - \Delta z_{bk})^2 &= \cos^2\phi \sum_{k=1}^{s} e_k^2 \cos^2\beta_k + \sin^2\phi \sum_{k=1}^{s} e_k^2 \sin^2\beta_k \\ &\quad + 2\cos\phi \sin\phi \sum_{k=1}^{s} e_k^2 \cos\beta_k \sin\beta_k.\end{aligned} \tag{12}$$

The first and second terms of the right side are defined positive, and only positive contributions are added up in their sums. Whereas the third (mixed) term may have either sign, and contributions of either sign occur in its sum. If the number s of surface points is sufficiently large, the mixed term will be small as compared to the sum of the two quadratic terms. Further on we make the assumption that it may be neglected. Equation (8) then can finally be written as

$$H_\phi = \sqrt{H_z^2 \cos^2\phi + H_h^2 \sin^2\phi} \tag{13}$$

where the two contributions, $(H_z \cos\phi)$ and $(H_h \sin\phi)$, add up quadratically.

Next, let the telescope be adjusted, with gravity, at some zenith distance θ. This adjustment moves each surface point by the amount $\Delta z_{bk}(\theta)$, the best-fit value of (10) at $\phi = \theta$. This means one must subtract for each contribution in (13) the corresponding contribution for $\phi = \theta$. Calling $H_{\phi\theta}$ the deviation from homology at angle ϕ, if adjusted at angle θ, this yields

$$H_{\phi\theta} = \sqrt{H_z^2(\cos\phi - \cos\theta)^2 + H_h^2(\sin\phi - \sin\theta)^2}. \tag{14}$$

Some interpretations of the quantity H_0 from (1) will now be given. If a telescope is adjusted without gravity and turned by a full 360° in elevation, defining a new best-fit paraboloid for any ϕ, then the rms deviation is

$$\text{rms}(H_\phi) = \sqrt{\tfrac{1}{2}(H_z^2 + H_h^2)} = H_0 \tag{15}$$

because $\langle\cos^2\phi\rangle = \langle\sin^2\phi\rangle = \frac{1}{2}$. This is why H_0 was called the standard deviation from homology of a structure. Taking the more realistic case of a telescope being adjusted with gravity at some zenith angle θ, then for $\phi = 90° - \theta$,

$$H_{90-\theta,\theta} = 2\sin(45° - \theta)H_0 \tag{16}$$

is the deviation from homology for observations done at a zenith angle of $\phi = 90° - \theta$. In particular, for $\theta = 15°$,

$$H_{75,15} = H_0. \tag{17}$$

Finally, values of the homology ratio q of (2) are given in Table I for two existing conventional telescopes, and for two NRAO designs optimized for homologous deformations. After this optimization, manufacturing tolerances were also taken into account in the computer analysis, by changing all joint coordinates by $\pm\frac{1}{4}$ in. (max) erection error, and by changing each bar area and each joint stiffness by ± 5 percent (max) of manufacturing tolerances, using equal-distributed random numbers. The resulting values of H_0 thus are considered to be realistic. (Both Δz_0 and H_0 use the unweighted deviations parallel to the telescope axis.)

TABLE I
STRUCTURAL RMS DEFORMATIONS Δz_0 OF SURFACE IN ZENITH AND HORIZON POSITION, AS COMPARED TO RMS DEVIATION H_0 FROM HOMOLOGY, FOR TWO CONVENTIONAL TELESCOPES AND TWO OPTIMIZED DESIGNS

	Δz_0 (mm)	H_0 (mm)	$q = \Delta z_0/H_0$	Weight (ton)
140-ft telescope	4.17	2.09	2.0	321
300-ft telescope	23.9	6.76	3.5	522
25-m VLA	4.06	0.399	10.2	70
65-m design	12.49	0.277	45.1	597
25-m design	2.36	0.0294	80.3	73

C. The Best Adjustment Angle

At which angle θ should a telescope actually be adjusted? An alt–azimuth mount, moving all 360° in azimuth, covers the whole sky if moving 90° in elevation, from zenith to horizon. The full elevation range thus is $0 \le \phi \le 90°$. If adjusted anywhere in between, the deviations will be largest at one of both limits, and a best angle θ could be defined by demanding equal deviations at both limits. Actually, however, it is better to replace the full range by a smaller "most useful range," $\phi_1 \le \phi \le \phi_2$. High surface accuracy is needed only for shortest wavelengths, where the atmospheric absorption and noise contribution becomes already important, making observations close to the horizon impossible; and $\phi_2 = 70°$ (20° above horizon) seems a reasonable choice. On the other side, close to the zenith there is only little sky, which furthermore is observed more conveniently before or after transit (avoiding fast azimuth movements); and a reasonable choice seems $\phi_1 = 10°$. Adopting these values,

$$10° = \phi_1 \le \phi \le \phi_2 = 70° \tag{18}$$

one then may demand that the telescope performs equally bad at both these limits:

$$H_{10,\theta} = H_{70,\theta}. \tag{19}$$

Equations (14) and (19) then yield, after some transformations,

$$g^2 = \frac{1 - A\cos\theta}{1 - B\sin\theta} \tag{20}$$

with

$$A = \frac{2}{\cos\phi_1 + \cos\phi_2} \quad \text{and} \quad B = \frac{2}{\sin\phi_1 + \sin\phi_2}. \tag{21}$$

For a given value of $g = H_h/H_z$, (20) can be solved for the best adjustment angle θ, either directly, numerically, or graphically. For the latter, Fig. 1 shows the angle θ as a function of g. Since the diameter of a telescope structure is considerably larger than its depth in the z direction, zenith deformations are usually larger than horizon deformations, and $g < 1$. The best angle θ is rather large, between 40° and 48.4° (see Fig. 1).

For a polar mount, one may select in the same way a range from 10° off the pole, to 20° above horizon. If the telescope is located at geographical latitude L, the most useful range is

$$L - 80° = \phi_1 \le \phi \le \phi_2 = 70°. \tag{22}$$

Green Bank, for example, is at $L = 38°$ latitude, giving $\phi_1 = -42°$.

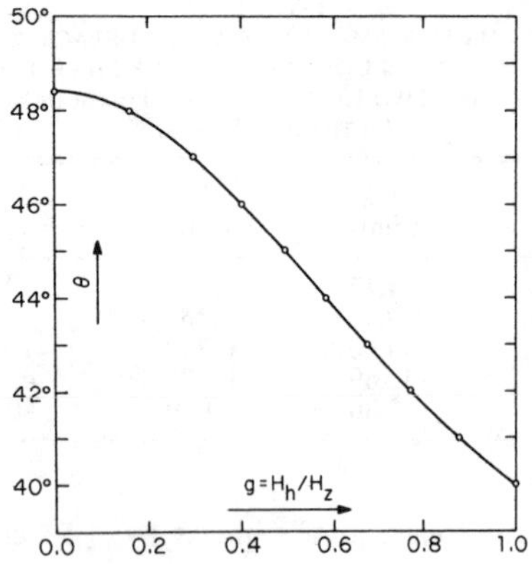

Fig. 1. Best zenith angle θ for adjustment of telescope, as function of telescope deformation parameter g of (3). Given for alt–azimuth mount, and demanding equal performance at 20° above horizon and at 10° below zenith.

However, a polar mount leads to an additional problem. Demand (19) with limits (22) would yield a large adjustment angle θ from (20), especially so for small g. If adjusted to this angle, $H_{\phi\theta}$ would be equally large at both limits as demanded, but it would show a still larger maximum in between, close to zenith. Since this maximum now occurs at a point important for a polar mount (except in the arctic), one must keep it well down, and good personal judgment is needed in selecting θ for a satisfying compromise. An example will be given later for the 140-ft telescope.

Regarding the application to a given telescope, it should be mentioned that the deformation parameters H_z, H_h, and g, and thus the best angle θ, can be obtained either from astronomical efficiency measurements or from a structural computer analysis. However, the actual adjustment needs either a point-by-point measurement of the surface with the telescope tilted by angle θ, or a surface measurement in zenith position plus a deformation analysis on the computer.

D. Astigmatism

For finding the amount of astigmatisms [4], and its variation with tilt angle ϕ, for a given telescope adjusted at angle θ, the structural deformations are first analyzed at angle θ, and a best-fit paraboloid is determined. Its values at each of the s surface points, $\Delta z_{bk}(\theta)$, are stored ($k = 1,\cdots,s$). At any other angle ϕ, a new analysis yields the deformations $\Delta z_k(\phi)$. Subtract the adjustment $\Delta z_{bk}(\theta)$ from the deformation, and call $\Delta z_{\theta k}(\phi) = \Delta z_k(\phi) - \Delta z_{bk}(\theta)$, which is the deviation of the adjusted surface from the design parabola. Next, find a best-fit paraboloid through these points $\Delta z_{\theta k}$. Call its values at each point $\Delta z_{pk}(\phi)$, and call

$$\Delta z_{dk}(\phi) = \Delta z_{\theta k}(\phi) - \Delta z_{pk}(\phi) \tag{23}$$

which is the deviation of the adjusted surface from its own best-fit paraboloid. The deviation from homology then is

$$H_{\phi\theta} = \sqrt{\frac{1}{s}\sum_{k=1}^{s} \{\Delta z_{dk}(\phi)\}^2}. \tag{24}$$

Astigmatic deviations from a paraboloid of revolution have the form

$$\Delta z_{ak}(\phi) = a(\phi)(r_k/R)^2 \cos(2\gamma_k) \tag{25}$$

where $a(\phi)$ is the amplitude of the astigmatism which may have either sign, r_k is the distance of point k from the telescope axis and γ_k its angle about the axis (with $\gamma = 0$ at $x = 0$ or north), and R is the telescope radius. Call

$$\text{residuals} = \rho_k(\phi) = \Delta z_{dk}(\phi) - \Delta z_{ak}(\phi). \tag{26}$$

A least squares fit for the amplitude then is obtained, by minimizing the sum of the squares of the residuals, as

$$a(\phi) = \frac{\sum_{k=1}^{s} \Delta z_{dk}(\phi)(r_k/R)^2 \cos(2\gamma_k)}{\sum_{k=1}^{s} (r_k/R)^4(\cos 2\gamma_k)^2} \tag{27}$$

and the remaining rms residual is

$$\rho(\phi) = \sqrt{\frac{1}{s}\sum_{k=1}^{s} \{\rho_k(\phi)\}^2}. \tag{28}$$

It was mentioned in the introduction that the astigmatism can be corrected with a deformable subreflector, being mechanically squashed by an elevation-dependent amount, or by two rotatable cylindrical lenses. For judging this correction, the remaining residual must be compared, on one side, with the uncorrected deviations from homology of (24). On the other side, the optimum telescope performance (at the adjustment angle, $\phi = \theta$) is determined by that part of the surface inaccuracy which cannot be adjusted away, meaning the internal bumpiness of the surface panels, which shall be called σ. In conclusion, a correction of the astigmatism is:

$$\begin{aligned} &\text{advisable,} && \text{if } \rho(\phi) \ll H_{\phi\theta} \\ &\text{sufficient,} && \text{if } \rho(\phi) \le \sigma \end{aligned} \tag{29}$$

where ϕ should be taken where it hurts most, either at the limits (18) for an alt–azimuth mount, or at the maximum of $H_{\rho\theta}$ near zenith for a polar mount. From the same reasoning as applied in the previous sections, $a(\phi)$ should follow a simple cosine law, with some amplitude and phase shift.

E. Weight Functions and Path Length

All average and rms procedures, as well as the program finding the best-fit paraboloid, may use weights w_k subscribed to each surface point. First, the structural joints where deformations are calculated may be not equally distributed along the surface, for example, crowded to the center and sparse at the rim; this is taken care of by measuring the area A_k supported by point k. If this area is measured along the curved surface, one should project it on the aperture plane:

$$w_k = \frac{A_k}{\sqrt{1 + (r_k/2F)^2}} \tag{30}$$

where F is the focal length. Second, for comparison with observations, it is necessary to consider the illumination pattern (of the feed) across the aperture. Mostly used is a "parabola with pedestal," for example

$$w_k = 1 - t(r_k/R)^2, \qquad \text{with } t = \begin{cases} 0.684, & \text{for 10 dB taper} \\ 0.822, & \text{for 15 dB taper.} \end{cases} \tag{31}$$

Third, for a more refined comparison, the deformations or deviations Δz in the previous formulas should be replaced by the path length difference Δp,

$$\Delta p = \frac{2\delta n}{\sqrt{1 + (r/2F)^2}} \tag{32}$$

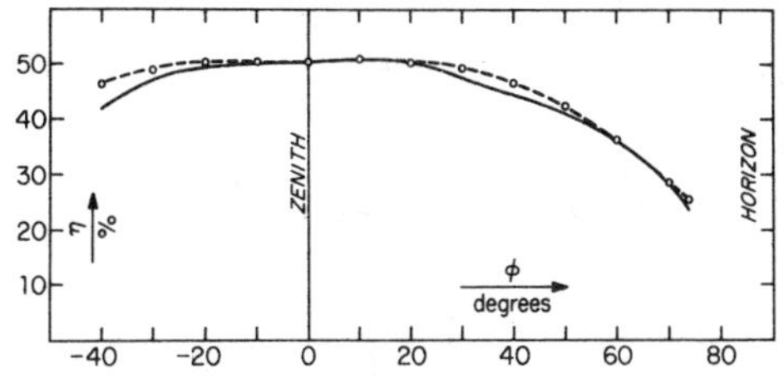

Fig. 2. Aperture efficiency η of 140-ft telescope, corrected for atmospheric absorption, at $\lambda = 2.8$ cm wavelength, as function of zenith distance ϕ along meridian. Full line: astronomical measurements. Circles and broken line: structural computer analysis (normalized to same efficiency at zenith).

where δn is the (three-dimensional) vector of the deformation, projected on the normal of the telescope surface.

Of all items mentioned, the illumination pattern is the most important one and will give considerable corrections, especially so for telescopes of conventional design where the deviations from homology are much larger at the rim than at the center. Whereas the areas A_k will not be too different for reasonable designs, and the use of path lengths instead of Δz will make a difference only for telescopes with short focal ratios F/D.

III. Applications to the 140-ft Telescope

A. Some General Checks

Although the 140-ft telescope has been studied in much more detail, the first application of one of the theoretical formulas occurred in 1970 for the 300-ft telescope. Astronomical measurements of the pointing error had shown a small hysteresis close to zenith, in contradiction to (7). Resolute insistence on the validity of this equation then actually led to the discovery of a partly loose joint in the feed support structure.

Attempts to measure with high accuracy the deformation of the 140-ft telescope as a function of tilt angle were started in 1970 [5] with a modulated light beam giving 0.08 mm rms accuracy. A detailed comparison is described by Findlay and Payne [6] where distance changes (during tilt) from the focus to five points on the surface, were measured by a radar technique of 0.08 mm rms accuracy, and where an improved very detailed computer model of the telescope was developed. Measurements and analysis agreed within 0.3 mm rms (1.2 mm max), which is 7 percent rms (13 percent max) of the deformations (and less than 10^{-4} of the distance). A further study showed that these deviations could be quantitatively explained by a slight off-axis position of the measuring instrument at the focus. This agreement between measurement and analysis was needed to give us confidence in the computer model, which is not trivial for a structure where many main members have a diameter comparable to their length.

Another check was done completely on the computer, about the validity of (14) regarding the neglected mixed term in (12). The structure is tilted through 180° in elevation, and every 10° a best-fit paraboloid is determined and the deviations from it are calculated, assuming adjustment at zenith angle $\theta = 30°$. In zenith and horizon position, the parameters H_z and H_h are calculated according to (8). Using these parameters, (14) then is used, and the results are compared with the directly calculated deviations. For the full range of 180°, both agree within 0.9 percent rms (2 percent max); in the range of (22), the agreement is 0.3 percent rms (0.5 percent max). These differences are so small that it is not clear whether they are real, or accumulated calculation errors during the many elaborate matrix operations. Either way, they are small enough to be neglected.

TABLE II
Gravitational Telescope Parameters, as Derived from Astronomical Observations and from Structural Analysis

		H_z (mm)	H_h (mm)	H_0 (mm)	g
Astronomical efficiency measurements		2.36	0.73	1.74	0.311
Structural analysis	no taper	2.95	0.25	2.09	0.085
	10 dB	2.62	0.37	1.87	0.141
	15 dB	2.51	0.38	1.80	0.150

B. Astronomical Observations and Structural Analysis

In 1972, K. Kellermann and M. Gordon (unpublished) calibrated the aperture efficiency (η) of the 140-ft telescope at $\lambda = 2.8$ cm wavelength. To their data, we apply a correction for atmospheric absorption, adopting (3 percent)/cos ϕ. The resulting curve along the meridian is shown in Fig. 2. The surface deviation at zenith was found by several observers, at $\lambda = 2.0$ and 1.3 cm, as

$$\sigma = 0.80 \text{ mm}. \tag{33}$$

Using the (somewhat unjustified) assumption that the correlation length of the deformations is small as compared to the telescope diameter, the deviations $H_{\phi 0}$ from homology for zenith adjustment, and the efficiency η then are connected by

$$\eta = 57.5\% e^{-(4\pi/\lambda)^2 (H_{\phi 0}{}^2 + \sigma^2)}. \tag{34}$$

The two gravitational parameters H_z and H_h then were determined, with (14), from two points of the observational curve, at $\phi = 50°$ and 70°. The result is given in Table II.

The telescope model was analyzed on the computer, using path-length differences of (32) and weights of (30); both illumination patterns of (31) were used, and "no taper" for completeness. For a comparison of observed and calculated parameters in Table II one should keep in mind that both determinations are completely independent, they have no input in common.

Table II shows the influence of the illumination on the averaging procedures. For the actual taper of 15 dB, H_z and H_0 give completely satisfying agreement between observation and calculation, whereas H_h (and thus g) differ by almost a factor two. This was first regarded as a serious discrepancy. But then a counter-test was made. Using H_z and H_h as determined by the computer analysis, $H_{\phi 0}$ is calculated from (14), to which $\sigma =$ 0.80 mm is added quadratically. The efficiency then is calculated from (34), and the result is entered in Fig. 2. There is a difference between the observed and the calculated curve, but it is only small, and the observers ascertain that it is well within the observing error, especially close to the pole (which is at $\phi = -52°$) where only very few radio sources are available. For further use, one may adopt the parameters from the last line of Table II.

C. Astigmatism

The procedure as described in Section II-D was applied to the 140-ft in a computer analysis, and the result is given in Fig. 3 for the amplitude $a(\phi)$. Comparison of curves a) and b) shows the influence of the adjustment angle. If $\theta \neq 0$, then $a(\phi)$ changes sign. However, it is not the actual astigmatism of the structure which changes sign: the astigmatism at $\theta = 30°$ has been removed by the adjustment, and thus shows up with opposite sign when the telescope is tilted back. Curve a) of Fig. 3 was found to follow a simple cosine law, with $a(\phi) = 13.3$ mm $(1 - \cos\phi)$, within 0.2 percent rms, thus confirming the statement at the end of Section II-D within the calculation accuracy. Comparison of curves a) and c) shows the influence of the illumination taper.

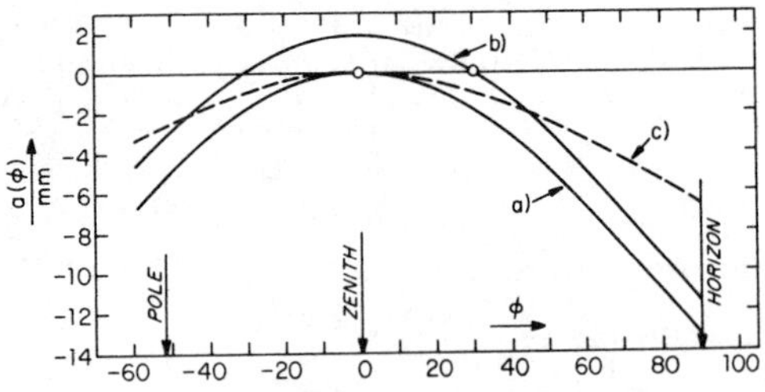

Fig. 3. Amplitude of astigmatism, $a(\phi)$, as function of zenith angle ϕ, for 140-ft telescope. (a) Adjusted at zenith ($\theta = 0$), no taper. (b) Adjusted at $\theta = 30°$, no taper. (c) Adjusted at zenith, 15 dB taper.

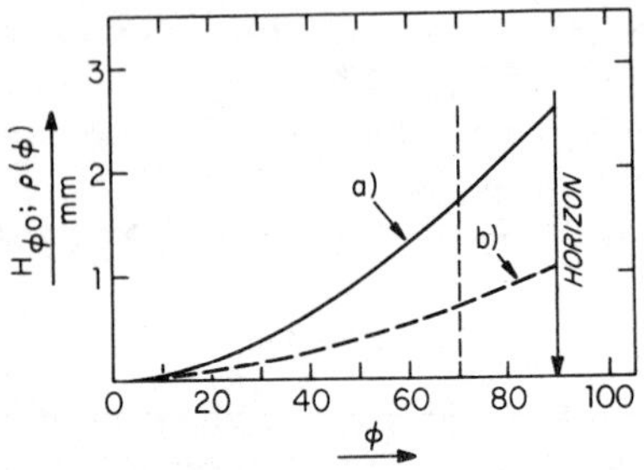

Fig. 4. Astigmatism correction, for zenith adjustment and 15 dB taper. (a) Uncorrected deviation from homology $H_{\phi 0}$. (b) Remaining residual $\rho(\phi)$ after correction is applied.

Since the values c) are smaller than a), most of the astigmatism must occur toward the rim of the telescope, as it might be expected.

Fig. 4 shows the (uncorrected) deviation $H_{\phi 0}$ from homology, and the remaining residual $\rho(\phi)$ after a correction for astigmatism has been applied. The residuals are about a factor 2.5 smaller than the uncorrected deviations.

Astigmatism means two different focal lengths in perpendicular planes. The focal length of the 140-ft is 720 in. For the three cases of Fig. 3, at the horizon ($\phi = 90°$), the difference is $\Delta F = 1.54$ in. for case a), $\Delta F = 1.33$ in. for case b), and $\Delta F = 0.77$ in. for case c).

IV. Suggested Improvements

Since it seems clear that no large short-wave telescope will be built in the foreseeable future, one should try to improve the 140-ft (and other existing telescopes) just as much as possible. Defining $\lambda = 16$ rms (Δz) as the shortest observational wavelength (where $\eta = \frac{1}{2}\eta_0$), the 140-ft is at present limited by its surface accuracy to $\lambda = 1.28$ cm at zenith, and limited by its deviation from homology to $\lambda = 3.0$ cm at 20° above horizon.

Improvements are suggested in three ways: a) reducing the surface accuracy by a new adjustment, b) adjusting for the best angle θ, and c) correcting the astigmatism.

It seems that the surface panels are somewhat more curved than the present focal length would demand, but two previous measurements did not agree on the amount. A new measurement is planned with a more accurate method being developed by Findlay and Payne [7]. If the previous results are confirmed, the rim of the telescope cannot be properly adjusted to the shorter focal length because the adjustment studs are too short at the rim, by about 2 in. It is suggested to adjust to the shorter focal length, but let the rim be down by 2 in.; and to correct for this with a secondary which again is at its rim by 2 in. below its hyperboloid. A preliminary estimate showed that the surface may be improved from its present value, $\sigma = 0.80$ mm, to

$$\sigma = 0.60 \text{ mm}. \tag{35}$$

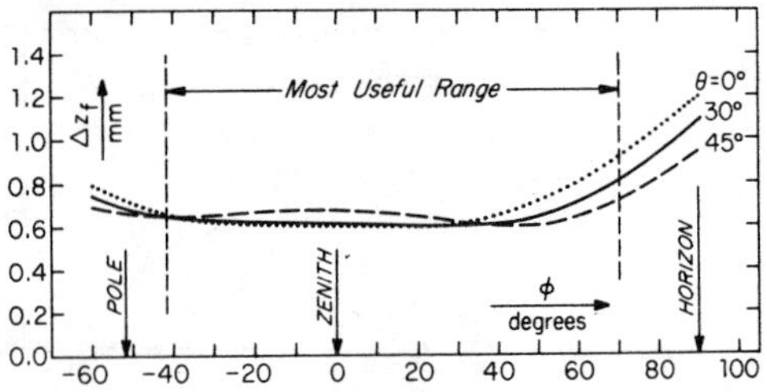

Fig. 5. Choice of best adjustment angle θ for polar mount. Shown is final surface accuracy $\Delta z_f(\phi,\theta)$ after astigmatic correction, in millimeters, as predicted for 140-ft telescope; ϕ = zenith distance along meridian. "Most useful range" starts at 10° off pole and goes to 20° above horizon. Different telescopes, and locations at different latitudes, will yield curves which are numerically different, but basically similar to curves shown.

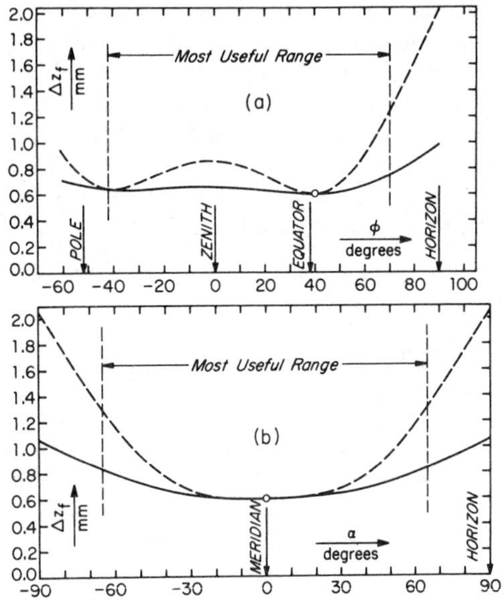

Fig. 6. Final surface accuracy Δz_f for best adjustment angle $\theta = 40°$; with astigmatic correction (full lines) and without it (broken lines). (a) Along meridian, ϕ = zenith distance. (b) Along zero declination α = hour angle.

Fig. 5 shows the performance after astigmatic correction, for several adjustment angles θ calculated for 15 dB taper and $\sigma = 0.6$ mm. We call $\Delta z_f(\phi,\theta)$ the final surface deviation from its best-fit paraboloid, when observing at ϕ and adjusted for θ, including the residuals ρ and the internal accuracy σ:

$$\Delta z_f(\phi,\theta) = \sqrt{\rho^2(\phi,\theta) + \sigma^2}. \tag{36}$$

For the polar mount, the best adjustment angle θ cannot be determined by the procedure used in Fig. 1, because the maximum close to zenith would then get very high. Discussions of Fig. 5 with several observers resulted in a compromise of

$$\theta \approx 40° \tag{37}$$

which, for a polar mount at Green Bank, means at declination $\delta \approx 0$. The best performance then is in a region centered on the celestial equator and thus containing most of the sky. Since the minimum of $\Delta z_f(\phi)$ at $\phi = \theta$ is fairly wide, one could as well shift θ off the equator either side by $\pm 10°$. Actually, the compromise means weighting the performance at zenith versus the performance at low elevation. The galactic center, for example, is at $\phi = 64°$, only 26° above the horizon; observers of galactic sources would prefer $\theta > 40°$, while $\theta < 40°$ would be preferred for extragalactic work.

Fig. 6 finally shows the expected performance for the adopted adjustment at $\theta = 40°$, with and without the astigmatic correction (calculated for 15 dB taper and $\sigma = 0.6$ mm). Since the previous discussion optimized the performance only along the meridian, other hour angles α must be checked, too. As the most important case, Fig. 6(b) gives the result for constant declination δ, along the equator ($\delta = 0$). As is to be expected from the symmetry of the structure, this result is very similar to the one of Fig. 5 for $\theta = 0$: a broad minimum with steeper flanks.

As to the conclusions suggested in (29), we find that $H/\rho = 2.5$ for large ϕ along the meridian, and 2.1 along the equator (both values at 20° above the horizon). The telescope performance can be improved by a good amount, and the astigmatic correction thus is *advisable*. The remaining residuals are $\rho = 0.43$ mm at 20° above the horizon on the meridian and 0.57 mm on the equator, $\rho = 0.26$ mm at the zenith, and still smaller for most of the sky. As compared to the internal surface accuracy of the panels, $\sigma = 0.6$ mm, the correction is also *sufficient*.

The shortest observational wavelength, $\lambda = 16$ rms (Δ_{zf}), then is $\lambda = 0.96$ cm in a fairly wide surrounding of $\alpha = \delta = 0$, and $\lambda = 1.06$ cm at the zenith. Along the meridian, it is $\lambda = 1.12$ cm at the galactic center and 1.18 cm at 20° above the horizon. Along the equator, $\lambda = 1.33$ cm at 20° above the horizon. These numbers show that the astigmatic correction almost removes all gravitational deformation for practical purposes.

References

[1] S. von Hoerner, *J. Struct. Div. Proc. Amer. Soc. Civil Eng.*, vol. 93, pp. 461–485, 1967.
[2] ——, *Astron. J.*, vol. 72, pp. 35–47, 1967.
[3] J. W. Findlay and S. von Hoerner, "A 65-m telescope for millimeter wavelength," National Radio Astron. Obs., Charlottesville, Va., 1972.
[4] M. Born and E. Wolf, *Principles of Optics*. New York: Pergamon, 1970.
[5] J. M. Payne, *Rev. Sci. Instr.*, vol. 44, pp. 304–306, 1973.
[6] J. W. Findlay and J. M. Payne, *IEEE Trans. Instrum. Meas.*, in press.
[7] ——, in preparation.

Microwave Holographic Metrology of Large Reflector Antennas

J. C. BENNETT, A. P. ANDERSON, PETER A. McINNES, SENIOR MEMBER, IEEE, AND A. J. T. WHITAKER

Abstract—**A microwave holographic technique for the determination of amplitude and phase of the principal and cross-polarized aperture fields of large reflector antennas is described. The hologram formation process utilizes the elevation over azimuth scanning system normally associated with these antennas, and, in this respect, appears to be unique among other proposed methods of field probing. The present work describes the means used to obtain vital information on the antenna structure such as *E*- and *H*-plane phase centers of the feed, and rms values of the reflector surface profile errors. Accurate prediction of *E*- and *H*-plane radiation patterns in the near- and far-field is also demonstrated.**

I. Introduction

REALIZATION and evaluation of the performance of an antenna are continuing requirements in the design of a communications system. Until recently, these requirements were met through the conventional recording of radiation patterns as angular power spectra and often the engineers' ability, through experience and intuition, to relate a poor pattern to the source of the problem within the antenna/feed structure.

The diffraction theory relating the radiated field to the fields in or near an antenna structure has been known for many years, although, in practice, a thorough vector formulation and solution of the problem may be prohibitively time consuming and costly. However, the traditional emphasis on recording the intensity of the radiation and neglecting the phase pattern does not permit proper utilization of the Fourier transform and its associated operations. Previous work [1] has indicated that near-field measurements made in amplitude and phase over a planar aperture can be used to yield far-field patterns to very good accuracy. This paper describes an alternative technique where the inherent scanning characteristics of the antenna are used to acquire data over a spherical recording aperture. A simple holographic recording technique permits utilization of a conventional microwave receiver which monitors received power.[1]

II. Hologram Formation Process

The basic microwave system used to form the hologram of a reflector antenna is shown in Fig. 1. The antenna elevation over azimuth positioning system is conveniently utilized for the scanning operation required. A static microwave horn provides the reference wave which is combined with the antenna signal via a hybrid-T. The hologram is obtained in raster format by performing continuous scans in elevation and small discrete increments in azimuth. The hologram signal is sampled at regular intervals during the elevation scan and the demodulated signal is quantized into 1000 levels and stored for subsequent processing. The function of the phase shifter is to synthesize an effective off-axis reference wave to ensure adequate separation of the reconstructed image from other interfering outputs. This is accomplished by incrementing the phase shifter by a constant amount between each elevation scan.

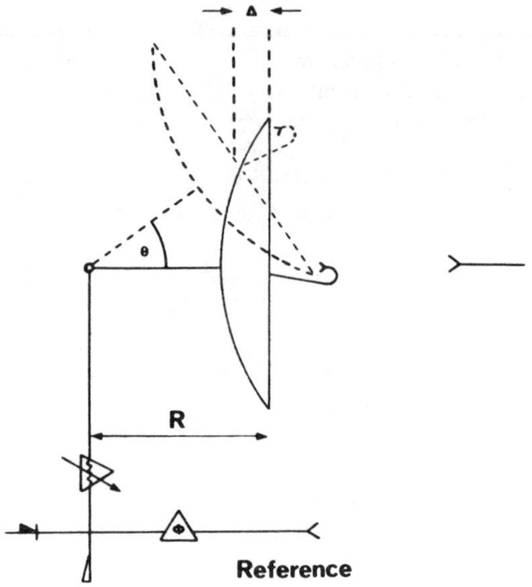

Fig. 1. Hologram formation geometry.

Since the reference horn is static, rotation of the antenna about an axis distant R behind its aperture will result in a relative phase change between the two received signals which is due solely to the changing spatial separation of the antenna and the transmitter. Consideration of the geometry of Fig. 1 shows that the separation Δ is given by

$$\Delta = R(1 - \cos\theta). \tag{1}$$

Thus, if θ is restricted to small angles,

$$\Delta = \frac{R\theta^2}{2}. \tag{2}$$

Hence, if the propagation constant is k, the reference wave also generates a quadratic phase term $\exp(-jk\Delta)$ which affects the focussing properties of the hologram.

In the receiving mode shown in Fig. 1, the hologram coordinates are $(p/(\lambda R_f), q/(\lambda R_f))$, where p and q are orthogonal distance variables on the focal surface in the

Manuscript received May 17, 1975; revised October 7, 1975. This work was supported by the U.K. Science Research Council.

The authors are with the Department of Electronic and Electrical Engineering, University of Sheffield, Sheffield, England.

[1] Some results have been briefly reported at professional group meetings [2], [3].

Reprinted from *IEEE Trans. Antennas Propagat.*, vol. AP-24, pp. 295–303, May 1976.

azimuth and elevation directions and R_f is the antenna focal length. Since $\theta = p/R_f$, (2) may be rewritten

$$\Delta = \frac{R\theta^2}{2} = \frac{Rp^2}{2R_f^2}. \tag{3}$$

Hence the spherical off-axis reference wave will be of the form

$$F_R = B \exp[-jk(p^2 + q^2)/(2R_f^2/R)] \exp[-jk\sigma p] \tag{4}$$

where B is constant and σ is a constant denoting the linear phase change in the azimuthal direction.

The focal surface distribution is given by

$$F_0 = D_F\left(\frac{p}{\lambda R_f}, \frac{q}{\lambda R_f}\right) \tag{5}$$

where F_0 is the Fourier transform of the aperture distribution and D_F is the focal surface distribution function.

The resulting hologram distribution is therefore

$$\begin{aligned} H(u,v) &= |F_0|^2 + |F_R|^2 + F_0^*F_R + F_0F_R^* \\ &= |F_0|^2 + |B \exp[-jk(p^2 + q^2)/(2R_f^2/R)] \\ &\quad \cdot \exp[-jk\sigma p]|^2 \\ &\quad + BF_0^* \exp[-jk(p^2 + q^2)/(2R_f^2/R)] \\ &\quad \cdot \exp[-jk\sigma p] \\ &\quad + BF_0 \exp[jk(p^2 + q^2)/(2R_f^2/R)] \\ &\quad \cdot \exp[jk\sigma p] \end{aligned} \tag{6}$$

where $u = p/(\lambda R_f)$ and $v = q/(\lambda R_f)$.

III. Image Reconstitution

The processing of data in the form of (6) is conveniently achieved by using the fast Fourier transform (FFT) algorithm. The second term of (6), which corresponds to hologram edge diffraction effects, may be removed computationally, giving

$$\begin{aligned} H(u,v) &= |F_0|^2 \\ &\quad + BF_0^* \exp[-jk(p^2 + q^2)/(2R_f^2/R)] \\ &\quad \cdot \exp[-jk\sigma p] \\ &\quad + BF_0 \exp[jk(p^2 + q^2)/(2R_f^2/R)] \\ &\quad \cdot \exp[jk\sigma p]. \end{aligned} \tag{7}$$

If an image distribution $\tilde{A}$ is required at the output, then the input to the FFT routine must be proportional to the Fourier transform of $\tilde{A}$ and be accompanied only by linear phase terms. The complex conjugate of the Fourier transform of $\tilde{A}$ is also an acceptable function since this merely produces a spatial reversal of $\tilde{A}$. Since F_0 has been defined as the Fourier transform of the aperture distribution then it is apparent that (7) may be converted into an acceptable form for the FFT by its premultiplication with either $\exp[\pm jk(p^2 + q^2)/2R_f^2/R)]$. Premultiplication by the negative exponent gives the new distribution

$$\begin{aligned} H_1(u,v) &= |F_0|^2 \exp[-jk(p^2 + q^2)/(2R_f^2/R)] \\ &\quad + BF_0^* \exp[-jk(p^2 + q^2)/(R_f^2/R)] \\ &\quad \cdot \exp[-jk\sigma p] + BF_0 \exp[jk\sigma p]. \end{aligned} \tag{8}$$

The contributions of these terms at the output plane (ξ,η) of the FFT will now be examined by considering their separate Fourier transforms.

If $H_1(u,v)$ is not bandlimited,

$$\begin{aligned} F\{H_1(u,v)\} &= \int_{-\infty}^{\infty}\int_{-\infty}^{\infty} \left|D_F\left(\frac{p}{\lambda R_f}, \frac{q}{\lambda R_f}\right)\right|^2 \\ &\quad \cdot \exp[-jk(p^2 + q^2)/(2R_f^2/R)] \\ &\quad \cdot \exp\left[j\frac{2\pi}{\lambda R_f}(p\xi + q\eta)\right] d\left(\frac{p}{\lambda R_f}\right) d\left(\frac{q}{\lambda R_f}\right) \\ &\quad + \int_{-\infty}^{\infty}\int_{-\infty}^{\infty} BD_F^*\left(\frac{p}{\lambda R_f}, \frac{q}{\lambda R_f}\right) \\ &\quad \cdot \exp[-jk(p^2 + q^2)/(R_f^2/R)] \\ &\quad \cdot \exp[-jk\sigma p] \\ &\quad \cdot \exp\left[j\frac{2\pi}{\lambda R_f}(p\xi + q\eta)\right] d\left(\frac{p}{\lambda R_f}\right) d\left(\frac{q}{\lambda R_f}\right) \\ &\quad + \int_{-\infty}^{\infty}\int_{-\infty}^{\infty} BD_F\left(\frac{p}{\lambda R_f}, \frac{q}{\lambda R_f}\right) \\ &\quad \cdot \exp[jk\sigma p] \\ &\quad \cdot \exp\left[j\frac{2\pi}{\lambda R_f}(p\xi + q\eta)\right] d\left(\frac{p}{\lambda R_f}\right) d\left(\frac{q}{\lambda R_f}\right) \end{aligned} \tag{9}$$

$$\begin{aligned} F\{H_1(u,v)\} &= \frac{-j}{\lambda R}[D(\xi,\eta) \circledast D^*(-\xi,-\eta)] \circledast \\ &\quad \cdot \exp\left(j\frac{\pi}{\lambda R}(\xi^2 + \eta^2)\right) \\ &\quad - j\frac{2B}{\lambda R}\left\{D(-\xi,-\eta) \circledast \right. \\ &\quad \left. \cdot \exp\left[j\frac{2\pi}{\lambda R}(\xi^2 + \eta^2)\right]\right\} \circledast \delta(\xi + \sigma R_f) \\ &\quad + BD(\xi,\eta) \circledast \delta(\xi - \sigma R_f) \end{aligned} \tag{10}$$

where $\circledast$ denotes the convolution process. Fig. 2 shows a typical ξ plane image distribution. The first term of (10) contributes a defocussed autocorrelation function of the aperture distribution which is centered about $\xi = 0$. The second term is a defocussed image of the aperture distribution situated at $\xi = -\sigma R_f$, whereas the desired in-focus image $D(\xi,\eta)$ is contained in the third term and positioned at $\xi = +\sigma R_f$.

In reality, the input plane spatial frequencies of the FFT are bandlimited. By application of the sampling theorem,

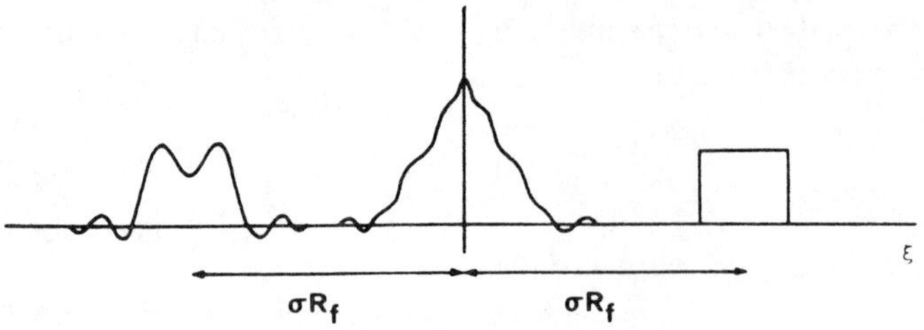

Fig. 2. ξ plane image distribution.

the required spatial range (2*S*) of the output plane is given by

$$2S = \frac{1}{\Delta f} \tag{11}$$

where Δf is the distance between spatial frequency samples. For the case of the microwave hologram,

$$\Delta f = \frac{\Delta p}{\lambda R_f} = \frac{\Delta \theta}{\lambda}. \tag{12}$$

Hence

$$2S = \frac{\lambda}{\Delta \theta} \tag{13}$$

defines the spatial range of the output plane in terms of the wavelength and the angle between samples.

It is important to consider the image positions in the output plane: these are determined by the choice of σ in (8). If the linear phase change between samples is $2\pi/T$ rad, then

$$k\sigma\Delta p = \frac{2\pi}{T} \tag{14}$$

or

$$\sigma = \frac{\lambda}{T\Delta p} = \frac{\lambda}{T\lambda R_f \Delta f} = \frac{2S}{TR_f}. \tag{15}$$

However, the image positions are given by

$$\xi = \pm \sigma R_f \tag{16}$$

$$\therefore \xi = \pm \frac{2S}{T}. \tag{17}$$

The significance of (13) and (17) becomes apparent by consideration of Fig. 3 where the effects of different values of T are demonstrated. In Fig. 3(a) the value of $T = T_1$ is large (i.e., the linear phase change is small) and the two images are degraded by the autocorrelation response. If a small value of $T (= T_2$, for example) is used, then violation of the sampling theorem occurs, and image degradation is incurred through interference with the adjacent output planes as shown in Fig. 3(b). A correct choice of $T = T_3$ ensures good image separation as in Fig. 3(c).

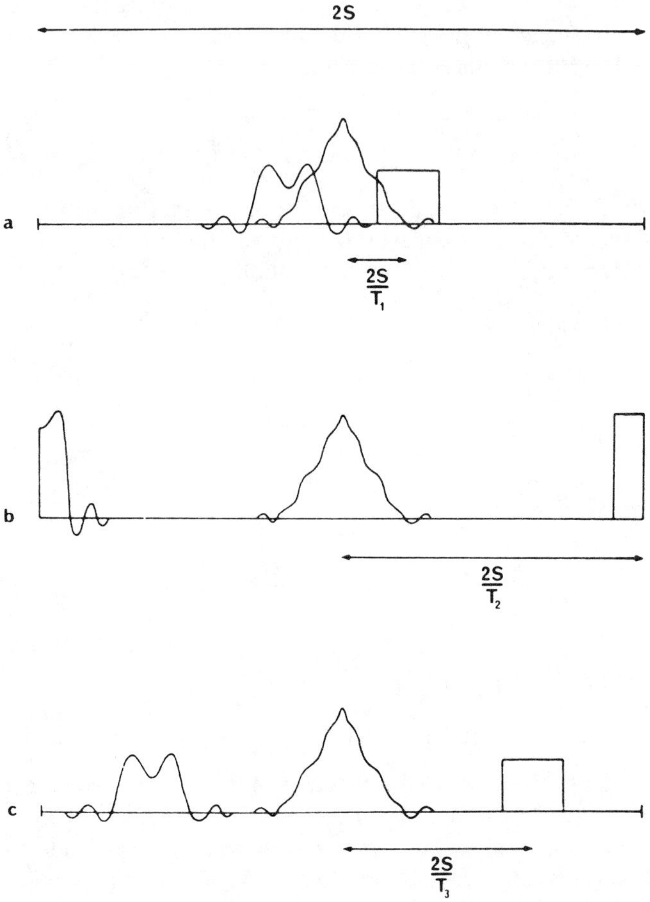

Fig. 3. Effect of T on image separation.

IV. Optimization of Data Logging

In order to optimize the fidelity of the recorded data, it is necessary to have an understanding of the effects of varying the hologram recording parameters. The choice of reference level determines the efficiency with which the quantization levels available in the data logging system are used.

Fig. 4 shows the structure of a section through a recorded antenna hologram where it can be seen, as predicted by the holographic equation (6), that the off-axis reference wave is modulated by the antenna pattern function. Here the peak of the main beam is assumed to possess unit field strength and the amplitude of the reference level is a. Suppose that a sidelobe of amplitude b is to be represented by N levels of an M level recording system. The maximum power level in the hologram is $(a + 1)^2$ and hence the power range of one quantization level is $(a + 1)^2/M$. Hence for the sidelobe in question,

$$(a + b)^2 = a^2 + N(a + 1)^2/M \tag{18}$$

or

$$b = [a^2 + N(a + 1)^2/M]^{1/2} - a. \tag{19}$$

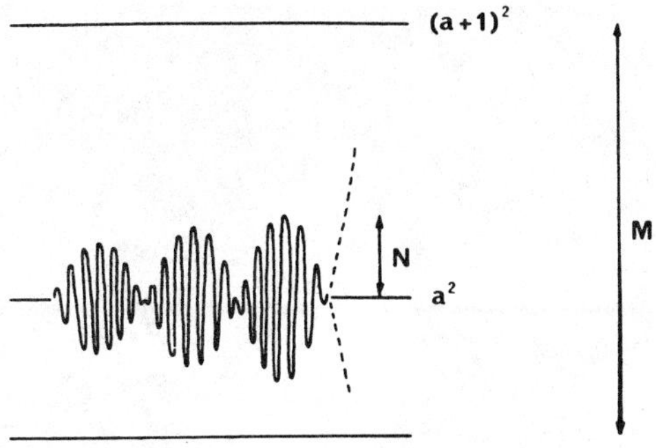

Fig. 4. Section through recorded antenna hologram.

In order to optimize the use of the N levels it is necessary to obtain the value of a which will provide a minimum value of b. Therefore,

$$\frac{db}{da} = \frac{1}{2}[a^2 + N(a+1)^2/M]^{-1/2}[2a + 2N(a+1)/M] - 1$$
$$= 0 \tag{20}$$

or

$$a = a_{\min} = \frac{1 - N/M}{1 + N/M}. \tag{21}$$

It can be seen that $a_{\min}$ need never be greater than the maximum signal level in the antenna pattern and, superficially, it may be said that as $a_{\min} \to 0$, greater recording fidelity is obtained. However, when $a_{\min} < 1$, the higher amplitude signals fail to fully modulate the reference, and information in this region is not properly recorded and, hence, in order to maximize information storage, the constraint $a_{\min} = 1$ should be applied. Having stipulated this condition, (19) may be computed for various values of N/M to produce the curve of Fig. 5, where

$$L = 20 \log_{10}\left(\frac{1}{b}\right) \text{ dB.} \tag{22}$$

Such a curve may be used to evaluate the capabilities of an M level recording system: that is, the interrelationship of N, M, and L may be deduced. For example, on considering the recording of a 30 dB sidelobe on a 1000 level recording system, it is found that 16 levels are available in the system for this purpose. It is of interest at this point to note the increase in sensitivity afforded by holographic recording. If only the antenna signal is recorded on a 1000 level system, one level alone is available for the description of a 30 dB sidelobe.

Under certain circumstances it may be permissible to reduce the value of $a_{\min}$ below unity. In this case a proportion of the lower spatial frequency component amplitude will be lost and greater definition of the high spatial frequency components (i.e., the outer sidelobes) will be obtained. If the components lost correspond to image information which has a long spatial period compared with the aperture size, then the image degradation due to this loss can be negligible.

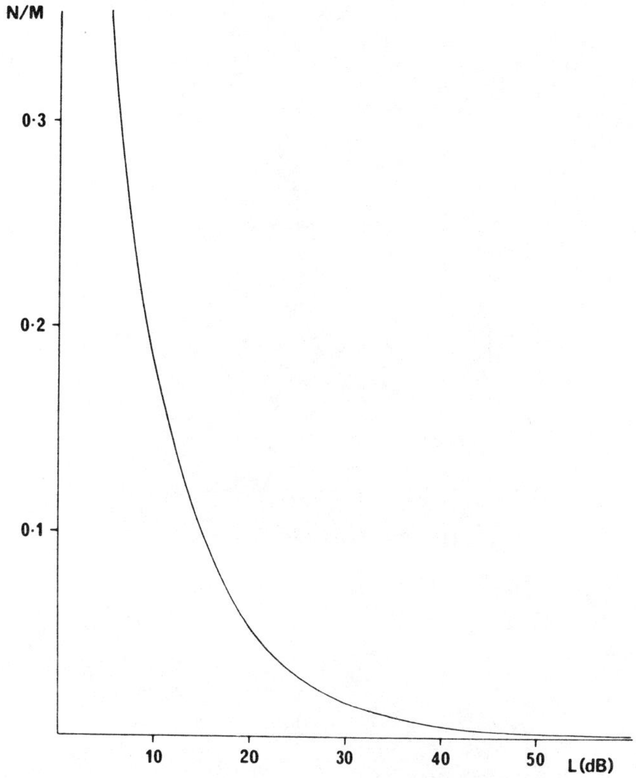

Fig. 5. Number of levels available in an M-level recording system to represent sidelobe of L dB when reference amplitude is equal to main beam amplitude.

Fig. 6. 3 m focal plane parabolic reflector.

V. Practical Results

A. Principally Polarized Holograms

Holograms of the 3 m focal plane dish with feed arm supports and the commercially produced reflector antenna (3.66 m diameter, $f/D = 0.33$) shown in Figs. 6 and 7 have been formed at 8.15 GHz and 11 GHz, respectively.

The principally polarized hologram of the focal plane dish is shown in Fig. 8 printed out on a 32-level intensity display system [4]. The FFT is then used to reconstruct

Fig. 7. 3.66 m parabolic reflector ($f/D = 0.33$).

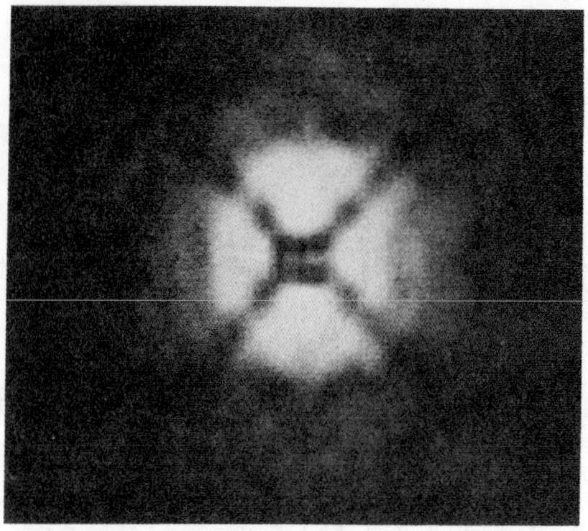

Fig. 9. Reconstructed aperture field amplitude distribution for 3 m paraboloid.

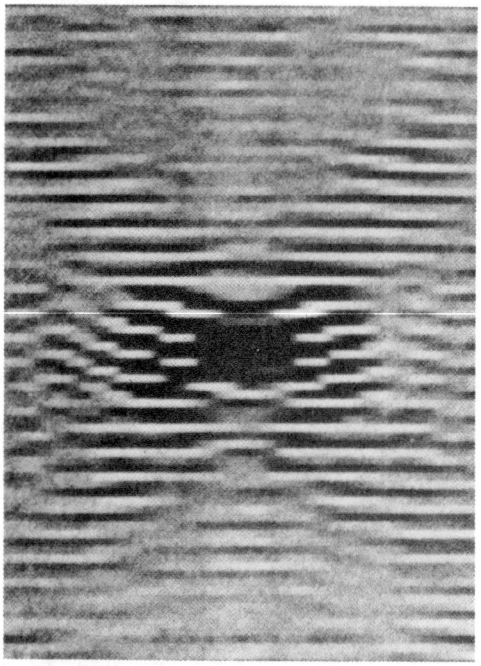

Fig. 8. *X* band hologram formed for 3 m paraboloid.

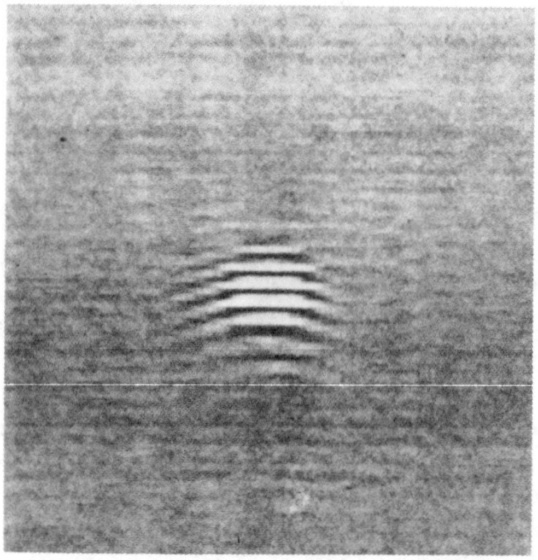

Fig. 10. Principally polarized hologram formed for 3.66 m paraboloid.

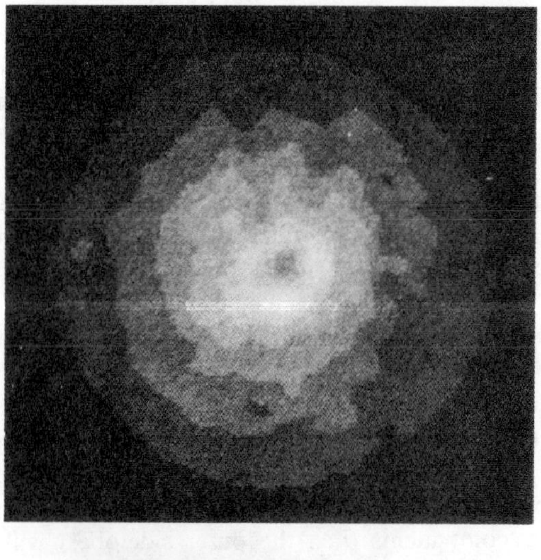

Fig. 11. 3.66 m paraboloid principally polarized aperture amplitude distribution (with 8 level legend).

the aperture field amplitude distribution, which is displayed in Fig. 9. This preliminary example is included to show how the feed support arms (width 1.38λ) and square feed box are clearly resolved. The dish edge does not appear on the display because of relatively low edge illumination.

For the holographic investigation of the commercially produced antenna, a near-field illuminating source was sited at a convenient distance ($0.23D^2/\lambda$) away along the test range. The antenna was scanned over a range of $\pm 5°$ in azimuth and elevation, and approximately 8000 samples of the hologram distribution were recorded. The hologram is shown in Fig. 10, and the reconstructed aperture field amplitude distribution is displayed with 8 levels in Fig. 11. Delineation of the aperture edge corresponds to the feed

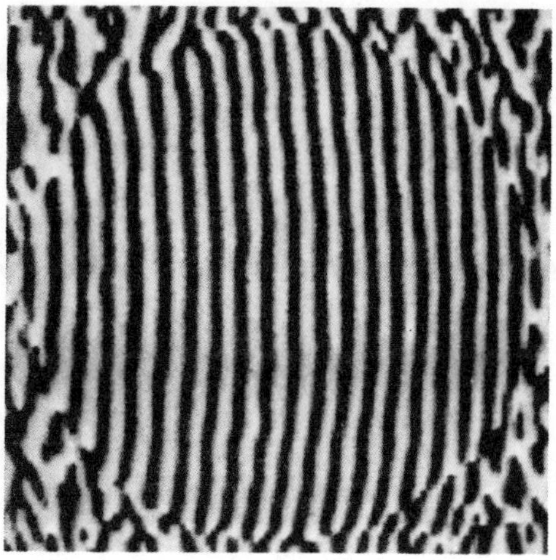

Fig. 12. Principally polarized interferogram for 3.66 m paraboloid.

illumination spill-over and the effect of feed blockage is seen as a perturbation of the contours.

Fig. 12 is a display of the aperture phase distribution produced by adding computationally an off-axis plane wave function to form an interferogram [5]. Delineation of the aperture edge is now observed as a breakup of the fringe pattern. The uniform visibility of fringes is due to the equal weighting given to all phase values in the computation. The fringe curvature indicates a quadratic-type phase variation which will be shown to be a combination of phase variations due to near-field illumination, axial feed position error, and different *E*- and *H*-plane phase centers.

Figs. 13(a) and (b) show the *E*- and *H*-plane phase distributions through the aperture center. From these it is possible to establish the positions of the *E*- and *H*-plane phase centers of the feed by removal of the known quadratic phase variation due to the near-field illuminating source. From the resulting phase differences between aperture center and edge, a geometric optics approach is used to locate the phase centers relative to the reflector focus. The reflector focus F_D and the location of the feed F_F specified by the antenna manufacturer are shown compared to the holographically derived phase centers in Fig. 13(c). It is seen that the separation of the two phase centers is 12.4 mm or 0.455λ. The production model of the feed is a design compromise so that the phase centers lie on each side of F_F, and this is clearly shown by the results.

A closer inspection of the aperture phase distribution is achieved by removing the phase components caused by the axial phase center displacements. Fig. 13(d) shows the effect of this operation on the *H*-plane distribution and reveals the cubic variation of phase indicative of lateral shift of the feed. (The linear phase variation was removed by setting the peak of the main beam to the zero degree position prior to recording.) Direct measurement of the reflector-feed geometry also indicated a lateral displacement of approximately 0.4λ in the *H*-plane. The theoretical aperture phase variation due to lateral displacement of the feed is shown in Fig. 13(e). Excluding errors due to the recording and reconstruction processes, any remaining

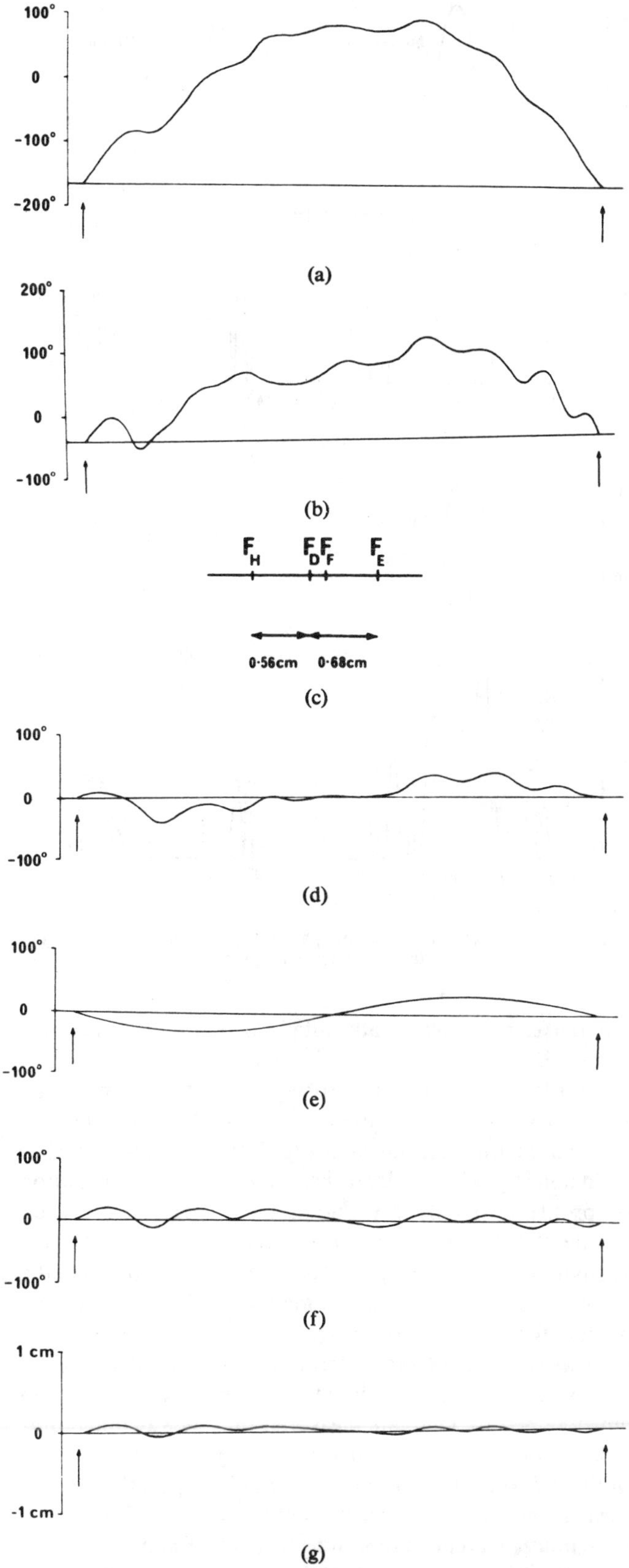

Fig. 13. (a) Reconstructed aperture phase distribution (*H*-plane). (b) Reconstructed aperture phase distribution (*E*-plane). (c) Phase center location. (d) Removal of phase errors due to axial displacement of *H*-plane phase center. (e) Phase variation due to directly measured lateral feed displacement. (f) Phase errors due to *H*-plane profile. (g) *H*-plane surface deformation profile. (Arrows denote aperture edge.)

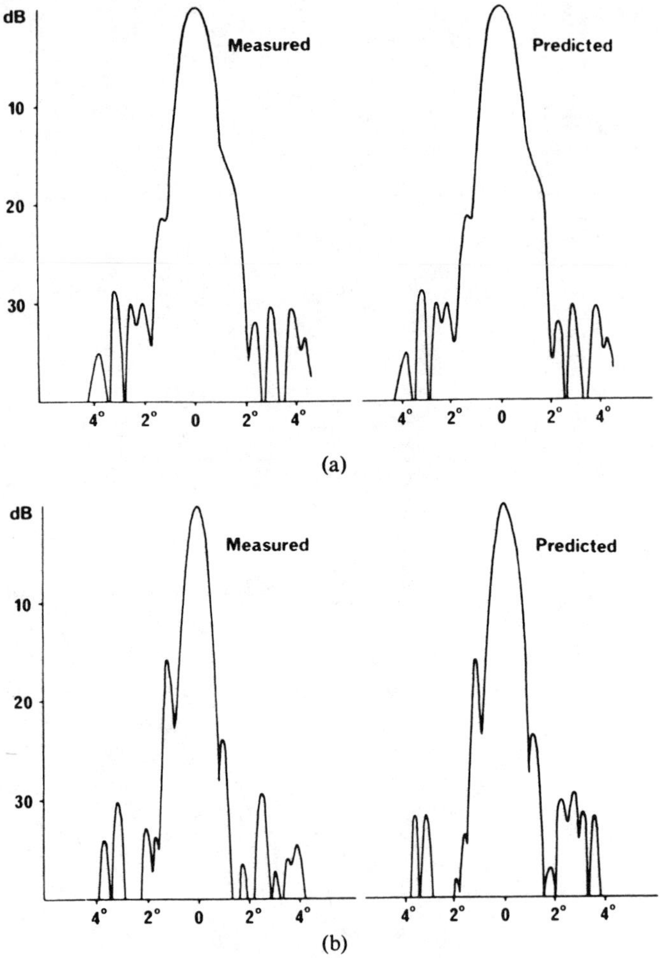

Fig. 14. (a) Near-field radiation patterns (H-plane). (b) Near-field radiation patterns (E-plane).

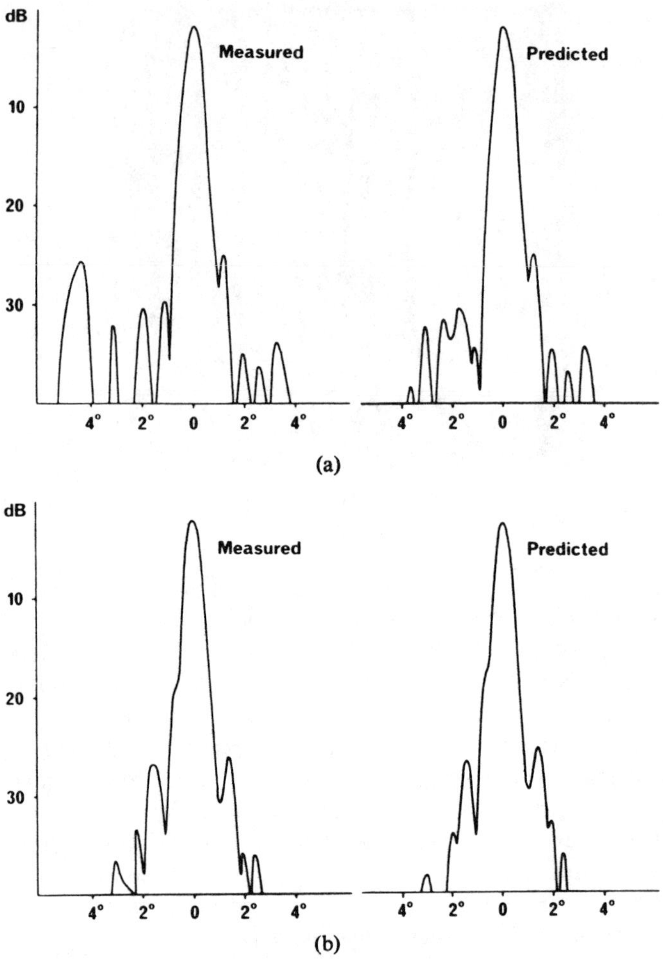

Fig. 15. (a) Far-field radiation patterns (H-plane). (b) Far-field radiation patterns (E-plane).

differences may now be attributed solely to reflector profile errors.

In order to examine these errors, the effect of the measured lateral feed shift is removed from Fig. 13(d) to give the residual distribution shown in Fig. 13(f). Using the analysis of Slater [6], which relates the aperture phase distribution to profile errors by ray optics, the surface deformation profile of Fig. 13(g) is obtained. This result is only for the H-plane profile through the aperture center. However, the whole reflector surface may be mapped in this way and then a value for the rms surface error obtained. By this method, the rms value of the profile error was found to be 0.093 cm, and is comparable with the manufacturer's result of 0.075 cm obtained by template measurements. It was not profitable to pursue these comparisons in greater detail since, in these particular experiments, the holographically derived measurements are known to contain small random errors in azimuth incrementation due to reading and setting the controls. These errors should be substantially reduced by automatic incrementation using synchro-digital converters.

An indication of the fidelity of the reconstructed amplitude and phase distributions is obtained by computing the Fourier transform of this data and comparing the result with the conventionally measured patterns. This process yields the distribution of Fig. 14(a) in the H-plane, where an excellent correspondence between the two patterns is seen. The computed and measured patterns for the E-plane (azimuth) direction are shown in Fig. 14(b). Excellent agreement is seen for the region encompassing the main beam and first sidelobes. However, the low-level sidelobe structure for the remainder of each pattern does not provide such good agreement and it is probable that these discrepancies are due to the azimuth positioning errors mentioned earlier.

It is important to note that by premultiplying the reconstructed aperture distribution by the appropriate quadratic phase term, the radiation pattern of the antenna may be obtained over any two-dimensional plane from the Fresnel zone to the far-field. As an example of this process, the quadratic term due to the near-field illumination function was removed from the reconstructed aperture distribution, and the E- and H-plane far-field patterns were predicted by carrying out a Fourier transform operation on this modified data. Figs. 15(a) and (b) show a comparison of the holographically derived far-field patterns and the conventionally measured results. Good agreement is seen between the predicted and experimental patterns except in the $-2°$ to $-4°$ region of the elevation patterns. The measured far-field patterns were performed in conditions of deep snow and the high sidelobes obtained in the above

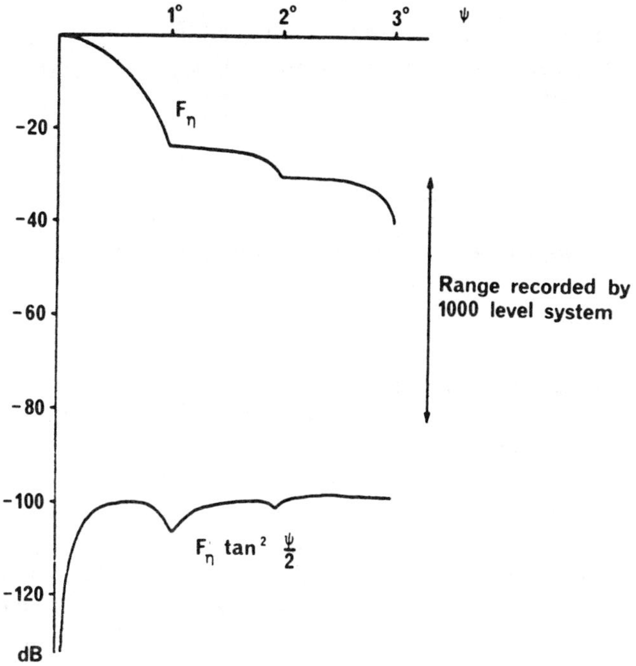

Fig. 16. Contribution of principal component to recorded data.

region are due to ground reflections. Small differences in the azimuth patterns are due to the positioning errors.

B. Crossed Polarized Hologram

This section describes a short investigation which was carried out to assess the feasibility of obtaining useful data from cross-polarized holograms.

Since the principal polarization components for this antenna-feed combination are as much as 30 dB greater than the cross-polarized components, it is necessary to ensure that their contribution to the hologram distribution is not significant.

It has been shown [7] that in the far-field of a radiating circular aperture, the measured orthogonal (or cross polarized) component[2] is given by

$$E_0(\psi,\phi_s) = j\,\frac{\exp(-jkR)}{\lambda R} \cdot \cos^2\frac{\psi}{2}\left[F_\varepsilon\left(1+\tan^2\frac{\psi}{2}\cos 2\phi_s\right) + F_\eta \tan^2\frac{\psi}{2}\sin 2\phi_s\right] \tag{23}$$

where F_ε and F_η are the spatial Fourier transforms of the orthogonal and principal components of the aperture distribution, respectively, and (ψ,ϕ_s,R) are the spherical coordinates of the far-field region. For the most significant contribution of F_η to the hologram distribution, $\phi_s = 45°$ and

$$E_0(\psi,45°) = j\,\frac{\exp(-jkR)}{\lambda R}\cos^2\frac{\psi}{2}\left[F_\varepsilon + F_\eta\tan^2\frac{\psi}{2}\right]. \tag{24}$$

[2] It is that component measured by conventional antenna-range techniques.

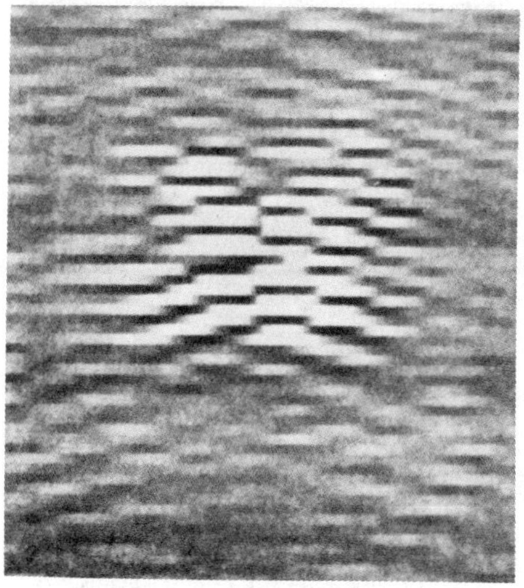

Fig. 17. Cross-polarized hologram formed for 3.66 m paraboloid.

Fig. 16 shows an envelope of F_η which has been obtained from the measured far-field radiation patterns of Fig. 15. Since the maximum principal and orthogonal components differ by 30 dB, then the range of signals detected by the 1000 level system may be inserted as shown. It is now apparent that the influence of the principal component may be ignored over the range for which the hologram was formed ($\psi = \pm 3°$), and (23) becomes

$$E_0(\psi,\phi_s) = j\,\frac{\exp(-jkR)}{\lambda R}\cos^2\frac{\psi}{2}\,[F_\varepsilon] \simeq j\,\frac{\exp(-jkR)}{\lambda R}\,F_\varepsilon. \tag{25}$$

Thus the far-field pattern is the Fourier transform of the aperture distribution and the method used for the formation and reconstruction process in Section V-A appears to be valid in this case.

For the cross-polarization measurement, the antenna was scanned over a range of $\pm 3°$ in azimuth and elevation, and approximately 3000 samples of the hologram distribution were recorded. The multilevel display of this data is shown in Fig. 17, and the computational reconstruction of the aperture cross-polarized amplitude distribution is shown in Fig. 18.

In order to interpret this reconstruction satisfactorily, it is first necessary to assess the character of the cross-polarized distribution which is to be expected for the feed used.

For a TE_{01} mode rectangular aperture situated at the focus, the orthogonal field component, $\varepsilon_0(\theta,\phi)$, in the antenna aperture may be approximated to

$$\varepsilon_0(\theta,\phi) = \frac{j}{2\lambda\rho}(1+\cos\theta)G(\theta,\phi)\left(\tan^2\frac{\theta}{2}\sin 2\phi\right) \tag{26}$$

where (ρ,θ,ϕ) are spherical polar coordinates with the origin at the focus, and $G(\theta,\phi)$ is the illumination function due to the feed.

This equation shows that the aperture illumination should consist, basically, of four quadrants defined by zero-field

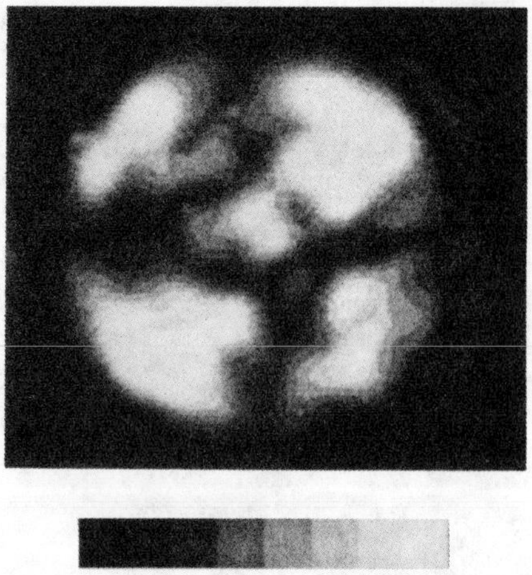

Fig. 18. 3.66 m paraboloid cross-polarized aperture amplitude distribution (with 8-level legend).

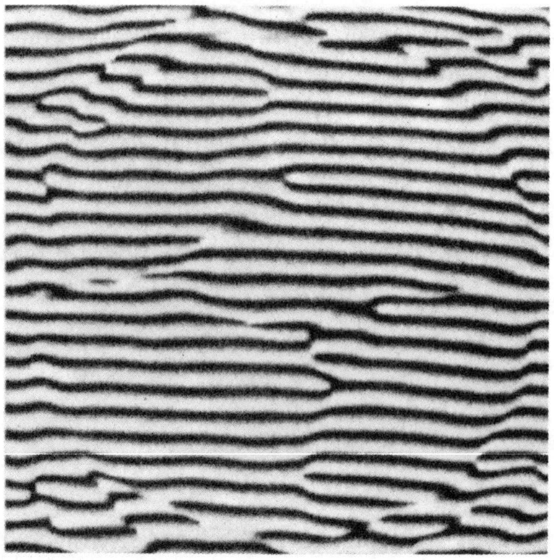

Fig. 19. Cross-polarized interferogram for 3.66 m paraboloid.

intensity along the axes determined by the condition

$$\sin 2\phi = 0. \tag{27}$$

The reconstruction shown in Fig. 18 displays these characteristics although some distortion of the expected distribution is present. Apart from any incorrect alignment of source and feed, it is suggested that this may be due, in part, to the presence of the feed in the aperture; reflections of the strong principal component from the feed structure may well exhibit significant cross-polarized components which contribute towards the finite fields which are seen in the center of the image of the aperture.

The interferometric representation of the aperture phase distribution is shown in Fig. 19. Equation (26) predicts that, for a focussed feed, adjacent quadrants in the aperture are in antiphase and, in fact, such rapid phase changes are apparent.

Conclusions

A comprehensive microwave holographic method for the recording and digital reconstruction of aperture field, near-field, and far-field patterns of large reflector antennas has been shown to yield excellent results. The accuracy appears to be at least comparable to other methods of measuring a single characteristic of the antenna, e.g., template or optical measurements of the profile and normal recordings of the radiated field in one region. The holographic technique has the added advantage that the pattern in any region can be generated using suitable transform operations.

The method does not require a separate probing and scanning system; it utilizes azimuth and elevation scanning of the antenna (or object) itself, which may be more convenient than planar scanning with a probe. Moreover, the spherical scan facility already exists in many systems. For example, the technique has been successfully applied to the Jodrell Bank Mk. IA 76 m reflector using the Mk. II reflector as the reference. In this case the antenna was illuminated by 31 cm radiation from a stellar radio source [8].

The limit to the proximity of the illumination source has not yet been tested. Our closest illumination distance has been $0.23D^2/\lambda$, which corresponded to an experimentally convenient transmitting site and is not a lower limit. The current interest in generating far-field distributions from near-field measurements should lead to the determination of a working limit.

References

[1] E. B. Joy and D. T. Paris, "Spatial sampling and filtering in near-field measurements," *IEEE Trans. Antennas Propagat.*, vol. AP-20, pp. 253–261, May 1972.

[2] J. C. Bennett, A. P. Anderson, P. A. McInnes, and A. J. T. Whitaker, "Investigation of the characteristics of a large reflector antenna using microwave holography," in *1973 IEEE G-AP Int. Symp.*, Boulder, Colo., Digest, pp. 298–301.

[3] ——, "Comprehensive measurement of aperture fields, near-fields and far-fields of a large reflector antenna from a microwave hologram," in *4th European Microwave Conf.*, Montreux, Switzerland, Proc., pp. 469–473.

[4] A. J. T. Whitaker and A. P. Anderson, "Improved optical images using a multilevel intensity display system for non-optical hologram distributions," *Laser Optics Technol.*, vol. 5, pp. 28–29, 1973.

[5] D. Gabor, G. W. Stroke, D. Brumm, A. Funkhouser, and A. Labeyrie, "Reconstruction of phase objects by holography," *Nature*, vol. 208, pp. 1159–1162, 1965.

[6] R. H. Slater, "Radiation pattern of imperfect paraboloidal reflectors," *Electron. Lett.*, vol. 6, pp. 796–798, 1970.

[7] A. W. Rudge and M. Shirazi, "Multiple beam antennas," Univ. of Birmingham, U.K., Final Rep. on ESRO/ESTEC Contract 1725/72, 1973.

[8] F. G. Smith, private communication.

Radiation from a Paraboloid with an Axially Defocused Feed

PAUL G. INGERSON AND WILLARD V. T. RUSCH

***Abstract*—The radiation characteristics of a paraboloid with an axially defocused idealized point-source feed are derived using numerical integration of the physical-optics currents. Results from the standard "linearized" analysis are compared with results obtained without approximating various factors in the integrand. In particular, the more complete analysis reveals that the defocusing curves are not symmetrical about the focus. At values of defocusing which cause deep nulls in the defocusing curves of less-tapered feeds, the angular beam exhibits bifurcation. Consequently, these less-tapered feeds are not generally suited to beam broadening applications.**

The experimental results of Landry and Chassé [1] for an axially defocused dipole-type feed illuminating a paraboloidal reflector confirm our analysis of the same geometry reported in [2]. Calculations made for $f/D = 0.35$ and $D/\lambda = 69.67$ yield results which are virtually identical to the data in [1, fig. 3], particularly with respect to the position of the displaced minima and the nonsymmetric maxima of the defocusing peaks. Because we feel that [2] may not have come to the general attention of the readers, the principal results are summarized in this communication.

A perfectly conducting paraboloidal reflector is illuminated by an idealized point source on the reflector axis (Fig. 1). The signed distance beyond the focus to the source is d. The vector from the focus to a point on the reflector is $\bar{\rho}$. The vector from the source to the same point is $\bar{\rho}'$. The difference in the magnitudes of these two vectors is primarily (but not exclusively) responsible for the antenna's defocusing characteristics. The physical-optics far-zone field is

$$\bar{E}_S = -\frac{j\omega\mu_0}{2\pi}\frac{\exp(-jkR)}{R}\int_{\text{front}} [\hat{n}\times\bar{H}_F(\rho,\theta',\phi')]_{\text{trans}} \cdot\exp(k\bar{\rho}\cdot\bar{a}_R)\,dS(\theta',\phi'). \quad (1)$$

The field of the feed is an idealized vector spherical wave

$$\bar{E}_F = \frac{\exp(-jk\rho')}{\rho'}\,\bar{e}(\theta'',\phi'')$$

$$\bar{H}_F = \left(\frac{\epsilon_0}{\mu_0}\right)^{1/2}(\bar{a}_{\rho'}\times\bar{E}_F). \quad (2)$$

The E- and H-fields are orthogonal to each other and to the direction of propagation. The constant-phase surfaces are spheres centered at the source. The directivity pattern is a function of the source angles θ'' and ϕ''. The distance relationship is

$$\rho' = \rho\left\{1+\left[\frac{d}{\rho}\left(\frac{d}{\rho}-2\cos\theta'\right)\right]\right\}^{1/2}. \quad (3)$$

The simplest line of analysis employs these approximations

$$\rho' \approx \rho - d\cos\theta' \quad \text{(in phase expressions)} \quad (4a)$$

$$\rho' \approx \rho \quad \text{(in amplitude expressions)} \quad (4b)$$

$$\bar{e}(\theta'',\phi'') \approx \bar{e}(\theta',\phi') \quad (4c)$$

$$\bar{a}_{\rho'} \approx \bar{a}_\rho. \quad (4d)$$

This set of approximations may be designated the "linear" analysis because the phase expression has been linearized. This linear analysis, or its equivalent, is used most commonly in [3], [4].

Manuscript received May 22, 1972; revised July 24, 1972. This paper was supported by the Joint Services Electronics Program through the Air Force Office of Scientific Research under Contract F44620-71-C-0067.

P. G. Ingerson is with Systems Group, TRW Inc., Redondo Beach, Calif. 90278.
W. V. T. Rusch is with the Department of Electrical Engineering, University of Southern California, Los Angeles, Calif. 90007.

Reprinted from *IEEE Trans. Antennas Propagat.*, vol. AP-21, pp. 104-106, Jan. 1973.

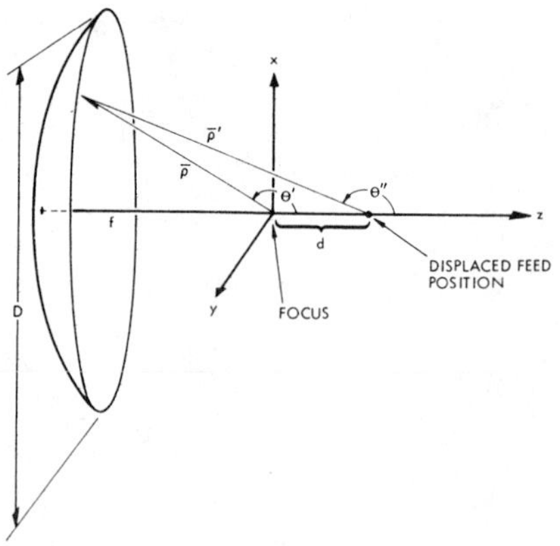

Fig. 1. Geometry of paraboloid with axially defocused feed.

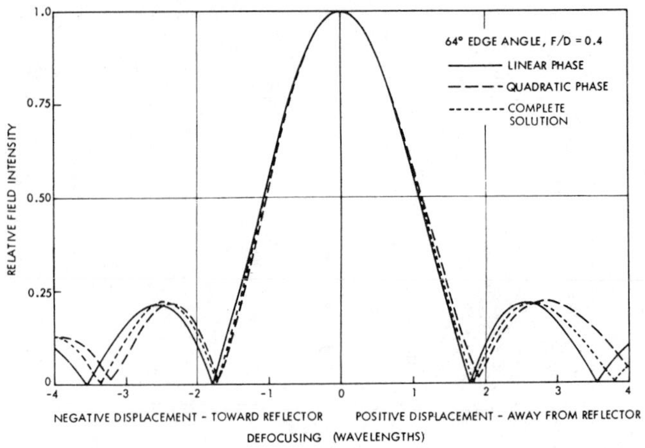

Fig. 2. Relative boresight field for defocused infinitesimal electric dipole.

The analysis of Kennaugh and Ott cannot be directly compared with the physical-optics analysis because their formulation of the problem uses spherical modes emanating from the focus. However, numerical evaluation has revealed that their results for axial defocusing are equivalent to the results of linear analysis, e.g., [3], their defocusing curve is symmetrical, their defocusing minima are regularly spaced, etc. Consequently, their analysis is, in effect, linearized. We suggest that this may have arisen from the approximations of (24) or from ignoring the second term in the integrand of [4, eq. (25a)].

Equation (1) may also be evaluated using numerical integration without making the approximations of (4). Since this procedure uses the complete incident fields of (2), for convenience we designate it the "complete" analysis. The disadvantage of the complete analysis is that certain closed-form expressions that are possible when the linear approximations are made are no longer feasible. However, the complete analysis allows the accuracy and range of validity of the "linear" analysis to be evaluated. In extreme cases of large defocusing, for example, the approximations of (4) become invalid, and the complete integration is necessary if results are desired.

If the source is an infinitesimal electric dipole of moment $\bar{p} = p\bar{a}_y$, the linear approximations permit (1) to be integrated, yielding the boresight radiated field:

$$\bar{E}_S(R, 0, 0) = -j\frac{k^2p}{4\pi\epsilon_0}(2kf)\bar{a}_y\frac{\exp(-jkR)}{R}\sin^2\left(\frac{\theta_0}{2}\right)$$

$$\cdot\exp(-j2kf)\left\{\exp\left(-jkd\cos^2\frac{\theta_0}{2}\right)\right\}\left\{\frac{\sin X}{X}\right\} \quad (5)$$

where θ_0 is the reflector edge angle and

$$X = \frac{2\pi(d/\lambda)}{1 + (4f/D)^2}.$$

Silver [5] has shown that using the linear approximation the Fresnel region field at a finite distance R' on the axis of a circular aperture varies as $\sin X'/X$, where $X' = ka^2/4R'$. Thus there is an obvious equivalence between "finite distance" and axial defocusing. The magnitude of this field is plotted as the solid curve in Fig. 2 versus d, for $f/D = 0.4$. For $X = \pm\pi$, $\pm 2\pi$, etc., the radiated field on axis is zero. At these points

$$\frac{d}{\lambda} = \pm\frac{m}{2}\left[1 + \left(\frac{4f}{D}\right)^2\right].$$

Here the main beam has widened and, in any plane, the field on either side of the axis is greater than the value on axis so that the beam appears bifurcated. It can also be shown [6] that for a defocused dipole feed and a distant axial field point, (5) satisfies the reciprocity theorem.

The results of the linear and complete analyses for a defocused dipole feed are compared in Fig. 2. Both curves coincide at the focus. The nulls of the linear analysis change to minima when the

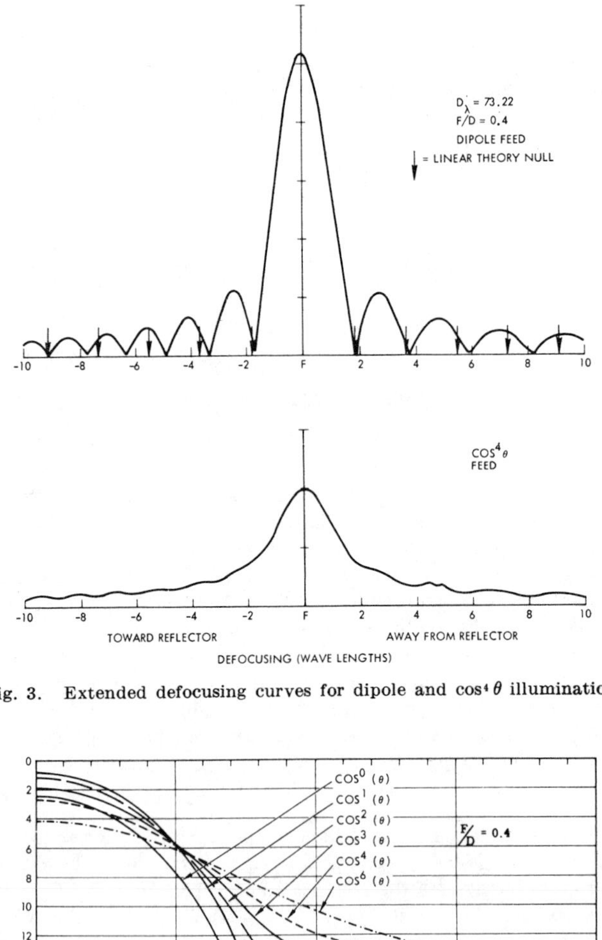

Fig. 3. Extended defocusing curves for dipole and $\cos^4\theta$ illumination.

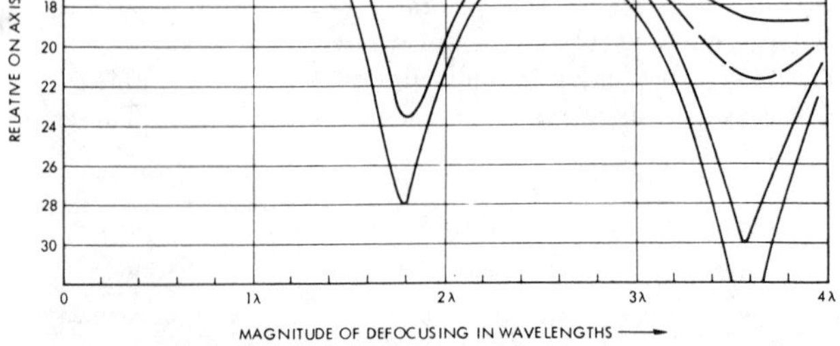

Fig. 4. Relative boresight gain as function of defocusing for $\cos^N\theta$ illumination.

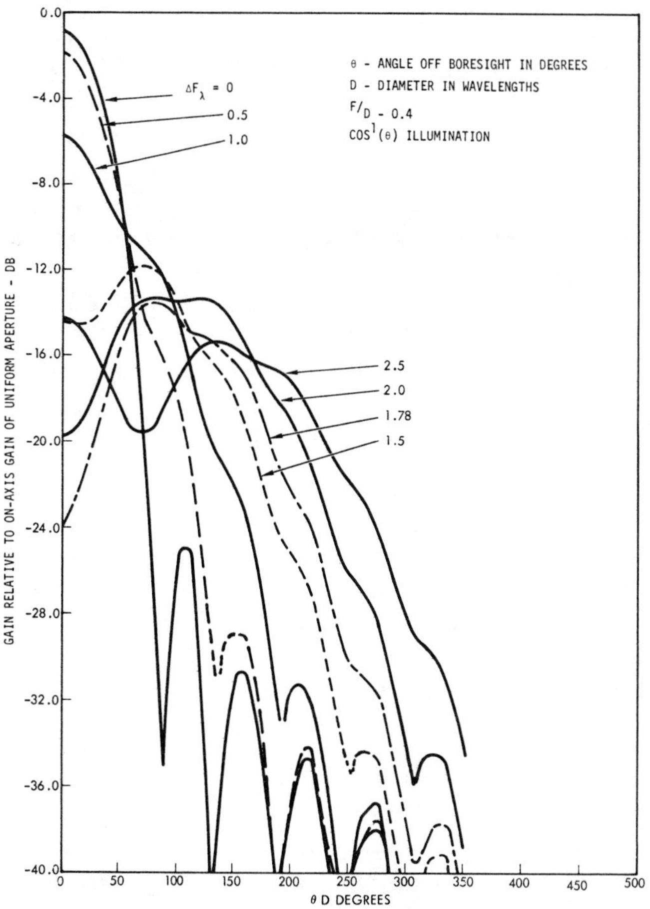

Fig. 5. Angular patterns for cos θ illumination with variable defocusing.

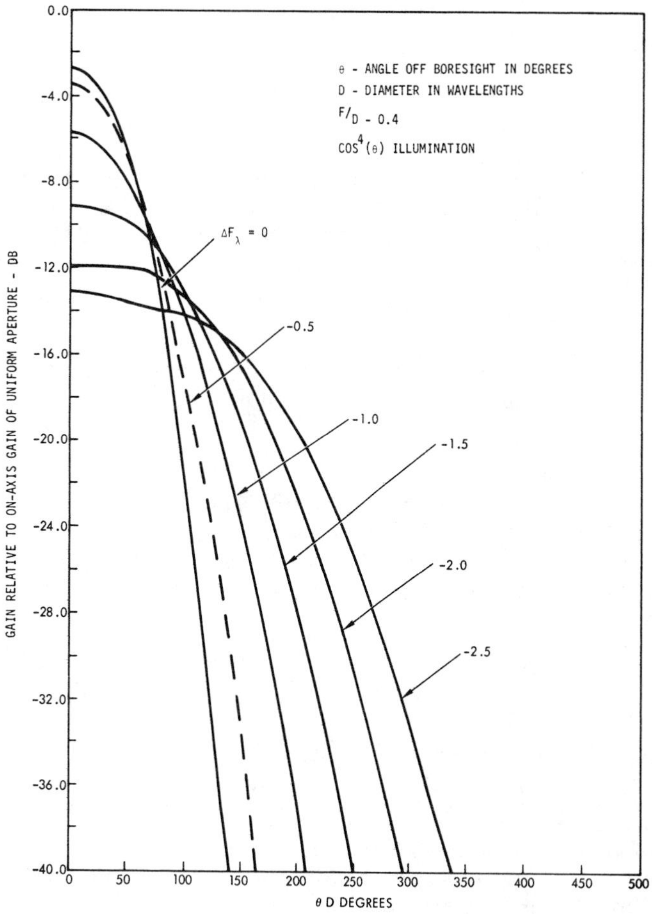

Fig. 6. Angular patterns for $\cos^4 \theta$ illumination with variable defocusing.

complete analysis is used. The minima shift *away* from the focus for positive defocusing and toward the focus for negative defocusing. The complete calculation reveals that the peak radiated field occurs when the dipole is slightly defocused toward the reflector. It has been shown that by simple power integration that this phenomenon occurs because, by moving toward the reflector, spillover is initially reduced faster than phase synchronism is lost by the defocusing. The curves closely resemble each other for $|d| \leq f/10$.

Extended defocusing curves for dipole and $\cos^4 \theta$ feeds with excursions of $\pm$ten wavelengths are plotted in Fig. 3. The amount of defocusing is sizeable compared to the focal length of 29 wavelengths. Consequently, only the complete formulation is used. The $\cos^4 \theta$ curve might be expected to have minima at large defocusing corresponding to its reduced effective aperture. However, the curve is generally monotonic, with no indication of nulls or minima, indicating that defocusing with highly tapered-illumination is more complex than simply a reduced effective aperture size.

Defocusing curves for the entire family of feed functions are plotted in Fig. 4. All of these curves were calculated using the linear approximation, so they are identical for positive and negative defocusing. The curves are normalized by the total feed power radiated so that spillover is included. Consequently, for example, cos θ is optimum for $f/D = 0.4$, etc.

The less tapered feeds exhibit deep minima. The dipole curve (not shown) has exact nulls at the same positions. As the taper is increased, the minima become less and less pronounced. At the deep minima in the less tapered defocusing curves, the angular beam exhibits bifurcation. *Consequently, these less-tapered feeds are not generally suited to beam-broadening applications.*

A family of angular patterns for a cos θ feed is plotted in Fig. 5 for different values of defocusing. Because of axial symmetry, only half of the pattern is shown. Linear theory is used to reduce computation expense. The beam becomes broken up and poorly defined with increased defocusing. At a defocusing distance of 1.78λ, where the deep null in the defocusing curves occurs in Fig. 4, the beam is bifurcated by 10 dB or so on axis. On the other hand, the angular patterns for a $\cos^4 \theta$ feed are plotted in Fig. 6. Bifurcation does not occur, and the beam remains well defined for all values of defocusing.

References

[1] M. Landry and Y. Chassé, "Measurement of electromagnetic field intensity in focal region of wide-angle paraboloid reflector," *IEEE Trans. Antennas Propagat.* (Commun.), vol. AP-19, pp. 539–543, July 1971.

[2] P. G. Ingerson and W. V. T. Rusch, "Studies of an axially defocused paraboloid," presented at the IEEE Symp. Antennas Propagat., Austin, Tex., Dec. 1969.

[3] E. M. Kennaugh and R. H. Ott, "Fields in the focal region of a parabolic receiving antenna," Ohio State Univ., Columbus, Ohio, Rep. 1223-16, Aug. 31, 1963.

[4] H. C. Minnett and B. M. Thomas, "Fields in the image space of symmetrical focusing reflectors," *Proc. Inst. Elec. Eng.*, vol. 115, no. 10, pp. 1419–1430, Oct. 1968.

[5] S. Silver, *Microwave Antenna Theory and Design.* New York: McGraw-Hill, 1949, sec. 6.9.

[6] W. V. T. Rusch and P. D. Potter, *Analysis of Reflector Antennas.* New York: Academic Press, 1971, ch. 4.

Bibliography for Part VIII

[1] J. Ruze, "The effect of aperture errors on the antenna radiation pattern," *Suppl. al Nuovo Cimento*, vol. 9, no. 3, pp. 364–380, 1952.

[2] P. G. Ingerson and W. V. T. Rusch, "Studies of an axially defocused paraboloid," in *IEEE AP-S Symp. Dig.*, Dec. 1969, pp. 62–68.

[3] D. K. Cheng and S. T. Moseley, "On-axis defocus characteristics of the paraboloidal reflector," *IRE Trans. Antennas Propagat.*, vol. AP-3, pp. 214–216, Oct. 1955.

[4] J. Robieux, "Influence de la précision de fabrication d'une antenne sur ses performances," *Ann. Radioelect.*, vol. 11, pp. 29–56, Jan. 1956.

[5] R. H. T. Bates, "Random errors in aperture distribution," *IRE Trans. Antennas Propagat.*, vol. AP-7, pp. 369–372, Oct. 1959.

[6] G. Swarup and K. S. Yang, "Monitoring paraboloidal reflector antennas," *Proc. IRE*, vol. 48, pp. 1918–1919, Nov. 1960.

[7] G. Swarup and K. S. Yang, "Phase adjustment of large antennas," *IRE Trans. Antennas Propagat.*, vol. AP-9, pp. 75–81, Jan. 1961.

[8] R. N. Bracewell, "Tolerance theory of large antennas," *IRE Trans. Antennas Propagat.*, vol. AP-9, pp. 49–58, Jan. 1961.

[9] C. Dragone and D. C. Hogg, "Wide-angle radiation due to rough phase fronts," *Bell Syst. Tech. J.*, vol. 42, pp. 2285–2296, Sept. 1963.

[10] F. V. Bale, J. A. Gourlay, and R. W. Meadows, "Measuring the shape of large reflectors by a simple radio method," *Electron. Lett.*, vol. 2, pp. 252–253, July 1966.

[11] E. Schanda, "The effect of random amplitude and phase errors of continuous apertures," *IEEE Trans. Antennas Propagat.*, vol. AP-15, pp. 471–473, May 1967.

[12] P. Miller, "Computation of polar diagrams from paraboloidal reflector aerials with systematic deformations," *Electron Lett.*, vol. 4, pp. 398–400, Sept. 20, 1968.

[13] D. E. N. Davies and A. W. Rudge, "Some results of electronic compensation for surface profile errors in parabolic reflectors," *Electron. Lett.*, vol. 4, pp. 433–434, Oct. 4, 1968.

[14] J. Ruze, "Feed support blockage loss in parabolic antennas," *Microwave J.*, vol. 11, pp. 76–80, Dec. 1968.

[15] Vu The Bao, "Influence of correlation interval and illumination taper in antenna tolerance theory," *Proc. Inst. Elec. Eng.*, vol. 116, pp. 195–202, Feb. 1969.

[16] H. L. Strachman and W. V. T. Rusch, "Loss-budget versus comprehensive analysis of gain loss for microwave reflector antennas," *IEEE Trans. Antennas Propagat.*, vol. AP-17, pp. 365–366, May 1969.

[17] G. Feix, "Focus broadening by astigmatism of large microwave parabolic antennas," *Appl. Opt.*, vol. 8, pp. 1631–1634, Aug. 1969.

[18] A. W. Rudge and D. E. N. Davies, "Electronically controllable primary feed for profile-error compensation of large parabolic reflectors," *Proc. Inst. Elec. Eng.*, vol. 117, pp. 351–358, Feb. 1970.

[19] M. Taylor, R. A. Keough, and A. W. Moeller, "Beam broadening of a monopulse tracking antenna by feed defocusing," *IEEE Trans. Antennas Propagat.*, vol. AP-18, pp. 622–627, Sept. 1970.

[20] R. H. Slater, "Radiation pattern of imperfect paraboloidal reflectors," *Electron. Lett.*, vol. 6, pp. 796–798, Dec. 10, 1970.

[21] J. H. Davis and J. R. Cogdell, "Reflector efficiency evaluation by frequency scaling," *IEEE Trans. Antennas Propagat.*, vol. AP-19, pp. 58–63, Jan. 1971.

[22] E. A. Parker and P. R. Cowles, "Techniques for compensating for reflector antenna surface errors with long correlation lengths," *Electron. Lett.*, vol. 8, pp. 366–367, July 13, 1972.

[23] P. R. Cowles and E. A. Parker, "Reflector surface error compensation in Cassegrain antennas," *IEEE Trans. Antennas Propagat.*, vol. AP-23, pp. 323–328, May 1975.

Part IX
Spherical Reflectors

The paraboloidal reflector is an ideal collimating device, but it has serious shortcomings for scanning. The spherical reflector, on the other hand, makes an ideal scanner if something can be done about its poor collimating properties. Perhaps the earliest effort in this direction was that reported by Ashmead and Pippard [1], who investigated the paraxial region of a spherical reflector and determined the optimum location for a point source feed in order to minimize path error in the aperture. This location, they found, is not at the half-radius point of the sphere, as taught in optics, but is displaced slightly toward the reflector. They showed that the correct location is at distance f from the spherical surface, such that the maximum value of the path error, over an aperture of diameter D, is given by

$$\Delta_{max} \simeq \frac{D^4}{2000 f^3}.$$

Essentially the same result is given by Li in the first paper of this part. He goes on to describe the design and test of a 10-ft-diameter hemispherical reflector at 11.2 GHz which had a useful scan range of ±70° in any direction.

If one draws a ray diagram for a plane wave incident on a spherical mirror, it is immediately evident that all rays cross the optical axis somewhere on a line which begins at the half-radius point (often called the paraxial focus) and extends along the axis toward the reflector. Thus, it is said that a spherical mirror possesses a line focus instead of a point focus. The first attempt to eliminate spherical aberration by use of a line source feed was reported in the classic paper of Spencer, Sletten, and Walsh [2]. They derived a geometrical optics solution for the design of a correcting line source and described two possible practical forms, one a dipole array fed by loaded waveguide, and the other an array of inclined slots in the narrow wall of loaded waveguide. Both were capable only of asymmetrical illumination in a manner analogous to that of an offset paraboloid. Because so much of this paper is devoted to discussions of these somewhat limited feed designs, it has not been included here. It does, however, form the basis for all subsequent line source feed development.

An account of some early line feed work is given by Love in [3] and, of course, the best known example of this technique is the 430-MHz line feed for the 1000-ft-diameter spherical reflector at Arecibo, Puerto Rico. The original feed design was described by Kay [4], and interest in the performance of this largest of reflectors stimulated a scalar diffraction analysis by Schell which forms the second paper in this part. He derived closed form expressions for the complex field on the axis of a spherical reflector due to an incident plane wave. This is another benchmark paper in the theory of spherical reflectors. It is followed (paper three) by a comprehensive and detailed analysis by McCormick of the full-blown vector diffraction problem in which expressions for all three field components in the region near the focal line are obtained. His treatment explains the problems of polarization mismatch and of spatial harmonics that can arise whenever the radiating sources of the line source are displaced from the axis. These problems, and a possible solution using a leaky cylindrical waveguide line source, were discussed in a short paper by Love and Gustincic [5], which is not included here.

Spherical aberration can be eliminated with a dual mirror configuration, referred to as a Gregorian-corrected system because the subreflector is usually concave. The initial suggestion to this effect is due to Head [6], and the method is explored from the point of view of geometrical optics in the fourth paper, by Holt and Bouche. The spherical Gregorian system is generally not too attractive because of the large amount of shadowing caused by the secondary mirror whose size does not decrease in relation to the main reflector, even at short wavelengths. The reader who is interested in optimized Gregorian designs should consult Phillips and Clarricoats [7].

The next four papers, five through eight, are all concerned with the fields near the axis when a plane wave is incident on a spherical reflector. A wealth of detail concerning the classical, or geometrical optics, nature of this region is given by Spencer and Hyde in paper five, followed by a detailed treatment of polarization effects in the sixth paper by the same authors. In paper seven, Hyde evaluates the field integrals by the method of stationary phase and presents graphical plots of the phase and amplitude of the various components. The same subject is handled in somewhat different but incisive fashion in the last of these four papers, by Thomas, Minnett, and Bao. They briefly indicate the aperture efficiencies to be expected when a corrugated waveguide feed is used that is capable of supporting higher hybrid modes.

In paper nine, a detailed description is given by LaLonde and Harris of the design and performance of a flat, slotted waveguide feed operating at 318 MHz in the Arecibo spherical reflector. With this flat feed, the aforementioned problems of polarization mismatch and spatial harmonics are eliminated because the sources (i.e., the slots) are very close to the axis; the penalty is that only one linear polarization can be handled. The untimely death of L. Merle LaLonde in the Spring of 1977 was saddening to all of us who have worked in this field, and, I am sure, represents a serious loss to the National Astronomy and Ionosphere Center at Cornell University and in Arecibo, Puerto Rico.

As an alternative to a line feed lying on the axis, spherical aberration may be corrected by means of an aperture-type feed lying on a surface that is transverse to the axis. This possibility seems first to have been examined seriously in a scalar analysis by Burrows and Ricardi [8]. Paper ten, by

Ricardi, follows up this approach in more detail, using a vector spherical wave analysis to determine the proper fields on an appropriate transverse surface. Such a practical system is considered and curves of aperture efficiency versus size of the transverse feed are given.

The final paper, by Love, returns to the problem of line feed correction and describes the scale model development of a line source for the Arecibo spherical reflector. Because the feed is required to support orthogonal polarizations of the dominant TE_{11} circular mode, it is a leaky cylindrical waveguide of appreciable diameter. As a consequence, the radiating sources are displaced from the axis, making it necessary to find solutions to the problems of polarization mismatch and spatial harmonics. Solutions were found, based on the concepts given in [5], and a successful 96-ft-long full-scale feed was built by simply scaling up from the model.

A Study of Spherical Reflectors as Wide-Angle Scanning Antennas*

TINGYE LI†

Summary—**A study is made of spherical reflectors for use as wide-angle scanning antennas. In order to keep the effects of spherical aberration within tolerable limits, the approach of using a restricted aperture is adopted. This approach is suitable for applications requiring very wide angles of scan.**

Experimental results show that the phase error over the illuminated aperture of a spherical reflector should not exceed one-sixteenth of a wavelength. This requirement determines the beamwidth of the primary source. A square-aperture horn with diagonal polarization is found to satisfy the requirements of a suitable feed for the reflector. Secondary patterns of a 10-foot-diameter hemispherical reflector illuminated by this horn at 11.2 kmc have a 3-db beamwidth of 1.76° and a relative sidelobe level of about −20 db throughout a total useful angle of scan of 140°. The measured gain is 39.4 db, which is equivalent to the gain of a uniformly illuminated circular aperture of 31-inch diameter.

Introduction

The suitability of spherical reflectors for use as wide-angle scanning antennas has been recognized for some time.[1] In order to minimize the effects of spherical aberration, two general approaches have been used. The first approach involves the use of a restricted aperture and a reflector of a sufficiently large radius.[1,2] This approach permits a simple design and is suitable for applications requiring very wide angles of scan. The second approach uses methods of compensation for spherical aberration. These methods include the use of phased line-source feeds,[3] multiple-source feeds,[4,5] auxiliary reflectors,[5,6] or correcting lenses.[7,8] With these methods of compensation a larger aperture can be effectively utilized. However, this is achieved at the expense of some complexity in design and of a reduction in the useful angle of scan.

In applications requiring very wide angles of scan, it is necessary to use a small illuminated aperture as compared with the reflector aperture. In such a case the first approach involving the use of a restricted aperture is applicable. The present study of the spherical reflector includes both theoretical and experimental investigations of this approach. In particular, the experimental work is devoted to the conception and realization of a suitable feed for the spherical reflector.

Theoretical Considerations

When a point source is placed at the focus of a spherical reflector, the reflected wave arising from it does not have a plane wave-front. This phenomenon is known as spherical aberration. The amount of departure of phase from a plane wave is a function of the aperture diameter and of the focal length. This functional relationship can be established by means of geometrical optics.

A hemispherical reflector and its geometry are shown in Fig. 1. A point source is assumed to be located at the focal point F. FB and BA are typical incident and reflected rays. The total path-length from the focal point F to the aperture plane at A is

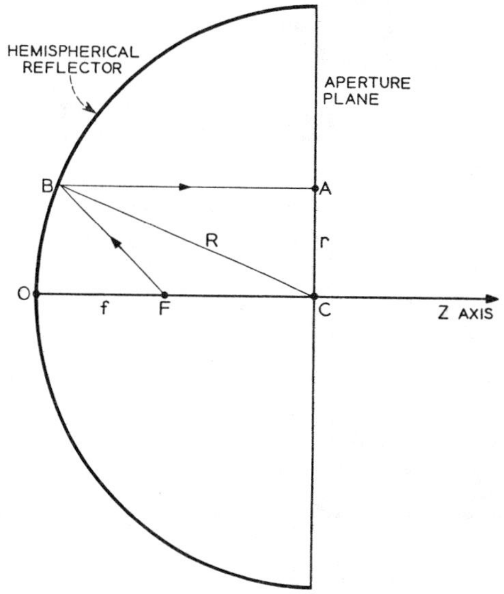

Fig. 1—Geometry of a spherical reflector.

$$FB + BA = d = \sqrt{R^2 - r^2} + \sqrt{r^2 + [\sqrt{R^2 - r^2} - (R - f)]^2} \quad (1)$$

where R = radius of the sphere, r = radial distance from the Z axis, and $f = OF$ = focal length. In (1) it is assumed that BA, the reflected ray, is parallel to the Z axis. This approximation is good for the range of variables given by $0 \leq r/R \leq 0.6$ and $0.45 \leq f/R \leq 0.50$.

* Manuscript received by the PGAP, February 10, 1959.
† Bell Telephone Labs., Inc., Holmdel, N. J.

[1] J. Ashmead and A. B. Pippard, "The use of spherical reflectors as microwave scanning aerials," *J. IEE*, vol. 93, pt. IIIA, pp. 627–632; 1946.

[2] A. S. Dunbar, "Applications of the Rayleigh criterion," Symp. on Microwave Optics, McGill Univ., Montreal, Can.; June, 1953.

[3] R. C. Spencer, C. J. Sletten, and J. E. Walsh, "Correction of spherical aberration by a phased line source," *Proc. NEC*, vol. 5, pp. 320–333; 1949.

[4] C. J. Sletten and W. G. Mavroides, "A Method of Side-Lobe Reduction," Naval Res. Lab., Washington, D. C., NRL No. 4043, pp. 1–12; April, 1952.

[5] W. Rotman, "Wide-angle scanning with microwave double-layer pillboxes," IRE Trans. on Antennas and Propagation, vol. AP-6, pp. 96–105; January, 1958.

[6] A. K. Head, "A new form for a giant radio telescope," *Nature*, vol. 179, pp. 692–693; April 6, 1957.

[7] H. B. Devore and H. Iams, "Microwave optics between parallel conducting sheets," *RCA Rev.*, vol. 9, pp. 730–732; December, 1948.

[8] H. N. Chait, "Wide-angle scan radar antenna," *Electronics*, vol. 26, pp. 128–132; January, 1953.

Reprinted from *IRE Trans. Antennas Propagat.*, vol. AP-7, pp. 223-226, July 1959.

The path-length difference between an axial ray *FOC* and a non-axial ray *FBA* is

$$\Delta = R + f - d. \tag{2}$$

Therefore, the phase error in wavelengths is

$$(\Delta/\lambda) = (R/\lambda)(2 - m - s - \sqrt{1 + m^2 - 2ms}) \tag{3}$$

where $m = 1 - f/R$ and $s = \sqrt{1 - (r/R)^2}$. A family of curves for $(\Delta/\lambda)/(R/\lambda)$ is plotted in Fig. 2.

When considering the phase error over a given aperture, the sum of the maximum absolute values of positive and negative phase errors is important. This sum is referred to as the total phase error and is plotted in Fig. 3 as a function of the focal length f for different aperture radii a. The optimum focal length for each aperture is determined by the minimum of each curve. For example, the optimum focal length for an aperture having a diameter equal to the radius of the sphere ($2a = R$) is equal to $0.4665R$. The corresponding total phase error is 0.004206 (R/λ) wavelength, or, 0.02643 (R/λ) radian.

It is of interest to note that the total phase error over a prescribed aperture is least when the phase error at the edge of the aperture is zero. Thus, equating (3) to zero and solving for f, the optimum focal length for an aperture of radius a is found to be

$$f_{op} = \frac{1}{4}(R + \sqrt{R^2 - a^2}). \tag{4}$$

Although this optimum focal length is one for which the total phase error is least over a prescribed aperture, it is not necessarily the focal length that yields the best radiation pattern when the aperture illumination is nonuniform. If the amplitude distribution is tapered, the focal length that yields the best radiation pattern would be somewhat longer. In practice, this can be determined easily by experiment.

It is clear from Fig. 3 that corresponding to a given aperture size there exists a minimum value of total phase error. This means that specification on phase-error tolerance sets the limit on the aperture size. The relationship between the maximum permissible aperture and the total allowable phase error is given by

$$\left(\frac{a}{R}\right)^4_{max} = 14.7 \frac{(\Delta/\lambda)_{total}}{(R/\lambda)}. \tag{5}$$

As an example, consider a spherical reflector having a 10-foot diameter. If the allowable phase error is $\lambda/16$ at 11.2 kmc, the maximum permissible aperture is found to have a diameter of 3.56 feet.

Experimental Studies

The experimental work was conducted using a hemispherical reflector of 10-foot diameter. The reflector is constructed of wood and has a mechanical tolerance of $\pm 1/32$ of an inch. The concave surface of the hemispherical bowl is covered with a coat of silver paint to render it reflecting. A photograph of the reflector is shown in Fig. 4.

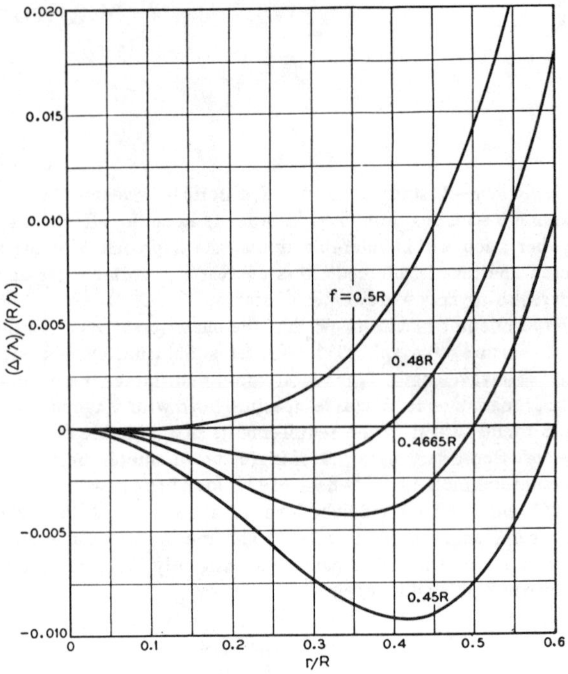

Fig. 2—Phase error over the aperture plane of a spherical reflector.

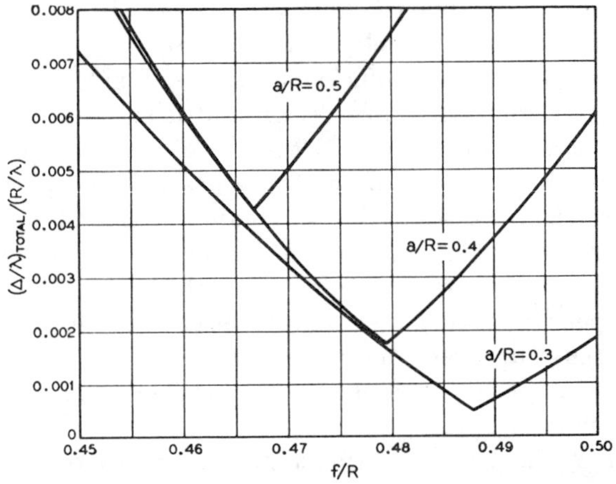

Fig. 3—Total phase error vs focal length for different aperture sizes.

Preliminary experimental studies consisted of taking radiation patterns of the reflector at *X*-band using several feed horns and different focal lengths. It was found that a suitable feed for the spherical reflector should have a 10-db beamwidth which produces an aperture over which the total phase error is less than $\lambda/16$. Furthermore, its radiation pattern should be rotationally symmetric and its sidelobes should be at least 25 db below peak intensity.

A square-aperture horn with diagonal polarization was found to satisfy the requirements. An exploded view of the horn is shown in Fig. 5. There are two

Fig. 4—10-foot diameter spherical reflector.

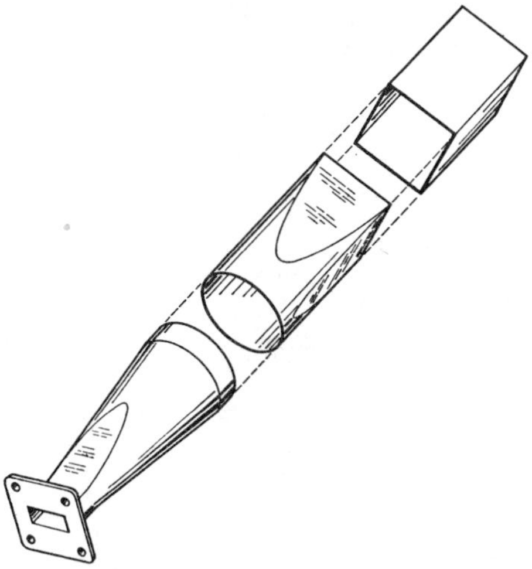

Fig. 5—The square-aperture horn.

transition sections in the horn; the first converts a TE_{10} wave in the rectangular waveguide to a TE_{11} wave in the circular waveguide, the second converts the TE_{11} wave in the circular waveguide to a wave polarized in the diagonal direction of the square waveguide. The diagonally polarized wave in the square waveguide is resolvable into a TE_{10} wave and a TE_{01} wave, both of which are in phase and of equal amplitude. The radiation characteristics of a square-aperture horn having a 3.9 cm by 3.9 cm aperture were measured at 11.2 kmc. Its *H*-, *E*-, and 45°-plane patterns were found to be essentially identical. Its 10-db beamwidth is 76° and all its sidelobe levels are at least 25 db below peak intensity.

Radiation characteristics of the 10-foot spherical reflector illuminated by the square-aperture horn were measured at 11.2 kmc. The radiation patterns in *E*-, *H*-, and 45°-planes for a focal length of 29.5 inches are plotted in Fig. 6. The measured gain is 39.4 db. This is

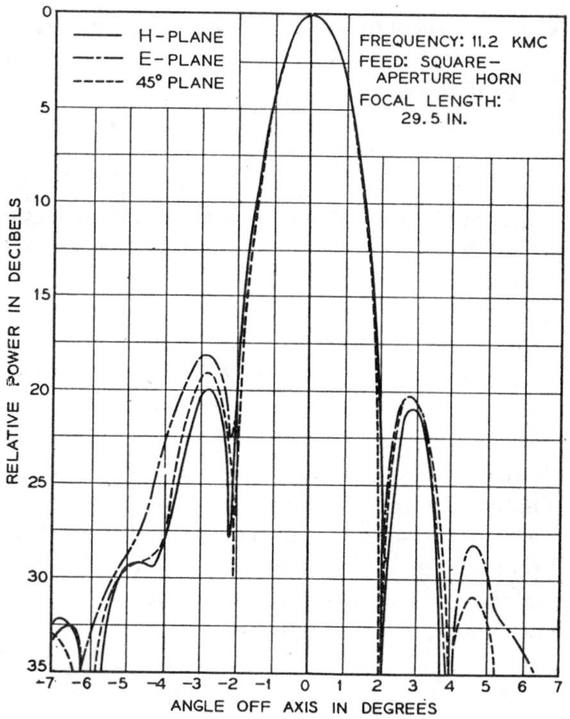

Fig. 6—Radiation patterns of 10-foot spherical reflector.

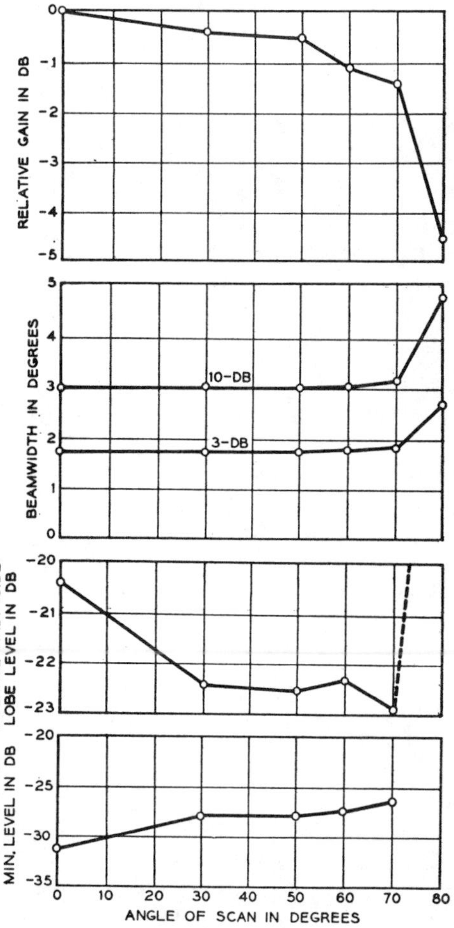

Fig. 7—*H*-plane scanning characteristics of 10-foot spherical reflector.

equivalent to the gain of a uniformly illuminated circular aperture of 31-inch diameter, or a typical paraboloid of 40-inch diameter.

Scanning characteristics of the spherical reflector with the square aperture horn were investigated. A summary of the results is shown graphically in Fig. 7. It is seen that the total useful angle of scan is about 140°. In order to achieve a wider scan, a smaller aperture or a larger reflector must be used.

Conclusion

The spherical reflector is well adapted for use as a wide-angle scanning antenna. By using a restricted aperture and choosing a proper focal point, the effects of spherical aberration can be kept within tolerable limits. The maximum allowable aperture corresponding to a prescribed phase error can be obtained from (5). The focal point that yields the most desirable radiation pattern is best determined experimentally.

Experimental results indicate that the total phase error over the illuminated aperture of the spherical reflector should be held within the $\lambda/16$ limit. This requirement sets an upper limit for the beamwidth of the primary source. A square-aperture horn with an aperture field polarized in the diagonal direction was found to be a suitable primary source. At 11.2 kmc, the radiation patterns of the spherical reflector illuminated by the square-aperture horn showed a 3-db beamwidth of 1.76° and a relative sidelobe level of about -20 db. An absolute gain of 39.4 db was measured and a total useful angle of scan of 140° was achieved. If a larger reflector or a smaller aperture is used, it should be possible to attain a lower sidelobe level and a wider angle of scan.

Acknowledgment

The author wishes to express his appreciation to A. B. Crawford and W. C. Jakes for many stimulating discussions during the course of this investigation. The assistance of W. E. Legg in the experimental phase of the work is also gratefully acknowledged. The hemispherical reflector was built by the carpenter shop personnel of the Holmdel Laboratory.

The Diffraction Theory of Large-Aperture Spherical Reflector Antennas*

A. C. SCHELL†, MEMBER, IEEE

Summary—**The field along the axis of a spherical reflector is determined from the geometry of the system rather than from each term of the aberration taken separately. The procedure shows how the field distribution changes from the case of small aberration, where there is a well-defined focus, to the geometric optics limit.**

A spherical reflector becomes an efficient antenna when a set of feed elements is located along the axis to reduce the effects of spherical aberration. The number and position of these elements is dictated by the size and curvature of the reflector and the allowable distortion of the wavefront.

I. Introduction

A SPHERICAL MIRROR used as a radio wave antenna leads to a system whose diffraction pattern can be steered without moving the reflector, thus providing a practical technique for constructing very large scanning antennas. By properly locating a number of radiating elements along the axis in the vicinity of the paraxial focus, feed structures for radio frequencies can be designed to nullify the effect of spherical aberration. With this modification, reflectors having a high degree of aberration have use as high-gain antennas.

A geometric optics treatment of the spherical reflector as an antenna, together with an experimentally verified method of correcting spherical aberration by a phased line-source feed, was published by Spencer, Sletten, and Walsh[1] in 1951. The antenna now being built for the Arecibo Ionospheric Observatory in Puerto Rico is based on a modification of that technique. The geometric optics analysis has a limited range of applicability, however, and does not provide an accurate measure of pattern deterioration.

In determining the effect of spherical aberration on the fields at the focus of a mirror or lens, the usual approach is to consider only the Seidel term corresponding to first-order spherical aberration when solving for the fields along the axis of the system. The resulting integrals cannot be simply evaluated, and asymptotic expressions are often used to obtain quantitative results.

A solution to the diffraction problem for all terms of spherical aberration along the axis is presented in this paper. The result is applicable to spherical mirrors or lenses of such size that edge effects can be neglected.

It is well known that a single feed located near the paraxial focus will illuminate the central portion of the reflector, the size of this portion being determined by the curvature of the reflector and the permissible phase error. If the reflector is large, additional feed elements along the axis will minimize phase errors in the aperture. The systematic method described has been developed for designing spherical reflector antenna systems with arbitrarily small amounts of apparent spherical aberration. The feed structures evolved represent a compromise between the single paraxial feed of small-aberration reflectors and the line-source feed dictated by geometric optics requirements. Very large aperture systems can be corrected by the use of two or three feed sections.

* Received October 31, 1962; revised manuscript received January 14, 1963. An earlier version of this paper was presented during Session 11—Radio Astronomy Antennas—of the 1961 WESCON Convention, San Francisco, Calif., August 22–25.

† Electromagnetic Radiation Laboratory, Air Force Cambridge Research Laboratories, Office of Aerospace Research (USAF), Bedford, Mass.

[1] R. C. Spencer, C. J. Sletten, and J. E. Walsh, "Correction of Spherical Aberration by a Phased Line Source," AF Cambridge Research Labs., Bedford, Mass., Rept. E 5069; May, 1951.

II. The Field Along the Axis of a Spherical Reflector

A cross section of a spherical reflector of radius of curvature R is shown in Fig. 1. The angle θ is measured from the center of curvature 0. The distance along the axis measured from the center of curvature is written as zR. A uniform plane wave traveling in the z direction, with the magnetic field intensity in the y direction, is assumed to be incident on the reflector. A reference plane for the incoming wave is located at $zR=0$. The surface current density on the reflector is

$$K = 2(\boldsymbol{n} \times \boldsymbol{H}_i),$$

which has two components:

$$K_x = 2H_i \cos\theta,$$

$$K_z = 2H_i \sin\theta \cos\phi.$$

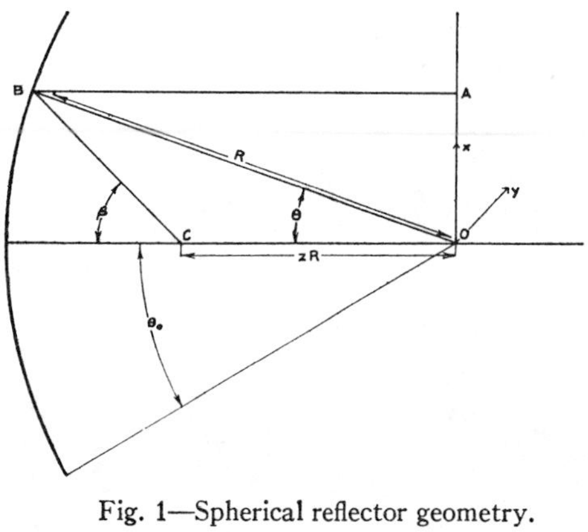

Fig. 1—Spherical reflector geometry.

Reprinted from *IEEE Trans. Antennas Propagat.*, vol. AP-11, pp. 428–432, July 1963.

It is clear that the z-directed current will not produce a field on the axis of the reflector because equal and opposite current elements are symmetrically disposed on the reflector surface.

Considering the current in the x direction, the field produced at a point C on the axis of the reflector is

$$E_x(z) = A\int_0^{2\pi}\int_0^{\theta_0} e^{-ikR\cos\theta}\cos\theta\frac{e^{-ik\overline{BC}}}{\overline{BC}}S(\beta,\phi)R^2\sin\theta d\theta d\phi,$$

where the factor $S(\beta, \phi)$ is the obliquity factor, or pattern, of each current element as viewed from the point C. From the geometry,

$$S(\beta, \phi) = 1 - \sin^2\beta\cos^2\phi,$$

which for the ϕ integration yields

$$\int_0^{2\pi}(1 - \sin^2\beta\cos^2\phi)d\phi = 2\pi(1 - \tfrac{1}{2}\sin^2\beta).$$

At the larger values of β the pattern of the elements produces a decrease in the contributions to E_x, but since the reduction for most β is not large enough to warrant the inclusion of this factor, $S(\beta, \phi)$ will be approximated by unity in this section.

The distance $\overline{BC}$ is found from the law of cosines

$$\overline{BC} = R\sqrt{1 + z^2 - 2z\cos\theta} = tR.$$

The integral to be evaluated is

$$E_x(z) = 2\pi R^2 A\int_0^{\theta_0}\frac{e^{-ikR(\cos\theta+\sqrt{1+z^2-2z\cos\theta})}}{\sqrt{1+z^2-2z\cos\theta}}\sin\theta\cos\theta d\theta, \quad (1)$$

and the constant A will be chosen later for appropriate normalization.

Changing the variables to

$$\sqrt{1 + z^2 - 2z\cos\theta} = t$$

gives

$$E_x(z) = 2\pi R^2 A\int_{t_0}^{t_s}\exp\left[-ikR\left(\frac{1+z^2}{2z} + t - \frac{t^2}{2z}\right)\right] \cdot\left(\frac{1+z^2-t^2}{2z^2}\right)dt,$$

and another change to

$$\frac{kR}{2z}(t - z)^2 = \frac{\pi}{2}u^2$$

yields

$$E_x(z) = \frac{\pi R^2 A}{z^2}\exp\left[-ikR\left(\frac{1+2z^2}{2z}\right)\right]\sqrt{\frac{\pi z}{kR}}\int_{u_0}^{u_1} \cdot e^{i(\pi/2)u^2}\left(1 - 2\sqrt{\frac{\pi z}{kR}}u - \frac{\pi z}{kR}u^2\right)du.$$

Carrying out the integration and using

$$C = \int\cos\frac{\pi}{2}u^2du, \qquad S = \int\sin\frac{\pi}{2}u^2du,$$

results in

$$E_x(z) = A\sqrt{\frac{\pi^3R^3}{z^3k}}\exp\left[-ikR\left(\frac{1+2z^2}{2z}\right)\right] \cdot\left\{\left(1 - \frac{iz}{kR}\right)[C(u_1) - C(u_0) + iS(u_1) - iS(u_0)] + 2i\sqrt{\frac{z^3}{\pi kR}}\left(e^{i(\pi/2)u_1^2} - e^{i(\pi/2)u_0^2}\right) + i\frac{zu_1}{kR}e^{i(\pi/2)u_1^2} - i\frac{zu_0}{kR}e^{i(\pi/2)u_0^2}\right\},$$

where the limits are

$$u_1 = \sqrt{\frac{kR}{\pi z}}(\sqrt{1 + z^2 - 2z\cos\theta_0} - z);$$

$$u_0 = \sqrt{\frac{kR}{\pi z}}(1 - 2z).$$

This expression can be evaluated for the field at any value z along the axis. The limits u_1 and u_0 are smooth functions of z, u_0 being zero at the paraxial focus $z=\frac{1}{2}$, and u_1 being zero at the crossing of the marginal rays given by $\cos\theta_0 = \frac{1}{2z}$.

For mirrors having a radius larger than $R=16\lambda$, the field in the vicinity of the paraxial focus can be determined with an error that is less than 0.5 per cent from the expression

$$E_x(z) = A\sqrt{\frac{\pi^3R^3}{z^3k}}\exp\left[-ikR\left(\frac{1+2z^2}{2z}\right)\right] \cdot\left\{C(u_1) - C(u_0) + iS(u_1) - iS(u_0) + i\sqrt{\frac{z}{\pi kR}} \cdot(\sqrt{1 + z^2 - 2z\cos\theta_0} + z) \cdot e^{i(\pi/2)u_1^2} - i\sqrt{\frac{z}{\pi kR}}e^{i(\pi/2)u_0^2}\right\}.$$

For very large mirrors the last two terms are negligible. The factor multiplying the Fresnel integrals is proportional to the geometric optics solution.

If all current elements were in such phase that their fields added in phase at the paraxial focus, the total field at that point would be

$$E_f = 2\pi R^2 A\int_{t_0}^{t_1}\left(\frac{1+z^2}{2z^2} - \frac{t^2}{2z^2}\right)dt$$

$$= 4\pi R^2 A\left[\left(\frac{5}{6} + \frac{\cos\theta_0}{3}\right)\sqrt{\frac{5}{4} - \cos\theta_0} - \frac{7}{12}\right]$$

$$= \pi R^2 A f(\theta_0).$$

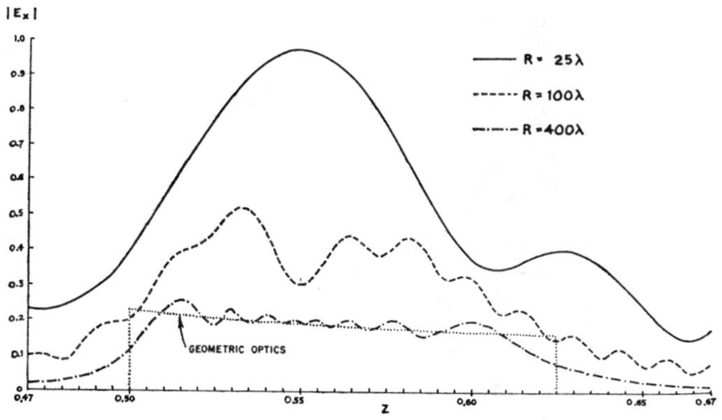

Fig. 2—Normalized magnitude of the field along the axis of a spherical reflector.

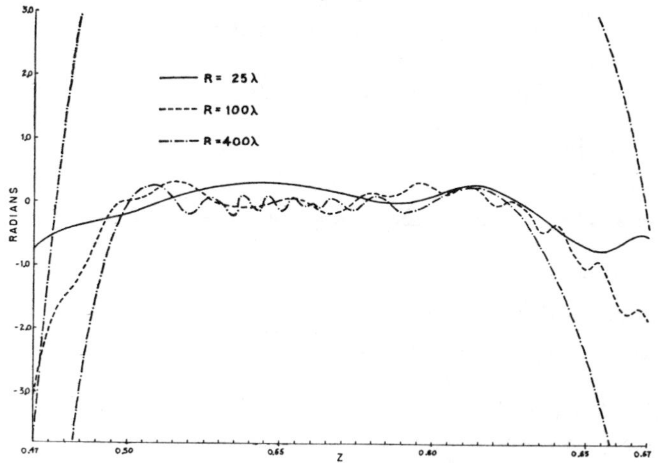

Fig. 3—Deviation from the geometric optical phase along the axis of a spherical reflector.

Normalizing the field expression by this quantity yields

$$E_n(z) = \left[\frac{1}{f(\theta)}\right]\left[\sqrt{\frac{\pi}{kR}}\right] \cdot \left[\frac{1}{z^{3/2}}\exp\left[-ikR\left(\frac{1+2z^2}{2z}\right)\right]\right] \cdot \left\{[C(u_1) - C(u_0) + iS(u_1) - iS(u_0)] + i\sqrt{\frac{z}{\pi kR}} \cdot (\sqrt{1+z^2-2z\cos\theta_0}+z) \cdot e^{i(\pi/2)u_1^2} - i\sqrt{\frac{z}{\pi kR}}\, e^{i(\pi/2)u_0^2}\right\}.$$

In this expression the first part is the normalization; the second shows that the normalized field decreases as the reflector size is increased; the third part is the geometric optics behavior; the fourth is the Fresnel integral behavior, and the rest contributes minor oscillations.

At the center of curvature the integral must be re-evaluated. Thus,

$$E_x(0) = 2\pi R^2 A \int_0^{\theta_0} e^{-ikR(\cos\theta+1)} \sin\theta\cos\theta d\theta$$
$$= 2\pi R^2 A e^{-2ikR}\left[\frac{1}{ikR}\right][1 - \cos\theta_0 e^{-ikR(\cos\theta_0-1)}].$$

The normalized field at $z=0$ is approximately

$$E_n(0) \cong \frac{2}{ikR}\,\frac{e^{-2ikR}}{f(\theta)}\,[1 - \cos\theta_0 e^{-ikR(\cos\theta_0-1)}].$$

The normalized field magnitude for spherical mirrors of half-angle $\theta_0 = \cos^{-1} 0.8 = 36.9°$ and radii of curvature $R = 25\lambda$, 100λ, and 400λ are plotted in Fig. 2. The curves for $R = 400\lambda$ are slightly smoothed. The geometric optics limit for $R = 400\lambda$ is also shown.

The deviation of the phase from the geometric optics phase

$$\exp\left[-ikR\left(\frac{1+2z^2}{2z}\right)\right]$$

is shown for the same cases in Fig. 3.

III. The Far-Field Pattern of the Spherical Reflector Antenna

The field intensity along the axis for an incident plane wave was determined in Section II. Section III deals with the far-field pattern of the antenna when the conjugate of the field distribution of (1) is placed on the axis of the reflector.

The current density on the reflector due to source element of strength $E_x^*(z)dz$ at a point z on the axis is approximately

$$K_x = S(\beta, \phi)E_x^*(z)dz\frac{e^{iktR}}{tR}\cos\theta.$$

The far-field pattern of a thin annulus of radius $R \sin\theta$ and width $R\cos\theta d\theta$ in the transverse plane is

$$R^2\sin\theta\cos\theta d\theta \int_0^{2\pi} e^{-ikR\sin\theta\sin\alpha\cos(\psi-\phi)} S(\beta, \phi)d\phi,$$

where α is the polar angle between the direction of observation and the axis of the reflector, and ψ is the azimuthal angle of the observation direction. The far-field pattern from all currents on the spherical reflector excited by the entire axial distribution is

$$F(\alpha, \psi) = \text{const} \int\int E_x^*(z)\frac{e^{-ik(tR+R\cos\theta)}}{tR} R^2\sin\theta\cos\theta \cdot \int e^{-ikR\sin\theta\sin\alpha\cos(\psi-\phi)}S(\beta, \phi)d\phi d\theta dz, \qquad (2)$$

where the effect of the obliquity factor of the current elements is included. Substituting the expression given for $E_x(z)$ in (1) and using the variables $u' = \cos\theta$ in (1) and $u = \cos\theta$ in (2) yields

$$F(\alpha, \psi) = \text{const}\int\int\int \frac{e^{-ikR(t-t'+u-u')}}{tt'}uu' \cdot \int e^{-ikR\sin\theta\sin\alpha\cos(\psi-\phi)}S(\beta, \phi)d\phi du du' dz.$$

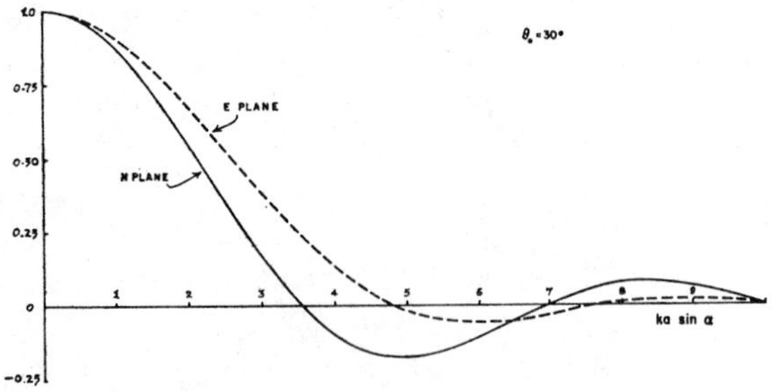

Fig. 4—Far-field radiation patterns of a spherical reflector with a line source feed.

Evaluating the u' and z integration by the method of steepest descent yields

$$F(\alpha, \psi) \approx \text{const} \int_0^{2\pi} \int_0^{\theta_0} e^{-ikR \sin\theta \sin\alpha \cos(\psi - \phi)} \cdot (1 - 4\sin^2\theta \cos^2\theta \cos^2\phi) \cos\theta \sin\theta d\theta d\phi.$$

With the indicated integration performed and $a = R \sin \theta_0$, the pattern in the E plane ($\psi = 0$) is given by

$$F(\alpha, 0) \approx \text{const} \Big\{ \Lambda_1(ka \sin \alpha_x) - 4 \sin^2\theta_0 \cdot \Big[\Lambda_1(ka \sin \alpha_x) - \frac{3}{4} \Lambda_2(ka \sin \alpha_x) \Big] + 4 \sin^4 \theta_0 \Big[\Lambda_1(ka \sin \alpha_x) - \frac{5}{4} \Lambda_2(ka \sin \alpha_x) + \frac{5}{12} \Lambda_3(ka \sin \alpha_x) \Big] \Big\},$$

where $\Lambda_p(x) = p!(2/x)^p J_p(x)$. The pattern in the H plane ($\psi = \pi/2$) is given by

$$F(\alpha, \pi/2) \approx \text{const} \{ \Lambda_1(ka \sin \alpha_y) - \sin^2 \theta_0 \Lambda_2(ka \sin \alpha_y) + \sin^4 \theta_0 [\Lambda_2(ka \sin \alpha_y) - \tfrac{1}{3}\Lambda_3(ka \sin \alpha_y)] \},$$

showing the effect of the element patterns of the currents on the reflector and the feed elements. The far-field radiation patterns for $\theta_0 = 30°$ are shown in Fig. 4.

IV. A Method of Feed Design

The axis of the spherical reflector contains a set of points at which the energy of the incoming plane wave is focused. For each of these points there is a region of the reflector that approximates a portion of a paraboloid. These regions can constitute a considerable portion of the total reflector. For example, a 40λ- to 100λ-diameter central portion of a typical spherical reflector can be illuminated without large phase errors by a feed element located near the paraxial focus. Another directive feed element located on the axis between the paraxial focus and the vertex can correctly illuminate a circularly symmetric ring of the reflector (see Fig. 5). A sufficient number of directive feed elements along the axis can, within a specified phase tolerance, illuminate all regions of the reflector.

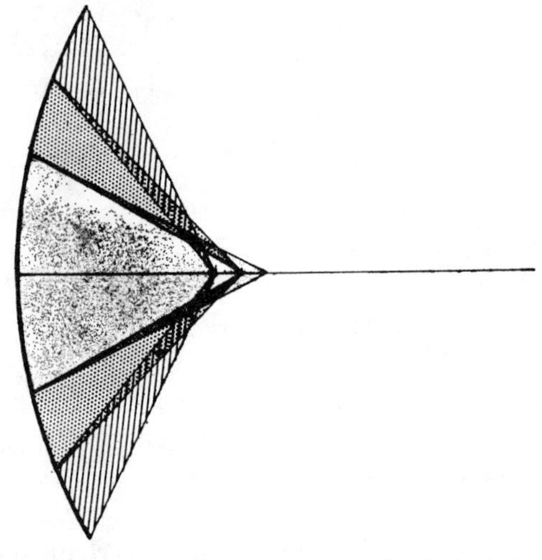

Fig. 5—Illumination of the reflector by feed sections.

For an incident on-axis plane wave, the path length error on the reflector due to a single feed at a point z_0 on the axis is

$$\Delta - \Delta_0 = \frac{2R}{z_0}\left(\sin^2\frac{\theta}{2} - \sin^2\frac{\theta_0}{2}\right)^2 + \text{higher-order terms,}$$

where $\cos\theta_0 = \frac{1}{2z_0}$ and Δ_0 is the path length from z_0 to the stationary phase point on the reflector. The mean-square path length error for a feed at z_0 illuminating the reflector from $\theta = 0$ to $\theta = \theta_1$ is

$$\overline{\delta^2} = \frac{1}{A}\int_0^{\theta_1} (\Delta - \Delta_0)^2 \sin\theta d\theta = \frac{8R^2}{Az_0^2}\int_0^{y_1} (y - y_0)^4 dy = \frac{8R^2}{5Az_0^2}[(y_1 - y_0)^5 + y_0^5],$$

where $y = \sin^2 \theta/2$ and

$$A = \int_0^{\theta_1} \sin\theta d\theta = 2y_1.$$

The mean path length is

$$\bar{\delta} = \frac{1}{A}\int_0^{\theta_1} (\Delta - \Delta_0)\sin\theta d\theta = -\frac{4R}{Az_0}\int_0^{y_1} (y - y_0)^2 dy = -\frac{4R}{3Az_0}[(y_1 - y_0)^3 + y_0^3].$$

The variance of the path length is

$$\sigma^2 = \overline{\delta^2} - \bar{\delta}^2 = \frac{8}{5}\frac{R^2}{Az_0^2}[(y_1 - y_0)^5 + y_0^5] - \frac{16R^2}{9A^2z_0^2}[(y_1 - y_0)^3 + y_0^3]^2.$$

Choosing y_0 to minimize this variance gives

$$y_0 = \frac{1}{2}y_1; \qquad z_0 = \frac{1}{2 - 2y_1},$$

and

$$y_1^2 = \sin^4 \frac{\theta_1}{2} = \frac{3\sqrt{5}}{R} z_0\sigma.$$

The amount of surface illuminated by the first feed is selective by specifying the tolerable phase deviation across the aperture.

The second feed section is located at z_2 and illuminates the reflector from θ_1 to θ_3. Choosing the parameters as before gives

$$(y_3 - y_1)^2 = \frac{3\sqrt{5}}{R} z_2\sigma; \qquad z_2 = \frac{\frac{1}{2}}{1 - y_1 - y_3}.$$

This procedure can be carried on until the entire reflector is illuminated by a number of feed sections. If the phase error is to be kept very small, a large number of feed sections must be used; each section must illuminate a small part of the reflector and thus each must be of considerable length. As the limit of very low phase errors is approached, the feed illumination discussed in Section III may be used.

As an example of this technique, consider a spherical reflector having a radius of curvature of 870 ft, or 380λ at 430 Mc. Assume that the aperture is uniformly illuminated, and that there is a 1-db decrease in gain due to the phase errors described above. The first feed element, which could be a horn, will be located at $z_0 = 0.513$ (11.4 ft below the paraxial focus of 435 ft) to illuminate the central portion of the reflector to $\theta_1 = 18.4°$, a total angle of 74° as viewed from z_0. The second array will have a phase center at $z_2 = 0.542$ (36.5 ft below the paraxial focus) to illuminate the reflector from $\theta_1 = 18.4°$ to $\theta_3 = 26.8°$. For the feed pattern to be substantially within an angle of 13.7°, the feed length must be at least 13 ft. The third feed section will be centered at $z_4 = 0.575$ (65.2 ft below the paraxial focus) to illuminate the reflector from $\theta_3 = 26.8°$ to $\theta_5 = 32.6°$. For the radiation pattern of this feed to be substantially contained within 9.8°, the feed length must be at least 16 ft. These three feeds will illuminate an aperture 936 ft in diameter. The total feed will be about 75-ft long measured from the paraxial focus. Two phase shifters located between the three feed sections will permit adjustment of relative phase at different frequencies.

V. Experimental Verification

A spherical reflector feed of the type described in Section IV was built and tested on the 10-ft diameter spherical reflector section at the Ipswich test site. A horn illuminated the central portion of the reflector, and a short waveguide array placed on the axis in front of the horn illuminated the outer portion of the reflector. The waveguide array blocked a considerable portion of the energy from the horn, yet the results were in substantial agreement with the calculated performance. The experimental X-band patterns (Fig. 6) show that as the relative amount of power supplied to the array is increased, the aperture illumination increases near the edges and the beamwidth decreases.

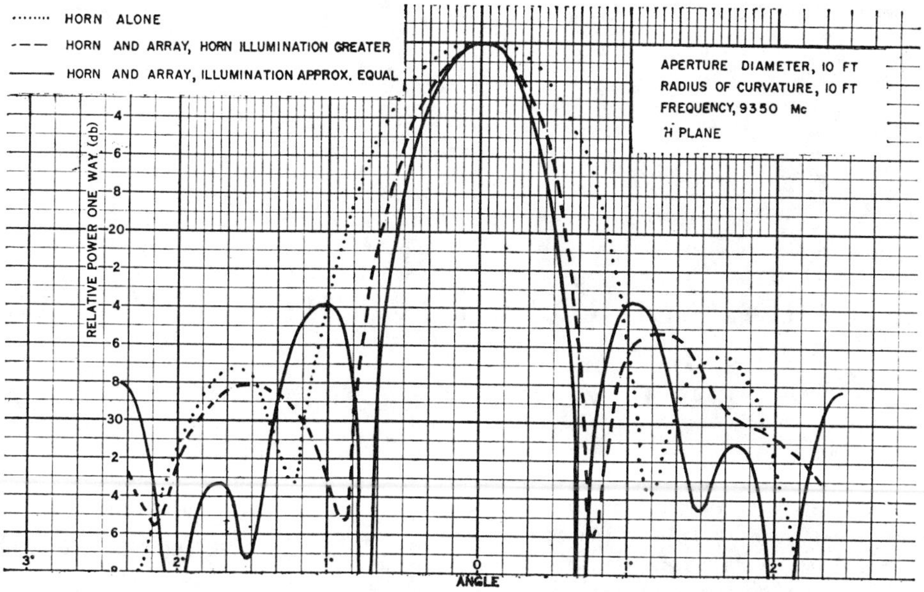

Fig. 6—Experimental secondary radiation patterns of a spherical reflector with horn and array.

A Line Feed for a Spherical Reflector

G. C. McCORMICK, MEMBER, IEEE

Abstract—**A line feed for a spherical reflector is considered on the basis of a plane-wave spectrum of radiation angles. It is shown that a feed excited by circumferential slots results in a gain deterioration of at least 3 dB. The correct excitation of the feed is indicated. Expressions for field components in the focal region are obtained.**

Manuscript received February 14, 1967; revised April 24, 1967.
The author is with the Radio and Electrical Engineering Division, National Research Council, Ottawa, Canada.

INTRODUCTION

BECAUSE of its symmetry, the spherical reflector permits scanning over many beamwidths by means of a translation and orientation of the feed. The feed may be a line source or have other spatial extension so as to correct for spherical aberration. The fields in the focal region are, therefore, of considerable interest and have been the subject of a number of recent papers [1]–[4].

The electromagnetic case (considered from the transmis-

Reprinted from *IEEE Trans. Antennas Propagat.*, vol. AP-15, pp. 639–645, Sept. 1967.

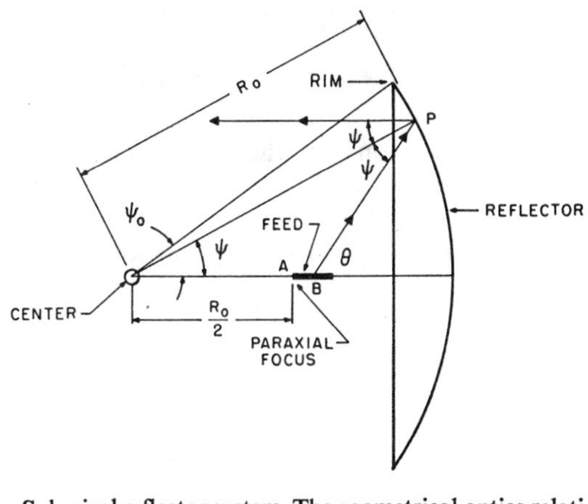

Fig. 1. Spherical reflector system. The geometrical optics relations are:

$$\theta = 2\psi$$
$$OB = BP = \frac{R_0}{2} \sec \psi$$
$$AB = \frac{R_0}{2} (\sec \psi - 1).$$

sion point of view) is complicated by the need to have a single polarization over the entire aperture in addition to the usual requirements of uniform phase and a reasonable amplitude taper. This paper deals with the polarization aspects of the problem and, in particular, with the basic limitations of a waveguide feed with circumferential slots. A more general feed capable of producing the ideal illumination is discussed. Finally, it is necessary to be mindful of the frequent need for using orthogonal polarizations, that is, for a feed having square or circular symmetry with no coupling between orthogonal excitations [5].

The basic system of a spherical reflector with feed is shown in Fig. 1, together with the geometrical optics relations. Consideration will first be given to an infinite cylindrical source lying along the reflector axis which leaks radiation at a constant angle θ. The discussion that follows is intended to show that the feed of Fig. 1 can be obtained by superimposing a spectrum of the infinite source solutions. The coordinate system used throughout has its origin at the center of curvature of the reflector, with the positive z direction pointing toward the reflector.

The Field Radiated by the Feed

A purely harmonic time dependence, $e^{j\omega t}$, will be assumed. The free-space Maxwell's equations can be reduced for cylindrical geometry by use of the z components of two Hertz vectors π and π^* [6].

Using circular cylinder coordinates r, ϕ, z ($\phi=0$ in the x-z plane), the quantities $f(r, \phi)$ and $g(r, \phi)$ will be defined by

$$\left.\begin{aligned} \pi_z &= f(r, \phi) e^{-jhz + j\omega t} \\ \pi_z^* &= g(r, \phi) e^{-jhz + j\omega t} \end{aligned}\right\} \tag{1}$$

where a common propagation constant h results from the excitation of the structure. The combination of the two solutions gives the following field components:

$$\left.\begin{aligned} E_r &= -j\left(h \frac{\partial f}{\partial r} + \frac{\mu_0 \omega}{r} \frac{\partial g}{\partial \phi}\right) e^{-jhz} \\ E_\phi &= j\left(-\frac{h}{r} \frac{\partial f}{\partial \phi} + \mu_0 \omega \frac{\partial g}{\partial r}\right) e^{-jhz} \\ E_z &= \kappa^2 f e^{-jhz} \\ H_r &= j\left(\frac{\epsilon_0 \omega}{r} \frac{\partial f}{\partial \phi} - h \frac{\partial g}{\partial r}\right) e^{-jhz} \\ H_\phi &= -j\left(\epsilon_0 \omega \frac{\partial f}{\partial r} + \frac{h}{r} \frac{\partial g}{\partial \phi}\right) e^{-jhz} \\ H_z &= \kappa^2 g e^{-jhz} \end{aligned}\right] \tag{2}$$

where $\kappa^2 = k^2 - h^2$, $k^2 = \mu_0 \epsilon_0 \omega^2$.

It will be supposed initially that radiation takes place from circumferential slots in a metal waveguide of radius a excited by a waveguide mode with the lowest angular distribution, say TE_{11} or TM_{11}, so that the angular field function is $\sin\phi/\cos\phi$. The slot will be idealized by supposing it to be continuous about the cylinder; then the structure can be regarded as a uniconducting cylinder, $\sigma_\phi = \infty$, $\sigma_z = 0$, which leaks radiation from the inside to the outside. This leads to the boundary condition at $r = a$, for x polarization,

$$\left.\begin{aligned} E_z &= A \cos \phi e^{-jhz} \\ E_\phi &= 0. \end{aligned}\right\} \tag{3}$$

It will be fruitful, however, to generalize the boundary conditions to

$$\left.\begin{aligned} E_z &= A \cos \phi e^{-jhz} \\ E_\phi &= B \sin \phi e^{-jhz} \end{aligned}\right\} \tag{4}$$

and eventually to

$$\left.\begin{aligned} E_z &= \sum A_n \cos n\phi e^{-jhz} \\ E_\phi &= \sum B_n \sin n\phi e^{-jhz}. \end{aligned}\right\} \tag{5}$$

Subsequent derivations will be based on (4)—and (4) reduces to (3) by putting $B = 0$, giving the ideal longitudinally polarized excitation. Additional effects due to the use of physical slots of appropriate length can be assessed by considering (5).

The exterior expansions for f and g are

$$f(r, \phi) = \sum_{-\infty}^{\infty} a_n e^{jn\phi} H_n^{(2)}(\kappa r) \tag{6}$$

$$g(r, \phi) = \sum_{-\infty}^{\infty} b_n e^{jn\phi} H_n^{(2)}(\kappa r). \tag{7}$$

From (2) and (4),

$$A \cos \phi = \sum_{-\infty}^{\infty} a_n e^{jn\phi} H_n^{(2)}(\kappa r),$$

from which

$$\left.\begin{aligned} a_1 = -a_{-1} &= \frac{A}{2\kappa^2 H_1^{(2)}(\kappa a)} \\ a_n &= 0 \quad \text{for } |n| \neq 1 \end{aligned}\right\} \tag{8}$$

$$f = 2a_1 \cos\phi H_1^{(2)}(\kappa r). \tag{9}$$

Also from (2) and (4)

$$B \sin\phi = j\left(-\frac{h}{a}\frac{\partial f}{\partial \phi} + \mu_0\omega \frac{\partial g}{\partial r}\right)_{r=a}.$$

As in the derivation of (8) the only nonzero terms in the expansions (6) and (7) are those for which $n = \pm 1$. Therefore,

$$B \sin\phi = j\left[-\frac{h}{a}(ja_1e^{j\phi} + ja_{-1}e^{-j\phi})H_1^{(2)}(\kappa a) + \mu_0\omega(-b_{-1}e^{-j\phi} + b_1e^{j\phi})\kappa H_1^{(2)\prime}(\kappa a)\right],$$

$$b_1 = b_{-1} = \frac{1}{2\mu_0\omega\kappa H_1^{(2)\prime}(\kappa a)}\left[\frac{jhA}{\kappa^2 a} - B\right] \tag{10}$$

and

$$g = 2jb_1 \sin\phi H_1^{(2)}(\kappa r). \tag{11}$$

The magnetic field components may be obtained from (2) by using (9) and (11)

$$H_r = 2j\kappa \sin\phi H_1^{(2)\prime}(\kappa r)(\epsilon_0\omega a_1 p(\kappa r) + jhb_1)e^{-jhz} \tag{12}$$

$$H_\phi = -2j\kappa \cos\phi H_1^{(2)\prime}(\kappa r)(\epsilon_0\omega a_1 + jhb_1 p(\kappa r))e^{-jhz} \tag{13}$$

$$\left.\begin{aligned} H_z &= 2j\kappa^2 b_1 \sin\phi H_1^{(2)}(\kappa r)e^{-jhz} \\ &= 2j\kappa \sin\phi H_1^{(2)\prime}(\kappa r) b_1 \kappa^2 r p(\kappa r)e^{-jhz} \end{aligned}\right\} \tag{14}$$

where

$$p(\kappa r) = \frac{H_1^{(2)}(\kappa r)}{\kappa r H_1^{(2)\prime}(\kappa r)}. \tag{15}$$

The electric field components may similarly be written

$$E_r = -2j\kappa \cos\phi H_1^{(2)\prime}(\kappa r)(ha_1 + j\mu_0\omega b_1 p(\kappa r))e^{-jhz} \tag{16}$$

$$E_\phi = 2j\kappa \sin\phi H_1^{(2)\prime}(\kappa r)(ha_1 p(\kappa r) + j\mu_0\omega b_1)e^{-jhz} \tag{17}$$

$$E_z = 2\kappa^2 a_1 \cos\phi H_1^{(2)}(\kappa r)e^{-jhz}. \tag{18}$$

Surface Currents on the Spherical Reflector

The incidence of the field given by (12) to (18) on the spherical reflector generates surface currents. It will be supposed that the reflector is of sufficient size so that the optical approximation is valid, viz., $J = 2(n \times H_i)$, where

$$H_i = r_1 H_r + \phi_1 H_\phi + z_1 H_z$$

$$n = -r_1 \sin\psi - z_1 \cos\psi$$

and r_1, ϕ_1, and z_1 are unit vectors. By using (12), (13), and (14), and the unit vectors

$$x_1 = r_1 \cos\phi - \phi_1 \sin\phi$$

$$y_1 = r_1 \sin\phi + \phi_1 \cos\phi$$

the following expressions for the current densities are obtained:

$$J_x = 2j\{-(F_2 + F_3) + (F_1 + F_3)\cos 2\phi\}e^{-j(hz+\kappa r)} \tag{19}$$

$$J_y = 2j \sin 2\phi(F_1 + F_3)e^{-j(hz+\kappa r)} \tag{20}$$

where

$$F_1 = -\kappa H_2^{(2)}(\kappa r)e^{j\kappa r}\cos\psi(jhb_1 - \epsilon_0\omega a_1) \tag{21}$$

$$F_2 = \kappa H_0^{(2)}(\kappa r)e^{j\kappa r}\cos\psi\,(jhb_1 + \epsilon_0\omega a_1) \tag{22}$$

$$F_3 = \kappa^2 H_1^{(2)}(\kappa r)e^{j\kappa r}\sin\psi b_1. \tag{23}$$

The coefficients appearing in (19) and (20) may be written

$$F_2 + F_3 = \kappa H_0^{(2)}(\kappa r)e^{j\kappa r} \cdot \cos\psi\left[\epsilon_0\omega a_1 + jhb_1\left(1 - \alpha\frac{H_2^{(2)}(\kappa r)}{H_0^{(2)}(\kappa r)}\right)\right] \tag{24}$$

$$F_1 + F_3 = \kappa H_2^{(2)}(\kappa r)e^{j\kappa r}\cos\psi[\epsilon_0\omega a_1 - jhb_1(1 + \alpha)] \tag{25}$$

where

$$\alpha = j\frac{\kappa}{h}\tan\psi\frac{H_1^{(2)}(\kappa r)}{H_2^{(2)}(\kappa r)}.$$

The analysis to this point has dealt with radiation from an infinite cylinder due to excitation with a fixed longitudinal propagation constant h. Radiation proceeds from the cylinder at an angle θ given by $\theta = \cos^{-1} h = \sin^{-1}\kappa$. The idea of a spectrum of radiation angles will now be introduced. Hence κ, h, a_1, and b_1 will be considered functions of θ, and radiation from many angles will be incident on every point of the reflector, all of which will contribute to the surface current. Therefore, J_x and J_y will be written as integrals using the relation

$$hz + \kappa r = kR_0 \cos(\theta - \psi). \tag{26}$$

$$J_x = 2j\int_C \{-(F_2 + F_3) + (F_1 + F_3)\cos 2\phi\}e^{-jkR_0\cos(\theta-\psi)}d\theta \tag{27}$$

$$J_y = 2j\sin 2\phi\int_C (F_1 + F_3)e^{-jkR_0\cos(\theta-\psi)}d\theta. \tag{28}$$

The range of integration is over the real angles 0 to π, and also over the imaginary angles 0 to $-i\infty$ and the complex angles π to $\pi + i\infty$ as shown by C in Fig. 2. The latter take into account the possible existence of surface waves and also reactive ripples along the surface of the radiator which are certain to be present if there are physical slots.

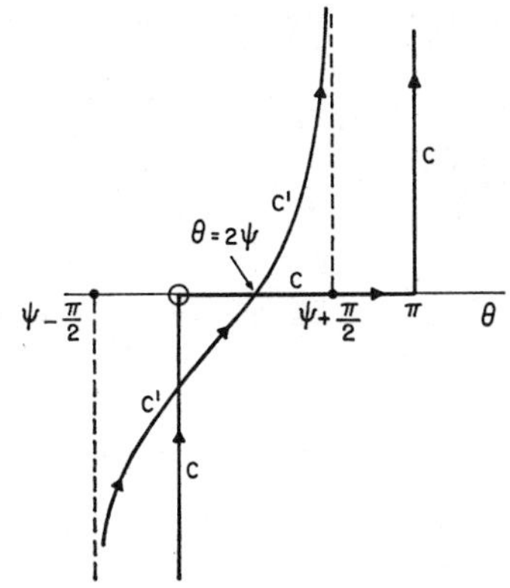

Fig. 2. Integration contours.

The Far Field and the Aperture Field

The far field is given by [7]

$$E' = \frac{-j\omega\mu_0}{4\pi R'} e^{-jkR'} \iint [J - (J\cdot R_1')R_1'] \cdot \exp[jk\boldsymbol{\varrho}\cdot R_1']ds \quad (29)$$

where the surface currents J are integrated over the reflector surface. Primed quantities E', R', θ', ϕ' refer to the distant field, and $\boldsymbol{\varrho}$ is the radius vector from the center to a point on the spherical surface as shown in Fig. 3; $|\rho| = R_0$. The radius vector to a point in the aperture is related to ρ by

$$\varrho_a = \varrho - m$$
$$m = R_0(\cos\psi - \cos\psi_0) \quad (30)$$

and

$$\varrho\cdot R_1' = \varrho_a \cdot R_1' - m + 2m\sin^2\frac{\theta'}{2}\cdot$$

If it be assumed that the system behaves like an antenna with a well-defined main beam and reasonable sidelobes, then (29) may be approximated by

$$\left.\begin{aligned} E_x' &= \frac{-j\omega\mu_0}{4\pi R'} e^{-jkR'} \iint J_x \sec\psi \cdot\exp[-jkm + jk\varrho_a\cdot R_1']td\phi dt \\ E_y' &= \frac{-j\omega\mu_0}{4\pi R'} e^{-jkR'} \iint J_y \sec\psi \cdot\exp[-jkm + jk\varrho_a\cdot R_1']td\phi dt \end{aligned}\right\} \quad (31)$$

where t is the distance from the axis to an element in the aperture as shown in Fig. 3. The neglected terms in (31) have order (λ/R_0), and the reflector will be assumed to have such size that they can properly be discarded. Equation (31) may be compared with the physical optics diffraction formulas

$$\left.\begin{aligned} E_x' &= \frac{j}{\lambda R'} e^{-jkR'} \iint e_x \exp[jk\varrho_a\cdot R_1']td\phi dt \\ E_y' &= \frac{j}{\lambda R'} e^{-jkR'} \iint e_y \exp[jk\varrho_a\cdot R_1']td\phi dt \end{aligned}\right\} \quad (32)$$

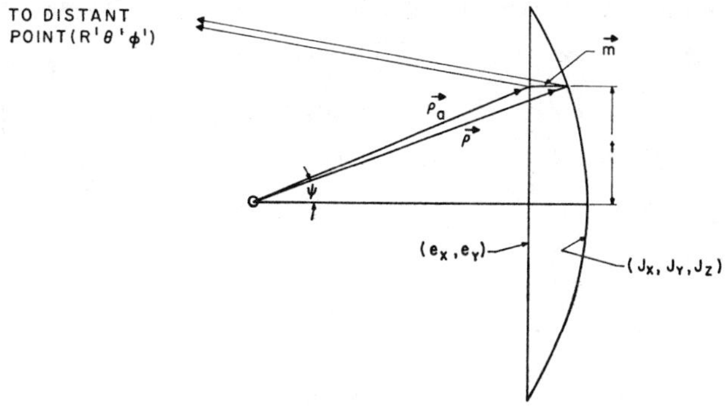

Fig. 3. Geometry of the aperture field and of the surface currents.

in which e_x and e_y are the aperture fields corresponding to the distant fields E_x' and E_y'. Equations (31) and (32) must be equally valid expressions over all significant portions of the far-field pattern, and this is possible only if the integrands are equal, or at most differ by terms that are reactive and do not propagate into the distant field. Hence

$$\left.\begin{aligned} e_x &= \frac{-\omega\mu_0}{2k} J_x \sec\psi\, e^{-jkm} \\ e_y &= \frac{-\omega\mu_0}{2k} J_y \sec\psi\, e^{-jkm}. \end{aligned}\right\} \quad (33)$$

Hence by comparison with (27) and (28),

$$\left.\begin{aligned} e_x &= e_0(\psi) + e_1(\psi)\cos 2\phi \\ e_y &= e_1(\psi)\sin 2\phi \end{aligned}\right\} \quad (34)$$

where

$$e_0(\psi) = \frac{j\omega\mu_0}{k}\sec\psi\, e^{-jkm}\int_C (F_2 + F_3) \cdot e^{-jkR_0\cos(\theta-\psi)}d\theta \quad (35)$$

$$e_1(\psi) = \frac{-j\omega\mu_0}{k}\sec\psi\, e^{-jkm}\int_C (F_1 + F_3) \cdot e^{-jkR_0\cos(\theta-\psi)}d\theta. \quad (36)$$

It would appear from (30) that e_0 and e_1 vary rapidly with ψ, a situation that is incompatible with their being suitable aperture functions. However, a rearrangement gives

$$e_0(\psi) = \frac{j\omega\mu_0}{k}\sec\psi \cdot e^{jkR_0\cos\psi_0}\int_C [(F_2 + F_3)e^{-2jkR_0\cos\theta/2}] \cdot e^{2jkR_0\cos\theta/2(1-\cos(\theta/2-\psi))}d\theta. \quad (37)$$

F_2 and F_3 are slowly varying in ψ; if the θ dependence of $a_1(\theta)$ and $b_1(\theta)$ is such that $[(F_2+F_3)e^{-2jkR_0\cos\theta/2}]$ is slowly varying in θ, the integral can be evaluated by stationary phase at the point $\theta = 2\psi$ and $e_0(\psi)$ becomes an acceptable function. The same rearrangement may be carried out for $e_1(\psi)$.

Therefore, the relative importance of $e_0(\psi)$ and $e_1(\psi)$ may be assessed by evaluating the terms of the expressions at $\theta = 2\psi$. The gain considerations of the following paragraph are not affected by the further approximation $\kappa r \gg 1$. At $\theta = 2\psi$ and $\kappa r \gg 1$, (24) and (25) reduce to

$$F_2 + F_3 = \kappa H_0^{(2)}(\kappa r) e^{j\kappa r} \cos\psi(\epsilon_0 \omega a_1 + jkb_1) \quad (38)$$

$$F_1 + F_3 = \kappa H_0^{(2)}(\kappa r) e^{j\kappa r} \cos\psi(-\epsilon_0 \omega a_1 + jkb_1) \quad (39)$$

and

$$e_x = \frac{e(t)}{2}(1 + \tau e^{j\sigma} + (1 - \tau e^{j\sigma}) \cos 2\phi) \quad (40)$$

$$e_y = \frac{e(t)}{2}(1 - \tau e^{j\sigma}) \sin 2\phi \quad (41)$$

where

$$\tau e^{j\sigma} = \frac{jkb_1}{\epsilon_0 \omega a_1} \quad (42)$$

$$e(t) = e_0(\psi) + e_1(\psi). \quad (43)$$

Equations (40) and (41) show the aperture fields in terms that can be directly calculated from the excitation fields on the cylindrical feed. In particular, the relative magnitudes of the cos 2ϕ term and of the cross-polarized term in (41) can be evaluated.

Gain Considerations

The gain of the antenna can be determined from (40) and (41). The power radiated through the aperture is given by

$$P = \frac{1}{2} Y_0 \int_0^{t_0} \int_0^{2\pi} (|e_x|^2 + |e_y|^2) t d\phi dt$$
$$= \frac{\pi Y_0}{2} \int_0^{t_0} e^2(1 + \tau^2) t dt \quad (44)$$

where $t_0 = R_0 \sin \psi_0$. The Poynting flux in the distant field is

$$S = \frac{1}{2} \frac{Y_0}{\lambda^2 R'^2} \left| \int_0^{t_0} \int_0^{2\pi} e_x t d\phi dt \right|^2$$
$$= \frac{\pi^2 Y_0}{2\lambda^2 R'^2} \left| \int_0^{t_0} (1 + \tau e^{j\sigma}) e t dt \right|^2. \quad (45)$$

The gain is given by

$$G = \frac{4\pi R'^2 |S|}{P} = \frac{4\pi^2}{\lambda^2} \frac{\left| \int_0^{t_0} (1 + \tau e^{j\sigma}) e t dt \right|^2}{\int_0^{t_0} (1 + \tau^2) e^2 t dt}.$$

If only a radial taper exists, the expression reduces to

$$G_0 = \frac{8\pi^2}{\lambda^2} \frac{\left| \int_0^{t_0} e t dt \right|^2}{\int_0^{t_0} e^2 t dt}.$$

Therefore, the deterioration in gain due to azimuthal variation and to cross polarization in the aperture is given by

$$\frac{G}{G_0} = \frac{1}{2} \frac{\left| \int_0^{t_0} (1 + \tau e^{j\sigma})^2 e t dt \right|^2}{\left| \int_0^{t_0} e t dt \right|^2} \frac{\int_0^{t_0} e^2 t dt}{\int_0^{t_0} (1 + \tau^2) e^2 t dt}. \quad (46)$$

If τ and σ are constant over the aperture,

$$\frac{G}{G_0} = \frac{1 + \tau^2 + 2\tau \cos \sigma}{2(1 + \tau^2)}. \quad (47)$$

If only circumferential slots are used, $B = 0$, and from (8), (10), (15), and (42),

$$\left.\begin{aligned} \tau e^{j\sigma} &= -\cos\theta p(\kappa a) \\ &\approx \frac{j \cos\theta}{\kappa a}. \end{aligned}\right\} \quad (48)$$

Hence $\sigma \approx \pi/2$, $G/G_0 \approx \frac{1}{2}$. A gain deterioration of 3 dB can be anticipated for this class of feed excitation.

With higher-order terms due to excitation by an even number of symmetrically disposed slots it can be shown that (44) should be replaced by

$$P = \frac{\pi Y_0}{2} \int_0^{t_0} [e^2(1 + \tau^2) + e_3^2(1 + \tau_3^2) + e_5^2(1 + \tau_5^2) + \cdots] t dt \quad (49)$$

while (45) remains unchanged. τ_n and e_n are given by

$$\tau_n = \cos\theta \left| \frac{H_{n-1}^{(2)}(\kappa a) + H_{n+1}^{(2)}(\kappa a)}{H_{n-1}^{(2)}(\kappa a) - H_{n+1}^{(2)}(\kappa a)} \right| \quad (50)$$

and

$$\frac{e_n(t)}{e(t)} = \frac{A_n}{A} \frac{H_1^{(2)}(\kappa a)}{H_n^{(2)}(\kappa a)} \quad (51)$$

where A_n is given by (5). It is seen that there is an increase in gain deterioration due to the azimuthal field variation which results from the use of slots with a circumferential spacing comparable with a wavelength. The value of the increase is typically 0.6 dB for a TE waveguide feed.

Field Calculations

It is evident that the gain deterioration of the preceding section is absent if $F_1 + F_3 = 0$. The remaining discussion will be confined to this case. The field in the vicinity of the feed cylinder necessary to achieve this desirable illumination will be examined. Substitution of $F_1 + F_3 = 0$ in (25) and (24) gives

$$F_2 + F_3 = 2\kappa h_0(\kappa r) \cos\psi \epsilon_0 \omega a_1(\theta) \quad (52)$$

where

$$h_0(\kappa r) = H_0^{(2)}(\kappa r) e^{j\kappa r} \left(1 - \frac{\alpha}{1 + \alpha} \frac{H_1^{(2)}(\kappa r)}{\kappa r H_0^{(2)}(\kappa r)}\right) \quad (53)$$

$$= H_0^{(2)}(\kappa r) e^{j\kappa r} + 0\left(\frac{1}{kR_0}\right) \quad (54)$$

$$\alpha = j \tan\theta \tan\psi \frac{H_1^{(2)}(\kappa r)}{H_2^{(2)}(\kappa r)}$$

$$\kappa r = kR_0 \sin\theta \sin\psi.$$

Substitution of (52) in (37) gives

$$2jk^2 \int_C h_0(\kappa r)\left[a_1(\theta)e^{-2jkR_0 \cos\theta/2}\right]e^{2jkR_0 \cos\theta/2(1-\cos(\theta/2-\psi))} \cdot \sin\theta d\theta = e_0(\psi)e^{-jkR_0 \cos\psi_0}$$

which may be regarded as an integral equation for $a_1(\theta)$. For purposes of evaluation, C will be deformed to the steepest descent path C', which cuts the real axis at $\theta=2\psi$. Then on putting $\theta-2\psi=\zeta e^{j\pi/4}$ the saddlepoint integration leads to

$$2jk^2 e^{j\pi/4}\int_{-\infty}^{\infty} h_0(S)\left[a_1(\theta)e^{-2jkR_0 \cos\theta/2}\right]e^{-(1/4kR_0 \cos\psi)\zeta^2}\sin\theta d\zeta \approx e_0(\psi)e^{-jkR_0 \cos\psi_0}$$

where $S=kR_0 \sin\theta \sin\theta/2$, and $e_0(\psi)$ is the design aperture function. It can be shown that a second saddlepoint exists on C at $\theta=\frac{1}{3}(\psi+\pi)$ but that its contribution to the integral is zero. Therefore,

$$a_1(\theta) = \frac{e^{-j(3\pi/4)}e_0(\theta/2)}{2k^2 h_0(S)\sin\theta}\sqrt{\frac{R_0 \cos\theta/2}{2\lambda}} \cdot e^{jkR_0(2\cos\theta/2-\cos\psi_0)}. \quad (55)$$

If $F_1+F_3=0$,

$$jhb_1 = \frac{\epsilon_0 \omega a_1}{1+\alpha}. \quad (56)$$

The field components at (r, ϕ, z), a point in the vicinity of the feed or elsewhere, may be obtained by substitution of (55) and (56) into (12) to (14), and into (16) to (18), followed by integration over the spectrum of radiation angles. Dropping terms of order $(1/kR_0)$

$$\begin{aligned} E_r &= -\cos\phi(M_0+M_2+M_{20}) \\ H_r &= Y_0 \sin\phi(M_0+M_2-M_{20}) \\ E_\phi &= \sin\phi(M_0-M_2+M_{20}) \\ H_\phi &= -Y_0\cos\phi(M_0-M_2-M_{20}) \\ E_z &= \cos\phi M_1 \\ H_z &= Y_0 \sin\phi M_1 \end{aligned} \quad (57)$$

where

$$M_0 = jC\int_0^{2\psi_0} \frac{e_0(\theta/2)}{m(S)}\cos^2\theta/2 H_0^{(2)}(\kappa r) \cdot e^{jk(2R_0\cos\theta/2-z\cos\theta)}\sin\theta d\theta \quad (58)$$

$$M_1 = C\int_0^{2\psi_0} \frac{e_0(\theta/2)}{m(S)}\sin^2\theta H_1^{(2)}(\kappa r) \cdot e^{jk(2R_0\cos\theta/2-z\cos\theta)}d\theta \quad (59)$$

$$M_2 = jC\int_0^{2\psi_0} \frac{e_0(\theta/2)}{m(S)}\sin^2\theta/2 H_2^{(2)}(\kappa r) \cdot e^{jk(2R_0\cos\theta/2-z\cos\theta)}\sin\theta d\theta \quad (60)$$

$$M_{20} = jC\int_0^{2\psi_0} \frac{e_0(\theta/2)}{m(S)}\sin^2\theta/2 H_2^{(2)}(\kappa r)\left(1-j\frac{H_1^{(2)}(S)}{H_2^{(2)}(S)}\right) \cdot e^{jk(2R_0\cos\theta/2-z\cos\theta)}\sin\theta d\theta \quad (61)$$

in which

$$Y_0 = \sqrt{\frac{\epsilon_0}{\mu_0}}$$

$$C = \frac{-\pi R_0}{2\lambda}e^{-jkR_0\cos\psi_0}$$

$$m(S) = \sqrt{\frac{\pi S}{2}}e^{-j\pi/4}H_0^{(2)}(S)e^{jS} \approx 1 \quad \text{for} \quad S \gg 1$$

and $e_0(\theta/2)$ equals the design aperture function $e_0(\psi)$.

A basic similarity will be observed between the solution derived above and those contained in the recent literature [1], [2], [4]. In these references, which deal with the diffraction rather than the transmission case, the Bessel function $J(\kappa r)$ appears instead of $H(\kappa r)^{(2)}$; also there are no terms corresponding to $m(S)$ and M_{20}.

Conclusions

The analysis presented here leads to the conclusion that a gain deterioration of 3 dB results from the use of a circular waveguide feed with circumferential slots. There is additional deterioration if the circumferential spacing of the slots is of the order of a wavelength or greater. It is further shown that the ideal illumination is possible using a feed that radiates both the longitudinal and the azimuthal field components given by (57). The ratio of the components at the excitation radius $r=a$ can be obtained approximately from (17), (18), and (56),

$$\frac{E_\phi}{E_z} \approx \frac{j\tan\phi}{\sin\theta}\left(\frac{\cos\theta}{\kappa a}+\frac{H_1^{(2)\prime}(\kappa a)}{H_1^{(2)}(\kappa a)}\right). \quad (62)$$

In designing a line feed for a particular spherical antenna, it may be sufficient to use (62) together with a calculation of the phase and radiation resistance of individual elements based on geometrical optics and other ad hoc considerations. However, a computer calculation of (57) would indicate the excitation required for a precise design and would be of particular interest near the ends of the feed where (62) breaks down.

Acknowledgment

The author wishes to thank R. F. Millar for reading the manuscript. His comments resulted in numerous improvements in content and presentation. Thanks are also extended to R. W. Breithaupt and R. A. Hurd with whom a number of points were discussed.

References

[1] A. C. Schell, "The diffraction theory of large-aperture spherical reflector antennas," *IEEE Trans. Antennas and Propagation*, vol. AP-11, pp. 428–432, July 1963.

[2] M. Kline and I. W. Kay, *Electromagnetic Theory and Geometrical Optics*. New York: Interscience, 1965, p. 350 ff.

[3] A. Boivin and E. Wolf, "Electromagnetic field in the neighbourhood of the focus of a coherent beam," *Phys. Rev.*, vol. 138, pp. B1561-B1565, June 21, 1965.

[4] M. Gravel and A. Boivin, "Electromagnetic field in focal region of wide angle spherical mirror," presented at the 1966 Meeting of the Optical Society of America.

[5] A. F. Kay, "A line source feed for a spherical reflector," AFCRL Rept. 529, May 1961.

[6] J. A. Stratton, *Electromagnetic Theory*. New York: McGraw-Hill, 1941, p. 349.

[7] S. Silver, *Microwave Antenna Theory and Design*. New York: McGraw-Hill, 1949, p. 88.

A Gregorian Corrector for Spherical Reflectors*

F. S. HOLT† AND E. L. BOUCHE‡

Summary—**The inherent spherical aberration of a spherical reflector antenna is corrected by using an auxiliary Gregorian reflector feed system that rotates about the center of curvature of the reflector. Tests at both *X*- and *K*-band frequencies demonstrate feasibility of the design for wide-angle scanning.**

Introduction

FOR ANTENNAS that must be capable of wide-angle scanning, paraboloidal reflectors of large aperture present serious problems of mechanical and structural design because of the difficulty of maintaining reflector surface tolerances. The use of spherical reflecting surfaces, explored in Air Force Cambridge Research Laboratories investigations since 1949, is feasible for wide-angle scanning if an appropriate feed system is rotated about the center of curvature of a spherical reflector. A system consisting of a spherical reflector having a 10-foot aperture and a Gregorian type of feed, designed and built to test the theory, showed good results at both *X*- and *K*-band frequencies.

Theory

Spherical reflectors, although not perfect focusing devices because of inherent spherical aberration, do nevertheless focus paraxial rays quite well. The paraxial focal point (F in Fig. 1) is located on the reflector axis at a distance half the reflector radius ($R/2$) from the reflector vertex V. Marginal rays do not focus at this point but intersect the reflector axis at some point between the paraxial focal point F and the reflector vertex V, the locus of intersection being determined by the f/D ratio of the reflector, that is, the ratio of the paraxial focal length f to the diameter D of the illuminated aperture.

For the phase error to be kept to $\pm\lambda/16$ in the reflected wavefront, it has been shown[1,2] that the diameter D of the illuminated aperture (KK in Fig. 1) of the spherical reflector must not exceed $256\lambda(f/D)^3$. Fig. 2 shows loss of gain from spherical aberration vs f/D and D/λ, as computed by Roy C. Spencer.[3]

To conserve space and keep the required feed motion to a minimum, most mechanically scanned microwave systems have f/D ratios of about 0.5 or less. If a spherical reflector has an f/D of 0.5 and the maximum phase error is limited to $\pm\lambda/16$, Ashmead and Pippard's relation yields an illuminated aperture of 32λ when a single point source feed is used. For systems whose f/D ratios are 0.5 or less and aperture diameters 32λ or more, some method of correction for spherical aberration is needed if satisfactory antenna gain and beam shape are to be maintained.

Corrector Feed Systems

Devices designed to correct spherical aberration include multiple point sources,[4] phased line sources,[5–8] correcting reflectors[9–11] and correcting lenses. The correcting device described in this paper is a small, shaped, auxiliary reflector with an associated feedhorn. Positioned in the radiating aperture of the spherical reflector so that the effect is that of a Gregorian optical system, it completely corrects the spherical aberration of the large reflector. Since the system is spherically symmetric, it is possible to scan the resulting beam by moving the correcting reflector and its feed as a unit. As this scanning is performed, different sections of the spherical reflector are illuminated. If the spherical reflector subtends a cone of full angle 2β, and the correcting reflector illuminates an area that subtends a cone of full angle 2α, both measurements made with respect to the center of curvature of the sphere (see Fig. 1), then the system has the capability of being scanned through a cone of full angle $2(\beta-\alpha)$ centered on the axis of symmetry of the spherical reflector.

Consider a family of incoming rays parallel to the reflector axis OV and incident on the spherical reflector

* Received April 22, 1963; revised manuscript received August 12, 1963.

† Microwave Physics Laboratory, Air Force Cambridge Research Laboratories, Office of Aerospace Research, USAF, Bedford, Mass.

‡ Physical Optics Group, Technical Operations Research, Inc., Burlington, Mass. Formerly with Air Force Cambridge Research Laboratories, Bedford, Mass.

[1] J. Ashmead and A. B. Pippard, "The use of spherical reflectors as microwave scanning aerials," *J.IEE*, vol. 93, pt. 111A, No. 4, pp. 627–632; 1946.

[2] J. F. Ramsey and J. A. C. Jackson, "Wide angle scanning performance of mirror aerials," *Marconi Rev.*, vol. XIX, pp. 119–140; 3rd Quarter, 1956.

[3] R. C. Spencer, "Theoretical Analysis of the Effect of Spherical Aberration on Gain," AF Cambridge Research Ctr., Bedford, Mass., AFCRC Tech. Rpt. No. E5082; December, 1951.

[4] RCA, private communication; 1960.

[5] R. C. Spencer, C. J. Sletten, and J. E. Walsh, "Correction of spherical aberration by a phased line source," *Proc. NEC*, vol. 5, pp. 320–333; 1950.

[6] R. R. MacMillan and A. W. Love, "Research Directed Toward Theoretical and Experimental Investigation of Corrected Line Source Feeds," Wiley Electronics Co., Phoenix, Ariz., A. F. Cambridge Research Ctr., Bedford, Mass., Rept. No. AFCRC-TR-60-122 (Rept. No. 93); February, 1960.

[7] E. E. Altshuler, "A Periodic Structure of Cylindrical Posts in a Rectangular Waveguide," A. F. Cambridge Research Labs., Bedford, Mass., Tech. Rept. No. AFCRL 175; April, 1961.

[8] A. F. Kay, "A Line Source Feed for a Spherical Reflector," AF Cambridge Research Labs., Bedford, Mass., Tech. Rept. No. AFCRL 529 (TRG, Somerville, Mass., Rept. No. 131); May, 1961.

[9] A. K. Head, "A new form of giant radio telescope," *Nature*, vol. 179, pp. 692–693; April, 1957.

[10] F. S. Holt and E. L. Bouche, "Spherical Cassegrain scanning antenna," *Proc. 9th Ann. Symp. USAF Antenna R & D Program*, Monticello, Ill.; October, 1959.

[11] Philco Corp., Palo Alto, Calif., private communication; June, 1961.

Reprinted from *IEEE Trans. Antennas Propagat.*, vol. AP-12, pp. 44–47, Jan. 1964.

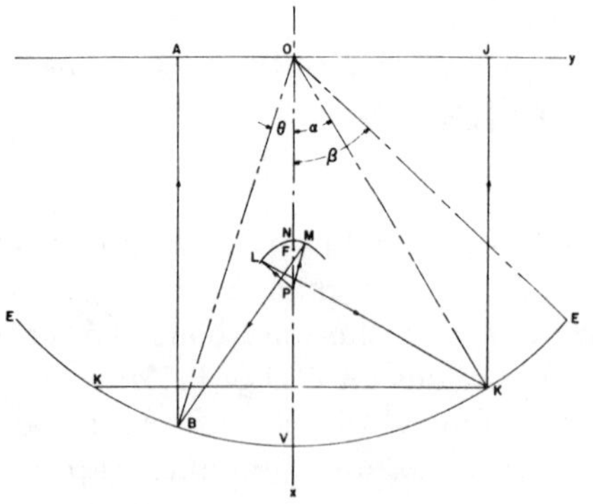

Fig. 1—Spherical Gregorian scanner.

F paraxial focus
V reflector vertex
EE reflector aperture
AOJ reference wavefront
$2(\beta-\alpha)$ total scan angle
P corrector focus
N corrector vertex
KK corrected aperture

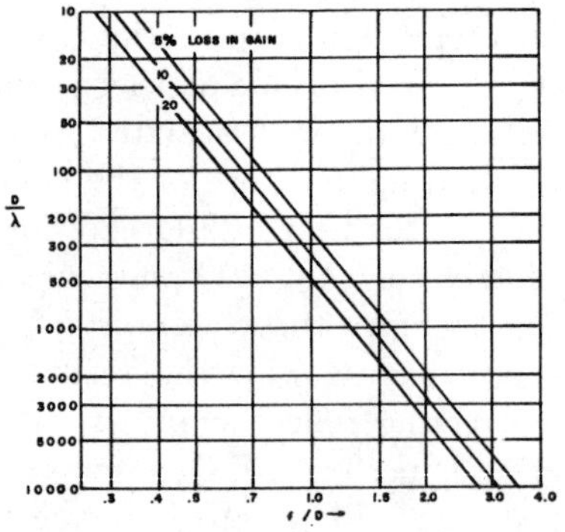

Fig. 2—Loss in gain for spherical reflectors.

(see Fig. 1). Because of symmetry it is sufficient to consider rays only in the xy plane. A typical ray is $\overline{AB}$. The general equation for the family of reflected rays $\overline{BM}$, expressed in terms of the parameters θ and R is:

$$(x - R\cos\theta)\tan 2\theta = (y - R\sin\theta). \tag{1}$$

The envelope of the family of rays reflected from the spherical surface is known as the caustic surface[12] and is given by

$$x_c = \frac{R(3\cos\theta - \cos 3\theta)}{4} \tag{2a}$$

$$y_c = \frac{R(3\sin\theta - \sin 3\theta)}{4}\cdot \tag{2b}$$

A ray trace of the caustic region is shown in Fig. 3. Energy density is high on the caustic surface and maximum near the paraxial focal point F. Inside the region bounded by the caustic surface and the cone of reflected marginal rays, two or more reflected rays pass through

[12] A degenerate caustic lies along the paraxial axis OV. When *line source* correctors are used as feeds with spherical reflectors they are placed along this axis.

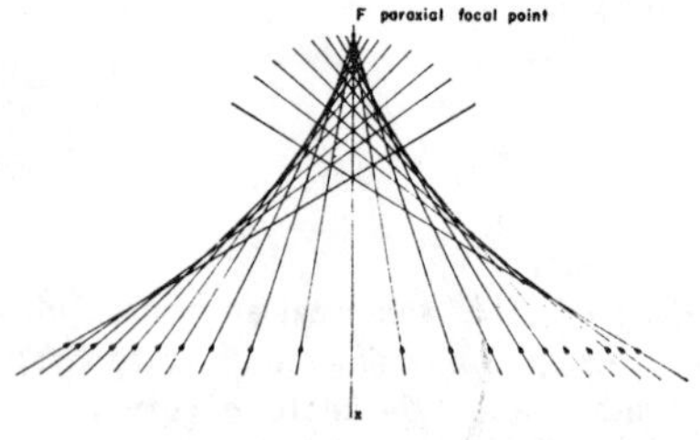

Fig. 3—Ray trace of the caustic region for a plane wave reflected from a spherical surface.

each point; outside this region, only one ray passes through each point. In ray-tracing design procedures each point of a reflecting surface permits only single-valued phase correction. Therefore, except under special conditions, the corrector must be located outside of the caustic region.

The equations for the correcting reflector surfaces are obtained by requiring all rays to have equal path lengths as measured from a reference wavefront passing through O (Fig. 1), that is,

$$\overline{AB} + \overline{BM} + \overline{MP} = \overline{OV} + \overline{VN} + \overline{NP}. \tag{3}$$

Defined with respect to a sphere of unit radius, the various coordinates and the surfaces or points they define are as follows:

$(c, s) = (\cos\theta, \sin\theta)$ spherical reflector

(u, v) corrector surface

$(n, 0)$ vertex N of correcting reflector

$(p, 0)$ focus P of corrector feedhorn.

The coordinates of the paraxial focus F become $(\frac{1}{2}, 0)$, and (3) can now be expressed as

$$\cos\theta + \left[(u-c)^2 + (v-s)^2\right]^{1/2} + \left[(u-p)^2 + v^2\right]^{1/2} = 2 - 2n + p. \tag{4}$$

Calculation of the slope of the ray $\overline{BM}$ yields

$$\tan 2\theta = \frac{v - s}{u - c}\cdot \tag{5}$$

Simultaneous solution of (4) and (5) for the coordinates (u, v) of the corrector surface yields

$$u = c - \frac{\left[c^2 - 4(1-n)c - (1+p^2) + (2+p-2n)^2\right](2c^2-1)}{4(pc^2 - c - n + 1)} \tag{6}$$

and

$$v = \pm\frac{(2cu - 1)s}{2c^2 - 1}, \tag{7}$$

where

$$n \le \frac{1}{2},$$

$$p \ge \frac{1}{2}.$$

Three special cases are of interest.

Case I: Given $n=\frac{1}{2}$, $p>\frac{1}{2}$, then

$$u = c - \frac{(c^2 - 2c + 2p)(2c^2 - 1)}{4(pc^2 - c) + 2}. \tag{8}$$

Case II: Given $n<\frac{1}{2}$, $p=\frac{1}{2}$, then

$$u=c-\frac{[c^2-4(1-n)c+(2.5-2n)^2-1.25](2c^2-1)}{2c^2+4(1-n-c)}. \tag{9}$$

Case III: Given $n=p=\frac{1}{2}$, then

$$u = (1 - c)c + 0.5. \tag{10}$$

The caustic and four specific examples of these cases are plotted in Fig. 4. Note the teardrop shapes of correctors A, B, and C. The points at which these surfaces intersect the caustic determine the geometric limits of usable corrector surface. Further restrictions on the extent of usable corrector surface are imposed by the practical difficulties of attaining proper illumination over solid angles greater than 2π. For a given feed position $p>\frac{1}{2}$ and a given angular aperture 2α on the sphere (Fig. 1), the corrector with minimum diameter and surface area is obtained by placing the corrector vertex at the paraxial focal point $F=(\frac{1}{2}, 0)$. Case I correctors (curve B), $n=\frac{1}{2}$, $p>\frac{1}{2}$, are of this type.

Case III (curve D), $n=p=\frac{1}{2}$, has serious disadvantages. Since the corrector surface lies within the caustic, incident rays are reflected from the outside of the corrector surface and antiphase components are present. Ray blocking by the corrector permits aperture correction only within approximately the range $20° \le \theta \le 30°$.

Experimental Model

The experimental model built to conform to Case I has a spherical reflector whose radius of curvature and aperture diameter are both 10 feet, that is, it is a 60° cap from the sphere. The correcting reflector has its vertex at the paraxial focus (to keep the required corrector diameter as small as possible) and has its focus at a point 1 foot from the paraxial focus (Figs. 5 and 6). This is equivalent to setting $n=0.5$ and $p=0.6$

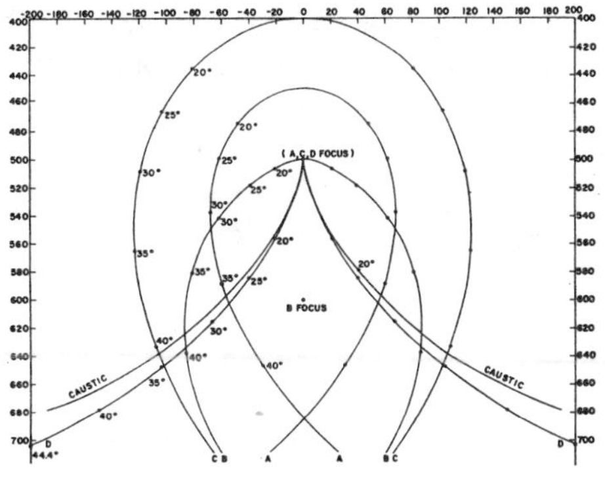

$$u = c - \frac{[c^2 - 4(1-n)c - (1+p^2) + (2+p-2n)^2](2c^2-1)}{4(pc^2-c-n+1)},$$

$$v = \pm\frac{(2cu-1)s}{2c^2-1};$$

where $c = \cos\theta$; $s = \sin\theta$; $R = 1000$

Curve	Corrector	
	Vertex n	Focus p
A	0.45	0.5
B	0.5	0.6
C	0.40	0.5
D	0.5	0.5

Fig. 4—Gregorian corrector shapes.

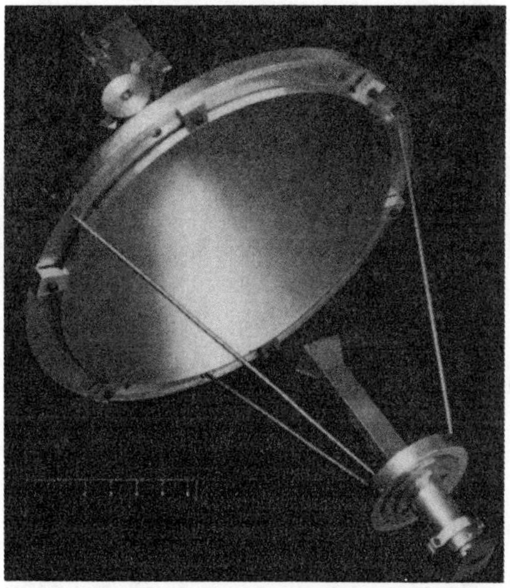

Fig. 5—The experimental Gregorian corrector showing the correcting reflector and X-band feedhorn ($n=0.5$; $p=0.6$). This corrector system illuminates the full aperture of a reflector whose radius of curvature and aperture diameter are 10 feet.

Fig. 6—The experimental 10 foot spherical reflector with the Gregorian correcting reflector and X-band feedhorn shown in Fig. 5.

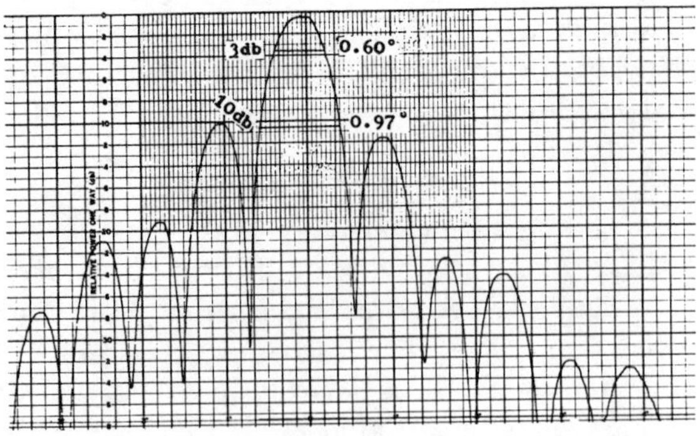

(a)

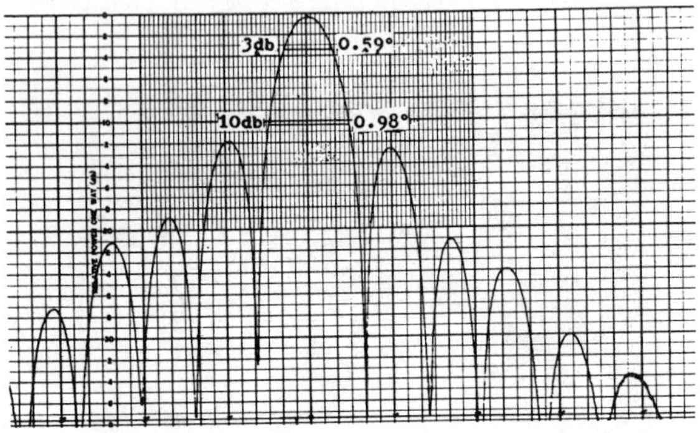

(b)

Fig. 7—Principal plane patterns of 10 foot spherical reflector fed by corrector and horn shown in Fig. 6 (9340 Mc; $n=0.5$, $p=0.6$). (a) H-plane. (b) E-plane.

(6). The corrector design equations, referenced to the center of curvature of the sphere, were

$$x = 120u$$

$$= 120\left[c - \frac{(c^2 - 2c + 1.6)(2c^2 - 1)}{4(0.6c^2 - c) + 2}\right] \text{ inches} \qquad (11)$$

$$y = \pm\, 120v = \pm\, 120\left[\frac{(2uc - 1)s}{2c^2 - 1}\right] \text{ inches.} \qquad (12)$$

The illumination cone was selected to have a full angle of $2\alpha = 60°$. Hence $f/D = 0.5$ and in (11) and (12),

$$-30° \leq \theta \leq 30°. \qquad (13)$$

For $\theta = \alpha = 30°$ (the marginal ray), (12) yields $y = Rv_{\max} = 7.5$ inches, and hence the diameter of the correcting reflector is 15 inches. Aperture blocking in this case is only 1.5 per cent. Since $\alpha = \beta$, it is evident that the entire 10-foot-diameter aperture of the spherical reflector is illuminated when the correcting reflector is aligned on the x axis. In this design, therefore, the gain and beamwidth are not preserved as the system is scanned.

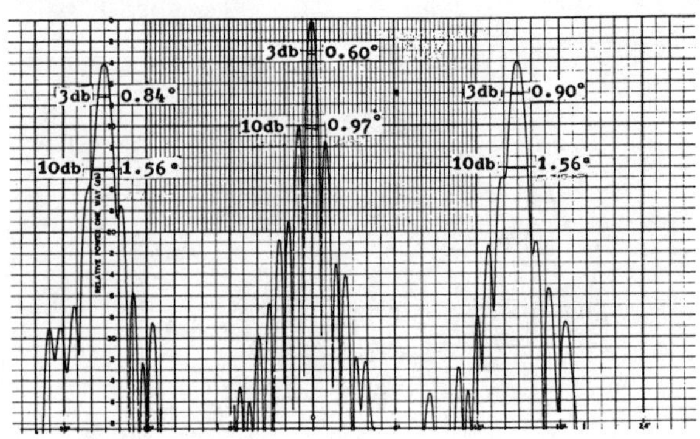

Fig. 8—H-plane scanning patterns obtained with system shown in Fig. 6 (9340 Mc; $n=0.5$, $p=0.6$; scan $= \pm 15°$).

Theoretical calculations for X-band predicted a half-power beamwidth of 0.54° and a sidelobe level of −11.6 db; experimental results showed a beamwidth of 0.59° for a 96λ aperture and first sidelobe levels in the range from −11 to −12 db. At K-band, experimental results showed a beamwidth of 0.22° for a 245λ aperture, with sidelobe levels again in the range from −11 to −12 db. (See Figs. 7(a), 7(b).)

It is interesting that the pattern holds up so well when the illumination from the corrector system is scanned off the sphere. The H-plane patterns in Fig. 8 show the loss in gain and rise in average sidelobe level as the beam is scanned in the H-plane to ±15° from its axial position.

Lower sidelobe levels are difficult to achieve with the spherical Gregorian system because the reflector geometry required to correct for spherical aberration causes a high inverse illumination taper across the antenna aperture. In most of the systems that have been investigated, this inverse taper is about +9 db. The resulting beamwidth is of course narrower than can be obtained with a uniformly illuminated aperture, but the increase in resolving power is obtained at the cost of a decrease in gain and a rise in sidelobe levels.

Studies of the Focal Region of a Spherical Reflector: Geometric Optics

ROY C. SPENCER, FELLOW, IEEE, AND GEOFFREY HYDE, MEMBER, IEEE

Abstract—**The classical optics of the spherical reflector is reviewed and extended with particular application to the focal region. The caustic surface, axial caustic, and circle of least confusion are described in detail. The manner in which the reflected rays combine in the focal region is then developed. In general, three reflected rays intersect at any field point P (P on the interior of the caustic surface, but off the axial caustic). The points of incidence of the three rays and their optical path lengths from a common reference plane are calculated, providing the necessary phase and divergence information for the focal region.**

Manuscript received October 17, 1967; revised January 15, 1968. This work was carried on mainly at the Missile and Surface Radar Division, Radio Corporation of America, Moorestown, N. J., and supported in part by the USAF Cambridge Research Laboratories, under Contract AF19(628)2758.

R. C. Spencer is a Consulting Physicist, 102 Devon Road, Cinnaminson, N. J.

G. Hyde is with the Missile and Surface Radar Division, Radio Corporation of America, Moorestown, N. J.

Introduction

ONE MAY ASK why the interest in spherical reflectors for large antennas, when the paraboloidal reflector, with or without the Cassegrainian subreflector, provides an excellent pencil beam antenna pattern. The answer is that the paraboloid is severely handicapped in angular scanning by the large mechanical moment of inertia of the rotating structure. Moreover, angular scanning by displacing the feed alone is thwarted by its large coma and astigmatism.

By contrast, the spherical reflector is perfectly symmetrical, though plagued by inherent spherical aberration. Ashmead and Pippard[1] showed that spherical reflectors could be used by illuminating them over the region which departed from the surface of a paraboloid by less than $\lambda/8$. Spencer[2] showed that the fractional loss in gain $\Delta G/G$ approximates

Reprinted from *IEEE Trans. Antennas Propagat.*, vol. AP-16, pp. 317–324, May 1968.

$\overline{\phi^2}$, the minimum mean-squared phase error across the aperture, with ϕ weighted by the amplitude function. For constant amplitude this is related to the Strehl definition (1894) and to Marechal's analysis (1944), discussed by Wolf.[3] Ruze[4] has reviewed antenna applications. Spencer[5] applied the criteria to a spherical reflector of diameter $d=N\lambda$. For uniform illumination

$$N = 1093.3\left(\frac{f}{d}\right)^3\left(\frac{\Delta G}{G}\right)^{\frac{1}{2}}, \quad f = R/2.$$

Thus for a 10 percent loss in gain, or $\frac{1}{2}$ dB, $N=340\,(f/d)^3$. The numerical constant changes slightly for tapered illumination.

Based on the idea that all parallel rays incident on a sphere cross the axis after reflection, Spencer, Sletten, and Walsh[6] proposed and tested a specially phased radial line source feed that corrects for spherical aberration. Kay[7] designed such a feed for the world's largest filled-aperture antenna, the 1000-foot diameter spherical reflector at Arecibo, Puerto Rico.[8] The reflector is fixed, but scanning in any direction out to 20° from the zenith is accomplished by moving the line source. Schell[9] reexamined the theory using physical optics.

Except for shadowing problems, a matching transverse feed system can be designed across any section of the focal region that intercepts most of the energy. Thus Sletten and Mavroides[10] partially corrected the amplitude and phase with a transverse cluster of feed horns, while Clasen[11] employed a transmitter approach with similar results. At the suggestion of Sletten of the USAF Cambridge Research Laboratories the transverse focal region was further studied by Spencer and Hyde,[12]–[14] who concluded that 1) polarization must be considered; 2) the method of stationary phase was adequate; and 3) for a field point within the caustic surface, but off the axis, the electric field is the sum of at most three contributions, corresponding to the three naturally reflected rays that intersect at the field point.

In view of the many contributions of classical optics to focal region studies, this paper reviews certain portions of geometric optics and Huygens' wave theory, applying and extending them as required.

Elementary Optics of the Sphere

Symmetry

Certain properties of the sphere are obvious, such as its angular symmetry, the constancy of its radius of curvature, and the fact that all normals pass through its center 0. In Fig. 1 it is convenient to choose an axis OV in the direction of the incoming parallel bundle of rays. A typical ray AB is separated from the axis by $OA=h=\sin\alpha$, for a normalized radius $R=1$. It is incident on the sphere at a point B at an angle of incidence $\alpha=\angle OBA$. The incident ray and normal at B define the plane of incidence which cuts the sphere in a great circle and includes both the reflected ray BA' and the axis OV. Therefore, reflection from a sphere reduces to the problem of reflection from a great circle.

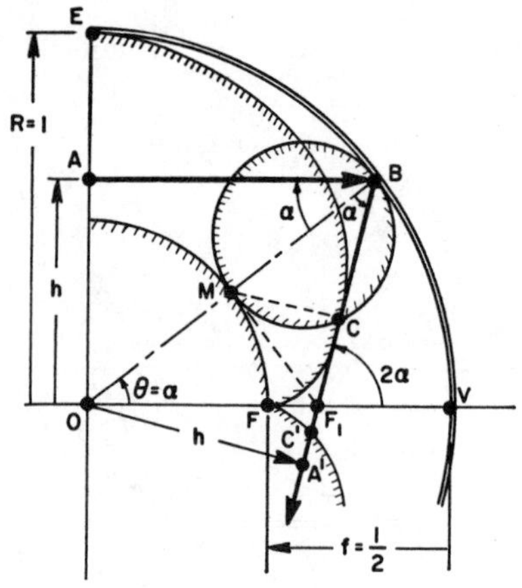

Fig. 1. Reflection of ray from sphere.

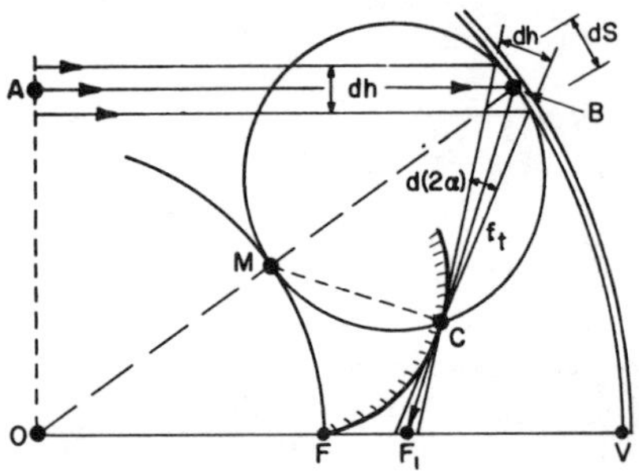

Fig. 2. Convergence of pencil beam.

Consequently, each reflected ray crosses the axis at some point F_1 so that there are no skew rays (rays that miss the axis). From the law of reflection $\angle OBA'=\angle OBA=\alpha$, and $OA'=OA=h$.

Classical Aberrations of the Sphere

Since triangle OF_1B in Fig. 1 is isosceles,

$$BF_1 = OF_1 = \tfrac{1}{2}\sec\alpha. \tag{1}$$

When α is small, this reduces to the *paraxial focal length*,

$$VF = OF = f = 1/2.$$

The shift in the intercept with the central ray

$$\Delta F = FF_1 = \tfrac{1}{2}(\sec\alpha - 1) \tag{2}$$

is the *longitudinal spherical aberration*.

Fig. 2 depicts a narrow bundle of parallel rays of differential height dh incident on the reflecting spherical surface element dS at B. This element exhibits classical astigmatism. Thus rays in the strip *in the plane of incidence* focus at C, determining the *tangential* focal length

$$f_t = BC = BM\cos\alpha = f\cos\alpha. \tag{3}$$

Equation (3) is true since $dh/f_t=d(2\alpha)$, and $dh/d\alpha=\cos\alpha$.

On the other hand, parallel rays in a strip *perpendicular* to this plane have the same h and pass through the same point F_1, determining the *sagittal focal length.*

$$f_s = BF_1 = f\sec\alpha. \qquad (1),(4)$$

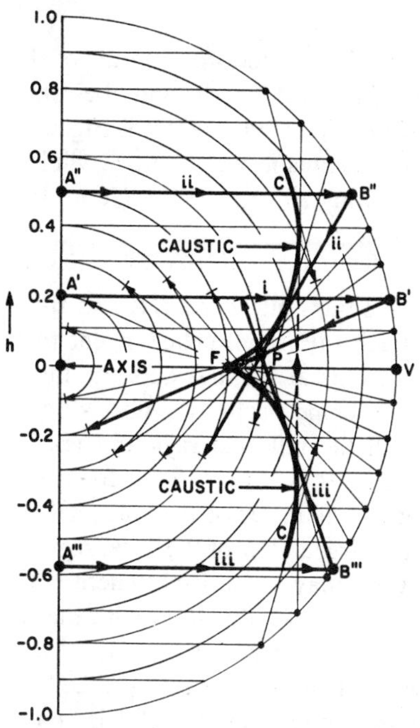

Fig. 3. Caustic by ray tracing.

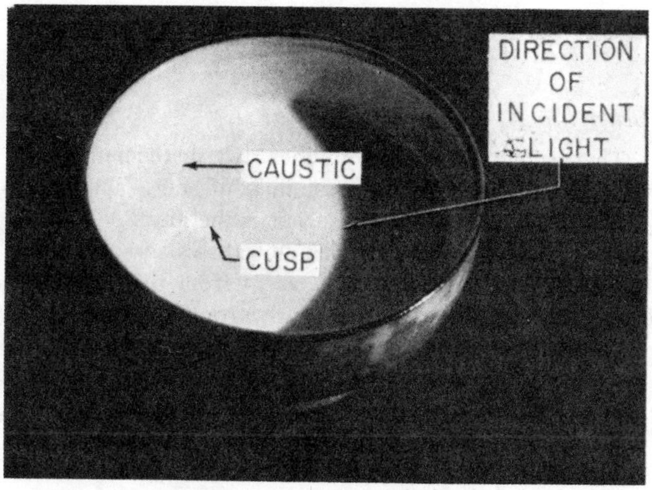

Fig. 4. Caustic in a bowl of milk.

The difference between the sagittal and tangential focal lengths is the *astigmatism* of the reflector surface element dS.

$$CF_1 = f_s - f_t = f(\sec\alpha - \cos\alpha). \tag{5}$$

From (3) and (4), $f=\sqrt{f_t f_s}$, the geometric mean of f_t and f_s.

Caustic Surfaces[15]–[17]

The locus of the tangential focus C for various values of h is the *caustic FCE* (see Fig. 1). Differential geometry states that it is the envelope of reflected rays. This is illustrated by drawing a series of parallel incident rays as in Fig. 3. Each individual reflected ray is tangent to the same concentric circle of radius h as is the incident ray. All reflected rays are then tangent to the caustic curve or envelope C. When the curve C is revolved about the axis OV there is formed the caustic surface of revolution. There is also the axial caustic OV.

Caustic surfaces may be observed in the reflections from the inner surface of a napkin ring or drinking glass. The "caustic in the coffee cup" is illustrated by the photograph of a bowl of milk in Fig. 4.

Kinematically, the caustic FCE in Fig. 1 is an epicycloid generated by a point C on the rim of a circle of diameter $BM=f$, rolling on a circle of radius $OM=f$. This particular epicycloid is called the *nephroid* because of its kidney shape. These concepts occur in the design of gears and gear teeth.

Related Properties of the Spherical Reflector

Use of Complex Number Notation

Complex number notation aids in deriving the optical properties of the caustic. In Fig. 5(a) the directions of the *incident ray, normal,* and *reflected ray* are, respectively, 1, $e^{i\alpha}$, and $-e^{2i\alpha}$. The position vector $\mathbf{OP}$ of a field point $P(r_p, \theta_p)$ on the reflected ray is then

$$\mathbf{r}_p = \mathbf{OB} + \mathbf{BP} = e^{i\alpha} - pe^{2i\alpha}, \quad p = BP. \tag{6}$$

The imaginary part of (6) is zero when p is at F_1, yielding

$$p = \tfrac{1}{2}\sec\alpha \tag{7}$$

which checks (1) for f_s. On differentiating (6) and arranging in components perpendicular to and parallel to the reflected ray

$$\frac{d\mathbf{r}}{d\alpha} = e^{2i\alpha}[i(\cos\alpha - 2p) + (\sin\alpha - dp/d\alpha)]. \tag{8}$$

When P is a point C on the caustic, $d\mathbf{r}$ is parallel to $e^{2i\alpha}$ and the imaginary part within the brackets of (8) is zero, whence

$$p = \tfrac{1}{2}\cos\alpha \tag{9}$$

which checks (3) for f_t. The real part is $ds/d\alpha$, the rate of change of s, the arc length along the caustic. On substituting the derivative of (9) for $dp/d\alpha$, we obtain

$$\frac{ds}{d\alpha} = \frac{3}{2}\sin\alpha. \tag{10}$$

Coordinates of the Caustic

The classical parametric equations for a point on the caustic are obtained from Fig. 5(b) by inspection,

$$\mathbf{r}_c = \mathbf{OB'} + \mathbf{B'C} = \tfrac{3}{4}e^{i\alpha} - \tfrac{1}{4}e^{3i\alpha} \tag{11a}$$

$$= z_c + ix_c. \tag{11b}$$

On equating real and imaginary parts

$$z_c = \tfrac{1}{4}[3\cos\alpha - \cos 3\alpha] = \tfrac{1}{2}\cos\alpha[1 + 2\sin^2\alpha] \tag{12}$$

$$x_c = \tfrac{1}{4}[3\sin\alpha - \sin 3\alpha] = \sin^3\alpha. \tag{13}$$

Also

$$r_c^2 = x_c^2 + y_c^2 = (1 + 3\sin^2\alpha)/4. \tag{14}$$

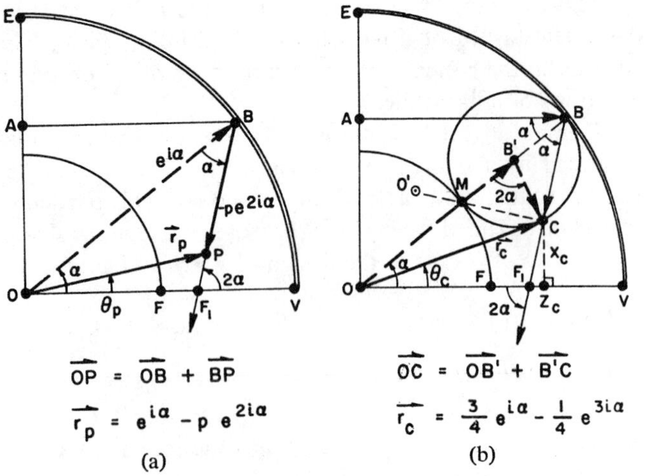

Fig. 5. Use of complex variable.

Arc Length

The arc length FC from the tip of the caustic is the integral of ds in (10).

$$s = \int_F^C ds = \tfrac{3}{2}(1 - \cos\alpha). \tag{15a}$$

The total arc length FCE is 3/2. If we let $s+s'=3/2$, then

$$s' = \tfrac{3}{2}\cos\alpha \tag{15b}$$

which is Whewell's equation.

Normal and Radius of Curvature

The normal to the caustic at C is perpendicular to BC and hence intersects M, the point of contact of the rolling circle of Figure 5(b). The radius of curvature ρ of the caustic at point C is the rate of change of arc length with the angle of the tangent, $\phi=2\alpha$.

$$\rho = \frac{ds}{d\phi} = \frac{1}{2}\frac{ds}{d\alpha} = \frac{3}{4}\sin\alpha. \tag{16}$$

This is $CO'=(3/2)CM$. One can now combine (15b) and (16) into the classical Cesaro equation

$$4\rho^2 + s'^2 = 9/4. \tag{17}$$

Anticaustic

The spherical reflector is the anticaustic of its caustic surface (the nephroid). Further, classical optics shows that

$$AB + BC + \widehat{CF} = OV + FV = 3/2 \tag{18}$$

where $\widehat{CF}=s$ (in Fig. 1).

Transformation of Variables

Because of the physical significance it is useful to express formulas in terms of h or s. Note that $h=\sin\alpha$ varies linearly across the aperture.

To transform to arc length s, transpose (15a)

$$\cos\alpha = 1 - \tfrac{2}{3}s = 1 - 2S \tag{19}$$

where $S=s/3$.

Table I lists various functions in terms of the variables α, h, and S. Note that focal lengths and aberrations are simple rational functions of the parameter S. Fig. 6 shows how the coordinates of the caustic vary with S, while Fig. 7 shows r_c and θ_c vs. S, using an expanded scale in the focal region.

Circle of Least Confusion

Fig. 8 features the central ray OV and the two marginal rays A_M, B_M. Each ray is reflected tangent to the caustic at C_M, crosses the axis at F_1, and cuts the far branch of the caustic at C_M'. Geometrically, $C_M'C_M'$ is the smallest circle that encloses all the reflected rays and is called the *circle of least confusion*. It is a likely area for a transverse feed. $C_MC_M=2h^3$ is the diameter of the marginal focus.

Fig. 9 is an enlarged view of the focal region. An exact expression for C_1C_1 may be derived, where C_1C_1 is the intersection of the sphere of radius OF_1 with the caustic surface. Now $C_1C_1=2h_1^3$. And it may be shown[12] that $h_1=t/(3)^{1/2}$, where $t=\tan\alpha$. Thus $C_1C_1=2t^3/(27)^{1/2}$. The diameter of the circle of least confusion $C_M'C_M'$ is somewhat smaller than C_1C_1. Again, $C_M'C_M'=2h'^3$. For small values of h, and thus of α and θ, $h' \doteq t/2$, whence $C_M'C_M' \doteq t^3/4$. But for $h=1/2$, h' is already about 3 percent larger than $t/2$.

Rays and Wavefronts

General

The techniques of both physics and mathematics have contributed to our understanding of rays and wavefronts.[18]–[20] In optics a wavefront is the locus of *constant phase* and in an isotropic medium crosses the rays at an angle of 90°. An equivalent concept from the differential geometry of surfaces is that the wavefront is the *orthogonal trajectory* of a set of straight lines (the rays). Such a bundle of rays is said to be orthotomic. If the bundle passes through a common point it is also *homocentric*.

The orthotomic property persists after reflection from, or refraction through, any number of smooth curved surfaces. However, the homocentricity is usually destroyed, because instead of passing through a single focal point, the rays are tangent to *two caustic surfaces* which are the *envelopes* of the rays. The Huygens wavefronts may be generated as involutes of the caustic surfaces. Consequently, wavefronts are *parallel curves* with a *common evolute* (the caustic surfaces).

One of the caustic surfaces of the sphere is the surface of revolution FCE about the axis of Fig. 1. The second surface has degenerated to the axis OV.

Three Images of a Spherical Reflector

Fig. 3 shows that as many as three naturally reflected rays i, ii, iii may intersect a field point $P(r_p, \theta_p)$ lying within the caustic. It follows that an observer at P would see light coming from reflector areas in the vicinity of the points of incidence B', B'', B''', which can be shown to be points of stationary phase.

Reversing the process, an observer at a distance will see images of P in line with the points B', B'', B'''. For example,

TABLE I

OPTICAL PROPERTIES OF SPHERICAL REFLECTOR IN TERMS OF α, h, OR PARAMETER S

Function	Angle, α	$h = \sin\alpha$	$S = s/3$
coordinates			
$OA = x$	$\sin\alpha$	h	$2[S(1-S)]^{\frac{1}{2}}$
$AB = z$	$\cos\alpha$	$(1-h^2)^{\frac{1}{2}}$	$(1-2S)$
focal lengths			
$BF_1 = OF_1 = f_s$	$\frac{1}{2}\sec\alpha$	$\dfrac{1}{2(1-h^2)^{\frac{1}{2}}}$	$\dfrac{1}{2(1-2S)}$
$BC = f_t$	$\frac{1}{2}\cos\alpha$	$\frac{1}{2}(1-h^2)^{\frac{1}{2}}$	$\frac{1}{2}(1-2S)$
aberrat ons			
FF_1 = spherical aberration	$\frac{1}{2}(\sec\alpha - 1)$	$\dfrac{1}{2}\left[\dfrac{1}{(1-h^2)^{\frac{1}{2}}} - 1\right]$	$\dfrac{S}{(1-2S)}$
$CF_1 = f_s - f_t$ = astigmatism	$\dfrac{\sin^2\alpha}{2\cos\alpha}$	$\dfrac{h^2}{2(1-h^2)^{\frac{1}{2}}}$	$\dfrac{2S(1-S)}{(1-2S)}$
caustic			
x_c	$\sin^3\alpha$	h^3	$8[S(1-S)]^{3/2}$
$2z_c$	$\cos\alpha(1+2\sin^2\alpha)$	$(1-h^2)^{\frac{1}{2}}(1+2h^2)$	$(1-2S)[1+8S(1-S)]$
$4r_c^2 = 4(x_c^2 + z_c^2)$	$(1+3\sin^2\alpha)$	$(1+3h^2)$	$[1+12S(1-S)]$
$\sin\theta_c = x_c/r_c$	$\dfrac{\sin^3\alpha}{(1+3\sin^2\alpha)^{\frac{1}{2}}}$	$\dfrac{h^3}{(1+3h^2)^{\frac{1}{2}}}$	$\dfrac{4[S(1-S)]^{3/2}}{[1+12S(1-S)]^{\frac{1}{2}}}$
ρ = radius of curvature	$\frac{3}{4}\sin\alpha$	$\frac{3}{4}h$	$\frac{3}{2}[S(1-S)]^{\frac{1}{2}}$
trigonometry	$\cos 2\alpha$	$(1-2h^2)$	$[1-8S(1-S)]^{\frac{1}{2}}$

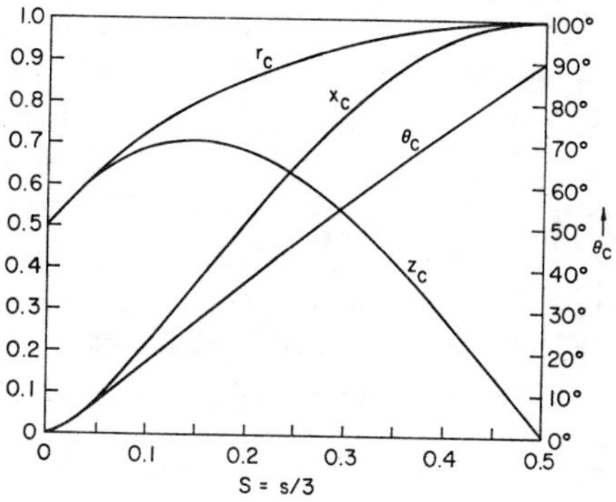

Fig. 6. Coordinates of caustic vs. parameter S.

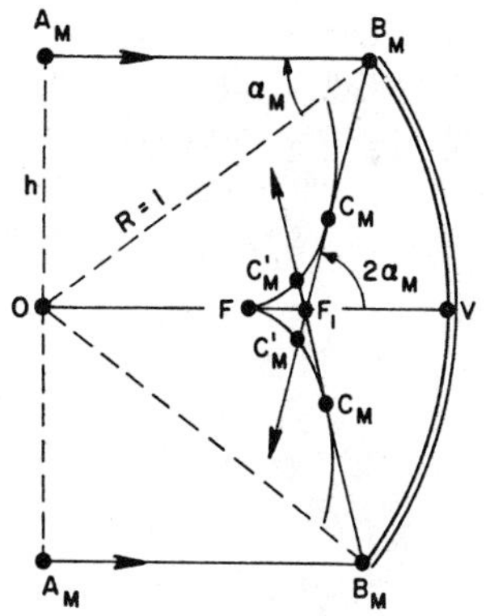

Fig. 8. Focal region of sphere.

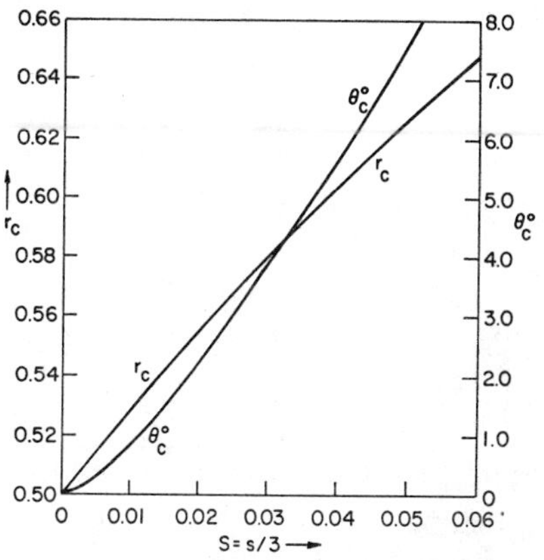

Fig 7. Polar coordinates of caustic (enlarged scale).

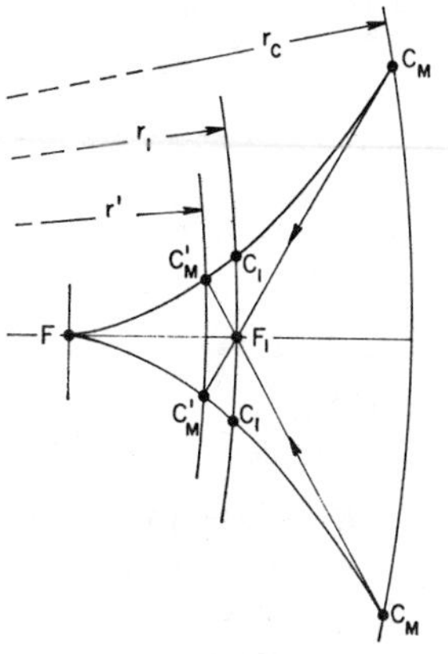

Fig. 9. Caustic region—circle of least confusion.

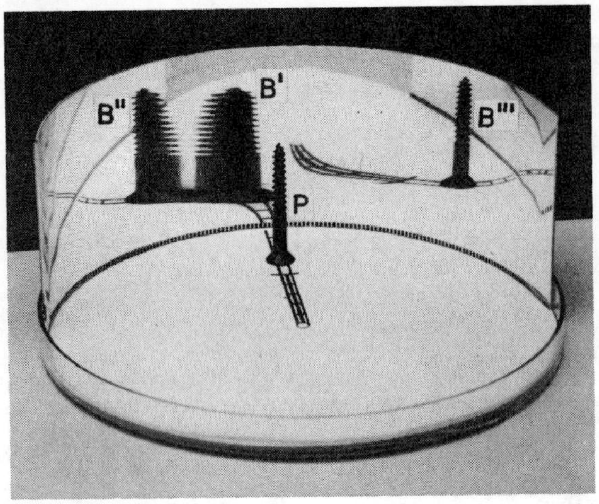

Fig. 10. Three images, B', B'', B''', of screw P in cylindrical reflector.

Fig. 10 is a photograph of the three images of a small wood screw in a cylindrical aluminum mirror. The variation in horizontal magnifications is marked, but the vertical magnification is unity because of the cylindrical mirror. The reader can examine such images by placing a pin inside the far rim of a cylindrical water glass or polished cylinder.

The amplitude of each of the contributions at P can be evaluated by the method of *stationary phase*, or by the method of *divergence*,[21],[22] or by the equivalent method of *optical image magnification*. An axiom in geometric optics states that "the brightness of a luminous source is unchanged when observed through a nonlossy optical system." Therefore, the *intensity* of illumination at P is proportional to the solid angle subtended at P by the image of a distant source, such as the sun. Two astigmatic images are formed at C and F_1 of Fig. 1. The solid angle magnification M is the product of the angular magnifications in the two principal planes,

$$M = \left(\frac{BC}{PC}\right)\left(\frac{BF_1}{PF_1}\right) = \left(\frac{h - x_c}{x_c - x_p}\right)\frac{h}{x_p}, \quad x_c = h^3. \tag{20}$$

For cylinders the second factor is unity. The intensity at P is the intensity at B times M. Equivalent formulas are obtained from divergence theory where the divergence $D = 1/M$. According to physical optics, the amplitude at P is the sum of three amplitudes with appropriate phase factors

$$\sum M^{\frac{1}{2}} e^{i\phi} \tag{21}$$

The M's are calculated using optical path lengths (20). The phases can also be calculated from optical path lengths, although a graphic method is presented next.

Phases from Wavefront Construction

The wavefronts of Fig. 11 were made by attaching an inking pen to a steel tape that was unrolled from a precision-made caustic curve. The increments of path length are $\Delta s = 0.003$ of the radius ($\Delta S = 0.001$). They provide a convenient linear scale along the caustic OC and the axis. The heavy triangular shaped wavefront for $S = 0.02$ illustrates the region of general interest. Note that the wavefront folds back on itself at the caustics. In fact, one can visualize the family of wavefronts as the loci of equidistant points of a straight rule that rolls without slipping about the caustic (a variety of *roulette*).

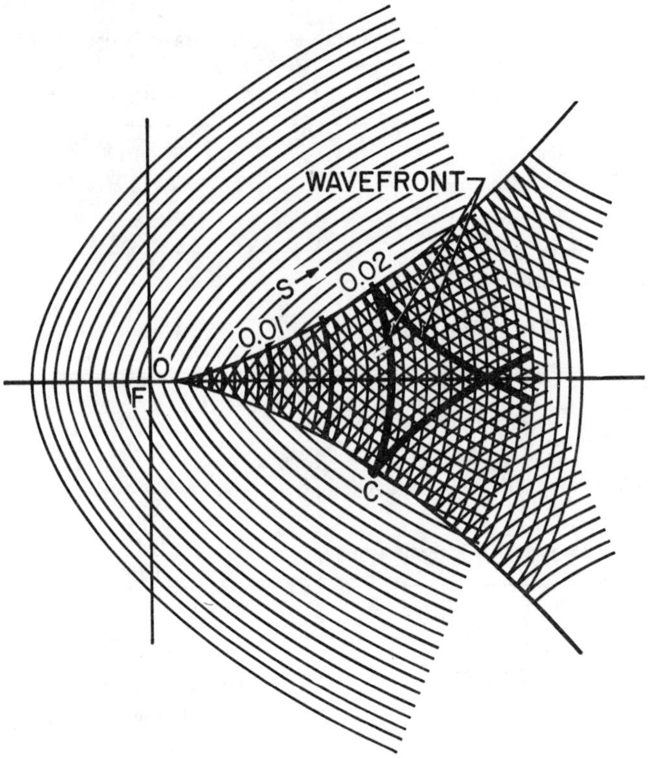

Fig. 11. Wavefronts generated with caustic template.

Such a chart is informative, and might have provided fairly good phase information. However, the more accurate computer calculation of path lengths was used, as described next.

Path Length and Coordinate Calculations

Fig. 12 shows a spherical reflector with unit radius $R = 1$, and a point $P(r_p, \theta_p)$ on a feed surface of constant radius r_p. We proceed to calculate the path length

$$\begin{aligned} l &= AB + BP = AB + BA' - PA' \\ &= 2AB - PA' = 2l_1 - l_p \end{aligned} \tag{22}$$

where

$$\begin{aligned} l_1 &= AB = (1 - h^2)^{\frac{1}{2}} \\ l_p &= (r_p{}^2 - h^2)^{\frac{1}{2}}. \end{aligned} \tag{23}$$

Fig. 13 is a plot of l vs. h with r_p as a parameter. Each curve is even and resembles a fourth degree function of h. There may be as many as four values of h with the same path length.

On taking the derivative of l with r_p constant

$$\frac{dl}{dh} = -h\left[\frac{2}{l_1} - \frac{1}{l_p}\right]. \tag{24}$$

The derivative is zero when $h = 0$ and when $l_p = l_1/2$. This latter condition indicates a point C on the caustic which is the tangential focal point by (3).

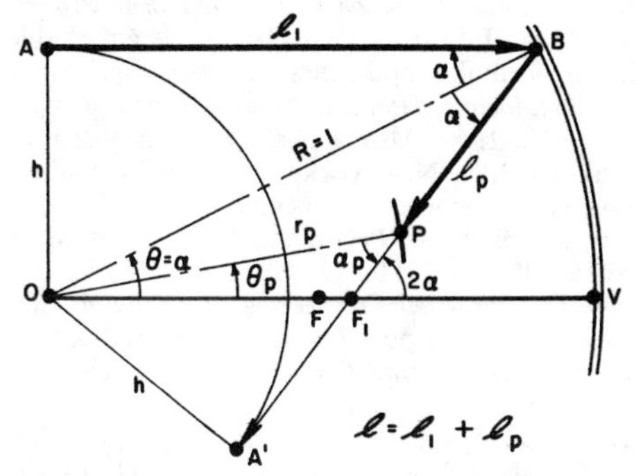

Fig. 12. Optical path length l.

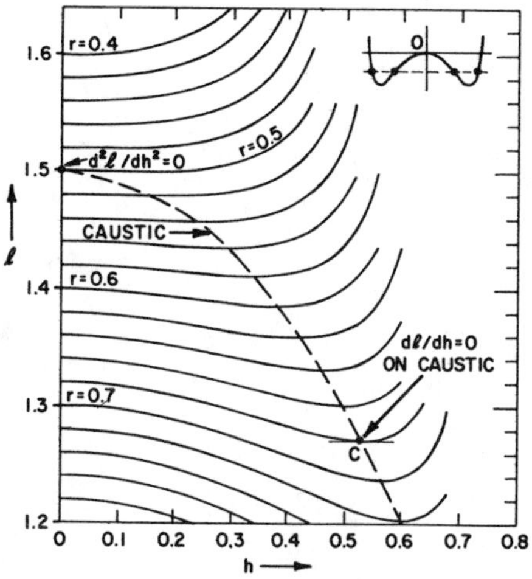

Fig. 13. Path length l vs. h.

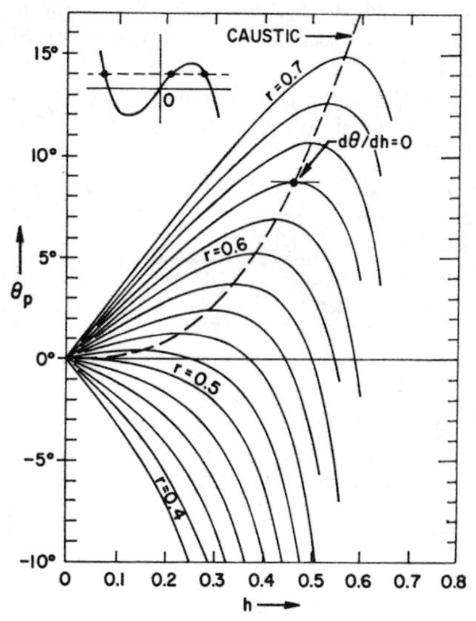

Fig. 14. Angle θ_p vs. h.

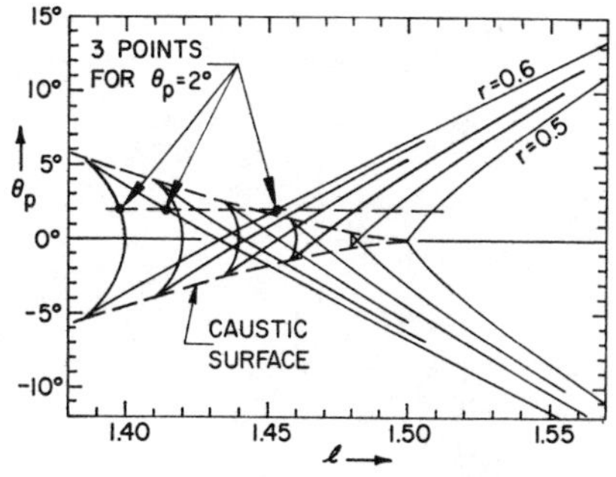

Fig. 15. The *fishtail*, angle θ_p vs. path length l.

The law of geodesics for an axially symmetric system holds along the optical path of Fig. 12.

$$R \sin \alpha = r_p \sin \alpha_p = h, \quad \text{a constant} \tag{25}$$

where α_p is the angle at which the ray cuts the radius vector r_p. From ΔOPF_1 we note that the angular coordinate of P is

$$\theta_p = 2\alpha - \alpha_p = 2 \sin^{-1} h - \sin^{-1}(h/r_p). \tag{26}$$

Fig. 14 shows θ_p vs. h, with r_p as a parameter. Each curve is odd and resembles a cubic. There may be three values of h for one value of θ_p.

Differentiation of θ_p yields

$$\frac{d\theta_p}{dh} = \frac{2}{l_1} - \frac{1}{l_p} \cdot \tag{27}$$

Consequently, $d\theta_p/dh$ is also zero when P is on the caustic. On comparing with (24) we see that

$$\frac{dl}{dh} = -h\frac{d\theta_p}{dh}, \quad \frac{dl}{d\theta_p} = -h. \tag{28}$$

Fig. 15, termed the *fishtail* is obtained from the two previous figures. With r_p as parameter, it indicates up to four values of θ_p for one value of l, and up to three values of l for one value of θ_p.

To evaluate (21) for the field at a point $P(r_p, \theta_p)$, one first obtains the three values of h for a particular r_p and θ_p, using Fig. 14. Next the corresponding values of the path lengths are obtained from Fig. 13. These are sufficient to make the evaluation.

REFERENCES

[1] J. Ashmead and A. P. Pippard, "The use of spherical reflectors as microwave scanning aerials," *J. IEE* (London), vol. 93, pt. IIIA, no. 4, p. 627, 1946.

[2] R. C. Spencer, "A least square analysis of the effect of phase errors on antenna gain," Cambridge Field Station, Air Material Command, Rept. E5025, January 1949.

[3] *Reports on Progress in Physics*, vol. 14. London: Physical Society, 1951, pp. 99, 100, 106.

[4] J. Ruze, "Antenna tolerance theory—a review," *Proc. IEEE*, vol. 54, pp. 633–640, April 1966.

[5] R. C. Spencer, "Theoretical analysis of the effect of spherical aberration on gain," USAF Cambridge Research Center, Cambridge, Mass., Tech. Rept. E5082, December 1951.

[6] R. C. Spencer, C. J. Sletten, and J. E. Walsh, "Correction of spherical aberration by a phased line source," *Proc. Nat'l Electronics Conf.* (Chicago, Ill., 1950), vol. 5. Also USAF Cambridge Research Center, Cambridge, Mass., Rept. E5069, May 1951.

[7] A. F. Kay, "A line source for a spherical reflector," USAF Cambridge Research Labs., Bedford, Mass., Tech. Rept. AFCRL-529 (TRG, Somerville, Mass., Rept. 131), May 1961.

[8] P. J. Klass, "New radar telescope to map ionosphere," *Aviation Week and Space Technology*, pp. 84–91, August 19, 1963.

[9] A. C. Schell, "The diffraction theory of large-aperture spherical reflector antennas," *IEEE Trans. Antennas and Propagation*, vol. AP-11, pp. 428–432, July 1963.

[10] C. J. Sletten and W. S. Mavroides, "A method of side lobe reduction," U. S. Naval Research Lab., Rept. 4043 on Sidelobe Conference, April 1952.

[11] C. Clasen, "The helisphere antenna" (A Study in Microwave Optics), M.S. thesis, Dept. of Physics, University of Pennsylvania, Philadelphia, 1959.

[12] R. C. Spencer and G. Hyde, "Transverse focal region of a spherical reflector, pt. I: geometric optics," Scientific Rept. 1 under AFCRL Contract AF 19(628)-2758, AFCRL-64-2921, March 1964.

[13] G. Hyde and R. C. Spencer, "Transverse focal region of a spherical reflector, pt. II: polarization," pt. II of Scientific Rept. 1 under AFCRL Contract AF 19(628)-2758, AFCRL-64-29211, May 1964.

[14] G. Hyde, "Transverse focal region fields of a spherical reflector," Scientific Rept. 2 under AFCRL Contract AF 19(628)-2758, AFCRL-66-48, January 1966.

[15] "Curves, special—the nephroid," in *Encyclopaedia Britannica*. The *nephroid* is so-named because of its resemblance to the kidney.

[16] R. C. Yates, *Handbook on Curves and their Properties*. Ann Arbor, Mich.: J. W. Edwards, 1947. This is a remarkable reference book with curves, formulas, and definitions (out of print).

[17] C. Zwikker, *Advanced Plane Geometry*. Amsterdam: North Holland Publishing Co., 1950. Also published in a paperback edition by Dover Publications, Inc., New York, N. Y. Zwikker discusses many types of curves using complex number notation.

[18] M. Born and E. Wolf, *Principles of Optics*, 1st ed. New York: Pergamon Press, 1959, pp. 111–113, 125–126, 168–170.

[19] C. R. Wylie, *Advanced Engineering Mathematics*, 1st ed. New York: McGraw-Hill, 1951, pp. 16, 310.

[20] D. J. Struik, *Differential Geometry*. Cambridge, Mass.: Addison-Wesley, 1950, pp. 39, 93–96, and Figs. 2–23.

[21] S. Silver, Ed., *Microwave Antenna Theory and Design*, Rad. Lab. Ser., vol. 12. New York: McGraw-Hill, 1949, pp. 138–143.

[22] M. Born and E. Wolf,[18] pp. 114–116: "The intensity at any point of a rectilinear ray is proportional to the Gaussian curvature $K=1/(R_1R_2)$, where R_1, R_2 are the two principal radii of curvature."

Studies of the Focal Region of a Spherical Reflector: Polarization Effects

GEOFFREY HYDE, MEMBER, IEEE, AND ROY C. SPENCER, FELLOW, IEEE

Abstract—**The polarization of the reflection of a uniform plane wave incident on a spherical reflector is analyzed using the current distribution method for scattered fields. The current distribution on the reflector is derived. For reflectors subtending about 60° or less, the radiation scattered in the direction of the circle of least confusion has essentially the same polarization as that reflected specularly from the tangent plane. The effective current, the component of surface current density radiating toward the focal region, is derived in several representations. Assuming $\hat{\imath}$ incident polarization, contour plots are provided for $\hat{\imath}$, $\hat{\jmath}$, and $\hat{k}$ components in spherical coordinates. Next, general formulas are derived for the $\hat{\imath}$, $\hat{\jmath}$, $\hat{k}$ components of the reflected fields, in terms of the direction cosines of the normal to the reflecting surface. These are displayed in terms of projections, and apply directly to the spherical reflector.**

Introduction

In an earlier paper the authors discussed the geometric optics[1] of the spherical reflector. They pointed out that for reflectors having large angles of incidence (such as Arecibo), the use of a purely scalar theory is precluded for the case of transverse feeds, and polarization must be considered. This paper derives the polarization components of the radiation scattered in the direction of the focal region when a linearly polarized plane wave is incident on a perfectly conducting spherical reflector.

The incident wave induces surface currents on the reflector which can be calculated by the current distribution method,[2] assuming that the radius $R \gg \lambda$, the conductivity $\rightarrow \infty$, and the edge discontinuities do not affect the currents. This method is equivalent to calculating the currents induced locally at the point of tangency in the infinite tangent plane.

The radiation scattered by the reflector to points in the focal region passes effectively through the small circle of least confusion,[1] in a direction approximately that of specular reflection. The polarization of this radiation is the same as the component of surface current radiating in that direction, i.e., the effective surface current density. The $\hat{\imath}$, $\hat{\jmath}$, $\hat{k}$ components of the effective current are calculated, as are the components of the field itself.

Manuscript received October 17, 1967; revised February 5, 1968. This work was carried on mainly at the RCA Missile and Surface Radar Division, Moorestown, N. J., and supported in part by the USAF Cambridge Research Laboratories, under Contract AF-19(628)-2758.

G. Hyde is with RCA Defense Microelectronics, Defense Electronic Products, Somerville, N.J. 08876

R. C. Spencer is a Consulting Physicist, 102 Devon Road, Cinnaminson, N. J.

[1] R. C. Spencer and G. Hyde, "Studies of the focal region of a spherical reflector: geometric optics," *IEEE Trans. Antennas and Propagation*, vol. AP-16, pp. 317–324, May 1968.

[2] S. Silver, *Microwave Antenna Theory and Design*, M.I.T. Radiation Lab. Ser., vol. 12. New York: McGraw-Hill, 1949, ch. 5.

Formulations in spherical and rectangular coordinates are compared. Since many formulas refer to the direction of the normal to the surface, they apply generally to smooth surfaces of infinite conductivity.

Currents on the Surface of a Spherical Reflector

Current Distribution Method[2]

Consider a uniform linearly polarized plane electromagnetic wave, with field vectors $\overline{E}_1$ and $\overline{H}_1$, incident on the perfectly conducting simply curved surface S at a point B as shown in Fig. 1(a). Let $\hat{n}_1$ be the unit vector in the direction of propagation of the incident wave; $\hat{e}_1$ and $\hat{h}_1$ the unit vectors for $\overline{E}_1$ and $\overline{H}_1$, respectively ($\hat{e}_1$, $\hat{h}_1$, and $\hat{n}_1$ forming a right-handed triad); and $\hat{n}_s$ the unit normal to S at B. Then the currents locally at B are assumed to be the same as those generated at B in an unbounded perfectly conducting plane tangent at B, provided B is remote from any discontinuities in S.

For reflection from the tangent plane, as shown in Fig. 1(b), the usual boundary conditions hold, whence the usual formulas for current and charge are obtained:

$$\overline{K} = \hat{n}_s \times \overline{H} = 2\hat{n}_s \times \overline{H}_1 = 2\left(\frac{\epsilon}{\mu}\right)^{\frac{1}{2}} \hat{n}_s \times (\hat{n}_1 \times \overline{E}_1) \qquad (1)$$

$$\eta = \epsilon \hat{n}_s \cdot \overline{E} = 2\epsilon \hat{n}_s \cdot \overline{E}_1. \qquad (2)$$

Noting that

$$\frac{E}{H} = (\mu/\epsilon)^{\frac{1}{2}} = \frac{E_1}{H_1}$$

then

$$\overline{K} = 2H_1 \hat{n}_s \times (\hat{n}_1 \times \hat{e}_1) \qquad (3)$$

and

$$\eta = 2\epsilon E_1 \hat{n}_s \cdot \hat{e}_1. \qquad (4)$$

Geometric optics has shown that the focal region of concern for transverse feed systems is near the circle of least confusion. This is the circle which encompasses all geometrically (specularly) reflected rays. It subtends less than 7° from any point on a spherical reflector that subtends 60° at its center (f/d ratio of 1/2) (see Fig. 2). Thus there is little error in taking the polarization of radiation to the focal region as that of the specularly reflected ray, i.e., that of radiation reflected from the tangent plane at the point of incidence. The direction $\hat{n}_2$ of the specularly reflected ray shown in Fig. 1(b) is given by the vector law of reflection:

$$\hat{n}_2 = \hat{n}_1 - 2\hat{n}_s(\hat{n}_1 \cdot \hat{n}_s). \qquad (5)$$

Reprinted from *IEEE Trans. Antennas Propagat.*, vol. AP-16, pp. 399-404, July 1968.

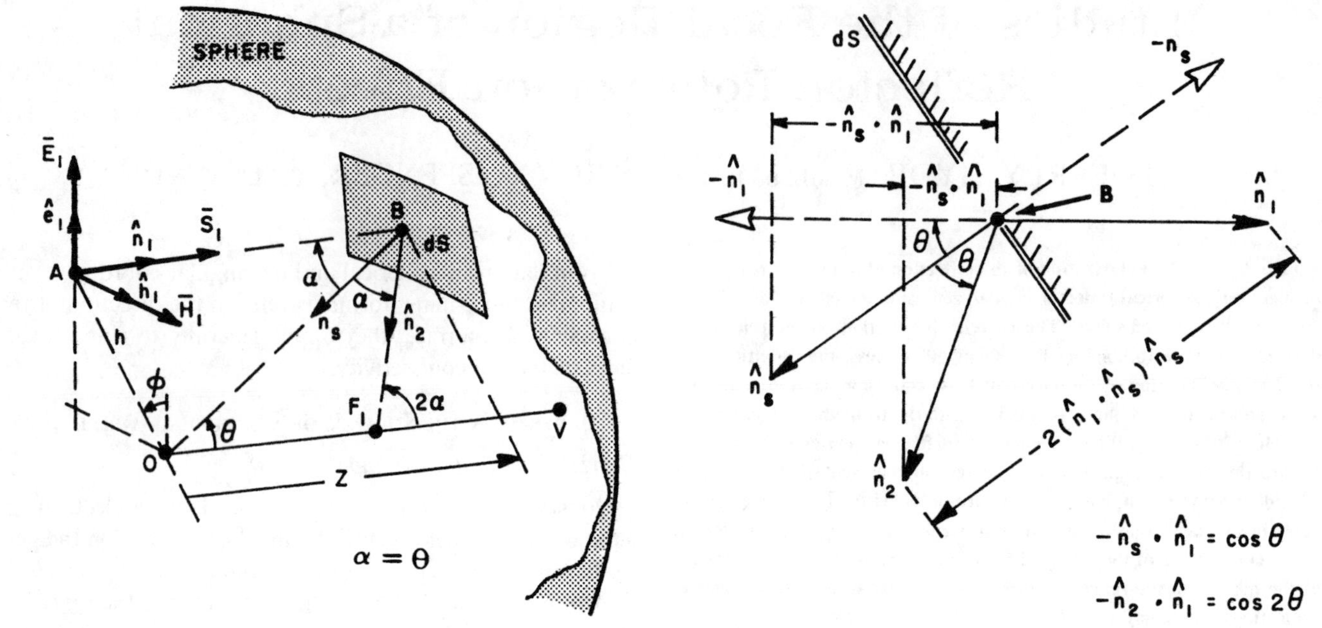

(a) (b)

Fig. 1. Plane wave reflected from a curved surface.

The component of surface current density $\overline{K}$ that radiates in the direction $\hat{n}_2$ is called the effective surface current density $\overline{K}_{2\perp}$ (or the effective current). It is perpendicular to $\hat{n}_2$ and is given by

$$\overline{K}_{2\perp} = \overline{K} - \hat{n}_2(\hat{n}_2 \cdot \overline{K}). \tag{6}$$

The reflected $\hat{h}_2$ and $\hat{e}_2$ vectors are the optical images of $\hat{h}_1$ and $\hat{e}_1$, respectively, except for reversal of the tangential component of $\hat{e}$. From the vector law of reflection,

$$\hat{h}_2 = \hat{h}_1 - 2\hat{n}_s(\hat{h}_1 \cdot \hat{n}_s) \tag{7}$$

$$-\hat{e}_2 = \hat{e}_1 - 2\hat{n}_s(\hat{e}_1 \cdot \hat{n}_s). \tag{8}$$

As yet, the particular case of interest has not been defined. There is no loss of generality if we choose $\hat{e}_1 = \hat{\imath}$ and $\hat{h}_1 = \hat{\jmath}$. Then $\hat{n}_1 = \hat{k}$. Further, the normal to the surface may be written as

$$\hat{n}_s = \frac{-\overline{OB}}{OB} = -\hat{\rho}, \quad \text{in spherical unit vectors}$$

$$= -(\hat{\imath}\ell + \hat{\jmath}m + \hat{k}n), \quad \text{in Cartesian unit vectors using direction cosines} \tag{9}$$

$$= -(\hat{\imath}\sin\theta\cos\phi + \hat{\jmath}\sin\theta\sin\phi + \hat{k}\cos\theta), \quad \text{using spherical coordinates.}$$

From the above, formulas for $\overline{K}_{2\perp}$, $\hat{h}_2$, and $-\hat{e}_2$ may be calculated.

The effective current $\overline{K}_{2\perp}$ and the unit reflected electric field $-\hat{e}_2$ are readily related using direction cosines. From (3) and (9),

$$\overline{K} = 2H_1(n\hat{\imath} - \ell\hat{k}) \tag{10}$$

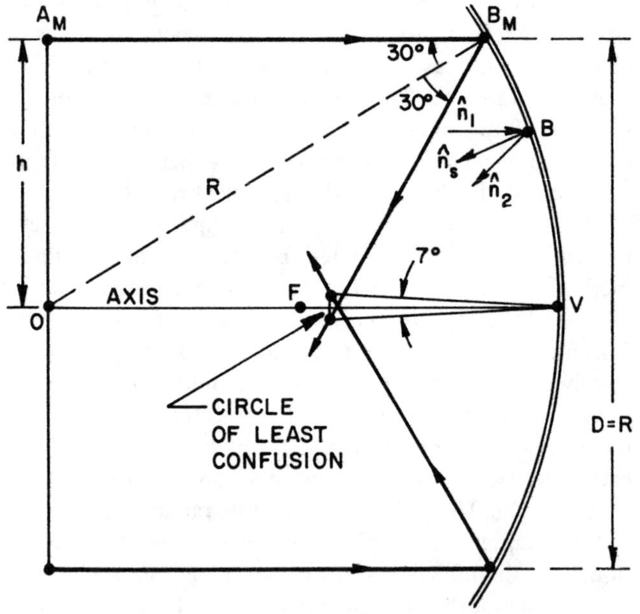

Fig. 2. Optics of a spherical reflector.

whence from (5), (6), and (8), the reflected $-\hat{e}_2$ is given by

$$\left(\frac{\overline{K}_{2\perp}}{2H_1}\right)\frac{1}{n} = \frac{\hat{k}_{2\perp}}{n} = -\hat{e}_2$$
$$= (1 - 2\ell^2)\hat{\imath} - 2\ell m\hat{\jmath} - 2\ell n\hat{k} \tag{11}$$

where $\overline{K}_{2\perp}/2H_1 = \hat{k}_{2\perp}$ is a normalization of $\overline{K}_{2\perp}$.

The factor $1/n$ can be considered to arise from the projection of an elemental area of the reflector dS on either the incident wavefront A_1, or the reflected wavefront A_2, since locally (using the tangent plane approximation) the elemental areas $dA_1 = dA_2 = -\hat{n}_1 \cdot \hat{n}_s dS = \hat{n}_2 \cdot \hat{n}_s dS = ndS$.

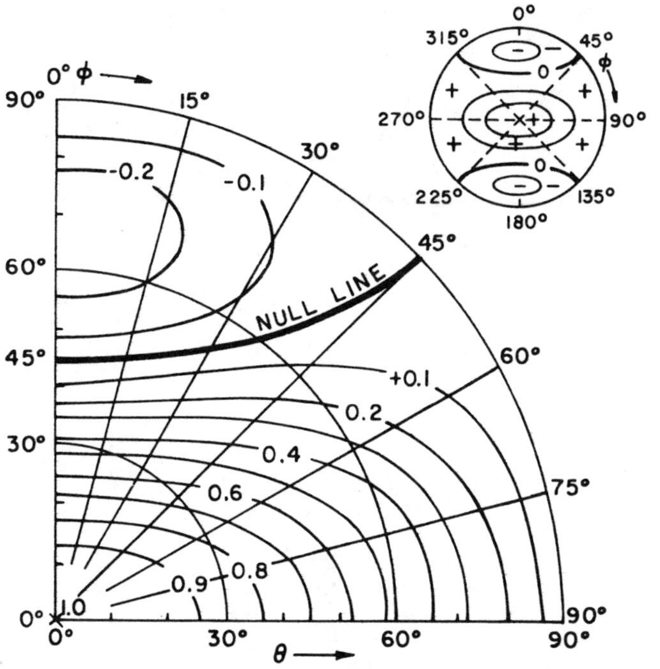

Fig. 3. The $\hat{\imath}$ component of the effective surface current density $\overline{K}_{2\perp}$.

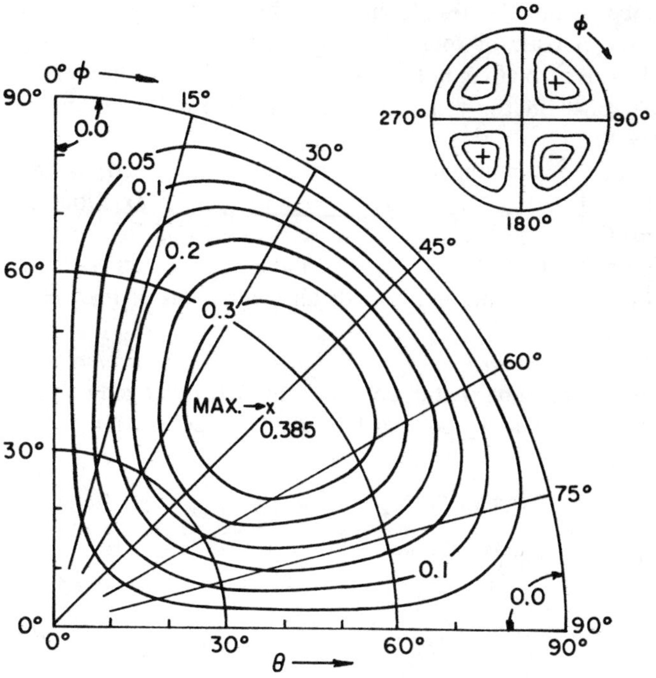

Fig. 4. The $\hat{\jmath}$ component of the effective surface current density $\overline{K}_{2\perp}$.

Effective Current on the Reflector

The effective current $\overline{K}_{2\perp}$ of (6), the component of $\overline{K}$ perpendicular to $\hat{n}_2$, is used when integrating over the reflecting surface to obtain the fields at a point in the focal region. From (5), (6), and (9), after substitution and manipulation, one obtains several representations for effective current:

$$\overline{K}_{2\perp} = \overline{K} - \hat{n}_2(\overline{K}\cdot\hat{n}_2)$$
$$= 2H_1[\hat{n}_s \times \hat{\jmath} - (\hat{\imath}\cdot\hat{n}_s)(\hat{k} - 2\hat{n}_s(\hat{n}_s\cdot\hat{k}))] \quad (12a)$$
$$= 2H_1[(n - 2n\ell^2)\hat{\imath} - 2\ell mn\hat{\jmath} - 2\ell n^2\hat{k}] \quad (12b)$$

using direction cosines

$$= 2H_1[\hat{\imath}\cos\theta(1 - 2\sin^2\theta\cos^2\phi) - \hat{\jmath}\sin\theta \cdot\sin 2\theta\sin\phi\cos\phi - \hat{k}\cos\theta\sin 2\theta\cos\phi] \quad (12c)$$

using spherical coordinates

where

$$\overline{K} = 2H_1[\hat{\imath}\cos\theta - \hat{k}\sin\theta\cos\phi] \quad (13a)$$

and

$$\overline{K}_{2\perp} = 2H_1[\hat{\rho}\cos\theta\sin\theta\cos\phi + \hat{\theta}\cos\phi(1 + \sin^2\theta) - \hat{\phi}\cos\theta\sin\phi] \quad (12d)$$

using spherical unit vectors

where

$$\overline{K} = 2H_1[\hat{\theta}\cos\phi - \hat{\phi}\cos\theta\cos\phi]. \quad (13b)$$

Components of Effective Current on a Reflector: The $\hat{\imath}$, $\hat{\jmath}$, and $\hat{k}$ components of the effective surface current density $\overline{K}_{2\perp}$ were computed from (12c) for an incident $\hat{\imath}$-polarized wave (with $2H_1 = 1$). Contours of constant levels of these components are graphed in Figs. 3, 4, and 5 with θ and ϕ as polar

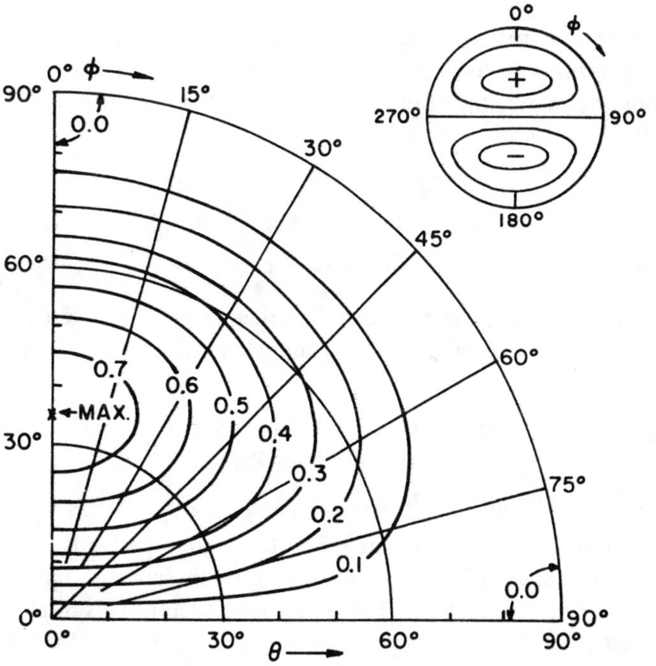

Fig. 5. The $\hat{k}$ component of the effective surface current density $\overline{K}_{2\perp}$.

coordinates. The graphs are analogous to what one would see if the contrours were drawn on a sphere which was then viewed from the center of the sphere.

The $\hat{\imath}$ component (Fig. 3) is unity at the origin, zero for $\theta = 90°$, and reverses sign along two null lines cutting through points ($\phi = 0, 180°$; $\theta = 45°$) and ($\phi = \pm 45°$, $\pm 135°$; $\theta = 90°$). Of course, in most practical situations, the null line lies outside the actual reflector, but its influence is felt in the elliptical squashed-down distribution of the $\hat{\imath}$ component, even for moderately deep reflectors. The $\hat{\jmath}$ component with its four-lobe symmetry results in the familiar cross-polarization characteristic of rotationally symmetric antennas (see Fig. 4). It is zero along both axes and has maxima (or minima) of ± 0.3849 in the four quadrants at $\phi = \pm 45°$, $\pm 135°$, and $\theta = 54.74°$.

The $\hat{k}$ component, which is usually neglected, is two-lobed (see Fig. 5). It is zero along the $\phi = 90°$ axis, but has maxima of ± 0.7698 at $\phi = 0°$, 180°, and $\theta = 35.26°$. Its magnitude of 0.6533 at $\phi = 0$, 180°, and $\theta = 22\frac{1}{2}°$ matches that of the $\hat{\imath}$ component, and exceeds all magnitudes of the $\hat{\jmath}$ component.

Current Distribution for an Incident Uniform Plane Wave of $\hat{\jmath}$ Polarization: Let us consider a linearly polarized uniform plane wave, now $\hat{\jmath}$-polarized, incident on a spherical reflector. Again

$$\hat{n}_1 = \hat{k}; \quad \text{but } \hat{e}_1 = \hat{\jmath}$$

whence from (3) and (9)

$$\begin{aligned} \overline{K} &= 2H_1(\hat{\jmath}n - \hat{k}m) \\ \hat{n}_2 &= -2\ell n\hat{\imath} - 2mn\hat{\jmath} + (1 - 2n^2)\hat{k}. \end{aligned} \tag{14}$$

Recalling $K_{2\perp} = \overline{K} - (\overline{K} \cdot \hat{n}_2)\hat{n}_2$,

$$\overline{K}_{2\perp} = 2H_1[-2\ell mn\hat{\imath} + n(1 - 2m^2)\hat{\jmath} - 2n^2m\hat{k}]. \tag{15}$$

Note that this has the same form as $\overline{K}_{2\perp}$ for an incident $\hat{\imath}$-polarized uniform plane wave, with the roles of $\hat{\imath}$ and $\hat{\jmath}$, ℓ and m interchanged. In spherical coordinates, we obtain for the $\hat{\jmath}$-polarized plane wave,

$$\begin{aligned} \overline{K}_{2\perp} = 2H_1[(&-\hat{\imath} \sin\theta \sin 2\theta \sin\phi \cos\phi \\ &+ \hat{\jmath} \cos\theta(1 - 2\sin^2\theta \sin^2\phi) \\ &- \hat{k} \sin 2\theta \cos\theta \sin\phi)] \end{aligned} \tag{16}$$

and again note the similarity in structure to (12c), with the roles of $\hat{\imath}$ and $\hat{\jmath}$, $\sin\phi$ and $\cos\phi$ interchanged, i.e., a 90° rotation.

Polarization of the Field Reflected in the Tangent Plane

Reflected Field from Image Theory

It was pointed out that reflection from a perfectly conducting plane takes place according to the laws of optics, except for reversal of the phase of the E vector. This implies that the reflected unit vector triad $(-\hat{e}_2, \hat{h}_2, \hat{n}_2)$ is precisely the mirror image of the incident triad $(\hat{e}_1, \hat{h}_1, \hat{n}_1)$. See Fig. 6. The angles α, β, γ that $\hat{e}_1, \hat{h}_1, \hat{n}_1$ make with the negative surface normal $-\hat{n}_s$ are their respective angles of incidence. If $\hat{e}_1, \hat{h}_1, \hat{n}_1$ are chosen parallel to the $\hat{\imath}, \hat{\jmath}, \hat{k}$ vectors, then the direction cosines of $-\hat{n}_s$ are

$$\ell = \cos\alpha, \quad m = \cos\beta, \quad n = \cos\gamma. \tag{17}$$

The reflected ray direction, in (5), reduces to

$$\hat{n}_2 = -2\ell n\hat{\imath} - 2mn\hat{\jmath} + (1 - 2n^2)\hat{k}. \tag{18a}$$

From symmetry considerations we can write down the equations for $-\hat{e}_2$ and $\hat{h}_2$ by permuting (18a), or by use of (7), (8), or (11).

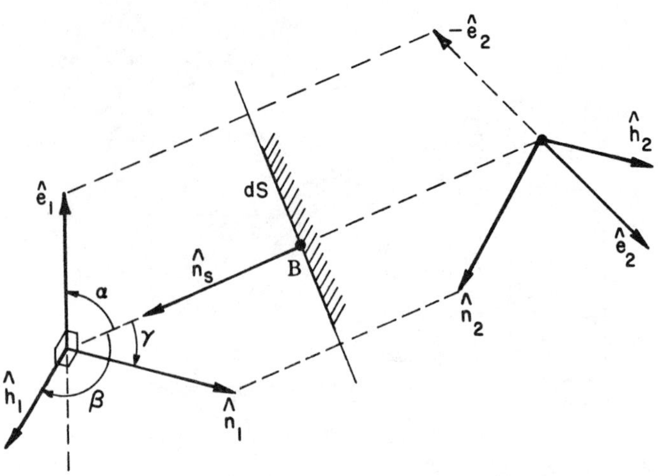

Fig. 6. Incident and reflected triads.

TABLE I
Components of Reflected Fields
(for $\hat{e}_1 = \hat{\imath}$, $\hat{h}_1 = \hat{\jmath}$, $\hat{n}_1 = \hat{k}$)

	$\hat{\imath}$	$\hat{\jmath}$	$\hat{k}$
$-\hat{e}_2$	$(1-2\ell^2)$	$-2\ell m$	$-2\ell n$
$\hat{h}_2$	$-2\ell m$	$(1-2m^2)$	$-2mn$
$\hat{n}_2$	$-2\ell n$	$-2mn$	$(1-2n^2)$

Table I combines these equations into a symmetrical matrix representing a rotational transformation of three unit orthogonal vectors. For example,

$$\hat{n}_2 = (\hat{n}_2 \cdot \hat{e}_1)\hat{e}_1 + (\hat{n}_2 \cdot \hat{h}_1)\hat{h}_1 + (\hat{n}_2 \cdot \hat{n}_1)\hat{n}_1. \tag{18b}$$

The one restriction is that each vector is reflected according to geometric optics. Therefore, the diagonal terms are respectively $-\cos 2\alpha$, $-\cos 2\beta$, $-\cos 2\gamma$. This is demonstrated in Fig. 1(b) for $\hat{n}_2$ where θ is the angle between n_s and $-\hat{n}_1$, and 2θ the angle between $\hat{n}_2$ and $-\hat{n}_1$. The corresponding projections are

$$\begin{aligned} -(\hat{n}_s \cdot \hat{n}_1) &= \cos\gamma = n \\ (\hat{n}_2 \cdot \hat{n}_1) &= -\cos 2\gamma = 1 - 2n^2. \end{aligned} \tag{18c}$$

Field Contours of Reflected $\hat{e}$, $\hat{h}$, and $\hat{n}$ Components

Table I contains but two basic types of coefficients which, for convenience, may be transformed by substitution into

$$\ell^2 + m^2 + n^2 = 1. \tag{19}$$

Typical of the diagonal terms is the $\hat{\imath}$ component of the negative reflected $\hat{e}$ vector

$$-e_{2i} = (1 - 2\ell^2) = -1 + 2(m^2 + n^2). \tag{20}$$

Typical of the off-diagonal terms are the $\hat{\jmath}$ component of $-\hat{e}_2$

$$-e_{2j} = -2\ell m \tag{21}$$

and its $\hat{k}$ component

$$-e_{2k} = -2\ell n = -2\ell(1 - \ell^2 - m^2)^{\frac{1}{2}}. \tag{22}$$

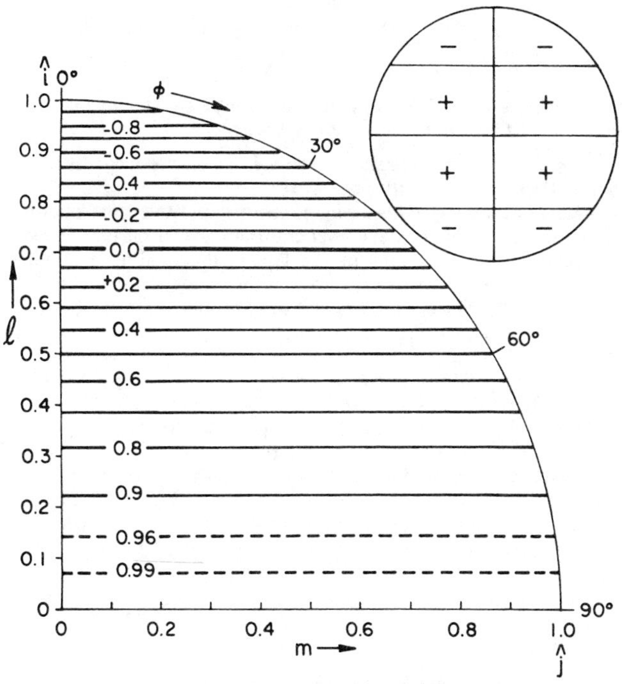

EQUIVALENT PROJECTIONS

	$-e_{2i}$	$-e_{2i}$	h_{2j}	h_{2j}	n_{2k}	n_{2k}
AXIS	k	-j	i	-k	j	-i
PLANE	i j	i k	j k	j i	k i	k j

Fig. 7. The $\hat{\imath}$ component of $-\hat{e}_2$, projected on $\hat{\imath}\hat{\jmath}$ plane.

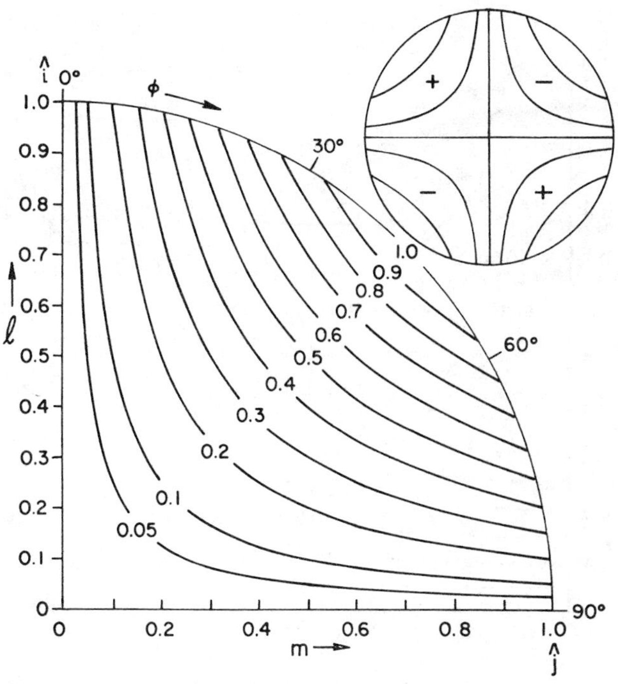

EQUIVALENT PROJECTIONS

	$-e_{2j}$	$-e_{2k}$	h_{2k}	h_{2i}	n_{2i}	n_{2j}
AXIS	k	j	i	k	j	i
PLANE	i j	k i	j k	i j	k i	j k

Fig. 9. The $\hat{\jmath}$ component of $-\hat{e}_2$, projected on $\hat{\imath}\hat{\jmath}$ plane.

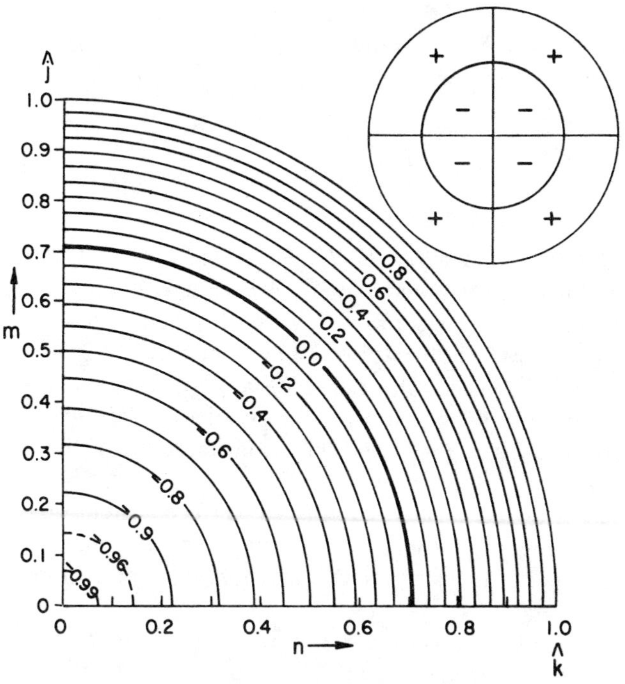

EQUIVALENT PROJECTIONS

	$-e_{2i}$	h_{2j}	n_{2k}
AXIS	i	j	k
PLANE	j k	k i	i j

Fig. 8. The $\hat{\imath}$ component of $-\hat{e}_2$, projected on $\hat{\jmath}\hat{k}$ plane.

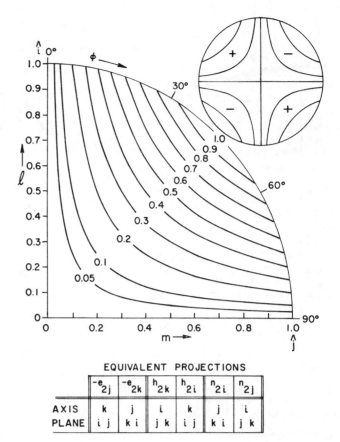

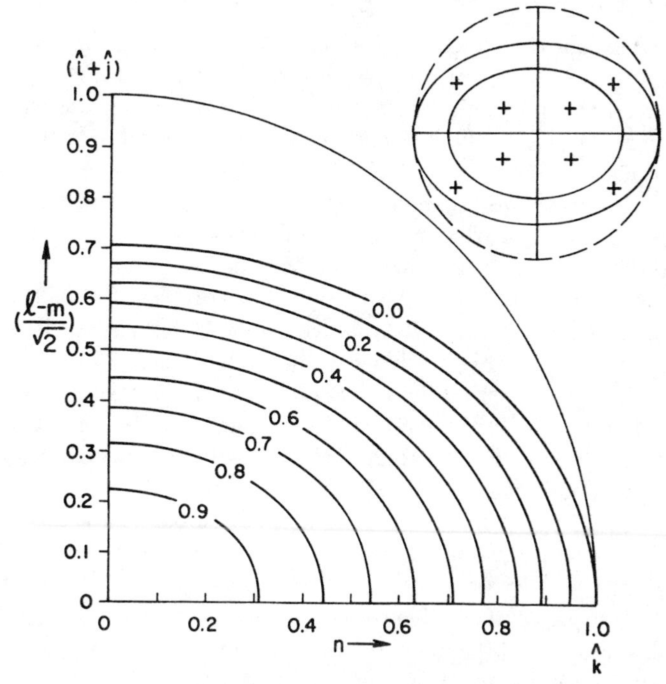

EQUIVALENT PROJECTIONS

	$-e_{2j}$	h_{2k}	n_{2i}
AXIS	-(i+j)	-(j+k)	-(k+i)
PLANE	(i-j), k	(j-k), i	(k-i), j

Fig. 10. The $\hat{\jmath}$ component of $-\hat{e}_2$, projected on $(\hat{\imath}-\hat{\jmath})$, $\hat{k}$ plane.

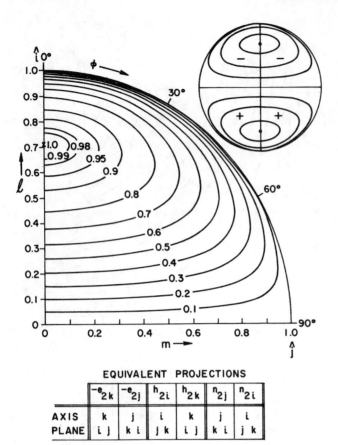

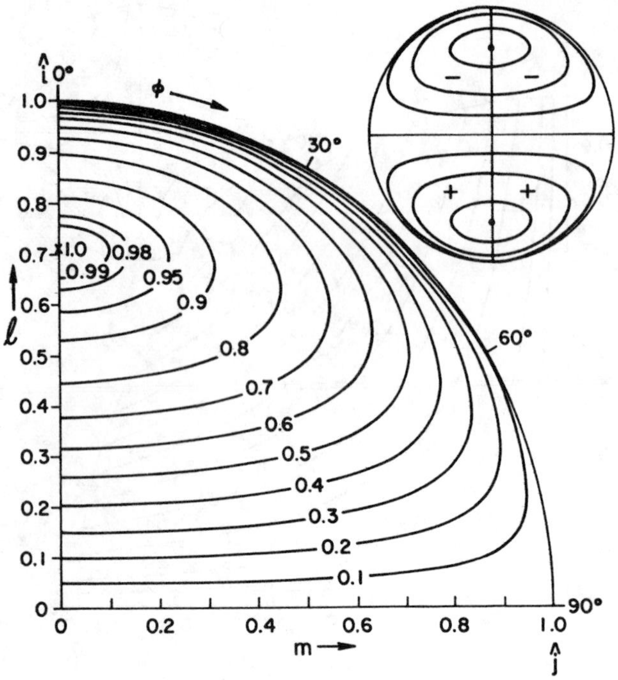

EQUIVALENT PROJECTIONS

	$-e_{2k}$	$-e_{2j}$	h_{2i}	h_{2k}	n_{2j}	n_{2i}
AXIS	k	j	i	k	j	i
PLANE	i j	k i	j k	i j	k i	j k

Fig. 11. The $\hat{k}$ component of $-\hat{e}_2$, projected on $\hat{i}\hat{j}$ plane.

Projections from the Unit Sphere onto Various Planes

The above changes in variables are effected merely by changes in direction of viewing, provided the contours are first drawn on a unit sphere. Thus, for example, when viewed along the $\hat{k}$ axis, the ℓ, m coordinates are projected onto the $\hat{i}\hat{j}$ plane.

Example of the $\hat{i}$ Component of $-\hat{e}_2$: Contours for constant $-e_{2i}$, in (20), are drawn as circles of constant $\ell=\cos\alpha$ about the $\hat{i}$ axis of the unit sphere, similar to circles of latitude on the earth. Projected on the $\hat{i}\hat{j}$ plane, they are a set of parallel lines as shown in Fig. 7. Projected on the $\hat{j}\hat{k}$ plane, they are a set of circles (Fig. 8). Now at normal incidence, $\ell=m=0$, $n=1$, and, for either figure, $-e_{2i}=1$, indicating reversal of phase. On the other hand, at grazing angle, $\ell=1$, $m=n=0$, and $-e_{2i}=-1$, indicating continuity with the incident e_1 vector, both being bounded on the top side by the reflector. There is a null for $\ell=\cos 45°$. If, in addition, $m=0$, then $n=\cos 45°$, and the reflected wave is complețly polarized along the $-\hat{k}$ axis.

Example of the $\hat{j}$ Component of $-\hat{e}_2$: The contours for this typical off-diagonal term, $-2\ell_m$, are the intersection with the unit sphere of a family of rectangular hyperbolic cylinders parallel to the $\hat{k}$ axis. These contours are closed ovals on the unit sphere, but, projected on the $\hat{i}\hat{j}$ plane as in Fig. 9, they are the rectangular hyperbolas $-2\ell m$ of (21). They are zero along the $\hat{i}$ and $\hat{j}$ axes but are ± 1 for $\ell=m=\cos 45°$. When projected on the $(\hat{i}-\hat{j})$ plane, as in Fig. 10, they are ellipses. This is because of the identity

$$1-2\ell m = m^2 + (\ell - m)^2. \tag{23}$$

The contraction of the projected $(\ell-m)$ scale produces an axial ratio of $\sqrt{2}$.

Example of the $\hat{k}$ Component of $-\hat{e}_2$: This component, $-2\ell n$, usually neglected in antenna design, attains a value of ± 1 for $n=\ell=\cos 45°$, coinciding with the null for the $\hat{i}$ component. It indicates complete polarization of the reflected wave along the $-\hat{k}$ axis. The contours on the unit sphere are similar to those of the $\hat{j}$ component, but with interchange of $\hat{j}$ and $\hat{k}$ axes. Fig. 11 shows the projection on the $\hat{i}\hat{j}$ plane. The closed contour ovals for $-e_{2k}$ were calculated by solving (22) for m:

$$m = \left[1 - \ell^2 - \frac{e_{2k}^2}{4\ell^2}\right]^{\frac{1}{2}}. \tag{24}$$

Spherical Coordinate Scales for ϕ and θ: The spherical coordinate scales ϕ and θ are easily included in any projection on the $\hat{i}\hat{j}$ plane (see Figs. 7, 9, and 11). Since $\tan\phi = m/\ell$, the ϕ scale is 0° in the $\hat{i}$ direction, swinging clockwise to 90° in the $\hat{j}$ direction. The radial distance in the $\hat{i}\hat{j}$ plane is $\sin\theta=(\ell^2+m^2)^{1/2}$, which coincides with the ℓ and the m scales, respectively, for $\phi=0°$ and 90°. Multiplication of the $-e_2$ components by $n=\cos\theta$, according to (11), yields the components of the normalized effective surface current density $\hat{k}_{2\perp}$ of Figs. 3 through 5. The resulting zero for $\theta=90°$ introduces at times an added maximum or minimum in the components of $\hat{k}_{2\perp}$.

Application to Components of $\hat{h}_2$ and $\hat{n}_2$: Figs. 7 through 11 consist of two projections of the basic diagonal term and three of the basic off-diagonal term of Table I. Although designed for components of the $-\hat{e}_2$ vector, each graph is accompanied by a chart showing how to apply it to components of $\hat{h}_2$ and $\hat{n}_2$. (See directions accompanying the figures.) One should keep in mind that the ϕ and θ scales apply only to projections on the $\hat{i}\hat{j}$ plane.

EXPLANATION OF FIGS. 7 TO 11; CONSTANT LEVEL CONTOURS OF COMPONENTS OF REFLECTED $\hat{e}$, $\hat{h}$, AND $\hat{n}$ VECTORS

Unit $\hat{e}_1$, $\hat{h}_1$, $\hat{n}_1$ vectors parallel to the $\hat{i}$, $\hat{j}$, $\hat{k}$ axes are incident on the tangent reflecting plane with normal $\hat{n}_s=-(\ell\hat{i}+m\hat{j}+n\hat{k})$. The $\hat{i}$, $\hat{j}$, $\hat{k}$ components of the reflected $-\hat{e}_2$, $\hat{h}_2$, $\hat{n}_2$ vectors of Table I are obtained from two basic sets of constant level contours drawn on unit spheres. Figs. 7 and 8 are projections of the typical diagonal term $(1-2\ell^2)$, while Figs. 9, 10, and 11 are projections of the typical off-diagonal term $(-2\ell m)$. Thus, by viewing the first set of contours along the $+\hat{k}$ axis, they are projected onto the $\hat{i}\hat{j}$ plane as in Fig. 7, depicting $-e_{2i}$, the $\hat{i}$ component of $-\hat{e}_2$.

Permutation of the indices of viewing axis and projection plane permits the same graph to be used for $\hat{h}_2$ or $\hat{n}_2$. For example, a single permutation of $\hat{i}$ to $\hat{j}$, $\hat{j}$ to $\hat{k}$, and $\hat{k}$ to $\hat{i}$, equivalent to two 90° rotations of the unit sphere, transforms its contours to those of $h_{2j}=(1-2m^2)$ of Table I. When viewed along the $\hat{i}$ axis, thus projecting the contours onto the $\hat{j}\hat{k}$ plane, they are identical with those of Fig. 7. Attached to Fig. 7 is a chart of components and their projections represented by the same figure. Similar charts accompany the other figures.

Studies of the Focal Region of a Spherical Reflector: Stationary Phase Evaluation

GEOFFREY HYDE, MEMBER, IEEE

Abstract—**The geometric optics and polarization properties of a spherical reflector are used to develop an integral representation of its focal region fields. These integrals are evaluated by the extended method of stationary phase for field points off the caustics, on the axial caustic, on the caustic surface, and at the paraxial focus. The contributions to the field at a field point are shown to arise respectively from three ordinary stationary points: a stationary ring and a stationary point at the vertex; an ordinary stationary point and a caustic type stationary point; and a fourth-order stationary point. The resulting formulas are used to compute the value of the focal region fields. The computed results are then compared to measured data.**

I. Introduction

THE DEVELOPMENT of a stationary phase approximation [1] for the fields in the transverse focal region of a spherical reflector builds upon results obtained earlier in geometric optics [2], [3] and polarization [4], [5] studies. Geometric optics was used to locate the focal region of a spherical reflector and map out many of its features, such as the caustic surface, the axial caustic, the paraxial focus, and the circle of least confusion, as shown in Fig. 1. In addition, path length from a reference plane to a point $P \equiv P(r_p, \theta_p)$ in the focal region was derived, giving the relative phase of different contributions.

The polarization analysis provided insight into the effect of a spherical reflector on an incident linearly polarized plane wave. The effective current density on the reflector $\overline{K}_{2\perp}$ was derived. It is the component of current that radiates from a point on the reflector in the direction of a field point in the focal region of the reflector. $\overline{K}_{2\perp}$ has a strong *longitudinal* or $\hat{k}$ component at off-axis points in the focal region.

From the currents on the spherical reflector, one may formulate an integral representation of the fields scattered to a point in the focal region [1], [6]. Because of the directional properties of feed systems, one may neglect the incident field, and consider only the scattered field in considering the energy focused by the reflector. For large reflectors $R \gg \lambda$ (in practice, for deep reflectors of the order of 140 λ), the method of stationary phase may be employed to evaluate the integrals, giving a leading-term approximation for the fields at a point in the focal regions. This method is a strong analog of the physical situation in geometric optics [2, Figs. 3 and 10], inasmuch as the stationary phase regions on the spherical reflector correspond closely to the multiple images formed in geometric optics. However, one now has a method for evaluation of fields at points on the caustics as well as at ordinary field points by considering higher-order terms in the Taylor expansion of the argument of the phase term, which is the three-dimensional formulation of the normalized path length of geometric optics.

Manuscript received March 11, 1968; revised July 16, 1968. This work was carried out in part at the Missile and Surface Radar Division, Radio Corporation of America, Moorestown, N. J., and supported in part by the Air Force Cambridge Research Laboratories under Contract AF 19(628)2758.

The author was with the Radio Corporation of America, Moorestown, N. J. He is now with COMSAT Laboratories, Communications Satellite Corporation, Washington, D. C.

Reprinted from *IEEE Trans. Antennas Propagat.*, vol. AP-16, pp. 646-656, Nov. 1968.

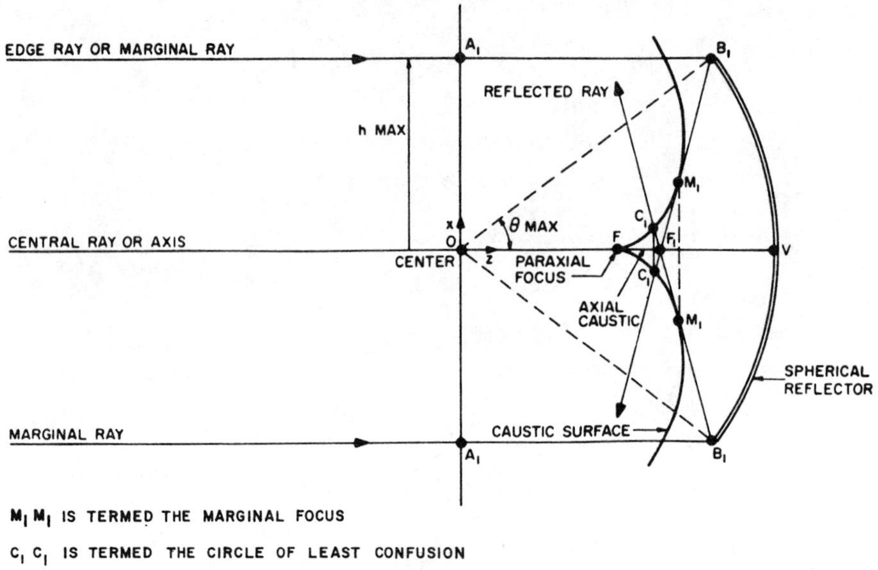

Fig. 1. Focal region of a spherical reflector.

In addition to the evaluation of the fields per se and their verification by experiment, there are three details of the result worth noting. First, there is a strong off-axis longitudinal component of field present in the focal regions. Then, there is the manner in which the fields at the caustic surface are built up not only by a contribution from a higher-order caustic-type stationary phase contribution, but also from an ordinary stationary point (which gets swamped out in the optical limit). Finally, there is the particular method for analysis of transverse focal regions which provides much insight into the physical situation while building on the results obtained from geometric optics and polarization analysis.

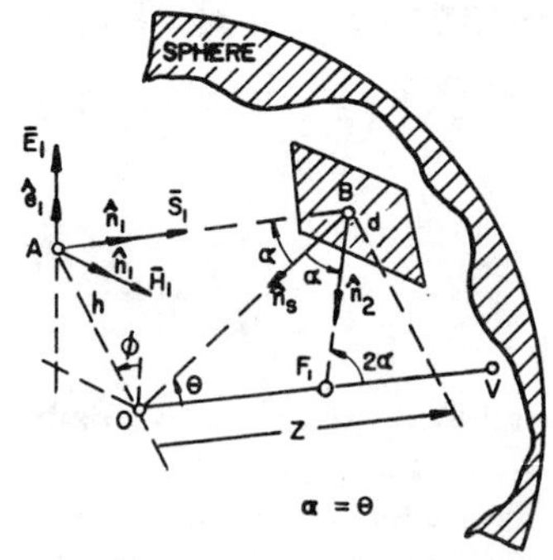

Fig. 2. Plane wave reflected from a curved surface.

II. Integral Representation [1]

For a uniform, linearly polarized plane wave ($\hat{\imath}$ polarized), incident upon the concave side of a large, smooth, perfectly conducting spherical reflector, as shown in Fig. 2, the current distribution method [7] yields an expression for $\bar{K}$, the current density on the reflector,

$$\bar{K} = 2H_1\hat{n}_s \times \hat{\jmath} = 2H_1(\hat{\imath}\cos\theta - \hat{k}\sin\theta\cos\phi). \quad (1)$$

Consider an element of current on the surface $I\,d\bar{l} = \bar{K}\,da$, where da is an element of surface area. The fields from such a current element at B are given by

$$d\bar{E} = \frac{u}{4\pi\rho} I\,dl\,e^{j(\omega(t-\rho/c)+\alpha)}(\hat{\xi}\sin\xi(j\omega + c/\rho + c^2/j\omega\rho^2) + \hat{\rho}2\cos\xi(c/\rho + c^2/j\omega\rho^2)) \quad (2)$$

where α is the phase of the current at B relative to some reference, and ρ and ξ are the element-centered spherical radial and latitudinal coordinates. Now if our reference is the phase of the plane wave as it passes by the center of curvature of the spherical reflector, the relative phases of currents on the reflector are given by $e^{j\frac{2\pi}{\lambda}L_1} = e^{j\frac{2\pi}{\lambda}Rl_1}$. Also the distance ρ to the field point from the element is $L_2 = Rl_2$ and $\hat{\rho} = \hat{n}_2$. With the time harmonic variation understood, noting that

$$\hat{\xi}\sin\xi K da = (\bar{K} - \hat{n}_2(\bar{K}\cdot\hat{n}_2))R^2\sin\theta\,d\theta\,d\phi$$

and that all induction and electrostatic terms are negligible at the distances involved, we may write

$$\bar{E} \approx \frac{j\omega\mu}{4\pi}\int_{\text{Reflector}} R\sin\theta\,d\theta\,d\phi\,\frac{\bar{K}_{2\perp}}{l_2}e^{-j(2\pi/\lambda)R(l_1+l_2)} \quad (3)$$

where $\bar{K}_{2\perp} = \bar{K} - \hat{n}_2(\hat{n}_2\cdot\bar{K}) = 2H_1[\hat{n}_s\times\hat{\jmath} - \hat{n}_2(\hat{n}_s\times\hat{\jmath})\cdot\hat{n}_2]$ from (1). $\bar{K}_{2\perp}$ is the component of $\bar{K}$ perpendicular to $\hat{n}_2$, the direction from the current element on the reflector to the field point in the focal region. $\hat{n}_2$ is not yet specified.

III. Stationary Phase Approximation for Focal Region Fields

The method of stationary phase [8]–[10] can be applied to integrals of the type

$$I(k) = \int_c g(x, y)e^{jkf(x,y)}dxdy \quad (4)$$

where k is a large number, $f(x, y)$ is independent of k and well behaved (say Taylor expandable) in C, and $g(x, y)$ is slowly varying and well behaved in C. The method hinges on the possibility, first exploited by Laplace [8], [10], that an integrand may behave in such a manner that the dominant contribution to $I(k)$ arises from a limited part of C (or several such parts). Thus integrating over the limited interval in C would yield a good approximation to $I(k)$. In this limited interval, hopefully one is able to represent or approximate the integrand in a more tractable manner. In the method of stationary phase, these hopes center on the oscillatory character of $e^{jkf(x,\,y)}$ as contrasted to the propriety expected of $g(x, y)$. Except near the stationary points of $f(x, y)$ in C, the rapid oscillations of $e^{jkf(x,\,y)}$ result in essentially zero net contribution to the integral, except possibly on the boundary of C. The similarity between this sort of thing and Fresnel's method of lunes is self-evident.

This mathematical procedure is an analog of the physical situation encountered in our spherical reflector problem for large reflectors, i.e., $R/\lambda \gg 1$. Our wave problem is exactly of the oscillatory nature required. The stationary phase regions correspond to the multiple images that geometric optics analysis has led us to expect [3, Fig. 19]. Also, the vector nature of (3) permits us to account for the polarization effects of the reflector.

Stationary Phase Conditions for a Spherical Reflector

Now

$$l = l_1 + l_2 \tag{5}$$

where

$$l_1 = \cos\theta \tag{6}$$

and

$$l_2 = \left[1 + r_p^2 - 2r_p(\sin\theta\sin\theta_p\cos(\phi-\phi_p) + \cos\theta\cos\theta_p)\right]^{1/2} \tag{7}$$

where $(1, \phi, \theta)$ are the coordinates of the current element on the reflector, (r_p, ϕ_p, θ_p) are the coordinates of the field point, l_2 is the distance between them, and the radial coordinate is normalized with respect to R. If we choose $2\pi R/\lambda$ to correspond to k, $l(\theta, \phi)$ to correspond to $f(x, y)$, and the rest of the integrands to correspond to $g(x, y)$, we see that for $R/\lambda \gg 1$, $2\pi R/\lambda$ is indeed large. Further, as seen from (5), (6), and (7), $l(\theta, \phi)$ is bounded, never zero, and is a well-behaved function of θ and ϕ in the region bounded by the paraxial focus, the marginal focus, and the caustic surface. In this region, $l_2(\theta, \phi)$ is also bounded and nonzero. Thus the functions corresponding to $g(x, y)$ are well behaved. Hence we may apply the stationary phase conditions

$$l_\phi(\theta, \phi) = 0 \tag{8}$$

$$l_\theta(\theta, \phi) = 0 \tag{9}$$

(where the subscripted variable denotes partial differentiation). These are the mathematical analogs of Fermat's principle, which states that ray-path length through an optical system is an extremum. Applying condition (8) to (5),

$$\frac{r_p\sin\theta_p\sin\theta\sin(\phi-\phi_p)}{l_2(\theta, \phi)} = 0. \tag{10}$$

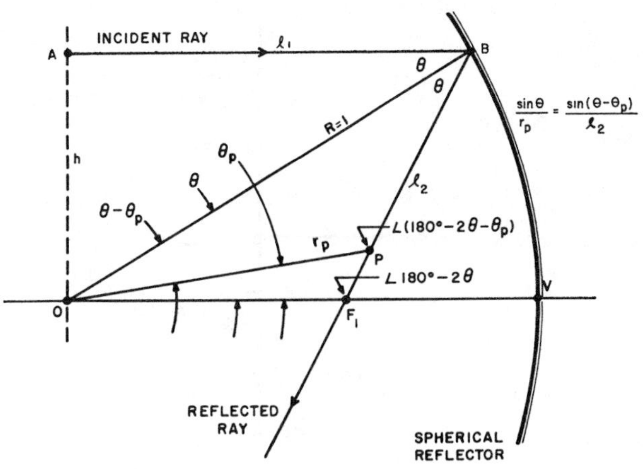

(a)

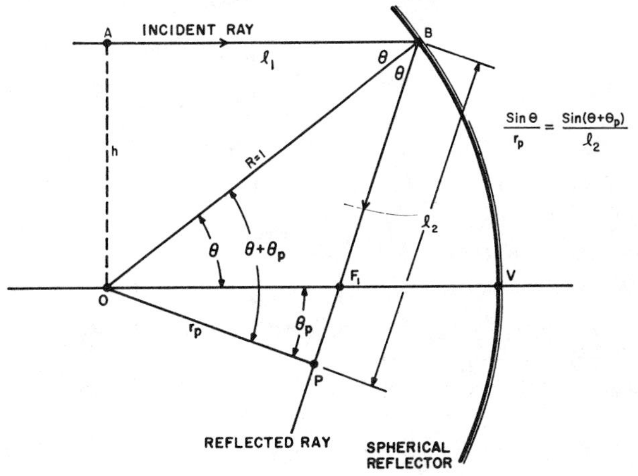

(b)

Fig. 3. (a) Interpretation of l_θ condition for $\phi-\phi_p=0$. (b) Interpretation of l_θ condition for $\phi-\phi_p=\pi$.

Since $l_2(\theta, \phi) \neq 0$, condition (8) is satisfied when $\theta_p=0$, $\theta=0$, or $(\phi-\phi_p)=0$ or π.

Similarly, condition (9) leads to

$$\frac{\sin\theta_p}{r_p} = \frac{-(\sin\theta_p\cos\theta\cos(\phi-\phi_p) - \cos\theta_p\sin\theta)}{l_2(\theta, \phi)}. \tag{11a}$$

If $\phi-\phi_p=0, \pi$, then (11a) becomes

$$\frac{\sin\theta}{r_p} = \frac{\sin(\theta \mp \theta_p)}{(1 + r_p^2 - 2r_p\cos(\theta\mp\theta_p))^{1/2}}. \tag{11b}$$

Geometrically, $\phi-\phi_p=0, \pi$ means that the stationary point on the reflector and the field point in the focal region are coplanar. Further, using (11b), one can show that the angle of incidence equals the angle of reflection. Thus the laws of geometric optics are implicit in conditions (8) and (9). Fig. 3(a) and (b) shows this for $\phi-\phi_p=0$ and $\phi-\phi_p=\pi$, respectively.

The other possible consequences of condition (8) and (10). can be shown to arise from (11b), as well. These are for $\theta_p=0$, field point on the axial caustic, which leads to the requirements that $r_p=l_2$ and $\theta \pm 0$, or that $\theta=0$ and $r_p=l_2=\frac{1}{2}$

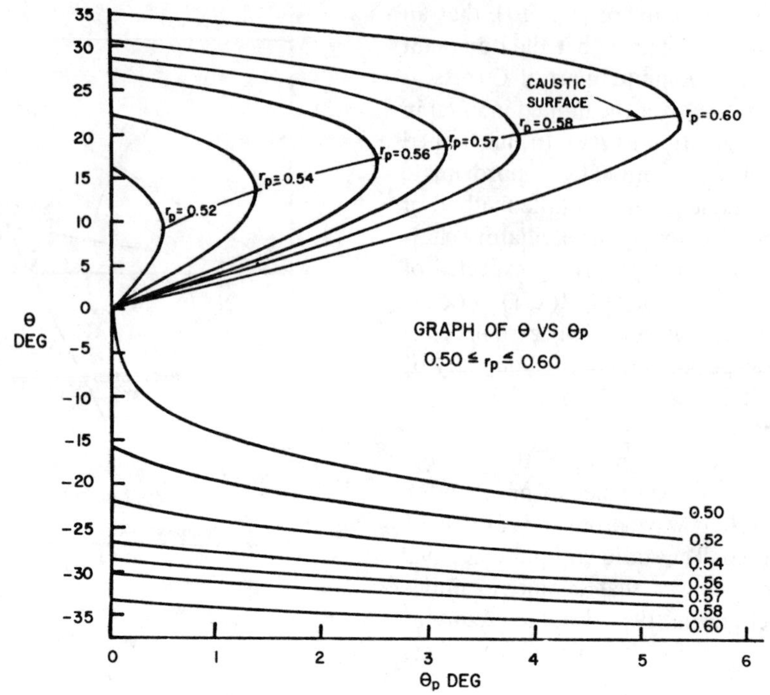

Fig. 4. Angular location of stationary points on spherical reflector.

(the paraxial focus). For the case $\theta=0$, $\theta_p=0$ and the field point is restricted to the axial caustic.

Thus

$$\phi - \phi_p = 0,\ \pi \tag{12}$$

is one general condition. The other general condition then is (11b). Conditions (11b) and (12) determine the direction $\hat{n}_2$ from the stationary point on the reflector $B(1, \theta, \phi)$ to the field point $P(r_p, \theta_p, \phi_p)$. Equation (11b) relates θ and θ_p as shown in Fig. 4. Equation (12) completes the restriction. Thus $\hat{n}_2$ is the vector reflection of $\hat{n}_1$ of Fig. 2 in the tangent plane at B. Whence

$$\hat{n}_2 = \hat{n}_1 - 2\hat{n}_s(\hat{n}_1 \cdot \hat{n}_s). \tag{13}$$

Noting from Fig. 2 that $\hat{n}_1 = \hat{k}$, and that

$$\hat{n}_s = -\frac{\overline{OB}}{OB} = -(\hat{\imath}\sin\theta\cos\phi + \hat{\jmath}\sin\theta\sin\phi + \hat{k}\cos\theta), \tag{14}$$

$$\overline{K}_{2\perp} = 2H_1(\hat{\imath}\cos\theta(1 - \sin^2\theta\cos^2\phi)) - \hat{\jmath}\sin\theta\sin 2\theta\sin\phi\cos\phi - \hat{k}\cos\theta\sin 2\theta\cos\phi) \tag{15}$$

as shown in [4], except that now restrictions on $\hat{n}_2$ are explicit, arising from the method of stationary phase.

Taylor Expansion of Path Length

Normalized path length

$$l = l_1 + l_2 = \cos\theta + (1 + r_p^2 - 2r_p\cos\delta)^{1/2} \tag{16}$$

where

$$l_1 = \cos\theta \tag{6}$$

$$l_2 = (1 + r_p^2 - 2r_p\cos\delta)^{1/2} \tag{17}$$

$$\cos\delta = \sin\theta\sin\theta_p\cos(\phi - \phi_p) + \cos\theta_p\cos\theta. \tag{18}$$

Let (θ_s, ϕ_s) be the coordinates of a stationary point on the reflector. Expanding $l(\theta, \phi)$ in a Taylor series about (θ_s, ϕ_s),

$$l(\theta, \phi) = l(\theta_s, \phi_s) + \sum_1^\infty \frac{1}{m!}\left[(\theta - \theta_s)\frac{\partial}{\partial\theta} + (\phi - \phi_s)\frac{\partial}{\partial\phi}\right]^m l(\theta, \phi)\bigg|_{\substack{\theta=\theta \\ \phi=\phi_s}} . \tag{19}$$

The integrals of (3) are evaluated asymptotically [1], representing $l(\theta, \phi)$ by the leading nonzero terms of its Taylor expansion. The zero-degree term $l(\theta_s, \phi_s)$ contributes a phase constant. The first degree term $(\theta-\theta_s)l_\theta(\theta_s, \phi_s)+(\phi-\phi_s)l_\phi(\theta_s, \phi_s)$ is set to zero by the stationary phase conditions (8) and (9). Thus the leading terms used are second degree or higher.

The coefficients of these terms all involve partial derivatives of $l(\theta, \phi)$. Since $l(\theta, \phi)=l_1(\theta)+l_2(\theta, \phi)$, one may consider separately the partial derivatives of $l_1(\theta)$ and $l_2(\theta, \phi)$ Because $l_1=l_1(\theta)$ only, only derivatives with respect to θ are nonzero, i.e.,

$$l_{1(\theta)^n} = (-1)^{2n+1-(-1)^n/4} \cdot [\cos\theta + \sin\theta + (-1)^n(\cos\theta - \sin\theta)]/2 \tag{20}$$

and

$$l_{1(\theta)^p(\phi)^q} = 0 \qquad q \neq 0. \tag{21}$$

Now

$$l_{2\gamma} = \frac{-r_p}{l_2}(\cos\delta)_\gamma, \qquad \gamma = \theta, \phi. \tag{22}$$

Thus higher-order derivatives with respect to ϕ and θ are expressible in terms of derivatives of $\cos\delta$ and $1/l_2$. In general,

$$\left[\left(\frac{1}{l_2}\right)^n\right]_\gamma = \frac{nr_p}{l_2^{n+2}}(\cos\delta)_\gamma, \qquad \gamma = \theta, \phi \tag{23}$$

and

$$\begin{aligned}(\cos\delta)_{(\phi)^m(\theta)^n} = {} & \frac{(-1)^{(2m+1-(-1)^m)/4}}{2} \\ & \cdot\left[\cos(\phi-\phi_p)+\sin(\phi-\phi_p) + (-1)^m(\cos(\phi-\phi_p)-\sin(\phi-\phi_p))\right] \\ & \cdot\frac{(-1)^{(2n-1+(-1)^n)/4}}{2} \\ & \cdot\left[\sin\theta+\cos\theta+(-1)^n(\sin\theta-\cos\theta)\right]\sin\theta_p \\ & +\left(\prod_{j=0}^{m}(1-j)\right)\frac{(-1)^{(2n+1-(-1))^n/4}}{4} \\ & \cdot\left[\cos\theta+\sin\theta+(-1)^n(\cos\theta-\sin\theta)\right]\cos\theta_p.\end{aligned} \tag{24}$$

All derivatives of $l_2(\theta, \phi)$ are terminating series of terms of the form $r_pD(\cos\delta)l^{-2n-1}$ where $D(\cos\delta)$ is a function of $\cos\delta$ and its derivatives. These are continuous, as is $l_2(\theta, \phi)$, and $l_2(\theta, \phi) \geq 0.3$ in the range of concern, whence the derivatives are continuous and the order of differentiation may be freely interchanged in all mixed derivatives of $l_2(\theta, \phi)$ and hence of $l(\theta, \phi)$. Using (20), (23), and (24), all the required derivatives of $l(\theta, \phi)$ may be formed.

Evaluation of Stationary Phase Integral Representation

Let

$$(\overline{K}_{2\perp}R\sin\theta)/l_2 = \overline{G}(\theta, \phi)H_1\lambda \tag{25}$$

where

$$\overline{G}(\theta, \phi) = \hat{\imath}G_i(\theta, \phi) + \hat{\jmath}G_j(\theta, \phi) + \hat{k}G_k(\theta, \phi) \tag{26}$$

and

$$G_i = 2R\sin\theta(\cos\theta - \sin\theta\sin 2\theta\cos^2\phi)/(\lambda l_2) \tag{27}$$

$$G_j = -R\sin^2\theta\sin 2\theta\sin 2\phi/(\lambda l_2) \tag{28}$$

$$G_k = -R\sin^2 2\theta\cos\phi/(\lambda l_2). \tag{29}$$

Then the electric field is given by

$$\begin{aligned}\frac{\overline{E}}{j\omega\mu/4\pi} = {} & \hat{\imath}H_1\lambda\int_{\text{Reflector}} G_i e^{j(2\pi/\lambda)Rl}d\theta\, d\phi \\ & + \hat{\jmath}H_1\lambda\int_{\text{Reflector}} G_j e^{j(2\pi/\lambda)Rl}d\theta\, d\phi \\ & + \hat{k}H_1\lambda\int_{\text{Reflector}} G_k e^{j(2\pi/\lambda)Rl}d\theta\, d\phi.\end{aligned} \tag{30}$$

Ordinary Points ($l_{\phi\phi}, l_{\theta\theta} \neq 0$): For points off the axial caustic and the caustic surface, the leading terms are second degree. At each such field point, from each of three stationary points on the reflector, for each of three polarizations, there is a contribution of the form

$$E_{\mu m} = \frac{H_1\lambda G_\mu(\theta_m, \phi_m)\epsilon_{\theta\theta}\epsilon_{\phi\phi}e^{j(2\pi/\lambda)Rl(\theta_m,\phi_m)}}{(R/\lambda)\,|\,l_{\theta\theta}(\theta_m, \phi_m)l_{\phi\phi}(\theta_m, \phi_m)\,|^{1/2}} \tag{31}$$

where

μ = polarization = i, j, k in turn,
m = designation of stationary point = 1, 2, 3,
$\epsilon_{\theta\theta}, \epsilon_{\phi\phi} = e^{\pm j\pi/4}$ according as $l_{\theta\theta}, l_{\phi\phi} \gtrless 0$ at (θ_m, ϕ_m).

Whence the field at a typical field point off the various caustics is given by

$$\begin{aligned}\frac{\overline{E}}{j\omega\mu/4\pi} \sim {} & \hat{\imath}(E_{i1}+E_{i2}+E_{i3})+\hat{\jmath}(E_{j1}+E_{j2}+E_{j3}) \\ & + \hat{k}(E_{k1}+E_{k2}+E_{k3}).\end{aligned} \tag{32}$$

This representation is valid for every field point in the focal region for which $l_{\theta\theta}, l_{\phi\phi} \neq 0$. For a field point $P(r_p, \theta_p, \phi_p)$ lying between the caustic surface and the axial caustic, (11b) yields three values of θ_m. Two are associated with the choice of conditions $\phi - \phi_p = 0$ and $(\theta_m - \theta_p)$, for which we assign $m = 1, 2$. These lie on the same side of the axial caustic as the field point, and may be interpreted as equivalent to rays shown in Fig. 3(a). The other ($m = 3$) is associated with the choice of conditions $(\theta - \theta_p) = \pi$ and $(\theta_m + \theta_p)$, and lies across the axial caustic (see Fig. 3(b)). $l_{\theta\theta}$ and $l_{\phi\phi}$ are given by

$$l_{\theta\theta}(\theta_m, \phi_m) = \pm\frac{\sin\theta_p\sin\theta_m}{l_2(\theta_m, \phi_m)} \tag{33}$$

$$\begin{aligned}l_{\phi\phi} = {} & -\cos\theta_m + \frac{r_p\cos(\theta_m \mp \theta_p)}{l_2(\theta_m, \phi_m)} \\ & - \frac{(r_p\sin(\theta_m \mp \theta_p))^2}{l_2^3(\theta_m, \phi_m)}\end{aligned} \tag{34}$$

$$\text{for } \theta_m \neq \theta_p,\ m = 1, 2, 3.$$

For $\theta_p = 0$, on the axial caustic $l_{\phi\phi} = 0$. This representation does not hold there. It can be shown [2] that on the caustic surface

$$x/r_p = \sin\theta_p = \sin^3\theta_m/r_p. \tag{35}$$

For the conditions $\phi - \phi_p = 0$ and $(\theta_m - \theta_p)$ for points on the caustic surface, substitution of (35) into (34) yields

$$l_{\theta\theta}(\theta_m, \phi_m) = 0.$$

The representation of (31) and (32) fails here too. Interestingly, the values of θ_m for the two stationary points for which $\phi_m - \phi_p = 0$ become the same, i.e., the two stationary points have coalesced, as seen in Fig. 4 near the maxima, where two values of θ approach each other for increasing θ_p.

Points on the Caustic Surface ($l_{\theta\theta} = 0$): The coalescing of the two stationary points brings about a higher-order stationary point, but it does not affect the third stationary point in the half-plane $(\phi_m-\phi_p)=\pi$. Thus the field on the caustic results from the interference of two different types of contribution. The ordinary contribution is given by (31). For the other, Kay [11] has shown how one may evaluate the fields across a caustic in terms of the Picht fields, given the geometry of the caustic. However, it is possible to use the method of stationary phase and asymptotic analysis directly. One may show that for the higher-order stationary point $B(\theta_0, \phi_0)$,

$$\begin{aligned} l(\theta, \phi) &= l(\theta_0, \phi_0) + \frac{1}{2!}(\phi - \phi_0)^2 l_{\phi\phi}(\theta_0, \phi_0) \\ &+ \frac{1}{3!}(\theta - \theta_0)(\phi - \phi_0)^2 l_{\theta\phi\phi}(\theta_0, \phi_0) \\ &+ \frac{1}{3!}(\theta - \theta_0)^3 l_{\theta\theta\theta}(\theta_0, \phi_0) \\ &+ \text{higher order terms.} \end{aligned} \tag{36}$$

Whence [1], [12], [13]

$$E_{\mu c} = H_1 \lambda G_\mu(\theta_0, \phi_0) \cdot \frac{4.98 e^{j(2\pi/\lambda)Rl(\theta_0,\phi_0)} \gamma_{\theta\theta\theta} \epsilon_{\phi\phi}}{\left(\frac{2\pi R}{\lambda}\right)^{5/6} |l_{\theta\theta\theta}(\theta_0, \phi_0)|^{1/3} |l_{\phi\phi}(\theta_0, \phi_0)|^{1/2}} \tag{37}$$

where

$$\gamma_{\theta\theta\theta} = e^{-j(\pi/3 \pm \pi/6)} \text{ according as } l_{\theta\theta\theta} \gtrless 0,$$

$$\mu = i, j, k, \text{ in turn,}$$

and

$$l_{\phi\phi} = (r_p \sin\theta_p \sin\theta_0)/l_2(\theta_0, \phi_0) \tag{38}$$

$$l_{\theta\theta\theta} = -(3\cos\theta_0 \sin\theta_0)/l_2(\theta_0, \phi_0). \tag{39}$$

Thus the total field at field points on the caustic surface is given by

$$\frac{\bar{E}}{j\omega\mu/4\pi} \sim \hat{\imath}[E_{ic} + E_{i3}] + \hat{\jmath}[E_{jc} + E_{j3}] + \hat{k}[E_{kc} + E_{k3}]. \tag{40}$$

This representation is valid for all points on the caustic surface except where $\theta_0=0$ and $\theta_p=0$ at the cusp on the caustic, i.e., at the paraxial focus.

In the limit of small wavelengths, the fields arising from the higher-order caustic will dominate because, from (31) and (37),

$$\frac{E_{\mu c}}{E_{\mu m}} = \frac{4.98(R/\lambda)^{1/6}}{(2\pi)^{5/6}} \left|\frac{G_\mu(\theta_0, \phi_0)}{G_\mu(\theta_3, \phi_3)}\right| \left|\frac{l_{\phi\phi}(\theta_3, \phi_3)}{l_{\phi\phi}(\theta_0, \phi_0)}\right|^{1/2} \cdot \frac{|l_{\theta\theta}(\theta_3, \phi_3)|^{1/2}}{|l_{\theta\theta\theta}(\theta_0, \phi_0)|^{1/3}} \propto (R/\lambda)^{1/6}. \tag{41}$$

However, for reflectors of the order considered computationally, these terms were of the same order of magnitude, and so a well-defined caustic in the geometric optics sense [2] is not observed. Thus while the asymptotic approximations are precise only in the limit of $R/\lambda \to \infty$, the analysis shows how the caustic surface loses its dominance for smaller reflectors.

Points on the Axial Caustic ($l_{\phi\phi}=0$): All points off the axial caustic have been treated. For those that remain, $\theta_p=0$ and there is no variation of $l(\theta, \phi)$ with ϕ. Hence now

$$l(\theta, \phi)\big|_{\theta_p=0} = l(\theta) = l(\theta_0) + \sum_{m=2}^{\infty} \frac{1}{m!}(\theta - \theta_0)^m l_{(\theta)^m}(\theta_0).$$

Examining (27) through (30) and integrating over ϕ, it is seen that the $\hat{\jmath}$ and $\hat{k}$ components vanish and

$$\frac{\bar{E}}{j\omega\mu/4\pi} = \hat{\imath} 2H_1\lambda \int_{\text{Reflector}} d\theta \sin\theta \cos\theta \frac{e^{-j(2\pi/\lambda)Rl}}{l_2} \cdot \int_0^{2\pi} d\phi(1 - 2\sin^2\theta\cos^2\phi). \tag{42}$$

(If in (42), the *obliquity factor*, the integrand with respect to ϕ, is set to unity, the resulting integral can be shown [1] equivalent to that evaluated by Schell [14], except for a phase constant.) Performing the ϕ integration,

$$\frac{\bar{E}}{j\omega\mu/4\pi} = \hat{\imath} H_1\lambda 4\pi R/\lambda \int_{\text{Reflector}} d\theta \sin\theta \cos^3\theta e^{-j(2\pi/\lambda)Rl(\theta)}. \tag{43}$$

Contribution from stationary points of $l(\theta)$ are of the form (for $l_{\theta\theta} \neq 0$)

$$E_{i\nu} = H_1\lambda 4\pi\sqrt{R/\lambda} \sin\theta_\nu \cos^3\theta_\nu \frac{e^{-j(2\pi/\lambda)Rl(\theta_\nu)}}{l_2(\theta_\nu)} \cdot \frac{\epsilon_{\theta\theta}}{|l_{\theta\theta}(\theta_\nu)|^{1/2}}. \tag{44}$$

For $\theta_p=0$,

$$l_{\theta\theta}(\theta) = -\cos\theta + \frac{r_p}{l_2}\cos\theta - \frac{r_p^2}{l_2^3}\sin^2\theta \tag{45}$$

$l_{\theta\theta} \neq 0$ except when $\theta=0$ *and* $r_p=l_2$, i.e., at the paraxial focus. When $\theta \neq 0$, $r_p=l_2$, whence $\cos\theta = 1/2r_p$

and

$$l_{\theta\theta} = -\sin^2\theta/r_p \tag{46}$$

and

$$E_{i\nu_1} = H_1\lambda 4\pi \frac{\sqrt{R/\lambda}}{r_p^{7/2}} e^{-j2\pi((1/2r_p + r_p)R/\lambda - 1/8)}. \tag{47}$$

When $\theta=0$, $r_p=l_2$, but $l_2=1-r_p$, and the integral of (43) must be reexamined. In the neighborhood of $\theta=0$, we have $\cos\theta \approx 1$, $\sin\theta \approx \theta$, and in the integrand denominator $l_2 \approx 1-r_p$, whence

$$\frac{\overline{E}}{j\omega\mu/4\pi} \approx \hat{\imath}H_1\lambda 4\pi \frac{R/\lambda}{1 - r_p}\int_{\text{Reflector}} d\theta\, \theta e^{-j(2\pi/\lambda)Rl}. \quad (48)$$

Applying Laplace's method, the integral to be evaluated is of the form

$$I(k) = \int_0^\infty dx\, xe^{-j\alpha x^2} \quad (49)$$

whence [15]

$$I(k) = 1/2e^{-(\pi/2)j}(\alpha)^{-1}\Gamma(1) \quad (50)$$

where

$$\alpha = \pi l_{\theta\theta}R/\lambda. \quad (51)$$

But

$$l_{\theta\theta}\bigg|_{\substack{\theta_p=0\\ \theta=0}} = \frac{2r_p - 1}{1 - r_p}.$$

Thus

$$E_{iv_2} = H_1\lambda \frac{2}{2r_p - 1} e^{-j2\pi((2-r_p)R/\lambda - 1/4)} \quad (52)$$

except when $l_{\theta\theta}=0$, which occurs for $r_p=0.5$, at the paraxial focus. Thus along the axial caustic

$$\frac{\overline{E}}{j\omega\mu/4\pi} \sim \hat{\imath}H_1\lambda\left[\frac{4\pi\sqrt{R/\lambda}}{r_p^{7/2}} e^{-j2\pi((1/2r_p+r_p)R/\lambda-1/8)} - \frac{2}{2r_p - 1} e^{-j2\pi((2-r_p)R/\lambda-1/4)}\right] \quad (53)$$

except at the paraxial focus.

The Paraxial Focus: The axial caustic joins the caustic surface at the cusp, the paraxial focus ($\theta_p=0$, $r_p=0.5$). The stationary conditions are fulfilled for $\theta=0$. Thus

$$\theta_p = \theta = 0; \qquad r_p = l_2 = 1/2.$$

There is no variation with ϕ. Variation with θ is slow. Indeed, of the first five derivatives of l, only $l_{(\theta)^4}$ is nonzero. Persisting with our stationary phase approach, we may now write

$$\frac{\overline{E}}{j\omega\mu/4\pi} \approx \hat{\imath}H_1\lambda 8\pi R/\lambda e^{-j2\pi(R/\lambda)l} \cdot \int_{\text{Reflector}} d\theta \sin\theta \cos^3\theta e^{-j(2\pi/\lambda)Rl_{(\theta)^4}\theta^4/4!}. \quad (54)$$

Then putting

$$\sin\theta \approx \theta, \quad \cos\theta \approx 1, \quad \text{and} \quad \frac{\pi}{12}R/\lambda l_{(\theta)^4} = \frac{\pi}{2}u^2 \quad (55)$$

one obtains

$$\frac{\overline{E}}{j\omega\mu/4\pi} \sim \hat{\imath}H_1\lambda \frac{4\pi\sqrt{6R/\lambda}}{\sqrt{l_{(\theta)^4}}} e^{-j(2\pi/\lambda)Rl} \cdot \int_{\text{Reflector}} du\, e^{-j(\pi/2)u^2} \quad (56)$$

and finally, noting $l_{(\theta)^4} = -3/2$, $l = +3/2$,

$$\frac{\overline{E}}{j\omega\mu/4\pi} \sim \hat{\imath}H_1\lambda 4\pi\sqrt{2R/\lambda}\, e^{-j3\pi(R/\lambda-1/4)}. \quad (57)$$

IV. Evaluation of Stationary Phase Representation

The formulas of the preceding sections were used in normalized form to calculate the fields in the focal region on a concentric spherical surface $r_p=0.563$, which is transverse to the axial caustic in the neighborhood of the circle of least confusion for a useful reflector section subtending about 60 degrees.

The range of θ_p, $0° \leq \theta_p \leq 3°$, was chosen to form a cap through the circle of least confusion that extends from the axial caustic cutting through the caustic surface. The choice of $R/\lambda = 140$ was dictated by measurement considerations. Calculations were made for $0 \leq \phi \leq 90°$ for the $\hat{\imath}, \hat{\jmath}$, and $\hat{k}$ components of normalized $\overline{E}$:

$$\overline{E}_n = \overline{E}\Big/\left(\frac{j\omega\mu}{4\pi}H_1\lambda\right). \quad (58)$$

In the neighborhood of the axial caustic and the caustic surface, the results on the caustics were calculated from the formulas, and interpolation was used to connect these to the calculations for ordinary field points. The results are shown in Fig. 5(a), (b), (c), and (d).

Experimental verification of the theoretical calculations has been made as part of a further study [16]. These experiments were conducted using a solid aluminum precision spherical reflector 10 ft in diameter, with a radius of curvature of 5.75 ft (140λ at 24 GHz, the measurement frequency), supplied by the Air Force Cambridge Research Laboratories. The transverse focal region fields were probed for $\hat{\imath}$ and $\hat{k}$ polarization at $r_p=0.56$. Fig. 6(a) and (b) show the comparison of measured and theoretical results. (The patterns have been adjusted relative to each other for best fit.) It is seen that the agreement is quite good. The angular correspondence of peaks and nulls is excellent. The relative strength at the peak of the lobes compares well. Deviation in the nulls and beyond the caustic surface is not unexpected, since it is in the regions of lower field strength where the limitations of the measurement setup and the theoretical model are most likely to be apparent. Overall, the agreement is adequate.

V. Conclusions

It has been demonstrated how the method of stationary phase may be used in extended form to find focal region fields for the case of the spherical reflector. In conjunction with the current distribution method, it can be used to account for polarization phenomena.

As expected, because both are precise in the limit of small wavelengths, a strong correspondence has been shown to exist between the geometric optics approach and the stationary phase method. However, the extended stationary phase method goes beyond geometric optics, permitting calculations on the axial caustic, caustic surface, and par-

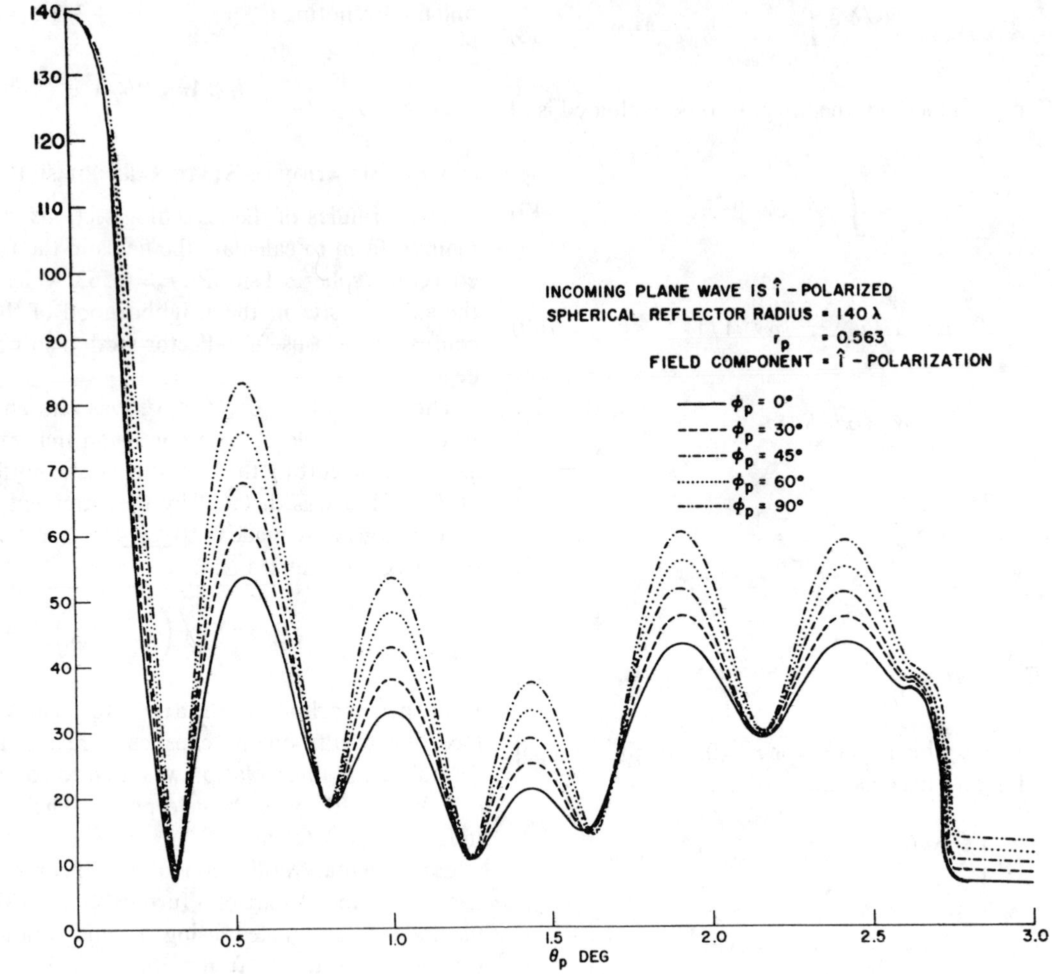

(a) $\hat{\imath}$ polarization

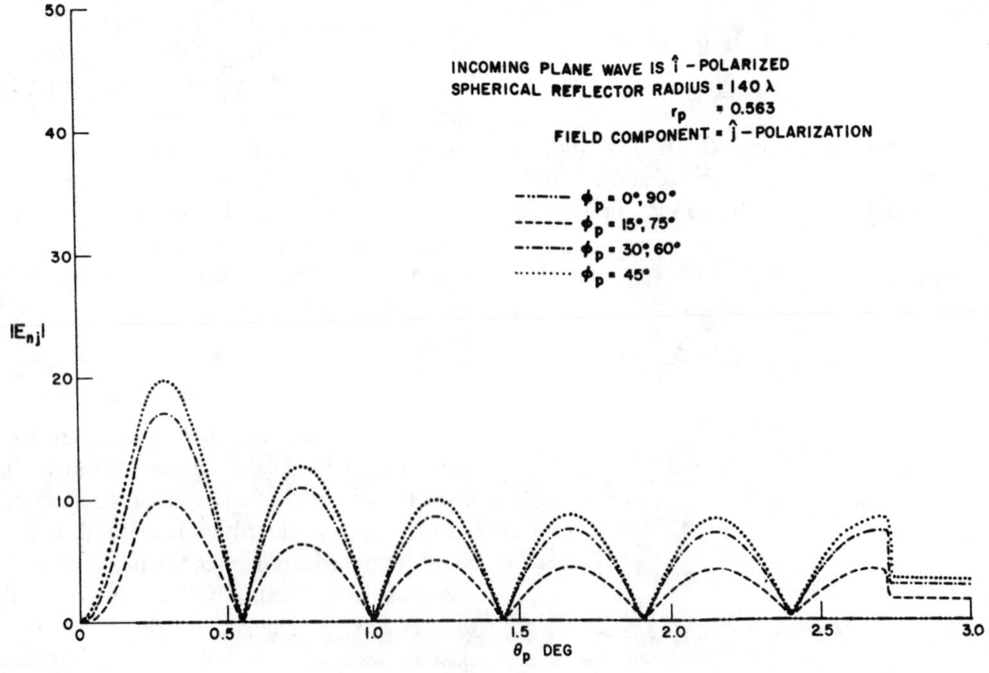

(b) $\hat{\jmath}$ polarization

Fig. 5. (a)–(d) Computed fields.

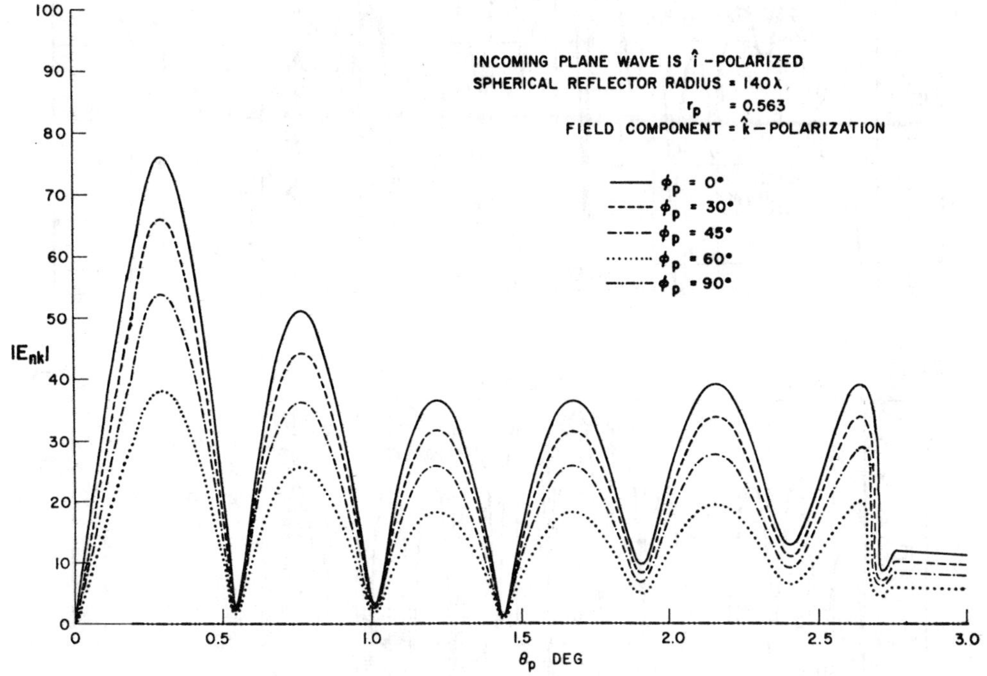

(c) $\hat{k}$ polarization

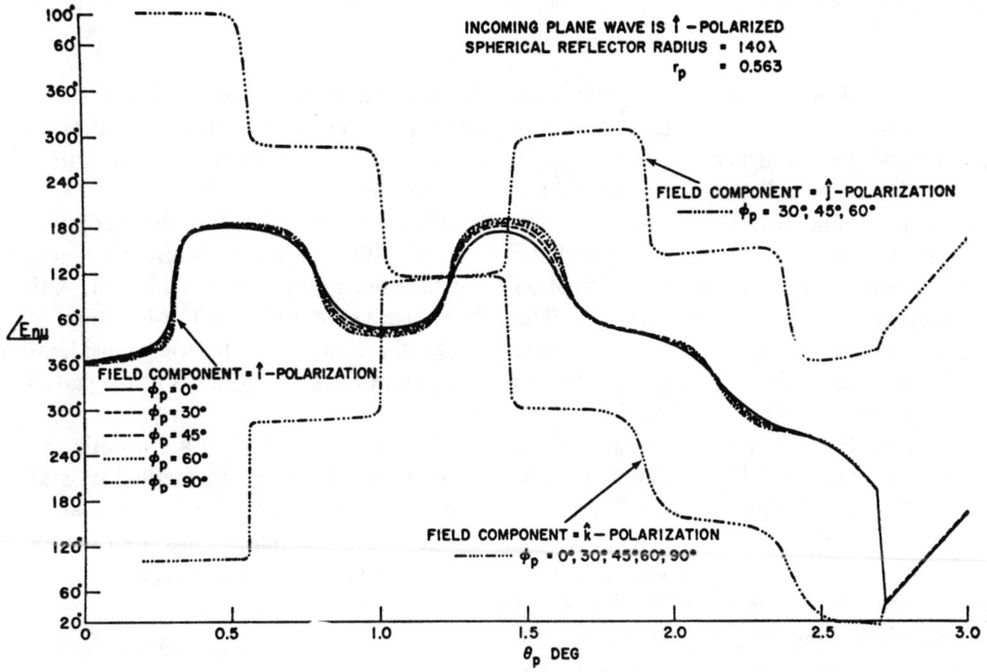

(d) phase

Fig. 5. (*cont'd*).

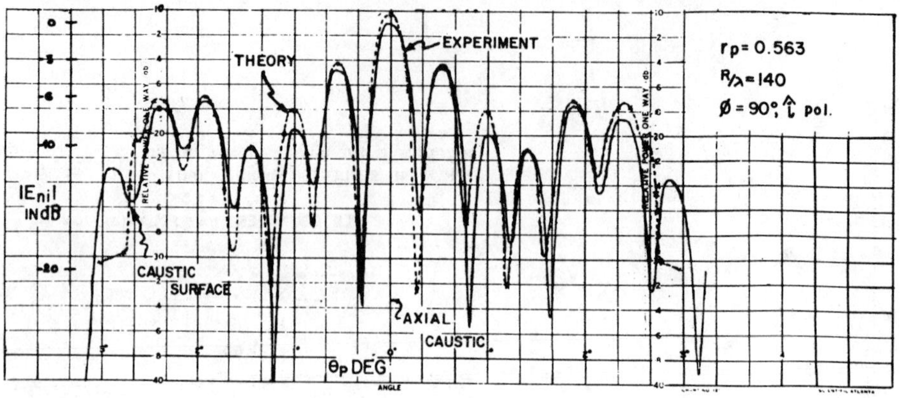

(a)

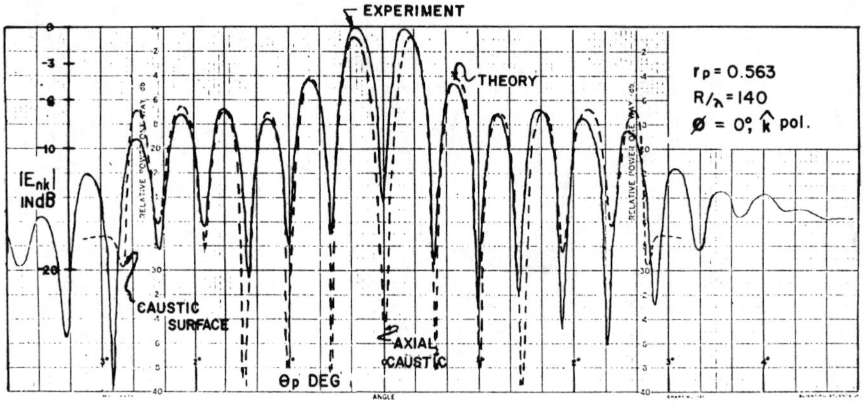

(b)

Fig. 6. Focal region fields, comparison of theoretical and experimental data. (a) $\hat{\imath}$ polarization. (b) $\hat{k}$ polarization.

axial focus, including polarization phenomena, and thus taking us a significant distance into the interregnum between geometric optics and electromagnetic fields.

The fields computed using the stationary phase model are shown in Fig. 5(a), (b), (c), and (d). The variation of the components of $\overline{E}$ with ϕ_p can be explained in terms of the variation with ϕ of the components of the effective current on the reflector $\overline{K}_{2\perp}$, keeping in mind that $\phi-\phi_p=0$, π. The variation with θ_p is the consequence of the interference of the three or less stationary regions on the reflector. Consider the case of ordinary field points. Contributions arise from three coplanar stationary points. Two lie on the same side of the axial caustic as the field point ($\phi-\phi_p=0$). One lies on the opposite side ($\phi-\phi_p=\pi$). The $\hat{\jmath}$ and $\hat{k}$ components of $\overline{E}$ resemble the interference pattern between two sources, especially for small θ_p. But for small θ_p, the inner of the stationary points for the same side ($\phi-\phi_p=0$) is in a region of small θ where $\overline{K}_{2\perp}$ is essentially $\hat{\imath}$ polarized. Thus the $\hat{\jmath}$ and $\hat{k}$ patterns arise from the interference of the two wide-angle contributors. The $\hat{\imath}$ pattern results from the interference of contributions of three stationary points.

At the caustic surface, the fields are the resultant of contributions from a higher order stationary point on the same side of the axial caustic as the field point, and an ordinary stationary point on the opposite side. On the axial caustic, the ϕ variation and cross polarization vanish, and the contributions come mainly from a stationary ring at $\theta=\cos^{-1}(1/2r_p)$, with a second-order contribution from a stationary point at the vertex.

Just off the axial caustic, the $\hat{k}$ component is particularly strong. This component is usually ignored. For transverse feeds, it must be taken into account. Although its existence is indicative of the non-TEM nature of the focal region fields, these fields result from consideration of the simple contributions of the currents at the stationary points on the reflector.

Thus it is seen that an asymptotic theory may be developed, using the extended method of stationary phase, that permits evaluation of fields in the focal region of a spherical reflector. This evaluation has been used as the basis for development of transverse feeds [16].

Acknowledgment

The author is indebted to Dr. R. C. Spencer, formerly with RCA, for his encouragement and physical insight freely given, and to C. J. Sletten, Chief of the Microwave Physics Laboratory, and O. Kerr, of the Air Force Cambridge Research Laboratories, under whose auspices much of this work was done.

References

[1] G. Hyde, "Focal region fields of a spherical reflector," Ph.D. dissertation, University of Pennsylvania, Philadelphia, 1967.

[2] R. C. Spencer and G. Hyde, "Studies of the focal region of a spherical reflector: geometric optics," *IEEE Trans. Antennas and Propagation*, vol. AP-16, pp. 317–324, May 1968.

[3] ——, "Transverse focal region of a spherical reflector, Part I: geometric optics," Air Force Cambridge Research Labs., Scientific Rept. 1, pt. 1, Contract AF19(628) 2758, AFCRL-64-292I, March 1964.

[4] G. Hyde and R. C. Spencer, "Studies of the focal region of a spherical reflector, Part II: polarization effects," *IEEE Trans. Antennas and Propagation*, vol. AP-16, pp. 399–405, July 1968.

[5] ——, "Transverse focal region of a spherical reflector, Part II: polarization," Air Force Cambridge Research Labs., Scientific Rept. 1, pt. 2, Contract AF19(628)2758, AFCRL-64-292II, May 1964.

[6] G. Hyde, "Transverse focal region fields of a spherical reflector," Air Force Cambridge Research Labs., Scientific Rept. 2, Contract AF19(628)2758, AFCRL-66-48, January 1966.

[7] S. Silver, *Microwave Antenna Theory and Design*, M.I.T. Radiation Lab. Ser., vol. 12. New York: McGraw-Hill, 1949, sec. 5.7.

[8] E. T. Copson, "The asymptotic expansion of a function defined by a definite integral or contour integral," Dept. of Scientific Research and Experiment, The Admiralty, London, England, Ref. S.R.E./ACS 106, 1946, sec. 3.

[9] A. Papoulis, *The Fourier Integral and Its Applications.* New York: McGraw-Hill, 1962, sec. 7-7.

[10] N. G. DeBruijn, *Asymptotic Methods in Analysis.* Amsterdam, The Netherlands: North-Holland Publishing Co., 1958, ch. 4–6.

[11] I. Kay, "Fields in the neighborhood of a caustic," *IRE Trans. Antennas and Propagation* (*Special Supplement*), vol. AP-7, pp. S255–S260, December 1959.

[12] M. V. Cerillo, "An elementary introduction to the theory of the saddlepoint method of integration," M.I.T. Research Lab. of Electronics, Cambridge, Mass., TR 55:2a, May 3, 1950.

[13] M. V. Cerillo and W. H. Kautz, "Properties and tables of the extended Airy–Hardy integrals," M.I.T. Research Lab. of Electronics, Cambridge, Mass., TR 144, November 15, 1951.

[14] A. C. Schell, "The diffraction theory of large-aperture spherical reflector antennas," *IEEE Trans. Antennas and Propagation*, vol. AP-11, pp. 428–432, July 1963.

[15] W. Grobner and N. Hofretter, "Integraltafel," in *Bestimmte Integrale*, vol. 2. Berlin: Springer, 1961, sec. 314.

[16] D. F. Bowman, "Transverse antenna feeds," Air Force Cambridge Research Labs., Final Rept., Contract AF19(628)2758, January 1966.

Fields in the Focal Region of a Spherical Reflector

Abstract—**An earlier analysis of the field structure and energy flow near the axis of any circularly symmetric focusing reflector is applied to the case of a spherical reflector. From the results an estimate is made of the attainable efficiency of corrugated-waveguide aperture-type feeds.**

Introduction

It has been shown that the fields in the focusing region near the axis of a circularly symmetric reflector illuminated by a linearly polarized wave incident normally on the aperture can be represented by a spectrum of axial hybrid waves [1]. The theory has been used to carry out a detailed study of the fields and energy flow near the focus of a paraboloid. It has also been shown that the hybrid waves can propagate as discrete modes inside cylindrical pipes with corrugated walls [2]. In the present communication some results of applying these concepts to spherical reflectors are summarized.[1]

Hybrid Axial Waves

The fields generated in the axial region of a spherical reflector illuminated by a linearly polarized normally incident plane wave **E,H** can be calculated from the equations given in [1]. For a reflector of radius R subtending a semiangle θ_0 at its center, the fields produced at a point ρ,ξ,z (origin at the center) by an annular zone $d\theta$ of the reflector become

$$
\begin{aligned}
E_\rho &= E_\rho(u)\sin\xi\\
E_\xi &= E_\xi(u)\cos\xi\\
E_z &= E_z(u)\sin\xi\\
H_\rho &= H_\rho(u)\cos\xi\\
H_\xi &= H_\xi(u)\sin\xi\\
H_z &= H_z(u)\cos\xi
\end{aligned}
\tag{1a}
$$

where

$$
\begin{aligned}
E_\rho(u) &= -\frac{1}{2}jkRE\kappa(\theta)[(\gamma+\Gamma)J_0(u) + (\gamma-\Gamma)J_2(u)]e^{-j\phi}\,d\theta\\
E_\xi(u) &= -\frac{1}{2}jkRE\kappa(\theta)[(\gamma+\Gamma)J_0(u) - (\gamma-\Gamma)J_2(u)]e^{-j\phi}\,d\theta\\
E_z(u) &= -kRE\kappa(\theta)\frac{\sin\theta}{\zeta}J_1(u)e^{-j\phi}\,d\theta\\
H_\rho(u) &= -j\frac{kRE}{2Z_0}\kappa(\theta)[(1+\gamma\Gamma)J_0(u) + (1-\gamma\Gamma)J_2(u)]e^{-j\phi}\,d\theta\\
H_\xi(u) &= +j\frac{kRE}{2Z_0}\kappa(\theta)[(1+\gamma\Gamma)J_0(u) - (1-\gamma\Gamma)J_2(u)]e^{-j\phi}\,d\theta\\
H_z(u) &= -\frac{kRE}{Z_0}\kappa(\theta)\gamma\frac{\sin\theta}{\zeta}J_1(u)e^{-j\phi}\,d\theta
\end{aligned}
\tag{1b}
$$

$$u = \frac{k\rho\sin\theta}{\zeta}$$

$$\kappa(\theta) = \frac{\sin\theta}{\zeta^2}\left[1-\frac{z}{R}\cos\theta\right]$$

$$\gamma = \frac{\zeta\cos\theta}{1-(z/R)\cos\theta}$$

$$\Gamma = \frac{\cos\theta - z/R}{\zeta}$$

$$\zeta = \sqrt{1-2\left(\frac{z}{R}\right)\cos\theta + \left(\frac{z}{R}\right)^2}$$

$$k = 2\pi/\lambda.$$

The phase $\phi = kR(\cos\theta+\zeta)$ is taken relative to the phase of the incident wave at the plane $z=0$. The field equations are accurate to a few percent out to radial distances $\rho/\lambda \le 0.35\sqrt{\zeta R/\lambda}$. For example, near the paraxial focus of the Arecibo 1000-foot reflector, $\rho/\lambda = 5.1$ at $\lambda = 70$ cm or $k\rho = 32$.

Equation (1) represents a hybrid wave (designated HE_{1n}) traveling along the axis away from the reflector. The parameter γ determines the ratio of the longitudinal fields of the TE_{1n} and TM_{1n} components of the wave. When $\gamma = 1$, the **F** and **H** fields are identical except for a rotation of 90 degrees in ξ, and the power flow is axially symmetrical [1].

Field on the Axis

The field on the axis $E(z)$ is linearly polarized in the plane of the incident field and is determined by integrating (1) with $u = 0$;

$$
\begin{aligned}
E(z) &= -\frac{1}{2}jkRE\int_0^{\theta_0}\kappa(\theta)(\gamma+\Gamma)e^{-j\phi}\,d\theta\\
&= -jkRE\int_0^{\theta_0}\frac{\sin\theta}{\zeta}\cdot\left[\cos\theta - \frac{(z/R)\sin^2\theta}{2\zeta^2}\right]\cdot e^{-jkR(\cos\theta+\zeta)}\,d\theta.
\end{aligned}
\tag{2}
$$

Schell [7] has also derived an expression for $E(z)$ which may be put in the following form:

$$E(z) = -jkRE\int_0^{\theta_0}\frac{\sin\theta\cos\theta}{\zeta}\cdot\left[1-\frac{\sin^2\theta}{2\zeta^2}\right]e^{-jkR(\cos\theta+\zeta)}\,d\theta. \tag{3}$$

The discrepancy between (2) and (3) arises because Schell has neglected the fields radiated by the z-directed components of the reflector current. These components are symmetrically disposed and produce zero resultant field in the z direction, but a resultant transverse field does exist and must be included.

For small θ_0 the second term containing $\sin^2\theta$ in each equation is insignificant, so that both equations then become identical. This approximation has been used by Schell to permit an analytical solution of (3). Fig. 1(a) compares (2) and (3) and also Schell's approximation of (3) when $\theta_0 = 36.9$ degrees and $R/\lambda = 400$.[2] The effect on $E(z)$ of varying θ_0 is shown in Fig. 1(b)–(e). For small values of θ_0 the field approaches that of a paraboloid of focal length f:

$$\frac{E(z)}{E} = 2k_0f\frac{\sin k_0z}{k_0z}$$

where $k_0 = k\sin^2\theta_0$ and the origin is transferred to the paraxial focus [1].

Fields in the Focal Region

To obtain the field components at a point in the focal region, the spectrum of axial waves given by (1) must be integrated from $\theta = 0$ to θ_0. In the region of most interest, between the paraxial and marginal foci, where the focused energy is concentrated, the deviation in γ from unity is greatest at the paraxial focus, where $\gamma \approx 1-0.75\theta^4$. Consequently, the maximum error in approximating γ by unity in (1b) in the focal region is 1 and 5 percent for $\theta_0 = 20$ and 30 degrees, respectively.

Manuscript received July 29, 1968; revised November 18, 1968.

[1] Following the submission of this communication, the authors' attention has been directed to some related developments in modern optical research. Using a diffraction formulation based on an integral of plane waves over the geometrical-optics ray directions [3], [4], expressions for the image fields of a spherical mirror have been derived by Kline and Kay [4] and by Gravel and Boivin [5]. These expressions differ in some interesting respects from our hybrid-wave solution, which is based essentially on the usual induced-current method of calculating the scattered fields [6].

[2] The method of normalization differs from that used by Schell. It should also be noted that the amplitude of the geometric optics limit ($\lambda \to 0$) shown in Schell's graph is arbitrarily scaled to fit the $R/\lambda = 400$ curve. The geometric optics limit of the hybrid-wave solution is in good agreement with that of a similar curve published by Gravel and Boivin [5] for $R/\lambda = 400$, but the details of the diffraction oscillations are different.

Reprinted from *IEEE Trans. Antennas Propagat.*, vol. AP-17, pp. 229–232, Mar. 1969.

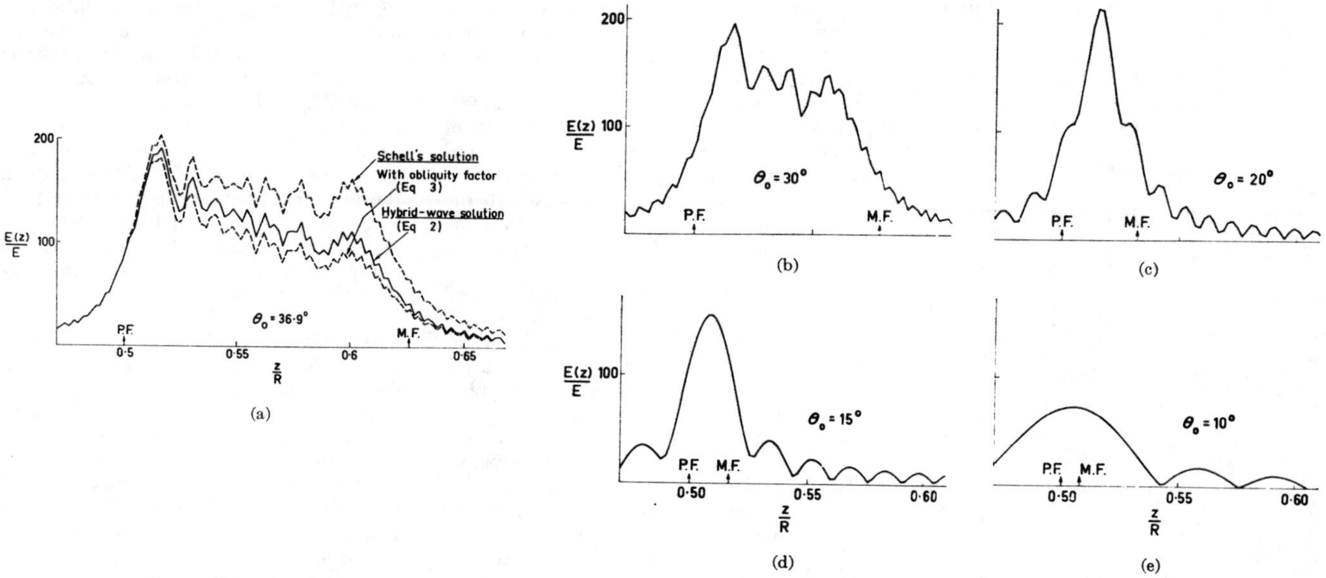

Fig. 1. Normalized field along axis of a spherical reflector, $R/\lambda = 400$ for various values of θ_0. Computation interval 0.002 in z/R. P.F.—paraxial focus; M.F.—marginal focus.

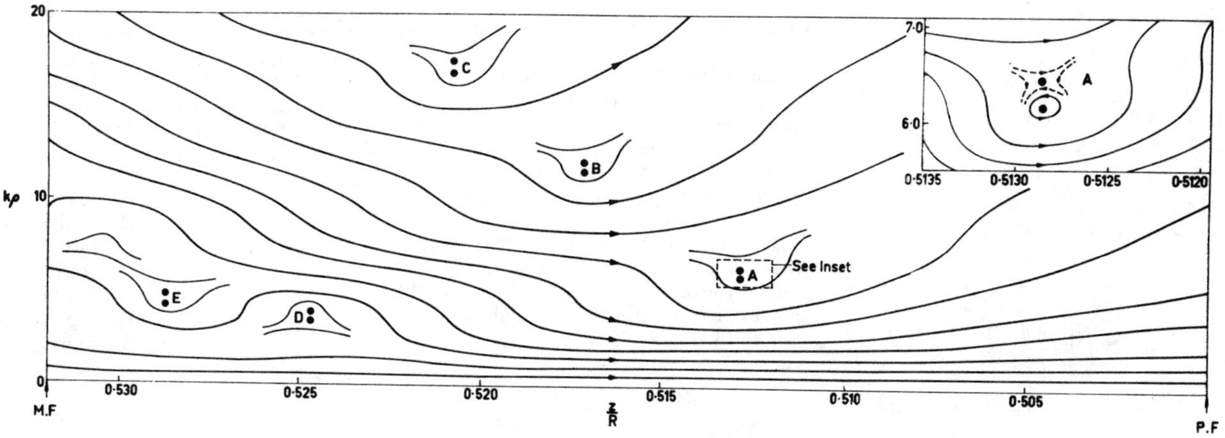

Fig. 2. Direction of energy flow in focal region of a spherical reflector, θ_0 = 20 degrees, R/λ = 400.

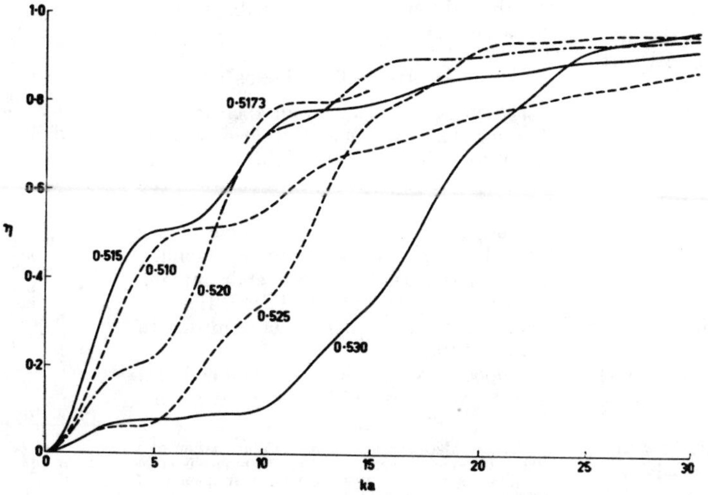

Fig. 3. Efficiency η of a spherical reflector, θ_0 = 20 degrees. R/λ = 400, with z/R as parameter.

The total field may then be written in the following form:

$$E_\rho = E_\rho(U) \sin \xi$$

$$E_\xi = E_\xi(U) \cos \xi$$

$$E_z = E_z(U) \sin \xi$$

$$H_\rho = H_\rho(U) \cos \xi$$

$$H_\xi = H_\xi(U) \sin \xi$$

$$H_z = H_z(U) \cos \xi \qquad (4a)$$

where

$$E_\rho(U) = -jkRE \sin^2 \theta_0[A(U) + B(U)]$$

$$E_\xi(U) = -jkRE \sin^2 \theta_0[A(U) - B(U)]$$

$$E_z(U) = -2kRE \sin^2 \theta_0 C(U)$$

$$H_\rho(U) = \frac{E_\rho(U)}{Z_0}$$

$$H_\xi(U) = -\frac{E_\xi(U)}{Z_0}$$

$$H_z(U) = \frac{E_z(U)}{Z_0} \qquad (4b)$$

$$U = k\rho \sin \theta_0$$

$$A(U) = \frac{1}{2} \csc^2 \theta_0 \int_0^{\theta_0} \kappa(\theta)(1 + \Gamma) \cdot J_0(u) e^{-j\phi} \, d\theta$$

$$B(U) = \frac{1}{2} \csc^2 \theta_0 \int_0^{\theta_0} \kappa(\theta)(1 - \Gamma) \cdot J_2(u) e^{-j\phi} \, d\theta$$

$$C(U) = \frac{1}{2} \csc^2 \theta_0 \int_0^{\theta_0} \frac{\sin \theta}{\zeta} \kappa(\theta) \cdot J_1(u) e^{-j\phi} \, d\theta. \qquad (4c)$$

Equations (4c) have been normalized so that $A(U) = 1$ at the paraxial focus when θ_0 is small.

Energy Flow in the Focal Region

The time average power density is circumferentially symmetrical about the axis when $\gamma = 1$ and is given by the real part of the complex Poynting vector:

$$S_z = -\frac{(kRE)^2}{2Z_0} \sin^4 \theta_0[(A_r^2 + A_i^2) - (B_r^2 + B_i^2)] \qquad (5)$$

$$S_\rho = \frac{(kRE)^2}{Z_0} \sin^4 \theta_0[(A_r - B_r)C_i - (A_i - B_i)C_r] \qquad (6)$$

where the subscripts r and i denote the real and imaginary parts, respectively, of the functions $A(U)$, $B(U)$, and $C(U)$.

Fig. 2 shows the lines of energy flow in a longitudinal plane between the paraxial and marginal foci for a spherical reflector with $\theta_0 = 20$ degrees and $R/\lambda = 400$.[3] Equations (5) and (6) show that the Poynting vector can be completely radial or completely longitudinal. There are also regions (labelled A, B, C, D, and E), where $A_r = B_r$, $A_i = B_i$, so that $S_z = S_\rho = 0$. These null points occur in pairs (indicated by heavy dots). In these regions a circulation or vortex of energy occurs, as shown in the inset enlargement of region A. Such vortices also occur with paraboloidal reflectors with the usual f/D values, but in this case they are centered in the focal plane near the dark bands of the Airy pattern [1].[4]

As indicated by Fig. 2, the focused energy approaches the axis substantially between the line of vortex loops A, B, C and the line D, E, and the concentration near the axis becomes greatest at about $z/R = 0.515$. The energy passing through an aperture of radius a centered on the axis increases rapidly with a in this region. This is demonstrated in Fig. 3, where the aperture energy, expressed as a fraction η of the energy incident on the reflector, is plotted against ka with z/R as parameter. It follows from (5) that

$$\eta = 2 \int_0^{U_a} [(A_r^2 + A_i^2) - (B_r^2 + B_i^2)] U \, dU \qquad (7)$$

where $U_a = ka \sin \theta_0$. The position along the axis for maximum η increases from approximately $z/R = 0.513$ for a small diameter aperture [see also Fig. 1(c)] to approximately $z/R = 0.52$, where almost 90 percent of the energy passes through a circle of radius $ka = 16$. A large increase in aperture size is necessary to achieve any significant increase in η beyond this.

It is of interest to compare these results with the geometric optics concept [9] of the "circle of least confusion." This is the smallest circle through which all the rays pass and has the following radius:

$$ka_c \approx 0.77 \frac{R}{\lambda} \tan^3 \theta_0. \qquad (8)$$

For $\theta_0 = 20$ degrees, $ka_c = 0.037R/\lambda$ at $z/R = 0.526$, so that $ka_c = 15$ when $R/\lambda = 400$. Fig. 3 shows that only about 70 percent of the energy passes through an aperture of this diameter, and that it is less than the maximum obtainable for this particular diameter.

Aperture-Type Feeds

The energy incident on an aperture feed A connected to a matched load will be completely absorbed if the fields over A are the conjugate of those generated when the feed is transmitting. In general, the fields do not exactly "match" and some of the focused energy, additional to that spilling past the feed, is lost. The overall antenna efficiency may be computed from the two fields using a transmission theorem due to Robieux [10].

Suppose now that A is the open end of a circular waveguide with appropriately corrugated walls [2]. During reception, the focused fields comprise an infinite spectrum of plane hybrid waves [see (4)] approaching A. When transmitting, a finite set of discrete hybrid modes propagates towards A within the guide. Assume as a first approximation that the aperture fields in each case are the incident fields, and that the fields elsewhere in the aperture plane, when transmitting, can be neglected. For maximum efficiency, the amplitudes and phases of the guided modes must be chosen to optimize the field match at A.

When A is a focal-plane feed for a paraboloid, the modes are cophased at A, and the fields can be precisely matched for discrete values of the guide radius a, corresponding to the "dark" rings of the Airy pattern.[5] The antenna efficiency for this optimum mode combination may be calculated by conventional methods from the radiation pattern of the feed. The value obtained agrees closely with η, the percentage of the focused energy incident on A, even for $a \approx \lambda$. Thus for the assumptions stated both methods of computing optimum efficiency have comparable validity.

This result suggests that the attainable efficiency with aperture-type feeds for spherical reflectors is given approximately by Fig. 3. In this case the waveguide modes must arrive at A with specific phase differences and the correct relative amplitudes. The best axial location z/R of the feed and its radius a are closely related to the positions of the outer sequence of dark rings A, B, and C in Fig. 2.

B. MacA. Thomas
H. C. Minnett
Vu The Bao
Div. of Radiophysics
C.S.I.R.O.
Epping, N.S.W.
Australia

[3] These values correspond to the illumination of a 600-foot-diameter section of the 1000-foot-diameter Arecibo dish (radius 870 feet) at a frequency of approximately 430 MHz.

[4] This interesting phenomenon has also been noted in recent analyses of aberration-free optical images [8].

References

[1] H. C. Minnett and B. MacA. Thomas, "Fields in the image space of symmetrical focusing reflectors," *Proc. IEE* (London), vol. 115, pp. 1419–1430, October 1968.
[2] ——, "A method of synthesizing radiation patterns with axial symmetry," *IEEE Trans. Antennas and Propagation* (*Communications*), vol. AP-14, pp. 654–656, September 1966.
[3] E. Wolf, "Electromagnetic diffraction in optical systems, I: an integral representation of the image field," *Proc. Roy. Soc.* (London), ser. A, vol. 253, pp. 349–357, December 1959.
[4] M. Kline and I. W. Kay, *Electromagnetic Theory and Geomerical Optics*. New York: Interscience, 1965, ch. 11.
[5] M. Gravel and A. Boivin, "Electromagnetic field in focal region of wide angle spherical mirror," presented at the meeting of the Optical Society of America, 1966 (Abstract in *J. Opt. Soc. Am.*, vol. 56, p. 1438, October 1966).
[6] S. Silver, *Microwave Antenna Theory and Design*. New York: McGraw-Hill, 1949, pp. 144–149.
[7] A. C. Schell, "The diffraction theory of large-aperture spherical reflector antennas," *IEEE Trans. Antennas and Propagation*, vol. AP-11, pp. 428–432, July 1963.
[8] A. Boivin, J. Dow, and E. Wolf, "Energy flow in the neighbourhood of the focus of a coherent beam," *J. Opt. Soc. Am.*, vol. 57, pp. 1171–1175, October 1967.
[9] R. C. Spencer and G. Hyde, "Studies of the focal region of a spherical reflector: geometric optics," *IEEE Trans. Antennas and Propagation*, vol. AP-16, pp. 317–324, May 1968.
[10] J. Robieux, "Lois générales de la liaison entre radiateurs d'ondes—application aux ondes de surface et à la propagation," *Ann. Radioélectricité*, vol. 14, pp. 187–229, July 1959.

[5] A two-hybrid-mode feed enclosing the central disk and first bright ring is in service at $\lambda = 6$ cm on the Parkes 210-foot radio telescope.

A High-Performance Line Source Feed for the AIO Spherical Reflector

L. MERLE LaLONDE AND DANIEL E. HARRIS

Abstract—An aberration-correcting line source feed has been designed, modeled, constructed, and tested in the Arecibo reflector. The feed is a linearly polarized flat waveguide 40-foot-long array illuminating 700 feet of the 1000-foot-diameter spherical reflector at 318 MHz. The antenna, illuminated by the new feed, yields an aperture efficiency of 70 percent with a peak gain near 56 dB, half-power beamwidth of 16.2 minutes of arc, and sidelobe levels below 4 percent of the on-axis gain. Vignetting losses are approximately 30 percent at the highest zenith angle.

I. Introduction

THE CONSTRUCTION of the Arecibo Ionospheric Observatory with a 1000-foot spherical reflector in 1960 generated a great interest in the problems of spherical aberration and their solution in the form of radio frequency line feeds. The problems were first attacked by Spencer, Sletten, and Walsh nearly two decades ago [1]. More recently, further theoretical investigation of line source feeds has been carried out [3]–[8].

Only two serious attempts have been made at feeding the AIO reflector with correcting line sources. The first was the original high-power dual circularly polarized feed [2]. The second was a linearly polarized receiving feed by Cohen and Perona [9] designed to illuminate 600 feet of the aperture. The first feed illuminates the full aperture with an efficiency of about 21 percent at 430 MHz. The second feed illuminates 600 feet of the aperture at 611 MHz with an efficiency of 30 percent. Though the loss due to reflector roughness is substantial at either of these frequencies, neither feed would produce aperture efficiencies on a perfect reflector comparable to those commonly attainable with parabolic reflectors and point source feeds.

Manuscript received April 1, 1969; revised June 16, 1969.

L. M. LaLonde is with the Center for Radiophysics and Space Research, Cornell University, Ithaca, N. Y. 14850.
D. E. Harris is with the Instituto Argentino de Radio Astronomia, Villa Elisa, Buenos Aires, Argentina.

Reprinted from *IEEE Trans. Antennas Propagat.*, vol. AP-18, pp. 41–48, Jan. 1970.

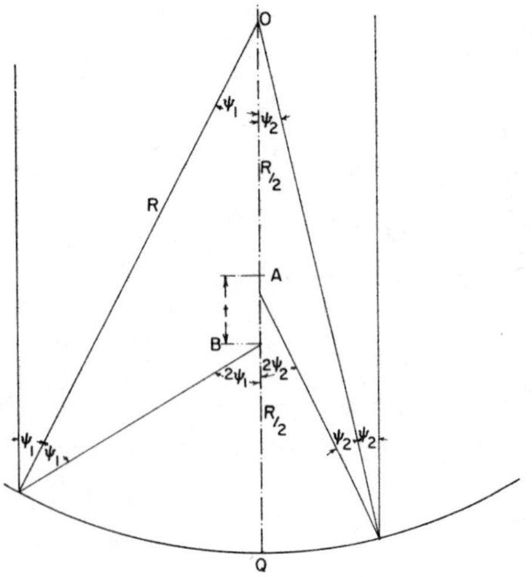

Fig. 1. Rays of a sphere.

TABLE I
1962 MHz Model Feed Dimensions

t (inches)	a (inches)	Cumulative Insertion Phase (degrees)
0	10.98	reference
8	10.98	469.7°
16	7.87	929.4°
24	6.52	1358.2°
32	5.74	1777.9°
40	5.20	2181.7°
48	4.83	2570.2°
56	4.53	2943.7°
64	4.31	3302.8°
72	4.12	3647.4°
80	3.98	3978.3°

Feed dimensions were calculated using a 3 percent correction for slowing effect of slots. Geometrical optics insertion phase length is shown as a function of distance below paraxial surface ($t = 0$).

As a consequence of the "Dicke Committee" report [10] the University obtained support from the NSF to 1) conduct a design study to investigate the feasibility of upgrading AIO to an operational capability at 10-cm wavelength, and 2) to develop under subcontract a receiving feed at 611 MHz to illuminate 750 feet of the reflector. It was also decided that an in-house attempt to develop a linearly polarized line feed at a lower frequency should be made, primarily to provide a feed at a frequency where surface errors caused a negligible reduction in gain. A secondary purpose was to have two groups independently working on similar problems, with the end results permitting valuable comparisons to be made. The in-house feed would be designed to illuminate 700 feet of the dish, some 200 feet more than possible with a point source illuminating the central Fresnel zone.

A somewhat arbitrary decision was made to design the feed at a frequency near 327 MHz, at which frequency the effects of reflector irregularities are trivial.

II. Design of the Feed

The design philosophy adopted for this feed was to keep the radiating elements as close to the axis as possible to avoid the problems in the square TRG feed. A radiating section of half-width ($b = \lambda/8$) guide had been checked at X band at Cornell in 1965 and found to be free of harmonics in the azimuthal pattern caused by the slots being displaced off-axis.

The use of waveguide with transverse slots as radiating elements as opposed to a string of dipoles has major advantages. It allows closer control over amplitude and phase of the elements, primarily because the slots are tightly coupled to the guide and more loosely coupled to each other. In the dipole string, mutual coupling is a severe problem. Furthermore, the power distribution and phasing of the dipole string must be done with power dividers and odd lengths of transmission line, subject to the adverse effects of humidity and weathering in general, unless very special precautions are taken; even with these, stability seems to be a problem. Also, the transmission lines and power dividers used in dipole feeds usually are much lossier than waveguide. Lastly, the waveguide feeds are mechanically simpler.

Thus we concluded that the feed should be designed in flat waveguide, holding the guide thickness to less than $\lambda/8$ so that the transverse slots cut in the broad face as radiators would be displaced less than $\lambda/16$ from the axis. In pattern testing, the primary illumination pattern measured at a distance from the feed corresponding to the position of the reflector would be used as a measure of proper performance of the feed.

The geometry of the sphere is shown in Fig. 1. Kay [2] and Schell [4] have considered correcting feeds by geometrical optics and diffraction theory. The design of this feed was based on geometrical optics, which leads to very nearly the same phase and amplitude requirements as diffraction theory [4]. Primary pattern measurements in the far field were used as a final measure of performance; empirically determined parameters for the waveguide and slot configuration were used to adjust the pattern to the desired illumination.

It can readily be shown [1], [2] from the geometry that

$$\cos\psi = \frac{f}{f+t} = \frac{1}{1+(t/f)} \tag{1}$$

where $f = R/2$, t is the distance below the paraxial surface along the axis of the feed, and

$$\lambda/\lambda_g = \cos 2\psi. \tag{2}$$

Using these equations and waveguide theory for the dominant mode, one calculates the a dimension of the waveguide as a function of t from

$$\frac{\lambda}{2a} = \left[1 - \left(\frac{1-2x-x^2}{1+2x+x^2}\right)^2\right]^{1/2} \tag{3}$$

where $x = t/f = t/435$ feet for the AIO reflector, and λ is obviously the free-space wavelength at the design frequency.

This equation allows for no perturbing effect of slots in the guide. Measurements made by TRG in 1.166-by-

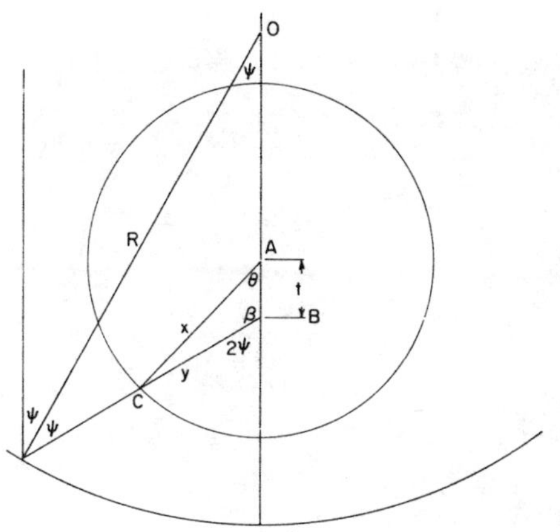

Fig. 2. Geometry appropriate for calculation of relative phase as a function of the angle θ at a distance X from top of the feed (rotation point). $\Delta\theta = AB + BC - X$.

0.583-inch guide would indicate that the slowing effect of slots of typical length would require a reduction in the a dimension of the guide of approximately 3 percent to adjust the phase velocity in the guide for the slot effects.

The model feed was designed to $\frac{1}{6}$ scale of the 327-MHz feed. The fixed b dimension of the feed was chosen at 4 inches full scale ($\lambda/9$), and the a dimension calculated using the 3 percent correction for the slot effect. Slots were cut in the broad faces at a uniform separation of a quarter of the free-space wavelength. The slots were cut purposely short so that it would be possible to lengthen them to achieve the desired amplitude taper.

Transitions from waveguide to coax were made for each end of the feed so that a measurement of the total insertion (radiation) loss by substitution could be made with ease.

A near field probe was devised, consisting of a dipole fed by a broad-band balun, mounted on a carriage which is capable of moving the dipole accurately along a line parallel to the axis of the feed, over its entire length.

The insertion phase of the feed is calculated from geometrical optics as

$$\Delta = (360/\lambda)[870t/(435 + t) - t] \text{ degrees} \qquad (4)$$

where t is in feet for the full scale feed.

Thus the design of the original model dictated those dimensions shown in Table I. The geometrical optics insertion phase is also noted in this table. The b dimension of the $\frac{1}{6}$ scale model is 0.75 inch, and all slots were initially cut 1 inch long and $\frac{1}{4}$ inch wide at a uniform separation of 1.5 inch, resulting in 53 slot pairs along the length of the feed.

The model feed was constructed entirely in brass for ease of machining and soft soldering of flanges. The wall thickness was chosen at 0.032 inch because a preliminary check of full scale thickness in aluminum for required rigidity indicated that 0.190-inch sheet must be used.

III. Model Measurements

First measurements on the model called for increasing the slot length uniformly until the radiation loss from the feed exceeded 10 dB, as measured by substitution. Then

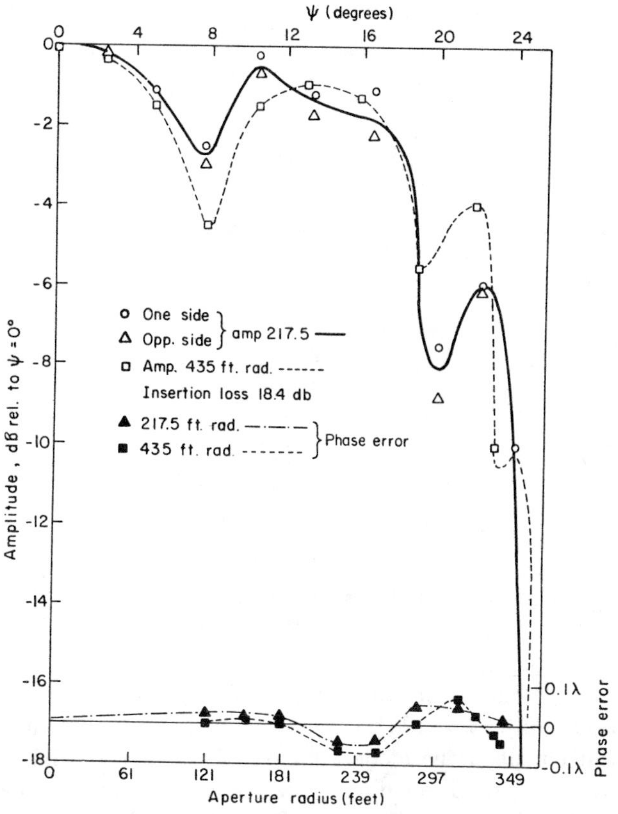

Fig. 3. Amplitude and phase error measured at half-paraxial and full-paraxial distance. E plane of $\frac{1}{6}$ scale model.

the near field probe was used to determine the phase distribution along the axis of the feed.

Using the near field probe and plotting phase distribution along the feed at several frequencies, we were able to determine that the bottom half of the feed was nearly correct in its insertion phase at a frequency of 1650 MHz, about 84 percent of the design frequency. In view of this, adjustable waveguide walls were placed in the narrow dimension of the feed and an empirical adjustment of the phase was made by moving the guide walls.

Since it is inconvenient to measure the pattern in the angle ψ, computations of required phase corrections were made as a function of the angle θ, the angle from the feed axis to a ray intersecting this axis at the paraxial surface consistent with the geometry shown in Fig. 2. Thus we were able to rotate the feed about its paraxial surface end and adjust the phase according to these computed values required for the proper correction for uniform phase across the aperture.

Fig. 3 shows the final results of amplitude and phase error obtained with the model. Note that some difference in amplitude is seen between the measurements at the two distances. This is not surprising since even the full paraxial distance is in the near field of the feed. The difference in amplitude is relatively insignificant however, and the difference in phase error completely so. The insertion loss of the feed in final form was 18 dB, showing that only 1.5 percent of the power inserted in the feed is radiated wastefully at the lower end of the feed.

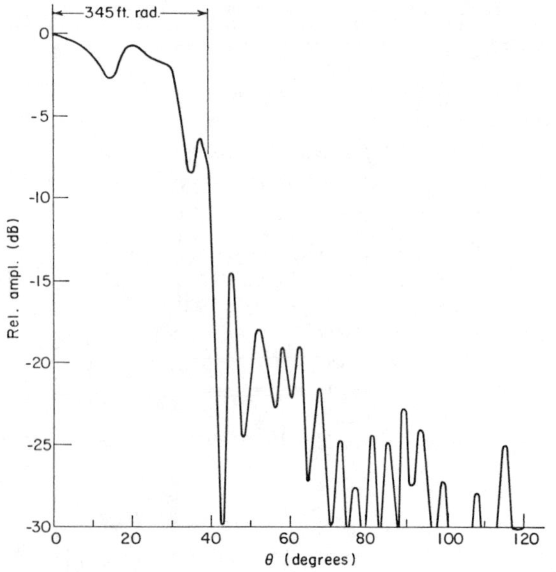

Fig. 4. E-plane amplitude pattern of $\frac{1}{6}$ scale model out to $\theta = 120°$. Radius of 500 feet is approximately at peak of first sidelobe.

Fig. 4 shows the amplitude of the primary E-plane pattern out to $\theta = 120°$, a limit imposed by the mechanical arrangement of the model support. Note the rapid falloff at the angle corresponding to the geometrical optics determined bottom end of the feed. Though a major lobe does appear between $\theta = 42°$ and $\theta = 70°$, its average amplitude is 20 dB below the main lobe. All other observable lobes appear to have an average level well under 25 dB below the main lobe.

The ripples in amplitude in the central region are thought to be due to standing waves in the feed. Attempts to remove them resulted only in their redistribution. In fact, these nulls have little effect on the antenna performance. They could probably be reduced by spacing slots by $\lambda_g/4$ rather than $\lambda/4$.

In addition to the pattern shown here, measurements were made by azimuthal rotation of the feed at constant angles of θ. These patterns showed no measurable distortion of the azimuthal field from spacial harmonics. This was as expected and, of course, a primary reason for keeping the b dimension of the guide to a minimum in the design.

A careful measurement of the model dimensions was made, and these were scaled by a factor of 6. The full scale feed was built in aluminum, precisely scaled to the model dimensions, although the mechanical details are somewhat different. The main difference between model and full scale feed is that the full scale feed has flanges, while the model did not. The flanges protrude from the face of the feed $1\frac{1}{2}$ inches ($\lambda/24$) and are spaced 12 feet apart. The effect of these flanges is thought to be negligible, but is in fact unknown.

The full scale feed is shown in Fig. 5. The feed was installed in the center of Carriage House no. 2 at AIO on December 12, 1968. An evaluation of the performance was initiated immediately by radio astronomy techniques observing "standard" radio sources.

Fig. 5. Feed being assembled at AIO, illustrating construction of the feed. Lines attached to flange in foreground are Mylar rope guys.

IV. Calculated Antenna Performance from Model Measurements

Measurements of the primary pattern of the full scale feed are impossible with facilities available in Ithaca, since handling the 40-foot-long 800-lb feed is quite impractical. It becomes very important, therefore, to compute parameters of the antenna performance in the form of characteristics of secondary pattern of the antenna when fed with the aperture illumination expected from the model feed measurements. Differences between the measured performance and the computed performance should be explicable and meaningful in terms of an evaluation of the accuracy of the electrical measurements of the model and the electrical scaling of the full scale feed to the model.

The efficiency of any antenna is computed from an on-axis gain measurement (peak gain) compared to the theoretical gain that would be realized from uniform amplitude over the aperture of the antenna with no phase error. It will be convenient therefore to discuss the factors affecting gain in some detail.

The major factors tending to reduce the gain of an antenna from the ideal are 1) "gain factor" from the taper in amplitude of the illumination, 2) phase errors in the illumination, 3) loss due to surface imperfections in the reflector, 4) spillover, 5) shadowing losses from physical blockage of the aperture by feed support structure, and 6) contingencies such as ohmic losses, polarization losses, opaqueness of the reflector etc., none of which will be considered here.

The losses attributable to items 1) and 2) are readily calculated with the aid of a computer program designed to compute antenna patterns from radial measurements of amplitude and phase. This is done by first calculating the on-axis gain with uniform amplitude and zero phase error, which must numerically equal $4\pi A\lambda^{-2}$. Second, we can compute the gain with the measured amplitude and zero phase error, and third, repeat the computation with measured amplitude and phase. Results of these three computations above are in order, 5.34×10^5, 5.02×10^5, and

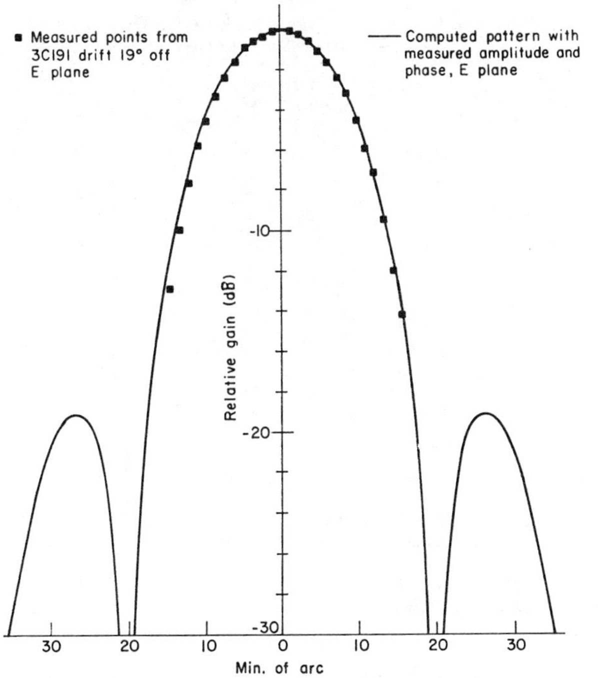

Fig. 6. *E*-plane pattern of feed from computer program with model measured illumination of Fig. 3. Plotted points are measured from drift curve of 3C-191, 19° from *E* plane.

4.74×10^5. The gain factor is just the ratio of the second to first, or 0.940. The loss factor due to illumination phase errors is the ratio of the third to second computation, or 0.944. The total gain reduction factor arising from nonuniform amplitude and phase errors is then 0.887, or the ratio of the third to first calculation. This would be the "aperture efficiency" in the absence of other losses.

The losses due to surface irregularities of the reflector are the largest and least well known of the individual cau es of reduction in gain. Using the results due to Ruze [11], we must consider the correlation interval of the reflector irregularities. The mesh support cables at AIO are spaced 3 feet apart and 100 feet apart in the N–S and E–W directions, respectively, or generally with a correlation interval $C \geq \lambda$. Two photogrammetric surveys of the AIO reflector would indicate that the rms error in the reflector surface is between 1 and 1.5 inches depending somewhat on temperature and perhaps more heavily on the accuracy of the measurements. Using a correlation interval of $C \geq \lambda$, and applying the 1- and 1.5-inch errors to Ruze's results gives a reduction of gain between 0.87 and 0.76 (−0.6 and −1.2 dB, respectively) from this cause.

Fig. 4 shows that the taper at the edge of the illuminated aperture (r = 350 feet) is very sharp, and the sidelobe structure outside the intended direction of radiation is relatively low. A reasonable average from Fig. 4 of the illumination outside the 350-foot radius is taken somewhat pessimistically at 23 dB below that on the aperture. Using these figures, we get a reduction in gain due to spillover of approximately 0.96.

Shadowing by the feed support structure is uncommonly low in the AIO system. Aperture blockage has been estimated at 1.5 percent of the full aperture; we will use 3 percent of half the aperture area and a gain reduction of 0.97 attributable to this cause.

Multiplication of the loss factors due to 1)–5) above result in an expected aperture efficiency of from 0.63 to 0.72, or a response in degrees of antenna temperature per flux unit of 8.1 to 9.3°K, depending upon the rms reflector error between the limits of 1.5 and 1 inch.

The secondary *E*-plane pattern of the antenna computed from the model illumination is shown in Fig. 6. This shows that we should expect a half-power beamwidth of approximately 16.2 minutes of arc (at 318 MHz) and a sidelobe level of approximately 1.2 percent of the peak gain. Superimposed on the computed pattern are data points from a drift curve of 3C-191, measured with the full scale feed in the Arecibo reflector.

V. Tests of the Full Scale Feed in the Spherical Reflector

The radio astronomical tests of the new line feed reported here were made during the latter part of December, 1968. The antenna was mounted at the center of Carriage House no. 2 and all other feeds were removed from the carriage house. The total power receiver operated initially at 327 MHz and consisted of a transistorized preamplifier of 10-MHz bandpass, a mixer, and IF amplifiers.

The backend of the receiver contained options for the following:

- 8-channel filter bank, Δf = 1 MHz, passive, IF ranging from 26.5 to 33.5 MHz;
- three individual passive filters centered at 30 MHz, Δf = 1, Δf = 3, and Δf = 8 MHz;
- six standard backends containing linear detectors;
- four square-law detectors and operational amplifiers;
- RC time constants, dc buckouts, and a Sanbourne 6 channel analog recorder.

Throughout these experiments the time constant employed was 0.1 second and the IF filter width was 1 MHz.

A. The Antenna Gain as a Function of Zenith Angle

The line feed was designed to illuminate 700 feet of the reflector. Vignetting should commence at a zenith angle of 11.3° for such an illuminated area.

Observations of 3C-47, 3C-264, 3C-334, and 3C-346 were made at 318 MHz. All sources were followed across the sky taking alternate declination (δ) and right ascension (α) scans, continuously correcting the coordinates for observed pointing errors ($<2'$).

Fig. 7 shows the normalized plot of the data. Each point should have an error considerably less than 5 percent (5 percent corresponds to a pointing error of $\sim$2ʹ.5). There is good agreement for all four sources.

B. Antenna Gain as a Function of Frequency

The peak antenna gain occurs at about 318 MHz and the preamplifier center frequency was shifted to that value in two steps. Fig. 8 shows the degrees per flux unit obtained for 14 different sources. These sources are all included in

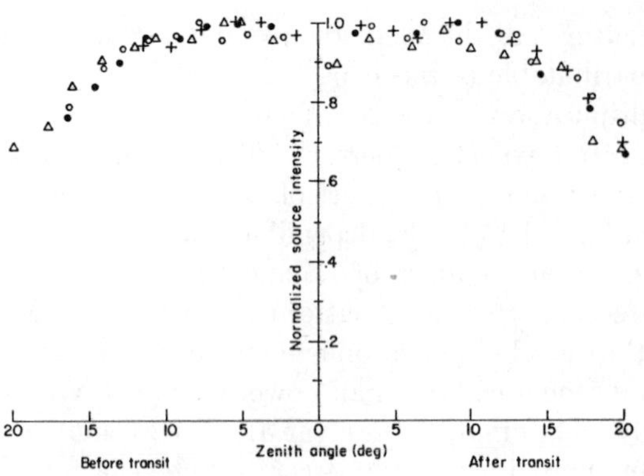

Fig. 7. Antenna gain dependence on zenith angle. Observations of 3C-47 (•), 3C-264 (+), 3C-334 (○), and 3C-346 (△) at 318 MHz.

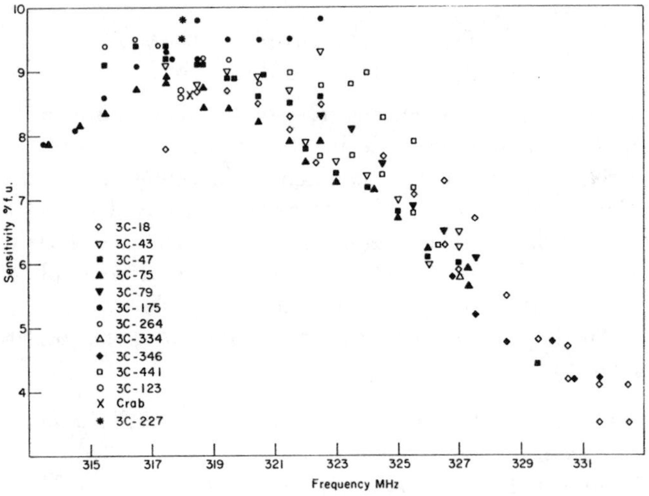

Fig. 8. Antenna gain dependence on frequency. Detector and zenith angle corrections are included.

the recent work on radio spectra by Kellermann *et al.* [12]. The sources we used were chosen from their Table 5 (calibration sources) or from the body of their Table 3 if their Table 4 indicated that the source had a straight spectrum. We have listed these sources and their assumed fluxes in Table II. We have not included errors since the large number of observations and sources used for Fig. 8 gives a scatter which includes all errors except for that from our calibration standard.

C. The Beamwidth and Sidelobes

The half-power beamwidth was measured for three small diameter sources. Table III gives the results.

The H plane is well determined at 16′.0. The E-plane width seems slightly larger at 16′.5 ± 0.3. The theoretical value derived from the model measurements is 16′.2.

Measurements indicate that the principal sidelobe which is ~25 minutes of arc from the center of the beam varies from ~4 percent to less than ~1 percent and is hardly ever equal on both sides.

The secondary lobes are less than 1 percent and usually of the order of ½ percent. The "grating lobe" of the reflector is present on all of the α scans. As suggested by F. D. Drake, the regular sag between the main N–S support cables of the AIO reflector could produce a grating lobe in the E–W direction. Since the spacing of these cables is 100 feet and the wavelength is 3 feet, the predicted feature would fall at 106 feet from the main beam axis. This lobe is at the 1 percent level and is always stronger than the secondary sidelobes.

A cut through one 45°-plane pattern is shown in Fig. 9 which shows the sidelobes, grating lobes, and typical symmetry of the 45°-plane patterns. E- and H-plane patterns are somewhat more asymmetrical and exhibit sidelobes with higher levels.

TABLE II
ASSUMED FLUX DENSITIES FOR SOURCES USED IN THE ANTENNA CALIBRATION

Source	Frequency (MHz)	S (flux units)
3C-18	320	12.8
3C-43	320	7.6
3C-47	327	15.0
	318	15.2
3C-75	327	17.4
	318	17.6
3C-79	327	16.5
3C-123*	318	137.0
CRAB Neb.†	318	1300.0
3C-175	320	9.9
3C-227	320	21.1
3C-264	320	16.8
M 87†	320	660.0
3C-334	327	6.5
3C-346	327	8.5
3C-441	327	7.8

Source intensities were derived from spectra drawn from [12].
* 10-dB pad inserted between antenna and receiver to minimize detector corrections.
† 20-dB pad inserted between antenna and receiver to minimize detector corrections.

D. Calculation of Aperture Efficiency

Radio astronomical measurements of standard radio sources may be used to calculate effective aperture from

$$\Delta T = (A' \times 10^{-26}/2k) \text{ degrees per flux unit (mks)} \quad (5)$$

where ΔT is the increase in antenna temperature, A' is the effective collecting area of the antenna, and k = Boltzmann's constant. For 100 percent aperture efficiency of a 700-foot-diameter aperture, we would get 12.9° per flux unit by numerical substitution into (5). The ratio of measured degrees per flux unit to this number for 100 percent efficiency is the aperture efficiency A'/A_0.

VI. CONCLUSIONS

The performance of this feed in the spherical reflector represents a substantial step forward in the technology of spherical reflector antennas. The most important implication, however, is that it proves the contention that spherical aberration can be made to work for, rather than against, efficient antenna systems. Feeds of this type can readily be built for higher or lower frequencies.

This new feed obtains an aperture efficiency of 70 percent on a reflector which is degrading the gain by a factor of 0.76 to 0.87. It is evident that if the reflector were perfect (as must be the case for comparable efficiency from a

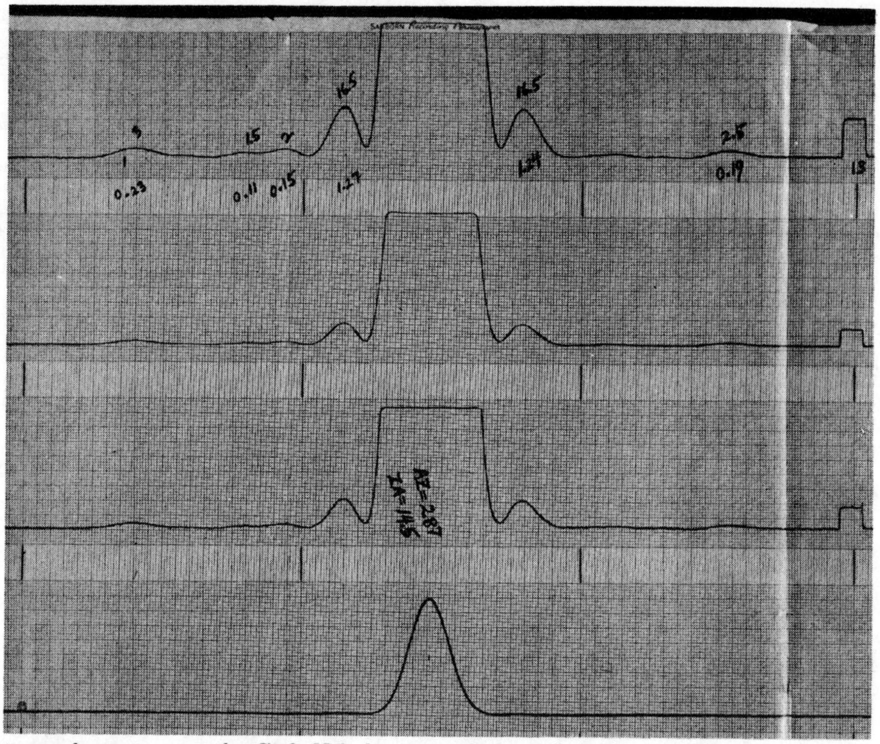

Fig. 9. Patterns of a cut across the Crab Nebula on a 45° plane of the feed. Calibration pulse is visible on all traces at right-hand side of the figure. Top two traces are linear detectors; bottom two, square-law. From left to center, observe grating lobe, third, second, and first sidelobe.

TABLE III
BEAMWIDTH MEASUREMENTS

Source	Zenith Angle (degrees)	Half-Power Beamwidth (minutes of arc)	Orientation	Type of Scan
3C-213	12.9	16.0	17° from H plane	δ
3C-123*	11.2	16.7	45° to principal planes (transit observation)	α
3C-228	7.5	16.0	15° from H plane	α
	6.4	16.5	10° from E plane	δ
3C-18	11.8	16.25	E plane	α
	12.7	16.0	5° from H plane	δ

* 10-dB pad between antenna and receiver to minimize detector corrections.

parabola), the measured efficiency would leap to 80–92 percent. Calculations from the model pattern substantiate this with a calculated efficiency of 88.7 percent without reflector error losses.

Aside from the strengthening of the obvious arguments in favor of fixed-reflector-type scanning antennas, the feed developed here puts the sphere in serious contention for consideration in designing high-performance fully steerable antennas. It is recognized, of course, that complex feeds for parabolic antennas may be developed so as to produce performance approaching that of a spherical antenna. However, in general this can not be done while retaining the flexibility of mounting many feeds simultaneously, which is inherent in the coma-free spherical reflector.

It is obvious that this feed does not represent the final solution to all of the problems involved in feeding spheres. We have now only one technique which is capable of producing linear polarization, though it is waveguide and therefore will handle high power. Remaining to be solved is the problem of handling high powers with polarization flexibility. Indeed, there have been a number of promising ways proposed in both line feed and surface feed approaches, some similar to the approach used here, though little has been done experimentally to check any proposed scheme. The feed described here shows the potential rewards for further development are great.

ACKNOWLEDGMENT

The authors gratefully acknowledge the assistance of D. M. Teeter and L. C. Tolliver in the task of building and testing both the model and full scale feeds. A. Niell assisted in obtaining the observations at AIO. It is a pleasure to acknowledge the helpful suggestions of G. Pettengill, F. S. Harris, A. Niell, Prof. V. H. Rumsey, and Dr. A. W. Love.

The Arecibo Ionospheric Observatory is operated by Cornell University with the support of the Advanced

Research Projects Agency and the National Science Foundation under contract with the Air Force Office of Scientific Research.

References

[1] R. C. Spencer, C. J. Sletten, and J. E. Walsh, "Correction of spherical aberration by a phased line source," *Proc. 1949 NEC* (Chicago, Ill.), vol. 5, p. 320.
[2] A. F. Kay, "A line source feed for a spherical reflector," Contract AF 19(604)-5532, AFCRL 529, May 1961.
[3] A. W. Love, "Spherical reflecting antennas with correcting line sources," *IEEE Trans. Antennas and Propagation*, vol. AP-10, pp. 529–537, September 1962.
[4] A. C. Schell, "Diffraction theory of large-aperture spherical reflector antennas," *IEEE Trans. Antennas and Propagation*, vol. AP-11, pp. 428–432, July 1963.
[5] F. S. Holt and E. L. Bouche, "A Gregorian corrector for spherical reflectors," *IEEE Trans. Antennas and Propagation*, vol. AP-12, pp. 44–47, January 1964.
[6] G. Hyde, "Focal region fields of a spherical reflector RCA," AFCRL-66-48, Contract AF 19(628)-2758, January 1966.
[7] G. C. McCormick, "A line feed for a spherical reflector," *IEEE Trans. Antennas and Propagation*, vol. AP-15, pp. 639–645, September 1967.
[8] A. W. Love and J. J. Gustincic, "Line source feed for a spherical reflector," *IEEE Trans. Antennas and Propagation* (Communications), vol. AP-16, pp. 132–134, January 1968.
[9] M. H. Cohen and G. E. Perona, "A correcting feed at 611 MHz for the AIO reflector," *IEEE Trans. Antennas and Propagation* (Communications), vol. AP-15, pp. 482–483, May 1967.
[10] "Report of the ad hoc advisory panel for large radio astronomy facilities," NSF, Washington, D. C., August 14, 1967.
[11] J. Ruze, in *Antenna Engineering Handbook*, H. Jasik, Ed. New York: McGraw-Hill, 1961, p. 2–39.
[12] K. I. Kellermann, I. I. K. Pauliny-Toth, and P. J. S. Williams, "The spectra of radio sources in the revised 3C catalog," *APJ*, vol. 157, pp. 1–34, July 1969.

Synthesis of the Fields of a Transverse Feed for a Spherical Reflector

LEÓN J. RICARDI, SENIOR MEMBER, IEEE

Abstract—The problem of designing a transverse feed for a spherical reflector is considered and a method is presented for synthesizing the fields on a surface of a sphere enclosing a feed that will produce a specified reflected field at the surface of a spherical reflector. The method identifies the reflector and a spherical surface enclosing the feed as a boundary value problem and uses a finite set of spherical waves to approximate the boundary conditions. A feed designed to excite this field will in turn produce the desired reflected field at the surface of the reflector, under the condition that that portion of the reflected field which is scattered by the feed may be neglected. It is shown that the feed need produce only a small part of the synthesized field to obtain an antenna efficiency of more than 70 percent. Some typical field distributions will be shown so as to indicate a method for designing a feed and to point out the correlation between the polarization of the synthesized field and the polarization of the reflected field at the surface of the reflector.

THE properties, practical application, and aberration of a spherical reflector are certainly not new to astronomers and microwave antenna designers. In fact, the spherical reflector was first studied by Christiaan Huygens [1] and Sir Issac Newton [2] in the middle of the seventeenth century. Both of them used the spherical mirror because the symmetry of its surface renders it easy to construct. Its contemporary popularity is primarily due to the large angle through which the radiated beam may be scanned by translation and orientation of the illuminating feed. This wide-angle property results from the symmetry of the surface; however, the resolving power and gain are usually compromised by the aberration that it introduces. Feeds for correcting and methods for determining this spherical aberration have been the topic of several papers published in the last two decades [4]–[16].

The first use of a spherical reflector as a microwave antenna occurred during World War II [3]; it was configured in its simplest form and certainly caused subsequent investigators to improve the theory, understanding, and performance of this device. The initial investigation of this device as a microwave antenna followed the methods laid down by the early researchers; they considered the microwave antenna as a receiving device and used primarily the principles of geometrical optics in their analysis. More recently, a method for synthesizing the fields of a line source located on the focal axis was derived [7], [16]. This paper addresses the similar problem associated with a transverse feed and presents a method for synthesizing the fields on a surface of a sphere enclosing a transmitting feed that will produce a specified reflected field at the surface of a spherical reflector of radius r_1. The method identifies the reflector and a spherical surface enclosing the feed as a boundary value problem and uses a finite set of spherical waves to approximate the boundary conditions. This leads to the determination of a field distribution $\boldsymbol{E}_{ft}$ tangent to the surface of a sphere, of radius r_2, which encloses the feed. A feed designed to excite this field will in turn produce the desired reflected field at the surface of the reflector, under the condition that we may neglect that portion of the reflected field which is scattered by the feed. It will be shown that the feed need produce $\boldsymbol{E}_{ft}$ only on a small portion of the surface $r = r_2$. Some typical field distributions will be shown so as to indicate a method for designing a feed and to point out the correlation between the polarization of $\boldsymbol{E}_{ft}$ and the polarization of the reflected field at the surface of the reflector.

Manuscript received June 29, 1970; revised August 27, 1970. This work was supported by the U. S. Air Force under Contract F19628-70-C-0230.

The author is with the M.I.T. Lincoln Laboratory, Lexington, Mass. 02173.

Derivation of Field Equations

In this section we will derive an equation for the synthesized field $\boldsymbol{E}_f(r_2,\theta,\phi)$ on the surface of a sphere enclosing a transmitting feed for a spherical reflector (Fig. 1), given the required reflected field $\boldsymbol{E}_1$ at the surface of the reflector. The latter has a circular aperture, it subtends an angle $2\theta_1$ measured at the center of the sphere (i.e., the origin of the coordinate system). $\boldsymbol{E}_f$ is expressed as an infinite sum of outward-propagating spherical waves; the relative amplitude of the waves is determined by letting the tangential component of $(\boldsymbol{E}_1 + \boldsymbol{E}_f)$ vanish at the surface of the reflector and the tangential component of $\boldsymbol{E}_f$ vanish over the remainder of the surface $r = r_1$. The latter requirement ensures that the reflector will intercept all the energy radiated by the feed. We will first derive a suitable expression for the field at $r = r_1$ in the absence of the reflector. Following this, a series expression for $\boldsymbol{E}_f$ and a method for evaluating the coefficients of each term will be derived. A justification for truncating the series representation of $\boldsymbol{E}_f$ will be given to complete the synthesis procedure.

Along the spherical surface $r = r_1$, we want

$$\boldsymbol{E}_{ft}(r_1,\theta,\phi) = \begin{cases} -\boldsymbol{E}_{1t}, & 0 < \theta < \theta_1 \\ 0, & \text{for all other } \theta \end{cases} \qquad (1)$$

Reprinted from *IEEE Trans. Antennas Propagat.*, vol. AP-19, pp. 310–320, May 1971.

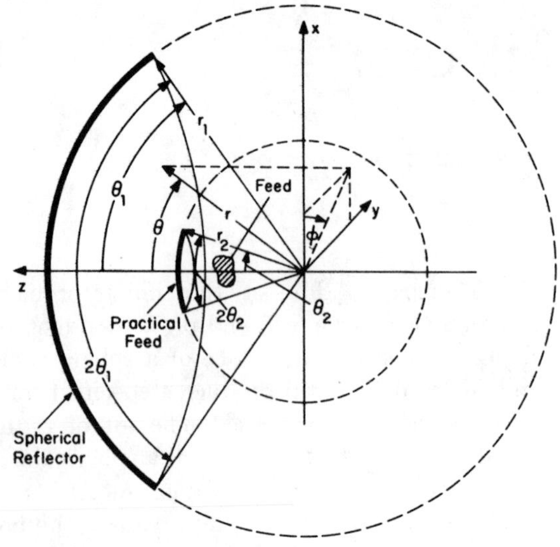

Fig. 1. Spherical reflector and feed geometry.

where the subscript t indicates the component tangent to the surface $r = r_1$. In the interest of mathematical simplicity and with, in principle, no loss in generality we will assume that the $\boldsymbol{E}_1$ is linearly polarized in the x direction (a more general $\boldsymbol{E}_1$ could be represented as a linear combination of an x and y directed polarization; the present development will produce the corresponding field equations). The antenna configuration and the coordinate system are shown in Fig. 1.

A classical solution [17] to boundary value problems uses Maxwell's equations as a departure point and assumes that the magnetic field is given by the curl of a vector potential $\boldsymbol{A}$, i.e.,

$$\nabla \times \boldsymbol{E} = -j\omega\mu \boldsymbol{H} \tag{2}$$

$$\nabla \times \boldsymbol{H} = j\omega\epsilon \boldsymbol{E} \tag{3}$$

$$\boldsymbol{H} = \nabla \times \boldsymbol{A} \tag{4}$$

where $\boldsymbol{E}$ and $\boldsymbol{H}$ are the time harmonic or complex representation of the instantaneous electric and magnetic fields, respectively. The instantaneous values are obtained through the relationship

$$\mathfrak{M} = \sqrt{2}\,\mathrm{Re}\,[\boldsymbol{M}\exp(j\omega t)] \tag{5}$$

where $\mathfrak{M}$ and $\boldsymbol{M}$ represent any of the vector fields, t is the time in seconds, ω the angular frequency in rad/s, and μ and ϵ are the permeability and the permittivity of the medium, respectively.[1] Equations (2)–(4) assume a source-free medium. Substitution of (4) into (2) and (3) and subsequent elimination of $\boldsymbol{E}$ from (2) and (3) results in the familiar vector wave or Helmholtz equation in $\boldsymbol{A}$

$$(\nabla^2 + k^2)\boldsymbol{A} = 0. \tag{6}$$

[1] These constitutive parameters are assumed to be isotropic and constant throughout the medium.

In a spherical coordinate system, (6) has the radiating modal solutions

$$\boldsymbol{A} = kr \sum_{m=0}^{\infty} \sum_{n=m}^{\infty} a_{mn} h_n^{(2)}(kr) P_n^m(\cos\theta) \begin{Bmatrix} \sin m\phi \\ \\ \cos m\phi \end{Bmatrix} \boldsymbol{\mu}_r. \tag{7}$$

Alternately, if we set $\boldsymbol{E} = -\nabla \times \boldsymbol{F}$, a second wave equation is obtained

$$(\nabla^2 + k^2)F = 0 \tag{8}$$

with the solution

$$\boldsymbol{F} = kr \sum_{m=0}^{\infty} \sum_{n=m}^{\infty} b_{mn} h_n^{(2)}(kr) P_n^m(\cos\theta) \begin{Bmatrix} \cos m\phi \\ \\ \sin m\phi \end{Bmatrix} \boldsymbol{\mu}_r. \tag{9}$$

In order to simplify the notation, we will use the modified spherical Bessel function [18]

$$\hat{H}_n^{(2)}(kr) = krh_n^{(2)}(kr) \tag{10}$$

where $h_n^{(2)}(kr)$, the spherical Hankel function of the second kind, is used to represent outward-traveling spherical waves. $k = 2\pi/\lambda$, λ being the wavelength. $P_n^m(\cos\theta)$ is the associated Legendre function of the first kind. It can be shown [17, pp. 267–268] that (7) and (9) give rise to two complete sets of modes, one which has no radial component of the electric field (TE) and one which has no radial component of the magnetic field (TM). A linear superposition of these modes is sufficient to describe any $\boldsymbol{E}$ or $\boldsymbol{H}$ in the assumed configuration; hence the total electric and magnetic fields are given by

$$\boldsymbol{E} = -\nabla \times \boldsymbol{F} - j\omega\mu\boldsymbol{A} + \frac{1}{j\omega\epsilon}\nabla(\nabla\cdot\boldsymbol{A}) \tag{11}$$

$$\boldsymbol{H} = \nabla \times \boldsymbol{A} - j\omega\epsilon\boldsymbol{F} + \frac{1}{j\omega\mu}\nabla(\nabla\cdot\boldsymbol{F}). \tag{12}$$

Substituting (7) and (9) into (11) and (12) gives the following expressions for the radial components of the fields:

$$E_r = \frac{1}{r^2}\sum_{m=0}^{\infty}\sum_{n=m}^{\infty} n(n+1)\hat{H}_n^{(2)}(kr)P_n^m(\cos\theta) \cdot [a_{mn}^e \cos m\phi + a_{mn}^o \sin m\phi] \tag{13}$$

$$H_r = \frac{1}{r^2}\sum_{m=0}^{\infty}\sum_{n=m}^{\infty} n(n+1)\hat{H}_n^{(2)}(kr)P_n^m(\cos\theta) \cdot [b_{mn}^e \cos m\phi + b_{mn}^o \sin m\phi] \tag{14}$$

where the superscripts e and o stand for even and odd, respectively. The same substitution will yield the transverse components of the fields. However, in the interest of clarity and simplification of notation we will use the spherical vector harmonics described by Morse and Feshbach [19], that is,

$$E_t = -\frac{1}{r}\sum_{m=0}^{\infty}\sum_{n=m}^{\infty}[n(n+1)]^{1/2} \cdot [j\delta \hat{H}_n^{(2)\prime}(kr)(a_{mn}^e \boldsymbol{B}_{mn}^e + a_{mn}^o \boldsymbol{B}_{mn}^o) + \hat{H}_n^{(2)}(kr)(b_{mn}^e \boldsymbol{C}_{mn}^e + b_{mn}^o \boldsymbol{C}_{mn}^o)] \quad (15)$$

$$H_t = \frac{1}{\delta r}\sum_{m=0}^{\infty}\sum_{n=m}^{\infty}[n(n+1)]^{1/2} \cdot [j\delta \hat{H}_n^{(2)}(a_{mn}^e \boldsymbol{C}_{mn}^e + a_{mn}^o \boldsymbol{C}_{mn}^o) + \hat{H}_n^{(2)\prime}(kr)(b_{mn}^e \boldsymbol{B}_{mn}^e + b_{mn}^o \boldsymbol{B}_{mn}^o)] \quad (16)$$

where

$$[n(n+1)]^{1/2}\boldsymbol{B}_{mn}^{\substack{e\\o}} = \begin{Bmatrix}\cos m\phi\\ \sin m\phi\end{Bmatrix}\frac{d}{d\theta}(P_n^m(\cos\theta))\,\boldsymbol{\mu}_\theta + \begin{Bmatrix}-m\sin m\phi\\ m\cos m\phi\end{Bmatrix}\frac{P_n^m(\cos\theta)}{\sin\theta}\,\boldsymbol{\mu}_\phi \quad (17)$$

$$[n(n+1)]^{1/2}\boldsymbol{C}_{mn}^{\substack{e\\o}} = \begin{Bmatrix}-m\sin m\phi\\ m\cos m\phi\end{Bmatrix}\frac{P_n^m(\cos\theta)}{\sin\theta}\,\boldsymbol{\mu}_\theta - \begin{Bmatrix}\cos m\phi\\ \sin m\phi\end{Bmatrix}\frac{d}{d\theta}(P_n^m(\cos\theta))\,\boldsymbol{\mu}_\phi \quad (18)$$

$$\boldsymbol{B}_{mn}^{\substack{e\\o}} = \boldsymbol{\mu}_r \times \boldsymbol{C}_{mn}^{\substack{e\\o}} \quad (19)$$

$$\boldsymbol{C}_{mn}^{\substack{e\\o}} = -\boldsymbol{\mu}_r \times \boldsymbol{B}_{mn}^{\substack{e\\o}} \quad (20)$$

$$\boldsymbol{B}_{mn}^{\substack{e\\o}}\cdot\boldsymbol{C}_{mn}^{\substack{e\\o}} = \boldsymbol{\mu}_r\cdot\boldsymbol{B}_{mn}^{\substack{e\\o}} = \boldsymbol{\mu}_r\cdot\boldsymbol{C}_{mn}^{\substack{e\\o}} = 0 \quad (21)$$

$$\delta = (\mu/\epsilon)^{1/2}$$

and $\boldsymbol{\mu}_r$, $\boldsymbol{\mu}_\theta$, and $\boldsymbol{\mu}_\phi$ are unit vectors along the r, θ, and ϕ directions, respectively.

Now let us consider the procedure for evaluating the a_{mn}^o, a_{mn}^e, b_{mn}^o, and b_{mn}^e. We assume that the reflector has been temporarily removed and require that

$$\boldsymbol{E}_{ft}(r_1,\theta,\phi) = -f_1(\theta)f_2(\phi)[\cos\theta\cos\phi\,\boldsymbol{\mu}_\theta - \sin\phi\,\boldsymbol{\mu}_\phi] \quad (22)$$

where $f_1(\theta)f_2(\phi) = |\boldsymbol{E}_1(r_1,\theta,\phi)|$ over the reflector. Note that $\boldsymbol{E}_1$ is assumed to be separable in r,θ,ϕ and that $f_1(\theta)$ can be assumed to be an even function of θ with no loss in generality.[2] At this point we introduce the temporary restriction that $|\boldsymbol{E}_1|$ must be an even function of ϕ. In most practical applications of the spherical reflector, the desired aperture distribution has this characteristic; the restriction is introduced here so that the fields of the feed can be represented by (13)–(16) with $a_{mn}^o = b_{mn}^e = 0$. This leads to a significant reduction in the subsequent mathematics; the solution is not less tractable without this restriction.

[2] This is because θ is limited to the range $0 < \theta < \pi$.

In order to satisfy the boundary condition at $r = r_1$ we use (15) and (22) to obtain

$$\left.\begin{matrix}-f_1(\theta)f_2(\phi)\boldsymbol{p}(\theta,\phi), & 0 \le \theta \le \theta_1\\ 0, & \text{for all other } \theta\end{matrix}\right\} = \sum_{m=0}^{\infty}\sum_{n=m}^{\infty}\frac{[n(n+1)]^{1/2}}{r_1} \cdot [a_{mn}\,j\delta\hat{H}_n^{(2)\prime}(kr_1)\boldsymbol{B}_{mn}^e + b_{mn}\hat{H}_n^{(2)}(kr_1)\boldsymbol{C}_{mn}^o] \quad (23)$$

where

$$\boldsymbol{p}(\theta,\phi) = \cos\theta\cos\phi\,\boldsymbol{\mu}_\theta - \sin\phi\,\boldsymbol{\mu}_\phi \quad (24)$$

with $a_{mn} = a_{mn}^e$ and $b_{mn} = b_{mn}^o$. Forming the dot product of both sides of (22) with $(\boldsymbol{B}_{pq}^o + \boldsymbol{C}_{rs}^e)\sin\theta\,d\theta\,d\phi$, integrating over the interval $0 < \theta < \pi$, $0 < \phi < 2\pi$ and invoking the orthogonality relations [19] (21), we have

$$a_{mn} = -W_{mn}\int_0^{2\pi}\int_0^{\theta_1} f_1(\theta)f_2(\phi)\boldsymbol{p}(\theta,\phi)\cdot\boldsymbol{B}_{mn}^e\sin\theta\,d\theta\,d\phi \quad (25)$$

$$b_{mn} = -W_{mn}'\int_0^{2\pi}\int_0^{\theta_1} f_1(\theta)f_2(\phi)\boldsymbol{p}(\theta,\phi)\cdot\boldsymbol{C}_{mn}^e\sin\theta\,d\theta\,d\phi \quad (26)$$

where

$$W_{mn} = -\frac{(2n+1)(n-m)!\,\epsilon_m r_1}{4\pi(n+m)!\,j\delta\hat{H}_n^{(2)\prime}(kr_1)[n(n+1)]^{1/2}} \quad (27)$$

$$W_{mn}' = \frac{W_{mn}\,j\delta\hat{H}_n^{(2)\prime}(kr_1)}{H_n^{(2)}(kr_1)} \quad (28)$$

and

$$\epsilon_0 = 1, \quad \text{for } m = 0$$

$$\epsilon_m = 2, \quad \text{for } m \ne 0.$$

An analytical method of performing the double integration indicated in (25) and (26) does not usually exist and standard methods of numerical integration usually depend on the character of $f_1(f_2)$. Alternatively, a method developed by Ricardi and Burrows [20] enables the use of the fast Fourier transform and a recurrence technique both of which are based on an existing formalism. Further discussion of this method is beyond the scope of this paper and can be obtained from [20].

Making use of the above simplifications, the tangential fields of the feed are given by

$$\boldsymbol{E}_{ft} = -\frac{1}{r}\sum_{m=0}^{M}\sum_{n=m}^{N}[n(n+1)]^{1/2} \cdot [a_{mn}\,j\delta\hat{H}_n^{(2)\prime}(kr)\boldsymbol{B}_{mn}^e + b_{mn}H_n^{(2)}(kr)\boldsymbol{C}_{mn}^o] \quad (29)$$

$$\boldsymbol{H}_{ft} = -\frac{-j}{\delta r}\sum_{m=0}^{M}\sum_{n=m}^{N}[n(n+1)]^{1/2} \cdot [a_{mn}\,j\delta\hat{H}_n^{(2)}(kr)\boldsymbol{C}_{mn}^e + b_{mn}H_n^{(2)\prime}(kr)\boldsymbol{B}_{nn}^o] \quad (30)$$

where a_{mn} and b_{mn} are given by (25) and (26) and the infinite sums have been truncated. Next, let us consider

N, the maximum value of n. In a study of the physical limitations of omnidirectional antennas, Chu [21] showed that the excitation of spherical waves, of order $n > kr$, requires the maintenance of a large amount of stored energy (principally in the immediate vicinity of the antenna) compared to the radiated energy. This gives rise to a highly reactive wave impedance and hence a highly reactive or high-Q antenna impedance. Consequently we will choose $N = kr_2$ to avoid the possibility of synthesizing fields which could only be produced by an undesirably high-Q or a supergain antenna [17, p. 309]. This restriction also sets $M = N = kr_2$ because $P_n^m(\cos\theta) = 0$ for $n < m$.

With the feed excited to produce $\mathbf{E}_{ft}(r_1,\theta,\phi)$, placing the reflector in its location will give rise to the reflected field $\mathbf{E}_1$ modified by the field scattered by the feed. However, at $r = r_2$ the strength of $\mathbf{E}_1$ is small compared to $\mathbf{E}_f$; hence the field scattered by the feed can be neglected when calculating $\mathbf{E}_f$ at the reflector. However, this scattered field can contribute to the far field of the antenna; this will be discussed in a later section.

In summary, $\mathbf{E}_{ft}$, the component of the field of the feed tangent to the surface of a sphere enclosing it, was set equal to two finite sums of Tesseral harmonics or vector wave functions [19] whose weights a_{mn}^e and b_{mn}^o were determined by a conventional expansion technique and a special recurrence technique [20]. The expansion was carried out such that $\mathbf{E}_{ft}(r_1,\theta,\phi) + \mathbf{E}_{1t}(r_1,\theta,\phi) = 0$ over the surface of the reflector and $\mathbf{E}_{ft}(r_1,\theta,\phi) = 0$ for all other values of θ and ϕ. Since its series representation has a finite number of terms, $\mathbf{E}_{ft}(r_1,\theta,\phi)$ can, in general, only approximate $\mathbf{E}_{1t}$; however, any truncation of this series ensures the least root-mean-square value of $(\mathbf{E}_{1t} - \mathbf{E}_{ft}(r_1,\theta,\phi))$ [2]. Furthermore, the truncation of the series is based on the desire that the feed not be required to radiate supergain waves [21] and consequently that it not be a high-Q antenna [17, p. 307]. It was also assumed that $\mathbf{E}_{1t}$ was an even function of ϕ; conversely, we could have chosen $\mathbf{E}_{1t}$ to be an odd function of ϕ and carried out the same procedure to determine a_{mn}^o and b_{mn}^e. The linear property of these fields permits a more general $\mathbf{E}_{1t}$ to be represented by a linear combination of both solutions.

Uniform Aperture Distribution—Focal Region Fields

In order to demonstrate an application of the method of synthesis presented in the previous section, let us consider the case when $\mathbf{E}_1$ is chosen to produce a linearly polarized field of unit amplitude over the aperture of the reflector. Our goal is to calculate $\mathbf{E}_f(r_2,\theta,\phi)$, for $0.5r_1 < r_2 < 0.62r_1$, in order to determine the probable size and location of a feed. Assuming that $\mathbf{E}_1$ is a plane wave propagating along the $-z$ axis, we have

$$\begin{aligned} \mathbf{E}_1 &= \exp(jkz)\,\boldsymbol{\mu}_z \\ &= (\sin\theta\cos\phi\,\boldsymbol{\mu}_r + \cos\theta\cos\phi\,\boldsymbol{\mu}_\theta - \sin\phi\,\boldsymbol{\mu}_\phi) \\ &\quad\cdot\exp(jkr\cos\theta). \end{aligned} \tag{31}$$

If the reflected field at the surface of the reflector is given by (31), an approximately uniform distribution will be produced over its aperture because of a slight amount of "spreading" as the reflected field propagates from the reflector to the aperture; nevertheless, the approximation is sufficiently accurate because the distance from the reflector's aperture to its vertex is less than the diameter of the reflector. Referring to (22), (23), and (31) we see that $f_1(\theta) = \exp(jkr\cos\theta)$ and $f_2(\phi) = 1$. It follows that only the $m = 1$ terms in (29) and (30) are not equal to zero. Making the appropriate substitutions in (25)–(28), it can be shown that

$$a_n = -\frac{f_n(kr_1)}{8k}\sum_{p=0}^{N} c_p\Big[(n+2)I_{p,n} + \frac{n}{2}(I_{p+2,n} + I_{p-2,n}) - (n+1)(I_{p+1,n-1} + I_{p-1,n-1})\Big] \tag{32}$$

$$b_n = -\frac{g_n(kr_1)}{8k}\sum_{p=0}^{N} c_p[(n+1)(I_{p+1,n} + I_{p-1,n}) - (2n+2)I_{p,n-1})] \tag{33}$$

where

$$f_n(kr_1) = \frac{(2n+1)kr_1}{(n(n+1))^2 j\delta \hat{H}_n^{(2)\prime}(kr_1)} \tag{34}$$

$$g_n(kr_1) = \frac{j\delta \hat{H}_n^{(2)\prime}(kr_1)}{\hat{H}_n^{(2)}(kr_1)} \tag{35}$$

$$I_{p,n} = \int_0^\pi \cos p\theta\, P_n^1(\cos\theta)\, d\theta \tag{36}$$

and a_{1n} and b_{1n} are replaced by a_n and b_n, respectively.

The $I_{p,n}$ can be evaluated using the readily determined values $I_{p,1}$ and $I_{p,2}$ and the recurrence relations [20]

$$I_{p,n+1} = \frac{2n+1}{n}(I_{p+1,n} + I_{p-n,n}) - \frac{n+1}{n}I_{p,n-1} \tag{37}$$

$$I_{0,n+1} = \frac{2n+1}{n}I_{1,n} - \frac{n+1}{n}I_{0,n-1}. \tag{38}$$

The Fourier coefficients are given by

$$c_p = \frac{\epsilon_p}{2\pi}\sum_{s=0}^{\infty}\epsilon_s j^s J_s(kr_1)\frac{\theta_1}{2}[\text{sinc}\,(p+s)\theta_1 + \text{sinc}\,(p-s)\theta_1] \tag{39}$$

where sinc $(x) = \sin x/x$ and $J_s(x)$ is the nth-order Bessel function of the first kind. The maximum value of s is determined by the weighing function $J_s(kr_1)$. When $s \geq kr_1$ the function decreases rapidly with increasing s. It is also true that the envelope of $|\text{sinc}(x)|$ decreases $\approx 1/x$ for $|x| \gg 1$. In view of this, the series in the right-hand member of (39) can be truncated at $s = S =$ integer value of kr_1, and the number of terms needed in the series for any given p could be somewhat less than S. Alternatively, the c_p can be determined by a fast Fourier transform technique, numerical integration, etc.; the series

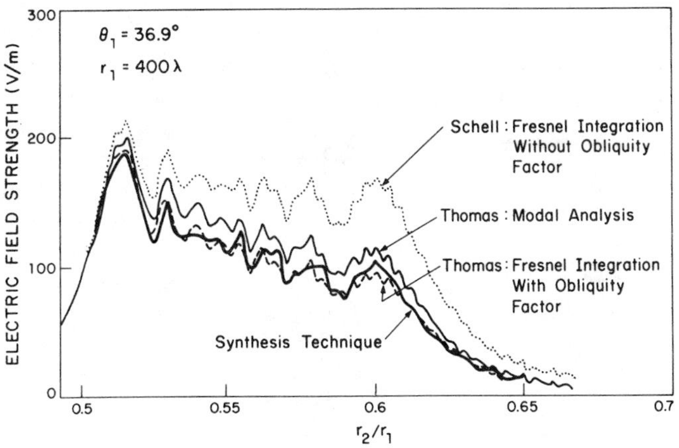

Fig. 2. Relative field strength on focal axis of spherical reflector; $r_1 = 400\lambda$, $\theta_1 = 36.9°$.

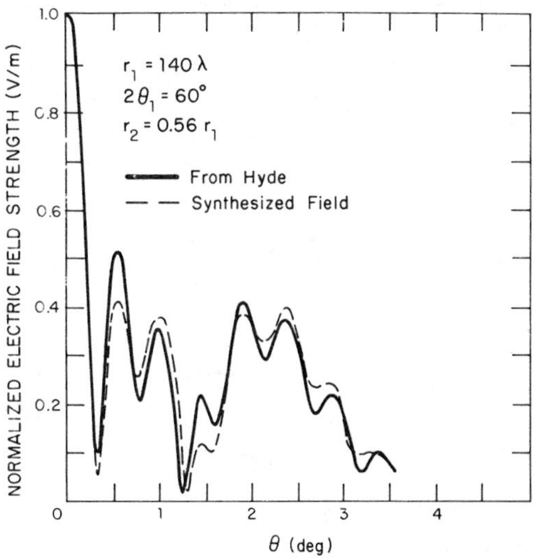

Fig. 3. Relative strength of θ component of electric field in focal region of spherical reflector.

evaluation shown here is presented merely to support the existence of a method.

Using (29), the field strength along the focal axis was calculated with $r_1 = 400\lambda$, $\theta_1 = 36.9$ by varying $0.5r_1 < r_2 < 0.65r_1$. These results are compared with those obtained by Schell [8] and Thomas *et al.* [16] in Fig. 2. The curve from Schell was computed neglecting the obliquity factor in the integral expression for the field produced by the current distribution on the reflector. The latter is determined by diffraction theory and an assumed incident plane wave. Thomas *et al.* repeated these calculations including the obliquity factor, and the results agree well with those obtained by his method of modal analysis and those obtained using the synthesis technique described here.

It is also interesting to compare the transverse field calculations of Hyde [24] with those obtained by the synthesis technique described here (Fig. 3). The former were obtained by a geometrical optics approximation to the integral of the surface current induced by an incident plane wave. Again the results are in reasonably good agreement. These figures are presented primarily to establish confidence in the various methods of analysis or synthesis.

As in the case of the receiving antenna, it is of interest to examine the character of the fields radiated by a feed located at various positions along the focal axis. In particular, we will vary r_2 from $0.5r_1$ to $0.62r_1$ ($r_1 = 250\lambda$ and $\theta_1 = 30°$) and plot both the amplitude and phase of $\boldsymbol{E}_{ft}$ as a set of contours in a two-dimensional space. This allows for the presentation of an interesting qualitative understanding of the character of the fields. In particular, refer to Fig. 4 where the strength of the θ component of the field of the feed is plotted versus z/r_1 and x/r_1 ($z = r_2 \cos\theta$; $x = r_2 \sin\theta$) with relative field strength as a parameter. The latter has a maximum value of 1.0 and the contours are plotted in steps of 0.1; this field has been normalized to a value of 144 V/m. The caustic surface [11] is indicated by the dashed line. The data might also be referred to as the E-plane near-field pattern of the feed; the far-field pattern is an approximation of the cosine θ variation required to produce a uniform field distribution in the aperture of the reflector.

Certainly of equal interest are the constant phase contours shown in Fig. 5. The dashed lines represent a phase of an odd multiple of π rad; the solid lines represent an even number of π rad. The contours in the upper left of the figure represent a wave that is propagating toward the right-hand lower corner of the figure and incident on the lower half of the reflector. A second wave propagating from the left and below the focal axis is incident on the upper half of the reflector. A third wave propagates along the focal axis toward the center of the reflector. In the vicinity of the circle of least confusion the relative strength of all three waves is such that a very "confused" picture of the phase of the field is presented. In fact, the scale of the plot shown is too small to permit a satisfactory representation of the data in the area near the circle of confusion. This area of confusion can be clarified by simply increasing the scale of the plot. The discontinuous phase "fronts" (contours) terminate at minima or nulls in the field strength where phase has little or no meaning. It will be shown later that the phase contours are valuable in the initial design of a transverse feed. Similar sets of amplitude and phase plots were obtained for the ϕ component (ϕ = 90°, or H-plane patterns) and the radial component (ϕ = 0°). Some observations concerning these contour representations of the field are as follows.

1) The caustic surface is a reasonable definition of the boundary between the focal region fields of significant and insignificant strength.

2) There exist three principal waves which correspond to the three stationary points of Spencer and Hyde's analysis [11].

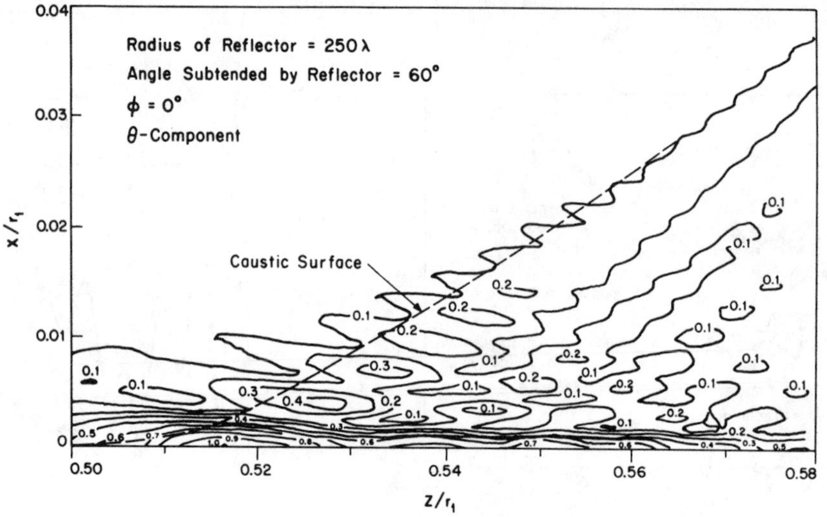

Fig. 4. Relative strength of θ component of synthesized field of correcting feed for spherical reflector; $r_1 = 250\lambda$, $\theta_1 = 30°$, $\phi = 0°$.

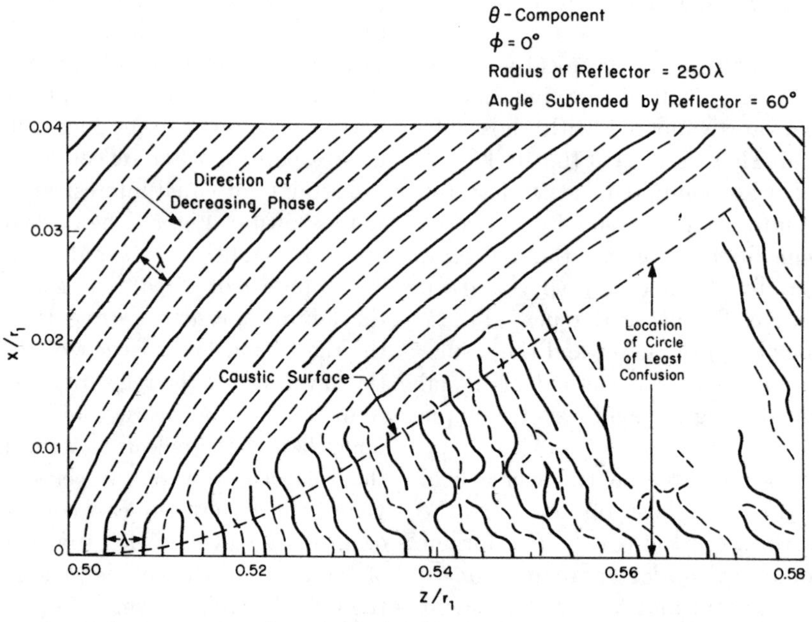

Fig. 5. Relative phase of θ component of synthesized field of correcting feed for spherical reflector; $r_1 = 250\lambda$, $\theta_1 = 30°$, $\phi = 0°$.

3) The size of the aperture of a transverse feed should not be significantly larger than that required to include the caustic surface within its aperture.

The latter observation gives rise to the very important engineering question: What is the relationship between antenna gain, or efficiency, and the size and location of the aperture of a transverse feed? In other words, let us define a practical feed as one whose aperture coincides with that section of a spherical surface of radius r_2 that subtends an angle $2\theta_2$ at the center of the spherical reflector (see Fig. 1). We will assume that the feed produces $\boldsymbol{E}_{ft}(r_2,\theta,\phi)$ only over its aperture and we wish to know the relationship between antenna directivity and θ_2 for various r_2. Toward this end let us consider the following method of analysis and the interesting results that can be derived from it.

Analysis of Practical Feed

As in the synthesis procedure, we will represent the field of the practical feed by the sum of $N = kr_2$ outward propagating spherical waves and calculate the current $\boldsymbol{K}$ induced by this field on the spherical reflector. In the far field of the reflector, the on-axis field strength $\boldsymbol{E}_0$ is then calculated by an appropriate integration of this current distribution. The antenna directivity can be determined from the well-known relationship

$$D = \frac{4\pi r_0^2 \boldsymbol{E}_0 \cdot \boldsymbol{E}_0^*}{P} \tag{40}$$

where the asterisk indicates the complex conjugate and P is the total input power required to produce $\boldsymbol{E}_0$ at r_0. Using the directivity obtained with a uniform field distribution over the aperture of the spherical reflector as the reference, the efficiency η of the antenna can be determined from

$$\eta = \frac{D\lambda^2}{(2\pi r_1 \sin \theta_1)^2}. \tag{41}$$

As in the derivation of $\boldsymbol{E}_f$, we will first express the component of the field tangent to the aperture of the practical feed as a finite Fourier series assuming that the field of the feed vanishes elsewhere on the surface $r = r_2$. This series characterized by the coefficients d_q and e_q can then be used to determine a new set of a_n and b_n, namely,

$$a_n' = -\frac{f_n(kr_2)}{4k} \sum_{q=0}^{N} d_q \left[\frac{n}{2} (I_{q+1,n} + I_{q-1,n}) - (n+1) I_{q,n-1} \right] + e_q I_{qn} \tag{42}$$

$$b_n' = -\frac{g_n(kr_2)}{4k} \sum_{q=0}^{N} d_q I_{q,n} + e_q \left[\frac{n}{2} (I_{q+1,n} + I_{q-1,n}) - (n+1) I_{q,n-1} \right] \tag{43}$$

where $I_{q,n}$ is given by (36).

Substituting these coefficients into (29) and (30), with $m = 1$, $\boldsymbol{K}$ can be calculated from the boundary condition

$$\boldsymbol{K} = -2\boldsymbol{\mu}_r \times \boldsymbol{H}_{ft}. \tag{44}$$

The far field of an electric-current distribution can be accurately represented by the second term in the right-hand member of (11); this is because $\boldsymbol{F} = 0$ and the other terms vary as $1/r^n$ where $n > 1$. Since we are interested in the x component of the radiated field, we have

$$E_{0x} = -j\omega\mu A_x \tag{45}$$

$$E_{0x} = -j\omega\mu \int_0^{\theta_1} \int_0^{2\pi} \frac{\boldsymbol{K} \cdot \boldsymbol{\mu}_x}{r_0 + r_1 \cos \theta} \cdot \exp\left[-jk(r_0 + r_1 \cos \theta)\right] r_1^2 \sin \theta \, d\theta \, d\phi. \tag{45a}$$

In the far field $r_0 \gg r_1$ and the denominator of the integrand can be accurately approximated by r_0. The previous procedure of expressing the integrand as a finite Fourier transform and then integrating analytically can be applied to the evaluation of (45a); however, this will not be described in this paper. The total power radiated by the feed is given by

$$P = \frac{\pi}{\delta} \sum_{n=1}^{N} \frac{n^2(n+1)^2}{2n+1} \{ |\delta a_n'|^2 + |b_n'|^2 \}. \tag{46}$$

Substituting this expression and (45) into (40) enables the determination of the antenna directivity D. It is important to point out that (40), (45), and (46) take "spill-over" and polarization losses into account and assume that the practical feed is free from all dissipative and aperture blocking losses. Substituting (40) into (41) gives the efficiency of the antenna [25]. The method of synthesis used and the analysis presented here characterize D and η as the maximum directivity and efficiency that can be obtained with a feed aperture defined by r_2 and θ_2 under the condition that the feed is not a high-Q antenna.

Maximum Efficiency of Spherical Reflector

The results of the previous section enable one to determine an upper bound on the efficiency that can be obtained by a transverse feed which does not excite high-Q spherical waves. This is a realistic restriction since it has been shown [21] that antennas which excite these waves are impractical. Furthermore, it has been shown [22] that the finite series used to represent the fields have the property that increasing the number of terms decreases the root mean square introduced by truncating them; consequently, the high-Q restriction determines the "best" series representation of the fields.

Using (40) and (41) the efficiency of the Aericebo spherical reflector operating at a frequency of 425 MHz (i.e., $r_1 = 379.6\lambda$, $\theta_1 = 35°$) was calculated for $\theta_2 = 2°$, 3°, 4°, 5°, and 6° with r_2/r_1 as a parameter. These results are plotted in Fig. 6 with a smooth curve drawn through the calculated points. First consider the curve for $r_2/r_1 = 0.5901$; the efficiency increases markedly from less than 20 percent to $\approx$95 percent with increasing θ_2 until $\theta_2 \approx 5°$. Larger values of θ_2 yield a relatively small increase in efficiency. The curve for $r_2/r_1 = 0.559$ has a similar character, except that the maximum efficiency is on the order of 83 percent and the knee of the curve occurs at $\theta_2 \approx 3°$. A review of a geometrical optics analysis [11] points out that the circle of least confusion is an appropriate site for a transverse feed since all of the incident rays must pass through the surface enclosed by this circle. The plausible argument that a transmitting feed located at this site would be equally capable of reversing the direction of propagation of energy is supported by the curves of Fig. 6 because the circle of least confusion is located at $r_2/r_1 = 0.594$ and it subtends an angle $2\theta_2 = 9.6°$. Furthermore, the intersection of the caustic surface and the sphere of radius $r_2 = 0.559r_1$ occurs at $\theta_2 = 2.5°$; qualitatively, the knee of the curves (Fig. 6) occurs very near that value of θ_2 at which the practical feed intersects the caustic surface. The curve for $r_2/r_1 = 0.518$ should have a knee at $\theta_2 = 0.45°$. The area of a practical feed which is located at $r_2 = 0.594r_1$ is less than one percent of the area of the radiating aperture; hence, neglecting aperture blockage in the synthesis and analysis is justified.

A similar set of data was obtained for $r_1 = 50\lambda$, $\theta_1 = 35°$. The results (Fig. 6) demonstrate that the data are somewhat independent of the frequency of operation; this is

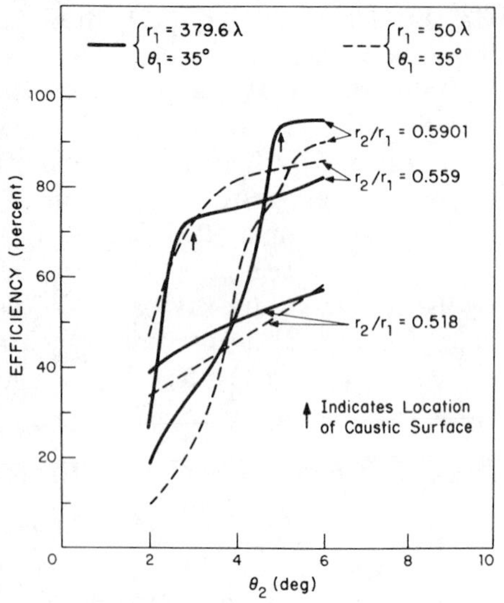

Fig. 6. Efficiency of practical feed with circular aperture on surface of sphere of radius r_2 and subtending angle $2\theta_2$; $\theta_1 = 35°$. Solid curve—$r_1 = 379.6\lambda$; dashed curve—$r_1 = 50\lambda$.

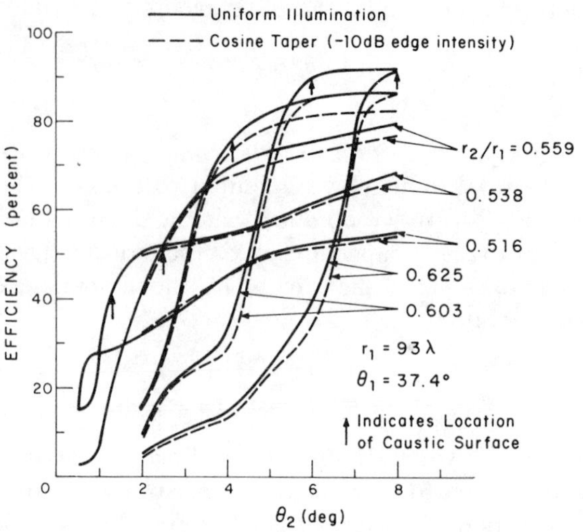

Fig. 7. Efficiency of practical feed with circular aperture on surface of sphere of radius r_2 and subtending angle $2\theta_2$; $r_1 = 93\lambda$, $\theta_1 = 37.4°$. Solid curve—uniform illumination; dashed curve—cos θ tapered illumination.

to be expected if there is to be any hope for a geometrical optics analysis. The knee of the curve is much more prominent for the larger reflector indicating the trend toward the predicted geometrical optics analysis.

Having established that the efficiency of the practical feed is relatively independent of the radius of the reflector, a more exhaustive examination of efficiency as a function of θ_2 and r_2 was carried out for $r_1 = 93\lambda$ and $\theta_1 = 37.4°$. The results are shown in Fig. 7 where vertical arrows indicate the values of θ_2 at which the spherical surface of radius r_2 intersects the caustic surface. The circle of least confusion is located at $r_2 = 0.61r_1$. Also shown is a similar set of data obtained when the synthesis and analysis procedures are carried out with a cosine tapered aperture distribution, that is,

$$\boldsymbol{E}_1 = \cos\theta(\sin\theta\cos\phi\ \boldsymbol{\mu}_r + \cos\theta\cos\phi\ \boldsymbol{\mu}_\theta - \sin\phi\ \boldsymbol{\mu}_\phi) \cdot \exp(jkr_1\cos\theta).$$

Since Figs. 6 and 7 indicate that the size of the aperture of the transverse feed determines its optimum location, the curves of Fig. 7 were used to construct a curve of efficiency versus r_2/r_1. In particular, r_2 is selected from Fig. 7 on the basis that the corresponding practical feed defined by r_2 and θ_2 produce the highest efficiency for a particular value of θ_2. This leads to a range of efficiency for a given value of r_2 (the range would decrease to a single value as the increment in r_2 is made infinitesimally small) which is plotted as a function of r_2 in Fig. 8. An estimate of the continuous curve that would be obtained if r_2 were a continuous variable is also shown. The corresponding values of θ_2 enable the width D of the aperture of the practical feed to be calculated. The curve D/r_1 (solid line) is obtained with the maximum value of θ_2; the curve $D(\theta_2)/r_1$ (dashed line) is based on the estimated value of θ_2. The aperture diameter D_c/r_1 that is defined by the intersection of the sphere $r = r_2$ and the caustic

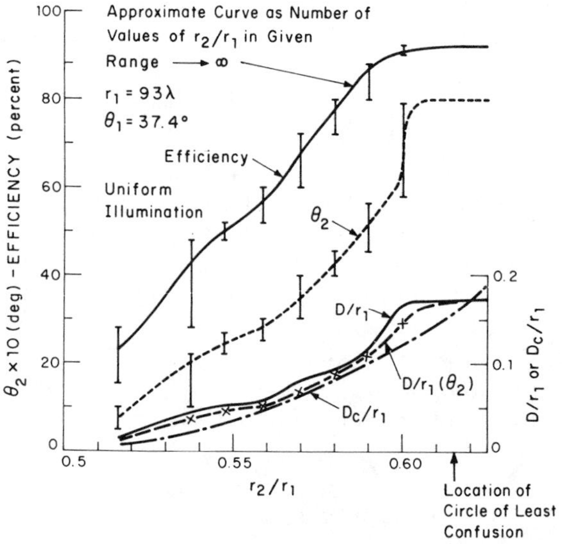

Fig. 8. Maximum efficiency for given practical feed with aperture D/r_1 as a function of r_2/r_1; $r_1 = 93\lambda$, $\theta_1 = 74.8°$. Data given in this plot are approximately correct for $50\lambda < r_1 < \infty$.

surface is given by the remaining curve. When the latter three curves are coincident, the aperture of the practical feed is very nearly equal to the diameter of the circle of least confusion. For somewhat smaller values of r_2 the aperture of the practical feed is significantly larger than the area defined by its intersection with the caustic surface. These results show that the classical geometrical optics analysis would in general define a feed aperture somewhat smaller than that required, except when the feed aperture is located at the circle of least confusion.

Also shown in Fig. 7 is the efficiency of a practical feed designed to produce an aperture illumination with a 10-dB cosine taper. Notice that these dashed curves are very nearly a vertical displacement of the solid curves. Consequently, the estimated curves shown in Fig. 8, although calculated for a uniformly illuminated aperture and $r_1 =$

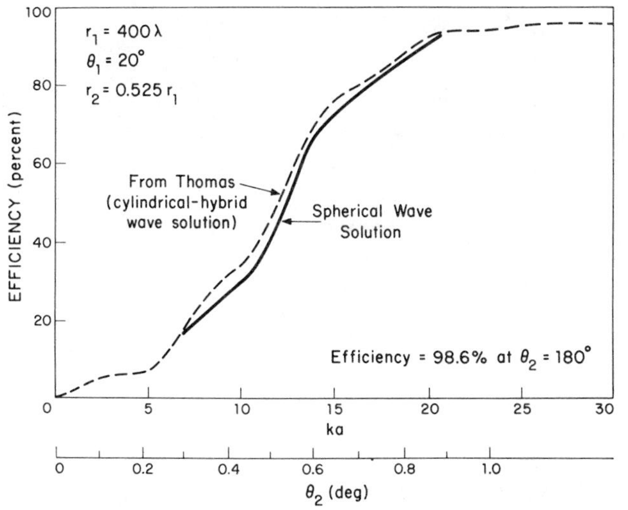

Fig. 9. Efficiency of practical feed with circular aperture of radius ka on spherical surface of radius r_2; $r_1 = 400\lambda$, $\theta_1 = 20°$, $r_2 = 0.525r_1$.

93λ, are a reasonably good approximation of the results that would be obtained with a tapered illumination.

On the basis of the foregoing, we can state that a practical feed designed to produce the synthesized fields $\boldsymbol{E}_f(r_2,\theta,\phi)$ over its aperture and located at, say, $r_2 = 0.55r_1$ will produce an efficiency of approximately 50 percent (see Fig. 8) when it illuminates a spherical reflector which subtends an angle of 74.8° and has a radius greater than, say, 50λ. The diameter of the practical feed aperture would have to be $\approx 0.05r_1$. Increasing the diameter of the practical feed to $0.1r_1$ requires that it be located at $r_2 = 0.585r_1$; the efficiency will be increased to ≈83 percent. These values of efficiency will be decreased by the aperture blockage introduced by the feed, losses due to inaccuracies in the reflector surface, and by any dissipative loss in the feed. Illumination taper and phase error, spillover, and loss due to energy radiated in the cross-polarized field have been taken into account.

Once again it is possible to check these results against those obtained by another author using a different method of analysis. In particular, Thomas *et al.* [16] calculated the efficiency of a hybrid mode exciter which presented a transverse aperture of radius ka illuminating a spherical reflector of radius $r_1 = 400\lambda$ and subtended angle $2\theta_1 = 40°$. The results obtained with the feed aperture located at $r_2 = 0.525r_1$ are plotted in Fig. 9. The corresponding results obtained by the synthesis and analysis procedure described in this paper are also plotted in Fig. 9 to demonstrate the agreement between the two methods.

Spherical Wave Feed

It was pointed out earlier that the field of the feed consists of three principal waves (see Fig. 5) whose strengths are approximately equal in the focal region. Since it is generally easier to build a feed that excites waves which propagate away from its aperture in a relatively perpendicular direction, it seems logical to investigate the properties of a feed whose aperture is approximately tangent

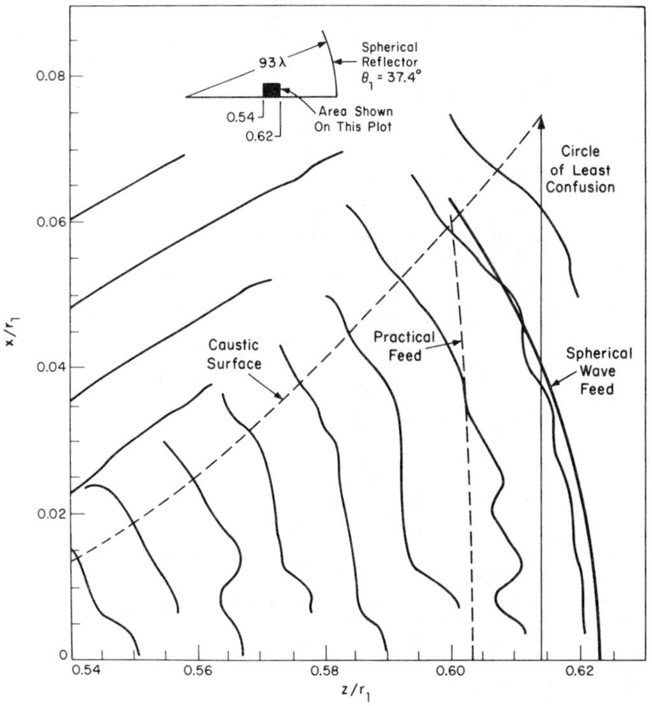

Fig. 10. Relative phase of θ component of synthesized field of correcting feed for spherical reflector; $r_1 = 93\lambda$, $\theta_1 = 37.4°$, $\phi = 0°$.

to the composite wavefront of the synthesized field. For example, consider the constant phase contours of the θ component of $\boldsymbol{E}_f(r,\theta,\phi)$ (see Fig. 10) which is synthesized to produce a uniform linearly polarized distribution over the aperture of the spherical reflector ($r_1 = 93\lambda$ and $\theta_1 = 37.4°$). Let us assume that the practical feed is located at $r_2 = 0.603r_1$ and extends out to $\theta_2 = 5.9°$ as indicated by the dashed curve. The practical feed can be replaced by a spherical wave feed[3] whose aperture is indicated by the solid line of radius r_e. The performance of the latter will be identical to that of the practical feed if the field tangent to the aperture of the spherical wave feed is equal to that component of $\boldsymbol{E}_{fp}(r,\theta,\phi)$ that is tangent to this aperture. This conclusion results when we recognize that the spherical wave feed completely encloses the aperture of the practical feed and then we apply the principle of uniqueness of electromagnetic fields [23].

The advantage of the foregoing can be illustrated by examining the fields tangent to the apertures of these feeds. For this purpose, refer to Fig. 11 where the θ component of the synthesized field that is tangent to the aperture of the practical feed is compared to the θ_e component of $\boldsymbol{E}_f$ that is tangent to the aperture of the spherical wave feed. The latter are referred to the eccentric coordinate system (r_e,θ_e,ϕ) where $r_e = 0.107r_1$. The magnitude of the fields is approximately the same for both feeds; however, the phase of the fields over the aperture of the spherical wave feed varies considerably less than

[3] The concept of the "spherical wave" feed was suggested to this author during a discussion in which Prof. L. J. Chu recommended expressing the field of the correcting feed as the sum of outward propagating spherical waves originating at a point on or near the paraxial focus.

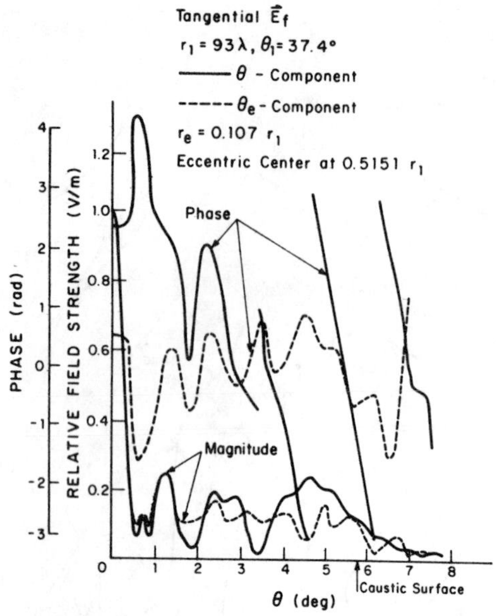

Fig. 11. Magnitude and phase of θ component of $E_f(0.603r_1,\theta,0°)$ and θ_e component of $E_f(r_e,\theta_e,0)$; $r_1 = 93\lambda$, $\theta_1 = 37.4°$. $r_e = 0.1070r_1$ with $r_e = 0$ located at $r = 0.5151r_1$, $\theta = 0$.

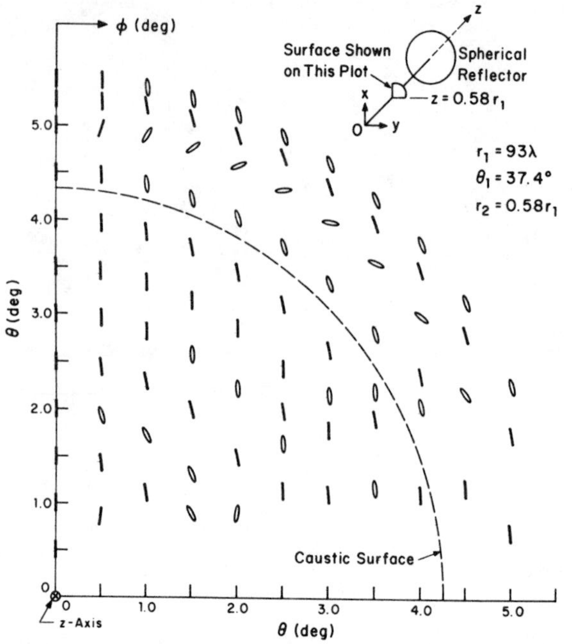

Fig. 12. Polarization of that component of $E_{fp}(0.61r_1,\theta,\phi)$ that is tangent to aperture of practical feed; $r_1 = 93\lambda$, $\theta_1 = 37.4°$.

the phase of the field over the aperture of the practical feed. This characteristic of E_f renders the design of a spherical wave feed somewhat easier than the design of the corresponding practical feed. Both feeds were designed to produce an antenna efficiency ≈89 percent (see Fig. 7). Smaller, less efficient practical feeds are somewhat easier to excite and an engineering compromise arises. The results shown in Figs. 7 and 8 enable an accurate assessment of the tradeoffs involved.

Polarization

The excellent correlation between the polarization of a transverse feed and the polarization of the field it produces at the aperture of the spherical reflector has often been considered one of the major advantages of this feed configuration. A quantitative illustration of the relationship is indicated in Fig. 12 where the polarization ellipse of E_{fp} has been plotted at several points on the surface of the practical feed. Each ellipse represents the trace of the component of E_f that is tangent to the feed aperture at the corresponding points on the surface $r = 0.58r_1$ ($r_1 = 93\lambda$, $\theta_1 = 37.4°$). As in the previous sections, E_f is synthesized to produce an x-directed linearly polarized field over the aperture of the reflector. Those points, where the polarization departs significantly from linear, correspond to nulls or minima in the field. The dashed line indicates the intersection between the caustic surface and the practical feed. These results substantiate the expected correlation.

Conclusions

A method of synthesis and analysis has been presented which enables the determination of the fields that a transmitting transverse feed must produce over its aperture to correct the inherent aberration introduced by a spherical reflector. The expressions for the synthesized field of the correcting feed are based on the assumptions that:

1) the required field distribution over the surface of the reflector is given,

2) the blockage effect of the feed is negligible,

3) the feed is restricted to a class of low-Q antennas which will not have the property of super gain [21].

The method of analysis gives a predictable efficiency of a practical feed. The calculated results shown indicate that the efficiency of a given size transverse feed can be optimized by appropriately choosing its location in the focal region of the reflector, or alternatively, given the desired efficiency, the smallest aperture transverse feed can be obtained by an appropriate choice of its location. These relationships are indicated in Fig. 8.

Acknowledgment

The author wishes to express his thanks and appreciations to Prof. L. J. Chu of the Massachussetts Institute of Technology for his guidance and help as the author's adviser in preparing his doctoral dissertation, from which many of the results presented here were derived. The many discussions and constructive criticisms given by Prof. H. R. Raemer of Northeastern University, Prof. J. I. Glaser of the Massachussetts Institute of Technology, Prof. M. Lowenthal of Northeastern University, and Dr. M. L. Burrows of the M.I.T. Lincoln Laboratory are also greatly appreciated. He is also indebted to Dr. F. S. Holt of the USAF Cambridge Research Center for his careful review of and suggested corrections to this paper.

References

[1] *Encyclopaedia Britannica*, 14 ed., vol. 11. New York: Encyclopaedia Britannica, 1929, see under Huygens, Christiaan, p. 905.

[2] *Encyclopaedia Britannica*, 14 ed., vol. 6. New York: Encyclopedia Britannica, 1929, see under Curves, Special, sec. 25 and 49, pp. 887–899.

[3] J. Ashmead and A. B. Pippard, "The use of spherical reflectors as microwave scanning aerials," *J. Inst. Elec. Eng.*, vol. 93, pt. 111a, 1946, pp. 627–632.

[4] R. C. Spencer, C. J. Sletten, and J. E. Walsh, "Correction of spherical aberration by a phased line source," *Proc. Nat. Elec. Conf.*, vol. 5, 1950, pp. 320–333.

[5] A. F. Kay, "A line source feed for a spherical reflector," Air Force Cambridge Research Labs., Bedford, Mass., Tech. Rep. AFCRL-529 (TRG, Somerville, Mass., Rep. 131), May 1961.

[6] A. W. Love, "Spherical reflecting antennas with corrected line sources," *IRE Trans. Antennas Propagat.*, vol. AP-10, Sept. 1962, pp. 529–537.

[7] G. C. McCormick, "A line feed for a spherical reflector," *IEEE Trans. Antennas Propagat.*, vol. AP-15, Sept. 1967, pp. 639–645.

[8] A. C. Schell, "The diffraction theory of large aperture spherical reflector antennas," *IEEE Trans. Antennas Propagat.*, vol. AP-11, July 1963, pp. 428–432.

[9] F. S. Holt and E. L. Bouche, "A Gregorian corrector for spherical reflectors," *IEEE Trans. Antennas Propagat.*, vol. AP-12, Jan. 1964, pp. 44–47.

[10] M. L. Burrows and L. J. Ricardi, "Aperture feed for a spherical reflector," *IEEE Trans. Antennas Propagat.*, vol. AP-15, Mar. 1967, pp. 227–230.

[11] R. C. Spencer and G. Hyde, "Transverse focal region of a spherical reflector, part I: geometric optics," RCA, AFCRL-64-922, AD-603788, Mar. 1964; also "Studies of the focal region of a spherical reflector: geometric optics," *IEEE Trans. Antennas Propagat.*, vol. AP-16, May 1968, pp. 317–324.

[12] G. Hyde and R. C. Spencer, "Transverse focal region of a spherical reflector, part II: polarization," RCA, AFCRL-64-292II, AD-607477, May 1964; also "Studies of the focal region of a spherical reflector: polarization effects," *IEEE Trans. Antennas Propagat.*, vol. AP-16, July 1968, pp. 399–404.

[13] G. Hyde, "Transverse antenna feeds focal region fields of a spherical reflector," RCA, AFCRL-66-48, Jan. 1966.

[14] D. F. Bowman, "Transverse antenna feeds," RCA, AFCRL-66-49, AD-628500, Jan. 1966.

[15] H. C. Minnett *et al.*, "Fields in the image space of symmetrical focusing reflectors," *Proc. Inst. Elec. Eng.*, vol. 115, Oct. 1968, pp. 1419–1430.

[16] B. MacA. Thomas, H. C. Minnett, and V. T. Bao, "Fields in the focal region of a spherical reflector," *IEEE Trans. Antennas Propagat.* (Commun.), vol. AP-17, Mar. 1969, pp. 229–232.

[17] R. F. Harrington, *Time Harmonic Fields*. New York: McGraw-Hill, 1961.

[18] S. A. Shelkunoff, *Electromagnetic Waves*. Princeton, N. J.: Van Nostrand Reinhold, 1943, pp. 51–52.

[19] P. M. Morse and H. Feshback, *Methods of Theoretical Physics*. New York: McGraw-Hill, 1953, pp. 1898–1900.

[20] L. J. Ricardi and M. L. Burrows, "A recurrence technique for expanding a function in spherical harmonics," *Math. Comput.*, to be published.

[21] L. J. Chu, "Physical limitations of omnidirectional antennas," *J. Appl. Phys.*, vol. 19, 1948, pp. 1163–1175.

[22] I. S. Sokolnikoff and R. M. Redheffer, *Mathematics of Physics and Modern Engineering.*, 2nd ed. New York: McGraw-Hill, 1966, pp. 76–78.

[23] J. A. Stratton, *Electromagnetic Theory*. New York: McGraw-Hill, 1941, p. 604.

[24] G. Hyde, "Studies of the focal region of a spherical reflector; stationary phase evaluation," *IEEE Trans. Antennas Propagat.*, vol. AP-16, Nov. 1968, pp. 646–656.

[25] *Definition of Antenna Terms*, IEEE Standard 145.

Scale Model Development of a High Efficiency Dual Polarized Line Feed for the Arecibo Spherical Reflector

A. W. LOVE

Abstract—The experimental development of a 1:6.535 scale model line feed for the Arecibo spherical reflector is described. The 14.7-ft long model at 2810 MHz simulates a 96.6-ft feed at 430 MHz capable of illuminating the full 1000-ft aperture of the reflector. The feed design requirements are discussed and an experimental program is outlined in which the necessary line source parameters were established using a number of leaky cylindrical test sections. Experimental measurements of both the near and far fields of the model feed are described and typical results are quoted and discussed. Finally, some results obtained after installation of a full size feed in the reflector are given and compared with predictions based on the model data.

I. Introduction

THE THEORY of aberration correcting feed systems for spherical reflectors has been significantly advanced in recent years. The requirements for line source feeds have been thoroughly analyzed [1]–[3], and transverse, or aperture type, feed systems have likewise received attention [4]–[6]. Much of the impetus for this work has stemmed, of course, from interest in the Arecibo Observatory's 1 000-ft aperture diameter spherical reflector. As a consequence, it has become possible to design and build linearly polarized line feeds that realize very high directivity, along with good polarization purity, reasonable sidelobe levels, and moderate bandwidth. One such feed, a flat waveguide slotted array described by LaLonde and Harris [7], achieved an aperture efficiency estimated to be in excess of 80 percent (after removing only those losses due to reflector surface roughness) when illuminating 700 ft of the aperture at 318 MHz.

Until recently, however, a comparably high level of performance has not been possible for dual polarized high power feeds which illuminate the full 1000 ft of aperture at 430 MHz. The slotted waveguide line source is very attractive when high power must be transmitted, and it certainly affords a most elegant solution to the problem of correcting spherical aberration. Yet its use, until recently, has been fraught with frustration and disappointment in the dual polarization case, largely for the following reason. The slotted guide must have a regular polygonal or circular cross section of sufficiently large radius "a" to permit propagation of two orthogonally polarized modes ($ka \gtrless 2$). But the radiation field of a slotted cylinder contains azimuthal phase and amplitude irregularities, due to the presence of spatial harmonics in the azimuth coordinate ϕ, of the form exp ($\pm jp\phi$), where the integer p exceeds unity. In addition, it turns out that neither axial nor circumferential slots alone can create the correct mixture of components E_θ and E_ϕ in the radiation field of the slotted cylinder. These two problems, of spatial harmonics and polarization mismatch, respectively, account for the failure of the original square waveguide feed [8] at Arecibo to produce an aperture efficiency comparable to that obtainable from a large paraboloid with a simple horn feed.

A new feed that avoids these difficulties has been developed in which radiation leaks out through rings of perforated holes in a tapered round waveguide carrying the dominant TE_{11} mode. Each ring has 6 identical rectangular holes spaced uniformly (60° apart) in the

Manuscript received February 1, 1973; revised April 17, 1973. This work was supported by the Advance Research Projects Agency, U.S. Department of Defence, under Office of Naval Research Contract N00014-70-C0017.

The author was with the Autonetics Division, North American Rockwell Corporation, Anaheim, Calif. He is now with the Space Division, North American Rockwell Corporation, Downey, Calif. 90241.

Reprinted from *IEEE Trans. Antennas Propagat.*, vol. AP-21, pp. 628–639, Sept. 1973.

Fig. 1. Photograph of model feed.

Fig. 2. New feed installed at Arecibo.

circumferential direction, the rings being separated by conducting disks that form short, radial transmission lines. A discussion of the underlying principles and the way in which both spatial harmonic and polarization mismatch losses are eliminated can be found in Love and Gustincic [9].

Optically, the line focus of the Arecibo reflector is 96.6-ft long. Because a model feed of such length would be too unwieldy for experimental work the system was scaled from 430–2810 MHz. This results in a waveguide model about 15-ft long that can be handled, albeit with some difficulty, on a conventional antenna positioner table, as is apparent in Fig. 1. Experimentally the development of the model took place in four stages. In the first, constant diameter sections representative of various positions along the feed were measured to determine the complex propagation constant $\gamma = \alpha + j\beta$ and the dependence of E_θ and E_ϕ in the radiated field upon hole size and shape and radial line length. In stage two the considerable body of data so obtained was analyzed, dimensions were established and a model constructed. Stage three consisted in measuring the tangential components of electric and magnetic fields very close to the surface of the model, followed by comparison with theoretically derived results. It is only in this stage that improper behavior of the feed can be localized and corrective action taken. The fourth stage comprised a thorough exploration, in both phase and amplitude, of the components E_θ and E_ϕ in the distant field of the model, followed by use of the experimentally measured values in a computer program to predict the secondary patterns and gain to be expected from the complete feed and reflector system.

A fifth stage, not properly part of this development program, occurred later when a scaled-up version of the model was built and installed in the Arecibo reflector; it can be seen in the photograph Fig. 2. At 430 MHz this feed produces a one-way gain of 60.0 dB for the 1000-ft aperture, an improvement of 3.8 dB over past performance [10]. The corresponding aperture efficiency is 53 percent in a reflector whose rms surface roughness (~1.2 in) causes a 1.3-dB reduction in gain. With an ideal reflector the aperture efficiency would reach 72 percent, a value exceeded only by the large Cassegrainian antennas used in the communication satellite terminals which achieve uniform illumination by the use of shaped reflectors. This is especially gratifying in view of the fact that the theoretical apodisation efficiency is 92 percent due to a −9.4-dB illumination taper designed into the new feed.

II. Design Theory

A. Near-Field Optimum Gain Theory

Use of a near-field testing procedure requires an intimate knowledge of the field $\bar{E}_b, \bar{H}_b$ which should exist close to the surface of the feed when it is transmitting. As shown by Rumsey [3], this field can be determined from a knowledge of the field $\bar{E}_a, \bar{H}_a$ in which the feed would be immersed when a plane wave is incident on the reflector, inducing currents which then reradiate. In the vicinity of the axis of the reflector this reradiated field can be decomposed into two parts,

$$\bar{E}_a = \bar{E}_{a1} + \bar{E}_{a2} \qquad \bar{H}_a = \bar{H}_{a1} + \bar{H}_{a2} \tag{1}$$

where the subscript 1 refers to waves traveling inward toward the axis and 2 refers to waves traveling outward, away from the axis. It is clear that $\bar{E}_{a1}, \bar{H}_{a1}$ is the field that would be created by the induced reflector currents if a perfect absorber were placed along the axis.

The field which should be created near the axis by a transmitting feed lying on the axis is then given by

$$\bar{E}_b = \bar{E}_{a1}^* \qquad \bar{H}_b = -\bar{H}_{a1}^* \tag{2}$$

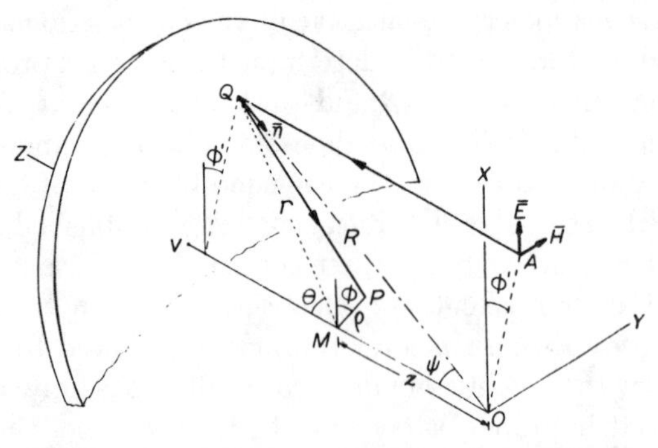

Fig. 3. Geometry of spherical reflector. (R = 870 ft; aperture diameter = 1000 ft; ψ_m = 35° 05′ at reflector rim).

and a feed which produces this field is optimum in the sense that, when receiving, it would extract maximum power from the incoming reflected plane wave. Conversely, when transmitting, it would yield maximum gain and aperture efficiency.

Expressions for the components of the vector "a" field have been derived by Thomas *et al.* [5]. The geometry is depicted in Fig. 3, in which a plane wave E_0, H_0, traveling in the z direction and linearly polarized with its electric field in the x direction, is incident on a spherical reflector of radius R. The field at a point $P(\rho\phi z)$ near the axis is obtained by summing the contributions from the reradiating current elements at all points such as Q on the reflector surface. The integration over ϕ' gives rise to Bessel functions of argument $\kappa\rho$, where $\kappa = k \sin\theta$. The final expressions for the field components at P then involve integrations over the single angle variable ψ, from 0 to its marginal value ψ_m.

It is convenient, however, to transform the independent variable ψ to a new variable u, and to define a position parameter c for the point P, by means of

$$u = \frac{r - z}{R} \quad \text{and} \quad c = \frac{R}{2z} \tag{3}$$

where $r \sin\theta = R \sin\psi$. The limits of integration then become

$$u_0 = 1 - \frac{1}{c}$$

$$u_1 = \left(1 + \frac{1}{4c^2} - \frac{1}{c}\cos\psi_m\right)^{1/2} - \frac{1}{2c}.$$

Following these changes the decomposition indicated by (1) can be performed upon the expressions given by Thomas *et al.* by substitution for the Bessel functions in the integrands by means of the well-known identity

$$J_n = \tfrac{1}{2}[H_n^{(1)} + H_n^{(2)}]$$

in which the Hankel functions of the first and second kind represent inward and outward going waves, respectively.

From (2) the optimum "b" field is then found to have tangential components given by

$$E_{bz} = kRE_0 \exp\left[jkR\left(c + \frac{1}{2c}\right)\right] \cos\phi \cdot e_z$$

$$E_{b\phi} = kRE_0 \exp\left[jkR\left(c + \frac{1}{2c}\right)\right] \sin\phi \cdot e_\phi$$

$$H_{bz} = kRH_0 \exp\left[jkR\left(c + \frac{1}{2c}\right)\right] \sin\phi \cdot h_z$$

$$H_{b\phi} = kRH_0 \exp\left[jkR\left(c + \frac{1}{2c}\right)\right] \cos\phi \cdot h_\phi \tag{4}$$

where $kR(c + 1/2c)$ is the geometrical optics phase.

It can be seen that the azimuthal dependence of the near axis fields displays a particularly simple form, expressed by $\cos\phi$ or $\sin\phi$. A leaky round waveguide will produce this same form of variation if it carries any TE_{1n} or TM_{1n} mode, and is therefore potentially suitable as a feed.

In (4) e and h are dimensionless near-field quantities given by

$$e_z = -c \int_{u_0}^{u_1} \cos(\theta - \psi) \sin\theta H_1^{(2)}(\kappa\rho) \exp(-jkRcu^2)\, du$$

$$e_\phi = -jc \int_{u_0}^{u_1} \left[\cos\psi H_0^{(2)}(\kappa\rho) - \frac{\sin(\theta - \psi)}{k\rho} H_1^{(2)}(\kappa\rho)\right] \cdot \exp(-jkRcu^2)\, du$$

$$h_z = -c \int_{u_0}^{u_1} \cos\psi \sin\theta H_1^{(2)}(\kappa\rho) \exp(-jkRcu^2)\, du$$

$$h_\phi = jc \int_{u_0}^{u_1} \left[\cos(\theta - \psi) H_0^{(2)}(\kappa\rho) - \frac{\sin\psi}{k\rho} H_1^{(2)}(\kappa\rho)\right] \cdot \exp(-jkRcu^2)\, du. \tag{5}$$

A check on the validity of these expressions for the optimum transmitting field can be made in the following way. Suppose a feed exists which creates the field $\bar{E}_b, \bar{H}_b$ at all points such as P, lying on a cylinder of radius ρ enclosing the feed. The radial component of the Poynting vector is then

$$S_\rho = E_{b\phi} H_{bz}^* - E_{bz} H_{b\phi}^* = k^2R^2E_0H_0(e_\phi h_z^* \sin^2\phi - e_z h_\phi^* \cos^2\phi).$$

All the power transmitted by the feed must flow through the cylindrical surface, i.e.,

$$P_t = \tfrac{1}{2} \operatorname{Re} \iint_{\text{cyl}} S_\rho \cdot \rho\, d\phi\, dz = \frac{\pi}{2} \rho k^2 R^2 E_0 H_0 \operatorname{Re} \int (e_\phi h_z^* - e_z h_\phi^*)\, dz. \tag{6}$$

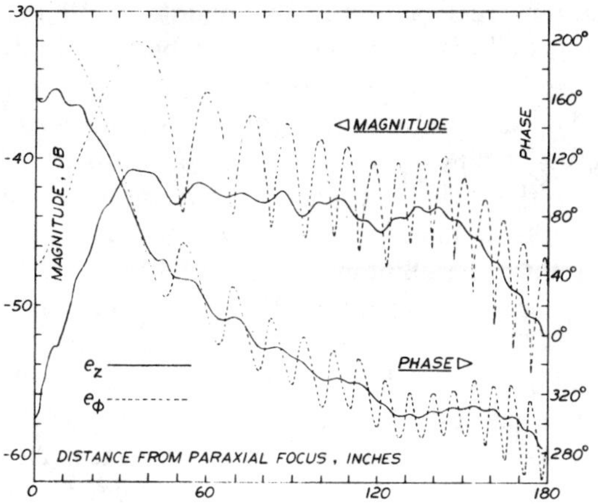

Fig. 4. Near fields e_z and h_z, for $k\rho = 2\pi$.

Now consider the optical limit of vanishing wavelength. In this case $kR \rightarrow \infty$ and a stationary phase evaluation of the integrals in (5) is exact. Then, because, $k\rho \rightarrow \infty$, the Hankel functions may be replaced by their asymptotic forms and in (6) the integration over z is found to be trivial, with the result that

$$P_t = \frac{\pi}{2} R^2 \sin^2 \psi_m E_0 H_0. \tag{7}$$

But this is exactly equal to the power in the incident plane wave that is intercepted by the aperture of the reflector. By reciprocity, then, the postulated feed would receive all the power incident on the reflector, yielding 100-percent aperture efficiency.

The introduction of a tapered illumination has been accomplished by assuming a radial tapering in the amplitude of the incident wave E_0,H_0 over the reflector aperture, with the inclusion of an appropriate apodization function under the integral sign in (5). The chosen function was simply cos 2ψ yielding an edge taper of -9.4 dB.

Due to singularities in the Hankel functions, the integrands in (5) may not be evaluated by the method of stationary phase. Consequently the quantities e and h were computer evaluated for the following parameter values appropriate to the Arecibo reflector at 430 MHz, or a 1/6.535 scale model at 2810 MHz; $kR = 2390$, $\psi_m = 35°05'$, and $k\rho = 2\pi, 2.5\pi$, corresponding to a field point P located at radial distances of 1λ and $5\lambda/4$ from the axis. Values were computed for points two inches apart along the model feed, i.e., from $t = 0$ to $t = 180$ in, where $t = z - R/2$ and $t = 0$ corresponds to the paraxial focus at $z = R/2$. Results for $k\rho = 2\pi$ are given in Fig. 4, which shows both the magnitude and phase angle for the electric field components e_z and e_ϕ. The same curves can be used for the magnetic field components for it happens that e_z and h_z are essentially identical, and that e_ϕ and $-h_\phi$ are similarly equal.

B. *Leaky Cylindrical Line Source*

The 42λ long feed can be imagined to be divided into constant diameter sections, each radiating a conical pattern at its own principal angle which satisfies the optical condition $\theta = 2\psi$. As discussed in [9], the external fields of a typical section are expressible in terms of 2-Hz potentials f and g, which are uniquely determined by the boundary conditions. Assuming a TE_{11} internal waveguide mode and a sufficient number of radiating holes per ring to suppress higher spatial harmonics, then the appropriate boundary conditions at the surface of the cylindrical section $\rho = a$ are

$$E_z = e_z \cos \phi \exp(-j\beta z) \qquad E_\phi = e_\phi \sin \phi \exp(-j\beta z) \tag{8}$$

where

$$\beta = k \cos \theta = 2\pi/\lambda_g \tag{9}$$

and e_z and e_ϕ are parameters that depend on the size and shape of the radiating holes, but which vary from section to section. So far as the feed as a whole is concerned, an optimum gain design would result if e_z and e_ϕ could be made to vary with position in just the manner prescribed by (5).

Once f and g are determined, the field components E_ρ, E_ϕ, and E_z can everywhere be deduced from the relation

$$\bar{E} = \nabla \times f\hat{z} + \nabla\left(\frac{\partial g}{\partial z}\right) + k^2 g\hat{z}$$

where $\hat{z}$ is a unit vector in the z direction. In the far field it is convenient to use spherical rather than cylindrical coordinates. With $\exp(-j\beta z)$ suppressed it is then found that

$$E_r = 0 \qquad E_\theta = A \cos \phi \qquad E_\phi = B \sin \phi \tag{10}$$

where

$$A = -\frac{k}{\kappa} e_z \frac{H_1^{(2)}(\kappa\rho)}{H_1^{(2)}(\kappa a)}$$

$$B = -\left(\frac{\beta e_z}{\kappa^2 a} + j e_\phi\right) \frac{H_1^{(2)}(\kappa\rho)}{H_1^{(2)\prime}(\kappa a)} \tag{11}$$

with

$$\kappa^2 = k^2 - \beta^2. \tag{12}$$

Now suppose this field is incident on the spherical mirror at the geometrical optics angle $\theta = 2\psi$. Then it can be shown that the reflected field in the aperture is of the form

$$E_x = -A \cos^2 \phi + B \sin^2 \phi$$
$$E_y = -(A + B) \sin \phi \cos \phi \tag{13}$$

so that polarization purity occurs for $A = -B$. This leads to the condition given in [9],

$$\frac{e_\phi}{e_z} = j \frac{k}{\kappa} \left[\frac{H_1^{(2)\prime}(\kappa a)}{H_1^{(2)}(\kappa a)} + \frac{\beta}{k\kappa a}\right]. \tag{14}$$

It is important to note that in the far field of any section of the feed E_θ and E_ϕ should have equal peak magnitudes in the principal direction of radiation.

If the large argument asymptotic forms are used for the Hankel functions, (14) becomes

$$\frac{e_\phi}{e_z} \simeq \frac{k}{\kappa}\left(1 + j\frac{\beta}{k\kappa a}\right) \simeq \frac{k}{\kappa}.$$

Thus to the extent that $\kappa a \gg 1$, e_ϕ and e_z are required to be in phase at the feed surface. But e_ϕ and e_z are created by the fields in the radiating holes and it has already been observed that these are in time phase quadrature. This can be seen by noting that the ratio of the circumferential to the axial wall current in a TE_{11} cylinder is $\kappa^2 a/j\beta$.

However, if the tangential field components are forced to propagate outward through a radial transmission line (formed by the conducting disks which separate each ring of holes) then e_ϕ and e_z can be brought into phase at some radius $\rho = b$ which then becomes the new radius of the feed cylinder.

The axial spacing $s = 0.47\lambda$ between disks, chosen to suppress multiple beams at small θ, actually results in a cutoff condition in the radial lines, as will be apparent in what follows.

The radial line is excited at $\rho = a$ by the fields e_z and e_ϕ due to the holes, and propagation occurs in a hybrid mixture of TM_{01} and TE_{11} modes. Thus, for the TM_{01} mode

$$E_z \sim H_1^{(2)}(k\rho) \qquad E_\phi = 0 \tag{15}$$

while for the TE_{11} mode,

$$E_z = 0 \qquad E_\phi \sim K_1'(h\rho) \tag{16}$$

where

$$h^2 = \left(\frac{\pi}{s}\right)^2 - k^2. \tag{17}$$

The modified Hankel function arises because, with $s < \lambda/2$, the radial wavenumber is jh, a pure imaginary.

At the new feed surface $\rho = b$, the tangential fields become

$$E_z|_{\rho=b} = e_z \frac{H_1^{(2)}(kb)}{H_1^{(2)}(ka)} \simeq e_z \left(\frac{a}{b}\right)^{1/2} \exp\left[-jk(b-a)\right] \tag{18}$$

$$E_\phi|_{\rho=b} = e_\phi \frac{K_1'(hb)}{K_1'(ha)} \tag{19}$$

in which ka and kb are both large enough to permit use of the asymptotic approximation in (18).

If the radial line is a quarter-wave long, i.e., $b - a = \lambda/4$, then (18) shows that E_z suffers a phase lag of 90° and a reduction in amplitude. On the other hand, E_ϕ suffers only a decrease in amplitude. Because the modified Hankel function in (19) is a real function of a real argument, E_ϕ undergoes no phase shift. This treatment is approximate, in that reflections at $\rho = b$ have been ignored, but it serves to show how the tangential fields, initially in quadrature at $\rho = a$, are brought into phase at $\rho = b$. The attenuation of the ϕ component predicted by (19) is very real; typically it amounts to 10 to 13 dB while that of E_z is somewhat smaller, being 5–10 dB.

A more complete theory was constructed in which the tangential fields were derived just above the surface of a conducting cylinder, $\rho = a$, due to a distribution of dipoles on the surface. A radial electric and two tangential magnetic dipole moments were calculated using the static polarizabilities for elliptically shaped holes, along with the fields appropriate to an internal TE_{11} waveguide mode. The tangential fields so derived were next transformed from the input $\rho = a$ to the output $\rho = b$ of the radial line, taking due account of reflections which occur at the interface between the radial line and free space. At $\rho = b$, these fields were then imposed as boundary conditions [cf. (8)] to determine the Hertz potentials and ultimately the complete radiation field of a cylindrical section of the feed. Theory and experiment were in good agreement for principal radiation angles exceeding $\theta \simeq 35°$, for which radial line length was predicted to be slightly longer than $\lambda/4$. Agreement was poor for small θ, however. The details will not be given here, although the theory did prove to be a useful guide over much of the feed.

C. Geometrical Optics

The use of ray optics to establish the proper variation of the complex propagation constant $\gamma = \alpha + j\beta$ along the leaky waveguide can be defended on the ground that the near field of the feed will later be tailored to fit the more rigorous requirements of the optimum gain theory given above. It has been shown [11] that spherical aberration can be eliminated and a collimated beam can be formed if the radiation from point z of the line source occurs at angle θ to the axis such that $\theta = 2\psi$, where

$$\sec\psi = \frac{2z}{R} = 1 + \frac{2t}{R} \tag{20}$$

and t is distance along the line source measured from the paraxial focus at $z = R/2$. But radiation occurs at the principal angle θ determined by $\cos\theta = \lambda/\lambda_g$, whence, providing the attenuation is not too large ($\alpha \ll k$),

$$\cos 2\psi = \frac{\lambda}{\lambda_g} = \frac{\beta}{k}. \tag{21}$$

Thus (20) and (21) determine the phase constant β as a function of position z in the waveguide line source.

By definition, the attenuation constant in the guide is given by

$$\alpha = -\frac{1}{2P}\frac{dP}{dt} \tag{22}$$

where P is the power at t in the guide for unit power input. So long as ohmic loss is negligible compared to radiation loss, P can be calculated in the following way. The power radiated in length dt of the line source is dP, and this power is distributed over an annular ring of radius ρ_A

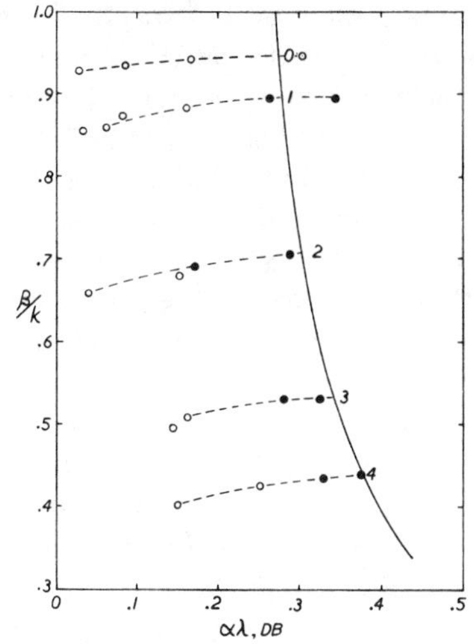

Fig. 5. α, β diagram (computed for −9.4 dB illumination taper and 4 percent unradiated power).

in the aperture, whose area is $dA = 2\pi\rho_A\, d\rho_A$. Let the apodization, i.e., the normalized field distribution over the aperture, be given by $F(\rho_A)$. Then

$$\frac{dP}{dA} = \frac{1}{2\pi\rho_A}\frac{dP}{d\rho_A} = K\,|\,F\,|^2 \tag{23}$$

and the total power radiated up to point t is

$$1 - P = 2\pi K \int_0^t |\,F\,|^2\, \rho_A\, d\rho_A.$$

But $\rho_A = R\sin\psi$, hence

$$P = 1 - \pi K R^2 \int_0^{\psi} |\,F\,|^2 \sin 2\psi\, d\psi. \tag{24}$$

If g is the fractional power which remains unradiated, then

$$g = 1 - \pi K R^2 \int_0^{\psi_m} |\,F\,|^2 \sin 2\psi\, d\psi \tag{25}$$

and this equation determines the constant K in terms of the design parameter g.

When (23)–(25) are substituted in (22) the resulting expression for α is

$$\alpha\lambda = \frac{2\lambda}{R}\,\frac{(1-g)\,|\,F\,|^2\cos^3\psi}{\int_0^{\psi_m}|\,F\,|^2\sin 2\psi\, d\psi - (1-g)\int_0^{\psi}|\,F\,|^2 \sin 2\psi\, d\psi}\ \text{nepers.} \tag{26}$$

Inspection of the denominator shows that extremely large values of α are required near the end of the feed if the fraction of unradiated power is to be negligible. As a compromise, a design value of 4 percent was used, so that $g = 0.04$ in (26). Largely for the sake of convenience in computing $\alpha\lambda$, an apodization of the form $F = \cos 2\psi$ was adopted; this leads to an illumination tapering to −9.4 dB at the edge of the reflector, where $\psi_m = 35°05'$.

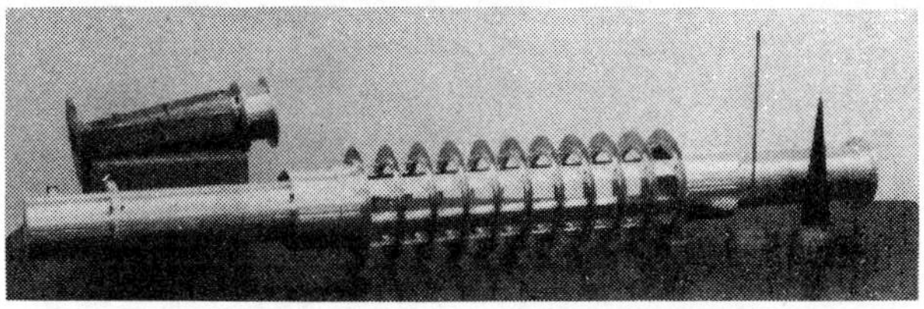

Fig. 6. Test Section I and ancillary pieces.

Equations (21) and (26) thus determine β/k and $\alpha\lambda$ in terms of the parameter ψ which, in turn, is related to position along the line source by (20). Values of $\alpha\lambda$ and β/k computed in this way are shown by the solid line in Fig. 5. Experimental values are indicated by circles, and will be explained later.

III. Experimental Data Acquisition

A. Test Sections

A successful feed design requires the synthesis of four different distributions, namely the phases and amplitudes of e_ϕ and e_z on a surface closely enclosing the feed. The difficulty is compounded by the fact that the four parameters are coupled in the sense that no one parameter can be altered without affecting, in some degree, the other three. To deal with this complexity in a rational way a less rigorous approach, based on geometrical optics, was initially used. In this case there are still four parameters to be dealt with but these are now the propagation constants, α and β, in the guide, along with the magnitude and phase of the ratio E_ϕ/E_θ in the principal direction of radiation in the distant field. Thus a constant diameter test section can be built, representative of some point along the feed which initially is not precisely known. The propagation constants, α and β, for this section can be measured in the laboratory, while E_θ and E_ϕ can be measured on a pattern range. If the peak magnitudes of E_θ and E_ϕ are not equal (in the direction $\theta = \cos^{-1}\beta/k$) then the hole shape is changed; at the same time the holes can be enlarged to cause an increase in α, the attenuation coefficient. The radial line length may also be adjusted if it is found that E_θ and E_ϕ are too far out of phase. This procedure is continued until the measurements show E_θ and E_ϕ to be in phase and equal in peak magnitude, while the attenuation coefficient $\alpha\lambda$ lies on the design curve shown in Fig. 5. The value of β/k is thus known and so is the location of the section in the feed.

In this work five such constant diameter test sections were built, intended to be representative of portions of the feed located at roughly equal distances from the paraxial end to the tip. Each test section consisted of ten constant diameter rings, as in Fig. 6, separated by fins which initially were all made $\lambda/4$ in radial length. The rings are bolted together in such a way as to clamp the

TABLE I
EXPERIMENTAL TEST SECTION DATA

Test Section No.	0	1	2	3	4
I. D. in inches	6.25	4.50	3.13	2.67	2.54
w = hole width, in.	1.480	1.398	1.155	1.090	1.100
l = hole length, in.	1.690	1.650	1.217	0.982	0.925
b-a = radial line length, in.	0.75	0.85	1.05	1.05	1.05
β/k	0.946	0.893	0.720	0.539	0.447
$\alpha\lambda$ in db	0.275	0.280	0.300	0.340	0.374
E_ϕ/E_θ in db	~ -5	<1.0	<1.0	< 0.5	< 0.5
δ = phase error, deg.	~-35	-3	-16	-15	-16
5th harmonic	2%	<1%	~1%	<1%	< 1%

fins securely between rings, ensuring good electrical contact not only from ring to ring, but to the fins as well. Use of ten rings results in a radiating section 4.7λ long, and this is sufficient to ensure that mutual coupling effects are included in the measurements. Appropriate tapered transitions to 3-in diameter round guide, coaxial adapters, absorbing terminations and sliding short circuits were also provided for use in the measurements.

Four such sections, designated 1, 2, 3, and 4, were initially built and a fifth was added later, called number zero. The radiating holes were all originally cut undersize and were $\frac{7}{8}$ in^2. The small circles in Fig. 5 represent experimentally measured values of α and β for the test sections; open circles refer to hole dimensions giving unequal magnitudes of E_θ and E_ϕ, while filled circles indicate that these far field components are balanced and very nearly in the correct phase relationship. That the experimental points do not always fall on the smooth dashed curves is not surprising, for a change in l, the axial hole length, is not equivalent to a similar change in w, the width in the circumferential direction. For test sections 0 and 1 interpolation was required in order to find hole dimensions which would lead to points falling exactly on the design curve for α,β. For test sections 2 and 3, short extrapolations were needed. The end result was a set of data representative of locations at distances t = 6, 12, 34, 62, and 77 ft along the full-size feed, corresponding to principal radiation angles of 18.9°, 26.7°, 43.9°, 57.4°, and 63.6°. A summary of this data is given in Table I.

Little difficulty was experienced in experiments involving sections 1, 2, 3, and 4. Different measurement techniques gave consistent and unambiguous values of α and β, while radiation pattern measurements gave results, with one exception, in good agreement with the theory mentioned in Section II-B. The phase error of 15° to 16° between E_θ and E_ϕ listed in Table I is a result of compromise. Experiments with radial line lengths slightly greater than $\lambda/4$ showed that the phase error could be eliminated, but that spatial harmonic content increased to unacceptably high levels. Surprisingly, a radial line length less than $\lambda/4$ appeared to be optimum for test section one.

Test section zero showed an anomalous behavior in its radiation pattern, giving an endfire beam (at $\theta = 0$) comparable in magnitude to the expected beam at $\theta = 19°$. There was also increased spatial harmonic content, a circumstance which introduced a sizeable uncertainty into the measurement of the phase angle between E_θ and E_ϕ. These components were unbalanced by 5 dB at $\theta = 19°$ although they were, of course, equal at endfire. Increasing the axial length of the holes would be expected to cause an increase in E_ϕ; this length, however, could not be increased beyond the value given in Table I without interfering with the fin clamping arrangement. Finally, the rather large phase error between E_θ and E_ϕ was again a compromise in favor of greatly reduced fifth-space harmonic. Some of the problems associated with this test section may have been due to higher modes. Due its large diameter it can support, at 2810 MHz, the TM_{01}, TE_{21}, TM_{11}, TE_{01}, and TE_{31} modes in addition to the dominant. Having regard to symmetry conditions and the method of excitation, the TM_{01}, TE_{01}, and TE_{21} modes would not be expected to occur. No radiation at angles appropriate to the TM_{11} or TE_{31} modes was evident in the patterns, however. The strong end-fire radiation could have been caused by a surface wave but the source of such a wave is unknown.

B. Experimental Techniques

Measurements of the phase constant β were initially made using three different methods, all of which gave results in good agreement. For example, β can be found from the principal angle of radiation, or, with a short-circuit termination, the standing wave pattern close to the surface of the test section can be explored and λ_g determined. By far the most accurate method, however, is to adopt a technique used in connection with periodic structures. The test sections become transmission cavities by the simple addition of input and output irises at each end of the section. It is evident that end effects are eliminated, since by image theory the structure appears infinitely long. Very small iris openings can be used so that the coupling is very light. By using different numbers of rings, e.g., 8, 9, and 10, a number of resonances at different frequencies can be observed and for each of these the value of β/k can be found from the relation

$$\frac{\beta}{k} = \frac{n\lambda_0}{2L}$$

where λ_0 is the resonant wavelength, L is the overall cavity length, and n is the number of integral half-wavelengths in the cavity at resonance. When β/k is plotted against resonant frequency a smooth curve can be drawn from which the value of β/k at 2810 MHz can be determined to an accuracy approaching 0.1 percent. This method was used exclusively in all later work.

The cavity measurements can also be used to determine α, the attenuation coefficient, by measuring the Q values at the resonant frequencies. Due to inability to measure small frequency differences with adequate accuracy this method was not used extensively. The preferred method consisted in determining the intrinsic insertion loss of a test section by measuring the input reflection coefficient as a function of position of a sliding short-circuit termination. Intrinsic loss, denoted by L_I, is defined as the attenuation suffered by a wave propagating in one direction only [12]. If ρ_1 and ρ_2 are the maximum and minimum VSWR's observed as the short circuit is moved, then L_I is given by, [13], $L_I = 10 \log_{10} (\bar{\rho} - 1)/(\bar{\rho} + 1)$ dB, where $\bar{\rho} = (\rho_1\rho_2)^{1/2}$. The method has two additional advantages; it gives the intrinsic VSWR of the section $\rho_0 = (\rho_1/\rho_2)^{1/2}$, and it permits identification of erroneous data which might be caused by possible higher mode resonances. Thus if the reflection coefficient data is plotted on a Smith chart, points which do not fall on an exact impedance circle can be rejected. A direct measurement of the insertion loss of a section by substitution is inaccurate because the losses are small, 1.3 to 1.8 dB, and mismatch loss is difficult to correct. Neither can adequate accuracy be obtained by measuring the ratio of the peaks of the forward and conjugate beams in the radiation pattern when a short-circuit termination is used.

The field components E_θ and E_ϕ were measured in the conventional way, with the test section mounted on a three-axis positioner table in the manner shown in Fig. 1. The lower axis (vertical) is used in recording E_θ and E_ϕ as functions of polar angle θ. To obtain the azimuthal distributions $E_\theta(\phi)$ and $E_\phi(\phi)$, the appropriate polar angle, θ, is set in by means of the middle or elevation axis (horizontal in Fig. 1). Rotation of the model about its own axis then generates either an E_θ or E_ϕ pattern, depending on the orientation of the far-field pickup antenna, a small pyramidal horn. It is an easy matter to see as little as 1 percent (in power) of the fifth-space harmonic in these cos ϕ, sin ϕ patterns.

The phase relationship between E_θ and E_ϕ in the principal direction of radiation was measured by what is thought to be a novel method. When the far-field pickup horn is oriented at 45° the received signal is proportional to $E_\theta \pm E_\phi$. If the two components have equal peak magnitudes and are in the desired antiphase condition (cf. (13)) then the $E_\theta \pm E_\phi$ patterns will display nulls at azimuth angles of $\phi = 45°$ or 135°. If the components are not exactly equal in magnitude then the nulls are displaced from the 45° or 135° azimuths. On the other hand, if the two components are not exactly in antiphase then the nulls become minima in which signal strength does not go to zero. Thus by measuring both the null depth and angular displacement, both the magnitude and phase of E_ϕ relative to E_θ can be determined. Interpretation becomes difficult if the spatial harmonic content exceeds about 2 percent.

A linear polarized TE_{11} waveguide mode was used in all these experimental investigations, but all measurements were repeated for the orthogonal linear mode to ensure that no problems would be encountered with use of circular polarization.

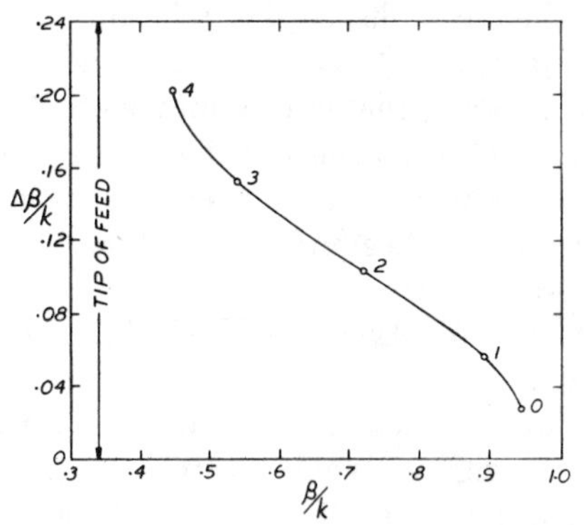

Fig. 7. Slowing factor $\Delta\beta/k$ versus β/k.

IV. Design of Model Field

A. *Interpretation of Experimental Data*

Table I lists hole dimensions, guide radius, and radial line length for five points that have α,β values falling exactly on the design curve in Fig. 5. All necessary dimensions for the complete model can thus be obtained by interpolation and extrapolation of this data. To reduce the risk of error, particularly in extrapolation, a search was made to find reasonably linear relationships between the various parameters. For this purpose it proved useful to use three derived quantities, the first of which is just the aspect ratio w/l, of the holes. Two more, termed slowing factors, are defined by

$$\frac{\Delta\beta}{k} = \frac{\beta}{k} - \frac{\beta_0}{k} \qquad \frac{\Delta\kappa}{k} = \frac{\kappa_0}{k} - \frac{\kappa}{k} \tag{27}$$

where β_0 is the propagation constant and κ_0 the cutoff wavenumber for an unperturbed cylinder of radius a.

Reasonably linear relationships were found to exist when the parameters $\Delta\beta/k$, w/l, and l were plotted against β/k. One such plot, that of $\Delta\beta/k$, is shown in Fig. 7, in which the numbered points refer to the test sections. Extrapolation into the region between test section 4 and the tip of the feed would clearly be risky; in fact, it happens that $\Delta\beta/k$ becomes complex because the wave slowing is so great that the unperturbed cylinder is beyond cutoff. Fortunately, the difficulty can be circumvented by plotting the parameter $\Delta\kappa/k$ against κ/k; in the tip region there is no anomalous behavior and extrapolation can safely be done. Feed cylinder radius at any point is then obtained from (27), along with the relation $\kappa_0 a = 1.84$.

Contrariwise, the slowing near the paraxial end is very slight. While extrapolation of $\Delta\beta/k$ is permissible, it leads to excessively large cylinder diameters. To avoid this situation a constant value of β/k (and hence of guide diameter) was used between test section 0 and the paraxial point, a distance of only 6 ft in the full size feed. This procedure results in incorrect phasing, of course, but the accumulated phase error is only 26°.

Having thus established all relevant dimensions as functions of β/k (or κ/k) the final step was to convert β/k into distance t along the feed by means of (20) and (21). There are a total of 90 rings in the model and the above procedure fixes the diameter of each ring at, say, its midpoint. For convenience in what follows, the rings are numbered serially, from $N = 1$ at the paraxial end, to $N = 90$ at the tip. As noted above, ring diameter was taken constant from $N = 1$ to $N = 5$. From $N = 6$ to $N = 20$, however, each ring was given a linear taper in diameter in order to avoid abrupt large steps between rings. In this region such steps would be close to $\lambda_g/2$ apart and a large reflection could build up. From $N = 21$ to $N = 90$ no taper was used; the changes in diameter are all small, the largest being 0.06 in between $N = 21$ and $N = 22$ in the model.

The hole dimensions change in an interesting way along the model. The length l simply decreases monotonically after $N = 5$. The width w, however, first decreases to a minimum at $N = 55$, then increases slightly. For ring $N = 39$, the holes are square, i.e., $w = l$; they are axially elongated for $N < 39$ and circumferentially elongated for $N > 39$. Radial line length was taken to be constant (0.75 in) from $N = 1$ to $N = 5$, then increased linearly with N to the value 1.05 in at $N = 21$ and remained constant thereafter.

B. Mechanical Design of Model

Each of the 90 ring sections was machined from aluminum tubing, with integral flanges at each end. The fins which form the radial lines were simply annular disks turned from 0.04-in thick sheet aluminum. Successive rings were joined together using 6-32 or 4-40 machine screws in such a way as to locate and capture a fin between the mating flanges. The assembled model was 178-in long, exclusive of input transition, and weighed 18 lb.

V. Experimental Measurements on Model

A. Near Field

For these measurements the model was cantilevered horizontally about 4 feet above the floor, sag being removed by strings dropped from the ceiling. A track, placed on the floor directly beneath the feed, guided a carriage actuated by motor-driven lead screws. The carriage supported a test antenna on the end of an adjustable boom in such a way that the test probe traveled at constant distance (to within 0.05 in) from the feed axis over the whole length of the feed. Two different distances were used, either 1λ or $5\lambda/4$, corresponding to $k\rho = 2\pi$ or 2.5π.

Two types of test antennas were used; a precision balun-fed [14] half-wave dipole for exploring tangential electric field, and a narrow slot for magnetic field. The slot was, in reality, the open end of a waveguide tapered to 0.1 in in the E plane. The dipole, aligned parallel to the axis of the model, was used to measure the component e_z along the azimuth line $\phi = 0$. Similarly, the slot responded

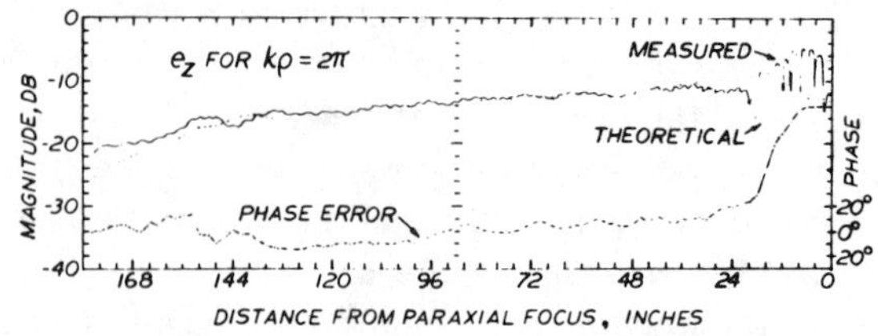

Fig. 8. Measured amplitude and phase of e_z along model at 2810 MHz.

to h_z along the azimuth line $\phi = \pi/2$, after rotation of the whole model through 90° around its own axis. Of course, e_ϕ and h_ϕ could equally well be explored using these probes but the considerably more oscillatory character of these components, particularly phase angle (see Fig. 4), tends to complicate matters.

Both magnitude and phase were measured with the aid of a network analyzer and were displayed on an antenna pattern recorder whose chart drive was servocontrolled to probe position. Fig. 8 shows such a record for the magnitude of e_z at the radial location $k\rho = 2\pi$. Superimposed on this record is the phase error of e_z, i.e., the departure of the measured phase from that shown in Fig. 4. The geometrical optics phase, $kR(c + 1/2c)$, undergoes nearly 27 complete revolutions of 2π radians over the length of the feed, and this variation has been removed in computing the phase error shown in the figure. Such magnitude and phase error distributions were investigated also for h_z at $k\rho = 2\pi$, and both e_z and h_z at $k\rho = 2.5\pi$. All these cases were again thoroughly explored for the orthogonal internal TE_{11} waveguide mode, so that a total of eight amplitude and eight phase distributions were recorded. Since the results given in Fig. 8 are quite typical, the other cases will not be presented here.

By way of explaining the large oscillations in the measured field shown in Fig. 8, it may be noted that at $k\rho = 2\pi$ the test probe passes within 0.3 in of the radial fins in the region near $t = 0$. Whenever the probe is centered under one of the fins its response drops to a low value. The effect is much less evident when the probe is farther away, e.g., at $k\rho = 2.5\pi$.

The dashed curve in Fig. 8 is the relative magnitude of e_z as computed theoretically from the first of (5), using the -9.4 dB illumination taper. It is clear that anomalous behavior occurs in both amplitude and phase over the first 20 in of the model feed. The relative magnitude of e_z is too high by many dB, and its phase is in error by 90° or more. Similar anomalies were observed for $k\rho = 2.5\pi$, for h_z and for the orthogonal mode of excitation.

However, from $t = 20$ in to the tip at $t = 178$ in, all the near-field measurements were in good agreement with theoretical values, although the observed magnitudes of both e_z and h_z were relatively somewhat too high near the tip. The all-important phase error was, in all cases, less than 45° from $t = 20$ in onward. In Fig. 8 there appears a slight linear decrease of 10 to 15° in the average level of the phase error, an effect also observed in the other near-field cases. It was found that this linear error could be elimi-

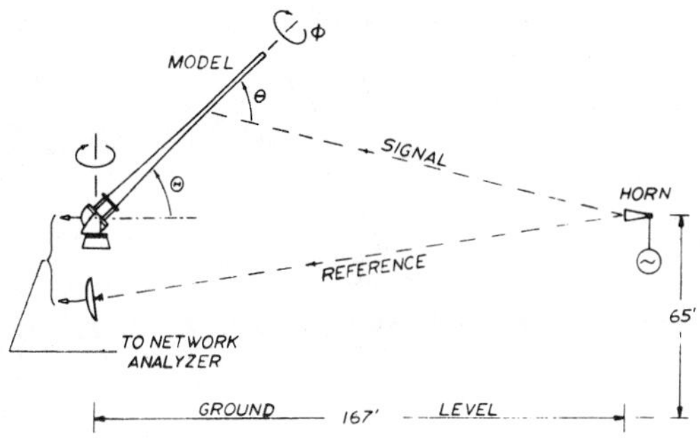

Fig. 9. Elevation view of pattern range.

nated by an upward shift in frequency to 2813 MHz, and that the maximum phase error was thereby decreased to 35° or less for all cases. It is worth noting that no adjustments had to be made, during these near field measurements, to any of the model feed dimensions in the region beyond $t = 20$ in.

Over the first 20-in, comprising rings $N = 1$ to $N = 10$, many attempts were made to suppress the anomalous behavior, but with little success. For example, the radiating holes in these rings were successively closed off by the use of adhesive-backed aluminum foil tape; in fact, in Fig. 8, all 6 holes in the first ring are covered. The only resulting benefit was the suppression of the improperly phased radiation in the vicinity of the closed rings. It is probable that the anomalies in this region are caused by higher modes in the guide, even though such modes did not appear to exist in test section zero. The tapered input transition would be expected to convert some of the TE_{11} mode to TM_{11}, and this latter mode can propagate freely up to about $t = 20$ in, at which point the guide diameter has decreased to about 5 in and cutoff occurs. The existence of the TM_{11} mode would, of course, create large phase errors since it propagates much faster than the TE_{11}. Attempts to detect the TM_{11} mode were inconclusive, largely due to difficulty in probing the internal field, and efforts to suppress the mode, if indeed it existed, were likewise futile.

A VSWR of 1.1 was measured at the coaxial connector of the input transition to the model. A substitution measurement of insertion loss through the feed (all rings radiating) gave 14.5 ± 0.5 dB, the uncertainty being largely due to the effects of mismatch at the tip of the feed. The corresponding fractional unradiated power is thus 3.6 ± 0.4 percent, in reasonably good agreement with the design value 4 percent.

B. Far Field

These patterns were recorded in essentially the same way as those for the test sections, described earlier, except that a network analyzer was used to measure phase as well as amplitude. Fig. 9 shows the model mounted on a positioner table located near the edge of a roof approximately 65 ft above ground level. Tensioning guys were used, as can be seen in Fig. 1, to ensure straightness of the model and concentricity of rotation about its axis. A small pyramidal horn, acting as a transmitting antenna, was located atop a 65-ft tower some 167-ft distant from the vertical axis of the positioner table. A reference signal for phase angle measurement was obtained by using a small reflector antenna mounted in a fixed position below the model feed. The signal and reference path lengths are roughly equalized in this way so that phase errors due to the long line effect are minimized.

Patterns are recorded in the following way. The feed is first positioned so that its axis points precisely at the center of the horn aperture. In this case, the angle Θ shown in Fig. 9 is initially zero. To obtain azimuthal patterns the feed is then elevated in a vertical plane through the table angle Θ, then rotated about its own axis through 360° in ϕ. The polar radiation angle θ is related to table angle Θ in a simple way which can easily be worked out from the geometry. For polar patterns, one again starts with the model axis pointing at the horn. However, the elevation drive is disabled and rotation takes place about a vertical axis, as indicated in Fig. 9. The feed now moves through table angle Θ in a horizontal, rather than a vertical, plane. In this way the field components E_θ and E_ϕ are measured on the surface of a distant sphere which, of course, does not correspond to the scaled reflector.

To relate the actual measured quantities to those that would be observed in the aperture of the reflector requires use of transformations based on geometrical optics that were first derived by Altshuler [15]. These will not be given here in detail. It turns out that the amplitude of E_θ or E_ϕ in the aperture differs negligibly (<1 dB) from the values determined in the above measurements. Fortuitously, table angle Θ is linearly proportional to radius in the aperture to an accuracy of better than 1 percent over the whole range in Θ. At the aperture edge Θ reaches the value 64°38′, while $\theta = 2\psi = 70°10'$. Due to these circumstances a measured polar pattern can be directly regarded as a radial aperture distribution.

Phase error as a function of aperture radius is obtained by subtracting, from the observed values, a correction factor (given by the Altshuler transformation) which varies from 0 at $\Theta = 0$ to nearly 29π radians at the aperture edge, $\Theta = 64°38'$. The observed phase angle changes very rapidly as Θ approaches its maximum value, reaching nearly 2π radians per degree at $\Theta = 64°38'$. For this reason the largest single source of inaccuracy in phase measurement is due to the readout error ($\sim$0.03°) in table angle. Averaging of several runs was resorted to in order to reduce this source of error to about $\pm 3°$ in phase angle. Azimuthal phase measurements do not suffer from this effect, but eccentricity of the feed as it rotates can contribute cyclic phase errors of up to 10°. Path length compensation plus use of a frequency source having stability of 1 in 10^8 renders negligible any error due to frequency drift. The remaining source of error is in the network analyzer; it amounts to about $\pm 2°$. Amplitude measurements are accurate to ± 0.5 dB over a 30 dB dynamic range.

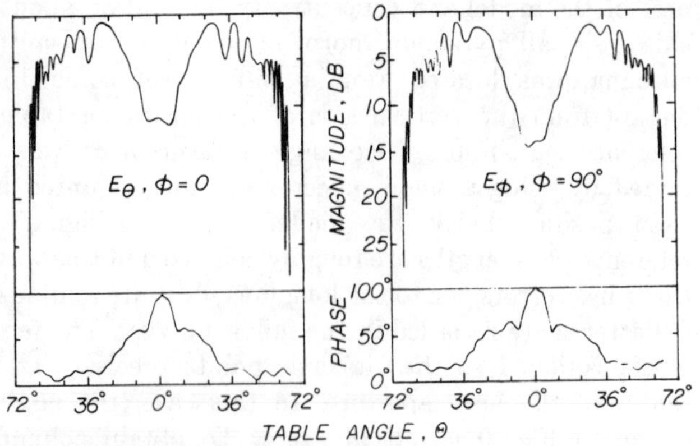

Fig. 10. Observed E and H plane patterns at 2813 MHz.

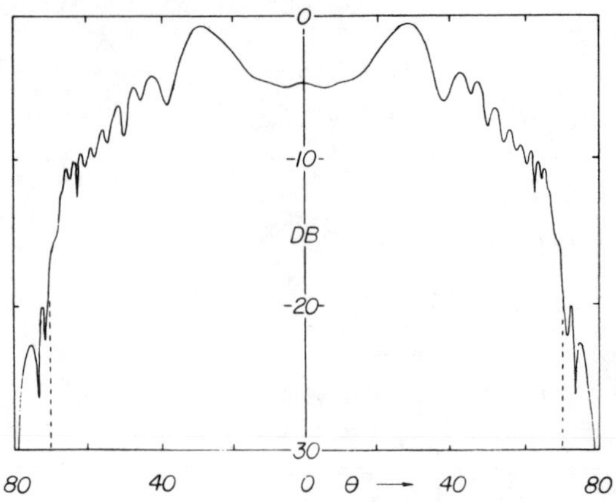

Fig. 12. Theoretical line feed pattern for sources on cylinder.

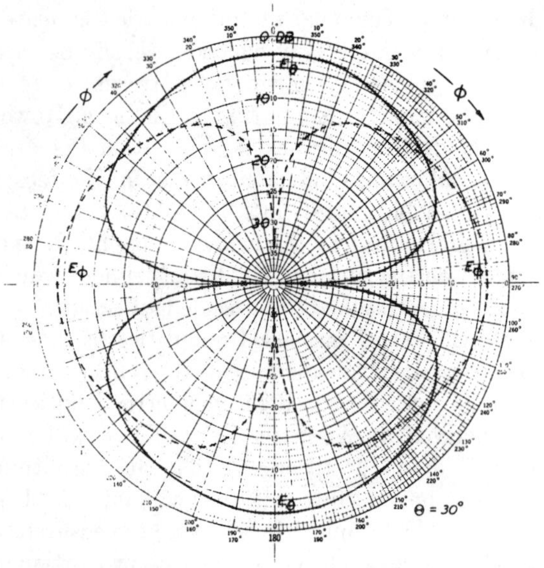

Fig. 11. Observed azimuthal distributions at polar angle 30°.

A typical set of observed E and H plane patterns taken at 2813 MHz is shown in Fig. 10. Other polar patterns, taken in the intercardinal (i.e., $\pm 45°$) planes, and for the orthogonal mode, showed very similar features, in particular, an underillumination in the region $0 < \Theta < 18°$, accompanied by a large phase error which is inconsequential because of the low amplitude level. Beyond $\Theta = 18°$ the observed phase error remained less than 45° in all cases.

As was done in the near field, the first few rings were successively closed off, using aluminum foil tape, and the polar phase and amplitude patterns were carefully examined to find the conditions likely to lead to least gain loss due to the underillumination. From these studies it was determined that radial phase error was least when no fewer than the first two, but no more than the first six, rings were closed off. This also appeared to yield the most acceptable amplitude patterns. In Fig. 10, rings $N = 1$ and $N = 2$ were closed.

Azimuthal phase and amplitude distributions were also thoroughly explored for both internal modes at polar angles $\Theta = 18°$, 30°, 42°, 51°, and 60.5°. A typical amplitude pattern is shown in Fig. 11 for $\Theta = 30°$. A small amount (not exceeding 1 percent in power) of the fifth-space harmonic is evident in these patterns and in those at $\Theta = 42$ and 60.5°. In general, there is an unbalance between the components with E_θ averaging 1–1.5 dB higher than E_ϕ. The phase distributions showed E_θ and E_ϕ to be either in phase or in antiphase to within 15° on the average, at all polar angles.

VI. Discussion and Conclusions

Because test section 0 radiated effectively at endfire, the deficiency on the part of the model feed came as a surprise and was first thought to be a manifestation of the anomalous behavior observed in the near field. It was suggested, however, that the highly reactive impedance presented by the $\lambda/4$ radial lines might inhibit surface waves and suppress endfire radiation. This implies, perhaps, that the near- and far-field anomalies are two separate phenomena. In any case, it was clear that the first six rings of the model did not contribute usefully and ought to be closed off. To calculate what effect this might have, a computer was programmed to sum the fields radiated by rings of six axial and six circumferential slots around a cylinder. Reflector currents were obtained by integrating the effects of all rings using appropriate excitation, with an empirical adjustment to the geometrical optics phase to account for the fact that the slots are not on axis. Finally, the program computed the far-field pattern and the gain of the system.

The interesting result was obtained that maximum gain occurs as a result of truncating the feed by closing off the first seven rings. The explanation is that this condition leads to decreased radiation in the $\theta = 180°$ direction, thereby reducing spillover loss more than enough to compensate for the loss in gain due to the decrease in central illumination. The computed aperture efficiency under these conditions was 85 percent, and the far-field pattern, shown in Fig. 12, showed a 5-dB deficiency in the illumination at the aperture center. The similarity to the observed patterns is striking, even to the number and the detail of the diffraction ripples. It should be noted that the experimental patterns were taken with the model terminated to absorb the small amount of unradiated power.

TABLE II
PREDICTED APERTURE EFFICIENCY

Apodisation efficiency (rings 1-6 closed)	.84
Effect of measured radial phase error	.94
Polarization mismatch (unbalance in E_θ and E_ϕ)	.99
Power unradiated at tip of feed	.96
Spatial harmonics in azimuth	.99
Spillover (estimated)	.96
Net efficiency with ideal reflector	.71
Arecibo reflector surface roughness (1.2 in. rms)	.74
Expected overall efficiency	.525

The experimental results reported above showed that, despite its deficiency in endfire radiation, a scaled-up version of the model feed was capable of high efficiency in the Arecibo reflector. An estimate of expected overall efficiency was provided by LaLonde [10] who used the experimentally observed aperture distributions in a computer program to predict gain and efficiency. The results are given in Table II. A major effect of the lack of central illumination is an increase in the sidelobes; the first minor lobe levels were predicted to occur at −16 dB.

The full size feed was built in the same way as the model, by simply scaling relevant dimensions, with the first six rings eliminated. It yielded a measured efficiency of 53% with −15.5 dB first sidelobes. The half power beam width is 8.8′ and bandwidth 12.5 MHz for a 3-dB gain loss. The feed is excited by a turnstile junction, allowing complete polarization flexibility, and it handles the 2.5-MW peak and 150-kW average power with no indication of breakdown. The improvement in gain over past performance is 3.8 dB, but the signal-to-noise ratio in the passive mode has actually increased by 5 dB because the reduced spillover results in lower antenna noise temperature. In the planetary radar mode the improvement in signal-to-noise ratio amounts to 9 dB.

ACKNOWLEDGMENTS

The author is indebted to his colleagues, J. P. Gressard for excellent mechanical design, R. L. Carlise and D. K. Waineo for helpful technical discussions, and C. A. Wiley for his long-time unflagging interest and stimulation. The advice and encouragement rendered throughout the course of the work by L. M. LaLonde of Cornell University were invaluable.

REFERENCES

[1] A. C. Schell, "Diffraction theory of large aperture spherical reflector antennas," *IEEE Trans. Antennas Propagat.*, vol. AP-11, pp. 428–432, July 1963.
[2] G. C. McCormick, "A line feed for a spherical reflector," *IEEE Trans. Antennas Propagat.*, vol. AP-15, pp. 639–644, Sept. 1967.
[3] V. H. Rumsey, "On the design and performance of feeds for correcting spherical aberration," *IEEE Trans. Antennas Propagat.*, vol. AP-18, pp. 343–351, May 1970.
[4] M. L. Burrows and L. J. Ricardi, "Aperture feed for a spherical reflector," *IEEE Trans. Antennas and Propagat.* vol. AP-15, pp. 227–230, May 1967.
[5] B. MacA. Thomas, H. C. Minnett, and V. T. Bao, "Fields in the focal region of a spherical reflector," *IEEE Trans. Antennas Propagat.*, vol. AP-17, pp. 229–231, Mar. 1969.
[6] L. J. Ricardi, "Synthesis of the fields of a transverse feed for a spherical reflector," *IEEE Trans. Antennas Propagat.*, vol. AP-19, pp. 310–320, May 1971.
[7] L. M. LaLonde and D. E. Harris, "A high-performance line source feed for the AIO spherical reflector," *IEEE Trans. Antennas Propagat.*, vol. AP-18, pp. 41–48, Jan. 1970.
[8] A. F. Kay, "A line source feed for a spherical reflector," A. F. Cambridge Res. Lab., Cambridge, Mass., Contract AF19(604)-5532, AFCRL rep. 529, ASTIA Doc. AD-261007, May 1971.
[9] A. W. Love and J. J. Gustincic, "Line source feed for a spherical reflector," *IEEE Trans. Antennas Propagat.* (Commun.), vol. AP-16, pp. 132–134, Jan. 1968.
[10] L. M. LaLonde and A. W. Love, "A new high power dual polarized line feed for the Arecibo reflector," presented at 1972 URSI Symp., Commission 5, Washington, D.C., April 15.
[11] R. C. Spencer, C. J. Sletten, and J. E. Walsh, "Correction of spherical aberration by a phased line source," in *Proc. Nat. Electron. Conf.*, vol. 5, pp. 320–333, 1950.
[12] E. L. Ginzton, *Microwave Measurements*. New York: McGraw-Hill, 1957, Sec. 11.5.
[13] K. Tomiyasu, "Intrinsic insertion loss of a mismatched microwave network," *IRE Trans. Microwave Theory Tech.*, vol. MTT-3, pp. 40–44, Jan. 1955.
[14] M. Gans, D. Kajfetz, and V. H. Rumsey, "Frequency independent baluns," *Proc. IEEE* (Lett.), vol. 53, pp. 647–648, June 1965.
[15] E. E. Altshuler, "Primary pattern measurement of a line source feed for a spherical reflector," *IRE Trans. Antennas Propagat.* (Commun.), vol. AP-10, pp. 214–215, Mar. 1962.

Bibliography for Part IX

[1] J. Ashmead and A. B. Pippard, "The use of spherical reflectors as microwave scanning aerials," *J. Inst. Elec. Eng.*, vol. 93, part III-A, pp. 627-632, 1946.

[2] R. C. Spencer, C. J. Sletten, and J. E. Walsh, "Correction of spherical aberration by a phased line source," in *Proc. Nat. Electron. Conf.*, vol. 5, 1949, pp. 320-333.

[3] A. W. Love, "Spherical reflecting antennas with corrected line sources," *IRE Trans. Antennas Propagat.*, vol. AP-10, pp. 529-537, Sept. 1962.

[4] A. F. Kay, "A line source feed for a spherical reflector," AFCRL Rep. 529 (ASTIA Doc. 261007), May 1961.

[5] A. W. Love and J. J. Gustincic, "Line source feed for a spherical reflector," *IEEE Trans. Antennas Propagat.*, vol. AP-16, pp. 132-134, Jan. 1968.

[6] A. K. Head, "A new form for a giant radio telescope," *Nature*, vol. 179, pp. 692-693, Apr. 6, 1957.

[7] C. J. E. Phillips and P. J. B. Clarricoats, "Optimum design of a Gregorian-corrected spherical reflector antenna," *Proc. Inst. Elec. Eng.*, vol. 117, pp. 718-734, Apr. 1970.

[8] M. L. Burrows and L. J. Ricardi, "Aperture feed for a spherical reflector," *IEEE Trans. Antennas Propagat.*, vol. AP-15, pp. 227-230, Mar. 1967.

[9] L. Ronchi and G. Toraldo di Francia, "An application of parageometrical optics to the design of a microwave mirror," *IRE Trans. Antennas Propagat.*, vol. AP-6, pp. 129-133, Jan. 1958.

[10] G. Toraldo di Francia, L. Ronchi, and V. Russo, "Experimental test of a stepped zone mirror for microwaves," *IRE Trans. Antennas Propagat.*, vol. AP-7, pp. S125-S131, Dec. 1959.

[11] V. Russo and G. Toraldo di Francia, "Correction of the astigmatism of a spherical diffraction reflector," *IRE Trans. Antennas Propagat.*, vol. AP-9, pp. 225-226, Mar. 1961.

[12] L. D. Bakrakh and I. V. Vavilova, "Spherical two-mirror antennas," *Radio Eng. Electron. Phys.*, vol. 6, pp. 1020-1028, July 1961.

[13] S. Kownacki, "High gain and high resolution antennas for space tracking and communication," *IEEE Trans. Antennas Propagat.*, vol. AP-11, pp. 594-595, Sept. 1963.

[14] P. M. Geruni, "Design problems of spherical dual-reflector antennas," *Radio Eng. Electron. Phys.*, vol. 9, pp. 1-8, Jan. 1964.

[15] R. F. Trainer and W. M. Young, "Research on an experimental corrected spherical reflector antenna using cosmic radio sources," *Microwave J.*, vol. 8, pp. 69-72, Jan. 1965.

[16] M. H. Cohen and G. E. Perona, "A correcting feed at 611 MHz for the AIO reflector," *IEEE Trans. Antennas Propagat.*, vol. AP-15, pp. 482-483, May 1967.

[17] T. Pratt, "Offset spherical reflector aerial with a line feed," *Proc. Inst. Elec. Eng.*, vol. 115, pp. 633-641, May 1968.

[18] T. Pratt and E. D. R. Shearman, "Sectoral hoghorn: A new form of line feed for spherical reflector aerials," *Electron. Lett.*, vol. 5, pp. 1-2, Jan. 9, 1969.

[19] P. J. B. Clarricoats and S. H. Lim, "Proposed high efficiency spherical reflector antenna," *Electron. Lett.*, vol. 5, pp. 709-711, Dec. 27, 1969.

[20] C. Ancona, "Focusing non-parabolic mirrors by using phase-corrected multiple beam sources," *Radio Sci.*, vol. 5, pp. 707-714, Apr. 1970.

[21] V. H. Rumsey, "On the design and performance of feeds for correcting spherical aberration," *IEEE Trans. Antennas Propagat.*, vol. AP-18, pp. 343-351, May 1970.

[22] H. S. Oranc and A. F. Fer, "Focused aperture antenna using a spherical main reflector with (Cassegrain) convex feed," *Electron. Lett.*, vol. 6, pp. 523-525, Aug. 6, 1970.

[23] J. I. Glaser, "Synthesis of the primary fields of a transmitting feed for a circular reflector," *IEEE Trans. Antennas Propagat.*, vol. AP-19, pp. 31-36, Jan. 1971.

[24] G. Poulton, "Efficiency of a stepped reflector when fed from an off-axis source," *Electron. Lett.*, vol. 7, pp. 666-667, Nov. 4, 1971.

[25] N. Amitay and H. Zucker, "Compensation of spherical reflector aberrations by planar array feeds," *IEEE Trans. Antennas Propagat.*, vol. AP-20, pp. 49-56, Jan. 1972.

[26] P. J. B. Clarricoats and D. C. Chang, "A proposed dielectric waveguide line feed for a millimetre wavelength radio-telescope," in *IEEE AP-S Symp. Dig.*, 1972, pp. 296-298.

[27] A. Ishimaru, I. Sreenivasiah, and V. K. Wong, "Double spherical Cassegrain reflector antennas," *IEEE Trans. Antennas Propagat.*, vol. AP-21, pp. 774-780, Nov. 1973.

[28] J. S. Shen and N. Brice, "Near-field gain calibration for large spherical antennas," *IEEE Trans. Antennas Propagat.*, vol. AP-21, pp. 787-792, Nov. 1973.

[29] R. H. Turrin, "A multibeam, spherical reflector satellite antenna for the 20- and 30-GHz bands," *Bell Syst. Tech. J.*, vol. 54, pp. 1011-1026, July-Aug. 1975.

[30] D. L. Doan and T. B. Vu, "Study of efficiency of spherical Gregorian reflector," *IEEE Trans. Antennas Propagat.*, vol. AP-23, pp. 819-824, Nov. 1975.

Author Index

Subject Index

R

S

T

W

X

Editor's Biography

A. W. Love (M'58-SM'75) was born and educated in Toronto, Ont., Canada, receiving the B.A. degree in mathematics and physics in 1938, the M.A. in physics in 1939, and the Ph.D. degree in physics in 1951, all from the University of Toronto. Following service as a Radar Officer in the U.K., the Middle East, and Australia in World War II. he then spent the years 1946–1948 as a Research Officer in the Commonwealth Scientific and Industrial Research Organization, Sydney, Australia. His work on noise standards in that organization's Radiophysics Laboratory led to interests in radio astronomy and microwave radiometry which he still maintains today.

Beginning in 1951 he spent six years in mining geophysical exploration activities with Newmont Exploration Limited, Jerome, AZ. He entered the aerospace field in 1957 with Wiley Electronics Company, Phoenix, AZ, where he developed the first successful airborne mapping radiometer, a passive sensor able to produce terrain images under all weather, day or night conditions. In 1963 he joined Rockwell International's Autonetics Division, Anaheim, CA, in charge of advanced antenna development. This was followed by two years in ECM and reentry systems work with National Engineering Science Company, Newport Beach, CA. Returning to Autonetics in late 1965, he engaged, for a time, in studies related to infrared and microwave emission processes in planetary atmospheres before he returned to the development of single and dual polarized line source feeds for large spherical reflectors. Since 1971 he has been with the Space Division of Rockwell International, Seal Beach, CA, pursuing the development of precision microwave radiometers and high beam efficiency antennas for the remote measurement of sea surface temperature from an orbiting satellite.

Dr. Love is listed in *American Men of Science*, has authored 17 papers in the above fields, and is the holder of five patents. He received the IEEE Antennas and Propagation Society's 1973 Best Paper Award for his paper, "Scale Model Development of a High Efficiency Dual Polarized Line Feed for the Arecibo Spherical Reflector," which appeared in the September 1973 issue of the TRANSACTIONS. He also edited the IEEE Press reprint volume *Electromagnetic Horn Antennas*, published in 1976.